T0337894

Entropy Theory and its Application in Environmental and Water Engineering

Entropy Theory and its Application in Environmental and Water Engineering

Vijay P. Singh

Department of Biological and Agricultural Engineering &
Department of Civil and Environmental Engineering
Texas A & M University
Texas, USA

WILEY-BLACKWELL

A John Wiley & Sons, Ltd., Publication

This edition first published 2013 © 2013 by John Wiley and Sons, Ltd

Wiley-Blackwell is an imprint of John Wiley & Sons, formed by the merger of Wiley's global Scientific, Technical and Medical business with Blackwell Publishing.

Registered office: John Wiley & Sons, Ltd, The Atrium, Southern Gate, Chichester, West Sussex, PO19 8SQ, UK

Editorial offices: 9600 Garsington Road, Oxford, OX4 2DQ, UK

The Atrium, Southern Gate, Chichester, West Sussex, PO19 8SQ, UK

111 River Street, Hoboken, NJ 07030-5774, USA

For details of our global editorial offices, for customer services and for information about how to apply for permission to reuse the copyright material in this book please see our website at www.wiley.com/wiley-blackwell.

Library of Congress Cataloging-in-Publication Data

Singh, V. P. (Vijay P.)
   Entropy theory and its application in environmental and water engineering / Vijay P. Singh.
      pages cm
   Includes bibliographical references and indexes.
   ISBN 978-1-119-97656-1 (cloth)
   1. Hydraulic engineering – Mathematics. 2. Water – Thermal properties – Mathematical models.
3. Hydraulics – Mathematics. 4. Maximum entropy method – Congresses. 5. Entropy. I. Title.
   TC157.8.S46 2013
   627.01′53673 – dc23

                                                                                    2012028077

A catalogue record for this book is available from the British Library.

Wiley also publishes its books in a variety of electronic formats. Some content that appears in print may not be available in electronic books.

Typeset in 10/12pt Times-Roman by Laserwords Private Limited, Chennai, India
Printed and bound in Singapore by Markono Print Media Pte Ltd

First Impression 2013

Dedicated to
My wife Anita,
son Vinay,
daughter-in-law Sonali
daughter Arti, and
grandson Ronin

Contents

Preface

Since the pioneering work of Shannon in 1948 on the development of informational entropy theory and the landmark contributions of Kullback and Leibler in 1951 leading to the development of the principle of minimum cross-entropy, of Lindley in 1956 leading to the development of mutual information, and of Jaynes in 1957–8 leading to the development of the principle of maximum entropy and theorem of concentration, the entropy theory has been widely applied to a wide spectrum of areas, including biology, genetics, chemistry, physics and quantum mechanics, statistical mechanics, thermodynamics, electronics and communication engineering, image processing, photogrammetry, map construction, management sciences, operations research, pattern recognition and identification, topology, economics, psychology, social sciences, ecology, data acquisition and storage and retrieval, fluid mechanics, turbulence modeling, geology and geomorphology, geophysics, geography, geotechnical engineering, hydraulics, hydrology, reliability analysis, reservoir engineering, transportation engineering, and so on. New areas finding application of entropy have since continued to unfold. The entropy theory is indeed versatile and its application is widespread.

In the area of hydrologic and environmental sciences and water engineering, a range of applications of entropy have been reported during the past four and half decades, and new topics applying entropy are emerging each year. There are many books on entropy written in the fields of statistics, communication engineering, economics, biology and reliability analysis. These books have been written with different objectives in mind and for addressing different kinds of problems. Application of entropy concepts and techniques discussed in these books to hydrologic science and water engineering problems is not always straightforward. Therefore, there exists a need for a book that deals with basic concepts of entropy theory from a hydrologic and water engineering perspective and then for a book that deals with applications of these concepts to a range of water engineering problems. Currently there is no book devoted to covering basic aspects of the entropy theory and its application in hydrologic and environmental sciences and water engineering. This book attempts to fill this need.

Much of the material in the book is derived from lecture notes prepared for a course on entropy theory and its application in water engineering taught to graduate students in biological and agricultural engineering, civil and environmental engineering, and hydrologic science and water management at Texas, A & M University, College Station, Texas. Comments, critics and discussions offered by the students have, to some extent, influenced the style of presentation in the book.

The book is divided into 16 chapters. The first chapter introduces the concept of entropy. Providing a short discussion of systems and their characteristics, the chapter goes on to discuss different types of entropies; and connection between information, uncertainty and entropy; and concludes with a brief treatment of entropy-related concepts. Chapter 2 presents the entropy theory, including formulation of entropy and connotations of information and entropy. It then describes discrete entropy for univariate, bivariate and multidimensional cases. The discussion is extended to continuous entropy for univariate, bivariate and multivariate cases. It also includes a treatment of different aspects that influence entropy. Reflecting on the various interpretations of entropy, the chapter provides hints of different types of applications.

The principle of maximum entropy (POME) is the subject matter of Chapter 3, including the formulation of POME and the development of the POME formalism for discrete variables, continuous variables, and two variables. The chapter concludes with a discussion of the effect of constraints on entropy and invariance of entropy. The derivation of POME-based discrete and continuous probability distributions under different constraints constitutes the discussion in Chapter 4. The discussion is extended to multivariate distributions in Chapter 5. First, the discussion is restricted to normal and exponential distributions and then extended to multivariate distributions by combining the entropy theory with the copula method.

Chapter 6 deals with the principle of minimum cross-entropy (POMCE). Beginning with the formulation of POMCE, it discusses properties and formalism of POMCE for discrete and continuous variables and relation to POME, mutual information and variational distance. The discussion on POMCE is extended to deriving discrete and continuous probability distributions under different constraints and priors in Chapter 7. Chapter 8 presents entropy-based methods for parameter estimation, including the ordinary entropy-based method, the parameter-space expansion method, and a numerical method.

Spatial entropy is the subject matter of Chapter 9. Beginning with a discussion of the organization of spatial data and spatial entropy statistics, it goes on to discussing one-dimensional and two-dimensional aggregation, entropy maximizing for modeling spatial phenomena, cluster analysis, spatial visualization and mapping, scale and entropy and spatial probability distributions. Inverse spatial entropy is dealt with in Chapter 10. It includes the principle of entropy decomposition, measures of information gain, aggregate properties, spatial interpretations, hierarchical decomposition, and comparative measures of spatial decomposition.

Maximum entropy-based spectral analysis is presented in Chapter 11. It first presents the characteristics of time series, and then discusses spectral analyses using the Burg entropy, configurational entropy, and mutual information principle. Chapter 12 discusses minimum cross-entropy spectral analysis. Presenting the power spectrum probability density function first, it discusses minimum cross-entropy-based power spectrum given autocorrelation, and cross-entropy between input and output of linear filter, and concludes with a general method for minimum cross-entropy spectral estimation.

Chapter 13 presents the evaluation and design of sampling and measurement networks. It first discusses design considerations and information-related approaches, and then goes on to discussing entropy measures and their application, directional information transfer index, total correlation, and maximum information minimum redundancy (MIMR).

Selection of variables and models constitutes the subject matter of Chapter 14. It presents the methods of selection, the Kullback–Leibler (KL) distance, variable selection, transitivity, logit model, and risk and vulnerability assessment. Chapter 15 is on neural networks comprising neural network training, principle of maximum information preservation, redundancy and diversity, and decision trees and entropy nets. Model complexity is treated in Chapter 16. The complexity measures discussed include Ferdinand's measure of complexity, Kapur's complexity measure, Cornacchio's generalized complexity measure and other complexity measures.

Vijay P. Singh
College Station, Texas

Acknowledgments

Nobody can write a book on entropy without being indebted to C.E. Shannon, E.T. Jaynes, S. Kullback, and R.A. Leibler for their pioneering contributions. In addition, there are a multitude of scientists and engineers who have contributed to the development of entropy theory and its application in a variety of disciplines, including hydrologic science and engineering, hydraulic engineering, geomorphology, environmental engineering, and water resources engineering – some of the areas of interest to me. This book draws upon the fruits of their labor. I have tried to make my acknowledgments in each chapter as specific as possible. Any omission on my part has been entirely inadvertent and I offer my apologies in advance. I would be grateful if readers would bring to my attention any discrepancies, errors, or misprints.

Over the years I have had the privilege of collaborating on many aspects of entropy-related applications with Professor Mauro Fiorentino from University of Basilicata, Potenza, Italy; Professor Nilgun B. Harmancioglu from Dokuz Elyul University, Izmir, Turkey; and Professor A.K. Rajagopal from Naval Research Laboratory, Washington, DC. I learnt much from these colleagues and friends.

During the course of two and a half decades I have had a number of graduate students who worked on entropy-based modeling in hydrology, hydraulics, and water resources. I would particularly like to mention Dr. Felix C. Kristanovich now at Environ International Corporation, Seattle, Washington; and Mr. Kulwant Singh at University of Houston, Texas. They worked with me in the late 1980s on entropy-based distributions and spectral analyses. Several of my current graduate students have helped me with preparation of notes, especially in the solution of example problems, drawing of figures, and review of written material. Specifically, I would like to express my gratitude to Mr. Zengchao Hao for help with Chapters 2, 4, 5, and 11; Mr. Li Chao for help with Chapters 2, 9, 10, 13; Ms. Huijuan Cui for help with Chapters 11 and 12; Mr. D. Long for help with Chapters 8 and 9; Mr. Juik Koh for help with Chapter 16; and Mr. C. Prakash Khedun for help with text formatting, drawings and examples. I am very grateful to these students. In addition, Dr. L. Zhang from University of Akron, Akron, Ohio, reviewed the first five chapters and offered many comments. Dr. M. Ozger from Technical University of Istanbul, Turkey; and Professor G. Tayfur from Izmir Institute of Technology, Izmir, Turkey, helped with Chapter 13 on neural networks.

My family members – brothers and sisters in India – have been a continuous source of inspiration. My wife Anita, son Vinay, daughter-in-law Sonali, grandson Ronin, and daughter Arti have been most supportive and allowed me to work during nights, weekends, and holidays, often away from them. They provided encouragement, showed patience, and helped in myriad ways. Most importantly, they were always there whenever I needed them, and I am deeply grateful. Without their support and affection, this book would not have come to fruition.

Vijay P. Singh
College Station, Texas

1 Introduction

Beginning with a short introduction of systems and system states, this chapter presents concepts of thermodynamic entropy and statistical-mechanical entropy, and definitions of informational entropies, including the Shannon entropy, exponential entropy, Tsallis entropy, and Renyi entropy. Then, it provides a short discussion of entropy-related concepts and potential for their application.

1.1 Systems and their characteristics

1.1.1 Classes of systems

In thermodynamics a system is defined to be any part of the universe that is made up of a large number of particles. The remainder of the universe then is referred to as surroundings. Thermodynamics distinguishes four classes of systems, depending on the constraints imposed on them. The classification of systems is based on the transfer of (i) matter, (ii) heat, and/or (iii) energy across the system boundaries (Denbigh, 1989). The four classes of systems, as shown in Figure 1.1, are: (1) Isolated systems: These systems do not permit exchange of matter or energy across their boundaries. (2) Adiabatically isolated systems: These systems do not permit transfer of heat (also of matter) but permit transfer of energy across the boundaries. (3) Closed systems: These systems do not permit transfer of matter but permit transfer of energy as work or transfer of heat. (4) Open systems: These systems are defined by their geometrical boundaries which permit exchange of energy and heat together with the molecules of some chemical substances.

The second law of thermodynamics states that the entropy of a system can only increase or remain constant; this law applies to only isolated or adiabatically isolated systems. The vast majority of systems belong to class (4). Isolation and closedness are not rampant in nature.

1.1.2 System states

There are two states of a system: microstate and macrostate. A system and its surroundings can be isolated from each other, and for such a system there is no interchange of heat or matter with its surroundings. Such a system eventually reaches a state of equilibrium in a thermodynamic sense, meaning no significant change in the state of the system will occur. The state of the system here refers to the macrostate, not microstate at the atomic scale, because the

Entropy Theory and its Application in Environmental and Water Engineering, First Edition. Vijay P. Singh.
© 2013 John Wiley & Sons, Ltd. Published 2013 by John Wiley & Sons, Ltd.

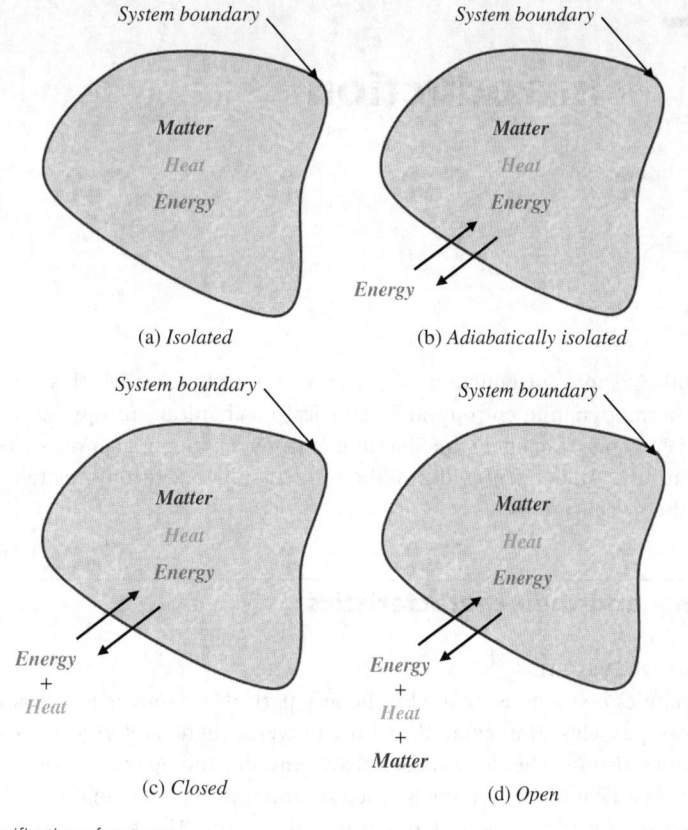

Figure 1.1 Classification of systems.

microstate of such a system will continuously change. The macrostate is a thermodynamic state which can be completely described by observing thermodynamic variables, such as pressure, volume, temperature, and so on. Thus, in classical thermodynamics, a system is described by its macroscopic state entailing experimentally observable properties and the effects of heat and work on the interaction between the system and its surroundings. Thermodynamics does not distinguish between various microstates in which the system can exist, and hence does not deal with the mechanisms operating at the atomic scale (Fast, 1968). For a given thermodynamic state there can be many microstates. Thermodynamic states are distinguished when there are measurable changes in thermodynamic variables.

1.1.3 Change of state

Whenever a system is undergoing a change because of introduction of heat or extraction of heat or any other reason, changes of state of the system can be of two types: reversible and irreversible. As the name suggests, reversible means that any kind of change occurring during a reversible process in the system and its surroundings can be restored by reversing the process. For example, changes in the system state caused by the addition of heat can be restored by the extraction of heat. On the contrary, this is not true in the case of irreversible change of state in which the original state of the system cannot be regained without making changes in the surroundings. Natural processes are irreversible processes. For processes to be reversible, they must occur infinitely slowly.

It may be worthwhile to visit the first law of thermodynamics, also called the law of conservation of energy, which was based on the transformation of work and heat into one another. Consider a system which is not isolated from its surroundings, and let a quantity of heat dQ be introduced to the system. This heat performs work denoted as dW. If the internal energy of the system is denoted by U, then dQ and dW will lead to an increase in U: $dU = dQ + dW$. The work performed may be of mechanical, electrical, chemical, or magnetic nature, and the internal energy is the sum of kinetic energy and potential energy of all particles that the system is made up of. If the system passes from an initial state 1 to a final state 2, then, $\int_1^2 dU = \int_1^2 dQ + \int_1^2 dW$. It should be noted that the integral $\int_1^2 dU$ depends on the initial and final states but the integrals $\int_1^2 dQ$ and $\int_1^2 dW$ also depend on the path followed.

Since the system is not isolated and is interactive, there will be exchanges of heat and work with the surroundings. If the system finally returns to its original state, then the sum of integral of heat and integral of work will be zero, meaning the integral of internal energy will also be zero, that is, $\int_1^2 dU + \int_2^1 dU = 0$, or $-\int_1^2 dU = -\int_2^1 dU$. Were it not the case, the energy would either be created or destroyed. The internal energy of a system depends on pressure, temperature, volume, chemical composition, and structure which define the system state and does not depend on the prior history.

1.1.4 Thermodynamic entropy

Let Q denote the quantity of heat. For a system to transition from state 1 to state 2, the amount of heat, $\int_1^2 dQ$, required is not uniquely defined, but depends on the path that is followed for transition from state 1 to state 2, as shown in Figures 1.2a and b. There can be two paths: (i) reversible path: transition from state 1 to state 2 and back to state 1 following the same path, and (ii) irreversible path: transition from state 1 to state 2 and back to state 1 following a different path. The second path leads to what is known in environmental and water engineering as hysteresis. The amount of heat contained in the system under a given condition is not meaningful here. On the other hand, if T is the absolute temperature (degrees kelvin or simply kelvin) (i.e., $T = 273.15 +$ temperature in $^\circ$C), then a closely related quantity, $\int_1^2 dQ/T$, is uniquely defined and is therefore independent of the path the system takes to transition from state 1 to state 2, provided the path is reversible (see Figure 1.2a). Note that when integrating, each elementary amount of heat is divided by the temperature at which it is introduced. The system must expend this heat in order to accomplish the transition and this heat expenditure is referred to as heat loss. When calculated from the zero point of absolute temperature, the integral:

$$S = \int_0^T \frac{dQ_{rev}}{T} \tag{1.1}$$

is called entropy of the system, denoted by S. Subscript of Q, rev, indicates that the path is reversible. Actually, the quantity S in equation (1.1) is the change of entropy ΔS ($= S - S_0$)

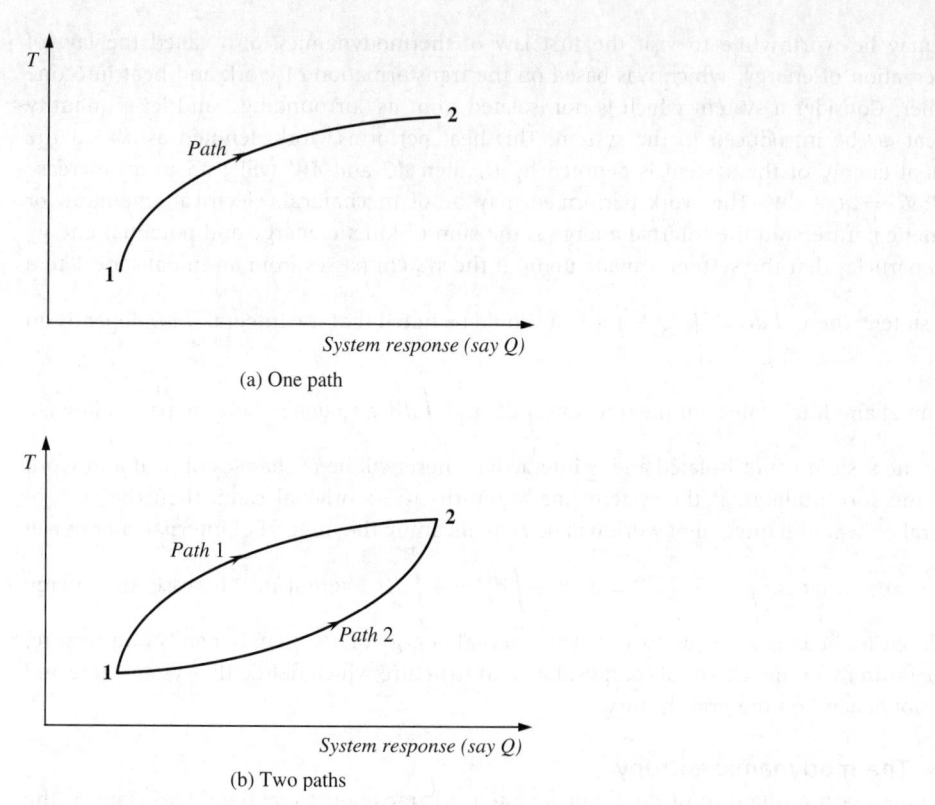

Figure 1.2 (a) Single path: transition from state 1 to state 2, and (b) two paths: transition from state 1 to state 2.

occurring in the transition from state 1 (corresponding to zero absolute temperature) to state 2. Equation (1.1) defines what Clausius termed thermodynamic entropy; it defines the second law of thermodynamics as the entropy increase law, and shows that the measurement of entropy of the system depends on the measurement of the quantities of heat, that is, calorimetry.

Equation (1.1) defines the experimental entropy given by Clausius in 1850. In this manner it is expressed as a function of macroscopic variables, such as temperature and pressure, and its numerical value can be measured up to a certain constant which is derived from the third law. Entropy S vanishes at the absolute zero of temperature. In 1865, while studying heat engines, Clausius discovered that although the total energy of an isolated system was conserved, some of the energy was being converted continuously to a form, such as heat, friction, and so on, and that this conversion was irrecoverable and was not available for any useful purpose; this part of the energy can be construed as energy loss, and can be interpreted in terms of entropy. Clausius remarked that the energy of the world was constant and the entropy of the world was increasing. Eddington called entropy the arrow of time.

The second law states that the entropy of a closed system always either increases or remains constant. A system can be as small as the piston, cylinder of a car (if one is trying to design a better car) or as big as the entire sky above an area (if one is attempting to predict weather). A closed system is thermally isolated from the rest of the environment and hence is a special kind of system. As an example of a closed system, consider a perfectly insulated cup of water in which a sugar cube is dissolved. As the sugar cube melts away into water, it would be logical to say that the water-sugar system has become more disordered, meaning its entropy has increased. The sugar cube will never reform to its original form at the bottom of the cup. However, that does not mean that the entropy of the water-sugar will never decrease. Indeed, if the system is made open and if enough heat is added to boil off the water, the sugar will recrystallize and the entropy will decrease. The entropy of open systems is decreased all the time, as for example, in the case of making ice in the freezer. It also occurs naturally in the case where rain occurs when disordered water vapor transforms to more ordered liquid. The same applies when it snows wherein one witnesses pictures of beautiful order in ice crystals or snowflakes. Indeed, sun shines by converting simple atoms (hydrogen) into more complex ones (helium, carbon, oxygen, etc.).

1.1.5 Evolutive connotation of entropy

Explaining entropy in the macroscopic world, Prigogine (1989) emphasized the evolutive connotation of entropy and laid out three conditions that must be satisfied in the evolutionary world: irreversibility, probability and coherence.

Irreversibility: Past and present cannot be the same in evolution. Irreversibility is related to entropy. For any system with irreversible processes, entropy can be considered as the sum of two components: one dealing with the entropy exchange with the external environment and the other dealing with internal entropy production which is always positive. For an isolated system, the first component is zero, as there is no entropy exchange, and the second term may only increase, reaching a maximum. There are many processes in nature that occur in one direction only, as for example, a house afire goes in the direction of ashes, a man going from the state of being a baby to being an old man, a gas leaking from a tank or air leaking from a car tire, food being eaten and getting transformed into different elements, and so on. Such events are associated with entropy which has a tendency to increase and are irreversible.

Entropy production is related to irreversible processes which are ubiquitous in water and environmental engineering. Following Prigogine (1989), entropy production plays a dual role. It does not necessarily lead to disorder, but may often be a mechanism for producing order. In the case of thermal diffusion, for example, entropy production is associated with heat flow which yields disorder, but it is also associated with anti-diffusion which leads to order. The law of increase of entropy and production of a structure are not necessarily opposed to each other. Irreversibility leads to a structure as is seen in a case of the development of a town or crop growth.

Probability: Away from equilibrium, systems are nonlinear and hence have multiple solutions to equations describing their evolution. The transition from instability to probability also leads to irreversibility. Entropy states that the world is characterized by unstable dynamical systems. According to Prigogine (1989), the study of entropy must occur on three levels: The first is the phenomenological level in thermodynamics where irreversible processes have a constructive role. The second is embedding of irreversibility in classical dynamics in which instability incorporates irreversibility. The third level is quantum theory and general relativity and their modification to include the second law of thermodynamics.

Coherence: There exists some mechanism of coherence that would permit an account of evolutionary universe wherein new, organized phenomena occur.

1.1.6 Statistical mechanical entropy

Statistical mechanics deals with the behavior of a system at the atomic scale and is therefore concerned with microstates of the system. Because the number of particles in the system is so huge, it is impractical to deal with the microstate of each particle, statistical methods are therefore resorted to; in other words, it is more important to characterize the distribution function of the microstates. There can be many microstates at the atomic scale which may be indistinguishable at the level of a thermodynamic state. In other words, there can be many possibilities of the realization of a thermodynamic state. If the number of these microstates is denoted by N, then statistical entropy is defined as

$$S = k \ln N \tag{1.2}$$

where k is Boltzmann constant (1.3806×10^{-16} erg/K or 1.3806×10^{-23} J/K (kg-m^2/s^2-K)), that is, the gas constant per molecule

$$k = \frac{R}{N_0} \tag{1.3}$$

where R is gas constant per mole (1.9872 cal/K), and N_0 is Avogadro's number (6.0221×10^{23} per mole). Equation (1.2) is also called Boltzmann entropy, and assumes that all microstates have the same probability of occurrence. In other words, in statistical mechanics the Boltzmann entropy is for the canonical ensemble. Clearly, S increases as N increases and its maximum represents the most probable state, that is, maximum number of possibilities of realization. Thus, this can be considered as a direct measure of the probability of the thermodynamic state. Entropy defined by equation (1.2) exhibits all the properties attributed to the thermodynamic entropy defined by equation (1.1).

Equation (1.2) can be generalized by considering an ensemble of systems. The systems will be in different microstates. If the number of systems in the i-th microstate is denoted by n_i then the statistical entropy of the i-th microstate is $S_i = k \log n_i$. For the ensemble the entropy is expressed as a weighted sum:

$$S = k \sum_{i=1}^{N} n_i \log n_i \tag{1.4a}$$

where N is the total number of microstates in which all systems are organized. Dividing by N, and expressing the fraction of systems by $p_i = n_i/N$, the result is the statistical entropy of the ensemble expressed as

$$S = -k \sum_{i=1}^{N} p_i \ln p_i \tag{1.4b}$$

where k is again Boltzmann's constant. The measurement of S here depends on counting the number of microstates. Equation (1.2) can be obtained from equation (1.4b), assuming the ensemble of systems is distributed over N states. Then $p_i = 1/N$, and equation (1.4b) becomes

$$S = -kN \frac{1}{N} \ln \frac{1}{N} = k \ln N \tag{1.5}$$

which is equation (1.2).

Entropy of a system is an extensive thermodynamic property, such as mass, energy, volume, momentum, charge, or number of atoms of chemical species, but unlike these quantities, entropy does not obey the conservation law. Extensive thermodynamic quantities are those that are halved when a system in equilibrium containing these quantities is partitioned into two equal parts, but intensive quantities remain unchanged. Examples of extensive variables include volume, mass, number of molecules, and entropy; and examples of intensive variables include temperature and pressure. The total entropy of a system equals the sum of entropies of individual parts. The most probable distribution of energy in a system is the one that corresponds to the maximum entropy of the system. This occurs under the condition of dynamic equilibrium. During evolution toward a stationary state, the rate of entropy production per unit mass should be minimum, compatible with external constraints. In thermodynamics entropy has been employed as a measure of the degree of disorderliness of the state of a system.

The entropy of a closed and isolated system always tends to increase to its maximum value. In a hydraulic system, if there were no energy loss the system would be orderly and organized. It is the energy loss and its causes that make the system disorderly and chaotic. Thus, entropy can be interpreted as a measure of the amount of chaos or disorder within a system. In hydraulics, a portion of flow energy (or mechanical energy) is expended by the hydraulic system to overcome friction, which then is dissipated to the external environment. The energy so converted is frequently referred to as energy loss. The conversion is only in one direction, that is, from available energy to nonavailable energy or energy loss. A measure of the amount of irrecoverable flow energy is entropy which is not conserved and which always increases, that is, the entropy change is irreversible. Entropy increase implies increase of disorder. Thus, the process equation in hydraulics expressing the energy (or head) loss can be argued to originate in the entropy concept.

1.2 Informational entropies

Before describing different types of entropies, let us further develop an intuitive feel about entropy. Since disorder, chaos, uncertainty, or surprise can be considered as different shades of information, entropy comes in handy as a measure thereof. Consider a random experiment with outcomes $x_1, x_2, \ldots, x_N$ with probabilities $p_1, p_2, \ldots, p_N$, respectively; one can say that these outcomes are the values that a discrete random variable X takes on. Each value of X, x_i, represents an event with a corresponding probability of occurrence, p_i. The probability p_i of event x_i can be interpreted as a measure of uncertainty about the occurrence of event x_i. One can also state that the occurrence of an event x_i provides a measure of information about the likelihood of that probability p_i being correct (Batty, 2010). If p_i is very low, say, 0.01, then it is reasonable to be certain that event x_i will not occur and if x_i actually occurred then there would be a great deal of surprise as to the occurrence of x_i with $p_i = 0.01$, because our anticipation of it was highly uncertain. On the other hand, if p_i is very high, say, 0.99, then it is reasonable to be certain that event x_i will occur and if x_i did actually occur then there would hardly be any surprise about the occurrence of x_i with $p_i = 0.99$, because our anticipation of it was quite certain.

Uncertainty about the occurrence of an event suggests that the random variable may take on different values. Information is gained by observing it only if there is uncertainty about the event. If an event occurs with a high probability, it conveys less information and vice versa. On the other hand, more information will be needed to characterize less probable or more uncertain events or reduce uncertainty about the occurrence of such an event. In a similar vein, if an event is more certain to occur, its occurrence or observation conveys less information and less information will be needed to characterize it. This suggests that the more uncertain an event the more information its occurrence transmits or the more information needed to characterize it. This means that there is a connection between entropy, information, uncertainty, and surprise.

It seems intuitive that one can scale uncertainty or its complement certainty or information, depending on the probability of occurrence. If $p(x_i) = 0.5$, the uncertainty about the occurrence would be maximum. It should be noted that the assignment of a measure of uncertainty should be based not on the occurrence of a single event of the experiment but of any event from the collection of mutually exclusive events whose union equals the experiment or collection of all outcomes. The measure of uncertainty about the collection of events is called entropy. Thus, entropy can be interpreted as a measure of uncertainty about the event prior to the experimentation. Once the experiment is conducted and the results about the events are known, the uncertainty is removed. This means that the experiment yields information about events equal to the entropy of the collection of events, implying uncertainty equaling information.

Now the question arises: What can be said about the information when two independent events x and y occur with probability p_x and p_y? The probability of the joint occurrence of x and y is $p_x p_y$. It would seem logical that the information to be gained from their joint occurrence would be the inverse of the probability of their occurrence, that is, $1/(p_x p_y)$. This shows that this information does not equal the sum of information gained from the occurrence of event x, $1/p_x$, and the information gained from the occurrence of event y, $1/p_y$, that is,

$$\frac{1}{p_x p_y} \neq \frac{1}{p_x} + \frac{1}{p_y} \tag{1.6}$$

This inequality can be mathematically expressed as a function $g(.)$ as

$$g\left(\frac{1}{p_x p_y}\right) = g\left(\frac{1}{p_x}\right) + g\left(\frac{1}{p_y}\right) \tag{1.7}$$

Taking g as a logarithmic function which seems to be the only solution, then one can express

$$-\log\left(\frac{1}{p_x p_y}\right) = -\log\left(\frac{1}{p_x}\right) - \log\left(\frac{1}{p_y}\right) \tag{1.8}$$

Thus, one can summarize that the information gained from the occurrence of any event with probability p is $\log(1/p) = -\log p$. Tribus (1969) regarded $-\log p$ as a measure of uncertainty of the event occurring with probability p or a measure of surprise about the occurrence of that event. This concept can be extended to a series of N events occurring with probabilities $p_1, p_2, \dots, p_N$, which then leads to the Shannon entropy to be described in what follows.

1.2.1 Types of entropies

There are several types of informational entropies (Kapur, 1989), such as Shannon entropy (Shannon, 1948), Tsallis entropy (Tsallis, 1988), exponential entropy (Pal and Pal, 1991a, b), epsilon entropy (Rosenthal and Binia, 1988), algorithmic entropy (Zurek, 1989), Hartley entropy (Hartley, 1928), Renyi's entropy (1961), Kapur entropy (Kapur, 1989), and so on. Of these the most important are the Shannon entropy, the Tsallis entropy, the Renyi entropy, and the exponential entropy. These four types of entropies are briefly introduced in this chapter and the first will be detailed in the remainder of the book.

1.2.2 Shannon entropy

In 1948, Shannon introduced what is now referred to as information-theoretic or simply informational entropy. It is now more frequently referred to as Shannon entropy. Realizing that when information was specified, uncertainty was reduced or removed, he sought a measure of uncertainty. For a probability distribution $P = \{p_1, p_2 \ldots, p_N\}$, where $p_1, p_2, \ldots, p_N$ are probabilities of N outcomes $(x_i, i = 1, 2, \ldots, N)$ of a random variable X or a random experiment, that is, each value corresponds to an event, one can write

$$-\log\left(\frac{1}{p_1 p_2 \cdots p_N}\right) = -\log\left(\frac{1}{p_1}\right) - \log\left(\frac{1}{p_2}\right) - \ldots - \log\left(\frac{1}{p_N}\right) \tag{1.9}$$

Equation (1.9) states the information gained by observing the joint occurrence of N events. One can write the average information as the expected value (or weighted average) of this series as

$$H = -\sum_{i=1}^{N} p_i \log p_i \tag{1.10}$$

where H is termed as entropy, defined by Shannon (1948).

The informational entropy of Shannon (1948) given by equation (1.10) has a form similar to that of the thermodynamic entropy given by equation (1.4b) whose development can be attributed to Boltzmann and Gibbs. Some investigators therefore designate H as Shannon-Boltzmann-Gibbs entropy (see Papalexiou and Koutsyiannis, 2012). In this text, we will call it the Shannon entropy. Equation (1.4b) or (1.10) defining entropy, H, can be re-written as

$$H(X) = H(P) = -K \sum_{i=1}^{N} p(x_i) \log[p(x_i)], \quad \sum_{i=1}^{N} p(x_i) = 1 \tag{1.11}$$

where $H(X)$ is the entropy of random variable $X: \{x_1, x_2, \ldots, x_N\}$, $P: \{p_1, p_2, \ldots p_N\}$ is the probability distribution of X, N is the sample size, and K is a parameter whose value depends on the base of the logarithm used. If different units of entropy are used, then the base of the logarithm changes. For example, one uses bits for base 2, Napier or nat or nit for base e, and decibels or logit or docit for base 10.

In general, K can be taken as unity, and equation (1.11), therefore, becomes

$$H(X) = H(P) = -\sum_{i=1}^{N} p(x_i) \log[p(x_i)] \tag{1.12}$$

$H(X)$, given by equation (1.12), represents the information content of random variable X or its probability distribution $P(x)$. It is a measure of the amount of uncertainty or indirectly the average amount of information content of a single value of X. Equation (1.12) satisfies a number of desiderata, such as continuity, symmetry, additivity, expansibility, recursivity, and others (Shannon and Weaver, 1949), and has the same form of expression as the thermodynamic entropy and hence the designation of H as entropy.

Equation (1.12) states that H is a measure of uncertainty of an experimental outcome or a measure of the information obtained in the experiment which reduces uncertainty. It also states the expected value of the amount of information transmitted by a source with probability distribution $(p_1, p_2, \ldots, p_N)$. The Shannon entropy may be viewed as the indecision of an observer who guesses the nature of one outcome, or as the disorder of a system in which different arrangements can be found. This measure considers only the possibility of occurrence of an event, not its meaning or value. This is the main limitation of the entropy concept (Marchand, 1972). Thus, H is sometimes referred to as the information index or the information content.

If X is a deterministic variable, then the probability that it will take on a certain value is one, and the probabilities of all other alternative values are zero. Then, equation (1.12) shows that $H(x) = 0$ which can be viewed as the lower limit of the values the entropy function may assume. This corresponds to the absolute certainty, that is, there is no uncertainty and the system is completely ordered. On the other hand, when all x_i s are equally likely, that is, the variable is uniformly distributed ($p_i = 1/N, i = 1, 2, \ldots, N$), that is, if all probabilities are equal, $p_i = p, i = 1, 2, \ldots, N$, then equation (1.12) yields

$$H(X) = H_{\max}(X) = \log N \tag{1.13}$$

This shows that the entropy function attains a maximum, and equation (1.13) thus defines the upper limit or would lead to the maximum entropy. This also reveals that the outcome has the maximum uncertainty. Equation (1.10) and in turn equation (1.13) show that the larger the number of events the larger the entropy measure. This is intuitively appealing because more information is gained from the occurrence of more events, unless, of course, events have zero probability of occurrence. The maximum entropy occurs when the uncertainty is maximum or the disorder is maximum.

One can now state that entropy of any variable always assumes positive values within the limits defined as:

$$0 \leq H(x) \leq \log N \tag{1.14}$$

It is logical to say that many probability distributions lie between these two extremes and their entropies between these two limits. As an example, consider a random variable X which takes on a value of 1 with a probability p and 0 with a probability $q = 1 - p$. Taking different values of p, one can plot $H(p)$ as a function of p. It is seen that for $p = 1/2$, $H(p) = 1$ bit is the maximum.

When entropy is minimum, $H_{\min} = 0$, the system is completely ordered and there is no uncertainty about its structure. This extreme case would correspond to the situation where $p_i = 1$, $p_j = 0$, $\forall j \neq i$. On the other hand, the maximum entropy $H_{\max}$ can be considered as a measure of maximum uncertainty and the disorder would be maximum which would occur if all events occur with the same probability, that is, there are no constraints on the system. This suggests that there is order-disorder continuum with respect to H; that is, more constraints on the form of the distribution lead to reduced entropy. The statistically most probable state corresponds to the maximum entropy. One can extend this interpretation further.

If there are two probability distributions with equiprobable outcomes, one given as above (i.e., $p_i = p, i = 1, 2, \ldots, N$), and the other as $q_i = q, i = 1, 2, \ldots, M$, then one can determine the difference in the information contents of the two distributions as $\Delta H = H_p - H_q = \log_2 p - \log_2 q = \log_2(p/q)$ bits, where H_p is the information content or entropy of $\{p_i, i = 1, 2, \ldots, N\}$ and H_q is the information content or entropy of $\{q_i, i = 1, 2, \ldots, M\}$. One can observe that if $q > p$ or $(M < N)$, $\Delta H > 0$. In this case the entropy increases or information is lost because of the increase in the number of possible outcomes or outcome uncertainty. On the other hand, if $q < p$ or $(M > N)$, then $\Delta H < 0$. This case corresponds to the gain in information because of the decrease in the number of possible outcomes or in uncertainty.

Comparing with H_{max}, a measure of information can be constructed as

$$I = H_{max} - H = \log n + \sum_{i=1}^{n} p_i \, \log p_i$$

$$= \sum_{i=1}^{n} p_i \, \log \left(\frac{p_i}{1/n} \right) = \sum_{i=1}^{n} p_i \, \log \left(\frac{p_i}{q_i} \right) \tag{1.15}$$

where $q_i = 1/n$. In equation (1.15), $\{q_i\}$ can be considered as a prior distribution and $\{p_i\}$ as a posterior distribution. Normalization of I by H_{max} leads to

$$R = \frac{I}{H_{max}} = 1 - \frac{H}{H_{max}} \tag{1.16}$$

where R is called the relative redundancy varying between 0 and 1.

In equation (1.12), the logarithm is to the base of 2, because it is more convenient to use logarithms to the base of 2, rather than logarithms to the base e or 10. Therefore, the entropy is measured in bits (short for binary digits). A bit can be physically interpreted in terms of the fraction of alternatives that are reduced by knowledge of some kind. These alternatives are equally likely. Thus, the amount of information depends on the fraction, not the absolute number of alternatives. This means that each time the number of alternatives is reduced to half based on some knowledge or one message, there will be a gain of one bit of information or the message has one bit of information. Consider there are four alternatives and this number is reduced to two, then one bit of information is transmitted. In the case of two alternative messages the amount of information $= \log_2 2 = 1$. This unit of information is called bit (as in binary system). The same amount of information is transmitted if 100 alternatives are reduced to 50, that is, $\log_2(100/50) = \log_2 2 = 1$. In general, one can express that $\log_2 x$ is bits of information transmitted or the message has if N alternatives are reduced to N/x. If 1000 alternatives are reduced to 500 (one bit of information is transmitted) and then 500 alternatives to 250 (another bit of information is transmitted), then $x = 4$, and $\log_2 4 = 2$ bits. Further, if one message reduces the number of alternatives N to N/x and another message reduces N to $N/2x$ then the former message has one bit less information than the latter. On the other hand, if one has eight alternative messages to choose from, then $\log_2 8 = \log_2 2^3 = 3$bits, that is, this case is associated with three bits of information or this defines the amount of information that can be determined from the number of alternatives to choose from. If one has 128 alternatives the amount of information is $\log_2 (2)^7 = 7$ bits.

The measurement of entropy is in nits (nats) in the case of natural logarithm (to the base e) and in logits (or decibles) with common logarithm. It may be noted that if $n^x = y$, then $x \log n = \log y$, meaning x is the logarithm of y to the base n, that is, $x \log_n n = \log_n y$. To be specific, the amount of information is measured by the logarithm of the number of choices. One can go from base b to base a as: $\log_b N = \log_b a \times \log_a N$.

From the above discussion it is clear that the value of H being one or unity depends on the base of the logarithm: bit (binary digit) for $\log_2$ and dit (decimal digit) for $\log_{10}$. Then one dit expresses the uncertainty of an experiment having ten equiprobable outcomes. Likewise, one bit corresponds to the uncertainty of an experiment having two equiprobable outcomes. If $p = 1$, then the entropy is zero, because the occurrence of the event is certain and there is no uncertainty as to the outcome of the experiment. The same applies when $p = 0$ and the entropy is zero.

In communication, each representation of random variable X can be regarded as a message. If X is a continuous variable (say, amplitude), then it would carry an infinite amount of information. In practice X is uniformly quantized into a finite number of discrete levels, and then X may be regarded as a discrete variable:

$$X = \{x_i, i = 0, \pm 1, \ldots, \pm N\} \tag{1.17}$$

where x_i is a discrete number, and $(2N + 1)$ is the total number of discrete levels. Then, random variable X, taking on discrete values, produces a finite amount of information.

1.2.3 Information gain function

From the above discussion it would intuitively seem that the gain in information from an event is inversely proportional to its probability of occurrence. Let this gain be represented by $G(p)$ or ΔI. Following Shannon (1948),

$$G(p) = \Delta I = \log\left(\frac{1}{p_i}\right) = -\log(p_i) \tag{1.18}$$

where $G(p)$ is the gain function. Equation (1.18) is a measure of that gain in information or can be called as gain function (Pal and Pal, 1991a). Put another way, the uncertainty removed by the message that the event i occurred or the information transmitted by it is measured by equation (1.18). The use of logarithm is convenient, since the combination of the probabilities of independent events is a multiplicative relation. Thus, logarithms allow for expressing the combination of their entropies as a simple additive relation. For example, if $P(A \cap B) = P_A P_B$, then $H(AB) = -\log P_A - \log P_B = H(A) + H(B)$. If the probability of an event is very small, say $p_i = 0.01$, then the partial information transmitted by this event is very large $\Delta I = 2$ dits if the base of the logarithm is taken as 10; such an outcome will not occur in the long run. If there are N events, one can compute the total gain in information as

$$I = \sum_{i=1}^{N} \Delta I_i = -\sum_{i=1}^{N} \log(p_i) \tag{1.19}$$

Each event occurs with a different probability.

The entropy or global information of an event i is expressed as a weighted value:

$$H(p_i) = -p_i \log p_i \tag{1.20}$$

Since $0 \le p_i \le 1$, H is always positive. Therefore, the average or expected gain in information can be obtained by taking the weighted average of individual gains of information:

$$H = E(\Delta I) = -\sum_{i=1}^{N} p_i(\Delta I_i) = -\sum_{i=1}^{N} p_i \log p_i \tag{1.21}$$

which is the same as equation (1.10) or (1.12). What is interesting to note here is that one can define different types of entropy by simply defining the gain function or uncertainty differently. Three other types of entropies are defined in this chapter.

Equation (1.21) can be viewed in another way. Probabilities of outcomes of an experiment correspond to the partitioning of space among outcomes. Because the intersection of outcomes is empty, the global entropy of the experiment is the sum of elementary entropies of the N outcomes:

$$H = H_1 + H_2 + \ldots + H_N = \sum_{i=1}^{N} H_i \tag{1.22a}$$

$$= -p_1 \log p_1 - p_2 \log p_2 - \ldots - p_N \log p_N = -\sum_{i=1}^{N} p_i \log p_i \tag{1.22b}$$

which is the same as equation (1.21). Clearly, H is maximum when all outcomes are equiprobable, that is, $p_i = 1/N$. This has an important application in hydrology, geography, meteorology, and socio-economic and political sciences. If a topology of data measured on nominal scales has classes possessing the same number of observations then it will transmit the maximum amount of information (entropy). This condition is not entirely true if by computing distances between elements one can minimize intra-class variance and maximize inter-class variance. This would lead to distributions with a smaller entropy but a higher variance value (Marchand, 1972).

Example 1.1: Plot the gain function defined by equation (1.18) for different values of probability: 0.1, 0.2, 0.3, 0.4, 0.5, 0.6, 0.7, 0.8, 0.9, and 1.0. Take the base of logarithm as 2 as well as e. What do you conclude from this plot?

Solution: The gain function is plotted in Figure 1.3. It is seen that the gain function decreases as the probability of occurrence increases. Indeed the gain function becomes zero when the probability of occurrence is one. For lower logarithmic base, the gain function is higher, that is, the gain function with logarithmic base of 2 is higher than that with logarithmic base e.

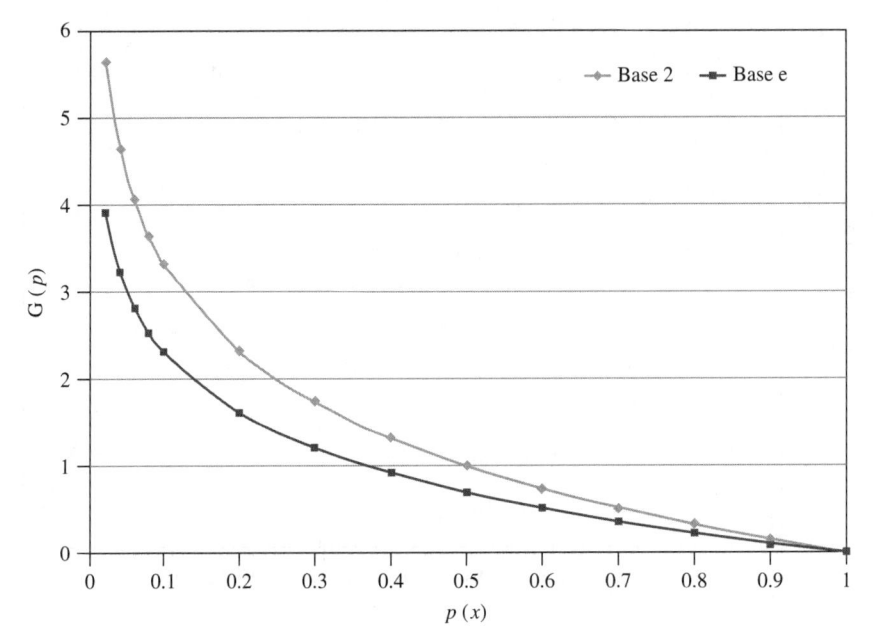

Figure 1.3 Plot of Shannon's gain function.

Example 1.2: Consider a two-state variable taking on values x_1 or x_2. Assume that $p(x_1) = 0.0$, 0.1, 0.2, 0.3, 0.4, 0.5, 0.6, 0.7, 0.8, 0.9, 1.0. Note that $p(x_2) = 1 - p(x_1) = $ 1.0, 0.9, 0.8, 0.7, 0.6, 0.5, 0.4, 0.3, 0.2, 0.1, and 0.0. Compute and plot the Shannon entropy. Take the base of the logarithm as 2 as well as e. What do you conclude from the plot?

Solution: The Shannon entropy for a two-state variable is plotted as a function of probability in Figure 1.4. It is seen that entropy increases with increasing probability up to the point where the probability becomes 0.5 and then decreases with increasing probability, reaching zero when the probability becomes one. A higher logarithmic base produces lower entropy and vice versa, that is, the Shannon entropy is greater for logarithmic base 2 than it is for logarithmic base e. Because of symmetry, $H(X_1) = H(X_2)$ and therefore graphs will be the same.

1.2.4 Boltzmann, Gibbs and Shannon entropies

Using theoretical arguments Gull (1991) has explained that the Gibbs entropy is based on the ensemble which represents the probability that an N-particle system is in a particular microstate and making inferences given incomplete information. The Boltzmann entropy is based on systems each with one particle. The Gibbs entropy, when maximized (i.e., for the canonical ensemble), results numerically in the thermodynamic entropy defined by Clausius. The Gibbs entropy is defined for all probability distributions, not just for the canonical ensemble. Therefore,

$$S_G \leq S_E$$

where S_G is the Gibbs entropy, and S_E is the experimental entropy. Because the Boltzmann entropy is defined in terms of the single particle distribution, it ignores both the internal energy and the effect of inter-particle forces on the pressure. The Boltzmann entropy becomes

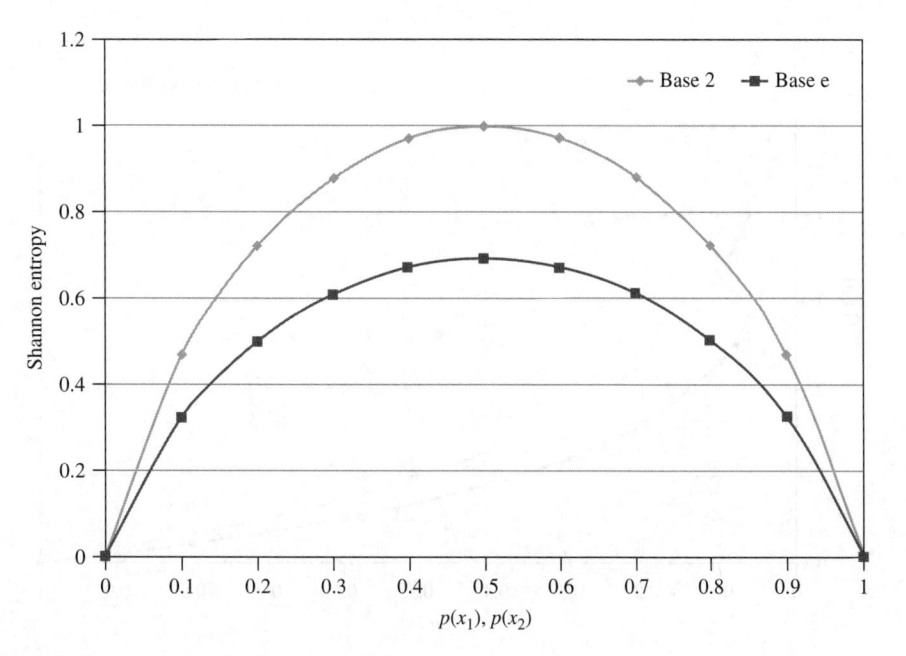

Figure 1.4 Shannon entropy for two-state variables.

the same as the Clausius entropy only for a perfect gas, when it also equals the maximized Gibbs entropy.

It may be interesting to compare the Shannon entropy with the thermodynamic entropy. The Shannon entropy provides a measurement of information of a system, and increasing of this information implies that the system has more information. In the canonical ensemble case, the Shannon entropy and the thermodynamic entropy are approximately equal to each other. Ng (1996) distinguished between these two entropies and the entropy for the second law of thermodynamics, and expressed the total entropy S of a system at a given state as

$$S = S_1 + S_2 \tag{1.23}$$

where S_1 is the Shannon entropy and S_2 is the entropy for the second law. The increasing of S_2 implies that the entropy of an isolated system increases as regarded by the second law of thermodynamics, and that the system is in decay. S_2 increases when the total energy of the system is constant, the dissipated energy increases and the absolute temperature is constant or decreases. From the point of view of living systems, the Shannon entropy (or thermodynamic entropy) is the entropy for maintaining the complex structure of living systems and their evolution. The entropy for the second law is not the Shannon entropy. Zurek (1989) defined physical entropy as the sum of missing information (Shannon entropy) and of the length of the most concise record expressing the information already available (algorithmic entropy), which is similar to equation (1.23). Physical entropy can be reduced by a gain of information or as a result of measurement.

1.2.5 Negentropy

The Shannon entropy is a statistical measure of dispersion in a set organized through an equivalent relation, whereas the thermodynamic entropy in a system is proportional to its ability to work, as discussed earlier. The second law of thermodynamics or Carnot's second principle is the degradation of energy from a superior level (electrical and mechanical energy) to a midlevel (chemical energy) and to an inferior level (heat energy). The difference in the nature and repartition of energy is measured by the physical energy. For example, if a system experiences an increase in heat, dQ, the corresponding increase in entropy dS can be expressed as

$$dS = \frac{dQ}{T} \tag{1.24}$$

where T is the absolute temperature, and S is the thermodynamic entropy.

Carnot's first principle of energy, conservation of energy, is

$$W - Q = 0 \tag{1.25}$$

and the second principle states

$$dS \geq 0 \tag{1.26}$$

where W is the work produced or output. This shows that entropy must always increase. Any system in time tends towards a state of perfect homogeneity (perfect disorder) where it is incapable of producing any more work, providing there are no internal constraints. The Shannon entropy in this case attains the maximum value. However, this is exactly the opposite of that in physics in that it is defined by Maxwell (1872) as follows: "Entropy of a system is the mechanical work it can perform without communication of heat or change of volume. When the temperature and pressure have become constant, the entropy of the system is exhausted."

Brillouin (1956) reintroduced the Maxwell entropy while conserving the Shannon entropy as negentropy: "An isolated system contains negentropy if it reveals a possibility for doing a mechanical or electrical work. If a system is not at a uniform temperature, it contains a certain amount of negentropy." Thus, Marchand (1972) reasoned that entropy means homogeneity and disorder, and negentropy means heterogeneity and order in a system:

Negentropy = −entropy

Entropy is always positive and attains a maximum value, and therefore negentropy is always negative or zero, and its maximum value is zero. Note that the ability of a system to perform work is not measured by its energy, since energy is constant, but by its negentropy. For example, a perfectly disordered system, with a uniform temperature contains a certain amount of energy but is incapable of producing any work because its entropy is maximum and its negentropy is minimum. It may be concluded that information (disorder) and negentropy (order) are interchangeable. Acquisition of information translates into an increase of entropy and decrease of negentropy; likewise decrease of entropy translates into increase of negentropy. One cannot observe a phenomenon without altering it and the information acquired through an observation is always slightly smaller than the disorder it introduces into the system. This implies that a system cannot be exactly reconstructed as it was before the observation was made. Thus, the relation between the information and entropy S in thermodynamics is: $S = k \log N$, k = Boltzmann's constant (1.3806×10^{-16} erg/K), and N = number of microscopic configurations of the system. The very small value of k means that a very small change in entropy corresponds to a huge change in information and vice versa.

Sugawara (1971) used negentropy as a measure of order in discussing problems in water resources. For example, in the case of hydropower generation, the water falls down and its potential energy is converted into heat energy and then into electrical energy. The hydropower station utilizes the negentropy of water. Another example is river discharge, which, with large fluctuations, has low negentropy or the smaller the fluctuation the higher the negentropy. In the case of a water treatment plant, input water is dirty and output water is clear or clean, meaning an increase in negentropy. Consider an example of rainwater distributed in time and space. The rainwater is in a state of low negentropy. Then, rainwater infiltrates and becomes groundwater and runoff from this groundwater becomes baseflow. This is in a state of high negentropy achieved in exchange of lost potential energy. The negentropy of a system can conserve entropy of water resources.

1.2.6 Exponential entropy

If the gain in information from an event occurring with probability p_i is defined as

$$G(p) = \Delta I = \exp[(1 - p_i)] \tag{1.27a}$$

then the exponential entropy, defined by Pal and Pal (1991a), can be expressed as

$$H = E(\Delta I) = \sum_{i=1}^{N} p_i \exp[(1 - p_i)] \tag{1.27b}$$

The entropy, defined by equation (1.27b), possesses some interesting properties. For example, following Pal and Pal (1991a), equation (1.27b) is defined for all p_i between 0 and 1, is continuous in this interval, and possesses a finite value. As p_i increases, ΔI decreases exponentially. Indeed, H given by equation (1.27b) is maximum when all p_i's are equal. Pal and Pal (1992) have mathematically proved these and other properties. If one were to plot the exponential entropy, the plot would be almost identical to the Shannon entropy. Pal and Pal (1991b) and Pal and Bezdek (1994) have used the exponential entropy in pattern recognition, image extraction, feature evaluation, and image enhancement and thresholding.

Example 1.3: Plot the gain function defined by equation (1.27a) for different values of probability: 0.1, 0.2, 0.3, 0.4, 0.5, 0.6, 0.7, 0.8, 0.9, and 1.0. What do you conclude from this plot? Compare this plot with that in Example 1.1. How do the two gain functions differ?

Solution: The gain function is plotted as a function of probability in Figure 1.5. It is seen that as the probability increases, the gain function decreases, reaching the lowest value of one when the probability becomes unity. Comparing Figure 1.5 with Figure 1.3, it is observed that the exponential gain function changes more slowly and has a smaller range of variability than does the Shannon gain function.

Example 1.4: Consider a two-state variable taking on values x_1 or x_2. Assume that $p(x_1) = 0.0,$ 0.1, 0.2, 0.3, 0.4, 0.5, 0.6, 0.7, 0.8, 0.9, and 1.0. Note that $p(x_2) = 1 - p(x_1) = 1.0,$ 0.9, 0.8, 0.7, 0.6, 0.5, 0.5, 0.4, 0.3, 0.2, 0.1, and 0.0. Compute and plot the exponential entropy. What do you conclude from the plot? Compare the exponential entropy with the Shannon entropy.

Solution: The exponential entropy is plotted in Figure 1.6. It increases with increasing probability, reaching a maximum value when the probability becomes 0.5 and then declines, reaching a minimum value of one when the probability becomes 1.0. The pattern of variation of the exponential entropy is similar to that of the Shannon entropy. For any given probability

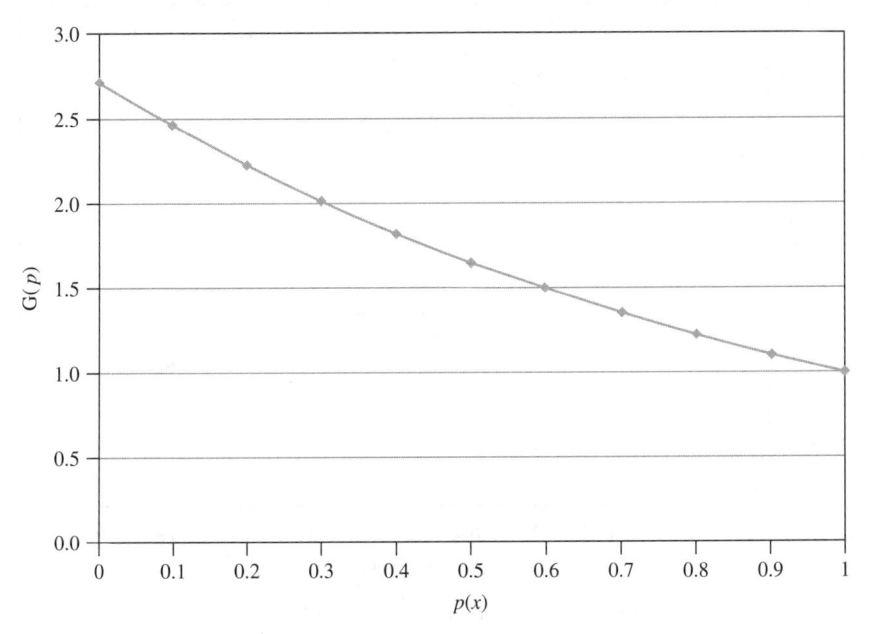

Figure 1.5 Plot of gain function of exponential entropy as defined by equation (1.27a).

value, the exponential entropy is higher than the Shannon entropy. Note that $H(X_1) = H(X_2)$; therefore graphs will be identical for X_1 and X_2.

1.2.7 Tsallis entropy

Tsallis (1988) proposed another formulation for the gain in information from an event occurring with probability p_i as

$$G(p) = \Delta I = \frac{k}{q-1}[(1 - p_i^{q-1})]$$ (1.28)

where k is a conventional positive constant, and q is any number. Then the Tsallis entropy can be defined as the expectation of the gain function in equation (1.28):

$$H = E(\Delta I) = \frac{k}{q-1}\sum_{i=1}^{N}p_i[(1 - p_i^{q-1})]$$ (1.29)

Equation (1.29) shows that H is greater than or equal to zero in all cases. This can be considered as a generalization of the Shannon entropy or Boltzmann–Gibbs entropy. The Tsallis entropy has some interesting properties. Equation (1.29) achieves its maximum when all probabilities are equal. It vanishes when $N = 1$; as well as when there is only one event with probability one and others have vanishing probabilities. It converges to the Shannon entropy when q tends to unity.

Example 1.5: Plot the gain function defined by equation (1.18) for different values of probability: 0.1, 0.2, 0.3, 0.4, 0.5, 0.6, 0.7, 0.8, 0.9, and 1.0. Take k as 1, and q as −1, 0,

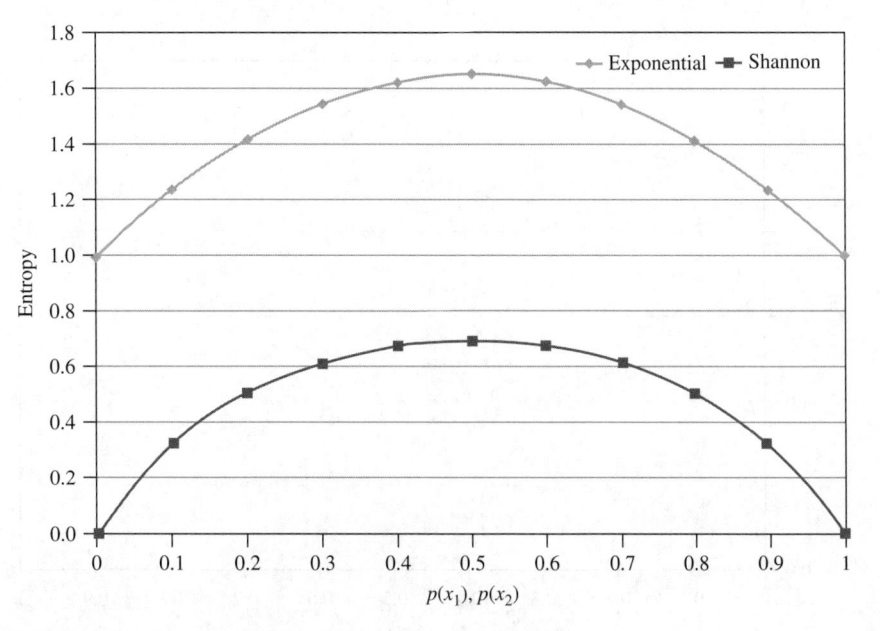

Figure 1.6 Plot of exponential and Shannon entropy for two-state variables.

1.1, and 2. What do you conclude from this plot? Compare the gain function with the gain functions obtained in Examples 1.1 and 1.3.

Solution: The Tsallis gain function is plotted in Figure 1.7. It is seen that the gain function is highly sensitive to the value of q. For $q = 1.1$, and $q = 2$, the gain function is almost zero; for $q = -1$, and 0, it declines rapidly with increasing probability – indeed it reaches a very small value when the probability is about 0.5 or higher. Its variation is significantly steeper than the Shannon and exponential gain functions, and its pattern of variation is also quite different.

Example 1.6: Consider a two-state variable taking on values x_1 or x_2. Assume that $p(x_1) = 0.0, 0.1, 0.2, 0.3, 0.4, 0.5, 0.6, 0.7, 0.8, 0.9, 1.0$. Note that $p(x_2) = 1 - p(x_1) = 1.0, 0.9, 0.8, 0.7, 0.6, 0.5, 0.4, 0.3, 0.2, 0.1$, and 0.0. Compute and plot the Tsallis entropy. Take q as 1.5 and 2.0. What do you conclude from the plot?

Solution: The Tsallis entropy is plotted in Figure 1.8. It increases with increasing probability reaching a maximum value at the probability of about 0.6 and then declines with increasing probability. The Tsallis entropy is higher for $q = 1.5$ than it is for $q = 2.0$.

1.2.8 Renyi entropy
Renyi (1961) defined a generalized form of entropy called Renyi entropy which specializes into the Shannon entropy, Kapur entropy, and others. Recall that the amount of uncertainty or the entropy of a probability distribution $P = (p_1, p_2, \ldots, p_n)$, where $p_i \geq 0$ and $\sum_{i=1}^{n} p_i = 1$, denotes the amount of uncertainty as regards the outcome of an experiment whose values have probabilities $p_1, p_2, \ldots, p_n$, measured by the quantity $H(p) = H(p_1, p_2, \ldots, p_n)$. $H(p, 1 - p)$ is a continuous function of p, $0 \leq p \leq 1$. Following Renyi (1961), one can also write: $H(wp_1, (1 - w)p_1, p_2, \ldots, p_n) = H(p_1, p_2, \ldots, p_n) + p_1 H(w, 1 - w)$ for $0 \leq w \geq 1$.

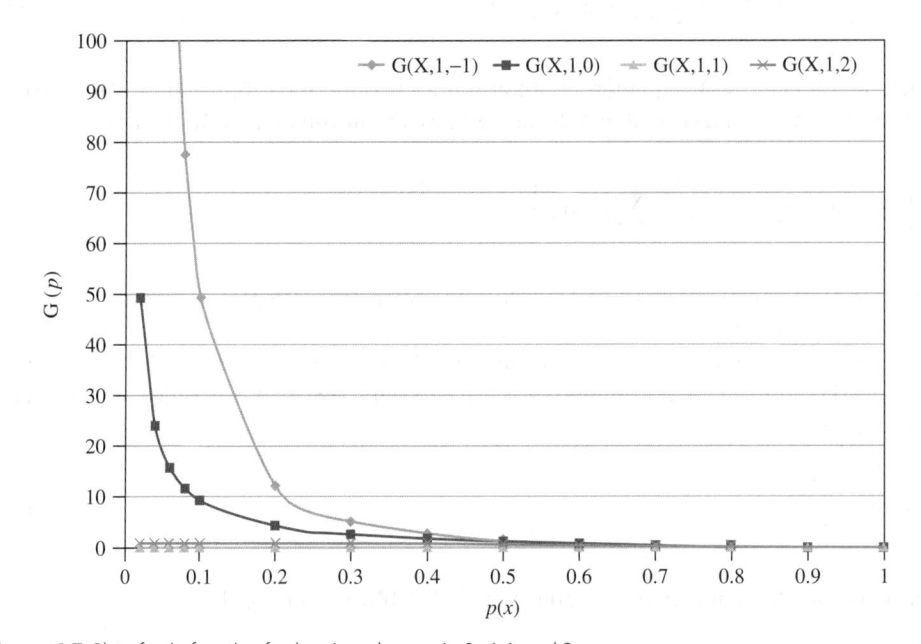

Figure 1.7 Plot of gain function for $k = 1$, and $q = -1$, 0, 1.1, and 2.

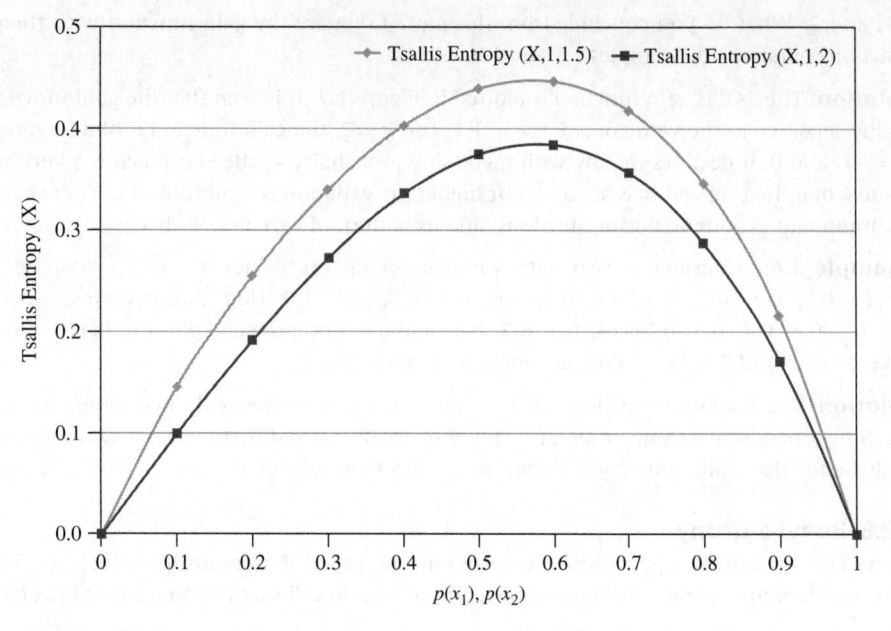

Figure 1.8 Plot of Tsallis entropy for two state variables with $q = 1.5$ and 2.

Renyi (1961) expressed

$$H_\alpha(p_1, p_2, \ldots, p_n) = \frac{1}{1-\alpha} \log_2 \left(\sum_{i=1}^{n} p_i^\alpha \right)$$ (1.30)

where $\alpha > 0$ and $\alpha \neq 1$. Equation (1.30) also is a measure of entropy and can be called the entropy of order α of distribution P. It can be shown from equation (1.30) that

$$\lim_{\alpha \to 1} H_\alpha(p_1, p_2, \ldots, p_n) = \sum_{i=1}^{n} p_i \log \frac{1}{p_i}$$ (1.31)

which is the same as equation (1.12). Thus, the Shannon entropy is a limiting case of the Renyi entropy given by equation (1.30) for $\alpha \to 1$.

Let $W(P)$ be the weight of the distribution P, $0 < W(P) < 1$. The weight of an ordinary distribution is 1. A distribution which has weight less than 1 is called an incomplete distribution:

$$W(P) = \sum_{i=1}^{n} p_i$$ (1.32)

For two generalized distributions P and Q, such that $W(P) + W(Q) \leq 1$,

$$H(P \cup Q) = \frac{W(P)H(P) + W(Q)H(Q)}{W(P) + W(Q)}$$ (1.33)

This is called the mean value property of entropy; the entropy of the union of two incomplete distributions is the weighted mean value of the entropies of the two distributions, where the entropy of each component is weighted by its own weight. This can be generalized as

$$H(P_1 \cup P_2 \cup P_3 \ldots \cup P_n) = \frac{W(P_1)H(P_1) + W(P_2)H(P_2) + \ldots W(P_n)H(P_n)}{W(P_1) + W(P_2) + \ldots W(P_n)} \tag{1.34}$$

Any generalized $P\{p_1, p_2, \ldots, p_n\}$ can be written as

$$P = \{p_1\} \cup \{p_2\} \cup \ldots \cup \{p_n\} \tag{1.35}$$

Thus, Renyi (1961) defined an entropy as

$$H(X) = \frac{1}{1-a} \log \frac{\sum_{i=1}^{N} p_i^a}{\sum_{i=1}^{N} p_i}, \ a \neq 1, \ a > 0 \tag{1.36}$$

This is an entropy of order a of the generalized distribution P. As $a \to 1$, equation (1.36) converges to the Shannon entropy. Thus, the Shannon entropy can be considered as a limiting case of Reny's entropy. The Kapur entropy is a further generalization of the Renyi entropy as

$$H(X) = \frac{1}{1-a} \ln \frac{\sum_{i=1}^{N} p_i^a}{\sum_{i=1}^{N} p_i^b}, \ a \neq 1, \ b > 0, \ a + b - 1 > 0 \tag{1.37}$$

If $b = 1$, equation (1.37) reduces to the Renyi entropy. For $b = 1$, and $a = 0$, equation (1.37) reduces to log N, if $p_i = 1/N$, which is Hartley's measure (Hartley, 1928).

1.3 Entropy, information, and uncertainty

Consider a discrete random variable $X : \{x_1, x_2, \ldots, x_N\}$ with a probability distribution $P(x) = \{p_1, p_2, \ldots, p_N\}$. When the variable is observed to have a value x_i, the information is gained; the amount of information I_i so gained is defined as the magnitude of the logarithm of the probability:

$$I_i = -\log p_i = |\log p_i|$$

One may ask the question: How much uncertainty was there about the variable before observation? The question is answered by linking uncertainty to information. The amount of uncertainty can be defined as the average amount of information expected to be gained by observation. This expected amount of information is referred to as the entropy of the distribution

$$H = \sum_{i=1}^{N} p_i I_i = -\sum_{i=1}^{N} p_i \log p_i = \sum_{i=1}^{N} p_i |\log p_i|$$

This entropy of a discrete probability distribution denotes the average amount of information expected to be gained from observation. Once a value of the random variable X has been observed, the variable has this observed value with probability one. Then, the entropy of the new conditional distribution is zero. However, this will not be true if the variable is continuous.

1.3.1 Information

The term "information" is variously defined. In Webster's *International Dictionary*, definitions of "information" encompass a broad spectrum from semantic to technical, including "the communication or reception of knowledge and intelligence," "knowledge communicated by others and/or obtained from investigation, study, or instruction," "facts and figures ready for communication or use as distinguished from those incorporated in a formally organized branch of knowledge, data," "the process by which the form of an object of knowledge is impressed upon the apprehending mind so as to bring about the state of knowing," and "a numerical quantity that measures the uncertainty in outcome of an experiment to be performed." The last definition is an objective one and indeed corresponds to the informational entropy. Semantically, information is used intuitively, that is, it does not correspond to a well-defined numerical quantity which can quantify the change in uncertainty with change in the state of the system. Technically, information corresponds to a well-defined function which can quantify the change in uncertainty. This technical aspect is pursued in this book. In particular, the entropy of a probability distribution can be considered as a measure of uncertainty and also a measure of information. The amount of information obtained when observing the result of an experiment can be considered numerically equal to the amount of uncertainty as regards the outcome of the experiment before performing it. Perhaps the earliest definition of information was provided by Fisher (1921) who used the inverse of the variance as a measure of information contained in a distribution about the outcome of a random draw from that distribution.

Following Renyi (1961), another amount of information can be expressed as follows. Consider a random variable X. An event E is observed which in some way is related to X. The question arises: What is the amount of information concerning X? To answer this question, let P be the probability (original, unconditional) distribution of X, and Q be the conditional distribution of X, subject to the condition that event E has taken place. A measure of the amount of information concerning the random variable X contained in the observation of event E can be denoted by $I(Q|P)$, where Q is absolutely continuous with respect to P. If $h = dQ/dP$, the Radon-Nikodym derivative of Q with respect to P, then a possible measure of the amount of information in question can be written as:

$$I_i(Q|P) = \int h \, \log_2 \, dQ = \int h \, \log_2 \, h dP \tag{1.38}$$

Assume X takes on a finite number of values: $X : \{x_1, \ x_2, \ \dots, x_n\}$ If $P(X = x_i) = p_i$ and $P(X = x_i|E) = q_i$, for $i = 1, \ 2, \ \dots, \ n$, then equation (1.38) becomes

$$I_1(Q|P) = \sum_{i=1}^{n} q_i \, \log_2 \, \frac{q_i}{p_i} \tag{1.39}$$

Also,

$$I_\alpha(Q|P) = \frac{1}{\alpha - 1} \, \log_2 \left(\sum_{i=1}^{n} \frac{q_i^\alpha}{p_i^{\alpha-1}} \right) \tag{1.40}$$

For $\alpha \to 1$,

$$\lim_{\alpha \to 1} I_\alpha(Q|P) = I_1(Q|P) \tag{1.41}$$

This measures the amount of information contained in the observation of event E with respect to the random variable X, or the information of order α obtained if the distribution P is replaced by distribution Q.

If $I(Q_1|P_1)$ and $I(Q_2|P_2)$ are defined, and $P = P_1 P_2$ and $Q = Q_1 Q_2$ and the correspondence between the elements of P and Q is that introduced by the correspondence between the elements of P_1 and Q_1 and those of P_2 and Q_2, then

$$I(Q|P) = I(Q_1|P_1) + I(Q_2|P_2) \tag{1.42}$$

If

$$W(P_1) + W(P_2) \le 1 \text{ and } W(Q_1) + W(Q_2) \le 1 \tag{1.43}$$

then

$$I(Q_1 U Q_2 | P_1 U P_2) = \frac{W(Q_1) I(Q_1|P_1) + W(Q_2) I(Q_2|P_2)}{W(Q_1) + W(Q_2)} \tag{1.44}$$

The entropies can be generalized as

$$I_1(Q|P) = \frac{\sum_{i=1}^{n} q_i \log_2 \frac{qi}{pi}}{\sum_{i=1}^{n} q_i} \tag{1.45}$$

Likewise,

$$I_\alpha(Q|P) = \frac{1}{\alpha - 1} \log_2 \left[\frac{\sum_{i=1}^{n} \frac{q_i^\alpha}{p_i^{\alpha-1}}}{\sum_{i=1}^{n} q_i} \right] \tag{1.46}$$

If P and Q are complete distributions then equation (1.45) will reduce to equation (1.39) and equation (1.46) to equation (1.40).

Information is a measure of one's freedom of choice when selecting an alternative or a message. Thus, it should not be confused with the meaning of the message. For example, two messages, one filled with meaning and the other with nonsense can be equivalent. Information relates not so much to what one does say as to what one could say. If there are two alternative messages and one has to choose one message then it is arbitrarily stated that the information associated with this case is unity which indicates the amount of freedom one has in selecting a message. Thus, the concept of information applies to the whole situation, not to individual messages. The messages can be anything one likes.

The measure of information is entropy. Entropy is a measure of randomness or shuffledness. Physical systems tend to become more and more shuffled, less and less organized. If a system is highly organized and it is not characterized by a large degree of randomness, then its information (entropy) is low.

If H is zero ($p_i = 1$, certainty) and ($p_j = 0$, $j \neq i$ impossibility) then information is zero and there is no freedom of choice. When one is completely free, H is maximum and reduces to zero when the freedom of choice is gone. Thus, H increases with the increasing number of alternatives or by equiprobability of alternatives if the number of alternatives is fixed. There is more information if the number of alternatives to choose from is more.

Entropy $H(X)$ permits to measure information and for that reason it is also referred to as informational entropy. Intuitively, information reduces uncertainty which is a measure of surprise. Thus, information I is a reduction in uncertainty $H(X)$ and can be defined as

$$I = H_I - H_O \tag{1.47}$$

where H_I is the entropy (or uncertainty) of input (or message sent through a channel), and H_O is the entropy (or uncertainty) of output (or message received). Equation (1.47) defines a reduction in uncertainty. Consider an input-output channel or transmission conduit. Were there no noise in the conduit, the output (the message received by the receiver or receptor) would be certain as soon as the input (message sent by the emitter) was known. This means that the uncertainty in output H_O would be 0 and $I = H_I$.

1.3.2 Uncertainty and surprise

The concept of information is closely linked with the concept of uncertainty or surprise. The quantity $- \log (1/p_i)$ can be used to denote surprise or unexpectedness (Watanabe, 1969). When all probabilities are equal, it is impossible to state that one possibility is more likely than another. This means there is complete uncertainty. Any information about the nature of an event under such conditions can be expected to shed more light than in any other condition. Maximum entropy is therefore a measure of complete uncertainty. Maximum uncertainty can be equated with a condition in which the expected information from actual events is also maximized. Now assume that $X = x_i$ and it occurs with probability one, $p_i = 1$; that is, the event occurs with certainty and hence there is no uncertainty. This means that $p_j = 0$, $j \neq i$. In this case, there is no surprise and therefore the occurrence of event $X = x_i$ conveys no information, since it is known what the event must be. One can state that the information content of observing x_i or the anticipatory uncertainty of x_i prior to the observation is a decreasing function of the probability $p(x_i)$. The more likely the occurrence of x_i, the less information its actual observation contains.

If x_i's occur with probabilities p_i's, $p_i \neq p_j$, $i = j = 0$, $\pm 1, \ldots, \pm N$, then there is more surprise and therefore more information that $X = x_i$ occurs with probability p_i than does $X = x_j$ with probability p_j where $p_j > p_i$. Thus, information, uncertainty and surprise are all related. Information is gained only if there is uncertainty about an event. Uncertainty suggests that the event may take on different values. The value that occurs with a higher probability conveys less information and vice versa. The probability of occurrence of a certain value is the measure of uncertainty or the degree of expectedness and hence of information. Shannon (1948) argued that entropy is the expected value of the probabilities of alternative values that an event may take on. The information gained is indirectly measured as the amount of reduction of uncertainty or of entropy.

The above discussion suggests that uncertainty can be understood to be a form of information deficiency or reflects information reduction, which may be because information is unreliable, biased, contradictory, vague, incomplete, imprecise, erroneous, fragmentary, or unfounded.

In many cases, information deficiency can be reduced and hence uncertainty. Consider, for example, prediction of a 100-year flood from a 20-year record. This prediction has uncertainty, say, u_1 (it can be referred to as a priori uncertainty). If the record length is increased to 50 years, the prediction will have less uncertainty, say u_2 (it can be referred to as posteriori uncertainty). The reduction in uncertainty due to a more complete record (or an action) is $u_1 - u_2$ which is equal to the information gain, that is, this is the amount of information realized as a result of uncertainty reduction. Klir (2006) refers to this uncertainty as uncertainty-based information, and reasons that this type of information does not encompass the concept of uncertainty in its entirety and is hence restricted somewhat. On the other hand, information is understood to reduce uncertainty or reflects uncertainty reduction. Klir (2006) calls this an information-based uncertainty.

1.4 Types of uncertainty

Uncertainty can appear in different forms. It can appear in both probabilistic and deterministic phenomena. In deterministic phenomena, it appears as a result of fuzziness about the phenomena, in data or in relations about the variables, and can be dealt with using the fuzzy set theory (plausibility, possibility, and feasibility). Probabilistic uncertainty is associated with the probability of outcomes and is entropy. This is also linked with arrow of time, meaning that it increases from past to present to future.

In environmental and water resources engineering models which express relations among states of given variables are constructed for a variety of purposes, including prediction, retrodiction, forecasting, diagnosis, prescription, planning, scheduling, control, simulation, detection, estimation, extrapolation, and design. Each of these purposes is subject to uncertainty. Depending on the purpose, unknown states of some variables are determined from the known states of other variables, using appropriate relation(s). If the relation is unique, the model is deterministic; otherwise it is nondeterministic and involves uncertainty. The uncertainty relates to the purpose for the construction of the model, and can thus be distinguished as predictive uncertainty, retrodictive uncertainty, forecasting uncertainty, diagnostic uncertainty, prescriptive uncertainty, planning uncertainty, scheduling uncertainty, control uncertainty, simulation uncertainty, detection uncertainty, estimation uncertainty, extrapolation uncertainty, and design uncertainty. It is logical that this uncertainty is incorporated into the model description. A decision is an action from a set of actions, based on the consequences of individual actions. Clearly, these actions are subject to anticipated uncertainty due to the uncertainty associated with consequences.

For probabilistic uncertainty, the value of $p(x)$ represents the degree of evidential support that x is the true alternative, $x \in X :\to [0, \ 1]$ set. Then the Shannon entropy measures the amount of uncertainty in evidence expressed by the probability distribution P on the finite set:

$$-c \sum p(x) \log_b p(x) \tag{1.48}$$

where c and b are constant, and $b \neq 1$. The choice of b and c determines the unit in which the uncertainty is measured. The most common measurement unit is a bit. If

$$-c \log_b \frac{1}{2} = 1 \tag{1.49}$$

then $b = 2$, $c = 1$, and it would imply $X = [x_1, x_2]$ and $p(x_1) = p(x_2) = 0.5$. This is often referred to as a normalization requirement. Thus, one bit is the amount of information gained or uncertainty removed when one learns the answer to a question whose two possible outcomes are equally likely. Thus, $H(p)$ is called the Shannon measure of uncertainty or Shannon entropy.

To gain further insight about the type of uncertainty measured by the Shannon entropy, one can write Shannon entropy as

$$H(p) = -\sum_{x \in X} p(x) \log_2 \left[1 - \sum_{y \neq x} p(y) \right] \tag{1.50}$$

Now consider the term

$$con(x) = \sum_{y \neq x} p(y) \tag{1.51}$$

which expresses the total evidence (sum) as a result of the alternatives that are different from x, that is, $y \neq x$. This evidence is in conflict with the one focusing on x. It is seen that $con(x) \in [0, 1]$ for each $x \in X$. The term $-\log_2 [1 - con(x)]$ in equation (1.51) increases monotonically with $con(x)$ and its range is extended from $[0, 1]$ to $[0, \infty]$. Thus, the Shannon entropy is the mean (expected value) of the conflict among evidences expressed by each probability distribution P.

Example 1.7: One way to gain further insight into the Shannon uncertainty is from equation $s(a) = c \log_b a$, where c and b are constants, and $b \neq 1$. The Shannon uncertainty here is analogous to the gain function defined by equation (1.18). Taking $c = -1$, and $b = 2, s(a) = c \log_2 a$. Plot this function taking $a = 0.0, 0.2, 0.4, 0.6, 0.8$, and 1.0. What do you conclude from this graph?

Solution: The function is plotted as a function of a in Figure 1.9. The function declines with increasing a and reaches zero when $a = 1$.

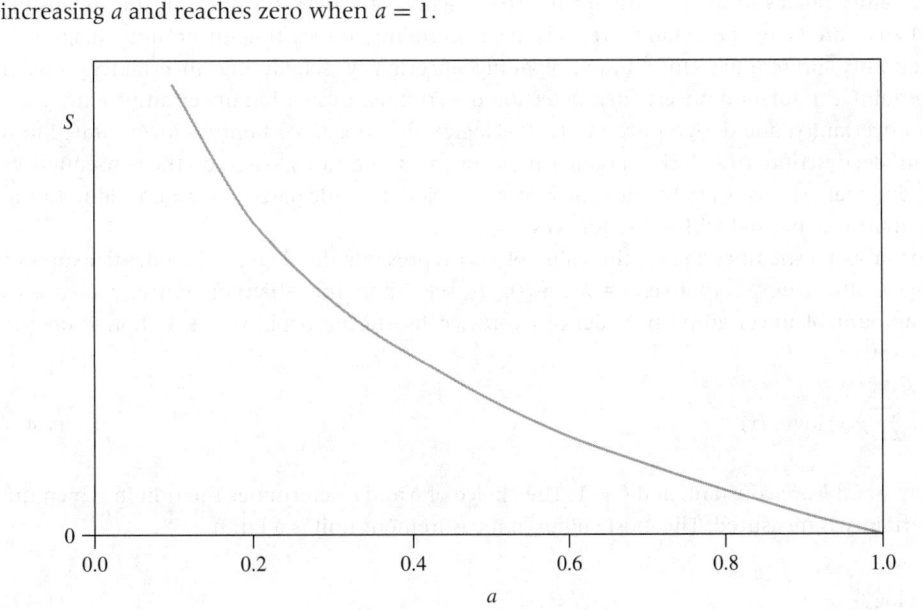

Figure 1.9 Plot of function $s(a) = -\log_2 a$ for different values of a.

Example 1.8: Consider $X = \{x_1, x_2\}$ with, $p(x_1) = a$, and $p(x_2) = 1 - a$, $a \in [0, 1]$; x_1 and x_2 represent two alternatives. The Shannon entropy depends only on a and is comprised of two components $S_1 = -a \log_2 a$ and $S_2 = -(1 - a) \log_2(1 - a)$; each component is analogous to the gain function. Compute the Shannon entropy as well as each of the two components, taking $a = 0.0, \ 0.2, \ 0.4, \ 0.6, \ 0.8,$ and 1.0, and graph them. What do you conclude from these graphs?

Solution: The Shannon entropy and each component thereof are plotted in Figure 1.10. The Shannon entropy graph is as before. The two components are mirror images of each other, as shown in Figure 1.10a. Graphs of S_1 and S_2 are shown in Figure 1.10a. Graph S_1 and S_2 are the same except for the change of scale.

1.5 Entropy and related concepts

Hydrologic and environmental systems are inherently spatial and complex, and our understanding of these systems is less than complete. Many of the systems are either fully stochastic or part-stochastic and part-deterministic. Their stochastic nature can be attributed to the randomness in one or more of the following components that constitute them: 1) system structure (geometry), 2) system dynamics, 3) forcing functions (sources and sinks), and 4) initial and boundary conditions. As a result, a stochastic description of these systems is needed, and the entropy theory enables the development of such a description.

Fundamental to the planning, design, development, operation, and management of environmental and water resources projects is the data that are observed either in field or experimentally and the information they convey. If this information can be determined, it can also serve as a basis for design and evaluation of data collection networks, design of sampling schemes, choosing between models, testing the goodness-of-fit of a model, and so on.

Engineering decisions concerning hydrologic systems are frequently made with less than adequate information. Such decisions may often be based on experience, professional judgment, thumb rules, crude analyses, safety factors, or probabilistic methods. Usually, decision making under uncertainty tends to be relatively conservative. Quite often, sufficient data are not available to describe the random behavior of such systems. Although probabilistic methods allow for a more explicit and quantitative accounting of uncertainty, their major difficulty occurs due to the lack of sufficient or complete data. Small sample sizes and limited information render the estimation of probability distributions of system variables with conventional methods difficult. This problem can be alleviated by the use of entropy theory which enables to determine the least-biased probability distributions with limited knowledge and data. Where the shortage of data is widely rampant as is normally the case in many countries, the entropy theory is particularly appealing.

1.5.1 Information content of data

One frequently encounters a situation in which to exercise freedom of choice, evaluate uncertainty or measure the information gain or loss. The freedom of choice, uncertainty, disorder, information content, or information gain or loss has been variously measured by relative entropy, redundancy, and conditional and joint entropies employing conditional and joint probabilities. As an example, in the analysis of empirical data, the variance has often been interpreted as a measure of uncertainty and as revealing gain or loss in information. However, entropy is another measure of dispersion – an alternative to variance. This suggests that it is possible to determine the variance whenever it is possible to determine the entropy

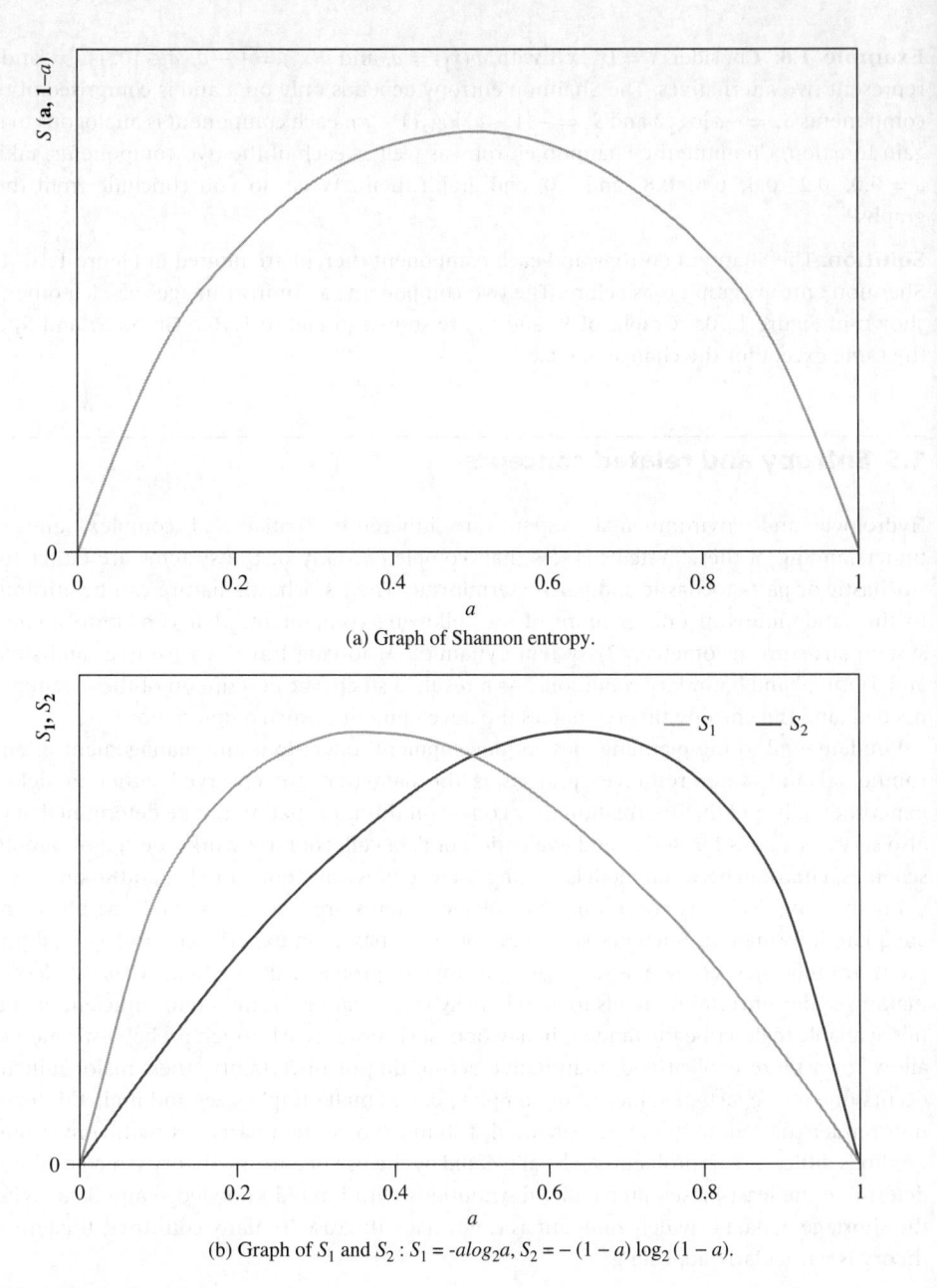

(a) Graph of Shannon entropy.

(b) Graph of S_1 and $S_2 : S_1 = -a\log_2 a$, $S_2 = -(1-a)\log_2(1-a)$.

Figure 1.10 Shannon entropy.

measure, but the reverse is not necessarily true. However, variance is not the appropriate measure if the sample size is small.

1.5.2 Criteria for model selection
Usually there are more models than one needs and a choice has to be made as to which model to choose. Akaike (1973) formulated a criterion, called Akaike Information Criterion (*AIC*),

for selecting the best model from amongst several models as

$$AIC = -2\log(maximized\ likelihood) + 2k \tag{1.52}$$

where k is the number of parameters of the model. AIC provides a method for model identification and can be expressed as minus twice the logarithm of the maximum likelihood plus twice the number of parameters used to find the best model. The maximum likelihood and entropy are uniquely related. When there are several models, the model giving the minimum value of AIC should be selected. When the maximum likelihood is identical for two models, the model with the smaller number of parameters should be selected, for that will lead to smaller AIC and comply with the principle of parsimony.

1.5.3 Hypothesis testing

Another important application of the entropy theory is in the testing of hypotheses (Tribus, 1969). With use of Bayes' theorem in logarithmic form, an evidence function can be defined for comparing two hypotheses. The evidence in favor of a hypothesis over its competitor is the difference between the respective entropies of the competition and the hypothesis under testing. Defining surprisal as the negative of the logarithm of the probability, the mean surprisal for a set of observations is expressed. Therefore, the evidence function for two hypotheses is obtained as the difference between the two values of the mean surprisal multiplied by the number of observations.

1.5.4 Risk assessment

There are different types of risk, such as business risk, social risk, economic risk, safety risk, investment risk, occupational risk, and so on. In common language, risk is the possibility of loss or injury and the degree of probability of such loss. Rational decision making requires a clear and quantitative way of expressing risk. In general, risk cannot be avoided and a choice has to be made between risks. To put risk in proper perspective, it is useful to clarify the distinction between risk, uncertainty, and hazard.

The notion of risk involves both uncertainty and some kind of loss or damage. Uncertainty reflects the variability of our state of knowledge or state of confidence in a prior evaluation. Thus, risk is the sum of uncertainty plus damage. Hazard is commonly defined as a source of danger and involves a scenario identification (e.g., failure of a dam) and a measure of the consequence of that scenario or a measure of the ensuing damage. Risk encompasses the likelihood of conversion of that source into the actual delivery of loss, injury, or some form of damage. Thus, risk is the ratio of hazard to safeguards. By increasing safeguards, risk can be reduced but it is never zero. Since awareness of risk reduces risk, awareness is part of safeguards. Qualitatively, risk is subjective and is relative to the observer. Risk involves the probability of scenario and its consequence resulting from the happening of the scenario. Thus, one can say that risk is probability and consequence. Kaplan and Garrick (1981) have analyzed risk using entropy. Luce (1960) has reasoned that entropy should be described as an average measure of risk, not of uncertainty.

Questions

Q.1.1 Assume that there are 256 possibilities in a particular case. These possibilities are arranged in such a way that each time an appropriate piece of information becomes available, the number of possibilities reduces to half. What is the information gain in bits if the number of possibilities is reduced to 128, to 64, to 32, to 16, to 8, to 4, and to 2?

Q.1.2 Assume that there are 10,000 possibilities in a particular case. These possibilities are arranged in such a way that each time an appropriate piece of information becomes available, the number of possibilities reduces to one tenth. What is the gain in information in decibels or dits if the number of possibilities is reduced to 1,000, to 100, and to 10?

Q.1.3 Consider that a random variable X takes on values x_i, $i = 1, 2, 3, 4, 5$, with probabilities $p(x_i) = 0.10, 0.20, 0.30, 0.25,$ and 0.15. Compute the gain in information for each value using the Shannon entropy, exponential entropy and Tsallis entropy with $q = 0.5$. Which entropy provides a larger gain?

Q.1.4 Consider the probabilities in Q.1.3. Order them in order of increasing surprise and relate the surprise to the gain in information computed in Q.1.3.

Q.1.5 Consider two distributions $P_i = p = 0.1$, $i = 1, 2, \ldots, 10$; $q_j = q = 0.05$, $j = 1,$ $2, 3, \ldots, 20$ having equiprobable outcomes. Compute the maximum entropy of each distribution in bits. Compare these two distributions by determining the difference in the information contents of these distributions. Is there a loss of information with the increase in the number of possible outcomes?

Q.1.6 Consider two distributions $P_i = p = 0.05$, $i = 1, 2, \ldots, 20$; $q_j = q = 0.10$, $j = 1,$ $2, 3, \ldots, 10$ having equiprobable outcomes. Compute the maximum entropy of each distribution in bits. Compare these two distributions by determining the difference in the information contents of these distributions. Is there a gain of information with the decrease in the number of possible outcomes?

Q.1.7 Consider that a discrete random variable X takes on 10 values with probability distribution $P : P : \{p_1, p_2, \ldots, p_{10}\}$ corresponding to $X : \{x_i, i = 1, 2, \ldots, 10\}$. What distribution P will yield the maximum and minimum values of the Shannon entropy?

Q.1.8 Consider an event A. The probability of the occurrence of event A can be regarded as a measure of uncertainty about its occurrence or non-occurrence. For what value of the probability will the uncertainty be maximum and why?

Q.1.9 Consider a coin tossing experiment. Let the probability of the occurrence of head be denoted as p and that of tail as q. Express the Shannon entropy of this experiment. Note $q = 1 - p$ or $p = 1 - q$. Plot a graph of entropy by taking different values of p. For what value of p does the entropy attain a maximum?

Q.1.10 Consider a six-faced dice throwing experiment. The dice is unbiased so the probability of the occurrence of any face is the same. In this case there are six possible events and each event is equally likely. Express the Shannon entropy of this experiment and compute its value. Now consider that the concern is whether an even-numbered or an odd-numbered face shows upon throw. In this case there are only two possible events: (even, odd). Express the Shannon entropy of this experiment and compute its value. Which of these two cases has higher entropy? Which case is more uncertain? Is there any reduction in uncertainty in going from case one to case two?

References

Akaike, H. (1973). Information theory and an extension of the maximum likelihood principle. *Proceedings, 2nd International Symposium on Information Theory*, B.N. Petrov and F. Csaki, eds., Publishing House of the Hungarian Academy of Sciences, Budapest, Hungary.

Batty, M. (2010). Space, scale, and scaling in entropy maximizing. *Geographical Analysis*, Vol. 42, pp. 395–421.

Brillouin, L. (1956). *Science and Information Theory*. Academic Press, New York.

Denbigh, K.G. (1989). Note on entropy, disorder and disorganization. *British Journal of Philosophical Science*, Vol. 40, pp. 323–32.

Fast, J.D. (1968). *Entropy: The Significance of the Concept of Entropy and its Application in Science and Technology*. Gordon and Breach, Science Publishers Inc., New York.

Fisher, R.A. (1921). On mathematical foundations of theoretical statistics. *Philosophical Transactions of the Royal Society of London*, Series A, Vol. 222, pp. 30–368.

Gull, S.F. (1991). Some misconceptions about entropy. Chapter 7 in: *Maximum Entropy in Action*, edited by B. Buck and V.A. McCauley, Oxford Science Publishers, pp. 171–86.

Hartley, R.V.L. (1928). Transmission of information. *The Bell System Technical Journal*, Vol. 7, No. 3, pp. 535–63.

Kaplan, S. and Garrick, B.J. (1981). On the quantitative definition of risk. *Risk Analysis*, Vol. 1, No. 1, pp. 11–27.

Kapur, J.N. (1989). *Maximum Entropy Models in Science and Engineering*. Wiley Eastern Ltd., New Delhi, India.

Klir, G. J. (2006). *Uncertainty and Information: Foundations of Generalized Information Theory*. John Wiley & Sons, New York.

Luce, R.D., (1960). The theory of selective information and source of its behavioral applications. In: *Developments in Mathematical Psychology*, edited by R.D. Luce, The Free Press, Glencoe.

Marchand, B., (1972). Information theory and geography. *Geographical Analysis*, Vol. 4, pp. 234–57.

Maxwell, J.C. (1872). *The Theory of Heat*. D. Appleton, New York (reproduced by University Microfilms).

Ng, S.K. (1996). Information and system modeling. *Mathematical and Computer Modeling*, Vol. 23, No. 5, pp. 1–15.

Pal, N.R. and Pal, S.K. (1991a). Entropy: A new definition and its applications. *IEEE: Transactions on Systems, Man, and Cybernetics*, Vol. 21, No. 5, pp. 1260–70.

Pal, N.R. and Pal, S.K. (1991b). Image model, Poisson distribution and object extraction. *International Journal of Pattern Recognition and Artificial Intelligence*, Vol. 5, No. 3, pp. 459–83.

Pal, N.R. and Pal, S.K. (1992). Some properties of the exponential entropy. *Informational Sciences*, Vol. 66, pp. 119–37.

Pal, N.R. and Bezdek, J.C. (1994). Measuring fuzzy uncertainty. *IEEE Transactions on Fuzzy Systems*, Vol. 2, No. 2, pp. 107–18.

Papalexiou, S.M. and Koutsoyiannis, D. (2012). Entropy based derivation of probability distributions: A case study to daily rainfall. *Advances in Water Resources*, Vol. 45, pp. 51–57.

Prigogine, I. (1989). What is entropy? *Naturwissenschaften*, Vol. 76, pp. 1–8.

Renyi, A. (1961). On measures of entropy and information. Proceedings, 4th Berkeley Symposium on Mathematics, *Statistics and Probability*, Vol. 1, pp. 547–561.

Rosenthal, H. and Binia, J. (1988). On the epsilon entropy of mixed random variables. *IEEE Transactions on Information Theory*, Vol. 34, No. 5, pp. 1110–14.

Shannon, C.E. (1948). A mathematical theory of communications, I and II. *Bell System Technical Journal*, Vol. 27, pp. 379–443.

Shannon, C.E. and Weaver, W. (1949). *The Mathematical Theory of Communication*. University of Illinois Press, Urbana, Illinois.

Sugawara, M. (1971). Water resources and negentropy. Proceedings of the Warsaw Symposium on Mathematical Models in Hydrology, Vol. 2, pp. 876–8, IAHS-UNESCO-WM, Warsaw, Poland.

Tsallis, C. (1988). Possible generalization of Boltzmann-Gibbs statistics. *Journal of Statistical Physics*, Vol. 52, No. 1/2, pp. 479–87.

Tribus, M. (1969). *Rational Description: Decision and Designs*. Pergamon Press, New York.

Watanabe, S. (1969). *Knowing and Guessing: A Quantitative Study of Inference and Information*. John Wiley & Sons, New York.

Zurek, W.H., 1989. Algorithmic randomness and physical entropy. *Physical Review A*, Vol. 40, No. 8, pp. 4731–51.

Additional References

Akaike, H. (1974). A new look at the statistical model identification. *IEEE Transactions on Automatic Control*, Vol. AC-19, No. 6, pp. 710–23.

Akaike, H. (1985). Prediction and entropy. Chapter 1 in: *A Celebration of Statistics*, edited by A.C. Atkinson and S.E. Fienberg, pp. 1–24, Springer-Verlag, Heidelberg, Germany.

Carnap, R. (1977). *Two Essays on Entropy*. University of California Press, Berkeley, California.

Cover, T.M. and Thomas, J.A. (1991). *Elements of Information Theory*. John Wiley & Sons, New York.

Harmancioglu, N.B. and Singh, V.P. (1998). Entropy in Environmental and Water Resources. pp. 225–241, Chapter in *Encyclopedia of Hydrology and Water Resources*, edited by D.R. Herschy, Kluwer Acdemic Publishers, Dordrecht.

Jaynes, E.T. (1958). *Probability Theory in Science and Engineering*. Colloquium Lectures in Pure and Applied Science, Vol. 4, Field Research Laboratory, Socony Mobil Oil Company, Inc., USA.

Khinchin, A.I. (1957). *The Entropy Concept in Probability Theory*. Translated by R.A. Silverman and M.D. Friedman, Dover Publications, New York.

Jaynes, E.T. (1957). Information theory and statistical mechanics, I. *Physical Review*, Vol. 106, pp. 620–30.

Jaynes, E.T. (1982). On the rationale of maximum entropy methods. *Proceedings of the IEEE*, Vol. 70, pp. 939–52.

Levine, R.D. and Tribus, M., eds., (1978). *The Maximum Entropy Formalism*, 498 p., The MIT Press, Cambridge, Massachusetts.

Singh, V.P. (1998). *Entropy-Based Parameter Estimation in Hydrology*. Kluwer Academic Publishers, Boston, Massachusetts.

Singh, V.P. (1998). The use of entropy in hydrology and water resources. *Hydrological Processes*, Vol. 11, pp. 587–626.

Singh, V.P. (1998). Entropy as a decision tool in environmental and water resources. *Hydrology Journal*, Vol. XXI, No. 1–4, pp. 1–12.

Singh, V.P. (2000). The entropy theory as tool for modeling and decision making in environmental and water resources. *Water SA*, Vol. 26, No. 1, pp. 1–11.

Singh, V.P. (2003). The entropy theory as a decision making tool in environmental and water resources. In: *Entropy Measures, Maximum Entropy and Emerging Applications*, edited by Karmeshu, Springer-Verlag, Bonn, Germany, pp. 261–97.

Singh, V.P. (2005). Entropy theory for hydrologic modeling. in: *Water Encyclopedia: Oceanography; Meteorology; Physics and Chemistry; Water Law; and Water History, Art, and Culture*, edited by J. H. Lehr, Jack Keeley, Janet Lehr and Thomas B. Kingery, John Wiley & Sons, Hoboken, New Jersey, pp. 217–23.

Singh, V.P. (2011). Hydrologic synthesis using entropy theory: Review. *Journal of Hydrologic Engineering*, Vol. 16, No. 5, pp. 421–33.

Singh, V.P. and Fiorentino, M., editors (1992). *Entropy and Energy Dissipation in Hydrology*. Kluwer Academic Publishers, Dordrecht, The Netherlands, 595 p.

White, H. (1965). The entropy of a continuous distribution. *Bulletin of Mathematical Biophysics*, Vol. 27, pp. 135–43.

2 Entropy Theory

Since the development of informational entropy in 1948 by Shannon, the literature on entropy has grown by leaps and bounds and it is almost impossible to provide a comprehensive treatment of all facets of entropy under one cover. Thermodynamics, statistical mechanics, and informational statistics tend to lay the foundation for what we now know as entropy theory. Soofi (1994) perhaps summed up best the main pillars in the evolution of entropy for quantifying information. Using a pyramid he summarized information theoretic statistics as shown in Figure 2.1, wherein the informational entropy developed by Shannon (1948) represents the vertex. The base of the Shannon entropy represents three distinct extensions which are variants of quantifying information: discrimination information (Kullback, 1959), mutual information (Lindley, 1956, 1961), and principle of maximum entropy (POME) or information (Jaynes, 1957, 1968, 1982). The lateral faces of the pyramid are represented by three planes: 1) the SKJ (Shannon-Kullback-Jaynes) minimum discrimination information plane, 2) the SLK (Shannon-Lindley-Kullback) mutual information plane, and 3) the SLJ (Shannon-Lindley-Jaynes) Bayesian information theory plane. Most of the information-based contributions can be located on one of the faces or in the interior of the pyramid. The discussion in this chapter on what we call entropy theory represents some aspects of all three faces but not fully.

The entropy theory may be comprised of four parts: 1) Shannon entropy, 2) principle of maximum entropy, 3) principle of minimum cross entropy, and 4) concentration theorem. The first three are the main parts and are most frequently used. One can also employ the Tsallis entropy or another type of entropy in place of the Shannon entropy for some problems. Before discussing all four parts, it will be instructive to amplify the formulation of entropy presented in Chapter 1.

2.1 Formulation of entropy

In order to explain entropy, consider a random variable X which can take on N equally likely different values. For example, if a six-faced dice is thrown, any face bearing the number 1, 2, 3, 4, 5, or 6 has an equal chance to appear upon throw. It is now assumed that a certain value of X (or the face of the dice bearing that number or outcome upon throw) is known to only one person. Another person would like to know the outcome (face) of the dice throw by asking questions to the person, who knows the answer, in the form of only yes or no. Thus, the number of alternatives for a face to turn up in this case is six, that is, $N = 6$. It can

Entropy Theory and its Application in Environmental and Water Engineering, First Edition. Vijay P. Singh.
© 2013 John Wiley & Sons, Ltd. Published 2013 by John Wiley & Sons, Ltd.

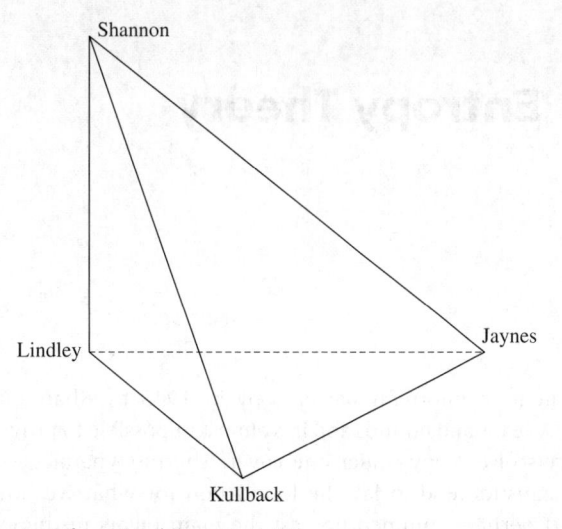

Figure 2.1 Pyramid showing informational-theoretic statistics.

be shown that the minimum number of questions to be asked in order to ascertain the true outcome is:

$$I = -\log_2 \frac{1}{N} = \log_2 N \tag{2.1}$$

where I represents the amount of information required to determine the certain value of X, $1/N$ defines the probability of finding the unknown value of X by asking a single question, when all outcomes are equally likely, and log is the logarithm to the base of 2. If nothing else is known about the variable, then it must be assumed that all values are equally likely in accordance with the principle of insufficient reason.

In general,

$$I_i = -\log_2 p_i \tag{2.2}$$

where p_i is the probability of outcome $i = 1, 2, \ldots, N$. Here I can be viewed as the minimum amount of information required to positively ascertain the outcome of X upon throw. Stated in another way, this defines the amount of information gained after observing the event $X = x$ with probability $1/N$. In other words I is a measure of information and is a function of N. The base of the logarithm is 2, because the questions being posed (i.e., questions admitting only either yes or no answers) are in binary form. The point to be kept in mind when asking questions is to gain information, not assent or dissent, and hence in many cases a yes is as good an answer as is a no. This information measure or equation (2.2) satisfies the following properties:

1 $I(x_i) = 0$, $p_i = 1$. This means that if the outcome of an event is certain, it provides no information or no information is gained by its occurrence.

2 $I(x_i) \geq 0$, $0 \leq p_i \leq 1$. This means that the occurrence of an event $X = x_i$ provides some or no information but does not lead to the loss of information.

3 $I(x_k) > I(x_i)$, $p_k < p_i$, $k \neq i$. This means that the less probable the event the more information one gains through its occurrence.

Example 2.1: If a six-faced dice is thrown, any face bearing the number 1, 2, 3, 4, 5, or 6 has an equal chance to appear. The outcome of the first throw is number 5 which is known to a person A. How many questions does one need to ask this person or how much information will be required to positively ascertain the outcome of this throw?

Solution: In this case, $N = 6$. Therefore, $I = -\log_2 \frac{1}{N} = \log_2 N = 2.585$ bits. This gives the minimum amount of information needed or the number of questions to be asked in binary form (i.e., yes or no). The number of questions needed to be asked is a measure of uncertainty. The questioning can go like this: Is the outcome between 1 and 3? If the answer is no then it must be between 4 and 6. Then the second question can be: Is it between 4 and 5? If the answer is yes, then the next question is: Is it 4 and the answer is no. Then the outcome has to be 5. In this manner entropy provides an efficient way of obtaining the answer. In vigilance, investigative or police work entropy can provide an effective way of interrogation. Another example of interest is lottery.

Example 2.2: Suppose that N tickets are sold for winning a lottery. In other words the winning ticket must be one of these tickets, that is, the number of chances are N. Let N be 100. Each ticket has a number between 1 and 100. One person, called Jack, knows what the winning ticket is. The other person, called Mike, would like to know the winning ticket by asking Jack a series of questions whose answers will be in the form of yes or no. Find the winning ticket.

Solution: The number of binary questions needed to determine the winning ticket is given by equation (2.1). For $N = 100$, $I = 6.64$. This says that it will take 6.64 questions to find the winning ticket. To illustrate this point, the questioning might go as shown in Table 2.1. The questioning in Table 2.1 shows how much information is gained simply by asking binary questions. The best way of questioning and finding an answer is by subdividing the class consisting of questions in half at each question. This is similar to the method of *regula falsi* in numerical analysis when determining the root of a function numerically.

Consider another case where two coins are thrown and a person knows what the outcome is. There are four alternatives in which head or tail can appear for the first and second coins, respectively: head and tail, head and head, tail and tail, and tail and head or one can simply write the number of alternatives N as 2^2. The number of questions to be asked in order to ascertain the outcome is again given by equation (2.1): $\log_2 4 = \log_2 2^2 = 2$.

Table 2.1 Questioning for finding the winning lottery ticket.

Question number	Question asked	Answer
1	Is the winning ticket between 50 and 100?	No
2	Is it between 25 and 49?	No
3	Is it between 1 and 12?	Yes
4	Is it between 7 and 12?	Yes
5	Is it between 7 an 9?	Yes
6	Is it between 7 and 8?	Yes
7	It is 7?	No
	Answer then is: 8	

Example 2.3: Consider that the probability of raining on any day in a given week is the same. In that week it rained on a certain day and one person knew the day it rained on. Another person would like to know the day it rained on by asking a number of questions to the person who knows the answer. The answers to be given are in binary form, that is, yes or no. What will be the minimum number of questions to be asked in order to determine the day it rained on?

Solution: In this case, $N = 7$. Therefore, the minimum number of questions to be asked is: $I = -\log_2 (1/7) = \log_2 7 = 2.807$

In the above discussion the base of the logarithm is 2 because of the binary nature of answers and the questioning is done such that the number of alternatives is reduced to half each time a question is asked. If the possible responses to a question are three, rather than two, then with n questions one can cover 3^n possibilities. For example, an answer to a question about weather may be: hot, cold, or pleasant. Similarly, for a crop farmers may respond: bumper, medium or poor. In such cases, the number of questions to cover N cases is given as

$$I = -\log_3 \frac{1}{N} = \log_3 N \tag{2.3}$$

Here the logarithm is to the base 3. The change from a base of 2 to a base of 3 corresponds to multiplication by a constant, that is,

$$\log_3 N = \log_3 2 \log_2 N \tag{2.4}$$

Example 2.4: Consider two days in a week: Monday and Tuesday. The weather on any of these two days can be hot (W), pleasant (J) or cold (C). What will be the minimum number of questions to be asked in order to determine what the weather would be on these two days? The weather man knows the answer and one would like to know the answer by asking the weather man a number of questions.

Solution: In this case, the number of possibilities, N, is: $3^2 = 9$. These include: WW, JJ, CC, WJ, WC, CJ, CW, JC, and JW. Therefore, the number of questions to cover nine cases is:

$$I = -\log_3 \frac{1}{9} = \log_3 9 = \log_3 (3)^2 = 2$$

The questioning can go like this: Were any of the two days hot? If the answer is no, then were any of the two days cold? If the answer was again no then it must be pleasant on both days. On the other hand, if the base of the logarithm is 2 then the questioning is done in a binary manner as before. The questioning can go like this: Was it hot on Monday? If the answer is no, then the next question is if it was cold. If the answer is again no then it clearly was pleasant on Monday. Was it hot on Tuesday? If the answer is no, then the next question is: Was it cold? If the answer is again no then it clearly was pleasant on Tuesday. In this manner, by asking four questions, one determined the weather on Monday and Tuesday. In this case, the information contained in the outcome of the experiment of questioning is:

$$I = -\log_2 \frac{1}{9} = \log_2 9 = 3.17 \, \text{bits}$$

In order to appreciate the informational value of entropy, one may ask a question: What is the information associated with an experiment whose outcome is certain? For example, the experiment may be whether the sun will set between midday (or noon) and midnight. In this case the outcome is certain and there is only one possible outcome. The number of questions to be asked to determine the outcome is 0. Therefore,

$$I = \log_2 1 = 0$$

In other words, the information contained in the outcome of the experiment is 0 bits.

To extend this point further, if the experiment entails flipping a coin then the number of possible outcomes is 2 and the information should be 1 bit. If the experiment involves throwing a dice as in Example 2.1, then there are six possible outcomes and the information of the outcome is $\log_2 6 = 2.585$ bits. If the experiment involves flipping three coins, then the number of possible outcomes would be $2^3 = 8$, and the information would be $\log_2 8 = 3$ bits. An interesting point to be noted here is that the information is the sum of information associated with flipping each individual coin, that is, $\log_2 2 + \log_2 2 + \log_2 2$. This shows that it does not matter whether the coins are flipped simultaneously or separately. In all of this discussion the underlying assumption is that outcomes are equally likely.

Can one extend the above concept of information to the outcomes of an experiment where they are not equally likely? To keep things simple, consider an experiment having N equally likely outcomes which are divided into two groups of N_1 and N_2 outcomes: $N = N_1 + N_2$. The question one asks is: Does a particular outcome belong to N_1 or N_2? One can write the probability, denoted as p_1, if the outcome belongs to group 1 as

$$p_1 = \frac{N_1}{N_1 + N_2} \tag{2.5a}$$

and similarly p_2 if it belongs to group 2 as:

$$p_2 = \frac{N_2}{N_1 + N_2} \tag{2.5b}$$

The question is: How much information is associated with the particular outcome? Remember the outcomes are not equally likely. The information associated with the outcome amongst N possible outcomes is: $\log_2 N$. Similarly, the information associated with the outcome amongst N_1 equally likely outcomes is: $\log_2 N_1$ and this happens only for a proportion of outcomes equal to N_1/N. Similarly for the outcome amongst N_2 equally likely outcomes the information is: $\log_2 N_2$ and this happens only for the proportion equal to N_2/N.

One way to determine the information associated with the particular outcome is to compute the information associated with N equally likely outcomes and subtract the excess information associated with N_1 or N_2 possible outcomes in the two groups. The information associated with the outcome of the experiment can now be computed by the arithmetic sum as

$$\begin{aligned} I &= \log N - \left[\frac{N_1}{N} \log N_1 + \frac{N_2}{N} \log N_2 \right] \\ &= -p_1 \log p_1 - p_2 \log p_2 \end{aligned} \tag{2.5c}$$

Since probabilities are less than unity, their logarithms are negative and hence I is positive.

The above discussion then leads to a form for the amount of information in the outcome of an experiment. Although it was assumed that all values of X were equally likely, that is, p_i's are equally or uniformly distributed, I holds for any kind of probability distribution

$P = \{p_1, p_2, \ldots, p_N\}$. Equations (2.5c) can be extended to define the self-information or information content of X.

Example 2.5: There was a music gathering of well-known musicians who charged for their performances. In order to break even, 500 (N) tickets, numbered from 1 to 500, were to be sold. There were five (n) sales persons (called A_i, $i = 1, 2, \ldots, 5$) who were tasked with selling the tickets. In order to encourage sales, a lottery ticket was included. One of the sales persons had the lottery ticket. The question was: what was the lottery ticket? Find the lottery ticket by asking binary questions.

Solution: In order to answer this question, one must first determine the sales person who had the winning ticket. The number of questions (n_1) required to determine the sales person having the lottery ticket is

$$n_1 = \log_2 n = \log_2 5 = 2.3219$$

Likewise the number of questions (n_2) required to determine the lottery ticket is

$$n_2 = \log_2 \left(\frac{N}{n}\right) = \log_2 \left(\frac{500}{5}\right) = \log_2 100 = 6.6439$$

The number of questions asked in the above two steps can be combined:

$$m = n_1 + n_2 = \log_2 n + \log_2 \left(\frac{N}{n}\right) = \log_2 N = \log_2 500 = 8.9658 \text{ bits}$$

$$n_1 + n_2 = 2.3219 + 6.6439 = 8.9658$$

Interestingly, this is the same as would be given directly by equation (2.1).

From the above discussion one can define the uncertainty U as the number of binary questions to be asked for determining the answer to a specific question where there is certain evidence. In the above example, the evidence is that there are N equally likely mutually exclusive possibilities or alternatives. Then the uncertainty is given by equation (2.1) and can be measured in bits. It should be noted that the questioning was done in two steps and at each step each question reduced the number of possibilities to half and the possibilities were equally likely. Let us now consider the case where possibilities are not equally likely.

Example 2.6: Consider the above example on the music gathering of musicians. In this case, sales person A_1 had to sell 75, A_2 75, A_3 250, A_4 50, and A_5 50 tickets. The question was: what was the lottery ticket? Find the lottery ticket by asking binary questions.

Solution: In this case, let m_i be the number of tickets the i-th sales person had to sell. Then

$$\sum_{i=1}^{5} m_i = N$$

Let A_i be the i-th sales person who sold the lottery ticket and this person had m_i tickets. There are two questions to be answered: 1) Which ticket is the lottery ticket? This question can be designated as Q_1. 2) Which sales person sold the lottery ticket? This question can be designated as Q_2. The uncertainty U associated with Q_1 can be expressed as

$$U(Q_1) = \log_2 \sum_{i=1}^{n} m_i = \log_2 \sum_{i=1}^{5} m_i = \log_2 500 = 8.9658 \text{ bits}$$

Likewise, the uncertainty associated with Q_2 can be expressed as

$$U(Q_2) = -\sum_{i=1}^{n} p(A_i) \log_2 [p(A_i)] = \sum_{i=1}^{5} \left(\frac{m_i}{N}\right) \log \left(\frac{m_i}{N}\right)$$

It may be noted that

$$p(A_1) = p(A_2) = 75/500 = 0.15; p(A_4) = p(A_5) = 50/500 = 0.1, p(A_3) = 250/500 = 0.5.$$

Therefore,

$$U(Q_2) = -\left[2\left(0.15 \log_2 0.15\right) + 0.50 \log_2 0.50 + 2\left(0.15 \log_2 0.15\right)\right] = 2.1422 \text{ bits}$$

Now the number of questions to be asked to determine the sales persons who sold the lottery ticket is computed as follows. 1) Did A_3 sell? If the answer is yes then the question Q_2 is answered in one attempt. The probability of getting the yes answer in one question is 1/2. Therefore the probability of having Q_2 answered in one question is $^1/_2$, because this person sold half the tickets. 2) If the answer is no, then the next question is did A_1 and A_2 sell? 3) If the answer is yes then the next question is if A_1 sold. 4) If the answer is yes then Q_2 is answered. This means that it takes two additional questions to answer Q_2. Thus, the expected number of questions needed to answer Q_2 therefore is expressed as $\frac{1}{2} \times 1 + \frac{1}{2} \times 4 = 2.5$, that is, 3.

2.2 Shannon entropy

Equations (2.1) and (2.2) denote the information content of a single outcome of a variable or process which may be of interest. However, of greater importance is the information content of the entire variable (or process) or the expected value of the information content of each value that the random variable X can take on. Expressed mathematically,

$$H = E[I] = \sum_{i}^{N} p_i I_i \tag{2.6}$$

or

$$H = E[I] = -\sum_{i=1}^{N} p_i \log_2 p_i \tag{2.7}$$

where N is the number of values X takes on, p_i is the probability of the i-th value of X occurring, I_i is the information content of the i-th value, and H is Shannon entropy of X

or $P:\{p_1, p_2, \ldots, p_N\}$. Equation (2.7) is a measure of the average uncertainty of the random variable and defines the entropy given by Shannon (1948). It can also be interpreted as the number of bits required, on average, to describe the random variable. For example, when a six-faced dice is thrown, any face bearing the number 1, 2, 3, 4, 5, or 6 has an equal chance to appear. Then the average uncertainty or information content of the random variable expressing the occurrence of the face upon throw is 2.585 bits as computed in Example 2.1. To see the validity of equation (2.7), consider the case where N outcomes are equally likely. Then, $p_i = 1/N$. One can write: $-\log_2 p_i = \log_2 N$. Substituting this in equation (2.7) gives $\log_2 N$, as it should.

If one imagines before carrying out an experiment N possible outcomes with probabilities $p_1, p_2, \ldots, p_N$, then equation (2.7) measures the uncertainty concerning the results of the experiment. However, if we imagine the outcomes after the experiment has already been done, then equation (2.7) measures the amount of information obtained from the experiment. It is intuitively plausible to say that the larger amount of information gained upon the knowledge of the outcome or required to determine the outcome means the greater a priori uncertainty of this outcome. In this sense, entropy is employed as an indicator of uncertainty. Thus, information can be considered as the opposite of uncertainty or vice versa.

Shannon (1948) developed the entropy theory for expressing information or uncertainty. To understand the informational aspect of entropy an experiment on a random variable X is performed. There may be N possible outcomes $x_1, x_2, \ldots, x_N$, with probabilities $p_1, p_2, \ldots, p_N$; $P(X = x_1) = p_1$, $P(X = x_2) = p_2$, $\ldots$, $P(X = x_N) = p_N$. These outcomes can be described by the probability distribution:

$$P(X) = (p_1, p_2, \ldots, p_N) ; \sum_i^N p_i = 1; p_i \geq 0, i = 1, 2, \ldots, N \tag{2.8}$$

If this experiment is repeated, the same outcome is not likely, implying that there is uncertainty as to the outcome of the experiment. Based on one's knowledge about the outcomes, the uncertainty can be more or less. For example, the total number of outcomes is a piece of information and the number of those outcomes with nonzero probability is another piece of information.

The probability distribution of outcomes, if known, provides a certain amount of information. Shannon (1948) defined a quantitative measure of uncertainty associated with a probability distribution or the information content of the distribution in terms of entropy as

$$H(X) = H(P) = -\sum_{i=1}^{N} p_i \log p_i = E\left[-\log p\right] \tag{2.9}$$

Quantity $H(P)$ or $H(X)$ is called Shannon entropy or informational entropy, and $H(X)$ is a measure of the average amount of information conveyed per outcome or message. It is a measure of prior uncertainty about X. It is assumed here that $0 \log 0 = 0$; this can be justified by continuity considerations, since $x \log x \to 0$ as $x \to 0$. This means that adding terms of zero probability does not change the value of entropy. Entropy depends on the probability distribution of X but not on the value of the variable itself. If X follows $p(x)$ and $g(x) = \log [1/p(x)] = -\log p(x)$, then the expected value of $g(x)$ is: $-\sum p(x) \log p(x) = E[\log (1/p(x))] = E[g(x)]$. Thus, entropy is the expected value of minus the logarithm of $p(x)$ or the opposite of information.

Entropy is bounded as $0 \leq H(X) \leq \log(N)$. $H(X) = 0$ if and only if the probability $p_i = 1$ for some i, and the remaining probabilities are all zero. This lower bound of entropy represents no uncertainty. When all events are equally probable and all p_i's are equal, then $H(X) = \log N$; this upper bound corresponds to the maximum uncertainty which increases with increasing N.

Entropy can also be considered as a statistic for measuring the spatial distribution of various geographic phenomena. Consideration of spatial dimension in which these phenomena are recorded is implicit rather than explicit in equation (2.9). The manner of partitioning of space has received little attention in water and environmental engineering. Different areal patterns can result in different conclusions. Thus, design of optimal spatial systems for geographical analysis or locational planning is desirable. Geometry of spatial systems in relation to the measurement of locational phenomena will be discussed in Chapters 9 and 10.

Example 2.7: Consider a coin tossing experiment in which the collection of events is comprised of head and tail. What is the entropy of the coin tossing experiment?

Solution: The probability of occurrence of head = the probability of occurrence of tail = 0.5, since there are only two mutually exclusive events ($N = 2$). Therefore, the entropy of the coin tossing experiment is:

$$H = -\frac{1}{2}\log_2\frac{1}{2} - \frac{1}{2}\log_2\frac{1}{2} = \log_2 2 = 1 \text{ bit}$$

Example 2.8: Consider an experiment throwing a six-faced dice with faces numbered as 1, 2, 3, 4, 5, and 6. What is the entropy of the dice throwing experiment? What is the amount of information gained about dice throwing? Now characterize the faces as odd (1, 3, 5) or even (2, 4, 6). When a dice is thrown, either an even numbered face or an odd numbered face shows up. Determine the entropy of the dice throwing experiment in this case and the information gained. What will be the loss of information by going from considering each face in experiment one to considering faces as even or odd in experiment two?

Solution: There are six faces of the dice, implying six mutually exclusive events. Therefore, the probability of any face showing upon throw is $p = 1/6$. Then, the entropy of the dice throw experiment one is:

$$H = -\frac{1}{6}\log_2\frac{1}{6} - \frac{1}{6}\log_2\frac{1}{6} - \frac{1}{6}\log_2\frac{1}{6} - \frac{1}{6}\log_2\frac{1}{6} - \frac{1}{6}\log_2\frac{1}{6} - \frac{1}{6}\log_2\frac{1}{6}$$

$$= -\log_2\frac{1}{6} = \log_2 6 = 2.585 \text{ bits}$$

The amount of information gained about the dice throwing experiment one $= \log_2 6 = 2.585$ bits.

In this case, when faces are characterized as odd or even, then there are only two events ($N = 2$). Therefore, the entropy of the dice throwing experiment two or the average uncertainty per event is

$$H = -\frac{1}{2}\log_2\frac{1}{2} - \frac{1}{2}\log_2\frac{1}{2} = -\log_2\frac{1}{2} = \log_2 2 = 1 \text{ bit}$$

The amount of information gained is $\log_2 2 = 1$ bit.

The amount of information gained in the first experiment is greater than that in the second experiment. The amount of uncertainty about the dice throwing experiment $= \log_2 6 - \log_2 2 = \log_2 3 = 1.585$ bits. Thus, the loss in information from the second experiment on characterizing faces as odd or even is $\log_2 3 = 1.585$ bits.

Example 2.9: There is a bag containing ten balls: five red, three black and two white. Choose one ball from the bag blindly and the ball may be red. If this experiment is repeated, the outcome of getting a red ball is not likely, implying that there is uncertainty as to the outcome of the experiment. How much uncertainty is associated with this experiment or what is the value of entropy in bits of this experiment?

Solution: In this experiment, $N = 3$, $x_1 = $ red, $x_2 = $ black, $x_3 = $ white, with corresponding probabilities, respectively, as: $p_1 = 5/10 = 0.5$, $p_2 = 3/10 = 0.3$, and $p_3 = 2/10 = 0.2$. The entropy of this experiment is:

$$H = E[I] = -\sum_i^N p_i \log_2 p_i = 1.485 \text{ bits}$$

2.3 Connotations of information and entropy

2.3.1 Amount of information

Consider a certain event that is going to occur. The event can occur in different ways leading to different outcomes, and one may determine how probable each outcome is from observations of similar events. However, what is not exactly known is which of these outcomes will actually happen. Consider an example to illustrate the point. A thirsty person would like to get a bottle of fresh water. Assume that there are 20 bottles available but except one all others do not have fresh water. Bottles are identifiable through color, shapes, marks on the top, and so on, and so is a fresh water bottle but not without seeing the bottle. Assume that bottles are identified by color marks on their tops. If the thirsty person grabs the bottle he can see what it contains. Clearly the event – grabbing the bottle – has 20 possible outcomes. Of course, if he is lucky the first bottle he picks up may be the fresh water bottle, and he can quench his thirst. However, it is not very likely, so that in order to pick the right bottle, he needs information. One can tell him that all the bottles having red marks are not fresh water bottles and there are five such bottles. With this piece of information, the number of alternatives or outcomes reduces from 20 to 15.

Similarly, if the person is told that all the bottles with green marks on the top are not fresh water bottles and such bottles are five bottles. Then, the number of alternatives reduces to 10. If he is simply told that the bottle with blue mark on the top is the fresh water bottle then he has all the information he needs to pick the fresh water bottle. In this case the number of alternatives is reduced from 20 to 1. In this example, the amount of information is measured by the reduction in alternatives or outcomes. Nothing is said about the truthfulness, value, believability, or understanding. Miller (1953) emphasized that information is not synonymous with "meaning." All that is being discussed is how much information there is, that is, the amount of information each piece of knowledge conveys. It should be noted however that the term information occurs in a particular way. In information theory, only the amount of information is measured or quantified, but it says nothing about the content, value, truthfulness, exclusiveness, history, or purpose of the information.

2.3.2 Measure of information

One way to measure the information in such examples as above may be to count the number of alternatives the supplied information eliminates. In that case, one unit of information is transmitted each time an alternative is eliminated. This however does not seem to be a good measure, at least intuitively. For example, it is more advantageous to eliminate two alternatives in case of five possible alternatives than in case of 20 alternatives. This would suggest the use of ratios or fractions of alternatives eliminated. In order to transmit the same information (i.e., $2/5 = 0.4$), the number of alternatives to be reduced would have to be 8 in case of 20 alternatives, that is, 20 to 12 alternatives.

The information is measured in bits. When the number of alternatives is reduced to half, one unit of information is obtained and this unit is called *bit*. If one message reduces n alternatives to n/x, then it possesses one bit less information than does a message that reduces n to $n/2x$, where x is factor. Thus, the amount of information in a message that reduces n to n/x, is $\log_2 x$ bits. In the case of water bottles, if the number of bottles is reduced from 20 to 5 then $x = 4$ and $\log_2 4 = 2$ bits of information. In this case, 20 has been halved twice: from 20 to 10 and from 10 to 5 alternatives.

Information is related to the freedom of choice and uncertainty to the choice made. Greater the freedom of choice, the greater the information and the greater the uncertainty that the alternative actually is a particular one. This connects greater freedom $\Rightarrow$ greater information $\Rightarrow$ greater uncertainty.

2.3.3 Source of information

When communicating information, three elements are involved: source (supplier of information), conduit or channel (transmitter or conveyor of information), and destination (receiver) of information. A schematic of information flow and components of a resulting communication system is given by Shannon (1948) as shown in Figure 2.2. The source generates messages (and hence information) by making a series of decisions among certain alternatives. The series of such decisions is often referred to as a message in a discrete system. The source selects a message from a set of n alternative messages. A message can be a chart, data, text, or table, because it provides information. The message flows through the channel and thus the channel is the connection between the source and the destination. The decisions made by the source must be translated into a form which is suitable for transmission through the channel. This may be accomplished by the introduction of a transmitter between the channel and the source. The transmission of the message is in the form of signals. Likewise,

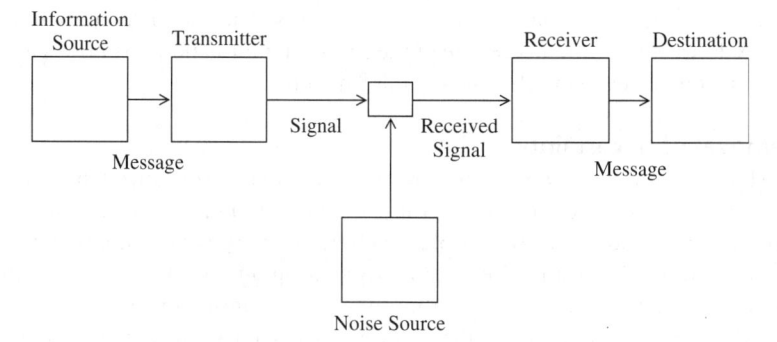

Figure 2.2 General communication system (adapted from Shannon, 1948).

the signals coming from the channel must be translated into a form acceptable to the destination. This entails introducing a receiver between the channel and the destination.

The receiver decodes the message for the destination. In this communication system, the transmitter accepts the message from a source which has entropy H, and encodes it, that is, it changes the message into a signal. The transmitter transmits $\log_2 n$ bits of information, where n is the number of signals. The transmitter has a finite capacity C which is defined in terms of the information it can transmit or more accurately in terms of its ability to transmit what comes out of the source of given information. This signal encodes the message. Then the receiver gets the information as to what was chosen and decodes it. The entropy theory yields the average of transmission over the channel as C/H. Irrespective of the coding, the transmission rate cannot exceed C/H and this equals the capacity of the channel.

Not all messages are equally likely. If some messages are more likely than others then a receiver can anticipate them and less information needs to be conveyed. Consider N messages each with probability $p = 1/N$ (equally likely) then the amount of information to be conveyed for this message is: $-\log_2 p = \log_2 N$. For unequally likely messages with probabilities p_1, $p_2, \ldots, p_N$, the amount of information associated with the ith message is: $-\log_2 p_i$.

Now consider an example of a raingage which measures rainfall. A transmitter changes the message into a signal. The hopper collects the rainwater and sends it to the tipping bucket. Receiver – an inverse transmitter – changes the transmitted signal into a message. The tipping bucket records the rainwater onto a graph which is the measurement. In the process of measurement, certain distortions or errors are added to the measurements. These are called noise.

There may be another source which transmits signal into a channel. This source is referred to as noise source and its signal as noise. The effect of noise is unpredictable beforehand, except statistically. One then may need to state the probability with which it alters one signal into another. Thus, loss of information occurs during the process of sending, transmitting, and receiving a message. For example, during transmission noise is introduced in the form of distortions, errors, or certain extraneous matter. The message received is not exactly the message sent. The received message has greater uncertainty and greater information, for the received signal is selected out of more varied set than the sent signal. This means that the information is enhanced during the process of transmission, and the enhancement may not be desirable. This enhancement or loss causes the amount of original information to decline. There is a multitude of factors that may affect the loss of information. For example, in the case of a chart, the language of the chart, its quality, amount of noise, experience of the reader, conditions on which the chart is read, and others affect the loss of information.

The connotation of information can be good or bad. For example, uncertainty due to the freedom of choice the sender has is desirable uncertainty, and the uncertainty due to errors or noise is undesirable uncertainty. Some of the received information is spurious and must be subtracted. Entropy is related to the missing information.

2.3.4 Removal of uncertainty

One can ask the question: How much uncertainty is removed? This uncertainty is referred to as entropy which is defined as an average value of information rate to eliminate uncertainty. Entropy represents a kind of uncertainty of the system before receiving information. Figure 2.3 shows a schematic for eliminating the uncertainty (entropy) from a system S. The uncertainty rate associated with the system is also the amount of information required to eliminate this uncertainty. The implication is that entropy determines the amount of information that must be added to the system. Consider, for example, a person as a receiver of the

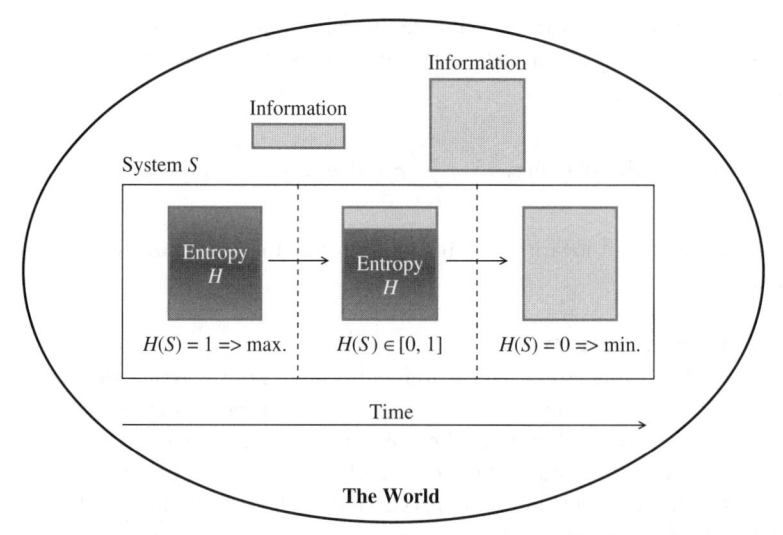

Figure 2.3 Schematic for eliminating uncertainty from a system (Adapted from Paszto et al., 2009, with permission).

information. Removal of uncertainty (entropy) can be insignificant, if this person has either a lack of knowledge or on the other hand complete knowledge/experience with the problem of interest. In the first instance, the person would not be able to comprehend the provided information. In the second instance, the person learns no new facts that he already does not have or know. This means a suitable person would be somewhere in between.

2.3.5 Equivocation

Entropy of one set may be related to another, that is, entropy of a message related to the signal. The message is generated by the source and the signal is what is actually received. The probabilities of these two are inter-related. The probability of receiving, say, a symbol (of a message) depends on the symbol that was sent. In the absence of noise, the received signal would correspond to the message symbols sent. In the presence of noise or errors, the probabilities of received symbols would be closely related to those of message symbols sent. Given the signals received, relative entropies are computed by averaging entropies. The relative entropy is referred to as equivocation. It measures the average uncertainty when the signal is known. In the absence of noise, there is no uncertainty as to the message for the known signal.

Thus, the capacity C of a channel equals the maximum rate (bits/s) at which useful information (i.e., total uncertainty minus noise uncertainty) can be transmitted over the channel. If C is equal to or greater than $H(x)$, then with appropriate coding the source message can be transmitted with little error. If C is less than $H(x)$, then there is always some uncertainty (noise) about what the message was after the signal was received. This undesirable uncertainty (equivocation) will be equal to or greater than $H(x) - C$.

2.3.6 Average amount of information

In practice, the interest is in the amount of information a source generates rather than the amount of information of a particular message. In general, different messages possess different amounts of information. It seems, therefore, logical to determine the average amount of information per message one can expect to get from the source, that is, the average for all the different messages the source may select. The expected value of information from source X is denoted as $H(X)$:

$$H(X) = \text{the mean value of} \left(-\log_2 p_i\right) = \sum_{i=1}^{N} p_i \left(-\log_2 p_i\right) \qquad (2.10)$$

$H(X)$ represents the mean logarithmic probability for all messages from source X. This is the Shannon entropy.

2.3.7 Measurement system

One may ask questions concerning a measurement system. How does one measure the amount of information? How does one measure the capacity of a measurement system? What are the characteristics of noise? How does the noise affect the accuracy of measurement? How can one minimize the undesirable effects of noise? To what extent can they be estimated? In hydrologic observations, problems seem to occur at three levels (Shannon and Weaver, 1949).

Level 1: The technical problem: How accurately can observations be made and transmitted? This entails accuracy of transference of observations from receiver to the sender.

Level 2: The semantic problem: How precisely do the transmitted observations convey the desired meaning? This entails interpretation of the meaning of information.

Level 3: The effectiveness problem: How effectively do the received measurements affect the conduct in the desired way. The success with which the meaning is conveyed to the receiver leads to the desired effect. Level 1 represents mostly engineering aspects, and level 2 and 3 philosophical aspects of the problem, but level 1 does impact levels 2 and 3.

2.3.8 Information and organization

Information offers a way to measure the degree of organization. If a system is well organized meaning its behavior is predictable, then one knows almost what the system is going to do before it does. This means that one learns little or acquires little information when the system does something. Extending this reasoning further, the behavior of a perfectly organized system is completely predictable and hence yields no information at all. On the other hand, the converse is true for a disorganized system. The more disorganized the system the more information its behavior yields and there is much to be learnt by observing the behavior. In this way, information, predictability, and organization are interrelated.

However, Denbigh (1989) argues that informational entropy has no bearing on the second law of thermodynamics. Neither is it related to orderliness, organization, or complexity as discussed by Denbigh and Denbigh (1985). Terms, such as order and organization and their negations, are broad and are subject to large variation, based on context, such as political, legal, etc. Changes in entropy, orderliness, and organization can occur independently. Denbigh (1989) argues that increased orderliness and increase in organization or complexity do not mean the same things.

2.4 Discrete entropy: univariate case and marginal entropy

Equation (2.7) can be generalized to bases of logarithm different from 2. For a random variable X which takes on values $x_1, x_2, \ldots, x_N$ with probabilities $p_1, p_2, \ldots, p_N$, respectively, entropy of X can be expressed as

$$H(X) = H(P) = -K \sum_{i=1}^{N} p(x_i) \log p(x_i) \tag{2.11}$$

where N is the sample size, and K is a parameter whose value depends on the base of the logarithm used. If different units of entropy are used, then the base of the logarithm changes. For example, one uses bits for base 2, Napier for base e, and decibels for base 10. In general, K can be taken as unity. $H(X)$, given by equation (2.11), represents the information content of random variable X and is referred to as the marginal entropy of X. It is a measure of the amount of uncertainty or indirectly the average amount of information content of a single value.

If X is a deterministic variable, then the probability that it will take on a certain value is one, and the probabilities of all other alternative values are zero. Then, equation (2.11) shows that $H(X) = 0$ which can be viewed as the lower limit of the values the entropy function may assume. This corresponds to the absolute certainty. On the other hand, when all x_is are equally likely, that is, the variable is uniformly distributed ($p_i = 1/N$, $i = 1, 2, \ldots, N$), then equation (2.11) yields

$$H(X) = H_{\max}(X) = \log N \tag{2.12}$$

This shows that the entropy function attains a maximum, and equation (2.12) thus defines the upper limit. This also reveals that the outcome has the maximum uncertainty. One can now state that entropy of any process/variable always assumes positive values within limits defined as:

$$0 \leq H(X) \leq \log N \tag{2.13}$$

One can now define relative entropy, H_R, or dimensionless entropy as the ratio of entropy to the maximum entropy:

$$H_R = \frac{H(X)}{H_{\max}(X)} = \frac{H(X)}{\log N} \tag{2.14}$$

Equation (2.14) expresses the comparative uncertainty of outcomes of X defined by $H(X)$ with respect to the entropy value if all outcomes were equally likely. Clearly, the more uncertain the outcome is, the closer the relative entropy (uncertainty) to unity is.

The marginal entropy satisfies the following properties:

1 It is a function of $p_1, p_2, \ldots, p_N$, meaning that $H = H(P) = H(p_1, p_2, \ldots, p_N)$.

2 It is a continuous function of $p_1, p_2, \ldots, p_N$, i.e., small changes in $p_1, p_2, \ldots, p_N$ would result in small changes in H. Often, this property is referred to as continuity.

3 It does not change when outcomes are rearranged among themselves, meaning thereby that H is a symmetric function about its arguments, $p_1, p_2, \ldots, p_N$. It does not change with numbering or ordering. This property is referred to as symmetry.

4 It is maximum when outcomes are equally likely or follow a uniform distribution which means that the maximum uncertainty is when:

$$p_1 = p_2 = \cdots = p_N = 1/N \tag{2.15}$$

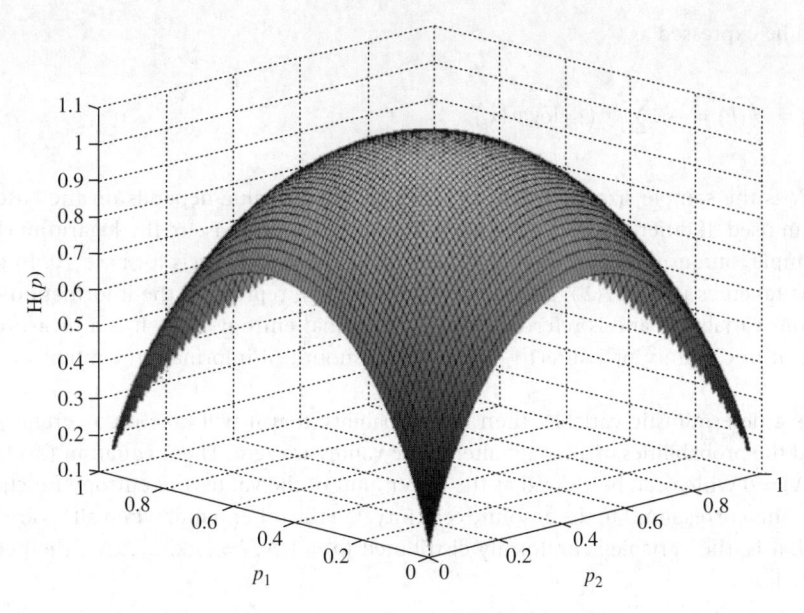

Figure 2.4 Entropy of a distribution $P : \{p_1, p_2, p_3\}$ as function of probabilities [Note: $p_1 + p_2 + p_3 = 1$].

The maximum value of H is $\log N$. This is the maximum property. Entropy is a smooth concave function of probabilities. For $N = 3$, Figure 2.4 shows the scaled entropy. The entropy of a distribution with infinite support may be infinite.

For the special case of $N = 2$, the entropy measured in bits is

$$0 \leq H(p) \leq 1$$

Maximum uncertainty or entropy can be considered to correspond to a condition where the expected information from actual events is also maximized.

Example 2.10: Show that the maximum of Shannon entropies for probability distributions with N elements is $\log_2 N$.

Solution: Let there be two probability distributions $P : \{p_1, p_2, \ldots, p_N\}$ and $Q : \{q_1, q_2, \ldots, q_N\}$. Using the inequality

$$-\sum_{i=1}^{N} p_i \log_2 p_i \leq -\sum_{i=1}^{N} p_i \log_2 q_i$$

for all probability distributions $\{p_1, p_2, \ldots, p_N\}$ and $\{q_1, q_2, \ldots, q_N\}$. The above equation is sometimes referred to as Gibbs theorem. If q's are uniformly distributed then

$$-\sum_{i=1}^{N} p_i \log_2 p_i \leq -\sum_{i=1}^{N} p_i \log_2 \frac{1}{N} = -\log_2 \frac{1}{N} \sum_{i=1}^{N} p_i = \log_2 N$$

Thus, entropy is bounded as $H(p_1, p_2, \ldots, p_N) \leq \log_2 N$, with the upper bound obtained for $p_i = 1/N$.

5 A corollary of the above property is that the uncertainty would increase with increasing N. If there are two positive integers, M and N, $M < N$, then $H(M) < H(N)$. This is also referred to as monotonicity.

6 If an impossible outcome is added, entropy or uncertainty does not change, that is,

$$H\left(p_1, p_2, \ldots, p_N, 0\right) = H\left(p_1, p_2, \ldots, p_N\right)$$

This property is referred to as expansibility.

7 Entropy is always positive and is equal to zero when the outcome is certain, that is, $p_k = 1, p_i = 0, i \neq k$. Thus, $0 \leq H \leq \log N$.

8 Any change toward equalization of probabilities increases entropy. For example, let us consider probabilities p_1 and p_2 of two events named 1 and 2. Assume that $p_1 < p_2$. If we increase p_1 and decrease p_2 by the same amount then entropy would increase which means uncertainty would increase.

Example 2.11: There are two bags numbered as 1 and 2. Number 1 bag contains ten balls: three red and seven black, and number 2 bag contains ten balls: four red and six black. The experiment is to choose one ball from number 1 bag without seeing. Compute its entropy. This experiment is repeated for number 2 bag. Compare the entropies of the two experiments on the two bags in bits.

Solution: For number 1 bag: $p_1 = 3/10$ for ten red balls, and $p_2 = 7/10$ for black balls.

$$H = E\left[I\right] = -\sum_i^N p_i \log_2 p_i = -\left(0.3 \times \log_2 0.3 + 0.7 \times \log_2 0.7\right) = 0.8813 \text{ bits}$$

For number 2 bag: $p_1 = 4/10$ for red balls, and $p_2 = 6/10$ for black balls.

$$H = E\left[I\right] = -\sum_i^N p_i \log_2 p_i = -\left(0.4 \times \log_2 0.4 + 0.6 \times \log_2 0.6\right) = 0.971 \text{ bits}$$

The entropy of number 2 bag is higher.

9 $H_N(P) = H(p_1, p_2, \ldots, p_N)$ is a convex function of $p_1, p_2, \ldots, p_N$. This means that the local maximum of $H_N(p)$ is equal to the global maximum.

10 Entropy also satisfies the property of subadditivity. A joint probability distribution is derived from marginal probability distributions. According to this property, the uncertainty of this joint distribution should not be larger than the sum of uncertainties of the corresponding marginal distributions. For two random variables, $X = \{x_1, x_2, \ldots, x_N\}$ with probability distribution as $P_X = \{p_1, p_2, \ldots, p_N\}$ and $Y = \{y_1, y_2, \ldots, y_N\}$ with probability distribution as $Q_Y = \{q_1, q_2, \ldots, q_N\}$, the sub-additivity property can be expressed as

$$H(p_1, p_2, \ldots, p_N; q_1, q_2, \ldots, q_N) \leq H\left(p_1, p_2, \ldots, p_N\right) + H(q_1, q_2, \ldots, q_N)$$

11 If the variables are independent or noninteractive, then the uncertainty of any joint distribution of variables becomes equal to the sum of uncertainties of the corresponding marginal distributions. This property is called additivity. For random variables X and Y defined as above, this property can be expressed as

$$H(p_1, p_2, \ldots, p_N; q_1, q_2, \ldots, q_N) = H\left(p_1, p_2, \ldots, p_N\right) + H(q_1, q_2, \ldots, q_N)$$

12 Entropy exhibits a recursive property defined as:

$$H_N(p_1, p_2, \ldots, p_N) = H_{N-1}(p_1 + p_2, p_3, p_4, \ldots, p_N) + (p_1 + p_2)H_2\left(\frac{p_1}{p_1 + p_2}, \frac{p_2}{p_1 + p_2}\right)$$

$$= H_{N-2}(p_1 + p_2 + p_3, p_4, \ldots, p_N) + (p_1 + p_2 + p_3)$$

$$\times H_3\left(\frac{p_1}{p_1 + p_2 + p_3}, \frac{p_2}{p_1 + p_2 + p_3}, \frac{p_3}{p_1 + p_2 + p_3}\right) \qquad (2.16)$$

The advantage of this property is that if $H_2(p_1, p_2)$ is known then one can successively determine $H_3(p_1, p_2, p_3), H_4(p_1, p_2, p_3, p_4), \ldots$ It then follows that

$$H_N(p_1, p_2, \ldots, p_N) \geq H_{N-1}(p_1 + p_2, p_3, p_4, \ldots, p_N) \geq H_{N-2}(p_1 + p_2 + p_3, p_4, p_5, \ldots, p_N)$$

$$\geq \ldots \geq H_2(p_1 + p_2 + \ldots + p_{N-1}, p_N) \qquad (2.17)$$

This means that uncertainty is reduced by combining outcomes and increased by decomposing outcomes. This property is also referred to as branching.

If X is divided into two groups A and B, defined as $A = \{x_1, x_2, \ldots, x_m\}$ and $B = \{x_{m+1}, x_{m+2}, \ldots, x_N\}$ and $p_A = \sum_{i=1}^{m} p_i$ and $p_B = \sum_{i=m+1}^{N} p_i$ then sometimes the branching property is also expressed as

$$H(p_1, p_2, \ldots, p_N) = H(p_A, p_B) + p_A H\left(\frac{p_1}{p_A}, \frac{p_2}{p_A}, \ldots, \frac{p_m}{p_A}\right) + p_B H\left(\frac{p_{m+1}}{p_B}, \frac{p_{m+2}}{p_B}, \ldots, \frac{p_N}{p_B}\right) \qquad (2.18)$$

This is essentially a grouping or consistency requirement.

Example 2.12: Assume that a probability distribution is given as:

$$P = \{p_1, p_2, \ldots, p_{10}\} = \{0.015, 0.035, 0.1, 0.15, 0.20, 0.25, 0.14, 0.06, 0.04, 0.01\}$$

Using these probability values show that the recursive property of entropy holds.

Solution: Entropy computations are shown in Table 2.2.

Entropy and the total entropy associated with the specified probability values are given in column 1 of Table 2.2. In columns 2 and 3, entropy is calculated as per the following formulae [equation (2.16)]:

Column 2:

$$H_{N-1}(p_1 + p_2, p_3, p_4, \ldots, p_N) + (p_1 + p_2)H_2\left(\frac{p_1}{p_1 + p_2}, \frac{p_2}{p_1 + p_2}\right)$$

Column 3:

$$H_{N-2}(p_1 + p_2 + p_3, p_4, \ldots, p_N) + (p_1 + p_2 + p_3)H_3\left(\frac{p_1}{p_1 + p_2 + p_3}, \frac{p_2}{p_1 + p_2 + p_3}, \frac{p_3}{p_1 + p_2 + p_3}\right)$$

Table 2.2 Computation for showing the recursive property of entropy.

Column 1			Column 2			Column 3		
P	P	H (bits)	P	p	H (bits)	P	P	H (bits)
p_1	0.015	0.0909						
p_2	0.035	0.1693	$p_1 + p_2$	0.05	0.2161			
p_3	0.1	0.3322	P_3	0.1	0.3322	$p_1 + p_2 + p_3$	0.1500	0.4105
p_4	0.15	0.4105	P_4	0.15	0.4105	p_4	0.1500	0.4105
p_5	0.2	0.4644	P_5	0.2	0.4644	p_5	0.2000	0.4644
p_6	0.25	0.5000	P_6	0.25	0.5000	p_6	0.2500	0.5000
p_7	0.14	0.3971	P_7	0.14	0.3971	p_7	0.1400	0.3971
p_8	0.06	0.2435	P_8	0.06	0.2435	p_8	0.0600	0.2435
p_9	0.04	0.1858	P_9	0.04	0.1858	p_9	0.0400	0.1858
p_{10}	0.01	0.0664	P_{10}	0.01	0.0664	p_{10}	0.0100	0.0664
	Σ	**2.8601**	$p_1/(p_1 + p_2)$	0.3	0.5211	$p_1/(p_1 + p_2 + p_3)$	0.1000	0.3322
			$p_2/(p_1 + p_2)$	0.7	0.3602	$p_2/(p_1 + p_2 + p_3)$	0.2333	0.4899
						$p_3/(p_1 + p_2 + p_3)$	0.666667	0.3900
				Σ	2.8601			
							Σ	2.8601

The total entropy as computed from individual probabilities is seen to be the same as that calculated in columns 2 and 3. It can therefore be deduced that entropy follows a recursive property.

If entropy is measured in bits, then one obtains:

$$H\left(\frac{1}{2}, \frac{1}{2}\right) = 1 \text{bit}$$

This is the normalization property, and can be amended if other measurement units are employed.

Example 2.13: Consider a random variable X which takes on two values, a value of 1 with a probability p and a value of 0 with a probability $q = 1m\,p$. Taking different values of p, plot $H(p)$ as a function of p. This is analogous to tossing a coin where the occurrence of head has probability p and that of tail has probability q. Compute the entropy by taking different values of p and graph it.

Solution: The entropy can be expressed as

$$H(P) = -p\log p - q\log q = -p\log p - (1 - p)\log(1 - p)$$

The graph of $H(P)$ versus p is shown in Figure 2.5. It is seen that for $p = 0.5$, $H(P) = 1$ bit is the maximum. The function $H(P)$ is symmetrical, convex and attains its maximum at $p = q = 0.5$. Also, $H(p = 0) = H(p = 1) = 0.0$.

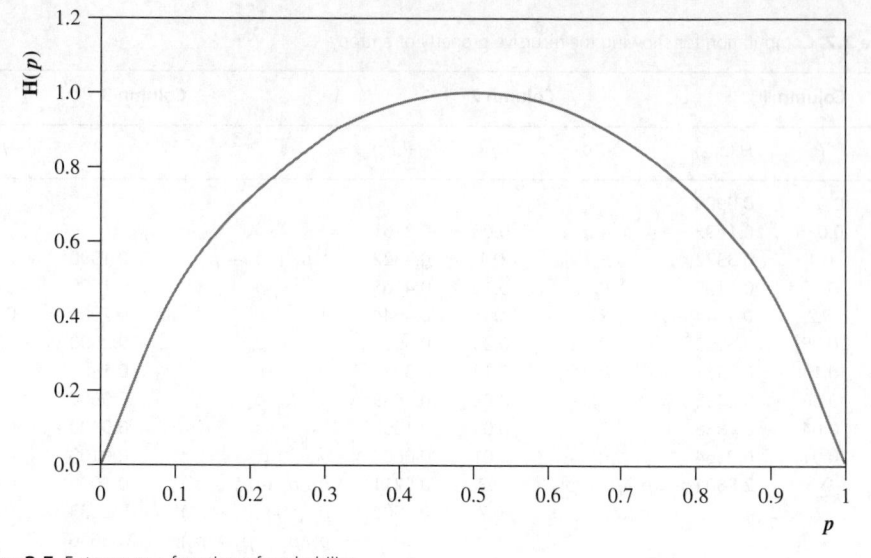

Figure 2.5 Entropy as a function of probability.

2.5 Discrete entropy: bivariate case

In general, when there are more sources generating information three cases arise: 1) The sources are independent, 2) the sources are perfectly correlated, and 3) the sources are partially correlated. To illustrate these cases, consider the power situation on a given day in a city. Let there be three people, labeled as A, B, and C, who make binary choices designated by P for power and F for failure. The question arises: What binary choices would these people make? This case entails triple-choice events and one may want to determine the outcome of the triple-choice event. Without further knowledge or information, there are eight possible outcomes $= 2^3$: PPP, PPF, PFF, FFF, FPP, FFP, PFP, FPF. If each outcome is equally likely, then $p_i = 1/8$, $i = 1, 2, \ldots, 8$, then one would need $\log_2 8 = \log_2 2^3 = 3$bits of information to determine the outcome. If one person (A) divulges what his choice would be then the number of outcomes reduces from 8 to 4: PP, PF, FF, FP. One bit of information then is gained from the disclosure by A. Extending this logic further, one can determine the amount of information to be gained if both B and C tell what they would do.

Case I: Both B and C make their choices independently. Knowledge about what B will choose conveys nothing about what C will do. If one bit of information is gained from B and one bit of information from C then the total amount of information gained from B and C together is two bits of information.

Case II: If B and C are perfectly correlated then knowledge about B also translates into knowledge about C, and vice versa. This means that once information is acquired from B then C adds nothing to the information already acquired. If both B and C always make the same choices then the number of outcomes reduces to four: PPP, FPP, PFF, FFF, for A makes a choice of either P or F. Then only two bits of information are needed to select the outcome.

Case III: In general, B and C are partially, not fully, correlated. If one knows what B will do, one can make a reasonable guess what C will do, and vice versa. The information gained from B is not duplicated by C. Therefore, this case falls between case I and case II.

2.5.1 Joint entropy

Consider two simultaneous experiments whose outcomes are represented by X and Y with probability distributions P_X and P_Y. For mathematically formulating the information associated with cases involving more than one variable, consider these two random variables X and Y. Let X take on values $x_1, x_2, x_3, \ldots, x_N$ with corresponding probabilities $P = \{p_1, p_2, p_3, \ldots, p_N\}$; and let Y take on values $y_1, y_2, y_3, \ldots, y_M$ with corresponding probabilities $Q = \{q_1, q_2, q_3, \ldots, q_M\}$, such that

$$\sum_{i=1}^{N} p_i = 1, \ p_i \geq 0; \quad \sum_{j=1}^{M} q_j = 1, \ q_j \geq 0 \tag{2.19}$$

Note that each value of a random variable represents an event. The joint probability of x_i and y_j can be denoted as $p(x_i, y_j) = p_{ij}, i = 1, 2, 3, \ldots, N; j = 1, 2, 3, \ldots, M$.

The joint entropy of X and Y can now be defined as

$$H(X, Y) = -\sum_{i=1}^{N} \sum_{j=1}^{M} p_{ij} \log p_{ij} \tag{2.20}$$

The joint entropy expresses the total amount of uncertainty contained in the merger or union of events representing X and Y, and depends on the joint probability distribution of X and Y.

If X and Y are stochastically independent, then $p(x_i, y_j) = p(x_i) \, p(y_j)$ or $p_{ij} = p_i p_j$ where $p(x_i) = p_i, p(y_j) = p_j$. This leads to the definition of joint entropy of independent variables as

$$H(X, Y) = H(X) + H(Y) \tag{2.21}$$

Clearly $H(X)$ depends on the probability distribution of X and the same for Y. $H(X, Y)$ in equation (2.21) indicates the total amount of uncertainty that X and Y entail or the total amount of information they convey, provided they are independent. Consider, for example, the flood process characterized by flood peak (X) and flood volume (Y). Then the uncertainty of the flood process would be represented by equation (2.20). A practical implication is that more observations on X and Y would reduce the uncertainty more than more observations of only X or Y.

2.5.2 Conditional entropy

Consider two simultaneous outcomes or events A and B occurring with probabilities P_A and P_B, respectively. If events A and B are correlated then the occurrence of event A gives some indication about the occurrence of event B. The probability of occurrence of B knowing the occurrence of A is $P(B|A)$ is different from the probability of occurrence of B, $P(B)$. This then explains the statistical dependence. The conditional entropy of $B|A$ is $H(B|A)$, that is, the entropy of B conditioned on A, and the global entropy of the two outcomes is

$$H(AB) = H(A) + H(B|A) \tag{2.22a}$$

Since $H(AB) = H(BA)$, one can write

$$H(A) + H(B|A) = H(B) + H(A|B) \tag{2.22b}$$

where $H(A|B)$ is the entropy of A conditioned on B or conditional entropy of B. Equation (2.22a) can be generalized for any number of outcomes, $A, B, C, \ldots, Z$, as

$$H(ABC \ldots Z) = H(A) + H(B|A) + H(C|AB) + \ldots + H(Z|ABC \ldots) \tag{2.23}$$

It is seen from equation (2.23) that the general entropy includes the contribution of each event after deduction of the contribution of what has already been accounted for by other events.

Consider an event $X = x_i$ and an event $X = m$. Then the conditional probability of x_i given m, denoted as $p(x_i|m)$, can be expressed as

$$p\left(x_i|m\right) = \frac{p\left(x_i, m\right)}{p\left(m\right)} \tag{2.24}$$

The conditional entropy becomes

$$H(X|m) = -\sum_{i=1}^{N} p\left(x_i|m\right) \log p(x_i|m) \tag{2.25}$$

This corresponds to uncertainty in X given m.

Now consider in the subsequence of trials X in which m occurs. Now consider another variable or sequence of trials $Y : \{y_j, j = 1, 2, \ldots, M\}$. Then, the conditional probability of y_j given $Y = w$, denoted as $p(y_j|w)$, can be expressed as

$$p\left(y_j|w\right) = \frac{p\left(y_j, w\right)}{p\left(w\right)} \tag{2.26}$$

The conditional entropy of X given y_j becomes

$$H\left(X|y_j\right) = -\sum_{i=1}^{N} p\left(x_i|y_j\right) \log p(x_i|y_j) \tag{2.27}$$

Now the conditional entropy of X given Y can be written as

$$H(X|Y) = -\sum_{j=1}^{M} p\left(y_j\right) H(X|y_j) \tag{2.28}$$

The conditional entropy of X given Y is the entropy $H(X|y_j)$ weighted over $j = 1, 2, \ldots, M$. This is the uncertainty about X if at each trial event y_j of Y is known to have occurred.

Example 2.14: Consider a six-faced dice: $X : \{1, 2, \ldots, 6\} = \{faces\}$. Let Y be defined as [even, odd], that is, it takes on two values as either even or odd. Compute the probability of an even face occurring upon throw, given an even face; probability of an even face occurring upon throw, given an odd face; probability of an odd face occurring upon throw, given an odd face; and probability of an odd face occurring upon throw, given an even face. Compute the entropy of dice throw given an even face, and entropy of the throw given an odd face. Also compute the entropy of the dice throw and then the reduction in uncertainty knowing that at each trial either odd or even face occurred.

Solution: Each face has a probability of occurrence as 1/6, that is, $p_i = 1/6$. However, the conditional probability $p(i - th\ even\ face|even\ face) = 1/3$ if i is even and $p(i - th\ even\ face|odd\ face) = 0$, i is odd. Likewise, $p(i - th\ odd\ face|odd\ face) = 1/3$, if i is odd, and $p(i - th\ odd\ face|even\ face) = 0$, if i is even. Then

$$H\ (X|even) = -\frac{1}{3}\log_2\frac{1}{3} - \frac{1}{3}\log_2\frac{1}{3} - \frac{1}{3}\log_2\frac{1}{3} = \log_2 3 = 1.585\ \text{bits}$$

$$H\ (X|odd) = -\frac{1}{3}\log_2\frac{1}{3} - \frac{1}{3}\log_2\frac{1}{3} - \frac{1}{3}\log_2\frac{1}{3} = \log_2 3 = 1.585\ \text{bits}$$

$$H\ (X|even) = H(X|odd)$$

Since even and odd faces constitute two events one of which occurs when the dice is thrown. Therefore, $p(even) = p(odd) = 0.5$, and $H(X) = \log_2 2 = 1\text{bit}$. Then,

$$H\ (X|Y) = 0.5\log_2 3 + 0.5\log_2 3 = \log_2 3 = 1.585\ \text{bits}$$

$$H\ (X) = -\frac{1}{6}\log_2\frac{1}{6} - \frac{1}{6}\log_2\frac{1}{6} - \frac{1}{6}\log_2\frac{1}{6} - \frac{1}{6}\log_2 - \frac{1}{6}\log_2\frac{1}{6} - \frac{1}{6}\log_2\frac{1}{6} = \log_2 6$$

$$= 2.585\ \text{bits}$$

This expresses the uncertainty about X. However, at each trial if it is known that either odd or even face occurred then the uncertainty is reduced from $\log_2 6 = 2.585$ bits to $\log_2 3 = 1.585$ bits.

Now the discussion is extended to two random variables X and Y. The conditional entropy $H(X|Y)$ can be defined as the average of entropy of X for each value of Y weighted according to the probability of getting that particular value of Y. Mathematically,

$$H\ (X|Y) = -\sum_{i,j} p\left(x_i, y_j\right)\log p(x_i|y_j) \tag{2.29a}$$

Similarly,

$$H\ (Y|X) = -\sum_{i,j} p\left(x_i, y_j\right)\log p(y_j|x_i) \tag{2.29b}$$

where $p(x_i, y_j)$, $i = 1, 2, 3, \ldots, N; j = 1, 2, \ldots, M$, are the joint probabilities; and $p(x_i|y_j)$ and $p(y_j|x_i)$ are conditional probabilities.

Let $X : \{x_1, x_2, \ldots, x_N\}$ and $Y : \{y_1, y_2, \ldots, y_M\}$.

$$P\left(X = x_i|Y = y_j\right) = p_{ji}/p_j = q_{ji} \tag{2.30}$$

$$H\left(X|y_j\right) = -\sum \frac{p_{ji}}{p_j}\log\frac{p_{ji}}{p_j} = -\sum_{i=1}^{N} q_{ji}\log q_{ji} \tag{2.31}$$

The conditional entropy $H(X|Y)$ of X given Y is the conditional entropy of $X : \{x_1, x_2, \ldots, x_N\}$ given $Y : \{y_1, y_2, \ldots, y_M\}$, that is,

$$H(X|Y) = -\sum_{j}^{M} p_j H\left(X|y_j\right) = -\sum_{i=1}^{N} \sum_{j=1}^{M} p_{ji} \log\left(\frac{p_{ji}}{q_j}\right) = -\sum_{i=1}^{N} \sum_{j=1}^{M} p_{ji} \log q_{ji} \qquad (2.32)$$

The conditional entropy $H(X|Y)$ in equation (2.29a) gives the amount of uncertainty still remaining in X after Y becomes known or has been observed, and is expressed in terms of conditional probabilities. With the knowledge of the values of Y, say y_j ($j = 1, 2, \ldots, M$), the uncertainty in X will be reduced. In other words, the knowledge of Y will convey information on X. Thus it can be stated that conditional entropy $H(X|Y)$ is a measure of the amount of uncertainty remaining in X even with the knowledge of Y; which is less than the same amount of information gained by observing X. The amount of reduction in uncertainty in X equals the amount of information gained by observing Y. It then follows that the conditional entropy of one variable, say X, with respect to the other, say Y, will be less than the marginal entropy of X:

$$H(X|Y) \leq H(X) \qquad (2.33a)$$

Likewise,

$$H(Y|X) \leq H(Y) \qquad (2.33b)$$

wherein equality holds if X and Y are independent of each other. It is noted that

$$H(Y|X) \neq H(X|Y) \qquad (2.34a)$$

but

$$H(X) - H(X|Y) = H(Y) - H(Y|X) \qquad (2.34b)$$

In the case of sending messages and consequent transmission of signals one can also state the average uncertainty as to the received signal for the message sent. Here $H(X)$ is the entropy or information of the source of messages, $H(Y)$ is the entropy or information of the received signals, $H(X|Y)$ is the equivocation or the uncertainty in the message source given the signals, and $H(Y|X)$ is the uncertainty in the received signal given the sent messages or the spurious part of the received signal information due to noise. The right side of the equation is useful information transmitted in spite of the noise.

When X and Y are dependent, as may frequently be the case, say, for example, flood peak and flood volume, then their joint (or total) entropy equals the sum of marginal entropy of the first variable (say X) and the entropy of the second variable (say Y) conditioned on the first variable (X) (that is, the uncertainty remaining in Y when a certain amount of information it conveys is already present in X):

$$H(X, Y) = H(X) + H(Y|X) \qquad (2.35a)$$

or

$$H(X, Y) = H(Y) + H(X|Y) \tag{2.35b}$$

Noting equations (2.35a) and (2.35b), the total entropy $H(X, Y)$ of two dependent variables X and Y will be less than the sum of marginal entropies of X and Y:

$$H(X, Y) \leq H(X) + H(Y) \tag{2.35c}$$

Note that one can also write

$$\log p(x, y) = \log p(x) + \log p(y|x) \tag{2.36a}$$

Expectation of both sides of this equation yields equation (2.35a). Furthermore, if Z is another variable exhibiting some degree of dependence on X and Y, then

$$H(X, Y|Z) = H(X|Z) + H(Y|X, Z) \tag{2.36b}$$

2.5.3 Transinformation

To develop an intuitive appreciation of transinformation, consider a river reach receiving inflow (as a function of time) at its upstream end and discharging flow at its downstream end (outflow as a function of time). It is assumed that the reach does not receive or discharge flow from its sides. Then the amount of association between the inflow hydrograph and the outflow hydrograph of the channel is measured by transmitted information which is referred to as transinformation. If inflow and outflow are perfectly correlated then all the inflow information is transmitted. Here the connotation is that the inflow hydrograph appears as outflow hydrograph undisturbed. No information is transmitted if inflow and outflow are independent. This, of course, does not occur in hydrology. The perfect correlation and independence represent two extremes. In nature most cases fall in between where some information is transmitted and some is not. Our interest is in the amount of information transmitted, not what the transmitted information is.

Transinformation of two different events A and B, $T(A, B)$, is the quantity of information common to the two events and is also called mutual information, $I(A, B)$, and can be expressed as

$$T(A, B) = I(A, B) = H(A) - H(A|B) \tag{2.37a}$$

Likewise,

$$T(B, A) = I(B, A) = H(B) - H(B|A) \tag{2.37b}$$

It is now noted that

$$T(B, A) = H(A) + H(B|A) - H(A|B) - H(B|A) = H(A) - H(A|B) = I(A, B) \tag{2.38}$$

This shows that the mutual information is symmetric and can be extended to any number of events. Consider three events A, B, and C. Then, one can write:

$$T(AB, C) = H(AB) - H(AB|C) \tag{2.39}$$

The concept of mutual information can be used in a variety of problems, such as network evaluation and design, flow regime classification, watershed clustering, and so on. In the network design the objective is to minimize mutual information between gages to avoid repetition of observed information. The bigger the mutual information between two gages A and B, the less useful the two are together. It can be said that the role of gage B, used after gage A, must be proportional to the new information it actually conveys after deduction of the mutual information already accounted for in A, that is, proportional to $H(B) - T(A, B) = H(B|A)$.

Now the discussion is extended to two discrete random variables X and Y. Transinformation represents the amount of information common to both X and Y or repeated in both X and Y, and is denoted as $T(X, Y)$. Since $H(X)$ represents the uncertainty about the system input (X) before observing the system output (Y) and the conditional entropy $H(X|Y)$ represents the uncertainty about the system input after observing the system output, the difference between $H(X)$ and $H(X|Y)$ must represent the uncertainty about the system input that is reduced by observing the system output. This difference is often called the average mutual information between X and Y, or transinformation T of X and Y. Transinformation is also referred to as mutual information and is a measure of the dependence between X and Y and is always non-negative. It is a measure of the amount of information random variable X contains about random variable Y. This means that $T(X, Y)$ defines the amount of uncertainty reduced in X when Y is known. It equals the difference between the sum of two marginal entropies and the total entropy:

$$T(X, Y) = H(X) + H(Y) - H(X, Y) \tag{2.40a}$$

or

$$T(X, Y) = -\sum_{i=1}^{N} p(x_1) \log p(x_i) - \sum_{j=1}^{M} p(y_j) \log p(y_j) + \sum_{i=1}^{N} \sum_{j=1}^{M} p(x_i, y_j) \log p(x_i, y_j) \tag{2.40b}$$

One can also write

$$T(X, Y) = \sum_{i=1}^{N} \sum_{j=1}^{M} p(x_i, y_j) \log \frac{p(x_i|y_j)}{p(x_i)} \tag{2.41a}$$

or

$$T(X, Y) = \sum_{i=1}^{N} \sum_{j=1}^{M} p(x_i, y_j) \log \frac{p(x_i, y_j)}{p(x_i) p(y_j)} \tag{2.41b}$$

$T(X, Y) \geq 0$.

When X and Y are independent, $T(X, Y) = 0$. Taking account of equations (2.37a) and (2.37b), equation (2.40a) can be written as

$$T(X, Y) = H(Y) - H(Y|X) \tag{2.42a}$$
$$T(X, Y) = H(X) - H(X|Y) \tag{2.42b}$$

Marginal entropy can be shown to be a special case of transinformation or mutual information, because

$$H(X) = T(X, X) \tag{2.43a}$$
$$H(Y) = T(Y, Y) \tag{2.43b}$$

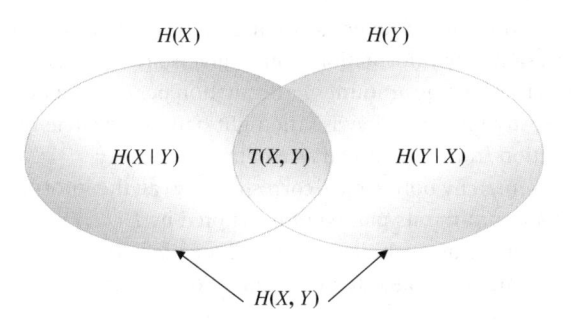

Figure 2.6 $T(X, Y)$: Information common to X and Y; $H(X \mid Y)$: information in X given Y; $H(Y \mid X)$: information in Y given X; and $H(X,Y)$: total information in X and Y together.

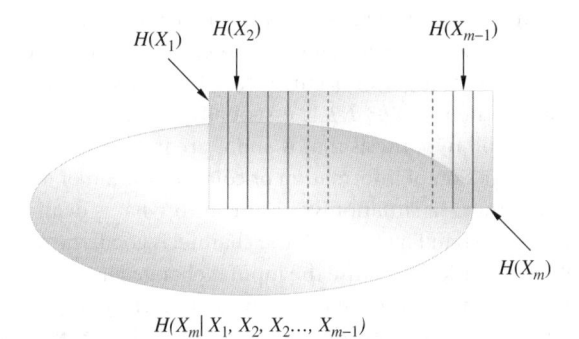

Figure 2.7 Marginal entropies $H(X_1)$, $H(X_2)$, $\cdots$, $H(X_m)$, and conditional entropy $H(X_m \mid X_1, X_2, \ldots, X_{m-1})$.

The concepts of marginal, conditional, and joint entropies and transinformation can be represented graphically as shown in Figures 2.6 and 2.7.

Let the entropy associated with X be represented by set S_1 and the entropy associated with Y by set S_2. Then union of sets S_1 and S_2 is equivalent to their joint entropy, the intersection of S_1 and S_2 equivalent to their transinformation or mutual entropy and the set difference of S_1 and S_2 equivalent to the conditional entropy. Noting equations (2.40a) and (2.40b) and referring to Figure 2.6, equations (2.42a) and (2.42b) show that the marginal entropy can be expressed as a sum of transinformation and conditional entropy:

$$H(X) = H(X|Y) + T(X, Y) \tag{2.44a}$$
$$H(Y) = H(Y|X) + T(X, Y) \tag{2.44b}$$

Equations (2.44a) and (2.44b) show that the marginal entropy of X is equal to the amount of uncertainty in X given the knowledge about Y plus transinformation or the amount of uncertainty that is reduced by the knowledge about Y.

Returning to the aforementioned three cases in Section 2.5.2, based on the degree of dependence, the situation in case III is graphed in Figure 2.6. The left circle corresponds to the information gained from B and the right circle corresponds to the information gained from C. The term $H(X)$ denotes the average amount of information in bits per event expected from source B. It is determined from probabilities of choices made by B. Likewise for $H(Y)$ and its computation. The overlap of two circles corresponds to the common information due to the correlation between B and C and its average amount in bits per event is denoted by T. The left half of the circle represents the information from B given the information from C, denoted as $H(X|Y)$ which defines the average amount of information per event that is to be obtained from source B after C is already known. Likewise the right half of the circle represents information from C given information from B, denoted as $H(Y|X)$.

The total area represented by both circles corresponds to all the information that both B and C contain. This total amount in bits per event is denoted by $H(X, Y)$. It is calculated from the joint probabilities of double (joint) choice of B and C together. The remainder of the quantities can be determined in a manner suggested by Figure 2.6. For example,

$$H(X|Y) = H(X, Y) - H(Y)$$
$$T = H(X) + H(Y) - H(X, Y)$$

T possesses the attributes of a measure of correlation (contingency, dependence) between B and C but it is not equivalent to correlation – rather it is more a measure of indepdence between B and C. If N is the number of occurrences of the event that are used to determine the probabilities involved, then $1.3863\ NT$ is essentially the same as the value of chi-square one calculates to test the null hypothesis that B and C are independent (Miller, 1953).

Now consider the transmission of information or communication of message again from one end of the conduit to another – a situation very similar to what is dealt with in input–output models in hydrology, as shown in Figure 2.8. If the channel is good, the output is closely related to the input but is seldom identical, because the input is changed in the process of transmission. If the change is random then it is referred to as noise reflecting the channel characteristics. The variables involved in the transmission can be associated with the various quantities of information shown in Figure 2.8. Let X be the source that generates the input information and Y be the source that generates the output information. $H(X)$ is the average amount of input information, and $H(Y)$ is the average amount of output information. The amount of information that is transmitted is the overlap or common information $T(X, Y)$, since X and Y are related sources of information. $T(X, Y)$ is the average amount of transmitted information. Sometimes T is referred to as the average amount of throughput information.

$H(X|Y)$ can be interpreted as the information that is put in but is not received. In other words, this information is lost in transmission. It is sometimes referred to as "equivocation" or lost information, because a receiver cannot determine if it was sent. Likewise, $H(Y|X)$ represents the information that is not put in but is not received; it is added during transmission. It is referred to as "noise" in the sense that it interferes with good communication.

$H(X, Y)$ represents the total amount of information one has when both input information and output information are known. One can say that $H(X, Y)$ is the sum of the lost information,

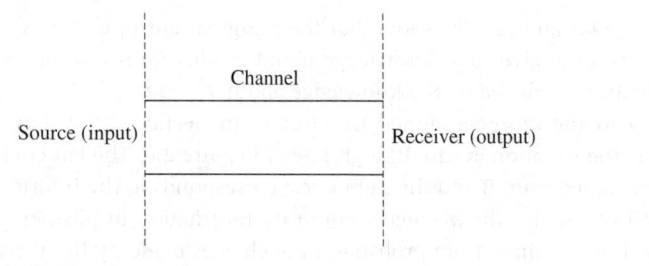

Figure 2.8 Transmission of information.

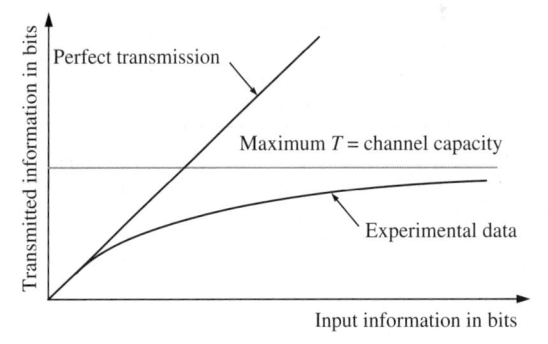

Figure 2.9 Amount of information transmitted as a function of the amount of input information. For perfect transmission, transmitted information = input information.

transmitted information and added information or noise:

$$H(X, Y) = H(X|Y) + T(X, Y) + H(Y|X) \tag{2.45}$$

where $H(X|Y)$ represents the lost information (or equivocation), $T(X, Y)$ represents the transmitted information, and $H(Y|X)$ represents the information not sent or system noise.

In communication, one wants to compute $T(X, Y)$, the amount of information transmitted by the channel. Each channel has an upper limit to the amount of information it transmits and this upper limit is referred to as channel capacity, C. As $H(X)$ increases, $T(X, Y)$ attains an upper bound and does not increase thereafter. This is depicted in Figure 2.9. Referring to the input-channel-output representation, $H(X)$ is the excitation or stimulus information, $H(Y)$ is the response information, and T measures the degree of dependence of responses on stimuli. T can also be considered as a measure of discrimination, and C is the basic capacity of the subject to discriminate among the given stimuli.

Consider student grading in a class. The purpose of the test is to discriminate among students with respect to some scholastic achievement criterion or dimension. Each student who takes the test achieves a certain score on this dimension. The result of the test will indicate this score. In the information parlance, the student can be considered as a channel. True test values are input information X, and test scores are output information Y. For a good test T is large and $H(Y|X)$ is small. The test score can discriminate accurately among students who take it. Transinformation helps define the number of classes students can be distinguished in. This analogy can be extended to see the similarity between any process of measurement and the transmission situation. Nature provides input, the process of measurement constitutes the channel, and measurements themselves constitute the output.

Consider the measurement of rainfall. The rain that falls is the input, the way it is measured is the channel, and the resulting measured values the output (see Figure 2.10).

In the preceding discussion it is assumed that all successive occurrences of the event are independent. In many cases this assumption is not tenable, and is not true in behavioral processes. Consider, for example, a sequence of rainfall and runoff or flood events in a watershed. Rainfall is the input, watershed is the channel and runoff is the output. The amount of information of runoff is not the same as the amount of information of rainfall, but there is some common information between them. This repeated information is "transinfomation." Let X be the rainfall source that generates input information and let Y be the runoff source that generates the output information. $H(X)$ is the average amount of information in bits in

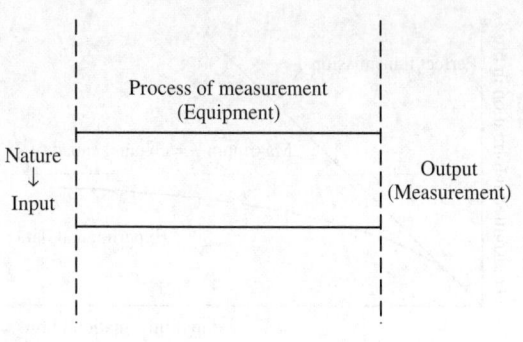

Figure 2.10 Measurement and transmission of information.

the N-rainfall sequence, and $H(Y)$ is the average amount of information in bits in the N-flood sequence. $H(Y|X)$ is the average amount of information per event in Y when X is known, and expresses the rate at which information is generated by the source. It measures the average number of bits per unit or event.

Transinformation and correlation

Although the joint information (transinformation) is a measure of dependence or independence, it is not equivalent to correlation. If event A is determined by event B then $p(A|B) = 1$, and $B \subset A$. The information in $(AB) = B$ is measured by $H(B)$, which may be small. The common information carried by two partially correlated variables is greater than the information of a group of two variables where one completely governs the other.

Calculation of transinformation

From the point of view of calculating T, consider a discrete input variable X (say, rainfall) taking on values x_i, $i = 1, 2, 3, \ldots, N$, and a discrete output variable Y (say, flood) taking on values y_j, $j = 1, 2, 3, \ldots, M$. It is assumed that when event x_i occurs, event y_j is caused. Thus, one can think of a joint input–output event (x_i, y_j) having a probability $p(x_i, y_j)$. Of course, it is true that

$$\sum_{i=1}^{N} p(x_i) = \sum_{j=1}^{M} p(y_j) = \sum_{i,j} p(x_i, y_j) = 1 \tag{2.46a}$$

The amount of information transmitted in bits per value (or signal) can be written from equation (2.40a) as

$$T(X, Y) = H(X) + H(Y) - H(X, Y)$$

where

$$H(X) = -\sum_{i=1}^{N} p(x_i) \log_2 p(x_i); \quad H(Y) = -\sum_{j=1}^{M} p(y_j) \log_2 p(y_j);$$

$$H(X, Y) = -\sum_{i,j} p(x_i, y_j) \log_2 p(x_i, y_j) \tag{2.46b}$$

One bit equals $-\log_2 (1/2)$ and denotes the information transmitted by a choice between two equally probable alternatives.

If a relation exists between X and Y then $H(X) + H(Y) > H(X, Y)$ and the size of the inequality defines $T(X, Y)$ which is a bivariate positive quantity measuring the association between X and Y. Now suppose that there are n observations of events (i, j) and n_{ij} denotes the number of times i occurred and j was caused. In other words,

$$n_i = \sum_j n_{ij}; \; n_j = \sum_i n_{ij}; \; n = \sum_{i,j} n_{ij} \tag{2.47}$$

where n_i denotes the number of times i occurred, n_j denotes the number of times j was caused, and n denotes the total number of observations. For doing calculations it is easier to use contingency tables where XY would represent cells and n_{ij} would be entries. In that case, the probabilities would be: $p(i) = n_i/n$, $p(j) = n_j/n$, and $p(i, j) = n_{ij}/n$; these are actually relative frequencies. Rather than using relative frequencies, one can also use a simpler notation for computing entropies in terms of absolute frequencies as follows:

$$s_{ij} = \frac{1}{n} \sum_{i,j} n_{ij} \log_2 n_{ij}; \; s_i = \frac{1}{n} \sum_i n_i \log_2 n_i; \; s_j = \frac{1}{n} \sum_j n_j \log_2 n_j; \; s = \log_2 n \tag{2.48a}$$

Then

$$T(X, Y) = s - s_i - s_j + s_{ij} \tag{2.48b}$$

The two-dimensional case of the amount of information transmitted can be extended to three or more dimensional cases (McGill, 1953, 1954). Consider three random variables: U, V, Y, where U and V constitute sources and Y the effect. In this case, X of the two-dimensional case has been replaced by U and V. Then, as shown in Figures 2.11a, and 2.11b, one can write

$$T(U, V; Y) = H(U, V) + H(Y) - H(U, V; Y) \tag{2.49}$$

Here X is divided into two classes U and V, with values of U as $k = 1, 2, 3, \ldots, K$; and values of V as $w = 1, 2, 3, \ldots, W$. The subdivision of X is made such that the range of values of U and V jointly constitute the values of X, with the implication that the input event i can be replaced by the joint event (k, w). This means that $n_i = n_{kw}$.

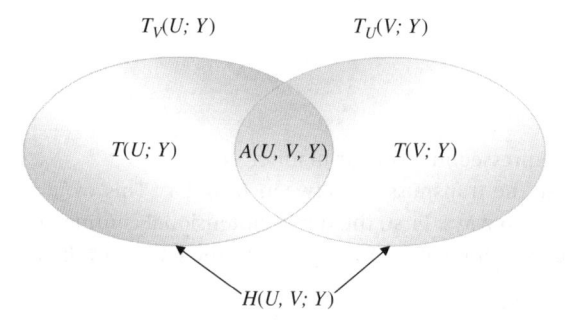

Figure 2.11a Schematic of the components of three-dimensional transmitted information. Three-dimensional information = part of bivariate transmission plus and interaction term.

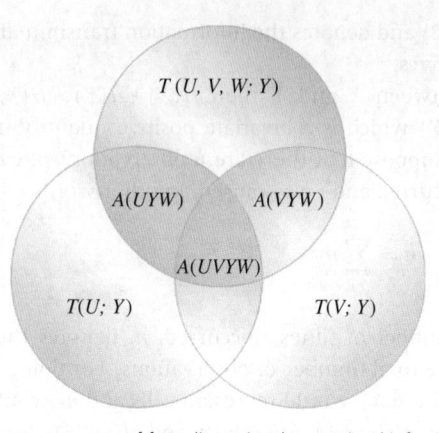

Figure 2.11b Schematic of the components of four-dimensional transmitted information with three sources and a single output.

Here $T(U, V; Y)$ measures the amount of information transmission that U and V transmit to Y. It can be shown that the direction of transmission is irrelevant, because

$$T(U, V; Y) = T(Y; U, V) \qquad (2.50)$$

This also implies that distinguishing sources from effect or transmitters from receivers do not gain anything, because the amount of information transmitted measures the association between variables and hence the direction in which information travels is immaterial. However, it is important to note that symbols cannot be permuted at will, because

$$T(U, Y; V) = H(U, Y) + H(V) - H(U, Y; V) \qquad (2.51)$$

but it is not necessarily equal to $T(U, V; Y)$.

For computing $T(U, V; Y)$, it should be noted that it can be expressed as a function of bivariate transmission between U and Y, and V and Y. Observations of the joint event (k, w, j) can be organized into a three-dimensional contingency table with UVY cells and n_{kwj} entries. Then one can compute the terms of

$$T(U, V; Y) = s - s_j - s_{kw} + s_{kwj} \qquad (2.52)$$

where

$$s_{kwj} = \frac{1}{n} \sum_{k,w,j} n_{kwj} \log_2 n_{kwj} \qquad (2.53)$$

Other terms can be expressed in a similar manner.

One can investigate the transmission between U and Y. This would involve eliminating V which can be done in two ways. First, the three-dimensional contingency table can be reduced to two dimensions by summing over V, resulting in the entries of the reduced table as

$$n_{kj} = \sum_w n_{kwj} \qquad (2.54)$$

The transmitted information between U and Y can be expressed as

$$T(U; Y) = s - s_k - s_j + s_{kj} \tag{2.55}$$

The second way for eliminating V is to compute the transmission between U and Y separately for each value of V and then average the transmitted values together. Designating $T_w(U;Y)$ as the information transmitted between U and Y for a single value of V, namely w, one can write the transmitted information $T_v(U; Y)$ as

$$T_v(U; Y) = \sum_w \frac{n_w}{n} \left[T_w(U; Y) \right] \tag{2.56}$$

It can be shown that

$$T_v(U; Y) = s_w - s_{kw} - s_{wj} + s_{kwj} \tag{2.57}$$

In a three-dimensional contingency table, three different pairs of variables occur. For transmission between V and Y, one can write

$$T(V; Y) = s - s_w - s_j + s_{wj} \tag{2.58}$$
$$T_u(V; Y) = s_k - s_{kw} - s_{kj} + s_{kwj} \tag{2.59}$$

The transmission between U and V can be expressed as

$$T(U; V) = s - s_k - s_w + s_{kw} \tag{2.60}$$
$$T_y(U; V) = s_j - s_{kj} - s_{wj} + s_{kwj} \tag{2.61}$$

Now, the information transmitted between U and Y can be reconsidered. If V affects the transmission between U and Y, that is, U and V are related, then $T_V(U;Y) \neq T(U;Y)$. This effect can be measured as

$$A(UVY) = T_V(U; Y) - T(U; Y) \tag{2.62}$$
$$A(UVY) = -s + s_k + s_w + s_j - s_{kw} - s_{kj} - s_{wj} + s_{kwj} \tag{2.63}$$

A little algebra shows

$$A(UVY) = T_v(U; Y) - T(U; Y) = T_u(V; Y) - T(V; Y) = T_y(U; V) - T(U; V) \tag{2.64}$$

Keeping this symmetry in mind, $A(UVY)$ can be regarded as the $U.V.Y$ interaction information, and is the gain (or loss) in information transmitted between any two of the variables due to the knowledge of the third variable.

Now the three-dimensional information transmitted from U, V to Y, that is, $T(U, V; Y)$, can be expressed as a function of its bivariate components:

$$T(U, V; Y) = T(U; Y) + T(V; Y) + A(UVY) \tag{2.65}$$
$$T(U, V; Y) = T_v(U; Y) + T_u(V; Y) - A(UVY) \tag{2.66}$$

Following these two equations, taken together, $T(U, V; Y)$ can be shown as in Figure 2.11a with overlapping circles. This figure assumes that there is a positive interaction between U, V, and Y, meaning that when one of the interacting variables is held constant, the amount of association between the remaining two increases, that is, $T_v(U;Y) > T(U;Y)$, and $T_u(V;Y) > T(V;Y)$.

For the three-dimensional case, one can write

$$H(Y) = H_{uv}(Y) + T(U; Y) + T(V; Y) + A(UVY) \tag{2.67}$$

where $H(Y) = s - s_j$ and $H_{uv}(Y) = s_{kwj} - s_{kw}$. This shows that the marginal information is partitioned into an error term and a set of correlation terms due to input variables. The error term is $H_{uv}(Y)$ and denotes the unexplained or residual variance in the output Y after the information due to inputs U and V has been removed. For the two-dimensional case, one can write

$$H(Y) = H_u(Y) + T(U; Y) \tag{2.68}$$

In this case H_u is the error term, because there is only one input variable U. Shannon (1948) has shown that

$$H_u(Y) \geq H_{uv}(Y) \tag{2.69}$$

This shows that if only U is controlled, the error term cannot be increased if V is also controlled. It can be shown that

$$H_u(Y) = H_{uv}(y) + T_{uv}(V; Y) \tag{2.70}$$

Now the issue of independence in three-dimensional transmission is considered. If the output is independent of the joint input then $T(U, V; Y) = 0$, that is,

$$n_{kwj} = \frac{n_{kw}n_j}{n} \tag{2.71}$$

Then it can be shown that

$$s_{kwj} = s_{kw} + s_j - s \tag{2.72}$$

This equation can be used to show that $T(U, V; Y) = 0$.

Now assume that $T(U, V; Y) > 0$, but V and Y are independent, that is,

$$n_{wj} = \frac{n_w n_j}{n} \tag{2.73}$$

This results in

$$s_{wj} = s_w + s_j - s \tag{2.74}$$

If s_{wj} from equation (2.58) is used in equation (2.74) then $T(V; Y) = 0$. Equation (2.74) does not lead to a unique condition for independence between V and Y.

If the input variables are correlated, then the question arises: How is the transmitted information affected? The three-dimensional transmitted information $T(U, V; Y)$ would account for only part of the total association in a three-dimensional contingency table. Let $C(U, V; Y)$ be the correlated information. Then one can write

$$C(U, V; Y) = H(U) + H(V) + H(Y) - H(U, V; Y) \qquad (2.75)$$

Adding to and subtracting from equation (2.75) $H(U, V)$ one obtains

$$C(U, V; Y) = T(U; V) + T(U, V; Y) \qquad (2.76)$$
$$C(U, V; Y) = T(U; V) + T(U; Y) + T(V; Y) + A(UVY) \qquad (2.77)$$

It is seen that $C(U, V; Y)$ can be employed to generate all possible components of the three correlated sources of information U, V, and Y.

Example 2.15: Using the s-notation method to compute the transinformation $T(A, B)$, $T(B, C)$, $T(A, C)$ and the interaction information $A(A, B, C)$. Data are given in Table 2.3.

Solution: Transinformation calculation:
Let us compute $T(A, B)$ first. From equation (2.48b) we know that we need to compute s, s_i, s_j and s_{ij}. All of the components can be obtained from equation (2.48a). In the following, by taking the transinformation $T(A, B)$ as an example, we compute all the components one by one.
a) Compute s: By dividing the ranges of variables A and B into five equal sized intervals and counting the number of occurrence in all combinations of these subintervals, the contingency table can be computed as

n Table

A \ B	1	2	3	4	5
1	15	6	1	1	0
2	10	8	4	3	1
3	2	4	0	0	1
4	0	0	0	1	1
5	1	0	1	0	0

From equation (2.47), we have

$$n = \sum_{i,j} n_{ij} = 15 + 10 + 2 + 0 + 1 + 6 + 8 + 4 + 0 + 0 + \ldots + 0 + 1 + 1 + 1 + 0 = 60$$

From equation (2.48a), we have

$$s = \log_2 n = \log_2 n = 5.9069$$

b) Compute s_i and s_j: By marginalizing out one of the two variables, the contingency of A and B can be obtained from the above contingency table.

Table 2.3 A Stream flow observations.

Year	Month	A	B	C
2000	1	61.21	2.54	6.86
2000	2	40.64	8.13	45.97
2000	3	122.68	22.10	15.75
2000	4	97.54	18.80	26.16
2000	5	179.83	23.88	80.52
2000	6	110.49	120.65	109.22
2000	7	12.45	12.45	7.87
2000	8	4.06	0.51	7.11
2000	9	62.23	16.51	48.00
2000	10	59.69	109.47	185.67
2000	11	347.98	84.07	140.46
2000	12	137.16	14.73	34.54
2001	1	120.14	30.48	64.77
2001	2	104.90	36.83	30.73
2001	3	175.01	33.27	59.44
2001	4	14.48	14.73	39.62
2001	5	89.15	60.20	72.39
2001	6	336.55	5.84	22.61
2001	7	41.66	12.45	25.15
2001	8	117.35	70.10	92.71
2001	9	168.66	52.32	67.31
2001	10	103.89	21.84	43.43
2001	11	72.39	77.98	126.49
2001	12	144.53	4.83	34.80
2002	1	49.78	10.67	6.60
2002	2	57.91	30.23	6.10
2002	3	82.55	41.91	27.69
2002	4	59.44	9.40	60.45
2002	5	101.85	32.00	50.55
2002	6	104.65	41.15	50.29
2002	7	90.68	73.15	416.56
2002	8	25.65	16.00	17.27
2002	9	29.72	32.78	144.78
2002	10	140.72	124.71	198.12
2002	11	99.82	13.21	23.88
2002	12	125.22	26.16	37.34
2003	1	18.54	4.06	19.05
2003	2	189.23	42.16	45.21
2003	3	51.56	35.31	41.15
2003	4	30.48	11.68	4.06
2003	5	56.64	38.86	23.62
2003	6	123.95	148.34	112.27
2003	7	111.25	16.00	182.12
2003	8	39.37	59.18	46.23
2003	9	90.17	68.83	119.38
2003	10	93.47	91.69	66.29
2003	11	116.33	20.57	23.11
2003	12	58.93	0	1.52
2004	1	110.24	35.81	57.91
2004	2	156.97	48.51	44.70
2004	3	75.95	48.26	95.25
2004	4	131.06	61.21	176.53
2004	5	105.16	22.86	31.24

(continued overleaf)

Table 2.3 *continued*

Year	Month	A	B	C
2004	6	212.34	105.41	232.16
2004	7	54.10	62.23	32.26
2004	8	105.66	119.38	59.44
2004	9	50.04	37.08	76.20
2004	10	179.07	150.37	77.72
2004	11	249.68	159.00	142.24
2004	12	66.80	10.41	5.84

From equation (2.47), we have

$$n_i(1) = \sum_j n_{1j} = 15 + 6 + 1 + 1 + 1 = 23$$

$$n_i(2) = \sum_j n_{2j} = 10 + 8 + 4 + 3 + 1 = 26$$

......

$$n_i(5) = \sum_j n_{5j} = 1 + 0 + 1 + 0 + 0 = 2$$

The results (contingency table for A) are shown in the shaded column in the following table.

n + Margin n Table

Similarly, we also have

$$n_j(1) = \sum_i n_{i1} = 15 + 10 + 2 + 0 + 1 = 28$$

$$n_j(2) = \sum_i n_{i2} = 6 + 8 + 4 + 0 + 0 = 18$$

......

$$n_j(5) = \sum_i n_{i5} = 0 + 1 + 1 + 1 + 0 = 3$$

Finally, from equation (2.48a) we have

$$s_i = \frac{1}{n} \sum_i n_i \log_2 n_i = \frac{1}{60} \left(23 \times \log_2 23 + 26 \times \log_2 26 + 7 \times \log_2 7 + 2 \times \log_2 2 + 2 \times \log_2 2 \right)$$

$$= 4.1651$$

Also we have

$$s_j = \frac{1}{n}\sum_j n_j \log_2 n_j = \frac{1}{60}\left(28 \times \log_2 28 + 18 \times \log_2 18 + 6 \times \log_2 6 + 5 \times \log_2 5 + 3 \times \log_2 3\right)$$
$$= 4.0256$$

c) Compute s_{ij}: From the joint contingency table and equation (2.48a) we can also compute s_{ij},

$$s_{ij} = \frac{1}{n}\sum_{i,j} n_{ij} \log_2 n_{ij}$$
$$= \frac{1}{60}\left(15 \times \log_2 15 + 10 \times \log_2 10 + 2 \times \log_2 2 + \ldots + 0 \times \log_2 0 + 1 \times \log_2 1 + \right.$$
$$\left. \ldots + 0 \times \log_2 0\right)$$
$$= 2.5681$$

Finally,

$$T(A, B) = s - s_i - s_j + s_{ij} = 5.9069 - 4.1651 - 4.0256 + 2.5681$$
$$= 0.2843$$

Similarly, $T(B, C)$ and $T(C, A)$ can be computed.

$$T(B, C) = 0.4890$$
$$T(C, A) = 0.1863$$

Interaction information calculation:
By dividing the ranges of the variables A, B and C into five equal sized intervals, the trivariate contingency table can be created. The resulting contingency table is

Counts,C:1.5-84.53

A	1	2	3	4	5
1	14	5	1	0	0
2	7	6	1	2	0
3	3	4	0	0	1
4	0	0	0	0	0
5	1	0	0	0	0

B

Counts,C:84.53-167.53

A	1	2	3	4	5
1	2	2	0	0	0
2	0	1	0	1	1
3	0	0	0	0	0
4	0	0	0	0	1
5	0	0	1	0	0

B

Counts,C:167.53-250.54

A	1	2	3	4	5
1	0	0	0	0	1
2	1	0	0	0	0
3	0	0	1	0	1
4	0	0	0	0	0
5	0	0	0	0	1

B

Counts,C:250.54-333.55

A	1	2	3	4	5
1	0	0	0	0	0
2	0	0	0	0	0
3	0	0	0	0	0
4	0	0	0	0	0
5	0	0	0	0	0

B

Counts,C:333.55-416

	1	2	3	4	5
1	0	0	0	0	0
2	0	0	0	0	0
3	0	0	1	0	0
4	0	0	0	0	0
5	0	0	0	0	0

A (rows), B (columns)

Summing up all the elements in the above trivariate joint contingency table, we have $n = 60$. Therefore,

$$s = \log_2 60 = 5.9069$$

The marginal contingency table can be obtained in the following way.
The marginal contingency table of A and B given $C \in (1.5, 84.53]$

Counts,C:1.5–84.53

	1	2	3	4	5	
1	14	5	1	0	0	→ 20
2	7	6	1	2	0	→ 16
3	3	4	0	0	1	→ 8
4	0	0	0	0	0	→ 0
5	1	0	0	0	0	→ 1
	↓	↓	↓	↓	↓	
	25	15	2	2	1	

A (rows), B (columns)

The marginal contingency table of A and B given $C \in (84.53, 167.53]$

Counts,C:84.53–167.53

	1	2	3	4	5	
1	2	2	0	0	0	→ 4
2	0	1	0	1	1	→ 3
3	0	0	0	0	0	→ 0
4	0	0	0	0	1	→ 1
5	0	0	1	0	0	→ 1
	↓	↓	↓	↓	↓	
	2	3	1	1	2	

A (rows), B (columns)

The marginal contingency table of A and B given $C \in (167.53, 250.54]$

Counts,C:167.53–250.54

A	1	2	3	4	5	
1	0	0	0	0	1	→ 1
2	1	0	0	0	0	→ 1
3	0	0	1	0	1	→ 2
4	0	0	0	0	0	→ 0
5	0	0	0	0	1	→ 1
	1	2	3	4	5	
	↓	↓	↓	↓	↓	
	1	0	1	0	3	

B

The marginal contingency table of A and B given $C \in (250.54, 333.55]$

Counts,C:250.54–333.55

A	1	2	3	4	5	
1	0	0	0	0	0	→ 0
2	0	0	0	0	0	→ 0
3	0	0	0	0	0	→ 0
4	0	0	0	0	0	→ 0
5	0	0	0	0	0	→ 0
	1	2	3	4	5	
	↓	↓	↓	↓	↓	
	0	0	0	0	0	

B

The marginal contingency table of A and B given $C \in (250.54, 333.55]$

Counts,C:333.55–416.

A	1	2	3	4	5	
1	0	0	0	0	0	→ 0
2	0	0	0	0	0	→ 0
3	0	0	1	0	0	→ 1
4	0	0	0	0	0	→ 0
5	0	0	0	0	0	→ 0
	1	2	3	4	5	
	↓	↓	↓	↓	↓	
	0	0	1	0	0	

B

Then the marginal contingency table for A can be

$$n_k(1) = \sum_{w,j} n_{1wj} = 20 + 4 + 1 + 0 + 0 = 25$$

$$n_k(2) = \sum_{w,j} n_{2wj} = 16 + 3 + 1 + 0 + 0 = 20$$

$$n_k(3) = \sum_{w,j} n_{3wj} = 8 + 0 + 2 + 0 + 1 = 11$$

$$n_k(4) = \sum_{w,j} n_{4wj} = 0 + 1 + 0 + 0 + 0 = 1$$

$$n_k(5) = \sum_{w,j} n_{5wj} = 1 + 1 + 1 + 0 + 0 = 3$$

The results are tabulated as:

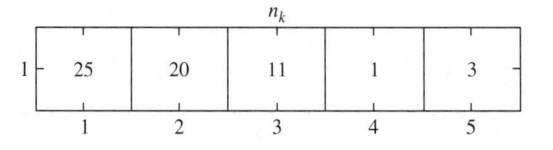

n_k

	25	20	11	1	3
	1	2	3	4	5

And the marginal contingency table for B can be

$$n_j(1) = \sum_{k,w} n_{kw1} = \underline{25} + \underline{2} + \underline{1} + \underline{0} + \underline{0} = 28$$

$$n_j(2) = \sum_{k,w} n_{kw2} = \underline{15} + \underline{3} + \underline{0} + \underline{0} + \underline{0} = 18$$

$$n_j(3) = \sum_{k,w} n_{kw3} = \underline{2} + \underline{1} + \underline{1} + \underline{0} + \underline{1} = 5$$

$$n_j(4) = \sum_{k,w} n_{kw4} = \underline{2} + \underline{1} + \underline{0} + \underline{0} + \underline{0} = 3$$

$$n_j(4) = \sum_{k,w} n_{kw4} = \underline{1} + \underline{2} + \underline{3} + \underline{0} + \underline{0} = 6$$

Tabulate the results as

n_j

	28	18	5	3	6
	1	2	3	4	5

The marginal contingency table for C can be computed as

$$n_w(1) = \underline{25} + \underline{15} + \underline{2} + \underline{2} + \underline{1} = 20 + 16 + 8 + 0 + 1 = 45$$

$$n_w(2) = 2 + \underline{3} + \underline{1} + \underline{1} + \underline{2} = 4 + 3 + 0 + 1 + 1 = 9$$

$$n_w(3) = \underline{1} + \underline{0} + \underline{1} + \underline{0} + \underline{3} = 1 + 1 + 2 + 0 + 1 = 5$$

$$n_w(4) = \underline{0} + \underline{0} + \underline{0} + \underline{0} + \underline{0} = 0 + 0 + 0 + 0 + 0 = 0$$

$$n_w(5) = \underline{0} + \underline{0} + \underline{1} + \underline{0} + \underline{0} = 0 + 0 + 1 + 0 + 0 = 1$$

Tabulate the results as

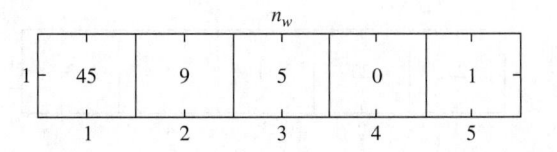

	n_w			
45	9	5	0	1
1	2	3	4	5

Therefore, we can have

$$s_k = \frac{1}{60}\left(25 \times \log_2 25 + 20 \times \log_2 20 + 11 \times \log_2 11 + 1 \times \log_2 1 + 3 \times \log_2 3\right) = 4.0891$$

$$s_j = \frac{1}{60}\left(28 \times \log_2 28 + 18 \times \log_2 18 + 5 \times \log_2 5 + 3 \times \log_2 3 + 6 \times \log_2 6\right) = 4.0256$$

$$s_w = \frac{1}{60}\left(45 \times \log_2 45 + 9 \times \log_2 9 + 5 \times \log_2 5 + 0 \times \log_2 0 + 1 \times \log_2 1\right) = 4.7879$$

Now s_{kw} s_{wj} and s_{kj} are computed. The bivariate contingency table can also be obtained from the trivariate contingency by marginalizing one of the three variables out. Using equation (2.54) we can have the bivariate joint contingency table of A and B:

			n_{kj}		
1	16	7	1	0	1
2	8	7	1	3	1
A 3	3	4	2	0	2
4	0	0	0	0	1
5	1	0	1	0	1
	1	2	3	4	5
			B		

Therefore s_{kj} can be computed as

$$s_{kj} = \frac{1}{60}\left(16 \times \log_2 16 + 8 \times \log_2 8 + 3 \times \log_2 3 + \ldots + 1 \times \log_2 1 + 1 \times \log_2 1\right) = 2.4802$$

Similarly, the bivariate joint contingency table of B and C is:

n_{wj}

	1	2	3	4	5
1	25	2	1	0	0
2	15	3	0	0	0
B 3	2	1	1	0	1
4	2	1	0	0	0
5	1	2	3	0	0

C

s_{wj} can be computed as

$$s_{wj} = \frac{1}{60} \left(25 \times \log_2 25 + 15 \times \log_2 15 + 2 \times \log_2 2 + \ldots + 0 \times \log_2 0 + 0 \times \log_2 0 \right) = 3.2035$$

The bivariate joint contingency table of A and C is:

n_{kw}

	1	2	3	4	5
1	20	4	1	0	0
2	16	3	1	0	0
A 3	8	0	2	0	1
4	0	1	0	0	0
5	1	1	1	0	0

C

and s_{kw} is computed as

$$s_{kw} = \frac{1}{60} \left(20 \times \log_2 20 + 16 \times \log_2 16 + 8 \times \log_2 8 + \ldots + 0 \times \log_2 0 + 0 \times \log_2 0 \right) = 3.1532$$

Using the trivariate contingency table and equation (2.53), s_{kwj} can be computed as

$$
\begin{aligned}
s_{kwj} &= \frac{1}{n} \sum_{k,w,j} n_{kwj} \log_2 n_{kwj} \\
&= \frac{1}{60} (14 \times \log_2 14 + 7 \times \log_2 7 + 3 \times \log_2 3 + \ldots + 0 \times \log_2 0 + 0 \times \log_2 0 \\
&\quad + 2 \times \log_2 2 + 0 \times \log_2 0 + 0 \times \log_2 0 + \ldots + 1 \times \log_2 1 + 0 \times \log_2 0 \\
&\quad + 0 \times \log_2 0 + 1 \times \log_2 1 + 0 \times \log_2 0 + \ldots + 0 \times \log_2 0 + 1 \times \log_2 1 \\
&\quad + 0 \times \log_2 0 + 0 \times \log_2 0 + 0 \times \log_2 0 + \ldots + 0 \times \log_2 0 + 0 \times \log_2 0 \\
&\quad + 0 \times \log_2 0 + 0 \times \log_2 0 + 0 \times \log_2 0 + \ldots + 0 \times \log_2 0 + 0 \times \log_2 0) \\
&= 1.9805
\end{aligned}
$$

So far we can use equation (2.63) to compute the interaction information

$$
\begin{aligned}
A(A, B, C) &= -s + s_k + s_w + s_j - s_{kw} - s_{wj} - s_{kj} + s_{kwj} \\
&= -5.9069 + 4.0891 + 4.7879 + 4.0256 - 3.1532 - 3.2035 - 2.4802 + 1.9805 \\
&= 0.1393
\end{aligned}
$$

It may be interesting to summarize this example now. The entire analysis is additive and hence is simple to calculate. The component measures of association (with drought duration and with inter-arrival time) plus the error or noise sum to the drought severity information. Following Garner and Hake (1951) and Miller (1953), the amount of information transmitted is approximately equal to the logarithm of the number of perfectly discriminated input classes.

More on mutual information

In many rural areas in developing countries of Asia and Africa, domestic water needs are met by fetching water from different sources of supply, such as open-dug wells, tanks, and ponds. Often there is no restriction as to who will fetch water from which source. To that end, consider the transport of a body of water from a location can be characterized as an event E. This event originates in an area denoted by a_i and the probability that this event occurs in area a_i is p_i ($i = 1, 2, \ldots, N$). The probability that event reaches the channel j, c_j, is p_j. The probability of the movement or flow between a_i and c_j is p_{ij}. If the information is received to the effect that the flow did occur between i and j, then each of these probabilities is raised to unity. The smaller the value of probability, the message about the actual movement is more informative. Let a priori transport probabilities be denoted as q_{ij}. Of course, the a posteriori probabilities of movement or transport of water are p_{ij}. Then one can write

$$
\sum_i \sum_j q_{ij} = 1 \tag{2.78a}
$$

$$
\sum_j q_{ij} = q_i \tag{2.78b}
$$

$$
\sum_i q_{ij} = q_j \tag{2.78c}
$$

$$
\sum_i q_{i.} = \sum_j q_{.j} = 1.0 \tag{2.78d}
$$

Equations (2.78b) and (2.778c) can be expressed in matrix form $Q : \{q_{ij}\}$ as:

$$
\begin{bmatrix}
q_{11} & q_{12} & \cdot & \cdot & \cdot & q_{1M} \\
q_{21} & q_{22} & \cdot & \cdot & \cdot & q_{2M} \\
\cdot & \cdot & \cdot & \cdot & \cdot & \cdot \\
\cdot & \cdot & \cdot & \cdot & \cdot & \cdot \\
\cdot & \cdot & \cdot & \cdot & \cdot & \cdot \\
q_{N1} & q_{N2} & \cdot & \cdot & \cdot & q_{NM}
\end{bmatrix}
$$

Analogous relations for p_{ij} can be written as

$$\sum_i \sum_j p_{ij} = 1.0 \tag{2.79a}$$

$$\sum_j p_{ij} = p_{i.} \tag{2.79b}$$

$$\sum_i p_{ij} = p_{.j} \tag{2.79c}$$

$$\sum_i p_{i.} = \sum_j p_{.j} = 1.0 \tag{2.79d}$$

Equations (2.79b) and (2.79c) can be written in matric form $P : \{p_{ij}\}$ as:

$$\begin{bmatrix} p_{11} & p_{12} & \cdot & \cdot & \cdot & p_{1M} \\ p_{21} & p_{22} & \cdot & \cdot & \cdot & p_{2M} \\ \cdot & & \cdot & \cdot & \cdot & \cdot \\ \cdot & & \cdot & \cdot & \cdot & \cdot \\ \cdot & & \cdot & \cdot & \cdot & \cdot \\ p_{N1} & p_{N2} & \cdot & \cdot & \cdot & p_{NM} \end{bmatrix}$$

These constitute matrices of probabilities. Note rows represent flow origins and columns flow destinations.

Comparison of the a priori and a posteriori matrices results in three issues. First, row and column entropies may or may not be maximized. If rows and columns are at the maximum of (log N, log M) then there is complete uncertainty about the origins and destinations. If the entropies are less than the maximum log N, then origins and destinations have some degree of concentration or organization reflecting systematic regularity. Second, if mutual entropy and expected mutual entropy or transinformation are the same, there is interchange reflecting movement within the bounds specified by the marginal probability distributions. The equality of mutual and expected mutual entropies reveals pure random movement and in that case marginal entropies are maximized. If the mutual and expected mutual entropies are not equal, movements depart from order or organization along margins. The transmitted information ultimately gets degraded to noise and thus the information entropy can be viewed in the same way as thermal entropy (thermodynamics). Of most value is the non-noisy or significant information, and this may be revealed by the differences between the a priori expectations and the a posteriori expectations.

Matrix P implies the joint, row, and column entropies:

$$H\left(p_{ij}\right) = -\sum_{i=1}^{N} \sum_{j=1}^{M} p_{ij} \log p_{ij} \quad \text{(Joint entropy)} \tag{2.80a}$$

$$H\left(p_{i.}\right) = -\sum_{i=1}^{N} p_{i.} \log p_{i.} \quad \text{(Row entropy)} \tag{2.80b}$$

$$H(p_{.j}) = -\sum_{j=1}^{M} p_{.j} \log p_{.j} \quad \text{(Column entropy)} \tag{2.80c}$$

Analogous expressions can be written for matrix Q as:

$$H(q_{ij}) = -\sum_{i=1}^{N}\sum_{j=1}^{M} q_{ij} \log q_{ij} \quad \text{(Joint entropy)} \tag{2.81a}$$

$$H(q_{i.}) = -\sum_{i=1}^{N} q_{i.} \log q_{i.} \quad \text{(Row entropy)} \tag{2.81b}$$

$$H\left(q_{.j}\right) = -\sum_{j=1}^{M} q_{.j} \log q_{.j} \quad \text{(Column entropy)} \tag{2.81c}$$

If movements are independent at origins and destinations the following measures of expected mutual entropy (transinformation) can be expressed:

$$T(q_{ij}) = \sum_{i=1}^{N}\sum_{j=1}^{M} q_{ij} \log \left(\frac{q_{ij}}{q_{i.}q_{.j}} \right) \tag{2.82a}$$

$$T(p_{ij}) = \sum_{i=1}^{N}\sum_{j=1}^{M} q_{ij} \log \left(\frac{p_{ij}}{p_{i.}p_{.j}} \right) \tag{2.82b}$$

One may also write

$$T(q_{ij}) = -\sum_{i=1}^{N} q_{i.} \log q_{i.} - \sum_{j=1}^{M} q_{.j} \log q_{.j} + \sum_{i=1}^{N}\sum_{j=1}^{M} q_{ij} \log q_{ij} \tag{2.83a}$$

Therefore,

$$T(q_{ij}) = H\left(q_{i.}\right) + H\left(q_{.j}\right) - H\left(q_{ij}\right) \tag{2.83b}$$

This means that the joint entropy of q_{ij} is less than the sum of marginal entropies of $q_{i.}$ and $q_{.j}$ because of the mutual information between them:

$$H(q_{ij}) < H\left(q_{i.}\right) + H\left(q_{.j}\right) \tag{2.84a}$$

Likewise,

$$T(p_{ij}) = H\left(p_{i.}\right) + H\left(p_{.j}\right) - H\left(p_{ij}\right) \tag{2.84b}$$

Transinformation is zero if the difference between joint and the sum of marginal entropies and the joint entropy (based on the margins) is zero. Transinformation increases with increasing deviations of q_{ij} and p_{ij} from their expectations $(q_{i.}q_{.j})$ and $(p_{i.}p_{.j})$. Furthermore, $\log(q_{ij}/q_{i.}q_{.j})$ is positive, zero or negative, depending on whether the probabilities of movement from i to j are greater than, equal to, or less than the independence level.

Three issues emerge from the a priori and a posteriori movement matrices. First, row and column matrices may or may not be maximized. Note rows represent origins of movement and columns destinations of movement. If rows and columns are at the maximum of log N then there is complete uncertainty about the origins and destinations. If their entropies are less than log N, then origins and/or destinations have some degree of concentration or organization reflecting systematic regularity. Second, if mutual and expected mutual entropy (transinformation) are the same, there is interchange reflecting movement within bounds specified by the marginal probability distributions. The equality of mutual and expected mutual entropies reveals pure random movement in the case where marginal entropies are maximized. If mutual and expected mutual entropies are not equal, flows depart from order or organization along margins.

Properties of mutual information

The mutual information or transinformation satisfies the following properties:

1 Transinformation between X and Y is symmetric, because it is a measure of common information between them, that is,

$$T(X, Y) = T(Y, X)$$

where the mutual information $T(X, Y)$ is a measure of the uncertainty about the system input X that is resolved by observing the system output, whereas the mutual information $T(Y, X)$ is a measure of uncertainty about the system output Y that is resolved by observing the system input X.

2 The mutual information $T(X, Y)$ between X and Y is always non-negative, that is,

$$T(X, Y) \geq 0$$

In other words, one does not lose information, on average, by observing system output Y. Furthermore, the mutual information vanishes only if the system input and output are independent.

3 The mutual information can be stated in terms of the system output entropy Y as:

$$T(X, Y) = H(Y) - H(Y|X)$$

where the right side represents the ensemble average of the information conveyed by the system output Y minus the ensemble average of the information conveyed by Y given that the system input is already known. The conditional entropy $H(Y|X)$ conveys information about the system noise, rather than about the system input X.

2.6 Dimensionless entropies

Sometimes it is useful to work with normalized or dimensionless entropies. Each type of entropy can be expressed in dimensionless terms by dividing the respective entropy by its maximum value. Thus, dimensionless marginal entropy, $H_*(X)$, can be written as:

$$H_*(X) = \frac{H(X)}{\log_2 |X|} \tag{2.85a}$$

where $|X|$ is the cardinality of X, meaning the number of elements or values of X. Clearly, when all values of X are equally likely, the entropy will be maximum. The joint entropy in dimensionless form can be expressed as

$$H_*(X, Y) = \frac{H(X, Y)}{\log_2(|X|\,|Y|)} \qquad (2.85b)$$

Likewise, the dimensionless conditional entropy can be written as

$$H_*(X|Y) = \frac{H(X|Y)}{\log_2 |X|} \qquad (2.85c)$$

These dimensionless entropies will vary within the range of 0 to 1. One can also define dimensionless transinformation as

$$T_*(X, Y) = \frac{T(X, Y)}{T_m(X, Y)} \qquad (2.86a)$$

where $T_m(X, Y)$ is the maximum value of transinformation which can be obtained as

$$T_m(X, Y) = \min\left\{\log_2 |X|, \log_2 |Y|\right\} \qquad (2.86b)$$

2.7 Bayes theorem

Bivariate entropies can also be expressed using set algebra. A set can be defined as a collection of objects considered for some purpose. The objects of the set are then called members or elements of the set. The usual convention is to denote sets by upper case letters and its elements by lower case letters. For example, if X denotes a set then x denotes its element. Symbolically, $x \in X$ means that x is a member of set X or x is contained in X. A set may contain a finite or an infinite number of elements. If the set contains a finite number of elements, say 1, 2, 5, and 7, then one way to define a set is to list its members explicitly within curly brackets as: $X = \{1, 2, 5, 7\}$. Another way to define a set is by specifying a characteristic that an object must possess in order to qualify for membership in the set. For example, if an object must be between 20 and 30, then $X = \{x|\ 20 \leq x \leq 30\}$. Symbol vertical bar stands for "such that." This manner of defining an object permits inclusion of an infinite number of objects if they can be included in the set.

A more common way to define a set is through a characteristic function which is like an indicator function. Here the concept of a universal set is invoked, wherein a set is analogous to a sample, and a universal set is analogous to a population. This means that members of any set are drawn from the universal set. Thus many sets can be drawn from the universal set. Now consider a set A, the characteristic function of set A, denoted by C_A, is then a function from the universal set to the set {0, 1} where

$$C_A = \begin{cases} 1 & \textit{if } x \textit{ is a member of } A \\ 0 & \textit{if } x \textit{ is not a member of } A \end{cases} \qquad (2.87)$$

for each $x \in X$.

Now Bayes' theorem is presented. This theorem is a powerful tool for computing conditional probabilities and for updating specified probabilities when new evidence becomes available. In other words, when given probabilities are expressed as prior probabilities and new evidence as conditional probabilities, then updated probabilities are posterior probabilities. To express Bayes' theorem, consider two sets A and B with probabilities $P(A)$ and $P(B)$, respectively, and $P(B)$ is not equal to zero. Then, the conditional probability of A given B is defined as

$$P(A|B) = \frac{P(A \cap B)}{P(B)} \tag{2.88}$$

Likewise, the conditional probability of B given A can be written as

$$P(B|A) = \frac{P(A \cap B)}{P(A)} \tag{2.89}$$

Combining equations (2.88) and (2.89), one obtains a relation between two conditional probabilities as

$$P(A|B)P(B) = P(B|A)P(A) \tag{2.90}$$

Equation (2.90) can also be expressed for one conditional probability in terms of another as

$$P(A|B) = \frac{P(B|A)P(A)}{P(B)} \tag{2.91}$$

Equation (2.91) is normally called the Bayes theorem. It may be noted that $P(B)$ can be expressed in terms of elementary mutually exclusive events A_i, $i = 1, 2, \ldots$, as

$$P(B) = \sum_i p(A_i \cap B) = \sum_i P(B|A_i)P(A_i) \tag{2.92}$$

Inserting equation (2.92) in equation (2.91), the Bayes theorem becomes

$$P(A|B) = \frac{P(B|A)P(A)}{\sum_i P(B|A_i)P(A_i)} \tag{2.93}$$

The Bayes theorem would be applied in several examples discussed in the chapter.

Example 2.16: Consider two variables rain (X) and wind (Y) on a given day in the month of August in north India. The states of these variables are 0 if the variable does not occur and 1 if it does. From empirical data it is found that

X	Y	$p(x, y)$
0	0	0.40
0	1	0.20
1	0	0.10
1	1	0.30

The marginal probabilities $p(x)$ and $p(y)$ are calculated from empirical data and are given as

X	$p(x)$	Y	$p(y)$
0	0.6	0	0.5
1	0.4	1	0.5

Compute the uncertainties associated with marginal, joint, and conditional probability distributions; and transinformation.

Solution: The uncertainties associated with these marginal probability distributions are:

$$H(X) = -0.6 \log_2 0.6 - 0.4 \log_2 0.4$$
$$= 0.971 \text{ bits}$$
$$H(Y) = -0.5 \log_2 0.5 - 0.5 \log_2 0.5$$
$$= 1.0 \text{ bit}$$

The uncertainty associated with the given joint probabilities can be expressed as

$$H(X, Y) = -0.40 \log_2 0.4 - 0.20 \log_2 0.20 - 0.10 \log_2 0.10 - 0.30 \log_2 0.30$$
$$= 1.846 \text{ bits}$$

The conditional uncertainties are calculated using equations (2.35a) and (2.35b) as:

$$H(X|Y) = H(X, Y) - H(Y) = 0.846 \text{ bits}$$
$$H(Y|X) = H(X, Y) - H(X) = 0.876 \text{ bits}$$

The transinformation or information transmission is calculated using equation (2.40a) as:

$$T(X, Y) = H(X) + H(Y) - H(X, Y) = 0.125 \text{ bits}$$

Example 2.17: Consider in the above example that variables X and Y are independent and their marginal probabilities are known. Compute the uncertainty associated with the joint distribution and compare it with the case in Example 2.16 where the actual joint distribution is known.

Solution: The joint entropy is calculated as

$$H(X, Y) = -0.6 \log_2 0.6 - 0.4 \log_2 0.4 - 0.5 \log_2 0.5 - 0.5 \log_2 0.5$$
$$= 1.971 \text{ bits}$$

Comparing it with that obtained in Example 2.16, where $H(X, Y) = 1.846$ bits, the gain in information is 0.125 bits when the actual probability distribution is known.

Example 2.18: A project work requires drainage work to be completed within the month of August. The contractor needs 15 rainless days, needs to rent earthmoving machinery, and needs to hire sufficient labor. Thus, there are three variables involved: occurrence of nonrainy days, availability of equipment for renting, and availability of labor for hiring. The first two variables can be represented by the set $X = Y = [0, 1]$, and the remaining variable by the set $Z = [0, 1, 2]$ corresponding to no labor, inadequate labor and enough labor. Empirical observations suggest that nonzero joint probabilities for different combinations of these three factors can be expressed as

X	Y	Z	$p(x, y, z)$
0	0	0	0.12
0	1	0	0.10
1	0	0	0.18
1	0	1	0.15
1	1	1	0.22
0	0	2	0.13
1	0	2	0.04
1	1	2	0.06

Observations lead to empirical marginal probabilities as

X	$p(x)$	Y	$p(y)$	Z	$p(z)$
0	0.35	0	0.62	0	0.40
1	0.65	1	0.38	1	0.37
				2	0.23

Compute joint, conditional, and marginal entropies. Note that there are 12 possibilities eight of which are given above and for the remaining four possibilities joint probabilities are zero. From this table one can compute $p(x, y)$, $p(y, z)$, and $p(x, z)$ which are tabulated below:

x	y	$p(x, y)$	x	z	$p(x, z)$	Y	z	$p(y, z)$
0	0	0.25	0	0	0.22	0	0	0.30
0	1	0.10	1	0	0.18	1	0	0.10
1	0	0.37	1	1	0.37	0	1	0.15
1	1	0.28	0	2	0.13	1	1	0.22
			1	2	0.10	0	2	0.17
						1	2	0.06

Solution: One can compute the uncertainties associated with different probability distributions as:

$$H(X, Y, Z) = -0.12 \log_2 0.12 - 0.10 \log_2 0.1 - 0.18 \log_2 0.18 - 0.15 \log_2 0.15$$
$$- 0.22 \log_2 0.22 - 0.13 \log_2 0.13 - 0.04 \log_2 0.04 - 0.06 \log_2 0.06$$
$$= 2.848 \text{ bits}$$

$$H(X, Y) = -0.25 \log_2 0.25 - 0.10 \log_2 0.1 - 0.37 \log_2 0.37 - 0.28 \log_2 0.28$$
$$= 1.877 \text{ bits}$$
$$H(X, Z) = -0.22 \log_2 0.22 - 0.18 \log_2 0.18 - 0.37 \log_2 0.37 - 0.13 \log_2 0.13$$
$$- 0.10 \log_2 0.10$$
$$= 2.171 \text{ bits}$$
$$H(Y, Z) = -0.30 \log_2 0.30 - 0.10 \log_2 0.10 - 0.15 \log_2 0.15 - 0.22 \log_2 0.22$$
$$- 0.17 \log_2 0.17 - 0.06 \log_2 0.06$$
$$= 2.423 \text{ bits}$$
$$H(X) = -0.35 \log_2 0.35 - 0.65 \log_2 0.65$$
$$= 0.934 \text{ bits}$$
$$H(Y) = -0.62 \log_2 0.62 - 0.38 \log_2 0.38$$
$$= 0.958 \text{ bits}$$
$$H(Z) = -0.40 \log_2 0.40 - 0.37 \log_2 0.37 - 0.23 \log_2 0.23$$
$$= 1.547 \text{ bits}$$

The conditional probabilities are calculated as

$$H(X|Y, Z) = H(X, Y, Z) - H(Y, Z) = 0.425 \text{ bits}$$
$$H(Y|X, Z) = H(X, Y, Z) - H(X, Z) = 0.676 \text{ bits}$$
$$H(Z|X, Y) = H(X, Y, Z) - H(X, Y) = 0.970 \text{ bits}$$
$$H(X, Y|Z) = H(X, Y, Z) - H(Z) = 1.300 \text{ bits}$$
$$H(X, Z|Y) = H(X, Y, Z) - H(Y) = 1.890 \text{ bits}$$
$$H(Y, Z|X) = H(X, Y, Z) - H(X) = 1.914 \text{ bits}$$

Now we can compute transinformation or information transmission as

$$T[(X, Y), Z] = H(X, Y) + H(Z) - H(X, Y, Z) = 0.577 \text{ bits}$$
$$T[(X, Z), Y] = H(X, Z) + H(Y) - H(X, Y, Z) = 0.282 \text{ bits}$$
$$T[(Y, Z), X] = H(Y, Z) + H(X) - H(X, Y, Z) = 0.509 \text{ bits}$$
$$T[(X, Y, Z] = H(X) + H(Y) + H(Z) - H(X, Y, Z) = 2.256 \text{ bits}$$

Example 2.19: Consider an urban area where in the month of July rainfall (X_1), flooding (X_2), wind (X_3), and power outage (X_4) frequently occur together. Empirical data suggest they occur with probabilities as $p(x_1) = p_1 = 0.45, p(x_2) = p_2 = 0.22$, $p(x_3) = p_3 = 0.18$, $p(x_4) = p_4 = 0.15$. Let the set $X = [x_1, x_2, x_3, x_4]$. There can be many combinations in which these events can occur, but for purposes of illustration consider only four combinations, as shown in Figure 2.12a. Compute the uncertainty or the probability distributions.

Solution: *Combination 1*: This is shown Figure 2.12a. Using equation (2.9), the entropy or uncertainty can be computed as:

$$H(P) = -0.45 \log_2 0.45 - 0.22 \log_2 0.22 - 0.18 \log_2 0.18 - 0.15 \log_2 0.15$$
$$= 1.855 \text{ bits}$$

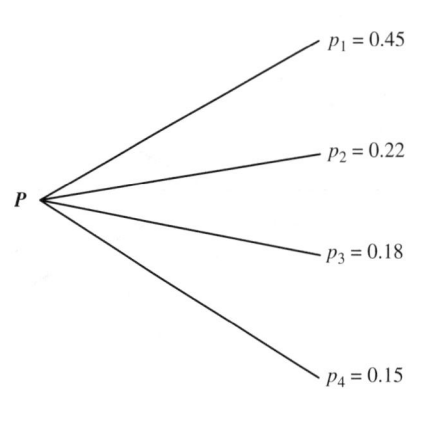

Figure 2.12a Combination 1.

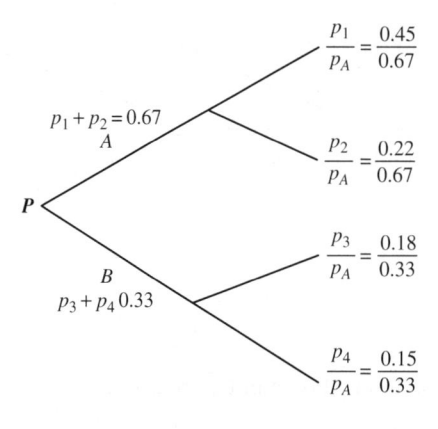

Figure 2.12b Combination 2.

Combination 2: This is shown in Figure 2.12b. It has two branches A and B to start with probabilities p_A which is made up of p_1 and p_2, and p_B which is made up of p_3 and p_4. The uncertainty can be computed using equation (2.16) or (2.18), the branching property given as:

$$H(P) = H(p_A, p_B) + p_A H\left(\frac{p_1}{p_A}, \frac{p_2}{p_A}\right) + p_B H\left(\frac{p_3}{p_B}, \frac{p_4}{p_B}\right)$$

$$= H(0.67, 0.33) + 0.67 H\left(\frac{0.45}{0.67}, \frac{0.22}{0.67}\right) + 0.33 H\left(\frac{0.18}{0.33}, \frac{0.15}{0.33}\right)$$

$$= 1.855 \text{ bits}$$

Combination 3: This is shown in Figure 2.12c. This has two branches p_1 and p_A which are made up of p_2, p_3 and p_4. The uncertainty therefore is computed as:

$$H(P) = H(p_1, p_A) + p_A H\left(\frac{p_2}{p_A}, \frac{p_3}{p_A}, \frac{p_4}{p_A}\right)$$

$$= H(0.45, 0.55) + 0.55 H\left(\frac{0.22}{0.55}, \frac{0.18}{0.55}, \frac{0.15}{0.55}\right)$$

$$= 1.855$$

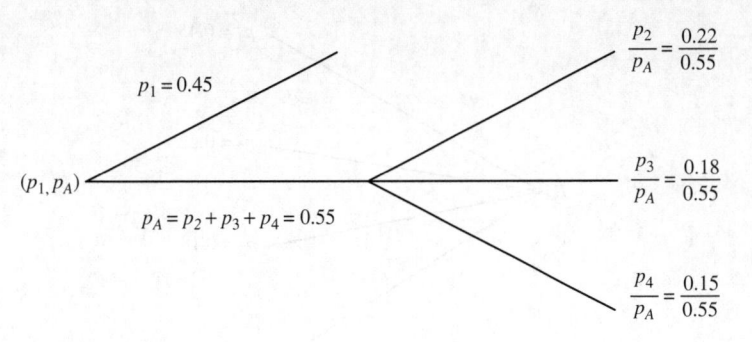

Figure 2.12c Combination 3.

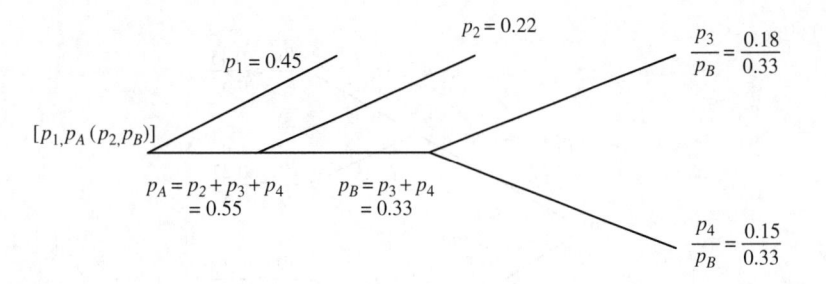

Figure 2.12d Combination 4.

Combination 4: This combination is shown in Figure 2.12d.

$$p_A = p_2 + p_3 + p_4$$
$$= 0.55$$
$$p_B = \frac{0.33}{0.55} \times 0.55 = 0.33$$
$$H(P) = H(p_1, p_A) + p_A H\left(\frac{p_2}{p_A}, \frac{p_B}{p_A}\right) + p_B H\left(\frac{p_3}{p_B}, \frac{p_4}{p_B}\right)$$
$$= H(0.45, 0.55) + 0.55 H\left(\frac{0.22}{0.55}, \frac{0.33}{0.55}\right) + 0.33 H\left(\frac{0.18}{0.33}, \frac{0.15}{0.33}\right)$$
$$= 1.855$$

These four combinations show that uncertainty is the same for all of them. Hence any combinations can be used for computing uncertainty.

Example 2.20: Consider a reservoir which can take on one of the three states: $X = [x_1, x_2, x_3]$, where x_1 is flood level, x_2 normal level, and x_3 low level. Transitions of states are assumed to occur at discrete times. Empirical data show that probabilities of these transitions are given as below:

Time t State			Time $t+1$ States			
	State	P	State	p	State	P
x_1	x_1	0	x_2	0.7	x_3	0.3
x_2	x_1	0.4	x_2	0.0	x_3	0.6
x_3	x_1	0.0	x_2	0.75	x_3	0.25

Assume that initially the reservoir is in state 1; this means that $p(x_1) = 1$, $p(x_2) = p(x_3) = 0$ at $t = 1$. Compute the uncertainty in predicting at time t sequences of states at $t+1$, $t+2$, and $t+3$.

Solution: The sequence of states is represented as a branch diagram shown in Figure 2.13. The diagram shows that there are eight sequences with nonzero probabilities. Let these sequences be denoted as S_1, S_2, ..., S_8, as indicated in the diagram. The probabilities of these sequences can be calculated in a general way as follows:

$$p(S_i) = p[x_i(t+1), \, x_j(t+2), \, x_k(t+3)], \; i, \, j, \, k = 1, 2, 3$$
$$= p[x_1(t)]p[x_i(t+1)|x_1(t)]p[x_j(t+2]|x_i(t+1))\, p[x_k(t+3)|x_j(t+2)]$$

More specifically,

$$p[S_1] = p[x_1(t)]p[x_2(t+1)|x_1(t)]p\big[x_1(t+2)|x_2(t+1)\big]p\big[x_2(t+3)|x_1(t+2)\big]$$
$$= 1 \times 0.7 \times 0.4 \times 0.7$$
$$= 0.196$$

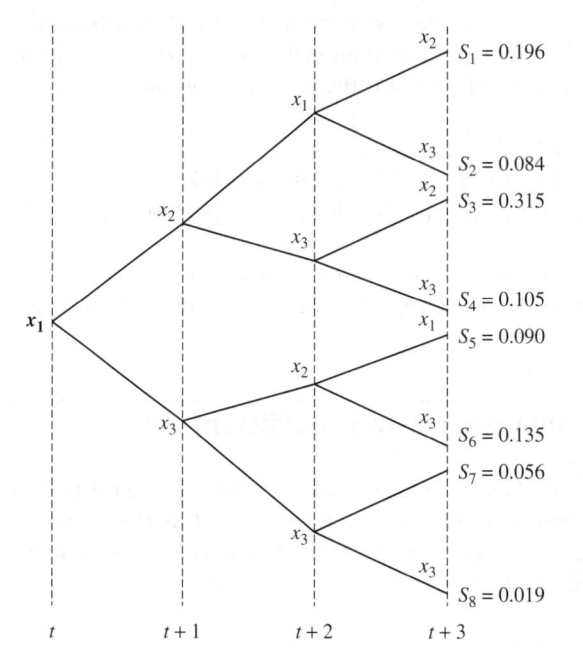

Figure 2.13 Branch diagram for the sequence of states.

$$p[S_2] = p[x_1(t)]p[x_2(t+1)|x_1(t)]p[x_1(t+2)|x_2(t)]p[x_3(t+2)|x_1(t+2)]$$
$$= 1 \times 0.7 \times 0.4 \times 0.3$$
$$= 0.084$$

$$p[S_3] = p[x_1(t)]p[x_2(t+1)|x_1(t)]p[x_3(t+2)|x_2(t+1)]p[x_2(t+3)|x_3(t+2)]$$
$$= 1 \times 0.7 \times 0.6 \times 0.75$$
$$= 0.315$$

$$p[S_4] = p[x_1(t)]p[x_2(t+1)|x_1(t)]p[x_3(t+2)|x_2(t+1)]p[x_3(t+3)|x_3(t+2)]$$
$$= 1 \times 0.7 \times 0.6 \times 0.25$$
$$= 0.105$$

$$p[S_5] = p[x_1(t)]p[x_3(t+1)|x_1(t)]p[x_2(t+2)|x_3(t+1)]p[x_1(t+3)|x_2(t+2)]$$
$$= 1 \times 0.3 \times 0.75 \times 0.4$$
$$= 0.09$$

Similarly,

$$p[S_6] = 1 \times 0.3 \times 0.75 \times 0.6 = 0.135$$
$$p[S_7] = 1 \times 0.3 \times 0.25 \times 0.75 = 0.056$$
$$p[S_8] = 1 \times 0.3 \times 0.25 \times 0.25 = 0.019$$

The amount of uncertainty associated with the prediction of a particular sequence is:

$$H(S_1) = 0.09H[0.0, 0.7, 0.3] + 0.567H[0.4, 0.0, 0.6] + 0.343H[0.0, 0.75, 0.25] = 0.908$$

We can also compute uncertainty associated with such a prediction. At time $t+3$, the prediction of states is based on the conditional probabilities of states at time $t+2$ and marginal probability at time t. Since $p[x_1(t)] = 1$, the relevant probabilities are:

$$p[x_1(t+3)]|x_1(t)] = 0.09$$
$$p[x_2(t+3)]|x_1(t)] = [0.196 + 0.315 + 0.056] = 0.567$$
$$p|x_3(t+3)]x_1(t)] = [0.084 + 0.105 + 0.135 + 0.019] = 0.343$$

At time $t+3$, state x_1 occurs one way, state x_2 can occur in three ways and state x_3 can occur in four ways. Therefore, considering these ways as independent, their probabilities need to be calculated.

2.8 Informational correlation coefficient

The informational correlation coefficient R_0 measures the mutual dependence between random variables X and Y and does not assume any type of distributional relationship between them. It is thus a measure of transferable information. It is a dimensional quantity and is expressed in terms of transinformation as

$$R_0 = \sqrt{1 - \exp(-2T_0)} \tag{2.94}$$

where T_0 is the transinformation or mutual information representing the upper limit of transferable information between two variables X and Y. If the values of probabilities in equation (2.41b) are computed from the corresponding sample frequencies (Harmancioglu et al., 1986), then the transinformation obtained represents the upper limit of transferable information between the two variables. When X and Y are normally distributed and linearly correlated, R_0 reduces to the classical Pearson correlation coefficient between X and Y, r_{xy}:

$$r_0 = r_{xy} = \frac{Cov(x, y)}{\sigma_x \sigma_y} \tag{2.95}$$

where $Cov(x, y)$ is the covariance between X and Y, σ_x is the standard deviation of X, and σ_y is the standard deviation of Y. These quantities can be computed from sample data as follows:

$$Cov(x, y) = \frac{1}{N} \sum_{i=1}^{N} (x_i - \bar{x})(y_i - \bar{y}) \tag{2.96}$$

where $\bar{x}$ is the mean of X, $\bar{y}$ is the mean of Y, and N is the sample size. The standard deviations are computed as:

$$\sigma_x = \sqrt{\frac{1}{N-1} \sum_{i=1}^{N} (x_i - \bar{x})^2}, \ \sigma_y = \sqrt{\frac{1}{N-1} \sum_{i=1}^{N} (y_i - \bar{y})^2} \tag{2.97}$$

Example 2.21: Let X and Y be two random variables that are normally distributed and linearly correlated. Show their informational correlation coefficient R_0 and compare it with the classical Pearson correlation coefficient between X and Y denoted as γ_{xy}^2.

Solution: For the normal distribution, it is known that

$$H(X) = \frac{1}{2}(\log(2\pi) + \log \sigma_x^2 + 1); \ \ H(Y) = \frac{1}{2}\left(\log(2\pi) + \log \sigma_y^2 + 1\right)$$

The joint entropy of X and Y can be shown to be:

$$H(X, Y) = \log(2\pi) + \frac{1}{2}(\log \sigma_x^2 + \log \sigma_y^2 + \log\left(1 - \gamma_{xy}^2\right) + 1$$

From equation (2.40), transinformation can be written as

$$T(X, Y) = H(X) + H(Y) - H(X, Y) = -\frac{1}{2}\log\left(1 - \gamma_{xy}^2\right)$$

Designating $T(X, Y)$ as T_0, it can be written as

$$R_0 = \sqrt{1 - \exp(-2T_0)} = \gamma_{xy}$$

Thus, the informational correlation coefficient R_0 between X and Y is γ_{xy} and it is equal to the classical Pearson correlation coefficient.

2.9 Coefficient of nontransferred information

In regression analysis, the coefficient of determination r^2 measures the amount of variance that can be explained by the regression relationship and thus is often used as a measure of transferred information by regression. Thus, $1-r^2$ represents the amount of unexplained variance. In a similar manner, R_0 can be used to measure the amount of information that the assumed relation between X and Y transfers with respect to the upper limit of transferable information. The coefficient of nontransferred information t_1 measures the percentage of information left in Y after transfer to X, and can be expressed as

$$t_1 = \frac{T_0 - T_1}{T_0} \tag{2.98a}$$

where T_1 is the transinformation computed for the relationship between X and Y. For example, if the relationship is described by the bivariate normal distribution then T_1 is given as

$$T_1 = -\frac{1}{2}\ln(1 - r^2) \tag{2.98b}$$

Otherwise, T_1 is the transinformation between X and Y and T_0 will be the marginal entropy of X. Quantity t_1 basically describes the relative portion, $T_0 - T_1$, of the untransferred information. Likewise, $1-t_1$ expresses the percentage of transferred information. Both R_0 and t_1 may be used to test the validity of the assumed relationship between two random variables X and Y.

Example 2.22: Annual rainfall and annual discharge runoff for San Antonio River at Elemndorf, Texas, is given in Table 2.4. Compute the correlation coefficient and the coefficient of determination between the selected two variables. Then compute the informational correlation coefficient. How different are these coefficients? Compute the coefficient of nontransferred information and compare it with the percentage of unexplained variance.

Solution: Annual rainfall and stream flow of San Antonio River near Elmendorf, Texas (USGS 08181800), covering a period of 44 years, are given in Table 2.4.

Mean rainfall $= 31.93$ in. and mean runoff $= 4.62$ in. The correlation coefficient is determined using equation (2.95) and the coefficient of determination by squaring it: $r^2 = r_{xy}^2$. Using the given data, $r_{xy} = 0.4391$ and $r^2 = 0.1928$. The informational correlation coefficient is obtained using equation (2.94) which requires calculation of mutual information or transinformation which involves computation of marginal and joint entropies. Here X and Y denote rainfall and discharge values. To that end, absolute frequency and probability tables for runoff and rainfall are prepared as shown in Tables 2.5 to 2.8.

From Tables 2.8 and 2.9, the transinformation value is obtained as:

$$T(X; Y) = H(X) + H(Y) - H(X, Y) = 1.53 + 1.96 - 2.86 = 0.63$$

The informational correlation coefficient R_0 between X and Y is

$$R_0 = \sqrt{1 - \exp(-2T_0)} = \sqrt{1 - \exp(-2 \times 0.63)} = 0.846$$

The coefficient of nontransferred information of X is

$$t_1 = \frac{T_0 - T_1}{T_0}$$

Table 2.4 Annual rainfall and stream flow of San Antonio River at Elemndorf, Texas.

Year	Rainfall (in.)	Runoff (in.)	Year	Rainfall (in.)	Runoff (in.)	Year	Rainfall (in.)	Runoff (in.)
1963	18.65	1.29	1978	35.99	5.00	1993	32.00	5.14
1964	31.88	1.69	1979	36.64	6.73	1994	40.43	3.09
1965	36.65	3.05	1980	24.23	2.78	1995	23.20	3.19
1966	21.44	2.02	1981	36.37	5.68	1996	17.80	1.66
1967	29.26	1.84	1982	22.96	3.56	1997	33.92	3.87
1968	30.42	3.98	1983	26.11	2.66	1998	42.05	3.22
1969	31.42	2.17	1984	25.95	1.90	1999	16.41	6.35
1970	22.74	2.50	1985	41.43	3.64	2000	35.85	2.05
1971	31.80	2.07	1986	42.73	4.67	2001	36.72	5.14
1972	31.49	4.93	1987	37.96	13.23	2002	46.27	12.86
1973	52.28	8.05	1988	19.01	2.89	2003	28.45	7.35
1974	37.00	6.17	1989	22.14	1.86	2004	45.32	6.54
1975	25.67	6.31	1990	38.31	3.16	2004	16.54	7.52
1976	39.13	4.32	1991	42.76	3.51	2006	21.34	1.79
1977	29.64	7.83	1992	46.49	13.92			

Table 2.5 Computation of marginal entropy of discharge.

Runoff	0–2	2–4	4–6	6–8	8–10	10–12	12–14	Total
Frequency	7	18	7	8	1	0	3	44
Probability	0.16	0.41	0.16	0.18	0.02	0.00	0.07	1.00
$H(X_i)$	0.29	0.37	0.29	0.31	0.09	0.00	0.18	1.53

Table 2.6 Computation of marginal entropy of rainfall.

Rainfall	15–20	20–25	25–30	30–35	35–40	40–45	45–50	50–55	Total
Frequency	5	7	6	7	10	5	3	1	44
Probability	0.11	0.16	0.14	0.16	0.23	0.11	0.07	0.02	1.00
$H(Y_i)$	0.25	0.29	0.27	0.29	0.34	0.25	0.18	0.09	1.96

Table 2.7 Absolute frequency contingency table for runoff and rainfall combinations.

Rainfall	Runoff 0–2	2–4	4–6	6–8	8–10	10–12	12–14	Marginal Runoff
15-20	2	1	0	2	0	0	0	5
20-25	2	5	0	0	0	0	0	7
25-30	2	1	0	3	0	0	0	6
30-35	1	4	2	0	0	0	0	7
35-40	0	3	4	2	0	0	1	10
40-45	0	4	1	0	0	0	0	5
45-50	0	0	0	1	0	0	2	3
50-55	0	0	0	0	1	0	0	1
Marginal Rain	7	18	7	8	1	0	3	

Table 2.8 Contingency table for computation of joint probability for runoff and rainfall combinations

Rainfall (in.)	Runoff (in.)							Marginal Runoff (in.)
	0–2	2–4	4–6	6–8	8–10	10–12	12–14	
15-20	0.05	0.02	0.00	0.05	0.00	0.00	0.00	0.11
20-25	0.05	0.11	0.00	0.00	0.00	0.00	0.00	0.16
25-30	0.05	0.02	0.00	0.07	0.00	0.00	0.00	0.14
30-35	0.02	0.09	0.05	0.00	0.00	0.00	0.00	0.16
35-40	0.00	0.07	0.09	0.05	0.00	0.00	0.02	0.23
40-45	0.00	0.09	0.02	0.00	0.00	0.00	0.00	0.11
45-50	0.00	0.00	0.00	0.02	0.00	0.00	0.05	0.07
50-55	0.00	0.00	0.00	0.00	0.02	0.00	0.00	0.02
Marginal Rain	0.16	0.41	0.16	0.18	0.02	0.00	0.07	

Table 2.9 Joint entropy calculation for runoff and rainfall combinations.

Number		7	7	1	2	3	Total:
Probability	$P(X_i, Y_i)$	0.05	0.02	0.11	0.07	0.09	1.00
Entropy	$H(X_i, Y_i)$	1.05	0.55	0.24	0.37	0.65	2.86

where $T_0 = H(X) = 1.53$ nats and $T_1 = T(X;Y) = 0.63$ nats. Therefore,

$$t_1 = \frac{1.53 - 0.63}{1.53} = 0.588$$

Similarly, the coefficient of nontransferred information of Y is

$$t_1 = \frac{T_0 - T_1}{T_0}$$

where $T_0 = H(Y) = 1.96$ nats and $T_1 = T(X;Y) = 0.63$ nats. Therefore,

$$t_1 = \frac{1.96 - 0.63}{0.63} = 0.679$$

The percentage of unexplained variance is $1 - r^2 = 1 - 0.1928 = 0.8072$.

2.10 Discrete entropy: multidimensional case

Consider n random variables $X_1, X_2, \ldots, X_n$. Each variable takes on values as x_{iN_i}, $i = 1$, $2, \ldots, n$; and N_i: $N_1, N_2, \ldots, N_n$. The multidimensional entropy of $X_1, X_2, \ldots, X_n$ can be expressed as

$$H(X_1, X_2, \ldots, X_n) = -\sum_{i_1=1}^{N_1} \sum_{i_2=2}^{N_2} \ldots \sum_{i_n=1}^{N_n} p(x_{i_1}, x_{i_2}, \ldots, x_{i_n}) \log\left[p\left(x_{i_1}, x_{i_2}, \ldots, x_{i_n}\right)\right] \tag{2.99}$$

in which $X_1 : \left\{ x_{i_1}, i_1 = 1, 2, \ldots, N_1 \right\}, \ldots, X_n : \left\{ x_{i_n}, i_n = 1, 2, \ldots, i_n = N_n \right\}$. The joint entropy, given by equation (2.99), represents the collective information or uncertainty of n random variables. In a similar manner, multidimensional conditional entropy of one random variable conditioned on all others can be defined as

$$H\left(X_n | X_1, \ldots, X_{n-1}\right) = -\sum_{i_1=1}^{N_1} \cdots \sum_{i_n=1}^{N_n} p\left(x_{i_1}, x_{i_2}, \ldots, x_{i_n}\right) \log\left[p\left(x_{i_n} \middle| x_{i_1}, x_{i_2}, \ldots, x_{i_{n-1}}\right) \right]$$

$$(2.100a)$$

It represents the uncertainty left in one random variable X_n when all others are known. The multidimensional transinformation can be defined as

$$T\left(X_1, X_2, \ldots, X_{n-1}; X_n\right) = H\left(X_n\right) - H\left(X_n | X_1, X_2, \ldots, X_{n-1}\right) \qquad (2.100b)$$

This represents the common information between random variable X_n and all other variables. Figure 2.7 graphically portrays these multivariate entropy concepts.

2.11 Continuous entropy

Most environmental and hydrologic processes are continuous in nature. The entropy concepts presented for discrete variables can be extended to continuous random variables. If the random variable X is continuous, the probability of a specific value of X is zero. That is the way the probability density function $f(x)$ of X is defined. Defining $F(x)$ as the probability that X has a value equal to or less than x, $f(x) = dF(x)/dx$. If $a \leq x \leq b$, where a and b are the lower and upper limits, the probability that X occurs between a and b can be defined as $\int_a^b f(x)\, dx$. If $b-a$ is small, then approximately, $\int_a^b f(x)\, dx = (b-a)f(a)$.

Now consider entropy as frequently used [defined by Shannon (1948)]:

$$H_s(X) = -\int f(x) \log f(x)\, dx \qquad (2.101a)$$

In this equation, H_s is not dimensionless, and depends on the coordinate system. To that end, consider a density function $q(w)$ in the new coordinate system where w and x are related. One can express that $f(x) = q(w)w'(x)$ and $dw = w'(x)dx$. Furthermore,

$$\begin{aligned}
H_s(W) &= -\int q(w) \log q(w)\, dw = -\int \frac{f(x)}{w'(x)} \log \frac{f(x)}{w'(x)}\, dw \\
&= -\int \frac{f(x)}{w'(x)} \left[\log f(x) - \log w'(x) \right] dx \\
&= -\int f(x) \log f(x)\, dx + \int f(x) \log w'(x)\, dx \\
&= H_s(X) + \int f(x) \log w'(x)\, dx = H_s(X) + E[\log w'(x)]
\end{aligned} \qquad (2.101b)$$

Equation (2.101b) shows that the entropy value depends on the coordinate system if it is used as defined by equation (2.101a).

Consider an example where coordinate systems x and w are linearly related as $w = mx + k$. Then, $dw = mdx$, and $w'(x) = m$. Then,

$$H_s(W) = H_s(X) + \int f(x) \log(m)\, dx = H_s(X) + \log m \qquad (2.102a)$$

Hence, $H_s(W) - H_s(X) = \log m$, a constant quantity. In this example, the difference between the entropies of the two distributions 1 and 2, H_{s1} and H_{s2} is the same in the two coordinate system x and w. However, in general this is not true, and both entropy H_s and the difference $H_{s1} - H_{s2}$ may depend on the coordinate system.

To illustrate, consider two populations of circles, with x denoting the radius and the area $w = \pi x^2$. One can write: $dw = 2\pi x dx$. Then, for two distributions

$$H_{s1}(W) - H_{s2}(W) = H_{s1}(X) - H_{s2}(X) + \int [f_1(x) - f_2(x)] \log(2\pi x)\, dx \qquad (2.102b)$$

In general, the integral will not be zero. It is seen from the above discussion that the meaning of entropy H_s is not the same as in the case of a discrete distribution, as remarked by Shannon (1948): "The scale of measurement sets an arbitrary zero corresponding to a uniform distribution over unit volume."

Equation (2.101a) is a special case of a more general definition of entropy defined as

$$H_s(X) = -k \int f(x) \log \left[\frac{f(x)}{m(x)} \right] dx \qquad (2.102c)$$

where k is constant depending on the base of logarithm, and $m(x)$ is a measure which guarantees the invariance of entropy to the choice of coordinate sustem. In practice, often both k and $m(x)$ are assumed unity. However, their physical import is not lost sight of.

2.11.1 Univariate case

A stochastic hydrologic process is represented by a continuous random variable X within a certain range and the probability density function $f(x)$ of the variable X is assumed known. The range within which the continuous variable assumes values is divided into N intervals of width Δx. One can then express the probability that a value of X is within the n-th interval as

$$p_n = P\left(x_n - \frac{\Delta x}{2} \leq X \leq x_n + \frac{\Delta x}{2}\right) = \int\limits_{x_n - (\Delta x/2)}^{x_n + (\Delta x/2)} f(x)dx \qquad (2.103)$$

For relatively small values of Δx, one approximates the probability p_n as

$$p_n \cong f(x_n)\, \Delta x \qquad (2.104)$$

First, consider the marginal entropy of X expressed by equation (2.9) which for given class interval Δx can be rewritten as

$$H(X; \Delta x) \cong -\sum_{n=1}^{N} p_n \log p_n = -\sum_{n=1}^{N} f(x_n) \log[f(x_n)\, \Delta x]\, \Delta x \qquad (2.105a)$$

This approximation would have an error whose sign would depend on the form of the function $-f(x)\log f(x)$. In order to reduce this approximation error, the Δx interval is chosen to be as small as possible. Let $p_i = p(x_i)\Delta x$ and $p(x_i) = f(x_i)$. Let the interval size Δx tend to zero. Then, equation (2.105a) can be expressed as

$$H(X; \Delta x) = -\lim_{\Delta x \to 0} \sum_{i=1}^{N} p(x_i)\,\Delta x \log\big[p(x_i)\,\Delta x\big] \qquad (2.105b)$$

Equation (2.105b) can be written as

$$H(X; \Delta x) = -\lim_{\Delta x \to 0} \sum_{i=1}^{N} p(x_i)\,\Delta x \log\big[p(x_i)\big] - \lim_{\Delta x \to 0} \sum_{i=1}^{N} p(x_i) \log(\Delta x)\Delta x \qquad (2.105c)$$

Equation (2.105c) can also be extended to the case where Δx_i varies with i, and shows that the discrete entropy of equation (2.9) increases without bound.

Equation (2.105c) converges to

$$H(X; \Delta x) = -\int_0^\infty f(x)\log f(x)\,dx - \lim_{\Delta x \to 0} \sum_{i=1}^{N} p(x_i)\log(\Delta x)\Delta x \qquad (2.106a)$$

Equation (2.106a) can be recast as

$$H(X; \Delta x) = -\int_0^\infty f(x)\log f(x)\,dx - \lim_{\Delta x \to 0} \log(\Delta x) \qquad (2.106b)$$

Equation (2.106b) is also written as:

$$H(X; \Delta x) = -\int_0^\infty f(x)\log f(x)\,dx - \log \Delta x \qquad (2.107a)$$

Moving $-\log \Delta x$ on the left side, equation (2.105c) can be written as

$$H(X; \Delta x) \cong -\sum_{n=1}^{N} p_n \log\left(\frac{p_n}{\Delta x}\right) = -\sum_{n=1}^{N} f(x_n)\log f(x_n)\,\Delta x \qquad (2.107b)$$

Equation (2.107b) is also referred to as spatial entropy (Batty, 2010) if x is a space dimension. The right side of equation (2.107b) can be written as

$$H(X) = -\int_0^\infty f(x)\log f(x)\,dx \qquad (2.108)$$

Equation (2.108) is the commonly used expression for the continuous Shannon entropy.

If a random variable X is continuous over the range $(0, \infty)$, then the Shannon entropy is expressed as

$$H(X) = -\int_0^\infty f(x)\log\big[f(x)\big]\,dx = -\int_0^\infty \log\big[f(x)\big]\,dF(x) = E\big[-\log f(x)\big] \qquad (2.109)$$

where $f(x)$ is the probability density function (PDF) of X, $F(x)$ is the cumulative probability distribution function of X, and $E[.]$ is the expectation of $[.]$. $H(X)$ is a measure of the uncertainty of random variable X of the system. It can also be understood as a measure of the amount of information required on average to describe the random variable. Thus, entropy is a measure of the amount of uncertainty represented by the probability distribution or of the lack of information about a system represented by the probability distribution. Sometimes it is referred to as a measure of the amount of chaos. If complete information is available, entropy $= 0$; otherwise, it is greater than zero. The uncertainty can be quantified using entropy taking into account all different kinds of available information.

Recall the discrete form of Shannon entropy given by equation (2.9) in which p_i is the probability of i-th event and N is the number of events. Here X is a discrete random variable taking on discrete values: x_i, $i = 1, 2, \ldots, N$, or $X:\{x_i, i = 1, 2, \ldots, N\}$; $X = x_i$ defines the i-th event. Of course, if X is a continuous random variable then the continuous form of entropy is expressed by equation (2.109) which measures the amount of information in the corresponding probability distribution. Equation (2.108) cannot be directly obtained from equation (2.9) by letting the interval size tend to zero and taking the limit.

From equation (2.106a) the continuous form of entropy can be expressed as

$$-\int_0^\infty f(x) \ln f(x)\, dx = -\sum_{i=1}^N p(x_i) \ln p(x_i) + \lim_{\Delta x \to 0} \sum_{i=1}^N p(x_i) \ln(\Delta x) \Delta x \tag{2.110a}$$

or

$$-\int_0^\infty f(x) \ln f(x)\, dx = -\sum_{i=1}^N p_i \ln p_i + \lim_{\Delta x \to 0} \sum_{i=1}^N p(x_i) \ln(\Delta x) \tag{2.110b}$$

Equation (2.110b) is not equal to the Shannon entropy and shows that the continuous entropy is the difference between the discrete entropy given by equation (2.9) and a term representing the logarithm of the interval of measurement. Therefore,

$$H = -\lim_{\Delta x \to 0} \sum_{i=1}^N p_i \ln\left(\frac{p_i}{\Delta x}\right) \quad \text{or} \quad H = -\lim_{\Delta x \to 0} \sum_{i=1}^N p_i \ln\left(\frac{p_i}{\Delta x_i}\right) \tag{2.111}$$

Equation (2.111) provides a discrete approximation of equation (2.110a) and constitutes the foundation for spatial analysis in geography, hydrology, climatology, ecosystem science, forestry, and watershed sciences.

Now consider a weighting function $g(x, y)$ such that

$$\int g(x, y)\, dx = \int g(x, y)\, dy = 1, \quad g(x, y) \geq 0 \tag{2.112a}$$

An averaged distribution $p(x, y)$ can be expressed using a generalized averaging operation:

$$p(y) = \int g(x, y) f(x)\, dx \tag{2.112b}$$

Then the entropy of $p(y)$ is equal to or greater than the entropy of the original distribution $f(x)$.

Example 2.23: A continuous random variable X with a normal probability density function of mean μ and standard deviation σ. What is the value of entropy of this random variable?

Solution: Note that $\int_{-\infty}^{\infty} f(x)\,dx = 1$; $\int_{-\infty}^{\infty} xf(x)\,dx = \mu$; $\int_{-\infty}^{\infty} f(x)\,(x-\mu)^2\,dx = \sigma^2$

Normal probability density function: $f(x) = \dfrac{1}{\sqrt{2\pi}\sigma}e^{-\frac{(x-\mu)^2}{2\sigma^2}}$

Substituting $f(x)$ into the Shannon entropy expression, one gets

$$H(X) = -\int_{-\infty}^{\infty} f(x)\log[f(x)]\,dx = -\int_{-\infty}^{\infty} f(x)\log\left[\frac{1}{\sqrt{2\pi}\sigma}e^{-\frac{(x-\mu)^2}{2\sigma^2}}\right]dx$$

$$= -\int_{-\infty}^{\infty} f(x)\log\left[\frac{1}{\sqrt{2\pi}\sigma}\right]dx - \int_{-\infty}^{\infty} f(x)\left[-\frac{(x-\mu)^2}{2\sigma^2}\right]dx$$

$$= -\log\left[\frac{1}{\sqrt{2\pi}\sigma}\right] + \frac{1}{2\sigma^2}\int_{-\infty}^{\infty} f(x)\,(x-\mu)^2\,dx$$

$$= \frac{1}{2}(\log 2\pi\sigma^2 + 1) = \frac{1}{2}(\log(2\pi) + \log\sigma^2 + 1) \tag{2.113}$$

If the variance is unity, then

$$H(X) = \frac{1}{2}\big[\log(2\pi) + 1\big] = 1.419 \text{ nats} \tag{2.114}$$

Equation (2.114) shows that the entropy of a normal variable depends only on the variance but not on the mean. This suggests that variance can also be considered as a measure of information.

2.11.2 Differential entropy of continuous variables

For a continuous random variable X with PDF $f(x)$,

$$h(X) = -\int_{-\infty}^{\infty} f(x)\log f(x)\,dx \tag{2.115}$$

is called the differential entropy of X, which is distinguished from the ordinary or absolute entropy. This can be seen as follows.

The discrete variable X takes on values $x_i = i\delta x$, where $i = 0, \pm 1, \pm 2, \ldots$, and δx approaches 0. The continuous variable takes on values in the interval $[x_i, x_i + \delta x]$ with probability $f(x_i)\delta x$. Letting δx go to 0, the ordinary entropy of the continuous random variable X can be expressed as:

$$H(X) = -\lim_{\delta x \to 0}\sum_{i=-\infty}^{\infty} f(x_i)\,\delta x\log[f(x_i)\delta(x)]$$

$$= -\lim_{\delta x \to 0}\left[\sum_{i=-\infty}^{\infty} f(x_i)\,\delta x\log[f(x_i)] + \log\delta x\sum_{i=-\infty}^{\infty} f(x_i)\,\delta x\right]$$

$$= -\sum_{i=-\infty}^{\infty} f(x)\log f(x)d(x) - \lim_{\delta x \to 0}\log\delta x\sum_{i=-\infty}^{\infty} f(x_i)\,\delta x = -h(x) - \lim_{\delta x \to 0}\log\delta x \tag{2.116}$$

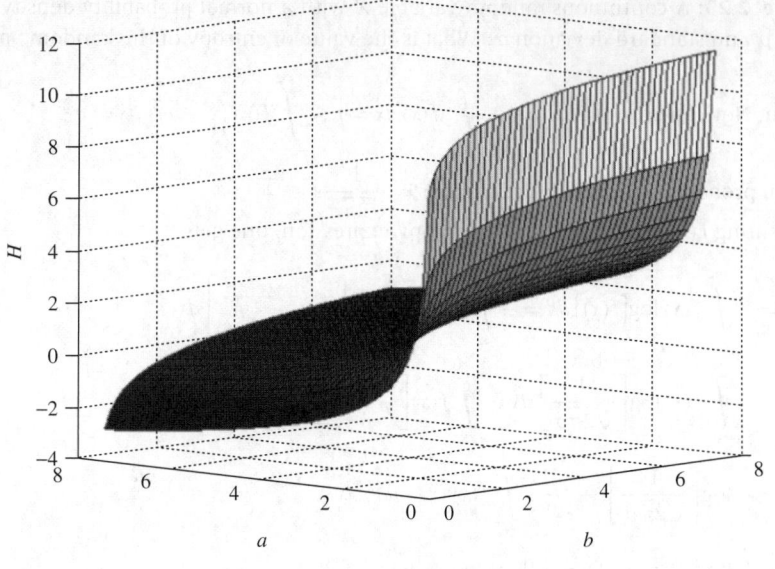

Figure 2.14 Entropy of the Pareto distribution as a function of parameters $a > $ and $b > 0$.

In the limit, as δx approaches 0, $-\log \delta x$ approaches infinity. This shows that the entropy of a continuous random variable is infinitely large. This is intuitive because X takes on many (an infinite number of) values in the interval $(-\infty, \infty)$, and uncertainty associated with the variable is of the order of infinity. Adopting $-\log \delta x$ as a reference, $h(X)$ is characterized as a differential entropy. For a dynamical system, the information processed is actually the difference between two entropy terms that have a common reference and will equal the difference between the corresponding differential entropy terms. In this text, the differential entropy $h(X)$ will be considered as entropy $H(X)$, keeping in mind that it is above the reference value of $-\log \delta x$.

For a continuous distribution, the differential entropy given by equation (2.115) shares some but not all the properties of discrete entropy. For continuous distributions, $H(X)$ is not scale independent, for $H(CX) = \log C + H(X)$ but is translation invariant, because $H(C + X) = H(X)$. If $f(.)$ is bounded then $H(X) > -\infty$. If $Var(x) < \infty$, then $H(X) < \infty$ (Ash, 1965, p. 237). But for a distribution with a finite entropy, the variance may not exist. Consider a Pareto distribution $f(x) = (a/b)(x/b)^{-(a+1)}, x > b, a > 0$, and $b > 0$. In this case, noting that $E[\ln x] = \ln b + (1/a)$, $H(X) = \log b - \log a + (1/a) + 1$ which is finite over the entire parameter space but $Var(x) = b^2 a(a-1)^{-2}(a-2)^{-2}$ is not defined if $a < 2$. Figure 2.14 shows the entropy of the Pareto distribution. This shows that variance cannot be employed for computing uncertainties of two Pareto distributions when $a < 2$, but the uncertainties based on the entropy difference can be computed: $H(X|a_1, b_1) - H(X|a_2, b_2)$.

If $\vec{X}$ is a random vector consisting of n random variables $X_1, X_2, \ldots, X_n$ then the differential entropy is given by the n-fold integral:

$$h(\vec{X}) = -\int f(\vec{X}) \log f(\vec{X}) \, d\vec{X} \tag{2.117}$$

where $f(\vec{X})$ is the joint PDF of $\vec{X}$.

2.11.3 Variable transformation and entropy

Consider a dimensionless variable X, say, river flow divided by mean flow having a probability density function $f(x)$, $x \in (a, b)$. Let there be a change of variable from X to Z:

$$x = g(z) \tag{2.118}$$

where $Z \in (a_0, b_0)$, $a_0 = g(a)$, $b_0 = g(b)$, and $z = g^{-1}(x)$ is a unique inverse. Let $q(z)$ be the PDF of Z. Also,

$$f(x)\, dx = f[g(z)] \left| \frac{dx}{dz} \right| dz = q(z)\, dz \tag{2.119}$$

where

$$q(z) = f[g(z)] \left| \frac{dx}{dz} \right| \tag{2.120}$$

where $q(z)$ is a PDF of Z:

$$\int_{a_0}^{b_0} q(z)\, dz = 1 \tag{2.121}$$

Now consider the Shannon entropy for Z:

$$H(Z) = - \int_{a_0}^{b_0} q(z) \ln q(z)\, dz \tag{2.122}$$

whereas for X:

$$H(X) = - \int_{a}^{b} f(x) \ln f(x)\, dx \tag{2.123}$$

Then,

$$H(Z) = - \int_{a}^{b} f(x) \left| \frac{dx}{dz} \right| \ln \left[f(x) \left| \frac{dx}{dz} \right| \right] dz = - \int_{a}^{b} f(x) \ln f(x)\, dx - \int_{a_0}^{b_0} f(x) \ln \left| \frac{dx}{dz} \right| dz$$

$$= H(X) - \int_{a}^{b} f(x) \ln \left| \frac{dx}{dz} \right| dz \tag{2.124}$$

This leads to

$$H(X) = H(Z) + \int_{a}^{b} f(x) \ln \left| \frac{dx}{dz} \right| dz \tag{2.125}$$

This shows that $H(Z) \leq H(X) + E[\ln |dx|]$.

2.11.4 Bivariate case

For two random variables, the bivariate entropy can be expressed as

$$H(X, Y) = -\int\limits_0^\infty \int\limits_0^\infty f(x, y) \log f(x, y)\, dx dy - \log(\Delta x \Delta y) \tag{2.126}$$

The conditional entropy of X with respect to Y can be expressed as

$$H(X|Y; \Delta x) = -\int\limits_0^\infty \int\limits_0^\infty f(x, y) \log[f(x|y)]\, dx\, dy - \log \Delta x \tag{2.127}$$

The range of variable Y is also divided using the same class interval Δx. It should be noted that the expressions for marginal entropy and conditional entropy include a subtractive constant term $\log \Delta x$. This means that the uncertainty is reduced as the class interval size is increased. Further, following (Papoulis, 1991), one can write

$$H(X|y) = -\int\limits_{-\infty}^\infty f(x|y) \log f(x|y)\, dx \tag{2.128}$$

$$H(X|Y) = -\int\limits_{-\infty}^\infty f(y)\, H(X|Y = y)\, dy = \int\limits_{-\infty}^\infty \int\limits_{-\infty}^\infty f(x, y) \log f(x|y)\, dx\, dy \tag{2.129}$$

$$H(X|Y = y) = E[-\log f(x|y)|Y = y] \tag{2.130a}$$

$$H(X|Y) = E[-\log f(x|y)] = E\{E[-\log f(x|y)|Y = y]\} \tag{2.130b}$$

Example 2.24: Consider the bivariate normal distribution of random variables X and Y:

$$f(x, y) = \frac{1}{2\pi \sigma_x \sigma_y \sqrt{1 - r^2}} \exp\left\{-\frac{1}{2(1 - r^2)}\left[\frac{(x - \bar{x})^2}{\sigma_x^2} - 2r\frac{(x - \bar{x})(y - \bar{y})}{\sigma_x \sigma_y} + \frac{(y - \bar{y})^2}{\sigma_y^2}\right]\right\}$$

where σ_x is the standard deviation of X, σ_y is the standard deviation of Y, and r is the coefficient of correlation between X and Y. Determine the entropy of this distribution.

Solution: The joint entropy $H(X, Y)$ can be expressed as

$$H(X, Y) = -\int\limits_{-\infty}^\infty \int\limits_{-\infty}^\infty f(x, y) \ln[f(x, y)]\, dx\, dy = E[-\ln f(x, y)]$$

Then,

$$\ln[f(x, y)] = \frac{-1}{2(1 - r^2)}\left[\frac{(x - \bar{x})^2}{\sigma_x^2} - 2r\frac{(x - \bar{x})(y - \bar{y})}{\sigma_x \sigma_y} + \frac{(y - \bar{y})^2}{\sigma_y^2}\right] - \ln\left[2\pi \sigma_x \sigma_y \sqrt{1 - r^2}\right]$$

Note that

$$E\left[\frac{(x-\bar{x})^2}{\sigma_x^2} - 2r\frac{(x-\bar{x})(y-\bar{y})}{\sigma_x\sigma_y} + \frac{(y-\bar{y})^2}{\sigma_y^2}\right] = 1 - 2r^2 + 1$$

Therefore,

$$E\left[-\ln f(x,y)\right] = 1 + \ln 2\pi\sigma_x\sigma_y\sqrt{1-r^2}$$

Thus, the entropy of the joint normal distribution is:

$$H(X, Y) = \ln 2\pi e\sqrt{\sigma_x^2\sigma_y^2 - r^2\sigma_x^2\sigma_y^2}$$

Example 2.25: Consider two random variables X and Y each having a normal distribution and they are jointly normal as well. Assume zero mean. Determine the conditional entropy and transinformation or mutual information.

Solution: The conditional probability density function $f(x|y)$ is normal with mean $r\sigma_x/\sigma_y$ and variance $\sigma_x^2(1-r^2)$, where r defines the degree of correlation. Then, the conditional entropy can be written as

$$H(X|Y) = E\left[-\ln f(x|y)\right] = \ln\left[\sigma_x\sqrt{2\pi e(1-r^2)}\right]$$

Since this is independent of Y, that is, $H(X|Y) = H(X)$, and $H(X) = \ln[\sigma_x\sqrt{2\pi e}]$, it then follows that

$$T(X, Y) = H(X) - H(X|Y) = -0.5\ln(1-r^2)$$

Further,

$$H(X|Y) + H(Y) = \ln 2\pi e\sqrt{\sigma_x^2\sigma_y^2 - r^2\sigma_x^2\sigma_y^2} = H(X, Y)$$

One can generalize this result.

Entropy $H(X)$ measures the relative information, with $-\log(\Delta x)$ serving as the datum, when Δx approaches zero (Lathi, 1969), and Δx is the division interval of the X domain. As a measure of the relative information, $H(X)$ can be positive, negative or zero, and therefore the conditional entropy connoting H_{Lost} can also be positive, negative or zero, since it is part of the information $H(X)$ and is bounded from above by $H(X)$. Negative $H(X)$ or negative H_{Lost} has no physical meaning. This difficulty occurs owing to the use of a relative coordinate system for which the origin is set at $-\log\Delta x$. In an absolute coordinate system where the origin is set at $-\infty$, and both H and H_{Lost} remain no longer negative and retain physical meaning.

An important difference between discrete and continuous cases is that entropy measures randomness of the discrete random variable in an absolute way, whereas for the continuous random variable entropy measures the randomness in relation to the coordinate system, which can be an assumed standard. If $X : \{x_1, x_2, \ldots, x_N\}$ is changed to coordinates $Y : \{y_1, y_2, \ldots, y_N\}$ then the new entropy is expressed as

$$H(Y) = \int \int \cdots \int f(x_1, x_2, \ldots, x_N) J\left(\frac{X}{Y}\right) \log\left[f(x_1, x_2, \ldots, x_N) J\left(\frac{X}{Y}\right)\right] dy_1 dy_2 \ldots dy_N$$

(2.131)

where $J(X/Y)$ is the Jacobian of the coordinate transformation. Equation (2.131) can be expressed as

$$H(Y) = H(X) - \int \int \cdots \int f(x_1, x_2, \ldots, x_N) J\left(\frac{X}{Y}\right) \log\left[J\left(\frac{X}{Y}\right)\right] dx_1 dx_2 \ldots dx_N \qquad (2.132)$$

Thus, the new entropy is defined as the old entropy minus the expected logarithm of the Jacobian. The coordinate system can be chosen with each small volume element $dx_1, dx_2, \ldots, dx_N$ given equal weight. In the changed coordinate system entropy measures randomness, wherein volume elements $dy_1, dy_2, \ldots, dy_N$ are assigned equal weight.

The entropy of a continuous distribution can be negative. The scale of measurement defines an arbitrary zero corresponding to a unique distribution over a unit volume. A distribution more confined than the uniform distribution has less entropy and it will be negative. The information rate and capacity will always be non-negative.

Now consider a case of linear transformation of coordinates:

$$y_j = \sum_i a_{ij} x_i$$

(2.133)

The Jacobian in this case is the determinant $\left|a_{ij}\right|^{-1}$. The entropy can then be written as

$$H(Y) = H(X) + \log\left|a_{ij}\right|$$

(2.134)

If the coordinates are rotated, then $J = 1$ and $H(X) = H(Y)$.

The rate of transmission of information can be expressed as

$$R = H(X) - H(X|Y)$$

(2.135)

where $H(X)$ is the entropy of the input and $H(X|Y)$ is the equivocation or the lost information. The minimum of R is obtained by varying the input X over all possible ensembles, that is, maximize

$$-\int f(x) \log f(x) \, dx + \int \int f(x, y) \log \frac{f(x, y)}{f(y)} \, dx \, dy$$

This can be expressed as

$$\int \int f(x, y) \log \frac{f(x, y)}{f(x) f(y)} \, dx \, dy$$

Noting that

$$\int \int f(x, y) \log \frac{f(x, y)}{f(x) f(y)} \, dx \, dy = \int f(x) \log(x) \, dx$$

(2.136)

If u is the message, x is the signal, y is the received signal perturbed by noise, and v is the recovered message, then

$$H(X) - H(X|Y) \geq H(U) - H(U|V) \tag{2.137}$$

irrespective of the operations performed on u to obtain x or on y to obtain v.

If the noise is added to the signal and is independent (in a probability sense), then $f(x|y)$ is a function of only the difference $z = y\text{-}x$:

$$f(y|x) = g(y - x) \tag{2.138}$$

For the received signal as the sum of the transmitted signal and the noise, the rate of transmission is

$$R = H(Y) - H(Y - X) \tag{2.139}$$

which is the entropy of the received signal minus the entropy of the noise. Since $y = x + z$,

$$H(X, Y) = H(X, Z) \tag{2.140}$$

Expanding the left side and considering X and Z as independent,

$$H(Y) + H(X|Y) = H(X) + H(Z) \tag{2.141a}$$

Thus,

$$R = H(X) - H(X|Y = y) = H(Y) - H(Z) \tag{2.141b}$$

Transinformation can be defined as

$$
\begin{aligned}
T(X, Y; \Delta x) &= -\int_0^\infty f(x) \log f(x)\, dx + \int_0^\infty \int_0^\infty f(x, y) \log f(x, y)\, dx\, dy \\
&= -\int_0^\infty \int_0^\infty f(x, y) \log f(x)\, dx\, dy + \int_0^\infty \int_0^\infty f(x, y) \log \frac{f(x, y)}{f(y)}\, dx\, dy \\
&= \int_0^\infty \int_0^\infty f(x, y) \log \frac{f(x, y)}{f(x)\, f(y)}\, dx\, dy
\end{aligned}
\tag{2.142}
$$

Equation (2.142) for transinformation shows that the information about X transferred by the knowledge of Y does not depend on the class interval selected.

Example 2.26: One would want to determine the effect of measurement interval size on the value of entropy. 1) Assume that the probability distribution given in Example 2.12 is for a random variable that has been measured at an interval of one unit. [Note the random variable has been normalized and is therefore dimensionless.] Compute the value of entropy using this distribution. 2) Now take the interval of measurement as two units. In this case,

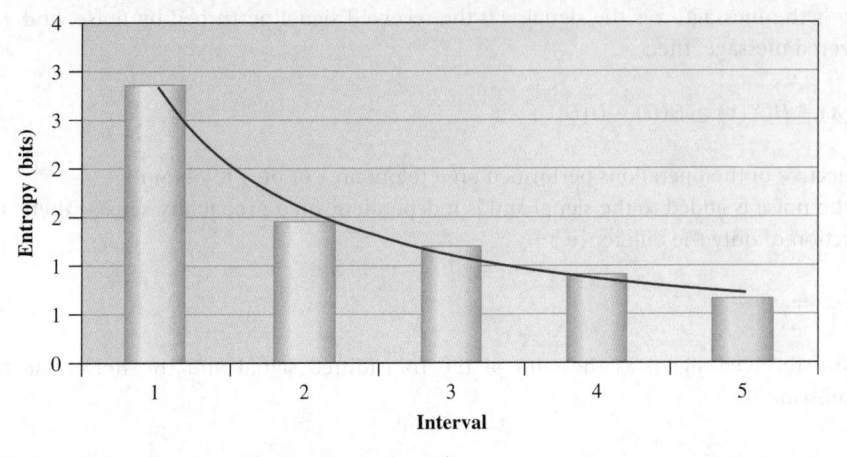

Figure 2.15 Variation of entropy with measurement interval.

probability values averaged over two class intervals can be used. Compute the entropy value for this measurement interval. 3) Increase the measurement interval to four times the original interval and then compute the entropy value. 4) Increase the measurement interval to five times the original interval and then compute the entropy value. 5) Compare entropy values and show how entropy varies with the measurement interval.

Solution:
a) Using the probabilities values from Example 2.12, a) $H(P) = 2.8601$ bits.
b) If the measurement interval is doubled, five probability values are obtained: $P = [0.05\ 0.25\ 0.45\ 0.20\ 0.05]$. Then $H(P) = 1.9150$ bits
c) If the measurement interval is increased four times the original value then three probabilities values are obtained as: $P = [0.3\ 0.65\ 0.05]$. Then, $H(P) = 1.1412$ bits
d) If the measurement interval is five times the original value then two probabilities values are obtained: $P = [0.5\ 0.5]$. Then, $H(P) = 1$ bit
e) The above entropy values are plotted in Figure 2.15. It is seen that there is a sharp drop in the magnitude of entropy, when the measurement interval increases multiplicatively. Intuitively, the decrease in entropy value with increasing measurement interval can be attributed to the reduction in the variability of observed values. Put another way, the probability of observing an event within the interval also increases when the measurement interval is increased. This leads to more certainty and hence less entropy.

Example 2.27: A random variable X of the normal probability distribution with a mean of zero and a variance of 2, compare the entropies of the variable for different values of interval Δx.

Solution: According to Example 2.23, the entropy value for the normally distributed continuous variable X is given as

$$H(X) = \frac{1}{2}\left(\ln(2\pi) + \ln\sigma_x^2 + 1\right) = 1.766 \text{ Napier}$$

Generate 1000 samples according to $X \sim N(0, 2)$ and count its frequencies for different intervals.

Table 2.10 Frequencies for different intervals.

$\Delta x = 1$	$-5\sim-4$	~-3	~-2	~-1	~0	~1	~2	~3	~4	~5	Total
Numbers of samples	5	19	75	138	192	238	180	91	48	14	1000
$\Delta x = 2$	$-5\sim-3$		~-1		~1		~3		~5		
Numbers of samples	24		203		430		271		62		1000

The result is listed in Table 2.10.

When $\Delta x = 1$, $H(X; \Delta x) \cong -\sum_{n=1}^{N} p_n \ln p_n - \ln \Delta x = 1.9602 - 0 = 1.9602$ napiers

When $\Delta x = 2$, $H(X; \Delta x) \cong -\sum_{n=1}^{N} p_n \ln p_n - \ln \Delta x = 1.3023 - 0.6931 = 0.6092$ napiers

2.11.5 Multivariate case

For multivariate case, the entropy of n random variables $X_1, X_2, \ldots, X_n$ can be expressed as

$$H(X_1, X_2, \ldots, X_n) = -\int_0^\infty \cdots \int_0^\infty f(x_1, x_2, \ldots, x_n) \log[f(x_1, x_2, \ldots, x_n)] \, dx_1 \, dx_2 \cdots dx_n - \log(\Delta x)^n$$

$$(2.143)$$

Similarly, the conditional entropy can be expressed as

$$H(X_n | X_1, X_2, \ldots, X_{n-1})$$
$$= -\int_0^\infty \cdots \int_0^\infty f(x_1, x_2, \ldots, x_n) \log[f(x_n | x_1, x_2, \ldots, x_{n-1}) \, dx_1 \, dx_2 \cdots dx_n - \log(\Delta x) \quad (2.144)$$

It can also be shown that

$$H(X_n) \geq H(X_n | X_{n-1}) \geq \cdots \geq H(X_n | X_{n-1}, X_{n-1}, \ldots, X_1) \quad (2.145)$$

Thus, as the degree of dependence increases and more multivariables interact, the conditional entropy decreases until some lower limit. The increase or decrease of information depends on the introduction or elimination of multivariables. This may be compared with the forward selection of variables in the regression analysis based on partial correlation.

2.12 Stochastic processes and entropy

Let there be random variables $X(t_1)$, $X(t_2), \ldots, X(t_m)$ having a joint probability density $f(x_1, x_2, \ldots, x_m)$. The joint entropy $H(X_1, X_2, \ldots, X_m) = E[-\ln f(x_1, x_2, \ldots, x_m)]$, referred to as the m-th order entropy of the stochastic process $X(t)$, denotes the uncertainty about those random variables and equals the information gained upon their observation. In general, the uncertainty about the values of $X(t)$ on the entire t-axis or an interval thereof is infinite. Therefore, only discrete-time processes are considered. Let X_n be a discrete-time stochastic process whose m-random variables can be expressed as $x_n, x_{n-1}, \ldots, x_{n-m+1}$. Then the m-th order entropy or joint entropy, $H(x_1, x_2, \ldots, x_m)$, indicates the uncertainty about m consecutive values of the process X_n. As an example, the first order entropy can be denoted by $H(X)$ representing the uncertainty about X_n for a specific value of n. This is a univariate entropy.

If the process is strictly white, that is, all random values $x_1, x_2, \ldots, x_m$ are independent then $H(x_1, x_2, \ldots, x_m) = mH(X)$. If the stochastic process is Markovian then the probability density function can be expressed as

$$f(x_1, x_2, \ldots, x_m) = f(x_m | x_{m-1}) \ldots f(x_2 | x_1) f(x_1) \tag{2.146}$$

Then the entropy can be written as

$$H(x_1, x_2, \ldots, x_m) = H(x_m | x_{m-1}) + \ldots + H(x_2 | x_1) + H(x_1) \tag{2.147}$$

Assuming X_n to be stationary, it follows that

$$H(x_1, x_2, \ldots x_m) = (m-1) H(x_1, x_2) - (m-2) H(X) \tag{2.148}$$

Thus, the m-order entropy of a Markovian stochastic process can be expressed in terms of first and second order entropies.

Conditional entropy: The conditional entropy of order m, $H(x_n | x_{n-1}, x_{n-2}, \ldots, x_{n-m})$, expresses the uncertainty about its present given its most recent observed values. It can then be shown that

$$H(x_n | x_{n-1}, x_{n-2}, \ldots, x_{n-m}) \leq H(x_n | x_{n-1}, x_{n-2}, \ldots, x_{n-m-1}) \tag{2.149}$$

This conditional entropy is a decreasing function of m. If $m \to \infty$ then

$$H_c(x) = \lim_{m \to \infty} H(x_n | x_{n-1}, x_{n-2}, \ldots, x_{n-m}) \tag{2.150}$$

measures the uncertainty about the present x_n given its entire past.

If x_n is strictly white, then

$$H_c(x) = H(x) \tag{2.151}$$

If x_n is Markovian, then

$$H(x_n | x_{n-1}, x_{n-2}, \ldots, x_{n-m}) = H(x_n | x_{n-1}) \tag{2.152}$$

Since x_n is a stationary stochastic process, then

$$H_c(x) = H(x_2 | x_1) = H(x_1, x_2) - H(x) \tag{2.153}$$

meaning that if x_{n-1} is obtained then the past has no influence on the uncertainty of the present.

Entropy rate: Let m be a block of consecutive samples. Then, $H(X_1, X_2, \ldots, X_m)/m$ expresses the average uncertainty per sample. For $m \to \infty$, the limit of this average, denoted as $H_r(X)$, expresses the entropy rate of the process X_n:

$$H_r(X) = \lim_{m \to \infty} \frac{1}{m} H(x_1, x_2, \ldots, x_m) \tag{2.154}$$

If X_n is strictly white then

$$H_r(X) = H(X) = H_c(X) \tag{2.155}$$

If X_n is Markovian then

$$H_r(X) = H(X_1, X_2) - H(X) = H_c(X) \tag{2.156}$$

The entropy rate of a process X_n is equal to its conditional entropy:

$$H_r(X) = H_c(X) \tag{2.157}$$

Let X_n be a normal stochastic process having variance σ^2 and power spectrum $S(w)$. Then the entropy rate can be expressed as

$$H_r(X) = \ln\sqrt{2\pi e} + \frac{1}{4\sigma} \int_{-\sigma}^{a} \ln\, S(w)\, dw \tag{2.158a}$$

Let Y_n be the output of a linear system $L(Z)$. Then the entropy rate $H_r(Y)$ of the output Y_n can be written as

$$H_{r(Y)} = H_r(X) + \frac{1}{2\sigma} \int_{-\sigma}^{\sigma} \ln|L[\exp(iwT)|\, dw \tag{2.158b}$$

If X_n is a normal process then Y_n is also a normal process. The entropy rate is expressed as before where

$$S(w) = S_y(w) = S_x(w)\left|L[\exp(iwT)]\right|^2 \tag{2.158c}$$

This leads to

$$H_{r(Y)} = \ln\sqrt{2\pi e} + \frac{1}{4\sigma} \int_{-\sigma}^{\sigma} \{\ln S_x(w) + \ln |L[\exp(iwT)|^2\}\, dw \tag{2.159}$$

2.13 Effect of proportional class interval

In the above discussion on measures of uncertainty, a fixed class interval has been assumed. In many cases, such as stream flow measurements, observation errors increase with the magnitude of flow and may indeed be approximately proportional to that flow. Consider, for example, an error of 0.1 cubic meter per second (cumec). It may be a large error at low flows but almost undetectable at high flows. In a similar vein, for a stream with flow above 50 cumecs a model predicting stream flow with an error of 0.1 cumec would be considered quite accurate but would not be considered satisfactory if observed stream flows were less than 0.2 cumec. This leads to the development of entropy concepts in terms of a class interval which is proportional to flow. This is tantamount to dividing the logarithms of flows into equal class intervals. Chapman (1986) addressed this issue and here his work, which is similar to that of Amorocho and Espildora (1973), is followed.

Consider a random variable Z, such that $Z = \ln X$. Let the range of Z be divided into N intervals of width Δz. One can now compute the probability that a value of X is within the i-th interval ($i = 1, 2, 3, \ldots, N$) as

$$p_i \cong p(z_i - \frac{1}{2}\Delta z \leq Z \leq z_i + \frac{1}{2}\Delta z) = p(\ln x_i - \frac{1}{2}\Delta z \leq \ln X \leq \ln x_i + \frac{1}{2}\Delta z)$$
$$= p[x_i \exp(-\Delta z/2) \leq X \leq x_i \exp(\Delta z/2)] \cong f(x_i)[x_i \exp(\Delta z/2) - x_i \exp(-\Delta z/2)] \qquad (2.160)$$

Expanding the exponential terms up to first order, one obtains:

$$p_i \cong f(x_i) x_i \Delta z \qquad (2.161)$$

if Δz is small. Then, the marginal entropy given by equation (2.9) can be expressed as

$$H(X; \Delta x/x) = H\ (X; \Delta x)\ = -\sum_i^N p_i \log p_i \cong -\sum_{i=1}^N x_i\, f(x_i) \ln\left[x_i\, f(x_i)\, \Delta z\right] \Delta z$$

$$\cong -\sum_{i=l}^N x_i\, f(x_i) \log[x_i\, f(x_i)]\Delta z - \log\Delta z \sum_{i=1}^N x_i\, f(x_i)\Delta z$$

$$\cong -\int_0^\infty f(x)\mathrm{In}[xf(x)]\, dx\ - \mathrm{In}(\Delta x/x) \qquad (2.162)$$

Note $dz = dx/x$ from the log transformation. Likewise, the conditional entropy can be written as

$$H(X|Y; \Delta x/x) \cong -\int_0^\infty \int_0^\infty f(x, y) \log[xf(x|y)]\, dx\, dy - \log(\Delta x/x) \qquad (2.163)$$

Transinformation can be derived by subtracting equation (2.163) from equation (2.162) as

$$T(X, Y; \Delta x/x) = -\int_0^\infty f(x) \log[xf(x)]\, dx + \int_0^\infty \int_0^\infty f(x, y) \log[xf(x|y)]\, dx\, dy \qquad (2.164)$$

Similar to equation (2.142), equation (2.164) shows that transinformation or the information about X transferred by the knowledge of Y does not depend on whether the class interval is constant or proportional.

Now the relation between expressions for marginal entropy and conditional entropy for a fixed class interval Δx [equations (2.109) and (2.144)] and those for a proportional class interval $\Delta x/x$ can be established [equations (2.162) and (2.163)] as:

$$[H(X; \Delta x) + \text{In } \Delta x] - [H(X; \Delta x/x) + \text{In}(\Delta x/x)]$$

$$= [H(X|Y; \Delta x) + \text{In}\Delta x] - [H(X|Y; \Delta x/x) + \text{In}(\Delta x/x)] = \int_0^\infty f(x) \, In(x) \, Inx dx \qquad (2.165)$$

For a lognormal distribution, the integral term in equation (2.165) defines the mean of z, μ_z, with the marginal entropy as: $H(X; \Delta x/x) = 0.5\ln(2\pi e\sigma_z^2) - \ln(\Delta x/x)$, where σ_z^2 is the variance of Z. Using $\Delta x/x = 0.05$ for daily stream flow data of Dry Creek, California, that Amorocho and Espildora (1973) used, Chapman (1986) found that marginal entropy was higher during low flow periods than during high flow periods and the cross over was where the mean flow was about $1/0.05 = 20$ cfs. He also noted that there was reduction in seasonal variation of the marginal entropy, with the ratio range/mean reduced from 1.59 for $\Delta x = 1$ cfs to 0.37 for $\Delta x/x = 0.05$. He also suggested that a value of $\Delta x/x = 0.05$ would be appropriate for stream flow data of good quality and as a general criterion of fit of hydrologic models to observed data.

Example 2.28: Take daily flows of Brazos River near Hempstead for the period of October 1950 through September 1979. Compute and compare marginal entropies for fixed and proportional intervals of time, respectively.

Solution: The fixed interval is taken as 1 cfs and the proportional interval as 0.05 cfs. Then, marginal entropy is computed for both intervals as shown in Figure 2.16. It is seen that the marginal entropy for the proportional interval is significantly less and varies less than that for the fixed intervals.

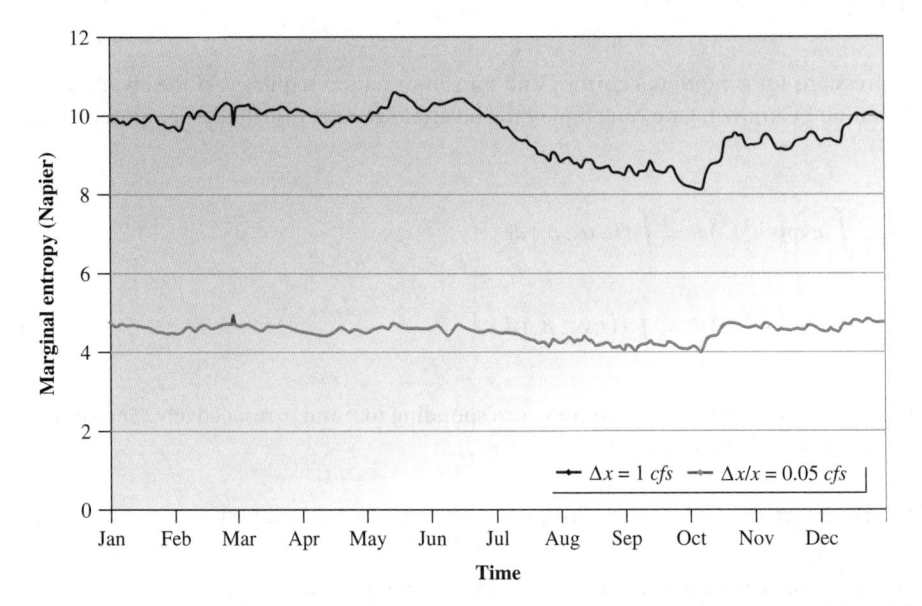

Figure 2.16 Marginal entropy of recorded daily flows of Brazos River near Hempstead, Texas (USGS 08111500). *Source*: US Geological Survey.

2.14 Effect of the form of probability distribution

The values of entropy and transinformation depend on the form of the probability distribution to be used. Frequently the lognormal distribution has been used for computing entropy. Consider the case of gamma distribution:

$$f(x; \alpha, \beta) = \frac{1}{\Gamma(\alpha)\,\beta^\alpha} x^{\alpha-1} \exp\left(-\frac{x}{\beta}\right) \tag{2.166}$$

where α and β are the form and scale parameters, respectively. Substitution of equation (2.166) in equation (2.162) yields

$$H(X; \Delta x/x) = \ln\Gamma(\alpha) + \frac{\mu_z}{\beta} - \alpha\psi(\alpha) - \ln(\Delta x/x) \tag{2.167}$$

where ψ is the digamma function (Abramowitz and Stegun, 1970) defined as

$$\psi(z) = \frac{d}{dz}\ln\Gamma(z) \tag{2.168}$$

The marginal entropy for equal class interval reduces to

$$H(X; \Delta x) = \ln[\beta\ln(\alpha)] + \frac{\mu_x}{\beta} - (\alpha - 1)\,\psi(\alpha) - \ln\Delta x \tag{2.169}$$

noting that the integral term in equation (2.166) is equal to

$$\int_0^\infty f(x)\ln x \, dx = \psi(\alpha) + \ln\beta \tag{2.170}$$

Expressions for conditional entropy and transinformation require that the bivariate gamma distribution be known. One can employ the bivariate gamma distribution proposed by Moran (1969):

$$\frac{1}{\sqrt{2\pi}} \int_{-\infty}^{z} \exp(-t^2/2)dt = \int_0^x f(t; \alpha_x, \beta_x)\, dt$$

$$\frac{1}{\sqrt{2\pi}} \int_{-\infty}^{w} \exp(-t^2/2)\, dt = \int_0^y f\left(t; \alpha_y, \beta_y\right) dt \tag{2.171}$$

where z and w are normalized variates corresponding to x and y, respectively. The conditional entropy becomes (Chapman, 1980):

$$H(X|Y; \Delta x/x) = \frac{1}{2}\ln\left(1 - \rho_{zw}^2\right) + \ln\Gamma\left(\alpha_x\right) + \frac{\mu_x}{\beta_x} - \alpha_x\psi\left(\alpha_x\right) - \ln(\Delta x/x) \tag{2.172}$$

where ρ_{zw} is the correlation coefficient between z and w.

The transinformation of X and Y is found to be

$$T(X, Y; \Delta x/x) = -\frac{1}{2}\ln\left(1 - \rho_{zw}^2\right) \tag{2.173}$$

Transinformation has the same form for any distribution which can be normalized as can be shown below. If $f(x, y)$ can be normalized to $g(x, y)$, then equation (2.142) becomes

$$T(X, Y) = \int\limits_{-\infty}^{\infty} \int\limits_{-\infty}^{\infty} g(z, w) \log\left[\frac{g(z, w)}{g(z)\, g(w)}\right] dz\, dw \tag{2.174}$$

If $g(z, w)$ is a bivariate normal distribution then equation (2.174) is the result. For such cases, conditional entropy can be obtained by adding transinformation to the marginal entropy. Numerical values of transinformation will be affected by the correlation coefficient ρ_{zw} which, in turn, depends on the form of transformation to normal.

If the probability distribution function is lognormal, as is frequently assumed, then one obtains the marginal entropy for a fixed class interval as

$$H(X; \Delta x) = \mu_z + \frac{1}{2}\ln\left(2\pi e\sigma_z^2\right) - \ln\Delta x \tag{2.175}$$

and conditional entropy as

$$H(X|Y; \Delta x) = \mu_z + \frac{1}{2}\ln\left[\left(2\pi e\sigma_z^2\right)\left(1 - \rho_{zw}^2\right)\right] - \ln\Delta x \tag{2.176}$$

The transinformation expression is the same as equation (2.173).

For a proportional class interval with the lognormal distribution, equation (2.162) yields

$$H(X; \Delta x/x) = \frac{1}{2}\ln\left(2\pi e\sigma_z^2\right) - \ln(\Delta x/x) \tag{2.177}$$

For recorded daily flows of Canadian River near Canadian (07228000), Texas, for a period of October 1950 through September 1979, assuming that daily flow is a random variable having a lognormal distribution, entropy was computed using a gamma distribution and the lognormal distribution. It was found that the marginal entropy for the recorded daily flows was less steady for the gamma distribution than for the lognormal distribution. This can be taken as an indication that the log-normal assumption was a better fit to the data, and suggests that calculations of marginal entropy can serve as an aid in selecting the appropriate form of the distribution for hydrologic data.

2.15 Data with zero values

In many instances time series may contain zero values. For example, flow in ephemeral and intermittent streams becomes zero some time after rainfall ceases. Consider a sequence of N values of a random variable X. Let n_x be the number of nonzero values. Then $N-n_x$ denotes the number of zero values and let the ratio of nonzero values be denoted as $k_x = n_x/N$. Considering

$f(x)$ as the probability density function for nonzero values only, the probability distribution for all X with class interval $\Delta z = \Delta x/x$ becomes

$$p(x_n) = (1 - k_x)\,\delta(x) + k_x f(x_n)\, x_n \Delta z \tag{2.178}$$

where $\delta(x)$ is the Dirac delta function. Substitution of equation (2.178) in equation (2.107a) yields the marginal entropy (Chapman, 1980):

$$H(X; \Delta x/x) = -(1 - k_x)\ln(1 - k_x) - k_x \ln k_x - k_x \int_0^\infty f(x)\ln[xf(x)]\,dx - k_x \ln(\Delta x/x) \tag{2.179}$$

Equation (2.179) reduces to equation (2.162) for $k_x = 1$ and to zero (certainty of no flow) for $k_x = 0$.

For the bivariate case, consider another random variable Y having the same number of observations N. With similar notations, let n_y be the number of nonzero values. Then $N - n_y$ denotes the number of zero values and let the ratio of nonzero values be denoted as $k_y = n_y/N$. Now let n_{xy} be the number of nonzero values common to both data sets and let the ratio of nonzero values be denoted as $k_{xy} = n_{xy}/N$. Then, for class intervals $\Delta z = \Delta x/x$ and $\Delta w = \Delta y/y$, one can write the joint probability and conditional probability as:

$$p(x_n, y_m) = \begin{cases} 1 - k_x - k_y + k_{xy} & x = 0,\ y = 0 \\ k_x - k_{xy} & x > 0,\ y = 0 \\ \\ k_y - k_{xy} & x = 0,\ y > 0 \\ k_{xy} x_n y_m f(x_n, y_m)\Delta z\,\Delta w & x > 0,\ y > 0 \end{cases} \tag{2.180}$$

and

$$p(x_n | y_m) = \begin{cases} \dfrac{1 - k_x - k_y + k_{xy}}{1 - k_y} & x = 0,\ y = 0 \\[2mm] \dfrac{k_x - k_{xy}}{1 - k_y} & x > 0,\ y = 0 \\[2mm] \dfrac{k_y - k_{xy}}{k_y} & x = 0,\ y > 0 \\[2mm] \dfrac{k_{xy}}{k_y} x_n f(x_n | y_m)\Delta z & x > 0,\ y > 0 \end{cases} \tag{2.181}$$

Substitution of equation (2.181) in equation (2.127) results in the expression for conditional entropy (Chapman, 1980):

$$\begin{aligned} H(X|Y; \Delta x/x) = &-(1 - k_x - k_y + k_{xy})\ln(1 - k_x - k_y + k_{xy}) \\ &-(k_x - k_{xy})\ln(k_x - k_{xy}) + (1 - k_y)\ln(1 - k_y) \\ &-(k_y - k_{xy})\ln(k_y - k_{xy}) + k_y \ln k_y - k_{xy}\ln k_{xy} \\ &-k_{xy}\int_0^\infty \int_0^\infty f(x, y)\ln[xf(x|y)]\,dx\,dy - k_{xy}\ln(\Delta x/x) \end{aligned} \tag{2.182}$$

Equation (2.182) reduces to equation (2.163) for $k_x = k_y = k_{xy} = 1$ and to zero for $k_x = k_y = k_{xy} = 0$. Note that transinformation is given by the difference between equations (2.179) and (2.182), and will be independent only if $k_x = k_{xy}$, that is, each flow occurrence is nonzero flow.

Example 2.29: In the recorded daily flows of May 14 of Canadian River near Canadian (07228000), Texas, for the period of October 1950 through September 1979, there are two zero values in the data set. Analyze the effect of zero values.

Solution: For equal class interval Δx, one uses the formula:

$$H(X; \Delta x) = -\int_{-\infty}^{\infty} f(x) \ln[f(x)] \, dx - \ln(\Delta x)$$

$$\int_{-\infty}^{\infty} f(x) \ln[f(x)] \, dx = 2.7007 \text{ Napier}$$

For unequal class interval Δx, one uses the formula:

$$H(X; \Delta x) = -(1 - k_x)\ln(1 - k_x) - k_x \ln k_x - k_x \int_0^{\infty} f(x) \ln[f(x)] \, dx - k_x \ln(\Delta x)$$

$$= 0.2449 + 0.9333 \times 2.7007$$

$$= 2.7656 \text{ Napier}$$

where $k_x = 0.9333$. The value of entropy is 2.7656 Napier.

2.16 Effect of measurement units

Transinformation depends on the correlation coefficient of the normalized variates and is therefore independent of their original units of measurement. For marginal entropy, let it be supposed that X is measured in units a or b and that $x_a = K x_b$ where K is just a constant. Taking its logarithm, $\ln x_a = \ln K + \ln x_b$. For the lognormal distribution $\ln z_a = \ln K + \ln z_b$ and $\sigma_{za} = \sigma_{zb}$. Thus equation (2.176) remains unchanged for a proportional class interval. Using equation (2.175) for a fixed class interval, one gets

$$H(X; \Delta x_a) = \mu_{za} + \frac{1}{2}\ln(2\pi e \sigma_{za}^2) - \ln \Delta x_a$$

$$= \mu_{za} + \ln K + \frac{1}{2}\ln(2\pi e \sigma_{zb}^2) - \ln \Delta x_b - \ln K = H(X; \Delta x_b) \tag{2.183}$$

Equation (2.183) shows that the marginal entropy remains unchanged, if the class interval remains the same absolute value.

For the gamma distribution defined by equation (2.166), the marginal entropy is given by equation (2.167) in terms of gamma parameters. These parameters can be estimated using the method of maximum likelihood as

$$\ln \alpha - \psi(\alpha) = \ln(\bar{x}/\bar{x}_G); \qquad \alpha\beta = \bar{x} \tag{2.184}$$

where $\bar{x}$ and $\bar{x}_G$ are the mean and geometric mean of X. For the scale change as defined above, the gamma parameters become

$$\alpha_a = \alpha_b, \quad \beta_a = K\beta_b \tag{2.185}$$

Equation (2.185) leads to

$$\left(\frac{\mu_x}{\beta}\right)_a = \left(\frac{\mu_x}{\beta}\right)_b \tag{2.186}$$

This means that equation (2.186) remains unchanged with change of units.

Example 2.30: Show if the change of units has any effect on entropy if the probability distribution is exponential, triangular, uniform, Gumbel, or two-parameter Pareto.

Solution: For an exponential distribution, the PDF is $f(x) = \lambda \exp(-\lambda x)$.

$$H(X, \lambda) = -\int_0^{\infty} f(x) \log[f(x)]\, dx = -\frac{1}{\lambda} \int_{\lambda}^{0} \log x\, dx = f(\lambda) = f[E(x)] \tag{2.187}$$

The above equation shows that there is no change in the marginal entropy provided $E(x)$ remains the same absolute value.

For a triangular distribution, the PDF is $f(x) = \frac{2}{\beta - \alpha}\left(\frac{x-\alpha}{\gamma-\alpha}\right)$, $\alpha < x < \gamma$, γ corresponding to the summit of the triangular distribution. The marginal entropy is expressed as

$$H(X; \alpha, \beta, \gamma) = -\int_{\alpha}^{\gamma} f(x) \log[f(x)]\, dx = \frac{\gamma - \alpha}{\beta - \alpha}\left[\log \frac{2}{\beta - \alpha} + \frac{1}{2}\right] \tag{2.188}$$

The above equation shows that there is no change in the marginal entropy, provided $(\gamma - \alpha)$ and $(\beta - \alpha)$ retain the same absolute values.

For a uniform distribution, the PDF is $f(x) = 1/(\beta - \alpha)$.

$$H(X; \alpha, \beta) = -\int_{\alpha}^{\beta} f(x) \log[f(x)]\, dx = -\log \frac{1}{\beta - \alpha} \tag{2.189}$$

The above equation shows that there is no change in the marginal entropy provided $(\beta - \alpha)$ retains the same absolute value.

For the Gumbel distribution, the PDF is:

$$f(x) = a \exp\{-a(x - b) - \exp[-a(x - b)]\} \tag{2.190}$$

where $a > 0$ and $-\infty < b < x$ are parameters. Entropy of this distribution is:

$$H(X; a, b) = -\ln a + a\bar{x} - ab \tag{2.191}$$

If $E(x) = \bar{x}$ remains the same, then there is no change in entropy.

For two-parameter Pareto distribution, the PDF is

$$f(x) = ba^b x^{-b-1} \qquad (2.192)$$

Entropy of this distribution is

$$H(X; a, b) = -\ln b - b \ln a + (b+1) E[\ln x] \qquad (2.193)$$

The marginal entropy depends on the average value of X which depends on the unit of measurement.

2.17 Effect of averaging data

Consider that daily flow data are available and entropies are calculated for these data. What happens if the data are averaged over shorter or longer time intervals? Or what happens if the data measurement interval is increased or decreased? This issue was analyzed by Chapman (1986). Increased interval would smoothen the data and that might lead to a decrease in marginal entropy and an increase in transinformation. Using lognormal distribution for daily, weekly, monthly, and yearly mean flows, Chapman (1986) found that marginal entropy exhibited little change for averaging intervals up to a month but was significantly less for mean annual flow. For increasing time interval, the conditional entropy decreased at a greater rate and as expected transinformation increased.

Example 2.31: Using daily, weekly, monthly, and annual average streamflow for Dry Creek, California, show the effect of averaging on entropies and transinformation.

Solution: Streamflow data are obtained for Dry Creek, California. Table 2.11 shows entropies and transinformation for daily, weekly, monthly, and annual mean flows. It is verified that a log normal distribution was appropriate for all these variates. The marginal entropy exhibits little change for averaging intervals up to a month, but is significantly less for mean annual flow. The conditional entropy decreases at a greater rate, resulting in the anticipated increase in transinformation as the averaging interval is increased.

Example 2.32: Using daily streamflow data form Brazos River near Hempstead, Texas, (USGS 08111500) for the period 1976 to 2006, investigate the effect of averaging on entropy. Consider several intervals as 1-day, 1-week (7-day), 15-day, 30-day (one month), 2-month, 3-month,

Table 2.11 Mean values of marginal and conditional entropy ($\Delta x/x = 0.05$) and transinformation, for Dry Creek, California, using a lognormal distribution.

	Day	Week	Month	Year
$H(X)$	4.45	4.46	4.27	3.64
$H(X|Y)$	3.66	3.57	3.28	2.07
$T(X, Y)$	0.79	0.89	1.09	1.57

Table 2.12 Effect of time interval on entropy.

Time interval	Marginal entropy (Napier) $\Delta x = 1$	Marginal entropy (Napier) $\Delta x / x = 0.05$
1 day	9.60	4.50
Weekly average	9.61	4.48
Fortnightly average	9.64	4.46
Monthly average	9.69	4.44
2 months average	9.73	4.40
3 months average	9.73	4.35
6 months average	9.73	4.21
Year average	9.68	4.00

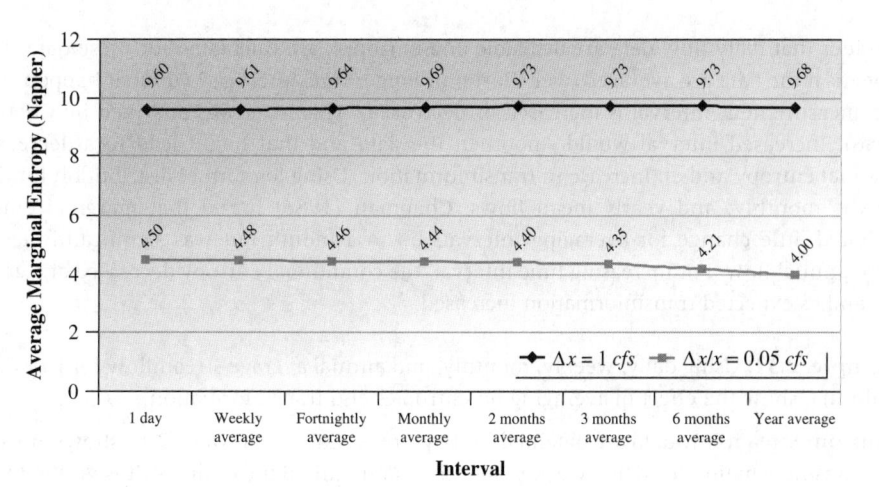

Figure 2.17 Effect of interval size on entropy.

6-month, and a year. Plot the entropy value against interval and see if the value increases or decreases.

Solution: Entropy calculations are done for data averaged over different time intervals. It is assumed that the daily stream flow follows a log-normal distribution. The corresponding average marginal entropy values are shown in Table 2.12 and a plot of average marginal entropy values is shown in Figure 2.17. It is seen that for the fixed class interval the entropy value increases slightly and then decreases, while for the proportional class interval a gradual decrease in the average entropy is observed.

2.18 Effect of measurement error

Let there be an observed value a of X and standard error s and let there be another value b. The value b is considered to differ significantly if $|a - b|$ is greater than a specified multiple of

s, cs, where c is a multiplication factor. The probability of a measurement of X which does not differ significantly from a can be stated as (White, 1965):

$$\int_{a-cs}^{a+cs} f(x)\, dx \cong 2csf(a)$$

Then, the absolute value of $\log[2csf(a)]$ defines the amount of information gained by the observation $I(x) = |\log[2csf(x)]|$. If $2cs$ can be viewed as an error factor, and f as the reciprocal of the extent of range of the variable, then the amount of information gained is the logarithm of ratio of the extent to the error. In general, both s and f vary with x and the product sf is dimensionless. Therefore, the information $I(x)$ per observation can be written as $I(x) = -\log[2cs(x)f(x)]$.

Consider the populations of circles again. Let it be assumed that the radius can be determined with an error that does not depend on the area. Then the error in the determination of area is proportional to the square root of area. In the above equation, since the product sf is dimensionless, this means its value does not depend on the coordinate system. For any change in coordinate system, s and f would change in a reciprocal manner. With this background, the Shannon entropy, analogous to the discrete case, can be defined as the average amount of information expected to be gained per observation as

$$H_s = \int f(x)\, I(x)\, dx = -\int f(x) \log\big[sf(x)\big] + K$$

where $K = -\log 2c$. If the interest is in only entropy changes then the value of K is irrelevant.

However, it is possible to assign a meaningful value to K. To that end, consider an extreme case, where the PDF, $f(x)$, has a small variance, and a measurement may unpredictably alter the variable. This suggests that less is known about the variable after the measurement than before. This means that one loses information by measurement, implying negative entropy. If, on average, information is neither gained nor lost, then entropy is zero. One can now define the zero point. H_s is zero if s is constant and $f(x)$ is normal with a standard deviation equal to s. Substitution into the above equation leads to $K = -\log\sqrt{2\pi e}$. Then, $c = \sqrt{2\pi e/2} = 2.066$. This corresponds to a confidence level of 3.9%.

Now consider two limitations on the accuracy with which the values of the variable are determined: 1) measurement error, and 2) random error. Both errors arise in case of hydrologic variables. In entropy, data may be investigated in two ways. First, considering $2cs$ as equal to the width of the interval for recording of data, the measurement error can be included in the error term. Approximating the integral by a sum, one can write

$$H_s = -\sum f_i \log(2cs_i f_i)\, \Delta x_i = -\sum (f_i \Delta x_i) \log(f_i \Delta x_i)$$

which is the entropy, H, of the distribution with the proportion of the value viewed as a discrete variable. The maximum entropy distribution will, of course, be uniform and then one can determine the difference between two distribution entropies. The difference will be the amount of information supplied by the new distribution.

2.19 Entropy in frequency domain

The previous discussion presents entropy either in time or space domain. Thus, the data available as a time series or space series can be employed to compute entropy. However, entropy can also be computed in the frequency domain as

$$H(\upsilon) = \int_{-\Delta t/2}^{\Delta t/2} \log[W(\upsilon)] \, d\upsilon \tag{2.194}$$

where υ is the frequency and $W(\upsilon)$ is the power spectrum related to the autocovariance function $\gamma(k)$ of the time series by the Wiener-Knitchine relationship:

$$\gamma(k) = 2\pi \int_{0}^{\pi} \cos(2\pi \upsilon k) \, W(2\pi \upsilon) \, d\upsilon \tag{2.195}$$

The entropy defined by equation (2.194) is known as the Burg entropy (Burg, 1975). Expanding equation (2.194) Burg (1975) derived

$$H(\upsilon) = \sum_{\upsilon} \log\left[\pi e \overline{W}(\upsilon)\right] \tag{2.196}$$

where $\overline{W}(\upsilon)$ is the normalized power spectrum. Both equations (2.194) and (2.196) are employed in univariate spectral analysis for reconstruction and extrapolation of time series. Equation (2.194) has been generalized in a manner similar to the form of the Shannon entropy (Johnson and Shore, 1984) and this will be discussed in detail in Chapter 11 on spectral analysis.

For multivariate spectral analysis, Burg (1977) suggested multivariate entropy as a function of the determinant of the power spectrum matrix $W(\upsilon)$:

$$H(\upsilon) = \int_{-\Delta t/2}^{\Delta t/2} \log\{\det[W(\upsilon)]\} \, d\upsilon \tag{2.197}$$

where det is the determinant.

2.20 Principle of maximum entropy

In search of an appropriate probability distribution for a given random variable, entropy should be maximized. In practice, however, it is common that some information is available on the random variable. The chosen probability distribution should then be consistent with the given information. There can be more than one distribution consistent with the given information. From all such distributions, one should choose the distribution that has the highest entropy. To that end, Jaynes (1957) formulated the principle of maximum entropy

(POME) a full account of which is presented in a treatise by Levine and Tribus (1978) and will be presented in the next chapter. According to POME, the minimally prejudiced assignment of probabilities is that which maximizes entropy subject to the given information, that is, POME takes into account all of the given information and at the same time avoids consideration of any information that is not given. The reasoning for POME comes from a simple practical need. Assume, for example, that we fit two probability density functions (PDF) to a histogram of data and that both PDFs fit equally well, that is, satisfy the same statistical criteria. Then the PDF with higher entropy should be the preferred one. In other words, for given information the best possible distribution that fits the data would be the one with the maximum entropy, since this contains the most reliable assignment of probabilities.

The information usually included in POME is specified as some statistics, including, for example, mean, variance, covariance, cross-variance, and so on, or linear combinations of these statistics. Since the POME-based distribution is favored over those with less entropy among those which satisfy the given constraints, according to the Shannon entropy as an information measure, entropy defines a kind of measure on the space of probability distributions. Intuitively, distributions of higher entropy represent more disorder, are smoother, are more probable, are less predictable, or assume less. The POME-based distribution is maximally noncommittal with regard to missing information and does not require invocation of ergodic hypotheses.

Constraints encode relevant information. POME leads to the distribution that is most conservative and hence most uninformative. If a distribution with lower entropy were chosen, it would mean that we would be assuming information that was not available, while a distribution with higher entropy would violate the known constraints. The maximum entropy leads to a probability distribution of particular macrostate occurring amongst all possible arrangements (or microstates) of the events under consideration.

2.21 Concentration theorem

The concentration theorem, formulated by Jaynes (1958), has two aspects. First, it shows the POME-based probability distribution best represents our knowledge about the state of the system by showing the spread of lower entropies around the maximum entropy value. Second, POME is the preferred method to obtain this distribution. The basis for these two aspects is contained in the Shannon inequality and the relation between entropy and the chi-square test.

The Shannon inequality says that the probability distribution given by POME has a greater entropy than any other distribution. Let $P = \{p_i, i = 1, 2, \ldots, N\}$ be the distribution given by maximization of the Shannon entropy. The entropy corresponding to this distribution obtained by the method of Lagrange multipliers is H_{max}. Let there be another distribution $Q = \{q_i, i = 1, 2, \ldots, N\}$ satisfying the same constraints as does P. Its entropy is H. For these two probability distributions, the Shannon inequality can be expressed as

$$\sum_{i=1}^{N} q_i \log\left[\frac{q_i}{p_i}\right] \geq 0 \tag{2.198}$$

where the equality holds only if $q_i = p_i$, $i = 1, 2, \ldots, N$. Thus, it can be written that

$$H_{max} - H \geq 0 \tag{2.199}$$

where $H_{max} - H = 0$ only if $Q = P$. If Q satisfies the same constraints as P that it can be shown that

$$H_{max} - H = \sum_{i=1}^{N} q_i \log\left[\frac{q_i}{p_i}\right] \tag{2.200}$$

Following Keshavan and Kapur (1992), an interesting result is obtained by letting $q_i = p_i(1 + \omega_i)$ where

$$\sum_{i=1}^{N} p_i \omega_i = 0 \tag{2.201}$$

where w_i is the weighting actor. Then, neglecting summation terms involving third or higher powers of ω, one can approximate equation (2.200) as

$$\Delta H = H_{max} - H = \sum_{i=1}^{N} q_i \log\left[\frac{q_i}{p_i}\right] = \sum_{i=1}^{N} p_i(1 + \omega_i)\log(1 + \omega_i) \tag{2.202}$$

Expanding the logarithmic term inside the summation, equation (2.202) can be approximated as

$$\Delta H \approx \sum_{i=1}^{N} p_i(1 + \omega_i)\left[\omega_i - \frac{\omega_i^2}{2} + \frac{\omega_i^3}{3} - \ldots\right] \approx \frac{1}{2}\sum_{i=1}^{N} p_i \omega_i^2$$

$$= \frac{1}{2}\sum_{i=1}^{N} p_i \frac{(q_i - p_i)^2}{p_i^2} = \frac{1}{2}\sum_{i=1}^{N} \frac{(q_i - p_i)^2}{p_i} = \frac{1}{2M}\sum_{i=1}^{N} \frac{(Mq_i - Mp_i)^2}{Mp_i} \tag{2.203}$$

where M is the total number of observations, and N is the number of class intervals. Then, Mq_i, $i = 1, 2, \ldots, N$, can be interpreted as the observed frequencies; and Mp_i, $i = 1, 2, \ldots, N$, as the frequencies computed from the POME-based distribution P which has the same moments that Q has. Let $O_i = Mq_i$ and $C_i = Mp_i$. Then, equation (2.203) can be written as

$$2M\Delta H = \sum_{i=1}^{N} \frac{(Mq_i - Mp_i)^2}{Mp_i} = \sum_{i=1}^{N} \frac{(O_i - C_i)^2}{C_i} = \chi^2 \tag{2.204}$$

where χ^2 is chi-square with n degrees of freedom defined by

$$n = N - m - 1 \tag{2.205}$$

where $(m+1)$ is the number of constraints. Because the chi-square distribution is known, one can determine the probability that the computed chi-square is less than the chi-square at 95% or 99% significance level:

$$P\left[(\chi^2) \leq \chi_n^2(0.95)\right] = 0.95; \; P\left[(\chi^2) \leq \chi_n^2(0.99)\right] = 0.99 \tag{2.206}$$

Table 2.13 Chi-square distribution.

N	2	3	4	5	6	8	10	15	20	25	30
$\chi_n^2(0.95)$	5.99	7.80	9.49	11.07	12.59	15.51	18.31	25.00	31.40	37.66	43.78
$\chi_n^2(0.99)$	9.21	11.35	12.28	15.09	16.81	20.09	25.21	30.58	37.67	44.31	50.89

For different degrees of freedom, n, the values of χ_n^2 (0.95) and χ_n^2 (0.99) are found in standard statistical textbooks or can be computed from the chi-square distribution as given in Table 2.13.

In order to determine the 95% and 99% entropy intervals, equation (2.206) with the use of equation (2.204) can be expressed as

$$P\left[H_{max} - H \le \frac{\chi_n^2(0.95)}{2M}\right] = 0.95 \tag{2.207a}$$

$$P\left[H_{max} - H \le \frac{\chi_n^2(0.99)}{2M}\right] = 0.99 \tag{2.207b}$$

Equations (2.207a) and (2.207b) can be expressed as

$$P\left[H_{max} \ge H \ge H_{max} - \frac{\chi_n^2(0.95)}{2M}\right] = 0.95 \tag{2.208a}$$

$$P\left[H_{max} \ge H \ge H_{max} - \frac{\chi_n^2(0.99)}{2M}\right] = 0.99 \tag{2.208b}$$

Thus, the 95% and 99% entropy intervals, respectively, are:

$$\left[H_{max} - \frac{\chi_n^2(0.95)}{2M}, H_{max}\right]; \left[H_{max} - \frac{\chi_n^2(0.99)}{2M}, H_{max}\right] \tag{2.209}$$

For a random variable X having a probability distribution that satisfies the same constraints as the POME-based distribution then its entropy will be greater than $\left[H_{max} - \frac{\chi_n^2(0.95)}{2M}\right]$ with a 95% probability. Likewise, its entropy with a 99% probability will be greater than $\left[H_{max} - \frac{\chi_n^2(0.99)}{2M}\right]$. If M is large, entropies of most probability distributions that satisfy a given set of constraints will be concentrated near the maximum entropy values. This means that the POME-based distribution is the best choice. It can also be noted that the length of the entropy interval increases with N and decreases with m, increases with confidence interval, and decreases rapidly with M. Thus, the concentration theorem states that for large M, $2M(H_{max}-H)$ is distributed as chi-square with $N-m-1$ degrees of freedom.

Thus, for the marginal entropy $H(x)$ of a random variable X, associated with its probability density function $f(x)$, the entropy for any PDF will be in the range given as

$$H_{max} - \Delta H \le H(x) \le H_{max} \tag{2.210a}$$

where H_{max} is given by POME as

$$H_{max} = \log(Z) + \sum_{k=1}^{m} \lambda_k a_k \tag{2.210b}$$

where m is the number of constraints, a_k, specified, Z is the potential function, and λ_k are Lagrange multipliers. If one considers N probabilities and observes M different realizations, the concentration of these probabilities near the upper bound H_{max} is given by the concentration theorem. Asymptotically $2M\Delta H$ is distributed over class C as χ^2 with $N-m-1$ degrees of freedom, independently of the nature of constraints. Denoting the critical value of χ^2 for k degrees of freedom at $100(1-\alpha)\%$ significant level as $\chi_c^2(\alpha)$, ΔH is given in terms of the upper tail area $1-F$ as:

$$\chi_c^2(1 - F) = 2M\Delta H \tag{2.211}$$

Example 2.33: Assume that a six-faced dice is being thrown 1000 times with no knowledge about its outcome. The problem is to determine frequencies or probabilities of the six faces of the dice and their distribution.

Solution: From POME it is known that the only noncommittal PDF is the uniform PDF for which probabilities are all equal:

$$f(x) = \frac{1}{6}, \quad f_i = \frac{1}{6}, \quad i = 1, 2, \ldots, 6$$

Can we make a more precise estimate of these frequencies knowing that the dice has been thrown 1000 times. It is known that $M = 1000$, $N = 6$, $m = 0$, and $N-m-1 = 5$ degrees of freedom. The maximum entropy is $H_{max} = \log(N) = \log(6) = 1.79176$ Napier. At a 95% significance level, from χ^2 tables, $\chi^2(0.95) = 11.07$. From equation (2.211),

$$2N\Delta H = 11.07$$

which yields $\Delta H = 0.0055$, and from equation (2.210a), 95% of all possible outcomes (confidence interval) will lie in the range:

$$1.786 \leq H(x) \leq 1.792$$

Thus, without invoking any empirical evidence, one is confident that the vast majority of the outcomes will have the PDF close to the uniform distribution.

2.22 Principle of minimum cross entropy

According to the Laplace principle of insufficient reason, all outcomes of an experiment should be considered equally likely unless there is information to the contrary. On the basis of intuition, experience or theory, a random variable may have an a priori probability distribution. Then, the Shannon entropy is maximum when the probability distribution of the random variable is that one which is as close to the a priori distribution as possible. This is referred to as the principle of minimum cross entropy (POMCE) which minimizes the Bayesian entropy (Kullback and Leibler, 1951). This is equivalent to maximizing the Shannon entropy.

2.23 Relation between entropy and error probability

There is a relation between entropy and prediction error probability. Intuitively, entropy H is a measure of the complexity or degree of randomness of a random variable. This means that if a random variable has a higher entropy then it is more difficult to predict or guess the value the random variable takes on and vice versa. Indeed this concept can be employed to assess the accuracy of schemes employed for winning in a gamble, achieving growth rates in stock market investments, and so on. The degree of difficulty in predicting the value of the random variable can be evaluated by the minimum possible error probability of any prediction scheme. Feder et al. (1992) showed that two random variables having the same entropy might have different minimum prediction error probabilities. In other words, the prediction error is not uniquely assessed by entropy (Feder and Merhav, 1994).

Consider a discrete random variable $X : \{x_i, i = 1, 2, \ldots, N\}$, with the probability distribution $p_i, i = 1, 2, \ldots, N$, for which the Shannon entropy is defined. If there are no constraints imposed on the random variable then the estimator of X is the value a with the highest probability. If $q = p(a) = \max_x p(x)$, then the minimum error probability in guessing the value of X is given by Feder and Merhav (1994) as

$$g(X) = \sum_{x \neq a} p(x) = 1 - q \tag{2.212}$$

For a uniform distribution with N outcomes, the maximum Shannon entropy is $\log N$ and highest possible minimum error probability is $(N-1)/N$. On the other hand, if the entire probability mass is concentrated on a single value, that is, the probability of that value is 1 and all other values have zero probabilities, then both entropy and minimum error probability for the random variable will be zero.

For a random variable X with entropy H, the upper and lower bounds of the minimum error probability $g(X)$ are given by Feder and Merhav (1994) as

$$\phi^{-1}(H) \geq g(X) \geq \Phi^{-1}(H) \tag{2.213}$$

where the maximum entropy is

$$\Phi(g) = H\big[p_{\max}(g)\big] = h(g) + g \log(N - 1) \tag{2.214a}$$

$$h(g) = -g \log g - (1 - g) \log(1 - g) \tag{2.214b}$$

and

$$\phi(g) = H[P_{\min}(g)]$$

$$= \begin{cases} h(g) & 0 \le g \le 1/2 \\ 2(1-g) + h(2g-1) & (1/2) \le g \le 2/3 \\ \quad \vdots & \quad \vdots \\ i \log i(1-g) + h(ig-(i-1)) & (i-1)/i \le g \le (i/(i+1)) \\ \quad \vdots & \quad \vdots \\ (N-1)\log(N-1)(1-g) + h((N-1)g - N + 2) & (N-2)/(N-1) \le g \le (N-1)/N \end{cases}$$

(2.215)

Equation (2.214b) is the binary entropy function. The quantity $P_{max}(g)$ can be expressed as

$$P_{\max}(g) = \left[1 - g, \frac{g}{N-1}, \cdots, \frac{g}{N-1}\right] \tag{2.216}$$

and $P_{min}(g)$ is

$$P_{\min}(g) = [p(1), p(2),, ..., p(N)] \tag{2.217}$$

where

$$\begin{aligned} p(1) &= 1 - g, \ p(2) = g, p(3) = \ldots p(N) = 0, & 0 \le g \le 1/2 \\ p(1) &= p(2) = 1 - g, p(3) = 2g - 1, p(4) = \ldots p(N) = 0, & 1/2 \le g \le 2/3 \\ p(1) &= \ldots p(N-1) = 1 - g, p(N) = 1 - (N-1)(1-g), & (N-2)/(N-1) \le g \le (N-1)/N \end{aligned} \tag{2.218}$$

Equation (2.214a) also implies that

$$H(X) \le h(g) + g \log (N - 1) \tag{2.219}$$

which is a special case of Fano's inequality. It may also be noted that

$$h(g) \ge H \ge 2g \tag{2.220}$$

or equivalently

$$\frac{1}{2}H \ge g \ge h^{-1}(H) \tag{2.221}$$

These bounds show that entropy and predictability do not have a one-to-one relationship.

Consider a random variable $X(i) = -\log p_i$, $i = 1, 2, \ldots, N$; p_i is the probability of $X(i)$. Let the Shannon entropy be $H(P)$ and empirical entropies from observations be $H(P_n)$. If $H(P_n)$ is appropriately normed then its asymptotic distribution is asymptotically normal with mean of $H(P)$, and the asymptotic variance $\sigma^2(P)$ is given (Feistauerova and Vajda, 1993) as

$$\sigma^2(P) = \sum_{i=1}^{N} p_i \log^2 p_i - H^2(P) \qquad (2.222)$$

This is the variance of the random variable $X(i) = -\log p_i$, defined above.

2.24 Various interpretations of entropy

Entropy can be viewed as an objective measure of some property of a system or as a subjective concept for use as a model building tool to maximize the use of available information. When entropy is used as a subjective concept it is associated not with the system itself but with the information about the system that is known. Wilson (1970) presented four different ways to view entropy: 1) Entropy as a measure of system property (such as order or disorder, reversibility or irreversibility, complexity or simplicity, etc.); 2) entropy as probability for measure of information, uncertainty, or probability; 3) entropy as a statistic of a probability distribution or measure of information or uncertainty; and 4) entropy as the negative of a Bayesian log-likelihood function for a measure of information. In addition to being a measure of uncertainty equation (2.9) has been associated with a multitude of interpretations (Kapur and Kesavan, 1992) some of which are outlined below.

2.24.1 Measure of randomness or disorder

Consider an experiment with N outcomes. Each outcome occurs with a certain probability $p_i, i = 1, 2, \ldots, N$. If all outcomes are equally likely, meaning all probability values are the same, that is, $p_1 = p_2 = \ldots = p_N = 1/N$, then the value of entropy would be maximum which is $\log N$ and it defines the upper bound. If one outcome occurs with certainty, that is, probability 1, then other outcomes will not occur at all and their probabilities of occurrence will be zero. In this case entropy will be zero, which also defines the lower bound of entropy, and there is no uncertainty or randomness. If probabilities of occurrence are not equal, that is, $p_i \neq p_j, i = 1, 2, \ldots, N; j = 1, 2, \ldots, N; i \neq j$, based on some knowledge about this experiment or the constraints that this experiment is designed to satisfy, then entropy will be less than the maximum and greater than zero. In this case the degree of randomness is less. This means that entropy can be considered as a measure of randomness. In environmental and water resources engineering, constraints may be expressed as laws of conservation of mass, momentum and energy, and flux laws. Each constraint reduces randomness and reduces entropy or disorder.

2.24.2 Measure of unbiasedness or objectivity

The argument employed here is that for a given set of constraints the distribution of probabilities yielding the maximum entropy is the most unbiased or objective distribution and this is the maximum entropy-based distribution. Any other distribution would lead to less than maximum entropy and would be biased or less than objective. Thus entropy can also be considered as a measure of objectivity or unbiasedness.

2.24.3 Measure of equality

As noted above, entropy varies from a maximum value to zero, depending on the equality or inequality of probabilities. The more unequal the probability the less the entropy value.

Thus entropy can be considered as a measure of equality or inequality. This is of interest in measuring economic, political or geographical inequalities of different types.

2.24.4 Measure of diversity

Watershed ecosystems may have different types of vegetation and animal species. One can then group, say vegetation, in different classes and determine the relative frequency of each class and hence entropy. Entropy will be higher when relative frequencies of different classes will be close to each other, and less when they would be more unequal. This means that greater diversity will lead to greater entropy. Thus entropy can be viewed as a measure of diversity or desegregation. In this case frequencies can simply be numbers or absolute frequencies.

2.24.5 Measure of lack of concentration

Since entropy is a measure of the degree to which M probabilities are equal to each other, it can be perceived as an inverse measure of concentration. For example, when $H = 0$, only one p_i is unity and the rest are 0, then there is the maximum concentration. If all probabilities are equal the concentration is minimum: $H = \log M$. Furthermore, as M increases and all p_is remain the same, $\log M$ increases, and the degree of concentration drops. Entropy is greater when probability values are spread out and declines when they are concentrated. This suggests that entropy can be used as a measure of dispersion or scatter or spread. Consider an example having M classes with absolute frequencies $f_1, f_2, \ldots, f_N$, where $f_1 + f_2 + \ldots + f_N = N$ (where N = total number of values). Entropy can be defined in the usual sense as

$$H(f) = -\sum_{i=1}^{M} \frac{f_i}{N} \log \frac{f_i}{N} = \log N - \frac{1}{N} \sum_{i=1}^{M} f_i \log f_i$$

In practical terms f_i can represent the average elevation of the i-th group of areas or points, the grade point average of the i-th group of students, the income of the i-th group of people, the height of the i-th group of trees, the weight of the i-th group of animals, and so on. $H(f)$ defines the spreading out of values.

2.24.6 Measure of flexibility

Flexibility emanates from the choice of options one has, the more options the more flexibility. Each option may be associated with a probability value. When all options are equally probable, entropy is maximum. This means one has maximum flexibility, and any option can be exercised. On the other hand, if probability values are unequal, the available options are not as many, and if the probabilities are highly unequal, the options become even more limited. In the extreme case, there may not be more than one option if all probability values, except one, are zero. In this way, entropy can be considered as a measure of flexibility.

2.24.7 Measure of complexity

The previous discussion shows that entropy can be used as a measure of uniformity or lack thereof (i.e., concentration). When probabilities are equal, entropy is maximum, and when they are unequal, entropy is less than maximum. The inequality of probabilities results from the constraints that are to be satisfied. Each constraint reduces entropy, decreases uniformity and thus introduces complexity. Hence the larger the number of constraints the smaller the entropy value and the larger the system complexity. Thus, the departure from the maximum entropy value can be regarded as a measure of complexity:

$$Complexity = H_{max} - H \qquad (2.223)$$

Often a normalized measure is more appropriate for measuring complexity:

$$Complexity(\%) = \frac{H_{max} - H}{H_{max} - H_{min}} \times 100 \qquad (2.224)$$

2.24.8 Measure of departure from uniform distribution

Consider two probability distributions $P = \{p_1, p_2, \ldots, p_N\}$ and $Q = \{q_1, q_2, \ldots, q_N\}$. A measure of directed divergence of P from Q can be expressed as

$$D_d = \sum p_i \log \frac{p_i}{q_i} \qquad (2.225)$$

If Q is uniformly distributed, that is, $q_1 = q_2 = \ldots = q_N = 1/N$, then

$$D_d = \sum p_i \log \frac{p_i}{(1/N)} = \sum p_i \log p_i + \log N = \log N - H \qquad (2.226)$$

Thus, a larger value of entropy H would result in a smaller value of directed divergence, meaning that P would be close to Q. Hence entropy can be viewed as a measure of departure of the given distribution from the uniform distribution.

2.24.9 Measure of interdependence

Consider n variables $X_1, X_2, \ldots, X_n$ with marginal probability distributions $g_1(x_1), g_2(x_2), \ldots, g_n(x_n)$ and the corresponding marginal entropies as $H_1, H_2, \ldots, H_n$. Let the joint probability distribution of these variables be $f(x_1, x_2, \ldots, x_n)$. The maximum value of entropy can be shown to be:

$$H_{max} = H_1 + H_2 + \ldots + H_n \qquad (2.227)$$

If the variables are independent then $H = H_{max}$ and if not then $H < H_{max}$. Thus the degree of interdependence of variables D_n can be defined as

$$D_n = H_{max} - H \geq 0 \qquad (2.228)$$

If the variables are discrete then it can be shown that

$$H_1 \leq H, H_2 \leq H, \ldots, H_n \leq H \qquad (2.229)$$

One can then write

$$\overline{D_n} = \frac{H_{max} - H}{(n-1)H}, \quad \overline{\overline{D_n}} = \frac{nH - H_{max}}{(n-1)H} \qquad (2.230)$$

When $H=H_{max}$ (for independent variables), $\overline{D_n} = 0, \overline{\overline{D_n}} = 1$; when $H=H_1$ (for perfect dependence), $H_{max} = nH_1$, then $\overline{D_n} = 1, \overline{\overline{D_n}} = 0$. Thus, both $\overline{D_n}$ and $\overline{\overline{D_n}}$ lie between 0 and 1, and can be used as measures of dependence. It may be noted that D_n, $\overline{D_n}$ and $\overline{\overline{D_n}}$ involve no assumptions of linearity or normality.

2.24.10 Measure of dependence

If there are two random variables X and Y, then one can construct a contingency table for computing joint probabilities. Let the random variable X have a range of values consisting of n categories (class intervals), while the random variable Y is assumed to have m categories (class intervals). The cell density or the joint frequency for (i, j) is denoted by f_{ij}, $i = 1, 2, \ldots, m$; $j = 1, 2, \ldots, n$, where the first subscript refers to the row and the second subscript to the column. The marginal frequencies are denoted by $f_{i.}$ and $f_{.j}$ for the row and the column values of the variables, respectively. For example, one may be measuring nitrate as an indicator of water quality at two wells. Their measurements would constitute a time series of nitrate at each well. Then the nitrate concentration at the two wells for the entire time series can be organized as a contingency table for purposes of computing the joint probability distribution of nitrate concentration at the two wells. Here nitrate at each well can be regarded as a random variable. The dimensions of the table would be $m \times n$ with elements f_{ij} such that

$$\sum_{j=1}^{n} f_{ij} = F_i, \quad \sum_{i=1}^{m} f_{ij} = G_j, \quad \sum_{i=1}^{m} F_i = \sum_{j=1}^{n} G_j = M \tag{2.231}$$

Entropy can be defined in the usual way as

$$H = -\sum_{j=1}^{n} \sum_{i=1}^{m} \frac{f_{ij}}{M} \log \frac{f_{ij}}{M}, \quad H_1 = -\sum_{i=1}^{m} \frac{F_i}{M} \log \frac{F_i}{M}, \quad H_2 = -\sum_{j=1}^{n} \frac{G_j}{M} \log \frac{G_j}{M} \tag{2.232}$$

If these entropies are maximized, using the constraints specified above, then one obtains

$$\frac{f_{ij}}{M} = \frac{F_i}{M} = \frac{G_j}{M}, \quad H_{max} = H_1 + H_2 \tag{2.233}$$

and

$$D = H_1 + H_2 - H \geq 0 \tag{2.234}$$

Here D vanishes if X and Y are independent. This means that D can be viewed as a measure of dependence of the two variables or two attributes. In the case of nitrate concentration, D can be viewed as a measure independence in the contingency table containing frequencies of nitrate observations.

2.24.11 Measure of interactivity

Consider the example of nitrate concentration in the two wells discussed in the above section. Then entropy H defined as above can be considered as a measure of interactivity between nitrate concentrations at the two wells. For higher interactivity the value of H would be larger. If the wells are independent and so are their nitrate concentrations, then the value of H would be zero.

2.24.12 Measure of similarity

Consider n m-dimensional vectors. These vectors can be obtained by organizing nitrate concentrations of the previous example as vectors where row represents time and column represents concentration:

$$
\begin{bmatrix}
c_{11} & c_{12} & \cdot & \cdot & & .c_{1n} \\
c_{21} & c_{22} & \cdot & \cdot & & .c_{2n} \\
\cdot & & & & & \\
\cdot & & & & & \\
\cdot & & & & & \\
c_{m1} & c_{m2} & \cdot & \cdot & & .c_{mn}
\end{bmatrix}
$$

Based on entropy, it is plausible to write a measure (Kapur and Kesavan, 1992) as

$$
S_r = \frac{2}{m(m-1)} \sum_{i=1}^{m} \sum_{j=1}^{n} \left[\sum_{k=1}^{n} c_{ik} \log \frac{c_{ik}}{c_{jk}} + \sum_{k=1}^{n} c_{jk} \log \frac{c_{jk}}{c_{ik}} \right]
\tag{2.235}
$$

which is always greater than or equal to 0 and it is 0 if the row vectors are identical. This means the nitrate concentration pattern is not changing with time. Thus, this can be regarded a measure of similarity.

2.24.13 Measure of redundancy

Consider an environmental phenomenon E controlled by $E_1, E_2, \ldots, E_N$ factors each giving rise to $i, j, k, \ldots, N$ outcomes. The global entropy of E will be maximum if 1) the outcomes of each factor are equiprobable, and 2) the factors are mutually and completely independent (completely, for three factors can be independent two by two, but dependent when three are considered together.) Thus, three cases can be considered: 1) equiprobable factors and complete independent factors, this will yield the maximum entropy $H_m(p)$; 2) not equiprobable outcomes but independent factors, this will lead to $H_d(p)$; and (3) not equiprobable outcomes, and not independent factors. The actual entropy $H(P)$ will be less than the maximum possible, because complete independence is unlikely.

One can compare the calculated entropy (or information) with the maximum value this entropy could have, using the ratio of the actual to the maximum entropy. This ratio is termed relative entropy. If the relative entropy of a source is, say, 0.6, then the source is 60% as free as it can be with these same alternatives. Redundancy is defined as one minus the relative entropy. The term redundancy used here is close to its meaning. If the fraction of alternatives were missing, the information would still be the same or could be made the same. Thus the fraction of alternatives is redundant. For example, the English language has 50% redundancy. This means that about half of the letters or words chosen for writing or speaking are subject to the freedom of choice and about half are governed by the statistical structure of the language.

Let E define a phenomenon, such as a watershed response as evapotranspiration, and $E_1, E_2, \ldots, E_N$ factors as temperature, radiation, wind velocity, and so on. controlling evapotranspiration. Because some of the factors are dependent, $H(E)$ or $H(P)$ tends to attain a limit as N tends to infinity, that is, as a new factor does not convey any new information. Then the ratio of $H(P)/H_m(P)$ expresses the constraint of the environment and subsequent loss of uncertainty or freedom of choice. Following Shannon and Weaver (1949), redundancy can be defined as

$$
R = 1 - \frac{H(P)}{H_m(P)}
\tag{2.236}
$$

in which

$$H(P) = H(E) = H(E_1) + H(E_2|E_1) + \ldots + H(E_N|E_1, E_2, \ldots, E_{N-1})$$ (2.237)

Considering a phenomenon caused by a number of factors each made of three outcomes, Marchand (1972) recognized tree types of entropies. First, the maximum entropy H_m will occur when the outcomes of factors are equally probable and the factors are completely and mutually independent. Second, entropy H_d will occur when the outcomes of each factor are not equally probable but factors are still mutually independent. Note that three factors can be independent two by two but dependent with the three together. Entropy H will occur when the factors are not independent and their outcomes are not equally probable. He then defined three types of redundancies:

1 Internal redundancy R_I :

$$R_I = 1 - \frac{H_d(P)}{H_m(P)}$$ (2.238)

Here the structure is more determined.

2 Structural redundancy R_S:

$$R_S = 1 - \frac{H(P)}{H_d(P)}$$ (2.239)

This measures the inner cohesion of the system and the degree to which it influences the human influence.

3 Global redundancy R:

$$R = 1 - \frac{H(P)}{H_m(P)}$$ (2.240)

This is equivalent to Shannon's definition which integrates internal constraints owing to the unequal distribution of the outcomes of each factor, and structural constraints which stem from the mutual dependencies among these factors. The concept of redundancy permits comparison of land use patterns of different areas with the same kind of development.

2.24.14 Measure of organization

Information theory can be employed to define organization in any point pattern in space. The spatial pattern of a distribution may encompass properties of shape, dispersion, relative location, density, and others. In other words, variations of density over an area can be analyzed. These may not follow any regular geometric pattern. The number and relative sizes of any nucleation may be analyzed. Likewise, the relative location of nucleations may also be analyzed. An example may be the areal distribution of human population in space. In these cases, there are variations in density, the number and sizes of cities or zones, and relative location of cities or zones.

Recall the Shannon entropy

$$H = \sum_{i=1}^{N} q_i \log \frac{1}{q_i}$$ (2.241)

where

$$I = K \log(1/q_i) \tag{2.242}$$

is the information. Equation (2.242) can be used to measure the amount of information gained between the first state where $I \neq 0$ and there are N equally likely outcomes $p_i = 1/N$, and the final state where $I = 0$, and $p_i = 1$, where the outcomes are no longer equally probable. The average expected information gain from a set of probabilities is given by equation (2.241) which applies to any number of probability sets. For two sets one before the other, one determines the average expected information gain between the first set and the second set, that is,

$$H = \sum_{i=1}^{N} p_i \left(\log \frac{1}{q_i} - \log \frac{1}{p_i} \right) = \sum_{i=1}^{N} p_i \log \frac{p_i}{q_i} \tag{2.243}$$

Term $\log(p_i/q_i)$ defines the amount of information gained from the first set $\{q_i\}$ to the second set $\{p_i\}$. Then equation (2.243) defines the weighted average information for the whole set. For the special case $p_i = 1$,

$$H = \log \frac{1}{q_i} \tag{2.244}$$

In this case p_i is selected from one of N expected outcomes and hence $q_i = 1/N$. This yields

$$I = K \log \left(\frac{1}{q_i} \right) = K \log N \tag{2.245}$$

Now consider areal distribution reflecting the density over an area in which the area is considered as a characteristic of the points, each point being surrounded by its area as in the case of Thiessen polygons for rainfall or soil moisture. The question one may ask is: What does this represent in the way of organization? To that end, a base or reference is defined using the maximum entropy. Note the total area and the amount of area for each individual point are defined, and fractions of the total area are treated as probabilities. Thus,

$$H = \sum_{i=1}^{N} p_i \log \frac{1}{p_i} \tag{2.246}$$

will be maximum when $p_i = 1/N$ or $a_i = A/N$, all individual areas are the same. This gives

$$H_{\max} = \log N \tag{2.247}$$

Equation (2.247) defines the base.

From the information gain, the degree of organization in the observed pattern can be expressed as

$$H = \sum_{i=1}^{N} p_i \log \frac{p_i}{1/N} \tag{2.248}$$

in which $q_i = 1/N$-all prior probabilities are equal.

Now divide the points and their areas into groups. For example, in the case of a state these can correspond to counties and let M be the number of these groups or counties or sets. If n_j denotes the number of objects, say people, in the j-th set s_j, $j = 1, 2, \ldots, M$, then for the j-th set or group $y_i = \sum_{i \varepsilon s_j} p_i$. Equation (2.243) can be expressed in disaggregated form as

$$H = \sum_{j=1}^{M} y_j \log \frac{y_j}{n_j/N} + \sum_{j=1}^{M} y_j \left[\sum_{i \varepsilon s_j} \frac{p_i}{y_j} \log \frac{p_i/y_j}{1/n_j} \right] \tag{2.249}$$

The entropy $\log N$ which represents the maximum entropy is known to be varying directly with N. The information gain is independent of N, thus allowing comparison of different sizes.

Organization can be related to the reduction in information from the case of maximum entropy. The uniform case, or equal areas, is the case of maximum order. This means that the maximum entropy, interpreted in a spatial sense, defines a highly significant regular pattern. Von Foerster (1960) defined order O as

$$O = 1 - \frac{H}{H_{max}} \tag{2.250}$$

which is the same as redundancy defined by Shannon. This makes matters muddy. The maximum entropy is being used here as an indication of order or the base for an exactly opposite definition of order.

Let us now consider organization in the relative sizes and number of nucleations. The entropy measure in the disaggregated form can be expressed as

$$H = \sum_j x_j \log \frac{1}{x_j} + \sum_j x_j \left[\sum_{i \varepsilon s_j} \frac{q_i}{x_j} \log \frac{1}{q_i/x_j} \right] \tag{2.251}$$

Equation (2.251) takes on a maximum value when all cities have the same population. This would mean there is no organization in their relative sizes which in the real world is not true. When sizes are different, the index will decrease and entropy will decrease.

Now consider effects of relative location from a central location or point of reference. For a watershed, this point can be the outlet, and in the case of a city it can be city center. The whole size range (of cities, say) can be laid out on each axis, with each row and each column represented by one city only. Entries along any one row may represent flow from city to all others. These flows can be computed as the product of two populations divided by their centers. These values are summed up for the whole matrix and then fractions of the summed values are re-entered and treated as probabilities. Then the entropy for the whole matrix can be expressed as

$$H = \sum p_i \sum \frac{p_{ij}}{p_i} \log \frac{p_i}{p_{ij}} \tag{2.252}$$

where p_{ij} is the flow from i-th to the j-th city, and $p_i = \sum_j p_{ij}$. For each row the maximum value is when all flow fractions have the same value. This would mean there is no flow preference between cities. As the value becomes smaller, the preference increases.

To determine the effect of distance another matrix is considered without dividing the product of populations by the square of the distance. The implication is that the distance has no effect – an ideal case. A measure of locational preference due to distance can be expressed by the index value of the first distance case as a percentage of the nondistance case.

2.25 Relation between entropy and variance

Although there does not appear to be an explicit relationship between entropy H and variance, this relationship can, by reparmeterization, be explicitly stated for certain distributions. By so doing these distributions can be compared as was done by Mukherjee and Ratnaparkhi (1986). Plotting H as a function of variance σ^2, assuming the variance is known, one can show those distributions that have higher entropy and those that have lower entropy. The close proximity of entropies may be reflected by the similarity of the shapes of distributions, whereas the difference may point to the longer tail of the distribution. The closeness of entropy may also point to the genesis of distributions.

Since entropy of a distribution is a function of its parameters, direct comparison of entropies is not simple or straightforward. However, it is plausible to construct an entropy-based measure of affinity or closeness for distributions having a common variance. Following Mukherjee and Ratnaparkhi (1986), the affinity A between two distribution functions f_1 and f_2 with common variance is defined as

$$A(f_1, f_2) = |H_1 - H_2| \tag{2.253}$$

where H_1 is entropy of f_1 and H_2 is entropy of f_2. Equation (2.253) can be written as

$$A(f_1, f_2) = \left| -E\left[\ln \frac{f_2}{f_1} \right] \right| \tag{2.254}$$

which is the expectation of the log-likelihood ratio and is not the same as the Kulback-Leibler minimum cross-entropy.

The affinity measure defined by equation (2.254) satisfies the following: 1) $A(f_1, f_2)=0$, if $f_1 = f_2$. That is, the affinity of a distribution with itself is zero. However, the converse is not necessarily true. 2) If $A(f_1, f_2) = A(f_2, f_1)$, then distributions are interchangeable. 3) $A(f_1, f_2)$ satisfies the triangular inequality. 4) It is a measure of the distance between two distributions having common variance.

Another measure, called similarity commonly used in cluster analysis, can be defined as

$$S(f_i, f_j) = 1 - \frac{A(f_i, f_j)}{\max\{A(f_i, f_j)\}} \tag{2.255}$$

for any i and j, and $i \neq j$. Thus, one can state that affinity is a monotonically decreasing function of similarity. The similarity measure can be employed for clustering distributions. Some distributions are more similar than others. Entropy may imply concentration of probability density toward location followed by long tails. Similarly, low values of entropy imply flatness of the distribution.

Consider $X_i, i = 1, 2, \ldots, N$, independent identically normally distributed random variables each having mean μ and variance σ^2. One can define

$$\overline{X} = \frac{1}{N} \sum_{i=1}^{N} \overline{X}_i, \quad S^2 = \frac{1}{N-1} \sum_{i=1}^{N} (X_i - \overline{X})^2 \tag{2.256}$$

where S^2 is an estimate of σ^2 and $\overline{X}$ is estimate of μ. An estimate of H, $\hat{H}$, for the normal distribution can be expressed as

$$\hat{H} = \frac{1}{2} \ln[2\pi e S^2] \tag{2.257}$$

It is known that the distribution of S^2 is chi-square with $(N-1)$ degrees of freedom. Then, from equation (2.257), it can be shown using transformation of variables that the distribution of $\hat{H}$ is log-chi-squared. If N is used as a scale factor in equation (2.257), then $\hat{H}$ becomes the maximum likelihood estimator of H.

Now consider two independent random samples, $X_{ij}, i = 1, 2; j = 1, 2, \ldots, N$, from normal distributions with mean and variance as μ_1, μ_2 and σ_1^2, σ_2^2, respectively. $S_i^2, i = 1, 2$, are sample variances. Then

$$D = \hat{H}_1 - \hat{H}_2 = \frac{1}{2} \ln\left(\frac{S_1^2}{S_2^2}\right) \tag{2.258}$$

has logarithmic F-distribution which can be approximated by the normal distribution. This follows from the observation that S_1^2/S_2^2 has F-distribution. Equation (2.258) further suggests that D constructed with other estimators may likely possess the desirable properties of the log-F distribution.

If $X_i, i = 1, 2, \ldots, N$, are exponentially distributed with parameter λ. Let $\overline{X}$ be the sample mean which is an unbiased estimator of σ then an estimate of H, $\hat{H}$, is expressed as

$$\hat{H} = 1 + \ln \overline{X} \tag{2.259}$$

It is known that $\overline{X}$ is gamma distributed with parameter $1/(N\lambda)$. Then, equation (2.259) leads to the distribution of $\hat{H}$ as log-gamma distribution. Furthermore, $\hat{H}$ is the maximum likelihood estimate of H, because $\overline{X}$ is the maximum likelihood estimate of σ.

If $X_{ij}, i = 1, 2; j = 1, 2, \ldots, N$, are two independent exponentially distributed random variables with parameter $\lambda_i, i = 1, 2$, and entropies $H_i, i = 1, 2$, then with sample means $\overline{X}_i, i = 1, 2$, the statistic

$$D = \hat{H}_1 - \hat{H}_2 = \ln\left(\frac{\overline{X}_1}{\overline{X}_2}\right) \tag{2.260}$$

has log-F distribution $(2N_1, 2N_2)$ degrees of freedom. This can be seen by recalling that the ratio of two gamma variables has an exact F distribution. Again, $D = \ln F$ can be approximated by the normal distribution.

In this manner, distributions having the same variance can be compared using entropy. High entropy indicates less extraneous information. Thus, entropy can also be used to select a distribution for a particular data set. In a similar manner, the affinity measure can be used to select a particular distribution.

2.26 Entropy power

Shannon (1948) introduced the concept of entropy power which is defined as the variance of the independent identically distributed (IID) components of an n-dimensional white Gaussian random variable having entropy $H(X)$. The entropy power $N(X)$ can be expressed as

$$N(X) = \frac{1}{2\pi e} \exp\left[\frac{2}{n} H(X)\right] \tag{2.261}$$

Since white noise has the maximum entropy for a given power, the entropy power of any noise (or any distribution) is less than or equal to its actual power. For two independent random variables X and Y, Shannon showed that

$$N(X + Y) \geq N(X) + N(Y) \tag{2.262}$$

Dembo (1989) has proved equation (2.262) and Zamir and Feder (1993) have presented generalizations of entropy power inequality.

2.27 Relative frequency

Consider N class intervals and in each class interval the number of values or absolute frequency is n_i, $i = 1, 2, \ldots, N$. Here $n_1 + n_2 + \ldots + n_N = N$. Let the probability of each class intervals be p_i, $i = 1, 2, \ldots, N$, where

$$p_i = \frac{n_i}{\sum\limits_{i=1}^{N} n_i} = \frac{n_i}{N}, \quad \sum_{i=1}^{N} n_i = N \tag{2.263}$$

For the relative frequency to be interpreted as probability, N must be sufficiently large. Then one can write

$$n_i = N p_i, i = 1, 2, \ldots, N \tag{2.264}$$

Sometimes, a sequence is referred to as typical, if $n_i \cong N p_i$; otherwise it is rare. Of course, $P = \sum\limits_{i=1}^{N} p_i = 1$ and $1\text{-}P = 0$. One can write

$$n_i \cong N p_i = \exp[N \ln p_i] \tag{2.265}$$

For all class interval sequences, one can express the probability of each typical sequence

$$P = \{p_1, p_2, \ldots, p_N\} = \exp[N p_1 \ln p_1 + N p_2 \ln p_2 + \ldots + N p_N \ln p_N = \exp[-NH(P)] \tag{2.266}$$

where $H(P)$ is the entropy of the sample. The number of typical class interval sequences, N_T, can be written as

$$N_T = \frac{P(U_t)}{P(t_s)} \approx \frac{1}{\exp[-NH(P)]} = \exp[NH(P)]$$

$$= \frac{Pr\,obability\ of\ union\ of\ all\ typical\ sequences}{Pr\,obability\ of\ each\ sequence} \tag{2.267a}$$

where U_t denotes the union of typical sequences, and t_s is a typical sequence. If all sequneces are equally likely then $H(P) = \ln N$ and $N_T = n^N$. Thus, for $n > 1$, one can express

$$N_T \approx \exp[NH(P)] << n^N \tag{2.267b}$$

For sufficiently large N, most sequences are rare.

2.28 Application of entropy theory

Applications of entropy can be grouped into two categories. The first category encompasses those applications which require determination of probability distributions or parameter estimation. Information is specified in the form of constraints and the determination is based on the application of the principle of maximum entropy (POME) or the principle of minimum cross entropy (POMCE) and entropy is maximized. Applications in this category are common in water engineering. The second category of applications entails the specification of (source) entropy and various random variables (code lengths) are constructed in order to minimize their expected values. The specification is based on the construction of optimum mappings (codes) of the random variables of interest into the given probability space.

Questions

Q.2.1 In the month of January it rained on a particular day in an area. The person who was interested to know the day it rained on was away from that area but he knew a friend who knew the day of rain. How many questions does that person need to ask his friend or how much information will be required to positively ascertain the day it rained on?

Q.2.2 Obtain two sets of daily rainfall data (only positive values) for two raingage stations. Compute the Shannon entropy for the two sets of data. On a relative basis, what do these entropy values say about the rainfall measurements of these two gages?

Q.2.3 Obtain two sets of data on runoff events for two small watersheds which are closely located. Compute the Shannon entropy for the two sets of data. On a relative basis, what do these entropy values say about runoff from these watersheds?

Q.2.4 Consider a set of rainfall amounts and the corresponding runoff amounts for a watershed. Compute marginal entropies of rainfall and runoff. Arrange these data as a two-dimensional contingency table and compute joint entropy of rainfall and runoff. Also compute transinformation. Compute the coefficient of correlation and informational correlation coefficient. How different are these two coefficients and what does the difference mean?

Q.2.5 Consider a set of runoff amounts and the corresponding sediment yields for a watershed. Compute marginal entropies of runoff and sediment yield. Arrange these data as a two-dimensional contingency table. Also compute transinformation. Compute the coefficient of correlation and informational correlation coefficient. How different are these two coefficients and what does the difference mean?

Q.2.6 Consider a set of rainfall amounts and the corresponding runoff amounts and sediment yields for a watershed. These data can be arranged as a three-dimensional contingency table. However, the three-dimensional contingency table can be portrayed as two sub-tables or two two-dimensional contingency tables one having sediment yield as output (or response) and rainfall as input (or stimulus) and the other having sediment yield as output (or response) and runoff as input (or stimulus). Compute the components of the sediment yield information.

Q.2.7 Consider a set of flood events with peak discharge, volume of flow, and duration for a watershed. Here peak discharge is considered as output (or response), and volume and duration as input variables (or stimuli). These data can be arranged as a three-dimensional contingency table. It is, however, more convenient to portray the three-dimensional contingency table as two sub-tables or two two-dimensional contingency tables one having peak discharge as output (or response) and volume as input (or stimulus) and the other having peak discharge as output (or response) and duration as input (or stimulus). Compute the components of the peak discharge yield information.

Q.2.8 Consider a set of rainfall-runoff events for a watershed. Compute marginal entropy of rainfall and of runoff. Compute transinformation between rainfall and runoff. For the watershed, rainfall can be considered as input and runoff as output. Determine the conditional entropy of runoff given rainfall and also conditional entropy of rainfall given runoff. What is the lost information? What is the noise? What is the transmitted information?

Q.2.9 For the data set in Q.2.8, change the class interval size and then compute entropy. Take at least three class interval sizes. How does entropy change with class interval size?

Q.2.10 Consider a stream flow time series for gaging station on a river. Compute the marginal entropy of stream flow using a constant class interval. Now consider a class interval proportional to flow. Compute entropy in this case and compare it with the entropy corresponding to the fixed class interval. What is the difference between two entropy values and what does the difference tell us?

Q.2.11 Obtain stream flow values from a gaging station where there are zero values. Compute entropy of such a record.

Q.2.12 Obtain daily rainfall values for a raingaging station. Fit a probability distribution to the rainfall values. Apply the concentration theorem to check if the fitted distribution is the right one.

References

Abramowitz, M. and Stegun, I.A. (1970). *Handbook of Mathematical Functions*. Dover, New York.

Amorocho, J. and Espildora, B. (1973). Entropy in the assessment of uncertainty in hydrologic systems and models. *Water Resources Research*, Vol. 9, No. 6, pp. 1511–22.

Ash, R.B. (1965). *Information Theory*. Dover, New York.

Batty, M.J. (2010). Space, scale, and scaling in entropy maximizing. *Geographical Analysis*, Vol. 42, pp. 395–421.

Burg, J.P. (1975). Maximum entropy spectral analysis. Unpublished Ph.D. thesis, Stanford University, 123 p., Palo Alto, California.

Burg, J.P. (1977). Maximum entropy spectral analysis. Paper presented at the 37th Annual Meeting of Society of Exploration Geophysics, Oklahoma City, Oklahoma.

Chapman, T.G. (1980). Equations for entropy in hydrologic time series. Department of Civil Engineering, Faculty of Military Studies, University of New South Wales, Duntroon, N.S.W., Australia.

Chapman, T.G. (1986). Entropy as a measure of hydrological data uncertainty and model performance. *Journal of Hydrology*, Vol. 85, pp. 111–26.

Dembo, A. (1989). Simple proof of the concavity of the entropy power with respect to added Gaussian noise. *IEEE Transactions on Information Theory*, Vol. 35, No. 4, pp. 887–8.

Denbigh, K.G. (1989). Note on entropy, disorder and disorganization. *British Journal of Philosophy and Science*, Vol. 40, pp. 323–32.

Denbigh, K.G. and Denbigh, J.S. (1985). *Entropy in Relation to Incomplete Knowledge*. Cambridge University Press, Cambridge, U.K.

Feder, M. and Merhav, N. (1994). Relations between entropy and error probability. *IEEE Transactions on Information Theory*, Vol. 40, No. 1, pp. 259–66.

Feder, M., Merhav, N. and Gutman, M. (1992). Universal prediction of individual sequences. *IEEE Transactions on Information Theory*, Vol. 38, pp. 1258–70.

Feisauerova, J. and Vajda, I. (1993). Testing system entropy and prediction error probability. *IEEE Transactions on Systems, Man, and Cybernetics*, Vol. 23, No. 5, pp. 1352–8.

Garner, W.R. and Hake, H.W. (1951). The amount of information in absolute judgments, *Psychology Review*, Vol. 58, pp. 446–59.

Harmancioglu, N.B., Yevjevich, V. and Obeysekara, J.T.B. (1986). Measures of information transfer between variables. in Proceedings of the Fourth International Hydrology Symposium on Multivariate Analysis of Hydrologic Processes, edited by H.W. Shen, pp. 481–99, Fort Collins, Colorado, July 1985.

Harmancioglu, N.B. and Singh, V.P. (1998). Entropy in environmental and water resources. *Encyclopedia of Hydrology and Water Resources*, R. W. Hershey and R. W. Fairbridge, eds., 225–41, Kluwer Academic Publishers, Boston, Massachusetts.

Jaynes, E.T. (1957). Information theory and statistical mechanics, I. *Physical Review*, 106, 620–30.

Jaynes, E.T. (1968). Prior probabilities. *IEEE Transactions on System Science and Cybernetics*, Vol. 70, pp. 939–52.

Jaynes, E.T. (1979). Concentration of distributions at entropy maxima. Paper presented at the 19th NBER-NSF Seminar on Bayesian Statistics, Montreal, October 1979. in E.T. Jaynes: Papers on Probability, Statistics and Statistical Physics, edited by R.D. Rosenkratz, pp. 315–36, D. Reidel Publishing Co., Boston, MA, 1983.

Jaynes, E.T. (1982). On the rationale of maximum entropy methods. *Proceedings of the IEEE*, Vol. 70, 939–52.

Johnson, R.W. and Shore, J.E. (1984). Which is the better entropy expression for speech processing:-SlogS or LogS? *IEEE Transactions on Acoustics, Speech, and Signal Proecssing*, Vol. ASSP-32, No. 1, pp. 129–37.

Kapur, J.N. and Kesavan, H.K. (1992). *Entropy Maximization Principles with Applications*. Academic Press, Inc., New York.

Kullback, S. (1959). *Information Theory and Statistics*. John Wiley & Sons, New York (reprinted in 1968 by Dover).

Kullback, S. and Leibler, R.A. (1951). On information and sufficiency. *Annals of Mathematical Statistics*, 22, 79–86.

Lathi, B.P. (1969). *An Introduction to Random Signals and Communication Theory*. International Textbook Company, Scanton, Pennsylvania.

Levine, R.D. and Tribus, M., eds. (1978). *The Maximum Entropy Formalism*, 498p., The MIT Press, Cambridge, Massachusetts.

Lindley, D.V. (1956). On a measure of information provided by an experiment. *The Annals of Mathematical Statistics*, Vol. 27, pp. 986–1005.

Lindley, D.V. (1961). The use of prior probability distributions in statistical inference and decision. Proceedings of the Fourth Berkeley Symposium, Vol. 1, pp. 436–8.

Marchand, B. (1972). Information theory and geography. *Geographical Analysis*, Vol. 4, pp. 234–57.

McGill, W.J. (1953). Multivariate transmission of information and its relation to analysis of variance. Report No. 32, Human Factors Operations Research Laboratories, MIT, Cambridge, Massachusetts.

McGill, W.J. (1954). Multivariate information transmission. *Psychometrica*, Vol. 19, No. 2, pp. 97–116.

Miller, G.A. (1953). What is information measurement? *American Psychologist*, Vol. 8, pp. 3–11.

Moran, P.A.P. (1969). Statistical inference with bivariate gamma distributions. *Biometrika*, Vol. 56, No. 3, pp. 627–34.

Mukherjee, D. and Ratnaparkhi, M.V. (1986). On the functional relationship between entropy and variance with related applications. *Communication in Statistics-Theory and Methods*, Vol. 15, No. 1, pp. 291–311.

Papoulis, A. (1991). *Probability, Random Variables and Stochastic Processes*, 3rd edition, McGraw Hill, New York.

Paszto, V., Tuček, P. and Voženílek, V. (2009). On spatial entropy in geographical data. GIS Ostrava 2009.

Shannon, C.E. (1948). A mathematical theory of communications, I and II. Bell System Technical Journal, Vol. 27, pp. 379–443.

Shannon, C.E. and Weaver, W. (1949). *The Mathematical Theory of Communication*. University of Illinois Press, Urbana, Illinois.

Soofi, E. (1994). Capturing the intangible concept of information. *Journal of the American Statistical Association*, Vol. 89, No. 428, pp. 1243–54.

Tribus, M. (1969). *Rational Description: Decision and Designs*. Pergamon Press, New York.

Von Foerster, H. (1960). On self-organizing systems and their environments. In: *Self-Organizing Systems*, edited by C. Yovits and C. Cameron, Pergamon Press, Oxford, pp. 31–50.

White, H. (1965). The entropy of a continuous distribution. *Bulletin of Mathematical Biophysics*, Vol. 27, pp. 135–43.

Wilson, A.G. (1970). *Entropy in Urban and Regional Modeling*. Pion, London, England.

Zamir, R. and Feder, M. (1993). A generalization of the entropy power inequality with applications. *IEEE Transactions on Information Theory*, Vol. 39, No. 5, pp. 1723–8.

Additional Reading

Akaike, H. (1973). Information theory and an extension of the maximum likelihood principle. *Proceedings, 2nd International Symposium on Information Theory*, B.N. Petrov and F. Csaki, eds., Publishing House of the Hungarian Academy of Sciences, Budapest, Hungary.

Bates, J.E. and Shepart, H.K. (1993). Measuring complexity using information fluctuation. *Physics Letters A*, Vol. 172, pp. 416–25.

Bevensee, R.M. (1993). *Maximum Entropy Solutions to Scientific Problems*. PTR Prentice Hall, Englewood Cliffs, New Jersey.

Carnap, R. (1977). *Two Essays on Entropy*. University of California Press, Berkeley, California.

Cover, T.M. and Thomas, J.A. (1991). *Elements of Information Theory*. John Wiley & Sons, New York.

Dalezios, N. R. and Tyraskis, P. A. (1989). Maximum entropy spectra for regional precipitation analysis and forecasting. *Journal of Hydrology*, Vol. 109, pp. 25–42.

Erickson, G.J. and Smith, C.R. (1988). *Maximum-Entropy and Bayesian Methods in Science and Engineering*. Kluwer Academic Publishers, Dordrecht, The Netherlands.

Fiorentino, M, Claps, P. and Singh, V.P. (1993). An entropy-based morphological analysis of river basin networks. *Water Resources Research*, Vol. 29, No. 4, pp. 1215–24.

Fougere, P.F., Zawalick, E.J. and Radoski, H.R. (1976). Spontaneous life splitting in maximum entropy power spectrum analysis. *Physics of the Earth and Planetary Interiors*, Vol. 12, pp. 201–7.

Gallager, R.G. (1968). *Information Theory and Reliable Communication*. Wiley, New York, N.Y.

Jaynes, E.T. (1958). Probability Theory in Science and Engineering. Colloquium Lectures in Pure and Applied Science, Vol. 4, Field Research Laboratory, Socony Mobil Oil Company, Inc., Dallas, Texas, USA.

Jaynes, E.T. (1982). On the rationale of maximum entropy methods. *Proceedings of the IEEE*, Vol. 70, pp. 939–52.

Kaplan, S. and Garrick, B.J. (1981). On the quantitative definition of risk. *Risk Analysis*, Vol. 1, No. 1, pp. 11–27.

Khinchin, A.I. (1957). *The Entropy Concept in Probability Theory*. Translated by R.A. Silverman and M.D. Friedman, Dover Publications, New York.

Krasovskaia, I. (1997). Entropy-based grouping of river flow regimes. *Journal of Hydrology*, Vol. 202, 173–1191.

Krstanovic, P.F. and Singh, V.P. (1991a). A univariate model for long-term streamflow forecasting: 1. Development. *Stochastic Hydrology and Hydraulics*, Vol. 5, 173–88.

Krstanovic, P.F. and Singh, V.P. (1991b). A univariate model for long-term streamflow forecasting: 2. Application. *Stochastic Hydrology and Hydraulics*, Vol. 5, 189–205.

Krstanovic, P.F. and Singh, V.P. (1992a). Evaluation of rainfall networks using entropy: 1. Theoretical development. *Water Resources Management*, Vol. 6, 279–93.

Krstanovic, P.F. and Singh, V.P. (1992b). Evaluation of rainfall networks using entropy: II. Application. *Water Resources Management*, Vol. 6, 295–314.

Krstanovic, P.F. and Singh, V.P. (1993a). A real-time flood forecasting model based on maximum entropy spectral analysis: 1. Development. *Water Resources Management*, Vol. 7, 109–29.

Krstanovic, P.F. and Singh, V.P. (1993b). A real-time flood forecasting model based on maximum entropy spectral analysis: 2. Application. *Water Resources Management*, Vol. 7, 131–51.

Moramarco, T and Singh, V. P. (2001). Simple method for relating local stage and remote discharge. *Journal of Hydrologic Engineering, ASCE*, Vol. 6, No. 1, 78–81.

Padmanabhan, G. and Rao, A.R. (1986). Maximum entropy spectra of some rainfall and river flow time series from southern and central India. *Theoretical and Applied Climatology*. Vol. 37, pp. 63–73.

Pachepsky, Y., Guber, A., Jacques, D., Simunek, J., Van Genuchten, M.T., Nicholson, T. and Cady, R. (2006). Information content and complexity of simulated soil water fluxes. *Geoderma*, Vol. 134, pp. 253–66.

Padmanabhan, G. and Rao, A.R. (1988). Maximum entropy spectral analysis of hydrologic data. *Water Resources Research*, Vol. 24, No. 9, pp. 1591–3.

Prigogine, I. (1989). What is entropy? *Naturwissenschaften*, Vol. 76, pp. 1–8.

Rao, A.R., Padmanabhan, G., and Kashyap, R.L. (1980). Comparison of recently developed methods of spectral analysis. Proceedings, Third International Symposium on Stochastic Hydraulics, pp. 165–75, Tokyo, Japan.

Shore, J.E. (1979). Minimum cross-entropy spectral analysis. NRL Memorandum Report 3921, Naval Research Laboratory, Washington D.C.

Singh, V.P. (1998). The use of entropy in hydrology and water resources. *Hydrological Processes*, Vol. 11, pp. 587–626.

Singh, V.P. (1998). Entropy as a decision tool in environmental and water resources. *Hydrology Journal*, Vol. XXI, No. 1–4, pp. 1–12.

Singh, V.P. (1998). *Entropy-Based Parameter Estimation in Hydrology*. Kluwer Academic Publishers, Boston, Massachusetts.

Singh, V.P. (2000). The entropy theory as tool for modeling and decision making in environmental and water resources. *Water SA*, Vol. 26, No. 1, pp. 1–11.

Singh, V.P. (2003). The Entropy theory as a decision making tool in environmental and water resources. In: *Entropy Measures, maximum Entropy and Emerging Applications*, edited by Karmeshu, Springer-Verlag, Bonn, Germany, pp. 261–97.

Singh, V.P. (2005). Entropy theory for hydrologic modeling. in: *Water Encyclopedia: Oceanography; Meteorology; Physics and Chemistry; Water Law; and Water History, Art, and Culture*, edited by J. H. Lehr, Jack Keeley, Janet Lehr and Thomas B. Kingery, John Wiley & Sons, Hoboken, New Jersey, pp. 217–23.

Singh, V.P. (2011). Hydrologic synthesis using entropy theory: Review. *Journal of Hydrologic Engineering*, Vol. 16, No. 5, pp. 421–33.

Singh, V.P. (2011). Entropy theory for earth science modeling, *Indian Geological Congress Journal*, Vol. 2, No. 2, pp. 5–40.

Singh, V.P. and Fiorentino, M., editors (1992). *Entropy and Energy Dissipation in Water Resources*. Kluwer Academic Publishers, Dordrecht, The Netherlands.

Ulrych, T.J. and Clayton, R.W. (1976). Time series modeling and maximum entropy. *Physics of the Earth and Planetary Interiors*, Vol. 12, pp. 188–99.

Wu, N. (1997). *The Maximum Entropy Method*. Springer, Heidelburg, Germany.

Yang, Y. and Burn, D. H. (1984). An entropy approach to data collection network design. *Journal of Hydrology*, Vol. 157, pp. 307–24.

3 Principle of Maximum Entropy

The principle of maximum entropy is one of the main pillars of entropy theory. It was briefly discussed in Chapter 2. The objective of this chapter is to discuss it more fully and illustrate its properties using examples.

3.1 Formulation

Consider that an experiment is conducted which produces N outcomes $(x_1, x_2, \ldots, x_N)$ with probabilities $(p_1, p_2, \ldots, p_N)$ which are not known; $x_1, x_2, \ldots, x_N$ are the values of the random variable X on which the experiment is conducted or are experimental outcomes, and N is the number of values or outcomes. The objective of the experiment is to determine the values of probabilities $p_1, p_2, \ldots, p_N$ or the probability distribution $P = \{p_1, p_2, \ldots, p_N\}$. Assume that nothing is known about X in terms of its moments (mean, variance, etc.). Of course, the total probability law will hold:

$$\sum_{i=1}^{N} p_i = 1, p_i \geq 0, \ i = 1, 2, \ldots, N \tag{3.1}$$

There can be a number of probability distributions that can satisfy equation (3.1). For example, one distribution could be: $P = \left\{\frac{1}{N}, \frac{1}{N}, \ldots, \frac{1}{N}\right\}$. Another distribution could be: $P = \left\{\frac{1}{2N}, \frac{3}{2N}, \ldots, \frac{1}{N}\right\}$. Likewise, one can construct an infinite number of probability distributions which can satisfy equation (3.1). Each of these distributions would have a particular value of entropy. There are also N degenerate distributions: $\{1, 0, 0, \ldots, 0\}$, $\{0, 1, 0, 0, \ldots, 0\}$, $\ldots, \{0, 0, 0, \ldots, 1\}$, each having zero entropy. Of the distributions having nonzero entropy there would be one distribution that would have the maximum entropy. It turns out that such a distribution would be the one with $p_i = 1/N$ and this distribution is the uniform (rectangular) distribution and its entropy is log N. The question arises: What would be the most appropriate probability distribution of X?

The principle of maximum entropy (POME) formulated by Jaynes (1957a, b), states that the most appropriate probability distribution would be the one having the maximum entropy or uncertainty. This is consistent with Laplace's principle of insufficient reason which states that

Entropy Theory and its Application in Environmental and Water Engineering, First Edition. Vijay P. Singh.
© 2013 John Wiley & Sons, Ltd. Published 2013 by John Wiley & Sons, Ltd.

all outcomes are equally likely when nothing is known about the random variable, except for what is described by equation (3.1). This makes sense, for it makes use of only the information that is given and avoids assuming anything that is not available or known. In that case the distribution would be the uniform distribution. According to POME, one should maximize uncertainty given by entropy. In other words, when only partial information in the form of constraints is given on X, that is, some information is missing, probabilities should be chosen that maximize the uncertainty about the missing information. This leads to the most random distribution subject to the specified constraints. The implication is that one should utilize whatever information one has and scrupulously avoid using any information that one does not have. This also means that one should be as noncommittal to the missing information as possible. This again is in accord with Laplace's principle of insufficient reason, that is, unless there is reason to believe, all outcomes of an experiment have equal probabilities of occurrence.

Now consider that an additional piece of information is given in the form of a constraint as:

$$p_1 x_1 + p_2 x_2 + \ldots + p_N x_N = \bar{x} \tag{3.2}$$

where $\bar{x}$ is the weighted mean of the N values of X. The probability distribution based on equation (3.1) may not satisfy equation (3.2). Thus, one needs a distribution that satisfies equation (3.1) as well as equation (3.2) and has maximum entropy, H_{max}. As before, there would be an infinite number of probability distributions each having a particular value of entropy, but there would be one distribution that would have the maximum entropy and there would be another distribution that would have the minimum entropy. It turns out that the distribution having the maximum entropy would be the exponential distribution. The entropy value of this distribution would be less than the entropy value of the uniform distribution log N, but would be greater than 0. This suggests that each new piece of information (in the form of constraint) that the distribution has to satisfy reduces (or at least does not increase) the maximum entropy and increases (or at least does not decrease) the minimum entropy, H_{min}. Indeed H_{max} is a monotonically decreasing (or at least nonincreasing) function of the number of constraints, and H_{min} increases (or at least does not decrease) as a function of the number of constraints. The reduction in uncertainty due to the knowledge of, say, average constraints quantifies the information content of the additional constraints. This idea of quantifying uncertainty reduction offers a useful tool for evaluating merits of explanatory variables in the class of probability distributions (Soofi, 1994).

Extending the above argument further, let there be M independent constraints on X such that a unique probability distribution (corresponding to the maximum entropy) would result, where $H_{max} = H_{min}$, as shown in Figure 3.1, and the uncertainty may be completely removed. For less than M constraints, there can be a number of distributions but these distributions would be biased. There would be only one distribution that would correspond to the maximum entropy and that distribution would be most random, most unbiased, and most objective. All these distributions would have H_{min} values (globally minimum) and would represent most biased distributions. Thus, POME can also be cast in other forms. For example, of all the distributions one chooses the distribution that has the maximum uncertainty, or chooses the distribution that is least committed to the information not known or not specified, or chooses the distribution that is most random or objective, or chooses the distribution that is most unbiased. Thus, POME can also be called the principle of maximum uncertainty. Van Campenhout and Cover (1981) have shown that all the well-known distributions are maximum entropy-based distributions, given appropriate moment constraints. Evans (1969)

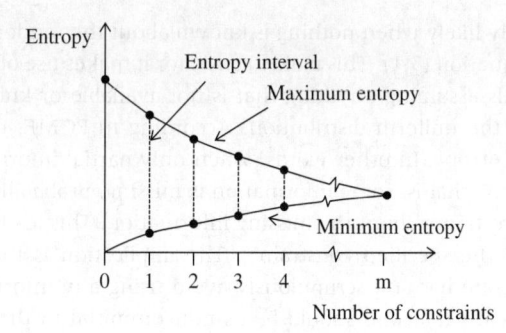

Figure 3.1 Variation of maximum and minimum entropy with the number of constraints.

called POME the principle of minimum information, since information is the negative of uncertainty, and discussed how to choose constraints. Kesavan and Kapur (1989) presented a generalization of POME.

As Jaynes (1957) has stated, the probability distribution based on the principle of maximum entropy is maximally noncommittal to missing information, agrees with what is known, and expresses maximum uncertainty with regard to other aspects and hence permits maximum freedom to accommodate the influence of subsequent data. POME is essentially a method of inductive inference. This means that different ways of using POME with the same information should yield consistent results. In other words, the results should be consistent for different methods of solving the same problem. To that end, Shore and Johnson (1980) formulated four consistency measures: 1) Uniqueness: The result should be unique. 2) Invariance: The result should be independent of the choice of coordinate system. 3) System independence: The result should not be dependent on whether systems are dependent or independent. 4) Subsystem independence: The result should not depend on whether a subset of a system is dealt with independently in terms of a separate conditional probability density function or full system probability density function.

Thus, one can now summarize entropy, information, and uncertainty. Information (or knowledge) is expressed in the form of constraints on the values of random variable and their associated probability values. Then, one does not have a unique value of uncertainty, rather one has a range of uncertainty defined by H_{max} and H_{min}, and as a result a definite statement on the value of uncertainty cannot be made. Any additional piece of information or constraint reduces the range of uncertainty, but not the value of uncertainty itself. As more and more constraints are specified, the range of uncertainty becomes smaller and smaller. It is plausible that with enough constraints the range may converge to a point or a definite value of uncertainty and when that equals 0, the probability distribution would be a degenerate one. In that case the experiment becomes deterministic. Otherwise it remains stochastic. This suggests that entropy provides a criterion for distinguishing between deterministic and stochastic phenomena. Thus, one may express: knowledge → information → constraints → entropy → uncertainty.

It may be noted that in practice constraints are derived from data and do not incorporate all the information provided by the data. This means that the POME formalism does not use all the information and may introduce errors when doing inference because sample averages may differ from statistical averages as emphasized by Feder (1986) who suggested a minimum description length (MDL) method introduced by Rissanen (1978, 1983, 1984). Feder (1986) showed POME to be a special case of MDL criterion.

3.2 POME formalism for discrete variables

POME involves determining probability distributions or parameters thereof, subject to given constraints. Often the problem is solved numerically. In a number of cases analytical solutions can be derived. POME states that one should maximize the Shannon entropy:

$$H(P) = -\sum_{i=1}^{N} p_i \log_e p_i = -\sum_{i=1}^{N} p_i \ln p_i \qquad (3.3)$$

subject to equation (3.1) and

$$\sum_{i=1}^{N} p_i g_j(x_i) = C_j, \quad j = 1, 2, ..., m \qquad (3.4)$$

where $g_j(x_i)$ is the j-th function of $X:\{x_i\}$ for expressing constraint C_j, and m is the number of constraints. In equation (3.3) the logarithm to the base e has been written as ln and this convention will be followed throughout the book. Equation (3.4) specifies m constraints.

Maximization of H can be done using the method of Lagrange multipliers where the Lagrangean function L can be expressed as

$$L = -\sum_{i=1}^{N} p_i \ln p_i - (\lambda_0 - 1)\left(\sum_{i=1}^{N} p_i - 1\right) - \sum_{j=1}^{m} \lambda_j \left(\sum_{i=1}^{N} p_i g_j(x_i) - C_j\right) \qquad (3.5)$$

where $\lambda_0, \lambda_1, \ldots, \lambda_m$ are the $(m+1)$ Lagrange multipliers corresponding to the $(m+1)$ constraints specified by equations (3.1) and (3.4). For convenience, in equation (3.5) one uses $(\lambda_0 - 1)$ as the first Lagrange multiplier instead of λ_0.

Differentiating L in equation (3.5) with respect to p_i and equating it to zero, one obtains

$$\frac{\partial L}{\partial p_i} = 0 \Rightarrow -\ln p_i - \lambda_0 - \sum_{j=1}^{m} \lambda_j g_j(x_i) = 0 \qquad (3.6)$$

Equation (3.6) gives

$$p_i = \exp\left[-\lambda_0 - \lambda_1 g_1(x_i) - \ldots - \lambda_m g_m(x_i)\right], \quad i = 1, 2, \ldots, N \qquad (3.7)$$

Equation (3.7) is the maximum entropy probability distribution with Lagrange multipliers $\lambda_0, \lambda_1, \ldots, \lambda_m$ which can be determined as follows.

Substituting equation (3.7) into equations (3.1) and (3.4), one, respectively, gets

$$\sum_{i=1}^{N} \exp\left[-\lambda_0 - \sum_{j=1}^{m} \lambda_i g_i(x_i)\right] = 1 \qquad (3.8)$$

and

$$\sum_{i=1}^{N} g_j(x_i) \exp\left[-\lambda_0 - \sum_{j=1}^{m} \lambda_i g_i(x_i)\right] = C_j, \quad j = 1, 2, \ldots, m \qquad (3.9)$$

The Lagrange multipliers are determined from equations (3.8) and (3.9). Equation (3.8) can be written as

$$\exp(\lambda_0) = Z = \sum_{i=1}^{N} \exp\left[-\sum_{j=1}^{m} \lambda_j g_j(x_i)\right] \tag{3.10}$$

or

$$\lambda_0 = \ln Z = \ln \sum_{i=1}^{N} \exp\left[-\sum_{j=1}^{m} \lambda_j g_j(x_i)\right] \tag{3.11}$$

Equation (3.10) is referred to as the partition function. Equation (3.11), also referred to as potential function, expresses λ_0 as a function of $\lambda_1, \lambda_2, \ldots, \lambda_m$. Differentiating equation (3.11) with respect to λ_j one obtains

$$G = \frac{\partial \lambda_0}{\partial \lambda_j} = \frac{\partial Z}{Z \partial \lambda_j} = \frac{\partial \ln Z}{\partial \lambda_j} = \frac{\sum_{i=1}^{N} \exp\left[-\sum_{j=1}^{m} \lambda_i g_i(x_i)\right] \frac{\partial}{\partial \lambda_j}\left[-\sum_{j=1}^{m} \lambda_j g_j(x_j)\right]}{\sum_{i=1}^{N} \exp\left[-\sum_{j=1}^{m} \lambda_j g_j(x_i)\right]} = -C_j \tag{3.12}$$

Equation (3.9) can be expressed as

$$C_j \exp(\lambda_0) = \sum_{i=1}^{N} g_j(x_i) \exp\left[-\sum_{j=1}^{m} \lambda_j g_j(x_i)\right], \ j = 1, 2, \ldots, m \tag{3.13}$$

Combining equations (3.10) and (3.13), one obtains

$$C_j = \frac{\sum_{i=1}^{N} g_j(x_i) \exp\left[-\sum_{j=1}^{m} \lambda_j g_j(x_i)\right]}{\sum_{i=1}^{N} \exp\left[-\sum_{j=1}^{m} \lambda_j g_j(x_i)\right]}; \ j = 1, 2, \ldots, m \tag{3.14}$$

Equation (3.14) expresses $C_1, C_2, \ldots, C_i$ as functions of $\lambda_1, \ldots, \lambda_m$. Thus, the Lagrange multipliers can be expressed in terms of constraints (the given information) and the resulting distribution is rendered as a nonparametric distribution. Agmon et al. (1979) developed an algorithm for determining the Lagrange multipliers using equation (3.14). Clearly, the algorithm entails a numerical solution and two stages. First, it verifies if the constraints are linearly independent, and then it checks if a feasible solution exists. These days, computer software packages are available for solving equation (3.14) for specified constraints.

Substitution of equation (3.7) in equation (3.3) gives the value of H_{max}:

$$H_{max} = -\sum_{i=1}^{N} p_i \left[-\lambda_0 - \lambda_1 g_1(x_i) - \lambda_2 g_2(x_i) - \ldots - \lambda_m g_m(x_i)\right]$$

$$= \lambda_0 + \lambda_1 C_1 + \lambda_2 C_2 + \ldots + \lambda_m C_m \tag{3.15}$$

It may be instructive to apply the POME formalism to derive a discrete distribution for illustrative purposes.

Example 3.1: Suppose a discrete random variable X takes on N values, but from equation (3.4), $C_i = 0$, $i = 1, 2, \ldots$. What should be the probability distribution of X?

Solution: From equation (3.7) one obtains

$$p_i = \exp\left[-\lambda_0\right], \ i = 1, 2, \ldots, N \tag{3.16}$$

Using equation (3.10), one gets

$$\exp\left(\lambda_0\right) = \sum_{i=1}^{N} \exp\left[0\right] = N \tag{3.17}$$

Inserting equation (3.17) in equations (3.7), one obtains:

$$p_i = \frac{1}{N}, \ i = 1, 2, \ldots, N \tag{3.18}$$

This is a uniform distribution. This means that if one knows nothing about the random variable or its values then for a discrete number they are all equally likely. This is consistent with Laplace's principle of insufficient reason.

Example 3.2: Consider a dice throwing experiment. Suppose one bets 2 dollars on the occurrence of an even-numbered face upon throw and one wins, on average, 50 cents per game. What will be the probability of occurrence of each face upon throw?

Solution: Let p_i define the probability of the i-th face, $i = 1, 2, \ldots, 6$. The probability of winning 0.50 dollar is cents/200 cents $= 0.25$. An average gain of 50 cents means that p (even numbered face) $- p$ (odd numbered face) $= 0.25$. Therefore, the constraint in this case is $p_2 + p_4 + p_6 = 0.75$, $p_1 + p_2 + p_3 = 0.25$. Maximizing entropy subject to this constraint yields

$$p_2 = p_4 = p_6 = 0.25, \quad p_1 = p_3 = p_5 = 0.0833.$$

This concurs with the principle of insufficient reason as to the outcome of events (even) and odd separately. In this example, the outcome depends on whether the dice is fair or not and it is not fair.

Example 3.3: Suppose a discrete random variable X (e.g., rainfall occurrence) takes on n states, each of which is characterized by either 1 or 0. Let S_i represent a specific ordering of these states. These states occur in many different ways and $\{S_i\}$ constitute an exhaustive mutually exclusive set of states. Let r_i represent the number of occurrences of 1 in S_i. It is assumed that the average value of r, $R = \bar{r}$, is known. Let $q = R/n$. Derive the probability distribution of S_i and the corresponding probability distribution of r_i.

Solution: The number of S_i having the same value of r with rainfall occurrence as 1 and non-occurrence as 0 each day a week (7 days) can be given as the binomial coefficient:

$$\binom{n}{r} = \frac{n!}{r!\,(n-r)!} \tag{3.19}$$

The probability distribution of S_i then becomes

$$p(S_i) = \exp(-\lambda_0 - \lambda_1 S_i) \tag{3.20}$$

From equation (3.4) one obtains

$$R = qn = \bar{r} = \sum_{i=1}^{n} r_i p(S_i) = \sum_{r=0}^{n} \binom{n}{r} p(S_i) \tag{3.21}$$

Then, from equation (3.10),

$$\exp(\lambda_0) = \sum_{i=1}^{n} \exp[-\lambda_1 r_i] = \sum_{r=0}^{n} \binom{n}{r} [\exp(-\lambda_1)]^r \tag{3.22}$$

It can be shown that

$$\exp(\lambda_0) = [1 + \exp(-\lambda_1)]^n; \quad \log[\exp(\lambda_0)] = n \log[1 + \exp(-\lambda_1)] \tag{3.23}$$

Therefore, from equation (3.14)

$$nq = R = -n \frac{\exp(-\lambda_1)}{1 + \exp(-\lambda_1)}; \quad \exp(-\lambda_1) = -\frac{q}{1-q} \tag{3.24}$$

Hence,

$$\exp(\lambda_0) = (1-q)^n \tag{3.25}$$

The probability distribution of states S_i is expressed as

$$p(S_i) = \frac{1}{(1-q)^{-n}} \left(\frac{q}{1-q}\right)^{r_i} = q^{r_i}(1-q)^{n-r_i} \tag{3.26}$$

Because $\binom{n}{r}$ states have the same value of r, the probability distribution of r is

$$p(r) = \binom{n}{r} q^r (1-q)^{n-r} \tag{3.27}$$

This is the binomial distribution.

Example 3.4: Consider grading students for their performance in a class as outstanding, excellent, very good, good, fair, or unsatisfactory. Thus, grading is assigned numbers from 6 to 1, respectively, with 6 denoting outstanding and 1 denoting unsatisfactory, with the six grades numbered as 1, 2, 3, 4, 5, and 6 from unsatisfactory to outstanding. Let a random variable X define the occurrence of a grade of student performance. Thus, X takes on values of 1, 2, 3, 4, 5, or 6, with probabilities of p_1, p_2, p_3, p_4, p_5, or p_6, respectively. It is assumed that the average grade of students is between very good and excellent:

$$\sum_{i=1}^{6} x_i p_i = \bar{x} = 3.5 \tag{3.28}$$

Determine the POME-based probability distribution of X.

Solution: Equation (3.1) with $N = 6$ is true. Using the method of Lagrange multipliers, one writes

$$L = -\sum_{i=1}^{6} p_i \ln p_i - (\lambda_0 - 1)\left(\sum_{i=1}^{6} p_i - 1\right) - \sum_{j=1}^{1} \lambda_j \left(\sum_{i=1}^{6} x_i p_i - \bar{x}\right) \tag{3.29}$$

Differentiating L with respect to p_i, one obtains:

$$\frac{\partial L}{\partial p_i} = 0 \Rightarrow -(1 + \ln p_i) - (\lambda_0 - 1) - x_i \lambda_1 = 0, \ i = 1, 2, \dots, 6 \tag{3.30}$$

This leads to

$$p_i = \exp\left[-\lambda_0 - \lambda_1 x_i\right] = a b^{x_i}, \ a = \exp(-\lambda_0), \ b = \exp(-\lambda_1) \tag{3.31}$$

The probabilities are in geometric progression and the probability distribution is geometric.

Parameters a and b can be determined with the aid of equation (3.1) and the information $\bar{x}$ given by equation (3.26). Thus,

$$\sum_{i=1}^{6} p_i = \sum_{i=1}^{6} a b^{x_i} = a \sum_{i-1}^{6} b^{x_i} = 1 \tag{3.32}$$

$$\sum_{i=1}^{6} x_i p_i = \sum_{i=1}^{6} x_i a b^{x_i} = a \sum_{i=1}^{6} x_i b^{x_i} = \bar{x} \tag{3.33}$$

Combining these two equations, one gets

$$\sum_{i=1}^{6} x_i b^{x_i} = \bar{x} \sum_{i=1}^{6} b^{x_i} \tag{3.34a}$$

or

$$(1 - \bar{x}) b + (2 - \bar{x}) b^2 + (3 - \bar{x}) b^3 + (4 - \bar{x}) b^4 + (5 - \bar{x}) b^5 + (6 - \bar{x}) b^6 = 0 \tag{3.34b}$$

Here $\bar{x}$ is 3.5. Thus, coefficients of the first three terms are negative and those of the last three terms are positive. By Descartes' rule of signs, this has only one positive root and that is $b = 1$ ($\bar{x} = 3.5$), and $a = 1/6$. This means that $p_i = 1/6$ and the probability distribution is uniform.

One can generate different probability distributions for different values of $\bar{x}$. Probabilities increase if $\bar{x} > (N + 1)/2$ and decrease if $\bar{x} < (N + 1)/2$. Kapur and Kesavan (1992) have noted that the probability distribution becomes $p_6, p_5, p_4, p_3, p_2, p_1$ if the mean is $(7 - \bar{x})$, and it becomes $p_1, p_2, p_3, p_4, p_5, p_6$ when the mean is $\bar{x}$.

Example 3.5: Consider different values of $\bar{x}$. in Example 3.4. Then, show that when the mean is $(7 - \bar{x})$ the probability distribution becomes $p_6, p_5, p_4, p_3, p_2, p_1$ and it becomes $p_1, p_2, p_3, p_4, p_5, p_6$ when the mean is $\bar{x}$, as noted by Kapur and Kesavan (1992).

Solution: For different values of $\bar{x}$. In Example 3.4, one gets different values of probabilities as shown below:

$\bar{x}$	p_1	p_2	p_3	p_4	p_5	p_6
1.0	1.000	0.000	0.000	0.000	0.000	0.000
2.0	0.4781	0.2548	0.1357	0.0723	0.03856	0.0205
3.0	...					
4.0	...					
5.0	0.0205	0.0385	0.0723	0.1357	0.2548	0.4781
6.0	1.000	0.000	0.000	0.000	0.000	1.000

This table shows that the probability distribution is indeed reversed.

In the above formulation it is noted that λ_0 is a convex function of $\lambda_1, \lambda_2, \ldots, \lambda_m$. This can be shown by taking the second derivative of λ_0 with respect to $\lambda_1, \lambda_2, \ldots, \lambda_m$ and determining that they are positive at every point in the interval. In this case the Hessian matrix will be positive definite. In a similar manner one can show that H_{max} is a concave function of $C_1, C_2, \ldots, C_m$. To that end, one can determine the Hessian matrix of the second order partial derivative H_{max} with respect to $C_1, C_2, \ldots, C_m$ and show that it is negative definite. An interesting result occurs when taking the derivatives of H_{max} with respect to $C_1, C_2, \ldots, C_m$, which are found to be equivalent to Lagrange multipliers.

Example 3.6: The Lagrange multiplier λ_0 is a function of $\lambda_1, \lambda_2, \ldots, \lambda_m$. Show that the corresponding Hessian matrix can be expressed as variance-covariance matrix. By so doing one can show that λ_0 is a convex function of $\lambda_1, \lambda_2, \ldots, \lambda_m$. One may note that

$$\frac{\partial^2 \lambda_0}{\partial \lambda_i \partial \lambda_j} = \text{cov}\left[g_i(x) g_j(x)\right] \tag{3.35a}$$

$$\frac{\partial^2 \lambda_0}{\partial \lambda_i^2} = \text{cov}[g_i(x) g_i(x)] = Var[g_i(x)] = E\left\{[g_i(x)]^2\right\} - \left\{E[g_i(x)]\right\}^2 \tag{3.35b}$$

Solution: To answer the question in this example, reference is made to Section 3.2. Differentiation of equation (3.10) with respect to λ_i yields

$$\exp\left(\lambda_0\right) \frac{\partial \lambda_0}{\partial \lambda_j} = \sum_{i=1}^{N} g_j\left(x_i\right) \exp\left[-\sum_{j=1}^{m} \lambda_j g_j\left(x_i\right)\right] \tag{3.36}$$

Differentiation of equation (3.36) with respect to λ_j yields

$$\exp\left(\lambda_0\right) \frac{\partial^2 \lambda_0}{\partial \lambda_j^2} + \exp\left(\lambda_0\right) \left(\frac{\partial \lambda_0}{\partial \lambda_j}\right)^2 = \sum_{i=1}^{N} g_j^2\left(x_i\right) \exp\left[-\sum_{j=1}^{m} \lambda_j g_j\left(x_i\right)\right] \tag{3.37}$$

Equation (3.37) can be written with the use of equations (3.7) as

$$\frac{\partial^2 \lambda_0}{\partial \lambda_j^2} + \left(\frac{\partial \lambda_0}{\partial \lambda_j}\right)^2 = \sum_{i=1}^{N} p_i g_j^2 (x_i) = E[g_j^2 (X)] \tag{3.38}$$

Equation (3.38) leads to

$$\frac{\partial^2 \lambda_0}{\partial \lambda_j^2} = -\left(\frac{\partial \lambda_0}{\partial \lambda_j}\right)^2 + \sum_{i=1}^{N} p_i g_j^2 (x_i) = -\left(\frac{\partial \lambda_0}{\partial \lambda_j}\right)^2 + E\left[g_j^2 (X)\right] \tag{3.39}$$

With the use of equation (3.35b) for the definition of variance, equation (3.39) can be expressed as

$$\frac{\partial^2 \lambda_0}{\partial \lambda_j^2} = E\left[g_j^2 (X)\right] - \left(E\left[g_j (X)\right]\right)^2 = Var\left[g_j (X)\right] \tag{3.40}$$

Differentiation of equation (3.36) with respect to λ_r yields

$$\exp (\lambda_0) \frac{\partial^2 \lambda_0}{\partial \lambda_j \partial \lambda_r} + \exp (\lambda_0) \left(\frac{\partial \lambda_0}{\partial \lambda_j}\right) \left(\frac{\partial \lambda_0}{\partial \lambda_r}\right) = \sum_{i=1}^{N} g_j (x_i) g_r (x_i) \exp \left[-\sum_{j=1}^{m} \lambda_j g_j (x_i)\right] \tag{3.41}$$

Equation (3.41) can be written as

$$\frac{\partial^2 \lambda_0}{\partial \lambda_j \partial \lambda_r} + \left(\frac{\partial \lambda_0}{\partial \lambda_j}\right) \left(\frac{\partial \lambda_0}{\partial \lambda_r}\right) = \sum_{i=1}^{N} p_i g_j (x_i) g_r (x_i) = E[g_j (X) g_r (X)] \tag{3.42a}$$

Equation (3.42a) leads to

$$\frac{\partial^2 \lambda_0}{\partial \lambda_j \partial \lambda_r} = -\left(\frac{\partial \lambda_0}{\partial \lambda_j}\right) \left(\frac{\partial \lambda_0}{\partial \lambda_r}\right) + \sum_{i=1}^{N} p_i g_j (x_i) g_r (x_i) = -\left(\frac{\partial \lambda_0}{\partial \lambda_j}\right) \left(\frac{\partial \lambda_0}{\partial \lambda_r}\right) + E\left[g_j (X) g_r (X)\right]$$

$$\tag{3.42b}$$

Recalling equations (3.12), (3.35a) and (3.35b), one can express equation (3.42b) as

$$\frac{\partial^2 \lambda_0}{\partial \lambda_j \partial \lambda_r} = E\left[g_j (X) g_r (X)\right] - E\left[g_j (X)\right] E\left[g_r (X)\right] = Cov\left[g_j (X), g_r (X)\right] \tag{3.43}$$

Constructing the Hessian matrix of λ_0, one gets

$$\begin{bmatrix} \dfrac{\partial^2 \lambda_0}{\partial \lambda_1^2} & \dfrac{\partial^2 \lambda_0}{\partial \lambda_1 \partial \lambda_2} & \cdots & \dfrac{\partial^2 \lambda_0}{\partial \lambda_1 \partial \lambda_m} \\[2mm] \dfrac{\partial^2 \lambda_0}{\partial \lambda_2 \partial \lambda_1} & \dfrac{\partial^2 \lambda_0}{\partial \lambda_2^2} & \cdots & \dfrac{\partial^2 \lambda_0}{\partial \lambda_2 \partial \lambda_m} \\[2mm] \cdots & \cdots & \cdots & \cdots \\ \cdots & \cdots & \cdots & \cdots \\ \cdots & \cdots & \cdots & \cdots \\ \dfrac{\partial^2 \lambda_0}{\partial \lambda_m \partial \lambda_1} & \dfrac{\partial^2 \lambda_0}{\partial \lambda_m \partial \lambda_2} & \cdots & \dfrac{\partial^2 \lambda_0}{\partial \lambda_m^2} \end{bmatrix}$$

Using the definitions of the derivatives of λ_0, the Hessian matrix can be written in terms of variances and covariances as

$$
\begin{bmatrix}
Var[g_1(X)] & Cov[g_1(X),g_2(X)] & \cdots & Cov[g_1(X),g_m(X)] \\
Cov[g_2(X),g_1(X)] & Var[g_2(X)] & \cdots & Cov[g_2(X),g_m(X)] \\
\cdots & \cdots & & \cdots \\
\cdots & \cdots & & \cdots \\
\cdots & \cdots & & \cdots \\
Cov[g_m(X),g_1(X)] & Cov[g_m(X),g_2(X)] & \cdots & Var[g_m(X)]
\end{bmatrix}
$$

Since variances and covariances are always positive, the Hessian matrix must be positive definite. This means that λ_0 is a convex function of $\lambda_1, \lambda_2, \ldots, \lambda_m$.

3.3 POME formalism for continuous variables

As in the case of discrete variables, the usual procedure for determining probability distributions or estimating their parameters in the case of continuous variables is based on the maximization of entropy which is accomplished using the method of Lagrange multipliers which will be presented here. However, in very simple cases entropy can also be maximized somewhat intuitively. To that end, it may be useful to recall for two arbitrary probability density functions $f(x)$ and $g(x)$:

$$
-\int_{-\infty}^{\infty} g(x) \log g(x)\, dx \leq -\int_{-\infty}^{\infty} g(x) \log f(x)\, dx \tag{3.44}
$$

Example 3.7: Consider a coin tossing experiment. The probability of the occurrence of head p can be regarded as a random variable. Determine the probability density of p, $f(p)$.

Solution: The objective is to apply POME, that is, one maximizes the entropy:

$$
H(p) = -\int_0^1 f(p) \log f(p)\, dp
$$

Since no prior information is available, $f(p)$ must be uniform over the interval $[0, 1]$. Therefore, $f(p) = 1$, leading to $H(p) = 0$. For any other density $g(p)$,

$$
-\int_0^1 g(p)\ \log\ g(p)\, dp \leq -\int_0^1 g(p)\ \log f(p)\, dp = 0
$$

3.3.1 Entropy maximization using the method of Lagrange multipliers

For a continuous variable X, one writes the Shannon entropy as

$$
H(f) = H(X) = -\int_a^b f(x)\ \ln f(x)\ dx \tag{3.45}
$$

where

$$\int_a^b f(x)\, dx = 1 \tag{3.46}$$

where a and b are upper and lower limits of the integral or the range of variable X. In order to maximize $H(f)$ subject to equation (3.46) and m linearly independent constraints C_r $(r = 1, 2, \ldots, m)$:

$$C_r = \int_a^b g_r(x)\, f(x)\, dx, \; r = 1, 2, \ldots, m \tag{3.47}$$

where $g_r(x)$ are some functions over an interval (a, b).

For maximization one employs the method of Lagrange multipliers and constructs the Lagrangean function L as:

$$L = -\int_a^b f(x)\, \ln f(x)\, dx - (\lambda_0 - 1)\left[\int_a^b f(x)\, dx - 1\right] - \sum_{r=1}^m \lambda_r\left[\int_a^b f(x)\, g_r(x)\, dx - C_r\right] \tag{3.48}$$

where $\lambda_1, \lambda_2, \ldots, \lambda_m$ are Lagrange multipliers.

In order to obtain $f(x)$ which maximizes or minimizes L, recall the Euler-Lagrange calculus of variations. Let there be a function G which is a function of random variable X and its probability density function (PDF) as well as the first derivative of the PDF. If the integral of G is written as

$$I = \int_a^b G\left(x, f(x), f'(x)\right) dx \tag{3.49}$$

then $f(x)$ can be obtained by differentiating G and equating the derivative to zero as

$$\frac{\partial G}{\partial f(x)} - \frac{d}{dx}\left(\frac{\partial G}{\partial f'(x)}\right) = 0 \tag{3.50}$$

The PDF, $f(x)$, so obtained maximizes or minimizes I. In our case L in equation (3.48) does not involve $f'(x)$ and is therefore a function of $f(x)$ alone. Therefore, one differentiates L with respect to $f(x)$ only and equates the derivative to zero:

$$\frac{\partial L}{\partial f} = 0 \Rightarrow -[1 + \ln f(x)] - (\lambda_0 - 1) - \sum_{r=1}^m \lambda_r g_r(x) = 0 \tag{3.51}$$

This yields

$$f(x) = \exp\left[-\lambda_0 - \lambda_1 g_1(x) - \ldots - \lambda_m g_m(x)\right] \tag{3.52}$$

Equation (3.52) is the POME-based probability distribution containing the Lagrange multipliers as parameters. Equation (3.52) is also written as

$$f(x) = \frac{1}{Z(\lambda_1, \lambda_2, \ldots, \lambda_m)} \exp\left[-\sum_{r=1}^{m} \lambda_r g_r(x)\right] \tag{3.53}$$

where Z is called the partition function. The Lagrange multipliers $\lambda_1, \lambda_2, \ldots \lambda_m$ can be determined by inserting equation (3.52) in equations (3.46) and (3.47):

$$\exp(\lambda_0) = \int_a^b \exp\left[-\sum_{r=1}^{m} \lambda_r g_r(x)\right] dx, \quad r = 1, 2, \ldots, m \tag{3.54}$$

and

$$C_r \exp(\lambda_0) = \int_a^b g_r(x) \exp\left[-\sum_{r=1}^{m} \lambda_r g_r(x)\right] dx, \quad r = 1, 2, \ldots, m \tag{3.55}$$

Equation (3.54) expresses the partition function Z as

$$Z(\lambda_1, \lambda_2, \ldots, \lambda_m) = \int_a^b \exp\left[-\sum_{r=1}^{m} \lambda_r g_r(x)\right] dx \tag{3.56}$$

Combining equation (3.54) and (3.55), one gets

$$C_r = \frac{\int_a^b g_r(x) \exp\left[-\sum_{r=1}^{m} \lambda_r g_r(x)\right] dx}{\int_a^b \exp\left[-\sum_{r=1}^{m} \lambda_r g_r(x)\right] dx}, \quad r = 1, 2, \ldots, m \tag{3.57}$$

Equation (3.56) expresses constraints as functions of the Lagrange multipliers or

$$\frac{\partial \log\left[Z(\lambda_1, \lambda_2, \ldots, \lambda_r)\right]}{\partial \lambda_r} = -C_r, \quad r = 1, 2, \ldots, m \tag{3.58}$$

Thus, the Lagrange multipliers can be determined from the given constraints.

Example 3.8: Assume that no constraints are given or no knowledge is available a priori for a random variable. Derive the probability distribution of this variable using POME.

Solution: In the case that all that is known is that equation (3.46) must be satisfied. Thus, entropy given by equation (3.45) is maximized subject to equation (3.46). Then, equation (3.53) becomes

$$f(x) = \frac{1}{Z(\lambda)} \exp\left[-\sum_{r=1}^{0} \lambda_r g_r(x)\right] \tag{3.59}$$

The expression inside the exponential is zero, since there are no constraints. Therefore,

$$f(x) = \frac{1}{Z(\lambda)} \tag{3.60}$$

Equation (3.56) yields

$$Z(\lambda) = \int_a^b dx = b - a \tag{3.61}$$

Therefore,

$$f(x) = \frac{1}{b-a} \tag{3.62}$$

which is a uniform probability density function on the interval $<a, b>$. According to the principle of insufficient reason, if nothing is known then the maximum noncommittal distribution is the uniform distribution as given by POME. The entropy of the uniform distribution is:

$$H(X) = -\int_a^b \frac{1}{b-a} \log\left[\frac{1}{b-a}\right] dx = \log(b-a) \tag{3.63}$$

and is measured in bits if the logarithm is with respect to the base of 2 or in Napier if the base is e.

It may be noted that if the limits of integration are extended as $a \to -\infty$ and $b \to \infty$, or $a \to 0$ and $b \to \infty$, POME would hold and the algebraic analysis would hold. When a random variable varies within these infinite limits, it cannot have a rectangular distribution if the only constraint is the total probability law.

Example 3.9: Assume that no constraints are given or no knowledge is available a priori for a random variable. Derive the probability distribution of this variable using POME if the limits of the variable are $[-\infty, \infty]$.

Solution: In the case that all that is known is that equation (3.46) must be satisfied. Thus, entropy given by equation (3.45) is maximized subject to equation (3.46). Then, equation (3.53) becomes

$$f(x) = \frac{1}{Z(\lambda)} \exp\left[-\sum_{r=1}^{0} \lambda_r g_r(x)\right] \tag{3.64}$$

The expression inside the exponential is zero, since there are no constraints. Therefore,

$$f(x) = \frac{1}{Z(\lambda)} \tag{3.65}$$

Equation (3.56) yields

$$Z(\lambda) = \int_{-\infty}^{\infty} dx = \infty + \infty \tag{3.66}$$

which is indeterminate. Therefore, in this situation, the probability distribution $f(x)$ cannot be determined.

Table 3.1 Stream flow (unit: cfs) of Brazos River at Richmond, Texas, for May from 1979 to 2007 [Note: – means the data is missing].

Year	Flow (cfs)	Year	Flow (cfs)	Year	Flow (cfs)
1979	26,330	1989	15,229	1999	3,752
1980	15,520	1990	39,470	2000	3,212
1981	–	1991	13,940	2001	7,968
1982	–	1992	23,060	2002	1,462
1983	–	1993	19,780	2003	1,883
1984	2,002	1994	11,050	2004	23,110
1985	7,581	1995	15,340	2005	–
1986	10,770	1996	818.60	2006	3,332
1987	4,994	1997	21,030	2007	29,800
1988	1,206	1998	2,771		

Example 3.10: Consider monthly (May) stream flow data for Brazos River at Richmond, Texas (gaging station USGS 08114000), as given in Table 3.1. Compute the mean and variance of the May-month stream flow data. Then, compute entropy for rectangular, exponential, and log-normal distributions that one may wish to fit the data. For the rectangular distribution, entropy is given as

$$H(X) = \ln(a - b) \tag{3.67}$$

where x is the value of stream flow, and $a < x > b$. For the exponential distribution, entropy is:

$$H(X) = \ln(e\bar{x}) \tag{3.68}$$

where e is the base of the logarithm. For the normal distribution, entropy is given as

$$H(Y) = \ln\left[S_y (2\pi e)^{0.5}\right] \tag{3.69a}$$

where S_y is the standard deviation of Y. Assume $X = \exp(Y)$, where $x \in (0, \infty)$, then Y has the log-normal distribution. The entropy for the log-normal distribution is:

$$H(X) = H(Y) + \bar{y} \tag{3.69b}$$

Show if entropy decreases as the number of constraints increases.

Solution: The mean, variance, and standard deviation of the given data are computed as

$$\bar{x} = \frac{1}{N}\sum_{i=1}^{N} x_i = 2216 \text{cfs}$$

$$Var(x) = E(X^2) - [E(X)]^2 = \frac{1}{N}\sum_{i=1}^{N}(x_i - \bar{x})^2 = 1.0595e + 008 (\text{cfs})^2$$

$$S_x = \sqrt{\frac{1}{N-1}\sum_{i=1}^{N}(x_i - \bar{x})^2} = 1050.60 \text{ cfs}$$

Now entropy is computed. For the rectangular distribution, $H(X) = \ln(a - b)$ in which $a = 39470$ cfs and $b = 818.6$ cfs then $H(X) = \ln(39470 - 818.6) = 10.56$ Napiers For the exponential distribution, $H(X) = \ln(e\bar{x}) = \ln(12216e) = 10.41$ Napiers For the normal distribution, $H(X) = \ln[S_x (2\pi e)^{0.5}] = \ln[1050.6 (2\pi e)^{0.5}] = 8.38$ Napiers. From these results it is seen that entropy decreases as the number of constraints increases. For the same set of data, the entropy for the rectangular distribution is the largest and the one for normal distribution is the smallest among the three distributions.

3.3.2 Direct method for entropy maximization

The entropy is maximized directly with respect to the specified constraints, that is, by differentiating entropy with respect to specified constraints. To illustrate the method, an example is considered.

Example 3.11: Consider a random variable $X \in (-\pi, \pi)$ Determine its probability density function $f(x)$ subject to the condition that the coefficients of its Fourier series expansion c_j are known for $|j| \le J$:

$$c_j = \frac{1}{2\pi} \int_{-\pi}^{\pi} f(x) \exp(ijx)\, dx, \quad |j| \le J$$

Solution: The PDF $f(x)$ can be expressed as a Fourier series:

$$f(x) = \sum_{j=-\infty}^{\infty} c_i \exp(-ijx), \quad -\pi \le x \le \pi$$

Entropy given as

$$H(X) = -\int_{-\pi}^{\pi} f(x)\, \log f(x)\, dx$$

is maximized subject to the specified constraints. Since H depends on c_j, its maximum is obtained as

$$\frac{\partial H}{\partial c_j} = \frac{\partial H}{\partial f}\frac{\partial f}{\partial c_j} = -\int_{-\pi}^{\pi}\big[\log f(x) + 1\big]\exp(ijx)\, dx = 0, \quad |j| > J$$

The term $\log f(x) + 1$ can be written in the interval $(-\pi, \pi)$ as:

$$\log f(x) + 1 = \sum_{j=-J}^{k=J} \gamma_k \exp(jkx)$$

and coefficients γ_k of the Fourier series expansion of $[\log f(x) + 1]$ should be zero for $|j| > J$. Thus,

$$f(x) = \exp\left[-1 + \sum_{k=-J}^{k=J} \gamma_k \exp(kjx)\right], \quad -\pi \le x \le \pi$$

Substitution of this equation into the constraint equation would lead to a system of nonlinear equations whose solution would yield parameters γ_k.

3.4 POME formalism for two variables

The POME formalism discussed for univariate cases in the preceding sections can now be extended to bivariate cases where one can maximize either joint entropy, conditional entropy, or transinformation (i.e., mutual entropy or joint information). Consider the case of two random variables X and Y for which one can write these three types of entropies, respectively, as:

$$H(X, Y) = -\int_0^\infty \int_0^\infty f(x, y) \ln f(x, y) \, dx \, dy \tag{3.70}$$

$$H(Y|X) = -\int \int f(x, y) \ln f(y|x) dx \, dy = -\int_0^\infty \int_0^\infty f(x, y) \ln \frac{f(x, y)}{f(x)} dx \, dy \tag{3.71}$$

$$T(X, Y) = H(X) + H(Y) - H(X, Y) = \int_0^\infty \int_0^\infty f(x, y) \ln \frac{f(x, y)}{f(x) \, g(y)} dx \, dy \tag{3.72}$$

First, the objective is to maximize $H(X, Y)$ given by equation (3.70). To that end, the following constraints are defined:

$$\int_0^\infty \int_0^\infty f(x, y) \, dy \, dx = 1 \tag{3.73}$$

$$\int_0^\infty \int_0^\infty x f(x, y) \, dy \, dx = \bar{x} \tag{3.74}$$

$$\int_0^\infty \int_0^\infty y f(x, y) \, dy \, dx = \bar{y} \tag{3.75}$$

$$\int_0^\infty \int_0^\infty xy f(x, y) \, dy \, dx = \overline{xy} = \sigma_{xy} \tag{3.76}$$

Using the method of Lagrange multipliers, one gets

$$f(x, y) = \exp\left(-\lambda_0 - \lambda_1 x - \lambda_2 y - \lambda_3 xy\right) \tag{3.77}$$

Inserting equation (3.77) in equation (3.73), one gets

$$\int_0^\infty \int_0^\infty f(x,y)\,dy\,dx = 1 = \exp(-\lambda_0) \int_0^\infty \int_0^\infty \exp(-\lambda_1 x - \lambda_2 y - \lambda_3 xy)\,dx\,dy$$

$$= \exp(-\lambda_0) \int_0^\infty \frac{\exp(-\lambda_1 x)}{\lambda_2 + \lambda_3 x}\,dx \qquad (3.78a)$$

One can also write

$$\int_0^\infty \int_0^\infty f(x,y)\,dy\,dx = 1 = \exp(-\lambda_0) \int_0^\infty \frac{\exp(-\lambda_2 y)}{\lambda_1 + \lambda_3 y}\,dy \qquad (3.78b)$$

Substituting equation (3.77) in equation (3.74), one obtains

$$\int_0^\infty \int_0^\infty x f(x,y)\,dy\,dx = \bar{x} = \exp(-\lambda_0) \int_0^\infty \int_0^\infty x \exp(-\lambda_1 x - \lambda_2 y - \lambda_3 xy)\,dx\,dy$$

$$= \exp(-\lambda_0) \left\{ \frac{1}{\lambda_3} \int_0^\infty \exp(-\lambda_1 x)\,dx - \frac{\lambda_2}{\lambda_3} \int_0^\infty \frac{\exp(-\lambda_1 x)}{\lambda_2 + \lambda_3 x}\,dx \right\}$$

$$= \frac{\exp(-\lambda_0)}{\lambda_3 \lambda_1} - \frac{\lambda_2}{\lambda_3} \qquad (3.79a)$$

Equation (3.79a) can be simplified as follows:

$$\int_0^\infty \int_0^\infty x f(x,y)\,dy\,dx = \bar{x} = \exp(-\lambda_0) \left[\int_0^\infty x \frac{\exp(-\lambda_1 x)}{\lambda_2 + \lambda_3 x}\,dx \right]$$

$$= \exp(-\lambda_0) \frac{1}{\lambda_3} \int_0^\infty \left(\frac{\lambda_2 + \lambda_3 x - \lambda_2}{\lambda_2 + \lambda_3 x} \right) \exp(-\lambda_1 x)\,dx$$

$$= \frac{\exp(-\lambda_0)}{\lambda_3} \left[\int_0^\infty \exp(-\lambda_1 x)\,dx - \lambda_2 \int_0^\infty \frac{\exp(-\lambda_1 x)}{\lambda_2 + \lambda_3 x}\,dx \right] \qquad (3.79b)$$

Taking advantage of equation (3.78b), equation (3.79b) can be simplified as

$$\int_0^\infty \int_0^\infty x f(x,y)\,dy\,dx = \bar{x} = \frac{\exp(-\lambda_0)}{\lambda_3} \int_0^\infty \exp(-\lambda_1 x)\,dx - \frac{\lambda_2}{\lambda_3} \left[\exp(-\lambda_0) \int_0^\infty \frac{\exp(-\lambda_1 x)}{\lambda_2 + \lambda_3 x}\,dx \right]$$

$$= \frac{\exp(-\lambda_0)}{\lambda_1 \lambda_3} - \frac{\lambda_2}{\lambda_3} \qquad (3.79c)$$

Substituting equation (3.77) in equation (3.75), one gets

$$\int_0^\infty \int_0^\infty y f(x,y)\,dy\,dx = \bar{y} = \exp(-\lambda_0) \int_0^\infty y \int_0^\infty \exp(-\lambda_1 x - \lambda_2 y - \lambda_3 xy)\,dx\,dy$$

$$= \exp(-\lambda_0) \int_0^\infty y \frac{\exp(-\lambda_2 y)}{(\lambda_1 + \lambda_3 y)}\,dy \qquad (3.80a)$$

Equation (3.80a) can be simplified as follows:

$$\int_0^\infty \int_0^\infty y f(x, y) \, dy \, dx = \bar{y} = \exp(-\lambda_0) \int_0^\infty \frac{(\lambda_1 + \lambda_3 y - \lambda_1) \exp(-\lambda_2 y)}{\lambda_3 (\lambda_1 + \lambda_3 y)} \, dy$$

$$= \frac{\exp(-\lambda_0)}{\lambda_3} \int_0^\infty \exp(-\lambda_2 y) \, dy - \frac{\lambda_1}{\lambda_3} \left[\exp(-\lambda_0) \int_0^\infty \frac{\exp(-\lambda_2 y)}{\lambda_1 + \lambda_3 y} \, dy \right]$$

(3.80b)

Equation (3.80b), with the use of equation (3.78b), simplifies to

$$\int_0^\infty \int_0^\infty y f(x, y) \, dy \, dx = \bar{y} = \frac{\exp(-\lambda_0)}{\lambda_2 \lambda_3} - \frac{\lambda_1}{\lambda_3}$$

(3.80c)

Substituting equation (3.77) in equation (3.76), one obtains

$$\int_0^\infty \int_0^\infty x y f(x, y) \, dy \, dx = \overline{xy} = \exp(-\lambda_0) \int_0^\infty x \exp(-\lambda_0 - \lambda_1 x) \, dx \int_0^\infty y \exp(-\lambda_2 y - \lambda_3 x y) \, dy$$

$$= \exp(-\lambda_0) \int_0^\infty \frac{x \exp(-\lambda_1 x)}{(\lambda_2 + \lambda_3 x)^2} \, dx$$

(3.81)

The marginal probability density function of X is obtained from equation (3.77) as

$$\int_0^\infty f(x, y) \, dy = f(x) = \frac{\exp(-\lambda_0 - \lambda_1 x)}{(\lambda_2 + \lambda_3 x)}$$

(3.82)

Similarly, the PDF of Y is obtained as

$$\int_0^\infty f(x, y) \, dx = g(y) = \frac{\exp(-\lambda_0 - \lambda_2 y)}{(\lambda_1 + \lambda_3 y)}$$

(3.83)

For conditional entropy $H(Y|X)$ one requires

$$\frac{f(x, y)}{f(x)} = \frac{\exp(-\lambda_0 - \lambda_1 x - \lambda_2 y - \lambda_3 xy)}{\exp(-\lambda_0 - \lambda_1 x)(\lambda_2 + \lambda_3 x)^{-1}} = (\lambda_2 + \lambda_3 x) \exp(-\lambda_2 y - \lambda_3 xy)$$

(3.84)

$$H(Y|X) = -\int_0^\infty \int_0^\infty f(x, y) \left[\ln(\lambda_2 + \lambda_3 x) - (\lambda_2 y + \lambda_3 xy) \right] dx \, dy$$

(3.85)

Equation (3.85) can be written as

$$H(Y|X) = -\int_0^\infty \int_0^\infty \exp(-\lambda_0 - \lambda_1 x - \lambda_2 y - \lambda_3 xy) \left[\ln(\lambda_2 + \lambda_3 x) - (\lambda_2 y + \lambda_3 xy) \right] dx \, dy$$

(3.86)

Equation (3.86) can be used to maximize the conditional entropy.

For transinformation, one requires

$$\frac{f(x, y)}{f(x) \, g(y)} = \frac{\exp\left(-\lambda_0 - \lambda_1 x - \lambda_2 y - \lambda_3 xy\right)}{\exp\left(-2\lambda_0 - \lambda_1 x - \lambda_2 y\right)\left[(\lambda_2 + \lambda_3 x)(\lambda_1 + \lambda_3 y)\right]^{-1}} \tag{3.87}$$

Transinformation can now be expressed as

$$T(X, Y) = \int_0^\infty \int_0^\infty f(x, y)\left[\ln(\lambda_1 + \lambda_3 y) + \ln(\lambda_2 + \lambda_3 x) + \lambda_0 - \lambda_3 xy\right] dx \, dy \tag{3.88}$$

Equation (3.88) can be expressed as

$$T(X, Y) = \int_0^\infty \frac{\ln(\lambda_1 + \lambda_3 y)}{(\lambda_1 + \lambda_3 y)} \exp\left(-\lambda_0 - \lambda_2 y\right) dy + \int_0^\infty \frac{\ln(\lambda_2 + \lambda_3 x)}{(\lambda_2 + \lambda_3 x)} \exp\left(-\lambda_0 - \lambda_1 x\right) dx$$

$$+ \lambda_0 - \lambda_3 \int_0^\infty \frac{x \exp\left(-\lambda_0 - \lambda_1 x\right)}{(\lambda_2 + \lambda_3 x)^2} dx \tag{3.89}$$

Equation (3.89) gives transinformation where the Lagrange multipliers are determined using constraint equations (3.73)–(3.76). One can then minimize the transinformation.

In the second case, the entropy given by equation (3.70) is maximized subject to the constraints given by equations (3.73) and (3.76) as well as the following:

$$\int_{-\infty}^\infty \int_{-\infty}^\infty x^2 f(x, y) \, dy \, dx = \overline{x^2} = \sigma_x^2 \tag{3.90}$$

$$\int_{-\infty}^\infty \int_{-\infty}^\infty y^2 f(x, y) \, dy \, dx = \overline{y^2} = \sigma_y^2 \tag{3.91}$$

Using the method of Lagrange multipliers, the entropy-based probability density function will be:

$$f(x, y) = \exp\left(-\lambda_0 - \lambda_1 x^2 - \lambda_2 y^2 - \lambda_3 xy\right) \tag{3.92}$$

Using equation (3.92) in equation (3.73), one gets

$$\exp(-\lambda_0) \int_{-\infty}^\infty \exp(-\lambda_1 x^2) dx \int_{-\infty}^\infty \exp\left[-\lambda_2\left(y^2 + \frac{\lambda_3}{\lambda_2 xy}\right)\right] dy$$

$$= \exp(-\lambda_0) \int_{-\infty}^\infty \exp\left[-\left(\lambda_1 - \frac{\lambda_3^2}{4\lambda_2}\right)\right] x^2 \frac{\sqrt{\pi}}{\sqrt{\lambda_2}} = 1 \tag{3.93}$$

This can be simplified as

$$\exp\left(-\lambda_0\right) \frac{\pi}{\sqrt{\lambda_2}} \frac{1}{\sqrt{\lambda_1 - \frac{\lambda_3^2}{4\lambda_2}}} = 1 = \exp\left(-\lambda_0\right) \frac{2\pi}{\sqrt{4\lambda_1 \lambda_2 - \lambda_3^2}} \tag{3.94}$$

This yields

$$\exp\left(\lambda_0\right) = \frac{2\pi}{\sqrt{4\lambda_1\lambda_2 - \lambda_3^2}} \tag{3.95}$$

Inserting equation (3.92) in equation (3.90), one obtains

$$\overline{x^2} = \exp\left(-\lambda_0\right) \frac{\sqrt{\pi}}{2\sqrt{\lambda_2}} \int_{-\infty}^{\infty} x^2 \exp\left[-\left(\lambda_1 - \frac{\lambda_3^2}{4\lambda_2}\right)x^2\right] dx$$

$$= \exp\left(-\lambda_0\right) \frac{\sqrt{\pi}}{2\sqrt{\lambda_2}} \frac{\sqrt{\pi}}{\left(\lambda_1 - \frac{\lambda_3^2}{4\lambda_2}\right)^{3/2}} = \frac{2\lambda_2}{\left(4\lambda_1\lambda_2 - \lambda_3^2\right)} = \sigma_x^2 \tag{3.96}$$

Similarly, substitution of equation (3.92) in equation (3.91) yields

$$\overline{y^2} = \sigma_y^2 = \frac{2\lambda_1}{\left(4\lambda_1\lambda_2 - \lambda_3^2\right)} \tag{3.97}$$

Substitution of equation (3.92) in equation (3.76) leads to

$$\sigma_{xy} = \overline{xy} = \exp\left(-\lambda_0\right) \int_{-\infty}^{\infty} x \exp\left(-\lambda_1 x^2\right) dx \int_{-\infty}^{\infty} \exp\left\{-\lambda_2\left[\left(y + \frac{\lambda_3}{2\lambda_2}x\right)^2 - \frac{\lambda_3^2}{4\lambda_2^2}x^2\right]\right\} dy \tag{3.98}$$

Taking $Y = y + \left(0.5\lambda_3/\lambda_2\right)x$, equation (3.98) can be simplified as

$$\sigma_{xy} = \exp\left(-\lambda_0\right) \left(-\frac{\lambda_3}{2\lambda_2}\right) \frac{\sqrt{\pi}}{\sqrt{\lambda_2}} \int_{-\infty}^{\infty} x^2 \exp[-\left(\lambda_1 - \frac{\lambda_3^2}{4\lambda_2}\right)x^2 dx = -\frac{\lambda_3}{\left(4\lambda_1\lambda_2 - \lambda_3^2\right)} \tag{3.99}$$

Thus, the joint probability distribution can be written as

$$f\left(x, y\right) = \frac{\sqrt{4\lambda_1\lambda_2 - \lambda_3^2}}{2\pi} \exp\left[-\left(\frac{4\lambda_1\lambda_2 - \lambda_3^2}{2}\right)\left(\sigma_y^2 x^2 + \sigma_x^2 y^2 - 2\sigma_{xy}xy\right)\right] \tag{3.100}$$

Note that

$$\sigma_x^2\sigma_y^2 - \sigma_{xy}^2 = \frac{1}{\left(4\lambda_1\lambda_2 - \lambda_3^2\right)} \tag{3.101}$$

Equation (3.100) becomes

$$f\left(x, y\right) = \frac{1}{2\pi\sqrt{\sigma_x^2\sigma_y^2 - \sigma_{xy}^2}} \exp[-\frac{\left(\sigma_y^2 x^2 + \sigma_x^2 y^2 - 2\sigma_{xy}xy\right)}{2\left(\sigma_x^2\sigma_y^2 - \sigma_{xy}^2\right)} \tag{3.102}$$

From equation (3.102) one can obtain

$$
f(x) = \int_{-\infty}^{\infty} f(x, y)\, dy = \frac{1}{2\pi\sqrt{\sigma_x^2\sigma_y^2 - \sigma_{xy}^2}} \exp\left[-\frac{\sigma_y^2 x^2}{2\left(\sigma_x^2\sigma_y^2 - \sigma_{xy}^2\right)}\right]
$$

$$
\times \int_{-\infty}^{\infty} \exp\left[-\frac{\sigma_x^2}{2\left(\sigma_x^2\sigma_y^2 - \sigma_{xy}^2\right)}\left(y^2 - \frac{2\sigma_{xy}}{\sigma_x^2}xy\right)\right] dy \tag{3.103}
$$

Equation (3.103) simplifies to

$$
f(x) = \frac{1}{\sqrt{2\pi}\,\sigma_x}\exp\left(-\frac{x^2}{2\sigma_x^2}\right) \tag{3.104}
$$

The conditional probability distribution can now be written as

$$
f(y|x) = \frac{f(x, y)}{f(x)} = \frac{1}{2\pi\sqrt{\sigma_x^2\sigma_y^2 - \sigma_{xy}^2}} \frac{\exp\left\{\left[-(4\lambda_1\lambda_2 - \lambda_3^2)/2\right]\left[\sigma_y^2 x^2 + \sigma_x^2 y^2 - 2\sigma_{xy}xy\right]\right\}}{\exp\left[-\frac{x^2}{2\sigma_x^2}\right]}\sqrt{2\pi}\,\sigma_x \tag{3.105}
$$

Equation (3.105) simplifies to

$$
f(y|x) = \frac{\sigma_x}{\sqrt{2\pi}\,(\sigma_x^2\sigma_y^2 - \sigma_{xy}^2)^{1/2}}\exp\left\{-\frac{\frac{\sigma_{xy}^2}{\sigma_x^2}x^2 + \sigma_x^2 y^2 - 2\sigma_{xy}xy}{2(\sigma_x^2\sigma_y^2 - \sigma_{xy}^2)}\right\} \tag{3.106}
$$

One can now write the joint entropy as well as marginal entropies.

One can also maximize $H(X, Y)$ subject to the constraints given by equations (3.90), (3.91) and (3.76), where $f(x)$ is also determined if only $\overline{x^2}$ is specified. In this case the joint probability distribution is given by equation (3.102), and $f(x)$ is given by equation (3.104). Thus, the same answer will be obtained in this case because of self-consistency conditions.

$$
H(X, Y) = \ln[2\pi\sqrt{\sigma_x^2\sigma_y^2 - \sigma_{xy}^2} + 1 \tag{3.107}
$$

The marginal entropies are

$$
H(X) = \ln\left(\sqrt{2\pi}\,\sigma_x\right) + \frac{1}{2} \tag{3.108}
$$

$$
H(Y) = \ln\left(\sqrt{2\pi}\,\sigma_y\right) + \frac{1}{2} \tag{3.109}
$$

Thus,

$$
T(X, Y) = -H(X, Y) + H(X) + H(Y) = \ln\left[\frac{\sigma_x\sigma_y}{\sqrt{\sigma_x^2\sigma_y^2 - \sigma_{xy}^2}}\right] = \frac{1}{2}\ln\left[\frac{\sigma_x^2\sigma_y^2}{\sigma_x^2\sigma_y^2 - \sigma_{xy}^2}\right] \tag{3.110}
$$

Example 3.12: Consider yearly rainfall and runoff for Brazos River for a number of years (equal to or greater than 30 years). Compute, mean, variance, standard deviation, and coefficient of variation of rainfall and runoff, and covariance between them. Then compute $f(x), f(y), f(x, y), T(X, Y), H(X), H(Y)$, and $H(X, Y)$

Solution: Stream flow data for Brazos River is obtained for a gaging station located near Bryan (USGS 08109000) from 1941–93 from the USGS website. Rainfall data for the Brazos County from 1941–93 is obtained from the Texas Water Development Board website [http://www.twdb.state.tx.us/]

$$\bar{x} = 39.5, \bar{y} = 5170.3, S_x = 76.3, S_y = 15741278, \sigma_x = 8.7, \sigma_y = 3967.5$$

$$C_{vx} = \frac{\sigma_x}{\bar{x}} = 0.22$$

$$C_{vy} = \frac{\sigma_y}{\bar{y}} = 0.77$$

$$Cov(x, y) = \sigma_{xy} = E((x - \bar{x})(y - \bar{y})) = 17709.9$$

$$f(x) = \frac{1}{\sqrt{2\pi}\sigma_x} \exp\left(-\frac{(x - \bar{x})^2}{2\sigma_x^2}\right) = \frac{1}{8.7\sqrt{2\pi}} \exp\left(-\frac{(x - 39.5)^2}{152.6}\right)$$

$$f(y) = \frac{1}{\sqrt{2\pi}\sigma_y} \exp\left(-\frac{(y - \bar{y})^2}{2\sigma_y^2}\right) = \frac{1}{3967.5\sqrt{2\pi}} \exp\left(-\frac{(y - 5170.3)^2}{31482556}\right)$$

$$f(x, y) = \frac{1}{2\pi\sqrt{\sigma_x^2\sigma_y^2 - \sigma_{xy}^2}} \exp\left[-\frac{\left(\sigma_y^2 x^2 + \sigma_x^2 y^2 - 2\sigma_{xy} xy\right)}{2\left(\sigma_x^2\sigma_y^2 - \sigma_{xy}^2\right)}\right]$$

$$= 5.34269 \times 10^{-6} \exp\left[-\frac{\left(3967.5x^2 + 8.7y^2 - 35419.9xy\right)}{1774803407}\right]$$

$$H(X) = \ln\left(\sqrt{2\pi}\sigma_x\right) + \frac{1}{2} = 3.59 \text{ Napier}$$

$$H(Y) = \ln\left(\sqrt{2\pi}\sigma_y\right) + \frac{1}{2} = 9.70 \text{ Napier}$$

$$T(X, Y) = \frac{1}{2}\ln\left[\frac{\sigma_x^2\sigma_y^2}{\sigma_x^2\sigma_y^2 - \sigma_{xy}^2}\right] = 0.15 \text{ Napier}$$

3.5 **Effect of constraints on entropy**

Let there be m constraints imposed on $f(x)$ and defined as

$$C_r = \int_a^b A_r(x) f(x)\, dx, \quad r = 1, 2, \ldots, m \tag{3.111}$$

in which $A_r(x)$ is known function of x for each $r = 1, 2, \ldots m$. Thus, C_r is the average of $A_r(x)$
The entropy-based PDF is

$$f(x) = \exp\left[-\lambda_0 - \sum_{r=1}^m \lambda_r A_r(x)\right] \tag{3.112}$$

where $\lambda_r, r = 1, 2, \ldots, m$, are Lagrange multipliers, determined from known constraints. The maximum entropy is then given as

$$H_m(f) = -\int_a^b f(x)\ln[f(x)]dx = -\lambda_0 - \sum_{r=1}^N \lambda_r C_r \tag{3.113}$$

Now suppose that $q(x)$ is another probability distribution obtained by satisfying n constraints, The maximum entropy is expressed as:

$$H_n(q) = -\int_a^b q(x)\ln q(x)\, dx \tag{3.114}$$

Note that n includes the previous m constraints plus some more. Then

$$H_n(q) \leq H_m(f), \quad n \geq m \tag{3.115}$$

The question is if equation (3.115) holds. To answer the question consider cross entropy,

$$H(q|f) = \int_a^b q(x)\ln\left[\frac{q(x)}{f(x)}\right]dx = \int_a^b q(x)\ln q(x)dx - \int_a^b q(x)\ln f(x)dx \tag{3.116}$$

It is known that $H(q|f) \geq 0$. Recalling Jensen's inequality

$$x - 1 \geq \ln x \geq 1 - \frac{1}{x} \quad \Rightarrow \quad x^2 - x \geq x\ln x \geq x - 1 \tag{3.117}$$

one can write

$$f(x) - f^2(x) \leq -f(x)\ln f(x) \leq 1 - f(x) \tag{3.118}$$

Multiplying by dx and integrating equation (3.118) with limits from a to b, one gets

$$1 - \int_a^b f^2(x)\, dx \leq H(f) \leq \int_a^b [1 - f(x)]dx = b - a - 1 \tag{3.119}$$

Using Jensen's inequality, equation (3.116) can be written as

$$H(q|f) \geq \int_a^b q(x) \left[1 - \frac{f(x)}{q(x)} \right] dx = 0 \tag{3.120}$$

From equation (3.116) and using equation (3.120) one can write

$$-\int_a^b q(x) \ln q(x) \, dx \leq -\int_a^b q(x) \ln f(x) \, dx \tag{3.121}$$

$$-\int_a^b q(x) \ln \left\{ \exp \left[-\lambda_0 - \sum_{r=1}^n \lambda_r A_r(x) \right] \right\} dx \leq -\int_a^b q(x) \ln \left\{ \exp \left[-\lambda_0 - \sum_{r=1}^m \lambda_r A_r(x) \right] \right\} dx \tag{3.122}$$

This yields

$$\lambda_0 + \sum_{r=1}^n \lambda_r C_r(x) \leq \lambda_0 + \sum_{r=1}^m \lambda_r C_r(x) \tag{3.123}$$

This shows that $H(q) \leq H(f)$.

Equation (3.116) can be written using expansion of the log term as

$$H(q|f) = -\int_a^b q(x) \ln \left(1 + \frac{q(x) - f(x)}{f(x)} \right) dx \geq -\frac{1}{2} \int_a^b q(x) \left[\ln(1 + \frac{q(x) - f(x)}{f(x)})^2 \right] dx \tag{3.124}$$

Therefore, using Jensen's inequality,

$$H(q|f) \geq \frac{1}{2} \int_a^b q(x) \left(\frac{f(x) - q(x)}{q(x)} \right)^2 \tag{3.125}$$

Because the first m constraints are the same, it may be noted that

$$-\int_a^b q(x) \ln q(x) \, dx = -\int_a^b f(x) \ln f(x) \, dx \tag{3.126}$$

Therefore,

$$H(q|f) = H_m(f) - H_n(q) \tag{3.127}$$

This shows that an increase in the number of constraints leads to a decrease in entropy or uncertainty as regards the system information.

Example 3.13: Consider the maximum yearly peak discharge values for Brazos River at College Station. Compute the mean, variance, standard deviation, and coefficient of variation.

Then, compute the probability distribution using only mean as the constraint and call it $f(x)$ Then using mean and variance in the logarithmic domain as the constraints compute the probability distribution and call it $q(x)$ which will be a log-normal distribution. Compute $H[f(x)]$ and $H[q(x)]$ and show that $H(q) < H(f)$.

Solution: $\bar{x} = 23741$, $s_x^2 = 397384148$, $\sigma_x = 19935$, $C_{vx} = \frac{\bar{x}}{\sigma_x} = 0.840$. With the mean as the constraint, we get the exponential distribution. For the exponential distribution, entropy is:

$$H(x) = \ln(e\bar{x})$$

where e is the base of the logarithm. When the variance is added, we can get the normal distribution. For the log-normal distribution, entropy is given as $I(x) = \ln[S_y (2\pi e)^{0.5}] + \bar{y}$, where $x = \exp(y)$, where $x \in (0, \infty)$, y is normally distributed. Then we can get:

$$H(f) = \ln(e\bar{x}) = 11.07$$
$$H(q) = \ln\left[S_y (2\pi e)^{0.5}\right] + \bar{y} = 11.01$$

Then it can be seen that $H(q) < H(f)$.

3.6 Invariance of total entropy

The Shannon entropy is a measure of information of the system described by a PDF. Let us suppose that different PDFs are proposed to describe the system. The total information of the system is fixed and different PDFs attempt to get the best estimate of this information. If these PDFs yield the same information content, then their parameters must be related to each other. To illustrate, consider that the system is described by an exponential distribution:

$$f(x) = a \exp(-ax), \ 0 \leq x < \infty, \ a \geq 0 \tag{3.128}$$

Now consider an extreme value type I distribution:

$$q(y) = a_0 \exp\left[-a_0 y - \exp(-a_0 y)\right], \quad -\infty < y < \infty \tag{3.129}$$

Entropy of equation (3.128) is

$$H(f) = \ln\left(\frac{e}{a}\right) \tag{3.130}$$

Likewise, entropy of equation (3.129) is

$$H(q) = \ln\left(\frac{e}{a_0}\right) - \lambda \tag{3.131}$$

where $\gamma = 0.57$ is Euler's constant. Equation (3.130) and (3.131) are related by the transformation:

$$ax = \exp(-a_0 y) \tag{3.132}$$

Thus,

$$f(x)dx = a \exp(-ax)dx = a_0 \exp[-\exp(-a_0 y)]dy = q(y)dy \tag{3.133}$$

If $f(x)$ and $q(y)$ yield the same information about the system then $H(f) = H(q)$ or equating equations (3.130) and (3.131), one obtains

$$a = a_0 \exp(\gamma) \tag{3.134}$$

This establishes a relation between parameters of the exponential and extreme value type I distributions. This means a_0 can be determined by knowing a.

Questions

Q.3.1 Consider an experiment on a random variable X with possible outcomes $X: \{x_1, x_2, x_3, x_4\}$ each with probability $P: \{p_1, p_2, p_3, p_4\}$. How many degenerate distributions can there be? Write these distributions. What would be the entropy of each of these degenerate distributions?

Q.3.2 Consider the experiment in Q.3.1. There can be many distributions having nonzero entropy. State the distribution that would have the maximum entropy amongst all non-zero entropy distributions. What would be the maximum entropy? What is the constraint on which this maximum entropy distribution is based?

Q.3.3 In the discrete POME formalism λ_0 is expressed as a function of Lagrange multipliers $\lambda_1, \lambda_2, \ldots, \lambda_m$ if there are m constraints. Show that this function is convex.

Q.3.4 In the discrete POME formalism, show if H_{max} is a concave function of constraints $C_1, C_2, \ldots, C_m$.

Q.3.5 Show that in the discrete POME formalism, derivatives of H_{max} with respect to constraints $C_1, C_2, \ldots, C_m$ are equivalent to Lagrange multipliers.

Q.3.6 Consider the continuous POME formalism. Show that partial derivatives of the partition function are equal to negative of constraints.

Q.3.7 Obtain data on the number of rainy days (n) for a number of years (say, 30 or more years) for College Station in Texas. Using the mean number of rainy days as a constraint, determine the discrete distribution that n follows. Fit this distribution to the histogram and discuss how well it fits. Compute the entropy of the distribution.

Q.3.8 Consider the number of rainy days as a continuous random variable. Use the data from Example 3.7. Using the mean number of rainy days as constraint, determine the continuous distribution that n follows. Fit this distribution to the histogram and discuss how well it fits. Compute the entropy of the distribution.

Q.3.9 Obtain the values of time interval between two successive rain events in College Station, Texas, for a number of years (say 30 or more years) and select the maximum value for each year. The maximum time interval between rainy days is considered here as a random variable. Plot a histogram of the maximum values and discuss what it looks like. Now compute the mean time interval in days. Using this statistic, determine the discrete distribution that the time interval follows. Fit this distribution to the histogram and discuss how well it fits. Compute the entropy of this distribution.

Q.3.10 Consider the time interval between two successive rain events and select the maximum value for each year for the data in Q.3.9. The maximum time interval is considered here as a continuous random variable. Using the mean time interval (maximum values) as a constraint, determine the continuous distribution that the time interval follows. Fit this distribution to the histogram and discuss how well it fits. Compute the entropy of the distribution.

Q.3.11 Obtain the values of time interval between two successive rain events in College Station, Texas, for a number of years (say 30 or more years) and select the minimum value for each year. The minimum time interval between rainy days is considered here as a random variable. Plot a histogram of the minimum values and discuss what it looks like. Now compute the mean time interval in days. Using this statistic, determine the discrete distribution that the time interval follows. Fit this distribution to the histogram and discuss how well it fits. Compute the entropy of this distribution.

Q.3.12 Consider the time interval between two successive rain events and select the minimum value for each year. Use the data from Q.3.11. The minimum time interval is considered here as a continuous random variable. Using the mean time interval (minimum values) as a constraint, determine the continuous distribution that the minimum time interval follows. Fit this distribution to the histogram and discuss how well it fits. Compute the entropy of the distribution.

Q.3.13 Consider yearly rainfall for a number of years for College Station, Texas. Consider yearly rainfall as a discrete random variable. Using the mean yearly rainfall as a constraint, determine the discrete distribution that yearly rainfall follows. Fit this distribution to the histogram and discuss how well it fits. Compute the entropy of the distribution.

Q.3.14 Consider yearly rainfall as a continuous random variable. Use the data from Q.3.13. Using the mean yearly rainfall as a constraint, determine the continuous distribution that yearly rainfall follows. Fit this distribution to the histogram and discuss how well it fits. Compute the entropy of the distribution.

Q.3.15 Obtain data on the number of days (n) having temperature above 36 $°C$ (100 $°F$) for a number of years (say, 30 or more years) for College Station in Texas. Using the mean number of days as a constraint, determine the discrete distribution that n follows. Fit this distribution to the histogram and discuss how well it fits. Compute the entropy of the distribution.

Q.3.16 Consider the number of days (n) having temperature above 36 $°C$ (100 $°F$) as a continuous random variable. Use the data from Q.3.15. Using the mean number of days as a constraint, determine the continuous distribution that n follows. Fit this distribution to the histogram and discuss how well it fits. Compute the entropy of the distribution.

Q.3.17 Obtain data on the number of days (n) having temperature equal to or below 0 $°C$ (32 $°F$) for a number of years (say, 30 or more years) for College Station in Texas. Using the mean number of days as a constraint, determine the discrete distribution that n follows. Fit this distribution to the histogram and discuss how well it fits. Compute the entropy of the distribution.

Q.3.18 Consider the number of days (n) having temperature equal to or less than 0 $°C$ (36 $°F$) as a continuous random variable. Use the data from Q.3.17. Using the mean number of days as a constraint, determine the continuous distribution that n follows. Fit this distribution to the histogram and discuss how well it fits. Compute the entropy of the distribution.

Q.3.19 Obtain the values of number of days without rainfall each year in College Station, Texas. The number of rainless days each year is considered here as a random variable.

Plot a histogram of the number of rainless days and discuss what it looks like. Now compute the mean number of rainless days. Using this statistic, determine the discrete distribution that the number of rainless days follows. Fit this distribution to the histogram and discuss how well it fits. Compute the entropy of this distribution.

Q.3.20 Consider the number of rainless days as a continuous random variable. Use the data from **Q.3.19**. Using the mean number of rainless days as a constraint, determine the continuous distribution that the number of rainless days follows. Fit this distribution to the histogram and discuss how well it fits. Compute the entropy of the distribution.

References

Agmon, N., Alhassid, Y. and Levine, R.D. (1979). An algorithm for finding the distribution of maximal entropy. *Journal of Computational Physics*, Vol. 30, No. 2, pp. 250–9.

Evans, R.A. (1969). The principle of minimum information. *IEEE Transactions on Reliability*, Vol. R-18, No. 3, pp. 87–90.

Feder, M. (1986). Maximum entropy as a special case on the minimum description length criterion. *IEEE Transactions on Information Theory*, Vol. IT-32, No. 6, pp. 847–9.

Jaynes, E.T. (1957a). Information theory and statistical mechanics, I. *Physical Reviews*, Vol. 106, pp. 620–30.

Jaynes, E.T. (1957b). Information theory and statistical mechanics, II. *Physical Reviews*, Vol. 108, pp. 171–90.

Jaynes, E.T. (1961). *Probability Theory in Science and Engineering*. McGraw-Hill Book Company, New York.

Jaynes, E.T. (1982). On the rationale of maximum entropy methods. *Proceedings of IEEE*, Vol. 70, pp. 939–52.

Kapur, J.N. (1989). *Maximum Entropy Models in Science and Engineering*. Wiley Eastern, New Delhi, India.

Kapur, J.N. and Kesavan, H.K. (1987). *Generalized Maximum Entropy Principle (with Applications)*. Sandford Educational Press, University of Waterloo, Waterloo, Canada.

Kapur, J.N. and Kesavan, H.K. (1992). *Entropy Optimization Principles with Applications*. 408 p., Academic Press, New York.

Kesavan, H.K. and Kapur, J.N. (1989). The generalized maximum entropy principle. *IEEE Transactions: Systems, Man and Cybernetics*, Vol. 19, pp. 1042–52.

Rissanen, J. (1978). Modeling by shortest data description. *Automatica*, Vol. 14, pp. 465–71.

Rissanen, J. (1983). A universal prior for integers and estimation by MDL. *Annals of Statistics*, Vol. 11, No. 2, pp. 416–32.

Rissanen, J. (1984). Universal coding, information, prediction and estimation. *IEEE Transactions on Information Theory*, Vol. IT-30, No. 4, pp. 629–36.

Shore, J.E. and Johnson, R.W. (1980). Properties of cross-entropy minimization. *IEEE Transactions on Information Theory*, Vol. 17, pp. 472–82.

Soofi, E. (1994). Capturing the intangible concept of information. *Journal the American Statistical Association*, Vol. 89, No. 428, pp. 1243–54.

Van Campenhout, J. and Cover, T.M. (1981). Maximum entropy and conditional probability. *IEEE Transactions on Information Theory*, Vol. IT-27, No. 4, pp. 483–9.

Additional Reading

Bevensee, R.M. (1993). *Maximum Entropy Solutions to Scientific Problems*. PTR Prentice Hall, Englewood Cliffs, New Jersey.

Campbell, L.L. (1970). Equivalence of Gauss's principle and minimum discrimination information estimation of probabilities. *Annals of Mathematical Statistics*, Vol. 41, No. 3, pp. 1011–15.

Carnap, R. (1977). *Two Essays on Entropy*. University of California Press, Berkeley, California.

Cover, T.M. and Thomas, J.A. (1991). *Elements of Information Theory*. John Wiley & Sons, New York.

Erickson, G.J. and Smith, C.R. (1988). *Maximum-Entropy and Bayesian Methods in Science and Engineering*. Kluwer Academic Publishers, Dordrecht, The Netherlands.

Goldszmidt, M., Morris, P. and Pearl, J. (1993). A maximum entropy approach to nonmonotonic reasoning. *IEEE Transactions on Pattern Analysis and Machine Intelligence*, Vol. 15, No. 3, pp. 220–32.

Khinchin, A.I. (1957). *The Entropy Concept in Probability Theory*. Translated by R.A. Silverman and M.D. Friedman, Dover Publications, New York.

Krstanovic, P.F. and Singh, V.P. (1988). Application of entropy theory to multivariate hydrologic analysis. Vol. 1, Technical Report WRR 8, Department of Civil Engineering, Louisiana State University, Baton Rouge, Louisiana.

Levy, W.B. and Delic, H. (1994). Maximum entropy aggregation of individual opinions. *IEEE Transactions on Systems, Man and Cybernetics*, Vol. 24, No. 4, pp. 606–13.

Li, X. (1992). An entropy-based aggregate method for minimax optimization. *Engineering Optimization*, Vol. 18, pp. 277–85.

Shore, J.E. and Johnson, R.W. (1980). Axiomatic derivation of the principle of maximum entropy and the principle of minimum cross entropy. *IEEE Transactions on Information Theory*, Vol. IT-26, No. 1, pp. 26–37.

Singh, V.P. (1998). The use of entropy in hydrology and water resources. *Hydrological Processes*, Vol. 11, pp. 587–626.

Singh, V.P. (1998). Entropy as a decision tool in environmental and water resources. *Hydrology Journal*, Vol. XXI, No. 1–4, pp. 1–12.

Singh, V.P. (2000). The entropy theory as tool for modeling and decision making in environmental and water resources. *Water SA*, Vol. 26, No. 1, pp. 1–11.

Singh, V.P. (2003). The entropy theory as a decision making tool in environmental and water resources. In: *Entropy Measures, Maximum Entropy and Emerging Applications*, edited by Karmeshu, Springer-Verlag, Bonn, Germany, pp. 261–97.

Singh, V.P. (2005). Entropy theory for hydrologic modeling. In: *Water Encyclopedia: Oceanography; Meteorology; Physics and Chemistry; Water Law; and Water History, Art, and Culture*, edited by J. H. Lehr, Jack Keeley, Janet Lehr and Thomas B. Kingery, John Wiley & Sons, Hoboken, New Jersey, pp. 217–23.

Singh, V.P. (2011). Hydrologic synthesis using entropy theory: Review. *Journal of Hydrologic Engineering*, Vol. 16, No. 5, pp. 421–33.

Singh, V.P. (1998). *Entropy-Based Parameter Estimation in Hydrology*. Kluwer Academic Publishers, Boston, Massachusetts.

Singh, V.P. and Fiorentino, M., editors (1992). *Entropy and Energy Dissipation in Hydrology*. Kluwer Academic Publishers, Dordrecht, The Netherlands, 595 p.

Stern, H. and Cover, T.M. (1989). Maximum entropy and the lottery. *Journal of the American Statistical Association*, Vol. 84, No. 408, pp. 980–5.

Wu, N. (1997). *The Maximum Entropy Method*. Springer, Heidelburg, Germany.

4 Derivation of Pome-Based Distributions

In Chapter 3, it has been discussed that the principle of maximum entropy (POME) leads to a least-biased probability distribution for a given set of constraints. This chapter extends this discussion further and derives a number of well-known probability distributions. The constraints are usually expressed in terms of moments (or averages of some kind), although it is not a necessary condition. Examples of such constraints are $E[x]$, $E[|x|]$, $E[x^2]$, $E[\ln x]$, $E[\ln(1-x)]$, $E[\ln(1+x)]$, $E[\{\ln(x)\}^2]$ and $E[\ln(1+x^2)]$. It must however be noted that in the case of a continuous variable the limits of integration for entropy and specification of constraints must be compatible, or else POME would not lead to a probability distribution or POME-based probability distribution would not exist.

4.1 Discrete variable and discrete distributions

Here, depending on the constraints some well-known discrete distributions are derived.

4.1.1 Constraint $E[x]$ and the Maxwell-Boltzmann distribution

Consider a random variable X which takes on values x_1, x_2, x_3, ..., x_N with probabilities p_1, p_2, p_3, ..., p_N where N is the number of values. It goes without saying that the total probability law holds:

$$\sum_{i=1}^{N} p_i = 1, \quad p_i \geq 0, \, i = 1, 2, \ldots, N \tag{4.1}$$

The expected value of the variable is known:

$$p_1 x_1 + p_2 x_2 + \cdots + p_N x_N = \sum_{i=1}^{N} x_i p_i = \bar{x} \tag{4.2}$$

Entropy Theory and its Application in Environmental and Water Engineering, First Edition. Vijay P. Singh.
© 2013 John Wiley & Sons, Ltd. Published 2013 by John Wiley & Sons, Ltd.

The objective is to derive the POME-based distribution $P = \{p_1, p_2, \ldots, p_N\}$, subject to equation (4.1) and (4.2). In other words, one maximizes the Shannon entropy:

$$H(p) = H(X) = -\sum_{i=1}^{N} p_i \ln p_i \tag{4.3}$$

subject to equation (4.1) and (4.2).

Following the POME formalism, one constructs the Lagrangean L:

$$L = -\sum_{i=1}^{N} p_i \ln p_i - (\lambda_0 - 1)\left(\sum_{i=1}^{N} p_i - 1\right) - \lambda_1 \left(\sum_{i=1}^{N} p_i x_i - \bar{x}\right) \tag{4.4}$$

Differentiating equation (4.4) with respect to p_i, $i = 1, 2, \ldots, N$, and equating each derivative to zero, one obtains:

$$\frac{\partial L}{\partial p_i} = 0 \Rightarrow -\ln p_i - \lambda_0 - \lambda_1 x_i = 0, \; i = 1, 2, \ldots, N \tag{4.5}$$

Equation (4.5) yields

$$p_i = \exp[-\lambda_0 - \lambda_1 x_i], \; i = 1, 2, \ldots, N \tag{4.6}$$

Equation (4.6) contains parameters λ_0 and λ_1 that are determined with the use of equation (4.1) and (4.2). Inserting equation (4.6) in equation (4.1), one gets

$$\exp(-\lambda_0) = \left[\sum_{i=1}^{N} \exp(-\lambda_1 x_i)\right]^{-1} \tag{4.7a}$$

or

$$\lambda_0 = \log\left[\sum_{i=1}^{N} \exp(-\lambda_1 x_i)\right] \tag{4.7b}$$

When equation (4.7a) is substituted in equation (4.6) the result is

$$p_i = \frac{\exp(-\lambda_1 x_i)}{\displaystyle\sum_{i=1}^{N} \exp(-\lambda_1 x_i)} \tag{4.8}$$

Equation (4.8) is called the Maxwell-Boltzmann (M-B) distribution used in statistical mechanics. Before discussing its application in different areas, one first determines its parameter λ_1 in terms of constraint $\bar{x}$. If the Lagrange multiplier for the average value of X is redundant, that is, $\lambda_1 = 0$, the exponential distribution reduces to a uniform distribution $p_i = 1/N$.

Inserting equation (4.6) in equation (4.2), one gets

$$\sum_{i=1}^{N} x_i \exp\left(-\lambda_0 - \lambda_1 x_i\right) = \bar{x} \tag{4.9}$$

Taking advantage of equation (4.7a), equation (4.9) yields

$$\frac{\sum_{i=1}^{N} x_i \exp\left(-\lambda_1 x_i\right)}{\sum_{i=1}^{N} \exp\left(-\lambda_1 x_i\right)} = \bar{x} \tag{4.10}$$

Equation (4.10) permits the estimation of λ_1 in terms of $\bar{x}$.

Kapur (1989), and Kapur and Kesavan (1987, 1992) have discussed mathematical details of the M-B distribution. If $\lambda_1 = 0$, clearly equation (4.6) would be a rectangular distribution with $p_i = 1/N$. If λ_1 is negative, then the probability increases as x_i increases, and if $\lambda_1 > 0$ then the probability decreases as x_i increases. In physics the M-B distribution has been employed to derive the microstates of a system on the basis of some knowledge about macroscopic data. For example, if a system had a large number of particles each with an energy level then the M-B distribution would be employed to determine the probability distribution of energy levels of particles, provided the expected energy of the system was somehow known. The M-B distribution, although developed in statistical mechanics, is applicable to a wide range of problems in environmental and water engineering. Fiorentino et al. (1993) employed the M-B distribution to describe the probability distribution of elevations of links in a river basin if the mean basin elevation was known. It is possible to employ the M-B distribution for representing the probability distribution of elevations, lengths, and drainage areas of channels of a given order, say, first order in a river basin.

The maximum entropy of the distribution becomes

$$H_{\max} = \lambda_0 + \lambda_1 \bar{x}$$

This shows that the maximum entropy is a function of the Lagrange multipliers and the constraint. It depends on the spread of the distribution. In this sense, entropy can be regarded as a system wide accessibility function where the partition and the x values relate to the probabilities across the system.

4.1.2 Two constraints and Bose-Einstein distribution

One can extend the discussion of the M-B distribution here by considering another moment-type constraint. Consider a case where values of a random variable X can be arranged into N categories. In each category X takes on k values (from 0 to ∞). The values of the random variable X can then be denoted as: $X : \{x_{jk}; j = 1, 2, \ldots, N; k = 1, 2, \ldots, M\}$, where the value of M can be as large as infinity. Thus, each value of random variable X can be associated with one of j attributes, where $j = 1, 2, \ldots, N$. In statistical mechanics x_{jk} represents the k number of particles associated with an energy level j; here k signifies the number of particles in energy level j. In hydrology, one considers the case of a river basin where there is a large number of links each associated with an elevation or elevation interval. It is possible to define a finite number of elevations or range of elevations, say N, in the basin. Then there can be k links ($k = 0, 1, 2, \ldots, \infty$) having elevations j ($j = 1, 2, \ldots, N$). Thus the number of links being of elevation j is a random variable. Corresponding to an elevation there can be k links between zero and infinity. Thus, one has two variables: 1) the number of links k and 2) the elevation j that k possesses. Similarly, consider a farm on which there are a large number of plants whose heights can be classified into a number of categories, and each plant is associated with a particular height. Then the number of plants of a particular height is a random variable. Here k represents the number of plants of a particular height j and N represents the categories of plant heights. For geomorphologic analysis, a river basin is ordered following the Horton-Strahler ordering scheme. Suppose a basin is a 6-th order basin using this scheme. Then, there can be a large number of channels for each order, except the highest order. Here, order is tantamount to category or energy level, denoted by j and number of channels in each order denoted by k is the random variable.

One considers p_{jk} as the probability of k links being in the j-th state or elevation from amongst N elevation states. The value x_j defines the value of variable X in the j-th state. Thus,

$$\sum_{k=0}^{\infty} p_{jk} = 1, j = 1, 2, \dots, N \tag{4.11}$$

It is known that

$$\sum_{j=1}^{N} \sum_{k=0}^{\infty} k p_{jk} = \bar{k} \tag{4.12}$$

and

$$\sum_{j=1}^{N} x_j \sum_{k=0}^{\infty} k p_{jk} = \bar{x} \tag{4.13}$$

where $\bar{k}$ is the average value of k (the number), and $\bar{x}$ is the average value of elevations of the basin.

The Shannon entropy for the j-th elevation category can be expressed as

$$H_j = -\sum_{k=0}^{\infty} p_{jk} \ln p_{jk}, j = 1, 2, \dots, N \tag{4.14}$$

The probability distribution of one elevation category is independent of the probability distribution of another elevation category. Therefore, the total entropy of the system is the sum of entropies of all the elevation categories:

$$H = H_1 + H_2 + \dots + H_N \tag{4.15}$$

or

$$H = -\sum_{j=1}^{N} \sum_{k=0}^{\infty} p_{jk} \ln p_{jk} \tag{4.16}$$

In order to determine p_{jk}, H in equation (4.16) is maximized, subject to equation (4.11) to (4.13). To that end, the Lagrange multiplier L is constructed as

$$L = -\sum_{j=1}^{N} \sum_{k=0}^{\infty} p_{jk} \ln p_{jk} - \sum_{j=1}^{N} (\lambda_0 - 1) \left(\sum_{k=0}^{\infty} p_{jk} - 1 \right) - \lambda_1 \left[\sum_{j=1}^{N} \sum_{k=0}^{\infty} k p_{jk} - \bar{k} \right]$$
$$- \lambda_2 \left[\sum_{j=1}^{N} x_j \sum_{k=0}^{\infty} k p_{jk} - \bar{x} \right] \tag{4.17}$$

Differentiating equation (4.17) with respect to p_{jk} and equating the derivative to zero, one obtains

$$\frac{\partial L}{\partial p_{jk}} = 0 \Rightarrow p_{jk} = \exp(-\lambda_0 j)\exp[-k(\lambda_1 + \lambda_2 x_j)]$$
$$= a_j \exp[-k(\lambda_1 + \lambda_2 x_j)], \, j = 1, 2, \dots, N; \, k = 0, 1, 2, \dots, \infty \tag{4.18}$$

where $a_j = \exp(-\lambda_0 j)$. Parameters λ_0, λ_1 and λ_2 are determined with the use of equations (4.11) to (4.13) Substituting equation (4.18) in equation (4.11), one gets

$$a_j \sum_{k=0}^{\infty} \exp[-k(\lambda_1 + \lambda_2 x_j)] = 1 \tag{4.19}$$

Equation (4.19) yields

$$a_j = 1 - \exp[-(\lambda_1 + \lambda_2 x_j)] \tag{4.20}$$

Coupling equation (4.20) with equation (4.18), one obtains

$$p_{jk} = \{1 - \exp[-(\lambda_1 + \lambda_2 x_j)]\}\exp[-k(\lambda_1 + \lambda_2 x_j)] \tag{4.21}$$

Substituting equation (4.21) in equation (4.12) one gets

$$\sum_{j=1}^{N}\sum_{k=0}^{\infty} k\{1 - \exp[-(\lambda_1 + \lambda_2 x_j)]\}\exp[-k(\lambda_1 + \lambda_2 x_j)] = \overline{k} \tag{4.22}$$

The inside summation designated as K_j in equation (4.22) can be simplified as

$$K_j = \sum_{k=0}^{\infty} k\{1 - \exp[-(\lambda_1 + \lambda_2 x_j)]\}\exp[-k(\lambda_1 + \lambda_2 x_j)]$$
$$= \{1 - \exp[-(\lambda_1 + \lambda_2 x_j)]\}\sum_{k=0}^{\infty} k\exp[-k(\lambda_1 + \lambda_2 x_j)]$$
$$= \frac{\exp[-\lambda_1 - \lambda_2 x_j]}{1 - \exp[-\lambda_1 - \lambda_2 x_j]} = \frac{1}{\exp(\lambda_1 + \lambda_2 x_j) - 1} \tag{4.23}$$

Substitution of equation (4.23) in equation (4.22) yields

$$\sum_{j=1}^{N} K_j = \sum_{j=1}^{N} \frac{1}{\exp(\lambda_1 + \lambda_2 x_j) - 1} = \overline{k} \tag{4.24}$$

Inserting equation (4.21) in equation (4.13) and taking advantage of equation (4.23) one gets

$$\sum_{j=1}^{N} K_j x_j = \sum_{j=1}^{N} \frac{x_j}{\exp(\lambda_1 + \lambda_2 x_j) - 1} = \overline{x} \tag{4.25}$$

Equation (4.24) and (4.25) are solved to determine λ_1 and λ_2 in terms of $\overline{k}$ and $\overline{x}$.

Equation (4.21) is the probability distribution of the number of links of different elevations. The distribution of the expected number of links in N states, K_j $(j = 1, 2, \ldots, N)$, is the Bose-Einstein (B-E) distribution. Note one can normalize the expected number of links of the j-the elevation as

$$p_j = \frac{K_j}{\bar{k}} \tag{4.26}$$

Then, $(p_1, p_2, \ldots, p_N)$ can be regarded as a probability distribution, wherein p_j can be viewed as the probability of a link having the j-th elevation. The B-E distribution was derived in statistical mechanics as the distribution of the expected number of particles in different states. This has wide ranging applications in environmental and water engineering as well as other disciplines.

4.1.3 Two constraints and Fermi-Dirac distribution

In this case the assumption is that each state contains either no particle or one particle. Thus, comparing with the B-E distribution, the number of particles (k) varies from 0 to 1 instead of ∞. As an example, a day can be considered as a state, and a day may contain a rainstorm or not. Thus, the number of rainstorms varies from 0 to 1. Therefore, constraint equation (4.11) to (4.13) become:

$$\sum_{k=0}^{1} p_{jk} = 1 \tag{4.27}$$

$$\sum_{j=1}^{N} \sum_{k=0}^{1} k p_{jk} = \bar{k} \tag{4.28}$$

$$\sum_{j=1}^{N} x_j \sum_{k=0}^{1} k p_{jk} = \bar{x} \tag{4.29}$$

Similarly, equation (4.16) becomes

$$H_j = -\sum_{j=1}^{N} \sum_{k=0}^{1} p_{jk} \ln p_{jk}, \ j = 1, 2, \ldots, N \tag{4.30}$$

As in the case of the B-E distribution, the method of Lagrange multipliers yields

$$p_{jk} = a_j \exp[-k(\lambda_1 + \lambda_2 x_j)] \tag{4.31}$$

Substituting equation (4.31) in equation (4.27), one obtains

$$a_j = \frac{1}{1 + \exp[-(\lambda_1 + \lambda_2 x_j)]} \tag{4.32}$$

Using equation (4.32) the expected number of particles in the j-th state can now be expressed as

$$K_j = \sum_{k=0}^{1} k p_{jk} = p_{j1} = \frac{1}{1 + \exp(\lambda_1 + \lambda_2 x_j)}, \ j = 1, 2, \ldots, N \tag{4.33}$$

Equation (4.33) is the Fermi-Dirac distribution.

4.1.4 Intermediate statistics distribution

This distribution occurs during transition from the B-E distribution to the F-D distribution. It is assumed that the number of particles in any state is constrained by a maximum value m_j (between ∞ in the B-E distribution and 1 in the F-D distribution). Therefore, constraint equation (4.11) to (4.13) and (4.16) become

$$\sum_{k=0}^{m_j} p_{jk} = 1, \ j = 1, 2, \ldots, N \tag{4.34}$$

$$\sum_{j=1}^{N} \sum_{k=0}^{m_j} k p_{jk} = \bar{k} \tag{4.35}$$

$$\sum_{j=1}^{N} x_j \sum_{k=0}^{m_j} k p_{jk} = \bar{x} \tag{4.36}$$

and

$$H = -\sum_{j=1}^{N} \sum_{k=0}^{m_j} p_{jk} \ln p_{jk} \tag{4.37}$$

Maximization of equation (4.37), subject to equation (4.34) to (4.36), using the method of Lagrange multipliers produces

$$p_{jk} = a_j \exp[-k(\lambda_1 + \lambda_2 x_j)], \quad j = 1, 2, \ldots, N; \ k = 0, 1, 2, \ldots, m_j \tag{4.38}$$

If

$$y_j = \exp(-\lambda_1 - \lambda_2 x_j) \tag{4.39}$$

Then, equation (4.38) becomes

$$p_{jk} = a_j y_j^k \tag{4.40}$$

Using equation (4.40) in equation (4.34), one gets

$$a_j(1 + y_j + y_j^2 + \ldots + y_j^{m_j}) = 1 \tag{4.41}$$

For the expected number of particles in the j-th state one can write

$$
\begin{aligned}
K_j &= \sum_{k=0}^{m_j} k p_{jk} = a_j \sum_{k=0}^{m_j} k y_j^k \\
&= \frac{y_j + 2y_j^2 + \ldots + m_j y_j^{m_j}}{1 + y_j + y_j^2 + \ldots + y_j^{m_j}}
\end{aligned} \tag{4.42}
$$

Equation (4.42) is referred to as the Intermediate Statistics Distribution (ISD). If $m_j = 1$ then equation (4.42) gives

$$K_j = \frac{y_j}{1 + y_j} = \frac{1}{1 + \exp(\lambda_1 + \lambda_2 x_j)} \tag{4.43}$$

which is the F-D distribution. If $m_j \rightarrow \infty$, equation (4.42) reduces to

$$K_j = \frac{y_j(1-y_j)^{-2}}{(1-y_j)^{-1}} = \frac{1}{\exp(\lambda_1 + \lambda_2 x_j) - 1} \tag{4.44}$$

which is the B-E distribution.

4.1.5 Constraint: *E[N]*: Bernoulli distribution for a single trial

Let a random variable be denoted by N which takes on anyone of two possible values – one associated with success and the other with failure. When one considers whether it is wet or dry, hot or cold, flooded or not, windy or tranquil, day or night, sunny or cloudy, clear or hazy, foggy or not foggy, urbanized or rural, rich or poor, tall or short, high or low, and so on, it is seen that only two mutually exclusive or collectively exhaustive events are possible outcomes. Such a variable is defined as the Bernoulli random variable. Now for the random variable N, one assigns a value of zero for ''nonoccurrence'' of the specified event and a value of one for its ''occurrence''. Let the probability of ''occurrence'' be p, and the probability of ''nonoccurrence'' be $q = 1 - p$. If a success is observed, $n = 1$ and if a failure is observed, $n = 0$. In the Shannon entropy the scale factor m is used as p.

Many everyday situations entail events which have just two possibilities. A highway bridge may or may not be flooded in the next year, an area may or may not get flooded this year, it may or may not rain today, it may be windy or may not be windy next week, it may snow or may not snow next week, it may be cloudy or sunny tomorrow, a car accident may or may not occur next week, a column may or may not buckle, an excavator may or may not cease to operate in the next week, and so on. For a Bernoulli variable, the probability of occurrence of the event in each trial is the same from trial to trial and the trials are statistically independent. The Bernoulli distribution is useful for modeling an experiment or an engineering process that results in exactly one of two mutually exclusive outcomes. The experiments involving repeated sampling of a Bernoulli random variable are frequently called Bernoulli trials, for example, tossing a coin repeatedly and observing their outcomes (head or tail).

The constraint equation can be specified as

$$\sum_{n=0}^{1} n p_n = E[n] = p \tag{4.45}$$

where p_n is the probability of occurrence of value n. Maximization of the Shannon entropy H then yields

$$p_n = \frac{1}{p \exp(\lambda_0)} \exp[-\lambda_1 n] \tag{4.46}$$

Substitution of equation (4.46) in equation (4.1) gives

$$\sum_{n=0}^{1} \frac{1}{p \exp(\lambda_0)} \exp[-\lambda_1 n] = 1 \tag{4.47}$$

Equation (4.47) yields

$$\exp(\lambda_0) = \frac{1 + \exp(-\lambda_1)}{p} \tag{4.48}$$

Equation (4.48) can also be expressed as

$$\lambda_0 = \ln[1 + \exp(-\lambda_1)] - \ln p \tag{4.49}$$

Differentiating equation (4.49) with respect to λ_1 one gets

$$\frac{\partial \lambda_0}{\partial \lambda_1} = -\frac{1}{1 + \exp(-\lambda_1)} \tag{4.50a}$$

Equation (4.50a) also equals

$$\frac{\partial \lambda_0}{\partial \lambda_1} = -p \tag{4.50b}$$

Therefore, equating equations (4.50a) and (4.50b) yields

$$\lambda_1 = \ln \frac{1-p}{p} \tag{4.51}$$

On substitution of equation (4.51) in equation (4.46), one obtains

$$p_n = p^n (1-p)^{1-n} \quad (n = 0, 1) \tag{4.52}$$

This is the Bernoulli distribution.

Several commonly used discrete distributions arise from examining the results of Bernoulli trials repeated several times. Three basic questions come to mind when one observes a set of Bernoulli trials: 1) How many successes will be obtained in a fixed number of trials? 2) How many trials must be performed until one observes the first success? 3) How many trials must be performed until one observes the kth success? To answer these three questions motivates the development of the binomial, geometric, and negative binomial distributions, respectively.

4.1.6 Binomial distribution for repeated trials

A binomial random variable represents the number of successes obtained in a series of N independent and identical Bernoulli trials, the number of trials is fixed and the number of successes varies from experiment to experiment. Consider a sequence of Bernoulli trials, where the outcomes of the experiment are mutually independent and the probability of success remains unchanged. For example, for a sequence of N years of flood data, the maximum annual flood magnitudes are independent and the probability of occurrence, p, of a flood in any year remains unchanged throughout the period of N years. If the random variable is whether the flood occurs or not, then the sequence of N outcomes is Bernoulli trials. Let the random variable be designated by K and its specific value by k. One wishes to determine the probability of exactly k occurrences (the number of successes) in N Bernoulli trials. Let the probability of success (say, occurrence of flood) be p.

When deriving the binomial distribution, it is useful to take the invariance measure function in the Shannon entropy as $m = \binom{N}{k}$, the binomial coefficient, which states the number of ways that exactly k successes can be found in a sequence of N trials. The constraints are given by equation (4.1) and

$$\sum_{k=0}^{N} k p_k = Np \tag{4.53}$$

Maximizing entropy subject to the constraints, one obtains the probability distribution as:

$$p_k = \frac{m}{\exp(\lambda_0)} \exp[-\lambda_1 k] \tag{4.54}$$

Substitution of equation (4.54) in equation (4.1) yields

$$\exp(\lambda_0) = \sum_{k=0}^{N} \binom{N}{k} [\exp(-\lambda_1)]^k 1^{N-k} = [\exp(-\lambda_1) + 1]^N \tag{4.55}$$

Taking the logarithm of equation (4.55) results in

$$\lambda_0 = N \ln[\exp(-\lambda_1) + 1] \tag{4.56}$$

Differentiating equation (4.56) with respect to λ_1 gives

$$\frac{\partial \lambda_0}{\partial \lambda_1} = -\frac{N \exp(-\lambda_1)}{1 + \exp(-\lambda_1)} = -Np \tag{4.57}$$

Equation (4.57) yields

$$\lambda_1 = \ln\left(\frac{1-p}{p}\right) \tag{4.58}$$

Substituting equation (4.58) in equation (4.55), one obtains

$$\exp(\lambda_0) = \left[\frac{p}{1-p} + 1\right]^N = \frac{1}{(1-p)^N} \tag{4.59}$$

Using equations (4.58) and (4.59) in equation (4.54), the probability distribution now is given as

$$P(K = k) = \binom{N}{k} p^k (1-p)^{N-k} \tag{4.60}$$

which is the binomial distribution.

   Here N must be an integer and $0 \leq p \leq 1$. This equation defines the distribution of K for given values of p and N. The binomial distribution has two parameters: the number of trials N and the probability of occurrence of the specified event in a single trial p. The probability of each sequence is equal to $p^k q^{N-k}$. With the use of the above equation, the probabilities that K will take on the values of 0, 1, 2, ..., N, which exhaust all possibilities, can be calculated.

4.1.7 Geometric distribution: repeated trials

The preceding discussion focuses on the number of successes occurring in a fixed number of Bernoulli trials. Here one focuses on the question of determining the number of trials when the first success would occur. For example, how many days would pass before the next rain if the probability of occurrence of rain on any day is p? What would be the year when a flood would occur if the probability of occurrence of flood in any year is p? When would the next accident occur? When would the next hurricane strike the Texas or Louisiana coast? When would the next earthquake hit the Los Angeles area? When would the next snowfall occur in

Denver? Thus a geometric random variable represents the number of trials needed to obtain the first success.

Assuming the independence of trials and a constant value of p, the distribution of N, the number of trials to the first success, can be found using the geometric distribution. The constraint equations for this distribution are given by equation (4.1) and

$$\sum_{n=0}^{\infty} np_n = m \tag{4.61}$$

In reality n must start at 1 but for purposes of summation, it is being taken as 0. Maximizing entropy subject to the given constraints, one gets the POME-based distribution:

$$p_n = \exp(-\lambda_0 - \lambda_1 n) = ab^n, \; a = \exp(-\lambda_0), \; b = \exp(-\lambda_1) \tag{4.62}$$

Inserting equation (4.62) in equation (4.1), one obtains

$$\sum_{n=0}^{\infty} \exp(-\lambda_0 - \lambda_1 n) = \sum_{n=0}^{\infty} ab^n = 1 \tag{4.63a}$$

Equation (4.63a) can be expressed as

$$\frac{a}{1-b} = 1, \; 0 < b < 1 \tag{4.63b}$$

Similarly from the second constraint, one gets

$$a\sum_{n=0}^{\infty} nb^n = m = a\frac{b}{(1-b)^2}, \; 0 < b < 1 \tag{4.64}$$

Equations (4.63a) and (4.64) yield

$$b = \frac{m}{1+m} = q, \quad a = \frac{1}{1+m} = 1 - q = p \tag{4.65}$$

Substituting equation (4.65) in equation (4.62) for the distribution function, one gets

$$p_n = \frac{1}{1+m}\left[\frac{m}{1+m}\right]^n = (1-q)q^n = p(1-p)^n \tag{4.66}$$

But n starts at 1. Therefore, for success at the n-th trial,

$$p_n = p(1-p)^{n-1} \tag{4.67}$$

This is the geometric distribution.

Example 4.1: In any given day the probability of rain event is $p = 0.1$; find the maximum entropy-based probability distribution.

Solution: Assume that the first rainfall event occurs after n days. The constraints for this problem are equations (4.1) and (4.61), and the geometric distribution derived by POME is given as equation (4.62). From equation (4.65), one can obtain:

$$a = 1 - q = p = 0.1, \; b = q = 0.9$$
$$p_n = ab^{n-1} = 0.1 \times (0.9)^{n-1}$$

4.1.8 Negative binomial distribution: repeated trials

The negative binomial random variable represents the number of trials needed to obtain exactly k successes. Here the number of successes, k, is fixed and the number of trials varies from experiment to experiment. Due to this reason it is thought of as a reversal of the binomial distribution, because the number of successes and number of trials are reversed. Each trial has two possible outcomes: success or failure and the probability of success is constant from one trial to another. In the binomial case the number of trials is fixed and the number of successes varies. Following the same procedure, one can derive the negative binomial distribution by maximizing entropy. To that end, it is useful to use the invariance measure as $\binom{N-1}{k-1}$. The distribution of k successes in N trials can be derived as

$$P_{N_k}(n) = \binom{N-1}{k-1}(1-p)^{n-k}p^k, \quad n = k, k+1, \ldots \tag{4.68}$$

where N_k is the trial number at which the k-th success occurs. This equation implies that $k-1$ successes in the preceding $n-1$ trials have already occurred. The probability of $k-1$ successes in $n-1$ trials is obtained from the binomial distribution.

This is the negative binomial distribution, also called as the Pascal distribution, with parameters k and p:

$$E(n) = \frac{k}{p} \tag{4.69}$$

$$Var(n) = \frac{k(1-p)}{p^2} \tag{4.70}$$

Example 4.2: In any given year, the probability of a city suffering a 100-year flood event as $p = 1/100 = 0.01$, find the maximum entropy probability distribution if the city will suffer k 100-year flood events in n years.

Solution: The probability that the city will suffer from k 100-year flood events in n years can be represented as negative binomial distribution. Here $p = 0.01$; $q = 1 - p = 0.99$. Based on equation (4.68), one can obtain that:

$$P(n, k, 0.01) = \binom{n-1}{k-1}(0.99)^{n-k}(0.01)^k, \quad n = k, k+1, \ldots$$

4.1.9 Constraint: $E[N] = n$: Poisson distribution

In hydrology, this distribution is used for rare events that occur randomly in time. Examples of such events include the time of start of rainfall. The binomial distribution is used when the random variable X is the number of times a specified event occurs in a fixed number of trials. When our interest is in the number of times a specified event occurs in a certain length of time, such as a given monitoring period, or how often the event is observed in a continuum of space, such as the length of a highway, an area of land, and so on, and the number of trials is not specified, then the binomial distribution cannot be used. In such cases, it is more appropriate to use the Poisson distribution. Instead of defining the probability of "occurrence" for the specified event in a single trial, as for the binomial distribution, what is defined here is the probability of occurrence per unit of time or of space. For example, the probability that lightning in New Orleans in the month of May will occur as, may be, 0.025 per day. It is assumed that the probability is the same for every day, every meter or every year. It is further assumed that the occurrences and the nonoccurrences are independent along the continuum. The difference between the binomial and Poisson distributions can be summarized by noting that both the occurrences and nonoccurrences can be specified for the binomial distribution, but not for the Poisson distribution.

The binomial and Poisson distributions share some similarities. The probability distribution of the number of occurrences X in a given continuum of time or space can be treated as a special case of the binomial distribution under two conditions: 1) The number of trials becomes infinitely large, and 2) the average number of occurrences defined by Np remains constant. By dividing the continuum into small intervals, the problem can be reduced to one of "occurrence" and "nonoccurrence" of the specified event in any of these intervals, provided these intervals are made so small that the probability of getting two or more "occurrences" in any interval is negligible.

The constraints are specified by equation (4.1) and

$$\sum_{n=0}^{\infty} np_n = E[n] = \bar{n} = \nu \tag{4.71}$$

where $\bar{n}$ is the mean of the distribution, n is the number of events, and p is the probability of n events to occur. Assume that in the Shannon entropy the scale factor $m = 1/n!$. With this scale factor, maximize H subject to the specified constraints. To that end, the Lagrange multiplier function, L, is constructed as

$$L = -\sum_{n=0}^{\infty} n!p_n \ln(n!p_n) - (\lambda_0 - 1)\sum_{n=0}^{\infty} p_n - \left[\lambda_1 \sum_{n=0}^{\infty} np_n - \nu\right] \tag{4.72}$$

Differentiating L in equation (4.72) with respect to p_n and equating to zero yields the entropy-based probability distribution:

$$p_n = m\exp[-\lambda_0 - \lambda_1 n] = \frac{1}{n!\exp(\lambda_0)}\exp(-\lambda_1 n) \tag{4.73}$$

where $z(\lambda_1) = \exp(\lambda_0)$ is the partition function. Substituting in equation (4.1), one obtains

$$Z(\lambda_1) = \exp(\lambda_0) = \sum_{n=0}^{\infty} \frac{1}{n!}\exp(-\lambda_1 n) = \exp[\exp(-\lambda_1)] \tag{4.74}$$

Equation (4.74) can be written as

$$\lambda_0 = \ln\left[\sum_{n=0}^{\infty} \frac{1}{n!}\exp(-\lambda_1 n)\right] = \exp(-\lambda_1) \tag{4.75}$$

Differentiating equation (4.75), one gets

$$\frac{\partial \lambda_0}{\partial \lambda_1} = -\nu \tag{4.76}$$

Also,

$$\frac{\partial \lambda_0}{\partial \lambda_1} = -\exp(-\lambda_1) \tag{4.77}$$

Therefore,

$$\exp(-\lambda_1) = v \tag{4.78}$$

Equation (4.78) yields

$$\lambda_1 = -\ln v \tag{4.79}$$

Substituting the values of λ_0 and λ_1 in the entropy-based probability distribution, one obtains

$$p_n = \frac{v^n e^{-v}}{n!}, \quad n = 0, 1, 2, \ldots, \infty \tag{4.80}$$

This is the Poisson distribution. This distribution has one parameter and is entirely specified by the average number of occurrences of the specified event over the interval of time or space in question.

Example 4.3: Consider that in a 40-year period, the total number of droughts is 70 with the longest drought lasting for 62 days in the growing season, find the maximum entropy distribution function if n droughts occur in the growing season.

Solution: From equation (4.71), one can obtain the average number of droughts per year as:

$$v = \sum_{n=0}^{\infty} n p_n = E(n) = \frac{70}{40} = 1.75$$

Then the entropy-based probability distribution can be expressed as:

$$p_n = \frac{1}{n!} 1.75^n e^{-1.75}$$

4.2 Continuous variable and continuous distributions

4.2.1 Finite interval [a, b], no constraint, and rectangular distribution

Let a random variable be X varying from a to b and having a probability density function $f(x)$. The Shannon entropy function is defined as

$$H = -\int_a^b f(x) \ln f(x) dx \tag{4.81}$$

Of course,

$$\int_a^b f(x) dx = 1 \tag{4.82}$$

Maximizing H in equation (4.81) subject to equation (4.82), one can determine the least-biased $f(x)$. To that end, the Lagrange multiplier L is constructed as:

$$L = -\int_a^b f(x) \ln f(x) dx - (\lambda_0 - 1) \left[\int_a^b f(x) dx - 1 \right] \tag{4.83}$$

where λ_0 is the Lagrange multiplier. Differentiating equation (4.83) with respect to $f(x)$ and setting the derivative equal to 0, one gets

$$\frac{\partial L}{\partial f(x)} = 0 \Rightarrow -[1 + \ln f(x)] - (\lambda_0 - 1) = 0 \tag{4.84}$$

Equation (4.84) yields

$$f(x) = \exp(\lambda_0) \tag{4.85}$$

Equation (4.85) is the POME-based probability distribution with λ_0 as parameter. Inserting equation (4.85) in equation (4.82), one obtains

$$\exp(\lambda_0) = \frac{1}{b-a} \tag{4.86}$$

Thus, equation (4.85) becomes

$$f(x) = \frac{1}{b-a} \tag{4.87}$$

which is the uniform distribution.

Example 4.4: Let a random variable vary from 1 to 100. Find the maximum entropy probability distribution.

Solution: Suppose the probability of random variable X is $f(x)$. For this problem, the only constraint is $\int_a^b f(x) dx = 1$. According to equation (4.87),

$$f(x) = \frac{1}{b-a} = \frac{1}{100-1} = \frac{1}{99} = 0.0101$$

4.2.2 Finite interval [a, b], one constraint and truncated exponential distribution

The constraint is given as

$$\int_a^b xf(x) dx = \bar{x} \tag{4.88}$$

The objective is to determine $f(x)$. Maximizing H, subject to equations (4.82) and (4.88), $f(x)$ is determined using the method of Lagrange multipliers. To that end, the Lagrangean L is constructed as

$$L = -\int_a^b f(x)\ln f(x)dx - (\lambda_0 - 1)\left[\int_a^b f(x)dx - 1\right] - \lambda_1\left[\int_a^b xf(x)dx - \bar{x}\right] \tag{4.89}$$

where λ_0 and λ_1 are parameters.

Differentiating L with respect to $f(x)$ and equating the derivative to zero, one gets

$$\frac{\partial L}{\partial f(x)} = 0 \Rightarrow -[1 + \ln f(x)] - (\lambda_0 - 1) - \lambda_1 x = 0 \tag{4.90}$$

Therefore,

$$f(x) = \exp(-\lambda_0 - \lambda_1 x) = a\exp(-\lambda_1 x), \quad a = \exp(-\lambda_0) \tag{4.91}$$

which is the POME-based distribution with λ_0 and λ_1 as parameters. These parameters can be determined with the aid of equation (4.82) and (4.88).

Inserting equation (4.91) in equation (4.82), one gets

$$a = \frac{\lambda_1}{\exp(-\lambda_1 a) - \exp(-\lambda_1 b)} \tag{4.92}$$

Therefore, equation (4.91) becomes

$$f(x) = \frac{\lambda_1}{[\exp(-\lambda_1 a) - \exp(-\lambda_1 b)]}\exp(-\lambda_1 x) \tag{4.93}$$

Inserting equation (4.93) in equation (4.88), one can determine λ_1 in terms of $\bar{x}$. Thus, equation (4.91) is the truncated exponential distribution.

A slightly different method for derivation of a distribution using POME entails expressing the zeroth Lagrange multiplier in two ways: 1) in terms of other Lagrange multipliers as an algebraic relation and 2) as logarithm of the integral of the probability density function minus the zeroth Lagrange multiplier. This will be illustrated for the case of beta distribution later in the chapter.

Example 4.5: Let a random variable vary from 0 to 2 with a mean value of 1, find the maximum entropy probability distribution.

Solution: Suppose the probability distribution of a random variable X is $f(x)$. For this problem, except for the constraint of $\int_0^2 f(x)dx = 1$, another constraint is $\int_0^2 xf(x)dx = 1$. From equations (4.93) and (4.88),

$$f(x) = \frac{\lambda_1}{1 - \exp(-2\lambda_1)}\exp(-\lambda_1 x);$$

$$\int_0^2 xf(x)dx = 1$$

Solving these equations, one gets $\lambda_1 = 0.008714$. Therefore,

$$f(x) = 0.5044\exp(-0.008714x)$$

4.2.3 Finite interval [0, 1], two constraints $E[\ln x]$ and $E[\ln(1 - x)]$ and beta distribution of first kind

In this case the constraint equations are given as

$$\int_a^b \ln x f(x)dx = E[\ln x] = \overline{\ln x} \tag{4.94}$$

and

$$\int_a^b \ln(1 - x)f(x)dx = E[\ln(1 - x)] = \overline{\ln(1 - x)} \tag{4.95}$$

The Shannon entropy is written as

$$H = -\int_0^1 f(x) \ln f(x)dx \tag{4.96}$$

and equation (4.82) as

$$\int_0^1 f(x)dx = 1 \tag{4.97}$$

For maximizing H in equation (4.96), subject to equations (4.94), (4.95), and (4.97), the Lagrange function L is constructed as

$$L = -\int_0^1 f(x) \ln f(x)dx - (\lambda_0 - 1)\left[\int_0^1 f(x)dx - 1\right] - \lambda_1 \left[\int_0^1 xf(x)dx - \overline{\ln x}\right]$$

$$-\lambda_2 \left[\int_0^1 \ln(1 - x)f(x)dx - \overline{\ln(1 - x)}\right] \tag{4.98}$$

Taking the derivative of L with respect to $f(x)$ and equating it to 0, one gets

$$\frac{\partial L}{\partial f(x)} = 0 \Rightarrow -[1 + \ln f(x)] - (\lambda_0 - 1) - \lambda_1 \ln x - \lambda_2 \ln(1 - x) = 0 \tag{4.99}$$

Therefore,

$$f(x) = ax^{-\lambda_1}(1 - x)^{-\lambda_2}, \quad a = \exp(-\lambda_0) \tag{4.100}$$

Equation (4.100) is the POME-based distribution with λ_0(or a), λ_2 and λ_1 as parameters. These parameters are determined with the use of equation (4.94), (4.95), and (4.97).

Substitution of equation (4.100) in equation (4.97) yields

$$\int_0^1 x^{-\lambda_1}(1 - x)^{-\lambda_2} dx = \frac{1}{a} \tag{4.101}$$

Equation (4.101) can be expressed in terms of the beta function as

$$\frac{1}{a} = \int_0^1 x^{m-1}(1-x)^{n-1}dx = B(m, n), \quad m = 1 - \lambda_1, \ n = 1 - \lambda_2 \tag{4.102}$$

Therefore, equation (4.100) becomes

$$f(x) = \frac{1}{B(m, n)}x^{m-1}(1-x)^{n-1} \tag{4.103}$$

which is the beta distribution of first kind. Parameters m and n are determined from:

$$\frac{1}{B(m, n)}\int_0^1 x^{m-1}(1-x)^{n-1}\ln x \, dx = \overline{\ln x} \tag{4.104}$$

$$\frac{1}{B(m, n)}\int_0^1 x^{m-1}(1-x)^{n-1}\ln(1-x) \, dx = \overline{\ln(1-x)} \tag{4.105}$$

Now a slightly different approach, alluded to in the beginning and discussed in Chapter 3, is considered. From equation (4.99), one writes

$$\exp(\lambda_0) = \int_0^1 \exp\left[-\lambda_1 \ln x - \lambda_2 \ln(1-x)\right] dx \tag{4.106}$$

The quantity $z(\lambda) = \exp(\lambda_0)$ is also called the partition function, and leads to

$$\lambda_0 = \ln \int_0^1 \exp\left[-\lambda_1 \ln x - \lambda_2 \ln(1-x)\right] dx \tag{4.107}$$

One can also simplify equation (4.106) as

$$\lambda_0 = \ln \Gamma(1 - \lambda_1) + \ln \Gamma(1 - \lambda_2) - \ln \Gamma(2 - \lambda_1 - \lambda_2) \tag{4.108}$$

Differentiating equation (4.107) with respect to λ_1 and λ_2, respectively, one gets

$$\frac{\partial \lambda_0}{\partial \lambda_1} = -E[\ln x] \tag{4.109}$$

$$\frac{\partial \lambda_0}{\partial \lambda_2} = -E[\ln(1-x)] \tag{4.110}$$

It should be emphasized that derivatives of the zeroth Lagrange multiplier with respect to other multipliers are equal to the constraints specified. Similarly, differentiating equation (4.108) with respect to λ_1 and λ_2 one obtains, respectively:

$$\frac{\partial \lambda_0}{\partial \lambda_1} = -\psi(m) + \psi(m+n), \quad m = 1 - \lambda_1, \ n = 1 - \lambda_2 \tag{4.111}$$

$$\frac{\partial \lambda_0}{\partial \lambda_2} = -\psi(n) + \psi(m+n) \tag{4.112}$$

where $\psi(m)$ and $\psi(m+n)$ are digamma functions defined as

$$\frac{\partial \ln(\Gamma(m))}{\partial \lambda_1} = -\psi(m) \qquad \frac{\partial \ln(\Gamma(m+n))}{\partial \lambda_2} = \psi(m+n) \tag{4.113}$$

Equating equations (4.109) and (4.110) to equations (4.111) and (4.112), respectively, one obtains:

$$E[\ln x] = \psi(m) - \psi(m+n) \tag{4.114}$$

$$E[\ln(1-x)] = \psi(m) - \psi(m+n) \tag{4.115}$$

The left side of equation (4.114) and of equation (4.115) is known and hence m and n can be found, and then in turn λ_1 and λ_2 and finally m and n.

Example 4.6: Let a random variate vary from 0 to 1 with a mean value of $E[\ln x] = -0.2$ and $E[\ln(1-x)] = -1.8$, find the maximum entropy probability distribution and graph it.

Solution: For this case, the distribution is the beta distribution of first kind. From equations (4.103), (4.104) and (4.105),

$$f(x) = \frac{1}{B(m,\ n)} x^{m-1}(1-x)^{n-1}$$

$$\frac{1}{B(m,\ n)} \int_0^1 x^{m-1}(1-x)^{n-1} \ln x\, dx = -0.2$$

$$\frac{1}{B(m,\ n)} \int_0^1 x^{m-1}(1-x)^{n-1} \ln(1-x)\, dx = -1.8$$

First the equations below are solved:

$$E[\ln x] = \psi(m) - \psi(m+n)$$

$$E[\ln(1-x)] = \psi(n) - \psi(m+n)$$

One gets: $m = 25.6925$, and $n = 5.5785$.
  Then one gets:

$$f(x) = \frac{1}{B(25.6925,\ 5.5785)} x^{24.6925-1}(1-x)^{4.4785-1}$$

The plot is shown in Figure 4.1.

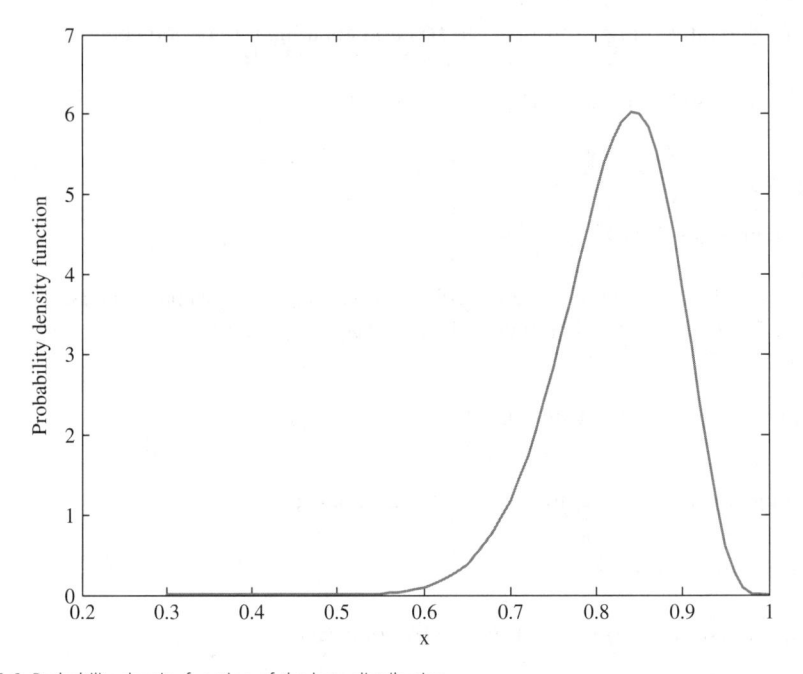

Figure 4.1 Probability density function of the beta distribution.

4.2.4 Semi-infinite interval $(0, \infty)$, one constraint $E[x]$ and exponential distribution

In this case the constraint equation is given as

$$\int_0^\infty x f(x) dx = E[x] = \bar{x} \tag{4.116}$$

The Shannon entropy is written as

$$H = -\int_0^\infty f(x) \ln f(x) dx \tag{4.117}$$

where

$$\int_0^\infty f(x) dx = 1 \tag{4.118}$$

The least-biased $f(x)$ is determined by maximizing equation (4.117), subject to equations (4.116) and (4.118). To that end, the Lagrangean L is constructed as

$$L = -\int_0^\infty f(x) \ln f(x) dx - (\lambda_0 - 1) \left[\int_0^\infty f(x) dx - 1 \right] - \lambda_1 \left[\int_0^\infty x f(x) dx - k \right] \tag{4.119}$$

Taking the derivative of L with respect to $f(x)$ and equating it to 0, one obtains

$$\frac{\partial L}{\partial f(x)} = 0 \Rightarrow -[1 + \ln f(x)] - (\lambda_0 - 1) - \lambda_1 x = 0 \tag{4.120}$$

Therefore,

$$f(x) = \exp(-\lambda_0 - \lambda_1 x) \tag{4.121}$$

Equation (4.121) is the POME-based distribution with λ_0 and λ_1 as parameters. Substituting equation (4.121) in equation (4.118), one obtains

$$\int_0^\infty \exp(-\lambda_0 - \lambda_1 x) dx = \lambda_1 \exp(\lambda_0) = 1 \tag{4.122}$$

Substituting equation (4.122) in equation (4.121), one gets

$$f(x) = \lambda_1 \exp(-\lambda_1 x) \tag{4.123}$$

Inserting equation (4.123) in equation (4.116), one gets

$$\int_0^\infty \lambda_1 x \exp(-\lambda_1 x) dx = k \ \ or \ \ \lambda_1 = \frac{1}{k} \tag{4.124}$$

Thus equation (4.123) becomes

$$f(x) = \frac{1}{k} \exp\left(-\frac{x}{k}\right), \ k = \bar{x} \tag{4.125}$$

which is the exponential distribution.

4.2.5 Semi-infinite interval, two constraints $E[x]$ and $E[\ln x]$ and gamma distribution

The constraint equations are expressed as equations (4.116) and (4.118) and

$$\int_0^\infty \ln x f(x) \, dx = \overline{\ln x} \tag{4.126}$$

To obtain the least-biased $f(x)$, H in equation (4.117) is maximized, subject to equations (4.118), (4.116), and (4.126). To that end the Lagrangean L is expressed as

$$L = -\int_0^\infty f(x) \ln f(x) dx - (\lambda_0 - 1) \left[\int_0^\infty f(x) dx - 1 \right] - \lambda_1 \left[\int_0^\infty x f(x) dx - \bar{x} \right]$$

$$- \lambda_2 \left[\int_0^\infty \ln x f(x) dx - \overline{\ln x} \right] \tag{4.127}$$

Differentiating equation (4.127) with respect to $f(x)$ and equating the derivative to 0, one gets

$$\frac{\partial L}{\partial f(x)} = 0 \Rightarrow f(x) = ax^{-\lambda_2} \exp(-\lambda_1 x), \ a = \exp(-\lambda_0) \tag{4.128}$$

Equation (4.128) is the least-biased POME-based distribution with a, λ_0 and λ_1 as parameters. Inserting equation (4.128) in equation (4.118), one obtains

$$\int_0^\infty x^{-\lambda_2} \exp(-\lambda_1 x) dx = \frac{1}{a} \tag{4.129}$$

Recall the definition of a gamma function:

$$\Gamma(n) = b^n \int_0^\infty x^{n-1} \exp(-bx) dx \tag{4.130}$$

Equation (4.129) can be recast as

$$\frac{1}{a} = \int_0^\infty x^{n-1} \exp(-bx) dx = \frac{\Gamma(n)}{b^n}, \quad n = 1 - \lambda_2; \ \lambda_1 = b \tag{4.131}$$

Thus, equation (4.128) becomes

$$f(x) = \frac{b^n}{\Gamma(n)} x^{n-1} \exp(-bx) \tag{4.132}$$

which is the gamma distribution.

Example 4.7: Let a random variate vary from 0 to ∞ with a mean value of $E[x] = 3$ and $E[\ln(x)] = 1$, find the maximum entropy probability distribution.

Solution: Suppose the probability of random variable X is $f(x)$. In this case, the distribution is the gamma distribution. Substituting equation (4.132) in the two constraint equations, one gets

$$f(x) = \frac{b^n}{\Gamma(n)} x^{n-1} \exp(-bx)$$

$$\int_0^\infty \frac{b^n}{\Gamma(n)} x^n \exp(-bx) \, dx = 2$$

$$\int_0^\infty \ln(x) \frac{b^n}{\Gamma(n)} x^{n-1} \exp(-bx) \, dx = 1$$

According to Singh (1998), we have the equation below for the parameter:

$$\frac{n}{b} = \bar{x} = 3$$

$$\Psi(n) - \ln n = E(\ln x) - \overline{\ln x} = -0.0986$$

Solving the above equations, one gets:

$$b = 1.744, \ n = 5.231$$

Then the maximum entropy probability distribution can be expressed as:

$$f(x) = \frac{1.744^{5.231}}{\Gamma(5.231)} x^{4.231} \exp(-1.744x)$$

4.2.6 Semi-infinite interval, two constraints $E[\ln x]$ and $E[\ln(1 + x)]$ and beta distribution of second kind

In this case the constraint equations are equation (4.126) and

$$\int_0^\infty \ln(1 + x) f(x) \, dx = E[\ln(1 + x)] = \overline{\ln(1 + x)} \tag{4.133}$$

The Shannon entropy given by equation (4.117) is maximized, subject to equation (4.126), (4.133) and (4.118), in order to obtain the least-biased $f(x)$. To that end, the Lagrangian L is expressed as

$$L = -\int_0^\infty f(x) \ln f(x) dx - (\lambda_0 - 1) \left[\int_0^\infty f(x) dx - 1 \right] - \lambda_1 \left[\int_0^\infty \ln x f(x) dx - \overline{\ln x} \right]$$

$$-\lambda_2 \left[\int_0^\infty \ln(1 + x) f(x) dx - \overline{\ln(1 + x)} \right] \tag{4.134}$$

Differentiating L with respect to $f(x)$ and equating the derivative to 0, one obtains

$$\frac{\partial L}{\partial f(x)} = 0 \Rightarrow f(x) = ax^{-\lambda_1}(1 + x)^{-\lambda_2}, \ a = \exp(-\lambda_0) \tag{4.135}$$

Equation (4.135) is the POME-based distribution with parameters a, λ_1 and λ_2. These parameters are determined using equations (4.126), (4.133) and (4.118).

Substituting equation (4.135) in equation (4.118), one gets

$$\int_0^\infty x^{-\lambda_1}(1 + x)^{-\lambda_2} dx = \frac{1}{a} \tag{4.136}$$

Recalling the definition of the beta function, equation (4.136) can be written as

$$\frac{1}{a} = \int_0^\infty x^{m-1}(1 + x)^{-(m+n)} dx = B(m, \ n), \ m - 1 = -\lambda_1; \ m + n = \lambda_2 \tag{4.137}$$

Therefore, equation (4.135) becomes

$$f(x) = \frac{1}{B(m, \ n)} \frac{x^{m-1}}{(1 + x)^{m+n}} \tag{4.138}$$

which is the beta distribution of the second kind with parameters m and n determined from equations (4.118) and (4.133).

Example 4.8: Let a random variate vary from 0 to ∞ with a mean value of $E[\ln x] = 1$ and $E[\ln(1 + x)] = 2$, find the maximum entropy probability distribution.

Solution: Let the probability distribution of the random variable X be $f(x)$. For this problem, the distribution is the beta distribution of the second kind. According to equations (4.138), (4.126) and (4.133),

$$f(x) = \frac{1}{B(m, n)} \frac{x^{m-1}}{(1 - x)^{m+n}} \tag{4.139}$$

$$\frac{1}{B(m, n)} \int_0^\infty x \frac{x^{m-1}}{(1 - x)^{m+n}} \, dx = 1 \tag{4.140}$$

$$\frac{1}{B(m, n)} \int_0^\infty \ln(x) \frac{x^{m-1}}{(1 - x)^{m+n}} \, dx = 2 \tag{4.141}$$

From equation (4.139)–(4.141), one can obtain the equations below:

$$\psi(m) - \psi(n) = -1$$

$$\psi(m) - \psi(m + n) = -2$$

where $m = \lambda_1 + \lambda_2 - 1$; $n = 1 - \lambda_1$. Then one gets $m = 0.3896$, $n = 0.5740$. The maximum entropy probability distribution is expressed as:

$$f(x) = \frac{1}{B(0.3896, 0.5740)} \frac{x^{-0.6104}}{(1 - x)^{0.9636}}$$

4.2.7 Infinite interval, two constraints $E[X]$ and $E[X^2]$ and normal distribution

In this case the constraint equations are

$$E[x] = m \tag{4.142}$$

$$E[x^2] = \sigma^2 + m^2 \text{ or } E[(x - m)^2] = \sigma^2 \tag{4.143}$$

where m is the mean of X, and σ^2 is the variance of X. Furthermore,

$$\int_{-\infty}^\infty f(x) \, dx = 1 \tag{4.144}$$

The Shannon entropy is expressed as

$$H = - \int_{-\infty}^\infty f(x) \ln f(x) \, dx \tag{4.145}$$

Equation (4.145) is maximized, subject to equations (4.142), (4.143) and (4.144), in order to obtain the least-biased $f(x)$. Hence the Lagrangean L is expressed as:

$$L = -\int_{-\infty}^{\infty} f(x)\ln f(x)dx - (\lambda_0 - 1)\left[\int_{-\infty}^{\infty} f(x)dx - 1\right] - \lambda_1\left[\int_{-\infty}^{\infty} xf(x)dx - m\right]$$
$$- \lambda_2\left[\int_{0}^{\infty} x^2 f(x)dx - \sigma^2 - m^2\right] \tag{4.146}$$

Differentiating equation (4.146) with respect to $f(x)$ and equating the derivative to 0, one gets

$$\frac{\partial L}{\partial f(x)} = 0 \Rightarrow f(x) = \exp(-\lambda_0 - \lambda_1 x - \lambda_2 x^2) \tag{4.147}$$

Equation (4.147) is the POME-based distribution with λ_0, λ_1 and λ_2 as parameters. These parameters are determined using equations (4.142)–(4.144) as:

$$\int_{-\infty}^{\infty} \exp(-\lambda_0 - \lambda_1 x - \lambda_2 x^2)dx = 1 \tag{4.148}$$

$$\int_{-\infty}^{\infty} x\exp(-\lambda_0 - \lambda_1 x - \lambda_2 x^2)dx = m \tag{4.149}$$

$$\int_{-\infty}^{\infty} x^2 \exp(-\lambda_0 - \lambda_1 x - \lambda_2 x^2)dx = \sigma^2 + m^2 \tag{4.150}$$

Note that equation (4.148) can be expressed as

$$\exp(\lambda_0) = \int_{-\infty}^{\infty} \exp\left(-\lambda_1 x - \lambda_2 x^2\right)dx = \exp\left(\frac{\lambda_1^2}{4\lambda_2}\right)\int_{-\infty}^{\infty} \exp-\left[\sqrt{\lambda_2}x + \frac{\lambda_1}{2\sqrt{\lambda_2}}\right]^2 dx \tag{4.151}$$

Taking $t = \sqrt{\lambda_2}x + \frac{\lambda_1}{2\sqrt{\lambda_2}}$, equation (4.151) can be written as

$$\exp(\lambda_0) = \frac{2\exp\left(\frac{\lambda_1^2}{4\lambda_2}\right)}{\sqrt{\lambda_2}}\int_{-\infty}^{\infty} \exp\left(-t^2\right)dt = \frac{\exp\left(\frac{\lambda_1^2}{4\lambda_2}\right)}{\sqrt{\lambda_2}}\sqrt{\pi} \tag{4.152}$$

A little algebraic manipulation yields (Singh, 1998):

$$\lambda_1 = -\frac{m}{\sigma^2}; \quad \lambda_2 = \frac{1}{2\sigma^2} \tag{4.153}$$

Finally, the resulting distribution is

$$f(x) = \frac{1}{\sqrt{2\pi}\,\sigma}\exp\left[-\frac{1}{2}\left(\frac{x-m}{\sigma}\right)^2\right] \tag{4.154}$$

which is the probability density function of the normal distribution, and corresponds to the case when mean and variance are known. It may be remarked that normal distribution can also be derived using only the variance as a constraint, because variance includes mean (Krstanovic and Singh, 1988).

Example 4.9: Let X be a random variable over the range from $-\infty$ to ∞ with a mean value of $E[x] = 0$ and $E[x^2] = 1$. Find the maximum entropy probability distribution of X and graph it.

Solution: Let the probability density function of X be denoted by $f(x)$. In this case, the distribution is the normal distribution. From equations (4.147)–(4.149) and (4.150) one obtains

$E[x] = m = 0$; $E[x^2] = \sigma^2 + m^2 = 1$. Thus, $m = 0$ and $\sigma = 1$. $f(x) = \exp(-\lambda_0 - \lambda_1 x - \lambda_2 x^2)$

$$\int_{-\infty}^{\infty} \exp(-\lambda_0 - \lambda_1 x - \lambda_2 x^2)dx = 1 \tag{4.155}$$

$$\int_{-\infty}^{\infty} x \exp(-\lambda_0 - \lambda_1 x - \lambda_2 x^2)dx = 0 \tag{4.156}$$

$$\int_{-\infty}^{\infty} x^2 \exp(-\lambda_0 - \lambda_1 x - \lambda_2 x^2)dx = 1 \tag{4.157}$$

According to $\lambda_1 = -\frac{m}{\sigma^2}$, $\lambda_2 = \frac{1}{2\sigma^2}$, one gets $\lambda_1 = 0$, $\lambda_2 = \frac{1}{2}$, $\lambda_0 = \frac{1}{2}\ln 2\pi + \frac{1}{2}\ln 2$.
The maximum entropy probability distribution is

$$f(x) = \frac{1}{\sqrt{2\pi}\,\sigma} \exp\left[-\frac{1}{2}\left(\frac{x-m}{\sigma}\right)^2\right] = \frac{1}{\sqrt{2\pi}} \exp\left[-\frac{1}{2}x^2\right] \tag{4.158}$$

Equation (4.158) can be obtained from equation (4.154) directly and it is graphed in Figure 4.2.

4.2.8 Semi-infinite interval, log-transformation $Y = \ln X$, two constraints $E[y]$ and $E[y^2]$ and log-normal distribution

Let there be a normal random variable Y over the interval $(-\infty, \infty)$ and be another random variable X related to Y as $Y = \ln X$. If Y is normally distributed then X would be log-normally distributed over the interval $(0, \infty)$. The constraint equations are equation (4.144) and

$$E[y] = m_y \text{ or } E[x] = E[\exp(y)] = m_x \tag{4.159}$$

$$E[y^2] = \sigma_y^2 + m_y^2 \tag{4.160}$$

Equation (4.145) is maximized, subject to equations (4.144), (4.159) and (4.160), in order to obtain the least-biased $f(x)$. Hence the Lagrangean L is expressed as:

$$L = -\int_{-\infty}^{\infty} f(x)\ln f(x)dx - (\lambda_0 - 1)\left[\int_{-\infty}^{\infty} f(x)dx - 1\right] - \lambda_1\left[\int_{-\infty}^{\infty}(\ln x)f(x)dx - m_x\right]$$

$$-\lambda_2\left[\int_{-\infty}^{\infty}(\ln x)^2 f(x)dx - \sigma_y^2 - m_y^2\right] \tag{4.161}$$

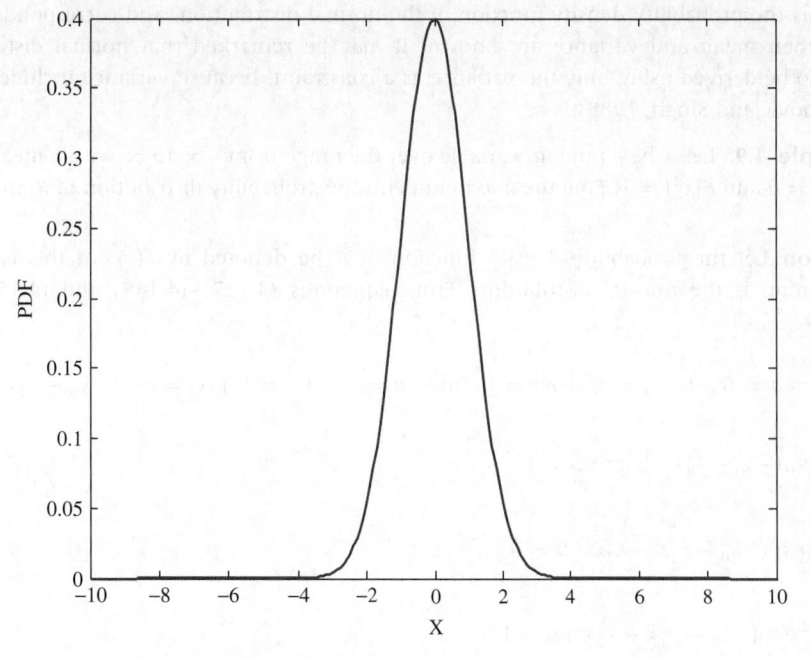

Figure 4.2 Probability density function of the maximum entropy distribution.

Differentiating equation (4.161) with respect to $f(x)$ and equating the derivative to 0, one gets

$$\frac{\partial L}{\partial f(x)} = 0 \Rightarrow f(x) = \exp[-\lambda_0 - \lambda_1 \ln x - \lambda_2 (\ln x)^2] \tag{4.162}$$

Equation (4.147) is the POME-based distribution with λ_0, λ_1 and λ_2 as parameters which can be determined by substitution of equation (4.162) in equations (4.144) and (4.159) and (4.160) as:

$$\int_{-\infty}^{\infty} \exp[-\lambda_0 - \lambda_1 \ln x - \lambda_2 (\ln x)^2] dx = 1 \tag{4.163}$$

$$\int_{-\infty}^{\infty} x \exp[-\lambda_0 - \lambda_1 \ln x - \lambda_2 (\ln x)^2] dx = m_y \tag{4.164}$$

$$\int_{-\infty}^{\infty} x^2 \exp[-\lambda_0 - \lambda_1 \ln x - \lambda_2 (\ln x)^2] dx = \sigma_y^2 + m_y^2 \tag{4.165}$$

A little algebraic manipulation yields (Singh, 1998):

$$\lambda_0 = \frac{1}{2} \ln \pi - \frac{1}{2} \ln \lambda_2 + \frac{(\lambda_1 - 1)^2}{4\lambda_2} \tag{4.166}$$

$$\lambda_1 = 1 - \frac{m_y}{\sigma_y^2} \tag{4.167}$$

$$\lambda_2 = \frac{1}{2\sigma_y^2} \tag{4.168}$$

Finally, the resulting distribution is

$$f(x) = \frac{1}{x\sqrt{2\pi}\,\sigma_y}\exp\left[-\frac{1}{2}\left(\frac{\ln x - m_y}{\sigma_y^2}\right)^2\right] \tag{4.169}$$

which is the probability density function of the lognormal distribution, and corresponds to the case when mean and variance are known.

4.2.9 Infinite and semi-infinite intervals: constraints and distributions

Following the above methodology, a number of probability distributions can be derived for given constraints. Without going through the algebra many of these distributions are summarized here.

Three-parameter lognormal distribution: constraints: $E[\ln(x - a)]$ and $E\{[\ln(x - a)]^2\}$, where $a > 0$ is a parameter. These constraints lead to a three-parameter lognormal distribution:

$$f(x) = \frac{1}{(x - a)\sigma_y\sqrt{2\pi}}\exp\left\{-\frac{[\ln(x - a) - m_y]^2}{2\sigma_y^2}\right\} \tag{4.170}$$

where $Y = \ln(x - a)$, m_y is the mean of Y, and σ_y^2 is the variance of Y.

Extreme value type I (or Gumbel) distribution: constraints: $E(x)$ and $E(\exp(-ax))$, where $a > 0$ and $-\infty < b < x$ are parameters. These constraints lead to extreme value type I (or Gumbel) distribution:

$$f(x) = a\exp\{-a(x - b) - \exp[-a(x - b)]\} \tag{4.171}$$

Log-extreme value type I distribution: constraints: $E(\ln x)$ and $E\{\exp[-a(\ln x - b)]\} = 1$, where a and b are parameters. These lead to the log-extreme value type I distribution:

$$f(x) = \frac{a}{x}\exp\{-a(\ln x - b) - \exp[-(\ln x - b)]\} \tag{4.172}$$

Extreme value type III distribution: constraints: $E[\ln(x - c)]$ and $E[(x - c)^a] = (b - c)^a$, where $a > 0$, $b > 0$ and c are parameters. These constraints yield the extreme value type III distribution:

$$f(x) = \frac{a}{b - c}\left(\frac{x - c}{b - c}\right)^{a-1}\exp\left[-\left(\frac{x - c}{b - c}\right)^a\right] \tag{4.173}$$

Here $(b - c)^a = E[(x - c)^a]$; $\psi(1) - \ln b = E[\ln(x - c)]$, where ψ is the digamma function.

Generalized extreme value distribution: constraints: $-E[\ln(1 - \frac{b}{a}(x - c)]$ and $E[(1 - \frac{b}{a}(x - c)]^{1/b}$, where $a > 0$, b and c are parameters. These constraints produce the generalized extreme value distribution:

$$f(x) = \frac{1}{a}\left[1 - \frac{b}{a}(x - c)\right]^{(1-b)/b}\exp\left\{-\left[1 - \frac{b}{a}(x - c)\right]^{1/b}\right\} \tag{4.174}$$

Weibull distribution: constraints: $E[\ln x]$ and $E[x^a] = b^a$, where $a > 0$ and $b > 0$ are parameters. These constraints yield the Weibull distribution:

$$f(x) = \frac{a}{b} \left(\frac{x}{b}\right)^{a-1} \exp\left[-\left(\frac{x}{b}\right)^a\right]$$

(4.175)

Here $b^a = E[\ln x^a]$ and $\psi(1) - \ln b = E[\ln x]$.

Pearson type III distribution: constraints: $E[\ln x] = ab + c$ and $E[\ln(x - c)]$, $a > 0$, $b > 0$, and $0 < c < x$ are parameters. These constraints lead to the Pearson type III distribution:

$$f(x) = \frac{1}{a\Gamma(b)} \left(\frac{x - c}{a}\right)^{b-1} \exp\left[-\left(\frac{x - c}{a}\right)\right]$$

(4.176)

Log-Pearson type III distribution: constraints: $E[\ln x]$ and $E[\ln(\ln x - c)]$: Also, $ab + c = E[\ln x]$ and $ba^2 = \sigma_y^2$. Here, $a > 0$ $b > 0$ and $c < 0 < \ln x$ are parameters and $Y = \ln X$. These constraints lead to the log-Pearson type III distribution:

$$f(x) = \frac{1}{ax\Gamma(b)} \left(\frac{\ln x - c}{a}\right)^{b-1} \exp\left[-\left(\frac{\ln x - c}{a}\right)\right]$$

(4.177)

Log-logistic distribution: constraints: $E[\ln x]$ and $E\{\ln[1 + (\frac{x}{a})^b]\}$, $a > 0$, and $b \geq 0$ are parameters. These constraints lead to the log-logistic distribution:

$$f(x) = \frac{(b/a)(x/a)^{b-1}}{[1 + (x/a)^b]^2}$$

(4.178)

Three-parameter log-logistic distribution: constraints: $E[\ln(\frac{x-c}{a})]$ and $E[1 + (\frac{x-c}{a})^b]$, a, b, and c are parameters. These constraints lead to the three-parameter log-logistic distribution:

$$f(x) = \frac{(b/a)[(x - c)/a]^{b-1}}{\{1 + [(x - c)/a]^b\}^2}$$

(4.179)

Two-parameter Pareto distribution: constraint: $E(\ln x)$. This constraint yields the two-parameter Pareto distribution:

$$f(x) = ba^b x^{-b-1}$$

(4.180)

Here $\frac{1}{b} + \ln a = E[\ln x]$, where $a > 0$ and $b > 0$ are parameters.

Two-parameter generalized Pareto distribution: constraint: $E[\ln(1 - a\frac{x}{b})]$, where a and b are parameters. This constraint leads to the two-parameter generalized distribution:

$$f(x) = \frac{1}{b} \left(1 - a\frac{x}{b}\right)^{\frac{1}{a}-1}, \quad a \neq 0$$

(4.181)

$$f(x) = \frac{1}{b} \exp\left(-\frac{x}{b}\right), \quad a = 0$$

(4.182)

Three-parameter generalized Pareto distribution: constraints: $E\left[\ln(1 - a\frac{x-c}{b})\right]$, where a, b, and c are parameters. This constraint yields the three-parameter generalized Pareto distribution:

$$f(x) = \frac{1}{b}\left[1 - a\frac{x-c}{b}\right]^{\frac{1}{a}-1}, \quad a \neq 0 \tag{4.183}$$

$$f(x) = \frac{1}{b}\exp\left(-\frac{x-c}{b}\right), \quad a = 0 \tag{4.184}$$

Laplace distribution: constraint: $E[|x|]$. This leads to the Laplace distribution:

$$f(x) = \frac{1}{\sigma}\exp\left[-\frac{|x|}{\sigma}\right] \tag{4.185}$$

Cauchy distribution: constraint: $E[\ln(1 + x^2)]$. This yields to the distribution:

$$f(x) = \frac{\Gamma(b)}{\sqrt{\pi}\Gamma\left(b - \frac{1}{2}\right)}\frac{1}{(1 + x^2)^b}, \quad b > \frac{1}{2} \tag{4.186}$$

where b is a parameter. When $b = 1$, the distribution specializes into the Cauchy distribution.

Rayleigh distribution: constraints: $E[\ln x]$ and $E[x^2]$. These constraints lead to the Rayleigh distribution:

$$f(x) = \frac{x}{\alpha^2}\exp\left[-\frac{x^2}{2\alpha^2}\right] \tag{4.187}$$

where $\alpha > 0$ is a parameter.

Chi-square distribution: constraints: $E[\ln x]$ and $-E[x]$. These constraints lead to the Chi-square distribution with k degrees of freedom:

$$f(x) = \frac{x^{(k-2)/2}\exp(-x/2)}{2^{k/2}\Gamma(k/2)} \tag{4.188}$$

Inverse normal (Gaussian) distribution: constraints: $E(x) = \mu$: $E(\frac{1}{x}) = \frac{1}{\lambda} + \frac{1}{\mu}$: These constraints lead to the inverse Gaussian distribution by letting $X = 1/Y^2$ in which Y has the probability density function as: $f(y) = \frac{2}{\sqrt{2\pi}\xi}\exp\left[-\frac{(y-v/y)^2}{2\xi^2}\right]$; $v = 1/\mu$, $\xi^2 = 1/\lambda$, the inverse Gaussian maximum entropy distribution can be obtained as:

$$f(x) = \left(\frac{\lambda}{2\pi x^3}\right)^{1/2}\exp\left\{-\frac{\lambda}{2\mu^2 x}(x - \mu)^2\right\} \tag{4.189}$$

Generalized gamma probability distribution: constraints: $E[\ln x]$ and $E[x^c]$: These constraints lead to

$$f(x) = \frac{c\lambda_2^{(1-\lambda_1)/c}}{\Gamma\left(\frac{1-\lambda_1}{c}\right)}x^{-\lambda_1}\exp(-\lambda_2 x^c) \tag{4.190}$$

Using the Maxwell-Boltzmann statistic, Lienhard (1964, 1972) derived equation (4.190) as a generalized probability distribution but his procedure is much more complicated. Equation (4.190) has three parameters: λ_1 λ_2 and c. Exponent c can be either specified or determined by trial and error or can be estimated using the entropy method. Equation (4.190) specializes into several distributions. For example, for $c = 1$, $\lambda_2 = 1/k$, and $(c - \lambda_1)/c = n$ it leads to a two-parameter gamma distribution. If x is replaced by $x - x_0$ then it would result in a Pearson type III distribution. If $y = \log x$ then equation (4.190) would lead to a log-Pearson type III distribution. If $c = 2$, $\lambda_2 = (m + 1)/(2k^2)$, and $\lambda_1 = -m$ then equation (4.190) becomes

$$h(t) = \frac{2}{k\Gamma\left(\frac{m+1}{2}\right)} \left(\frac{m+1}{2}\right)^{(m+1)/2} \left(\frac{x}{k}\right)^m \exp\left[-\frac{m+1}{2}\left(\frac{x}{k}\right)^2\right] \tag{4.191}$$

Equation (4.191) is the Lienhard equation. For $m = 2$, equation (4.191) reduces to

$$h(t) = \frac{1}{k\Gamma\left(\frac{3}{2}\right)} \left(\frac{3}{2}\right)^{3/2} \left(\frac{x}{k}\right)^2 \exp\left[-\frac{3}{2}\left(\frac{x}{k}\right)^2\right] \tag{4.192}$$

Lienhard (1964) used equation (4.192) for representing the unit hydrograph.

If $c = 2$, $\lambda_2 = (a/b)$, and $\lambda_1 = 1 - 2a$ then equation (4.190) becomes

$$h(t) = \frac{2}{\Gamma(a)} \left(\frac{a}{b}\right)^a x^{2a-1} \exp\left[-\frac{a}{b}x^2\right] \tag{4.193}$$

Equation (4.193) is the Nakagami–m distribution function and is a slightly different form of equation (4.190) and has received some attention in recent years (Rai et al., 2010). Equation (4.193) is another form of the IUH (instantaneous unit hydrograph) equation.

If $c = 1$, $\lambda_1 = 0$, and $\lambda_2 = 1/k$, equation (4.190) reduces to an exponential distribution. For $\lambda_1 = 1 - c$, and $\lambda_2 = (1/k^c)$, equation (4.190) reduces to the Weibull distribution. If $c = 2$, $\lambda_1 = -1$ and $\lambda_2 = (1/k^c)$ then equation (4.190) becomes the Rayleigh distribution used in reliability analysis. If $c = 2$, $\lambda_2 = (1/k^2)$, and $\lambda_1 = 0$ then equation (4.190) becomes

$$h(t) = \frac{1}{k}\frac{2}{\sqrt{\pi}} \exp\left[-\left(\frac{t}{k}\right)^2\right] \tag{4.194}$$

Equation (4.194) is the Maxwell molecular speed distribution which is used in quantum physics.

Extended Burr III (EBIII) distribution: constraints: $E[\ln(\frac{b}{x})]$ and $E[\ln[1 - a(\frac{b}{x})^c]]$. These constraints lead to the extended Burr III distribution:

$$f(x) = cb^{-1}\left(\frac{b}{x}\right)^{c+1}\left[1 - a\left(\frac{b}{x}\right)^c\right]^{\left(\frac{1}{a}-1\right)}, \quad a \neq 0 \tag{4.195a}$$

$$= cb^{-1}\left(\frac{b}{x}\right)^{c+1}\exp\left(-\left(\frac{b}{x}\right)^c\right), \quad a = 0 \tag{4.195b}$$

For $a \leq 0$, the range of x is from 0 to infinity, or $0 \leq x \leq \infty$; for $a > 0$, $x \geq ba^{1/c}$; for $a = 0$ this distribution corresponds to the Fréchet Distribution (Shao et al., 2008). For low flow analysis, one can get the EBIII distribution by applying the transformation $x \rightarrow 1/x$ to the extended Burr XII distribution (Shao et al., 2008; Hao and Singh, 2009a).

Extended three-parameter Burr XII distribution: constraints: $E[\ln(\frac{x}{b})]$ and $E[\ln(1 - a(\frac{x}{b})^c)]$: These constraints lead to the extended three-parameter Burr XII distribution:

$$f(x) = cb^{-1} \left(\frac{x}{b}\right)^{c-1} \left[1 - a\left(\frac{x}{b}\right)^c\right]^{(\frac{1}{a}-1)}, \quad a \neq 0 \tag{4.196a}$$

$$= cb^{-1} \left(\frac{x}{b}\right)^{c-1} e^{-(\frac{x}{b})^c}, \quad a = 0 \tag{4.196b}$$

for $a \leq 0$, $0 \leq x \leq \infty$, and for $a > 0$, $0 \leq x \leq ba^{-1/c}$; for $a = 0$ it corresponds to the Weibull distribution; b is a scale parameter; and c is a shape parameter. The location parameter is determined by c and a. The probability density function is unimodal at $\text{mod}(x) = [(c-1)/(c-a)]^{1/c}$ if $c > 1$ and $a < 1$; L-shaped if $c \leq 1$ and $a \geq 1$; and J-shaped otherwise (Shao et al., 2004; Hao and Singh, 2009b).

Triangular distribution: constraint: $E[x]$: Here $0 \leq x \leq a$. This constraint leads to a triangular probability distribution:

$$f(x) = \frac{2x}{a}, \quad 0 \leq x \leq a \tag{4.197a}$$

If $a \leq x \leq 1$, then the constraint becomes $E[1 - x]$ and the triangular probability density function becomes

$$f(x) = \frac{2(1-x)}{1-a}, \quad a \leq x \leq 1 \tag{4.197b}$$

Questions

Q.4.1 The M-B distribution is a one-parameter distribution determined from the average value of the attribute under consideration. Plot the distribution for different values of the average and discuss how its shape changes with changes in the mean value. Based on the shape, can this distribution be applied to any problem in water and environmental engineering that you can think of? Derive the entropy of the M-B distribution and plot it as a function of its parameter(s). Discuss the graph and reflect on its physical import.

Q.4.2 Consider a third-order river basin. Obtain elevations of all the links in the basin and graph the probability distribution of link elevations. Determine the mean elevation. Then, use the M-B distribution to represent the probability distribution of elevations of links and comment on its adequacy.

Q.4.3 For the basin in Q.4.2, obtain lengths of channels of all orders and graph the probability distribution of channel lengths. Determine the average channel length. Then, use the M-B distribution to represent the probability distribution of channel lengths and comment on its adequacy.

Q.4.4 For the basin in Q.4.2, obtain drainage areas of channels of all orders and graph the probability distribution of channel areas. Determine the average channel drainage area. Then, use the M-B distribution to represent the probability distribution of areas and comment on its adequacy.

Q.4.5 For the basin in Q.4.2, obtain the distances of all first and second order channels from the basin outlet and graph the probability distribution of these distances corresponding to first order channels. Determine the mean distance. Then, use the M-B distribution to represent the probability distribution of channel distances and comment on its adequacy.

Q.4.6 The B-E distribution is a two-parameter distribution determined from the average values: number and magnitude of the attribute under consideration. Plot the distribution for different values of the average number and magnitude and discuss how the distribution shape changes with changes in the mean values. Based on the shape, can this distribution be applied to any problem in water and environmental engineering that you can think of? Derive the entropy of the B-E distribution and plot it as a function of its parameter(s). Discuss the graph and reflect on its physical import.

Q.4.7 For the basin in Q.4.2, obtain elevations of all the links in the basin. For each elevation (category or range) there can be a number of links. Determine the mean elevation and the mean number of links. The number of links of a particular elevation is a random variable. Graph the probability distribution of link numbers. Then, use the B-E distribution to represent the probability distribution of number of links and comment on its adequacy.

Q.4.8 For the basin in Q.4.2, obtain lengths of channels of all orders. Determine the average channel lengths and average number. Then, use the B-E distribution to represent the probability distribution of the number of channels and comment on its adequacy.

Q.4.9 For the basin in Q.4.2, obtain drainage areas of channels of all orders and graph the probability distribution of channel areas. Determine the average drainage area and number of channels. Then, use the B-E distribution to represent the probability distribution of the number of channels and comment on its adequacy.

Q.4.10 For the basin in Q.4.2, obtain the distances of channels of all orders from the basin outlet and graph the probability distribution of these distances. Determine the mean distance and mean number of channels. Then, use the B-E distribution to represent the probability distribution of number of channels and comment on its adequacy.

Q.4.11 The F-D distribution is a two-parameter distribution determined from the average values of the number of, say, particles and magnitude of, say, energy of particles in the system under consideration. This gives the distribution of the expected number of particles in each of, say, N states. Plot the distribution for different values of the average number and magnitude and discuss how the distribution shape changes with changes in the mean values. Based on the shape, can this distribution be applied to any problem in water and environmental engineering that you can think of? Derive the entropy of the F-D distribution and plot it as a function of its parameter(s). Discuss the graph and reflect on its physical import. Note that in this case each state contains either zero or one event or particle.

Q.4.12 The ISD is a two-parameter distribution determined from two constraints: the average values of the number and magnitude of the attribute in the system under consideration. This gives the distribution of the expected number of, say, particles in the i-th state. Plot the distribution for different values of the average number and magnitude and discuss how the distribution shape changes with changes in the mean values. Based on the shape, can this distribution be applied to any problem in water and environmental engineering that you can think of? Derive the entropy of the ISD and plot it as a function of its parameter(s). Discuss the graph and reflect on its physical import.

Q.4.13 The binomial distribution is a one-parameter distribution determined from one constraint, and gives the distribution of the number of k successes in N Bernoulli trials.

Plot the distribution for different values of k and magnitude and probability of success p and discuss how the distribution shape changes with changes in parameter values. Based on the shape, can this distribution be applied to any problem in water and environmental engineering that you can think of? Derive the entropy of this distribution and plot it as a function of its parameter(s). Discuss the graph and reflect on its physical import.

Q.4.14 The geometric distribution is a one-parameter distribution determined from one constraint, and gives the distribution of the number of trials k before the first success would occur in N Bernoulli trials. Plot the distribution for different values of k and probability of success p and discuss how the distribution shape changes with changes in the parameter values. Based on the shape, can this distribution be applied to any problem in water and environmental engineering that you can think of? Derive the entropy of this distribution and plot it as a function of its parameter(s). Discuss the graph and reflect on its physical import.

Q.4.15 The negative binomial distribution is a one-parameter distribution determined from one constraint, and gives the distribution of the number of N Bernoulli trials needed to achieve k successes. Plot the distribution for different values of k and magnitude and probability of success p and discuss how the distribution shape changes with changes in parameter values. Based on the shape, can this distribution be applied to any problem in water and environmental engineering that you can think of? Derive the entropy of this distribution and plot it as a function of its parameter(s). Discuss the graph and reflect on its physical import.

Q.4.16 The Poisson distribution is a one-parameter distribution determined from one constraint, and gives the probability distribution of the average number of occurrences per unit of time or space. Plot the distribution as a function of its parameter and discuss how the distribution shape changes with changes in the parameter values. Based on the shape, can this distribution be applied to any problem in water and environmental engineering that you can think of? Derive the entropy of this distribution and plot it as a function of its parameter(s). Discuss the graph and reflect on its physical import.

Q.4.17 Consider the exponential distribution and its truncated form. Both are one-parameter distributions determined from one constraint, and give the probability distribution of the random variable. Plot each distribution as a function of its parameter and discuss how the distribution shape changes with changes in parameter values. Based on the shape, can this distribution be applied to any problem in water and environmental engineering that you can think of? Derive the entropy of this distribution and plot it as a function of its parameter(s). Discuss the graph and reflect on its physical import.

Q.4.18 Consider the beta distributions of first kind and second kind. Both are two-parameter distributions determined from two constraints, and give the probability distribution of the random variable. Plot each distribution as a function of its parameters and discuss how the distribution shape changes with changes in parameter values. Based on the shape, can this distribution be applied to any problem in water and environmental engineering that you can think of? Derive the entropy of this distribution and plot it as a function of its parameters. Discuss the graph and reflect on its physical import.

Q.4.19 Consider the gamma distribution which is a two-parameter distribution determined from two constraints and gives the probability distribution of the random variable. Plot the distribution as a function of its parameters and discuss how the distribution shape changes with changes in parameter values. Based on the shape, can this distribution be applied to any problem in water and environmental engineering that you can think of? Derive the entropy of this distribution and plot it as a function of its parameters. Discuss the graph and reflect on its physical import.

Q.4.20 Consider the normal distribution which is a two-parameter distribution determined from two constraints and gives the probability distribution of the random variable. Plot the distribution as a function of its parameters and discuss how the distribution shape changes with changes in parameter values. Based on the shape, can this distribution be applied to any problem in water and environmental engineering that you can think of? Derive the entropy of this distribution and plot it as a function of its parameters. Discuss the graph and reflect on its physical import.

Q.4.21 Consider the two-parameter log-normal distribution which is a two-parameter distribution determined from two constraints and gives the probability distribution of the random variable. Plot the distribution as a function of its parameters and discuss how the distribution shape changes with changes in parameter values. Based on the shape, can this distribution be applied to any problem in water and environmental engineering that you can think of? Derive the entropy of this distribution and plot it as a function of its parameters. Discuss the graph and reflect on its physical import.

Q.4.22 Consider the three-parameter log-normal distribution which is a three-parameter distribution determined from two constraints and gives the probability distribution of the random variable. Plot the distribution as a function of its parameters and discuss how the distribution shape changes with changes in parameter values. Based on the shape, can this distribution be applied to any problem in water and environmental engineering that you can think of? Derive the entropy of this distribution and plot it as a function of its parameters. Discuss the graph and reflect on its physical import.

Q.4.23 Consider the extreme value type I distribution (also called the Gumbel distribution) which is a two-parameter distribution determined from two constraints and gives the probability distribution of the random variable. Plot the distribution as a function of its parameters and discuss how the distribution shape changes with changes in parameter values. Based on the shape, can this distribution be applied to any problem in water and environmental engineering that you can think of? Derive the entropy of this distribution and plot it as a function of its parameters. Discuss the graph and reflect on its physical import.

Q.4.24 Consider the log-extreme value type I distribution which is a two-parameter distribution determined from two constraints and gives the probability distribution of the random variable. Plot the distribution as a function of its parameters and discuss how the distribution shape changes with changes in parameter values. Based on the shape, can this distribution be applied to any problem in water and environmental engineering that you can think of? Derive the entropy of this distribution and plot it as a function of its parameters. Discuss the graph and reflect on its physical import.

Q.4.25 Consider the generalized extreme value distribution which is a three-parameter distribution determined from three constraints and gives the probability distribution of the random variable. Plot the distribution as a function of its parameters and discuss how the distribution shape changes with changes in parameter values. Based on the shape, can this distribution be applied to any problem in water and environmental engineering that you can think of? Derive the entropy of this distribution and plot it as a function of its parameters. Discuss the graph and reflect on its physical import.

Q.4.26 Consider the extreme value type III distribution which is a three-parameter distribution determined from two constraints and gives the probability distribution of the random variable. Plot the distribution as a function of its parameters and discuss how the distribution shape changes with changes in parameter values. Based on the shape, can this distribution be applied to any problem in water and environmental engineering that you can think of? Derive the entropy of this distribution and plot it as a function of its parameters. Discuss the graph and reflect on its physical import.

Q.4.27 Consider the Weibull distribution which is a two-parameter distribution determined from two constraints and gives the probability distribution of the random variable. Plot the distribution as a function of its parameters and discuss how the distribution shape changes with changes in parameter values. Based on the shape, can this distribution be applied to any problem in water and environmental engineering that you can think of? Derive the entropy of this distribution and plot it as a function of its parameters. Discuss the graph and reflect on its physical import.

Q.4.28 Consider the Pearson type III distribution which is a three-parameter distribution determined from two constraints and gives the probability distribution of the random variable. Plot the distribution as a function of its parameters and discuss how the distribution shape changes with changes in parameter values. Based on the shape, can this distribution be applied to any problem in water and environmental engineering that you can think of? Derive the entropy of this distribution and plot it as a function of its parameters. Discuss the graph and reflect on its physical import.

Q.4.29 Consider the log-Pearson type III distribution which is a three-parameter distribution determined from two constraints and gives the probability distribution of the random variable. Plot the distribution as a function of its parameters and discuss how the distribution shape changes with changes in parameter values. Based on the shape, can this distribution be applied to any problem in water and environmental engineering that you can think of? Derive the entropy of this distribution and plot it as a function of its parameters. Discuss the graph and reflect on its physical import.

Q.4.30 Consider the logistic distributions: two-parameter log-logistic and log-logistic three parameter. Plot each distribution as a function of its parameters and discuss how the distribution shape changes with changes in parameter values. Based on the shape, can this distribution be applied to any problem in water and environmental engineering that you can think of? Derive the entropy of this distribution and plot it as a function of its parameters. Discuss the graph and reflect on its physical import.

Q.4.31 Consider the Pareto distributions: two-parameter, two-parameter generalized and three parameter. Plot each distribution as a function of its parameters and discuss how the distribution shape changes with changes in parameter values. Based on the shape, can this distribution be applied to any problem in water and environmental engineering that you can think of? Derive the entropy of this distribution and plot it as a function of its parameters. Discuss the graph and reflect on its physical import.

Q.4.32 Consider the Cauchy distribution which is a one-parameter distribution determined from one constraint and gives the probability distribution of the random variable. Plot the distribution as a function of its parameter and discuss how the distribution shape changes with changes in parameter values. Based on the shape, can this distribution be applied to any problem in water and environmental engineering that you can think of? Derive the entropy of this distribution and plot it as a function of its parameters. Discuss the graph and reflect on its physical import.

Q.4.33 Consider the Raleigh distribution which is a one-parameter distribution determined from one constraint and gives the probability distribution of the random variable. Plot the distribution as a function of its parameter and discuss how the distribution shape changes with changes in parameter values. Based on the shape, can this distribution be applied to any problem in water and environmental engineering that you can think of? Derive the entropy of this distribution and plot it as a function of its parameters. Discuss the graph and reflect on its physical import.

Q.4.34 Consider the Chi-square distribution which is a one-parameter distribution determined from one constraint and gives the probability distribution of the random variable. Plot the distribution as a function of its parameter and discuss how the distribution shape changes with changes in parameter values. Based on the shape, can this distribution be applied to any problem in water and environmental engineering that you can think of? Derive the entropy of this distribution and plot it as a function of its parameters. Discuss the graph and reflect on its physical import.

References

Fiorentino, M., Claps, P. and Singh, V.P. (1993). An entropy-based morphological analysis of river-basin networks. *Water Resources Research*, Vol. 29, No. 4, pp. 1215–24.

Hao, Z. and Singh, V. P. (2009a). Entropy-based parameter estimation for extended three parameter Burr III distribution for low-flow frequency analysis. *Transactions of the ASABE*, Vol. 52, No. 4, pp. 1–10.

Hao, Z. and Singh, V.P. (2009b). Entropy-based parameter estimation for extended Burr XII Distribution. *Stochastic Environmental Research and Risk Analysis*, Vol. 23, pp. 1113–22.

Kapur, J.N. (1989). *Maximum Entropy Models in Science and Engineering*. Wiley Eastern, New Delhi, India.

Kapur, J.N. and Kesavan, H.K. (1987). *Generalized Maximum Entropy Principle (with Applications)*. Sandford Educational Press, University of Waterloo, Waterloo, Canada.

Kapur, J.N. and Kesavan, H.K. (1992). *Entropy Optimization Principles with Applications*. 408 p., Academic Press, New York.

Krstanovic, P.F. and Singh, V.P. (1988). Application of entropy theory to multivariate hydrologic analysis, Vol. 1, Technical Report WRR8, 269 pp., Water Resources Program, Department of Civil Engineering, Louisiana State University, Baton Rouge, Louisiana.

Lienhard, J.H. (1964). A statistical mechanical prediction of the dimensionless unit hydrograph. *Journal of Geophysical Research*, Vol. 69. No. 24, pp. 5231–8.

Lienhard, J.H. (1966). A physical basis for the generalized gamma distribution. *Quarterly of Applied Mathematics*, Vol. 25, pp. 330–4.

Lienhard, J.H. (1972). Prediction of the dimensionless unit hydrograph. *Nordic Hydrology*, Vol. 3. pp. 107–9.

Rai, R.K., Sarkar, S., Upadhyay, A. and Singh, V.P. (2010). Efficacy of Nakagami-m distribution function for deriving unit hydrograph. *Water Resources Management*, Vol. 24, pp. 563–75.

Shao, Q.X., Wong, H., Xia, J., Wai-Cheung, I.P. (2004) Models for extremes using the extended three-parameter Burr XII system with application to flood frequency analysis. *Hydrologic Sciences Journal*, Vol. 49, No. 4, pp. 685–701.

Shao, Q.X., Chen, Y.D. and Zhang, L. (2008). An extension of three-parameter Burr III distribution for low-flow frequency analysis. *Comp. Statistics and Data Analysis*, Vol. 52, No. 3, pp. 1304–14.

Singh, V.P. (1998). *Entropy-Based Parameter Estimation in Hydrology*. Kluwer Academic Press, Boston.

Additional Reading

Agmon, N., Alhassid, Y. and Levine, R.D. (1979). An algorithm for finding the distribution of maximal entropy. *Journal of Computational Physics*, Vol. 30, pp. 250–8.

Arora, K. and Singh, V.P. (1987a). A comparative evaluation of the estimators of commonly used flood frequency models: l. Monte Carlo simulation. Completion Report, Louisiana Water Resources Research Institute, Louisiana State University, Baton Rouge, Louisiana.

Arora, K. and Singh, V.P. (1987b). A comparative evaluation of the estimators of commonly used flood frequency models: 2. Computer programs. Completion Report, Louisiana Water Resources Research Institute, Louisiana State University, Baton Rouge, Louisiana.

Arora, K. and Singh, V.P. (1987c). An evaluation of seven methods for estimating parameters of the EVl distribution. In *Hydrologic Frequency Modeling* edited by V.P. Singh, D. Reidel Publishing Company, Boston, pp. 383–94.

Arora, K. and Singh, V.P. (1987d). On statistical intercomparison of EV 1 estimators by Monte Carlo simulation. *Advances in Water Resources*, Vol. l0, No. 2, pp. 87–l07.

Arora, K. and Singh, V.P. (1989a). A comparative evaluation of the estimators of log-Pearson type (LP) 3 distribution. *Journal of Hydrology*, Vol. l05, pp. 19–37.

Arora, K. and Singh, V.P. (1989b). A note on the mixed moment estimation for the log-Pearson type 3 distribution. *Journal of the Institution of Engineers*, Civil Engineering Division, Vol. 69, Part CI5, pp. 298–30l.

Arora, K. and Singh, V.P. (1990). A comparative evaluation of the estimators of the log Pearson type (LP) 3 distribution - A reply. *Journal of Hydrology*, Vol. 117, pp. 375–6.

Basu, P.C. and Templeman, A.B. (1984). An efficient algorithm to generate maximum entropy distributions. *International Journal of Numerical Methods in Engineering*, Vol. 20, pp. 1039–55.

Ciulli, S., Mounsif, M., Gorman, N. and Spearman, T.D. (1991). On the application of maximum entropy to the moments problem. *Journal of Mathematical Physics*, Vol. 32, No. 7, pp. 1717–19.

Collins, R. and Wragg, A. (1977). Maximum entropy histograms. *Journal of Physics*, A: Math. Gen., Vol. 10, No. 9, pp. 1441–64.

Dowson, D.C. and Wragg, A. (1973). Maximum-entropy distributions having prescribed first and second moments. *IEEE Transactions on Information Theory*, Vol. IT-19, pp. 689–93.

Fiorentino, M., Arora, K. and Singh, V.P. (1987a). The two-component extreme value distribution for flood frequency analysis: another look and derivation of a new estimation method. *Stochastic Hydrology and Hydraulics*, Vol. l, pp. 199–208.

Fiorentino, M., Singh, V.P. and Arora, K. (1987b). On the two-component extreme value distribution and its point and regional estimators. In *Regional Flood Frequency Analysis* edited by V.P. Singh, D. Reidel Publishing Company, Boston, pp. 257–72.

Frontini, M. and Tagliani, A. (1994). Maximum entropy in the finite Stieltjes and Hamburger moment problem. *Journal of Mathematical Physics*, Vol. 35, No. 12, pp. 6748–56.

Griffeath, D.S. (1972). Computer solution of the discrete maximum entropy problem. *Technometrics*, Vol. 14, No. 4, pp. 891–7.

Guo, H. and Singh, V.P. (1992a). A comparative evaluation of estimators of extreme-value type III distribution by Monte Carlo simulation. Technical Report WRR24, Water Resources Program, Department of Civil Engineering, Louisiana State University, Baton Rouge, Louisiana.

Guo, H. and Singh, V.P. (1992b). A comparative evaluation of estimators of Pareto distribution by Monte Carlo simulation. Technical Report WRR25, Water Resources Program, Department of Civil Engineering, Louisiana State University, Baton Rouge, Louisiana.

Guo, H. and Singh, V.P. (1992c). A comparative evaluation of estimators of log-logistic distribution by Monte Carlo simulation. Technical Report WRR26, Water Resources Program, Department of Civil Engineering, Louisiana State University, Baton Rouge, Louisiana.

Guo, H. and Singh, V.P. (1992d). A comparative evaluation of estimators of two-component extreme-value distribution by Monte Carlo simulation. Technical Report WRR27, Water Resources Program, Department of Civil Engineering, Louisiana State University, Baton Rouge, Louisiana.

Hao, Z. and Singh, V.P. (2009a). Entropy-based parameter estimation for extended Burr XII distribution. *Stochastic Environmental Research and Risk Analysis*, Vol. 23, pp. 1113–22.

Hao, Z. and Singh, V.P. (2009b). Entropy-based parameter estimation for extended three-parameter Burr III distribution for low-flow frequency analysis. *Transactions of ASABE*, Vol. 52, No. 4, pp. 1193–202.

Jain, D. and Singh, V.P. (1986). Estimating parameters of EVl distribution for flood frequency analysis. *Water Resources Bulletin*, Vol. 23, No. l, pp. 59–72.

Jain D. and Singh, V.P. (1987). Comparison of some flood distributions using empirical data. In *Hydrologic Frequency Modeling* edited by V.P. Singh, D. Reidel Publishing Company, Boston, pp. 467–86.

Kociszewski, A. (1986). The existence conditions for maximum entropy distributions having prescribed the first three moments. *Journal of Physics, A: Math. Gen.*, Vol. 19, pp. L823–7.

Landau, H.J. (1987). Maximum entropy and the moment problem. *Bulletin, American Mathematical Society*, Vol. 16, No. 1, pp. 47–77.

Li, Y., Singh, V.P. and Cong, S. (1987). Entropy and its application in hydrology. In *Hydrologic Frequency Modeling* edited by V.P. Singh, D. Reidel Publishing Company, Boston, pp. 367–82.

Lisman, J.H.C. and van Zuylen, M.C.A. (1972). Note on the generation of most probable frequency distributions. *Statistica Neerlandica*, Vol. 26, No.1, pp. 19–23.

Liu, I.S. (1972). Method of Lagrange multipliers for exploration of the entropy principle. *Archive for Rational Mechanics and Analysis*, Vol. 46, No. 2, pp. 131–48.

Mead, L.R. and Papanicolaou, N. (1984). Maximum entropy in the problem of moments. *Journal of Mathematical Physics*, Vol. 25, No. 8, pp. 2404–17.

Mukherjee, D. and Hurst, D.C. (1984). Maximum entropy revisited. *Statistica Neerlandica*, Vol. 38, No. 1, pp. 1–12.

Naghavi, N., Singh, V.P. and Yu, F.X. (1993a). Development of 24-hour rainfall frequency maps for Louisiana. *Journal of Irrigation and Drainage Engineering*, ASCE, Vol. 119, No. 6, pp.1066–80.

Naghavi, B., Yu, F.X. and Singh, V.P. (1993b). Comparative evaluation of frequency distributions for Louisiana extreme rainfall. *Water Resources Bulletin*, Vol. 29, No. 2, pp. 211–19.

Ormoneit, D. and White, H. (1999). An efficient algorithm to compute maximum entropy densities. *Econometric Reviews*, Vol. 18, No. 2, pp. 127–40.

Siddall, J.N. and Diab, Y. (1975). The use in probabilistic design of probability curves generated by maximizing the Shannon entropy function constrained by moments. *Journal of Engineering for Industry*, Vol. 97, pp. 843–52.

Singh, V.P. (1986). On the log-Gumbel (LG) distribution. *Hydrology*, Vol. VIII, No. 4, pp. 34–42.

Singh, V.P. (1987a). Reply to Comments by H.N. Phien and V.T.V. Nguyen on Derivation of the Pearson type III distribution by using the principle of maximum entropy (POME). *Journal of Hydrology*, Vol. 90, pp. 355–7.

Singh, V.P. (1987b). On application of the Weibull distribution in hydrology. *Water Resources Management*, Vol. l, No. l, pp. 33–43.

Singh, V.P. (1987c). On the extreme value (EV) type III distribution for low flows. *Hydrological Sciences Journal*, Vol. 32, No. 4/l2, pp. 52l–33.

Singh, V.P. (1987d). *Hydrologic Modeling Using Entropy*. The VII IHP Endowment Lecture, Centre of Water Resources, Anna University, Madras, India.

Singh, V.P. (1989). Hydrologic modeling using entropy. *Journal of the Institution of Engineers, Civil Engineering Division*, Vol. 70, Part CV2, pp. 55–60.

Singh, V.P. (1992). Entropy-based probability distributions for modeling of environmental and biological systems. Chapter 6, pp. 167–208, in *Structuring Biological Systems: A Computer Modeling Approach*, edited by S. S. Iyengar, CRC Press, Inc., Boca Raton, Florida.

Singh, V.P. (1996). Application of entropy in hydrology and water resources. Proceedings, International Conference on From Flood to Drought, IAHR-African Division, Sun City, South Africa, August 5-7.

Singh, V.P. (1997). The use of entropy in hydrology and water resources. *Hydrological Processes*, Vol. 11, pp. 587–626.

Singh, V.P. (1997c). Effect of class interval size on entropy. *Stochastic Hydrology and Hydraulics*, Vol. 11, pp. 423–31.

Singh, V.P. (1998a). *Entropy-Based Parameter Estimation in Hydrology*. Kluwer Academic Publishers, Boston, 365 pp.

Singh, V.P. (1998b). Entropy as a decision tool in environmental and water resources. *Hydrology Journal*, Vol. XXI, No. 1-4, pp. 1–12.

Singh, V.P. (2000). The entropy theory as tool for modeling and decision making in environmental and water resources. *Water SA*, Vol. 26, No. 1, pp. 1–11.

Singh, V.P. (2002a). Entropy theory in environmental and water resources modeling. In: *Advances in Civil Engineering: Water Resources and Environmental Engineering*, edited by J. N. Bandhopadhyay and D. Nagesh Kumar, Indian Institute of Technology, Kharagpur, India, pp. pp. 1–11, January 3-6.

Singh, V.P. (2002b). Statistical analyses design. In: *Encyclopedia of Life Support Systems*, edited by A. Sydow, EOLSS Publishers Co., Ltd., Oxford, UK.

Singh, V.P. (2003). The entropy theory as a decision making tool in environmental and water resources. In: *Entropy Measures, Maximum Entropy and Emerging Applications*, edited by Karmeshu, Springer-Verlag, Bonn, Germany, pp. 261–97.

Singh, V.P. (2005). Entropy theory for hydrologic modeling. In: *Water Encyclopedia: Oceanography; Meteorology; Physics and Chemistry; Water Law; and Water History, Art, and Culture*, edited by J. H. Lehr, Jack Keeley, Janet Lehr and Thomas B. Kingery, John Wiley & Sons, Hoboken, New Jersey, pp. 217–23.

Singh, V.P. (2010). Entropy theory for hydrologic modeling. *Beijing Normal University Journal of Research*, Vol. 46, No. 3, pp. 229–40.

Singh, V.P. and Ahmad, M. (2004). *A comparative evaluation of the estimators of the three-parameter generalized Pareto distribution. Statistical Computation and Simulation*, Vol. 74, No. 2, pp. 91–106.

Singh, V.P. and Chowdhury, P.K. (1985). On fitting gamma distribution to synthetic runoff hydrographs. *Nordic Hydrology*, Vol. l6, pp. 177–92.

Singh, V.P., Cruise, J.F. and Ma, M. (1989). *A comparative evaluation of the estimators of two distributions by Monte Carlo method*. Technical Report WRR13, 126 p., Water Resources Program, Department of Civil Engineering, Louisiana State University, Baton Rouge, Louisiana.

Singh, V.P., Cruise, J.F. and Ma, M. (1990a). A comparative evaluation of the estimators of the three-parameter lognormal distribution by Monte Carlo simulation. *Computational Statistics and Data Analysis*, Vol. l0, pp. 71–85.

Singh, V.P., Cruise, J.F. and Ma, M. (1990b). A comparative evaluation of the estimators of the Weibull distribution by Monte Carlo simulation. *Journal of Statistical Computation and Simulation*, Vol. 36, pp. 229–41.

Singh, V.P. and Deng, Z.Q. (2003). Entropy-based parameter estimation for kappa distribution. *Journal of Hydrologic Engineering*, ASCE, Vol. 8, No. 2, pp. 81–92.

Singh, V.P. and Fiorentino, M., editors (1992a). *Entropy and Energy Dissipation in Hydrology*. Kluwer Academic Publishers, Dordrecht, The Netherlands, 595 p.

Singh, V.P. and Fiorentino, M. (1992b). A historical perspective of entropy applications in water resources. *Entropy and Energy Dissipation in Water Resources*, edited by V.P. Singh and M. Fiorentino, Kluwer Academic Publishers, Dordrecht, The Netherlands, pp. 21–61.

Singh, V.P. and Guo, H. (1995a). Parameter estimation for 2-parameter Pareto distribution by POME. *Water Resources Management*, Vol. 9, pp. 81–93.

Singh, V.P. and Guo, H. (1995b). Parameter estimation for 3-parameter generalized Pareto distribution by POME. *Hydrological Science Journal*, Vol. 40, No. 2, pp. 165–81.

Singh, V.P. and Guo. H. (1995c). Parameter estimation for 2-parameter log-logistic distribution (LLD) by POME. *Civil Engineering Systems*, Vol. 12, pp. 343–57.

Singh, V.P. and Guo, H. (1997). Parameter estimation for 2 - parameter generalized Pareto distribution by POME. *Stochastic Hydrology and Hydraulics*, Vol. 11, No. 3, pp. 211–28.

Singh, V.P., Guo, H. and Yu, F.X. (1993). Parameter estimation for 3-Parameter log-logistic distribution. *Stochastic Hydrology and Hydraulics*, Vol. 7, No. 3, pp. 163–78, 1993.

Singh, V. P. and Hao, L. (2009). Derivation of velocity distribution using entropy. 33rd IAHR Congress 2009, August 9–14, Vancouver, Canada.

Singh, V.P. and Harmancioglu, N.B. (1997). Estimation of missing values with use of entropy. In: *Integrated Approach to Environmental Data Management Systems*, edited by N. B. Harmancioglu, N. Alpaslan, S.D. Ozkul and V.P. Singh, pp. 267–75, Kluwer Academic Publishers, Dordrecht, The Netherlands.

Singh, V.P., Jain, S.K. and Tyagi, A.K. (2007). *Risk and Reliability Analysis in Civil and Environmental Engineering*. 783 pp., ASCE Press, Reston, Virginia.

Singh, V.P. and Krstanovic, P.F. (1986). A stochastic model for sediment yield. In: *Multivariate Analysis of Hydrologic Processes*, edited by H. W. Shen, J. T. B. Obeysekera, V. Yevjevich, and D. G. DeCoursey, pp. 755–67, Colorado State University, Fort Collins, Colorado.

Singh, V.P. and Krstanovic, P.F. (1987). A stochastic model for sediment yield using the principle of maximum entropy. *Water Resources Research*, Vol. 23, No. 5, pp. 781–93.

Singh, V.P. and Krstanovic, P.F. (1988). A stochastic model for water quality constituents. Proceedings, Sixth Congress of the IAHR-APD, Kyoto, Japan, July 20–22.

Singh, V.P., Rajagopal, A.K. and Singh, K. (1986). Derivation of some frequency distributions using the principle of maximum entropy (POME). *Advances in Water Resources*, Vol. 9, No. 2, pp. 91–106.

Singh, V.P. and Rajagopal, A.K. (1987a). Some recent advances in application of the principle of maximum entropy (POME) in hydrology. *IAHS Publication* No. 164, pp. 353–64.

Singh, V.P. and Rajagopal, A.K. (1987b). A new method of parameter estimation for hydrologic frequency analysis. *Hydrological Science and Technology*, Vol. 2, No. 3, pp. 33–40.

Singh, V.P. and Singh, K. (1985a). Derivation of the gamma distribution by using the principle of maximum entropy (POME). *Water Resources Bulletin*, Vol. 21, No. 6, pp. 941–52.

Singh, V.P. and Singh, K. (1985b). Derivation of the Pearson type (PT) III distribution by using the principle of maximum entropy. *Journal of Hydrology*, Vol. 80, pp. 197–214.

Singh, V.P. and Singh, K. (1985c). Pearson Type III Distribution and the principle of maximum entropy. Proceedings of the Vth World Congress on Water Resources held June 9–15, 1985, in Brussels, Belgium, Vol. 3, pp. 1133–46.

Singh, V.P. and Singh, K. (1987). Parameter estimation for TPLN distribution for flood frequency analysis. *Water Resources Bulletin*, Vol. 23, No. 6, pp. 1185–92.

Singh, V.P. and Singh, K. (1988). Parameter estimation for log-Pearson type III distribution by POME. *Journal of Hydraulic Engineering*, ASCE, Vol. 114, No. 1, pp. 112–22.

Singh, K. and Singh, V.P. (1991). Derivation of bivariate probability density functions with exponential marginals. *Stochastic Hydrology and Hydraulics*, Vol. 5, pp. 55–68.

Singh, V.P., Singh, K. and Rajagopal, A. K. (1985). Application of the principle of maximum entropy (POME) to hydrologic frequency analysis, Completion Report 06, 144 p., Louisiana Water Resources Research Institute, Louisiana State University, Baton Rouge, Louisiana.

Taglani, A. (1993). On the application of maximum entropy to the moments problem. *Journal of Mathematical Physics*, Vol. 34, No. 1, pp. 326–37.

Tagliani, A. (2003). Entropy estimate of probability densities having assigned moments: Hausdorff case. *Applied mathematics Letters*, Vol. 15, pp. 309–14.

Verdugo Lazo, A.C.G. and Rathie, P.N. (1978). On the entropy of continuous probability distributions. *IEEE Transactions on Information Theory*, Vol. IT-24, No. 1, pp. 120–2.

Wragg, A. and Dowson, D.C. (1970). Fitting continuous probability density functions over $(0,\infty)$ using information theory ideas. *IEEE Transactions on Information Theory*, Vol. IT-16, pp. 226–30.

Zellner, A. and Highfield, R.A. (1988). Calculation of maximum entropy distributions and approximation of marginal posterior distributions. *Journal of Econometrics*, Vol. 37, pp. 195–209.

5 Multivariate Probability Distributions

Multivariate distributions arise frequently in hydrologic science and water engineering. This chapter extends the discussion of the previous chapter on univariate distributions to multivariate distributions using POME. Let X_1 and X_2 be two random variables, such as rainfall depth and duration in case of rainfall, flood peak and volume in case of floods, drought duration and inter-arrival time in case of droughts, sediment load and concentration in case of sediment transport, and so on. In a similar manner there can be more than two variables, such as rainfall depth X_1, duration X_2, and inter arrival time X_3 in case of rainfall; drought duration, severity, and areal extent in case of droughts; and flood peak, volume, and duration in case of floods. These random variables are different; they may be dependent or independent of each other. Each individual variable can have Markovian dependencies. Consider, for example, annual stream flow of a river as a random variable. Now consider the same annual stream flow series but lagged by say five years. Then the five-year lagged values can be considered to represent another random variable. Likewise, 10-year lagged values would represent another random variable. Thus, the zero-lag series (original), five-year lag series, and 10-year lag series are three time series leading to three random variables from the same stream flow process; the latter two are obtained from the same present random series. In such cases, even one variable with Markovian dependency will have to be treated in the same way as two or more variables with cross-correlations.

The objective of this chapter is to derive multivariate normal and multivariate exponential distributions using the principle of maximum entropy (POME). Derivation of multivariate probability distribution function (PDF) entails 1) specification of constraints, 2) maximization of entropy using the method of Lagrange multipliers, 3) formulation of partition function, 4) relation between the Lagrange multipliers and constraints, 5) derivation of the PDF, and 6) determination of entropy of the PDF.

5.1 Multivariate normal distributions

5.1.1 One time lag serial dependence

First, a variable with one-lag serial dependency is treated. The values of the variable with one lag-one dependency are the values of the same process $X(t)$ but at one time lag part. This then

Entropy Theory and its Application in Environmental and Water Engineering, First Edition. Vijay P. Singh.
© 2013 John Wiley & Sons, Ltd. Published 2013 by John Wiley & Sons, Ltd.

reduces to a case involving two random variables, X_1 and X_2, one time lag apart: $X = \{X_1, X_2\}$. It is assumed that X_1 and X_2 have some cross-correlation. The objective is to derive a joint probability density function (PDF) of X_1 and X_2.

Specification of constraints

The constraints to be imposed on these variables X: $\{X_1, X_2\}$ are variance σ^2 and autocovariance γ:

$$\sigma_{X_i}^2 = \int\limits_{-\infty}^{\infty} \int\limits_{-\infty}^{\infty} (x_i - \mu)^2 f(x_1, x_2) dx_1 dx_2, i = 1, 2 \tag{5.1}$$

$$\gamma(1) = \int\limits_{-\infty}^{\infty} \int\limits_{-\infty}^{\infty} (x_1 - \mu)(x_2 - \mu) f(x_1, x_2) dx_1 dx_2 \tag{5.2}$$

where $f(x_1, x_2)$ is the joint PDF of X_1 and X_2 where their values, x_1 and x_2 represent two consecutive values of the same hydrologic process $X(t)$, and μ is the mean of X_1 and X_2. Both σ_X^2 and $\gamma(1)$ are approximated by their sample values, respectively, as:

$$S_{X_i}^2 = \frac{1}{N} \sum_{t=0}^{N-1} (x_{t+i} - \bar{x})^2, i = 1, 2 \tag{5.3}$$

$$c(1) = \frac{1}{N} \sum_{t=0}^{N-1} (x_t - \bar{x})(x_{t+1} - \bar{x}) \tag{5.4}$$

where N denotes the number of observations and $t \in (0, T)$, and $\bar{x}$ is the sample mean. For the same hydrologic process second order stationarity is assumed: $S_{X_1}^2 = S_{X_2}^2 = S_X^2$.

The total probability constraint can be expressed as

$$\int\limits_{-\infty}^{\infty} \int\limits_{-\infty}^{\infty} f(x_1, x_2) dx_1 dx_2 = 1 \tag{5.5}$$

Maximization of entropy

Entropy of two variables, joint entropy, can be written as

$$H(X_1, X_2) = -\int\limits_{-\infty}^{\infty} \int\limits_{-\infty}^{\infty} f(x_1, x_2) \ln[f(x_1, x_2)] dx_1 dx_2 \tag{5.6}$$

Maximizing the joint entropy $H(X_1, X_2)$ using the method of Lagrange multipliers, subject to equations (5.1), (5.2) and (5.5), one obtains the POME-based PDF:

$$f(x_1, x_2) = \exp[-\lambda_0 - \lambda_1 (x_1 - \mu)^2 - \lambda_2 (x_2 - \mu)^2 - \lambda_3 (x_1 - \mu)(x_2 - \mu)] \tag{5.7a}$$

or

$$f(x_1, x_2) = \frac{1}{Z(\lambda_1, \lambda_2, \lambda_3)} \exp[-\lambda_1 (x_1 - \mu)^2 - \lambda_2 (x_2 - \mu)^2 - \lambda_3 (x_1 - \mu)(x_2 - \mu)] \tag{5.7b}$$

where $\lambda_0, \lambda_1, \lambda_2$ and λ_3 are the Lagrange multipliers, and $Z(\lambda_1, \lambda_2, \lambda_3)$ is the partition function.

Formulation of partition function

Using equation (5.7b) in equation (5.5), the partition function can be written as

$$Z(\lambda_1, \lambda_2, \lambda_3) = \exp(\lambda_0)$$

$$= \int_{-\infty}^{\infty} \int_{-\infty}^{\infty} \exp[-\lambda_1(x_1 - \mu)^2 - \lambda_2(x_2 - \mu)^2 - \lambda_3(x_1 - \mu)(x_2 - \mu)]dx_1 dx_2 \quad (5.8a)$$

For simplicity, one can write $x_1 - \mu = x_1$ and $x_2 - \mu = x_2$, keeping in mind that these values are mean-corrected. Equation (5.8a) can be written as

$$Z(\lambda_1, \lambda_2, \lambda_3) = \exp(\lambda_0) = \int_{-\infty}^{\infty} \int_{-\infty}^{\infty} \exp[-\lambda_1 x_1^2 - \lambda_2 x_2^2 - \lambda_3 x_1 x_2]dx_1 dx_2 \quad (5.8b)$$

Taking the logarithm of equation (5.8b), one obtains

$$\lambda_0 = \ln \int_{-\infty}^{\infty} \int_{-\infty}^{\infty} \exp[-\lambda_1 x_1^2 - \lambda_2 x_2^2 - \lambda_3 x_1 x_2]dx_1 dx_2 \quad (5.9)$$

The exponential terms in equation (5.9) can be expressed as error functions, noting that

$$\int_{-\infty}^{\infty} \exp[-a^2 x^2]dx = \frac{\sqrt{\pi}}{a} \quad (5.10)$$

Separating the variables of integration, equation (5.8b) can be written as

$$\exp(\lambda_0) = \int_{-\infty}^{\infty} \exp[-\lambda_1 x_1^2]dx_1 \int_{-\infty}^{\infty} \exp\left[-\lambda_2\left(x_2 + \frac{\lambda_3}{2\lambda_2}x_1\right)^2 + \frac{4\lambda_3^2}{4\lambda_2}\right]dx_2$$

$$= \int_{-\infty}^{\infty} \exp\left[-x_1^2\left(\lambda_1 - \frac{\lambda_3^2}{4\lambda_2}\right)\right]dx_1 \int_{-\infty}^{\infty} \exp\left[-\lambda_2\left(x_2 + \frac{\lambda_3}{2\lambda_2}x_1\right)^2\right]dx_2$$

$$= \sqrt{\frac{\pi}{\lambda_1 - \frac{\lambda_3^2}{4\lambda_2}}}\sqrt{\frac{\pi}{\lambda_2}} = \frac{2\pi}{\sqrt{4\lambda_1\lambda_2 - \lambda_3^2}} \quad (5.11)$$

Equation (5.11) is valid whenever $4\lambda_1\lambda_2 - \lambda_3^2 > 0$, that is, for $\lambda_2 > 0$ and $4\lambda_1\lambda_2 > \lambda_3^2$.
Taking the logarithm of equation (5.11), one can write λ_0 as:

$$\lambda_0 = \ln(2\pi) - \frac{1}{2}\ln[4\lambda_1\lambda_2 - \lambda_3^2] \quad (5.12)$$

Relation between Lagrange multipliers and constraints

Differentiating equation (5.9) with respect to the Lagrange multiplier λ_1 and using equation (5.1) yield

$$\frac{\partial \lambda_0}{\partial \lambda_1} = -\sigma_X^2 \quad (5.13)$$

Differentiating equation (5.9) with respect to Lagrange multiplier λ_2 and using equation (5.1) yield

$$\frac{\partial \lambda_0}{\partial \lambda_2} = -\sigma_X^2 \qquad (5.14)$$

Differentiating equation (5.9) with respect to Lagrange multiplier λ_3 and using equation (5.2) yield

$$\frac{\partial \lambda_0}{\partial \lambda_3} = -\gamma(1) \qquad (5.15)$$

Differentiating equation (5.12) with respect to $\lambda_1, \lambda_2,$ and λ_3 separately, one obtains, respectively:

$$\frac{\partial \lambda_0}{\partial \lambda_1} = -\frac{2\lambda_2}{4\lambda_1\lambda_2 - \lambda_3^2} \qquad (5.16)$$

$$\frac{\partial \lambda_0}{\partial \lambda_2} = -\frac{2\lambda_1}{4\lambda_1\lambda_2 - \lambda_3^2} \qquad (5.17)$$

$$\frac{\partial \lambda_0}{\partial \lambda_3} = \frac{\lambda_3}{4\lambda_1\lambda_2 - \lambda_3^2} \qquad (5.18)$$

Equating equation (5.13) to equation (5.16), equation (5.14) to equation (5.17), and equation (5.15) to equation (5.18), one obtains

$$\sigma_X^2 = \frac{2\lambda_2}{4\lambda_1\lambda_2 - \lambda_3^2} \qquad (5.19)$$

$$\sigma_X^2 = \frac{2\lambda_1}{4\lambda_1\lambda_2 - \lambda_3^2} \qquad (5.20)$$

and

$$-\gamma(1) = \frac{\lambda_3}{4\lambda_1\lambda_2 - \lambda_3^2} \qquad (5.21)$$

From equations (5.19), (5.20) and (5.21), one obtains

$$\lambda_1 = \lambda_2 = \frac{1}{2} \frac{\sigma_X^2}{\sigma_X^4 - [\gamma(1)]^2} \qquad (5.22)$$

and

$$\lambda_3 = \pm \frac{\gamma(1)}{\sigma_X^4 - [\gamma(1)]^2} \qquad (5.23)$$

Substituting equations (5.22) and (5.23) in equation (5.11), the partition function becomes:

$$\exp(\lambda_0) = 2\pi \sqrt{\sigma_X^4 - [\gamma(1)]^2} \qquad (5.24)$$

Probability density function

Inserting the partition function and Lagrange multipliers in equation (5.7b), the POME-based PDF becomes:

$$f(x_1, x_2) = \frac{1}{2\pi\sqrt{\sigma_X^4 - [\gamma(1)]^2}} \exp\left\{-\frac{1}{2}\frac{\sigma_X^2[(x_1 - \mu)^2 + (x_2 - \mu)^2] \pm 2\gamma(1)[x_1 - \mu)(x_2 - \mu)]}{\sigma_X^4 - [\gamma(1)]^2}\right\}$$

(5.25)

which is the bivariate normal probability density function.

The expression inside the exponential in equation (5.25) can be re-arranged as

$$-\frac{1}{2}\left[\frac{(x_1 - \mu)^2\sigma_X^2}{\sigma_X^4 - [\gamma(1)]^2} \pm \frac{(x_1 - \mu)(x_2 - \mu)\gamma(1)}{\sigma_X^4 - [\gamma(1)]^2} \pm \frac{\gamma(1)(x_1 - \mu)(x_2 - \mu)}{\sigma_X^4 - [\gamma(1)]^2} + \frac{(x_2 - \mu)^2\sigma_X^2}{\sigma_X^4 - [\gamma(1)]^2}\right]$$

or

$$-\frac{1}{2}\left[\frac{(x_1 - \mu)\sigma_X^2}{\sigma_X^4 - [\gamma(1)]^2} \pm \frac{(x_2 - \mu)\gamma(1)}{\sigma_X^4 - [\gamma(1)]^2} \pm \frac{\gamma(1)(x_1 - \mu) + \sigma_X^2(x_2 - \mu)}{\sigma_X^4 - [\gamma(1)]^2}\right]\begin{bmatrix} x_1 - \mu \\ x_2 - \mu \end{bmatrix}$$

$$= -\frac{1}{2}\begin{bmatrix} x_1 - \mu & x_2 - \mu \end{bmatrix}\begin{bmatrix} \dfrac{\sigma_X^2}{\sigma_X^4 - [\gamma(1)]^2} & \pm\dfrac{\gamma(1)}{\sigma_X^4 - [\gamma(1)]^2} \\ \pm\dfrac{\gamma(1)}{\sigma_X^4 - [\lambda(1)]^2} & \dfrac{\sigma_X^2}{\sigma_X^4 - [\gamma(1)]^2} \end{bmatrix}\begin{bmatrix} x_1 - \mu \\ x_2 - \mu \end{bmatrix}$$

(5.26)

The joint PDF can now be written in compact form as

$$f(X) = \frac{1}{2\pi|\gamma_a|^{1/2}} \exp\left[-\frac{1}{2}(X - \mu)\gamma_a^{-1}(X - \mu)^*\right]$$

(5.27)

where $(X - \mu) = [x_1 - \mu, x_2 - \mu]$ is the mean corrected vector, superscript * denotes the transpose matrix; γ_a is the autocovariance matrix:

$$\gamma_a = \begin{bmatrix} \sigma_X^2 & \gamma(1) \\ \gamma(1) & \sigma_X^2 \end{bmatrix}$$

(5.28)

and γ_a^{-1} is the inverse of the autocovariance matrix given as:

$$\gamma_a^{-1} = \begin{bmatrix} \dfrac{\sigma_X^2}{\sigma_X^4 - [\gamma(1)]^2} & -\dfrac{\gamma(1)}{\sigma_X^4 - [\gamma(1)]^2} \\ -\dfrac{\gamma(1)}{\sigma_X^4 - [\gamma(1)]^2} & \dfrac{\sigma_X^2}{\sigma_X^4 - [\gamma(1)]^2} \end{bmatrix}$$

(5.29)

Equation (5.29) is valid only if λ_3 in equation (5.23) is defined with negative sign. Equation (5.27) is equivalent to the bivariate normal distribution when the autocovariance matrix γ_a is replaced by the cross-covariance matrix γ_c of the same dimension and the mean-corrected vector of the univariate process, (μ, μ) or $(\bar{x}, \bar{x})$, by the mean corrected vector of the bivariate process, (μ_x, μ_y) or $(\bar{x}, \bar{y})$.

The determinant of the autocovariance matrix given by equation (5.28) is:

$$|\gamma_a| = \sigma_X^4 - [\gamma(1)]^2$$

(5.30)

From the condition on Lagrange multipliers mentioned earlier, it is concluded that

$$|\gamma_a| > 0 \ \ if \ \ \lambda_1 = \lambda_2 \tag{5.31}$$

The autocovariance matrix must always be nonsingular, which is a necessary condition for the PDF to exist. For its real solution from equation (5.21), one gets

$$|\gamma_a| = 4\lambda_1^2 - \lambda_3^2 > 0 \tag{5.32}$$

or

$$\lambda_1 = \lambda_2 > \frac{\lambda_3}{2} \ \ for \ \lambda_1 > 0 \tag{5.33}$$

Thus, the two principal minors [A principal minor of order n is obtained by deleting all rows and columns except for the first n row and columns] of γ_a must be positive. Term γ_a^{-1} can now be re-written as

$$\gamma_a^{-1} = \begin{bmatrix} 2\lambda_1 & \lambda_3 \\ \lambda_3 & 2\lambda_2 \end{bmatrix} \tag{5.34}$$

underscoring the important connection between the Lagrange multipliers $\lambda_1, \lambda_2, \lambda_3$ and matrix γ_a^{-1}. Because of the identical values on the diagonal, equation (5.34) can be written as

$$\gamma_a^{-1} = \begin{bmatrix} 2\lambda_1 & \lambda_3 \\ \lambda_3 & 2\lambda_1 \end{bmatrix} \tag{5.35}$$

Equations (5.22), (5.23) and (5.24) relate the Lagrange multipliers to elements σ_X^2 and $\gamma(1)$ of the autocovariance matrix.

For the independent hydrologic process $X(t)$, there is no transfer of information from one lag to another, that is, $\gamma(1) = 0$. From equation (5.28), the off-diagonal Lagrange multiplier dependent on $\gamma(1) = 0$ is zero. Therefore, an independent hydrologic process has diagonal matrix Λ as

$$\Lambda = \begin{bmatrix} \lambda_1 & 0 \\ 0 & \lambda_1 \end{bmatrix} \tag{5.36}$$

Dependent hydrologic processes have two carriers of dependencies or information: one represented by Lagrange multiplier λ_1 proportional to the variance and another by λ_2 proportional to the autocovariance of the first lag.

Example 5.1: Plot $f(x_1, x_2)$ for various values of $\sigma^2 X$ and $\gamma(1)$.

Solution: Assume the mean of X_1 and the mean of X_2 are zero. The joint probability $f(x_1, x_2)$ for various values of σ_X^2 and $\gamma(1)$ are computed and plotted, as shown in Figure 5.1.

Example 5.2: Compute σ_X^2 and $\gamma(1)$ for the data series of monthly discharges from Brazos River at the gaging station at Waco, Texas (USGS gaging station number: 08096500), for the period from 1970–2001 to 2006–2012. Plot $f(x_1, x_2)$ as a function of σ_X^2 and $\gamma(1)$.

Solution: Monthly discharge data are first log-transformed beforehand. Then, the variance and co-variance are computed: $\sigma_x = 9.34 \times 10^3$ (cubic meters per second or cms in short) and $\gamma(1) = 5.25 \times 10^3$ (cms). The graph of $f(x_1, x_2)$ is constructed, as shown in Figure 5.2.

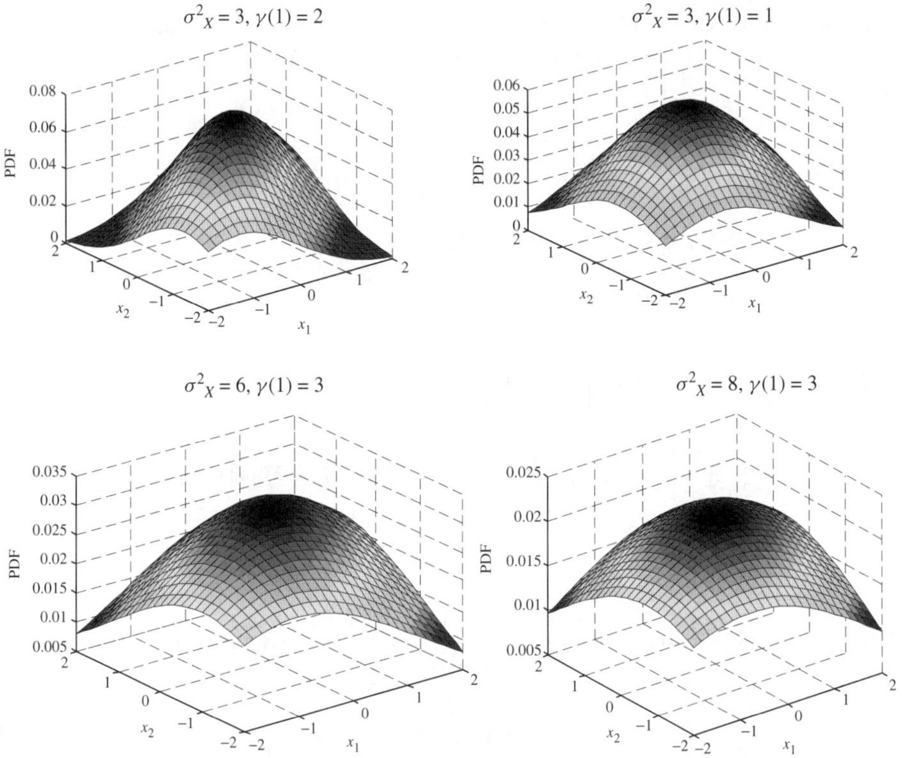

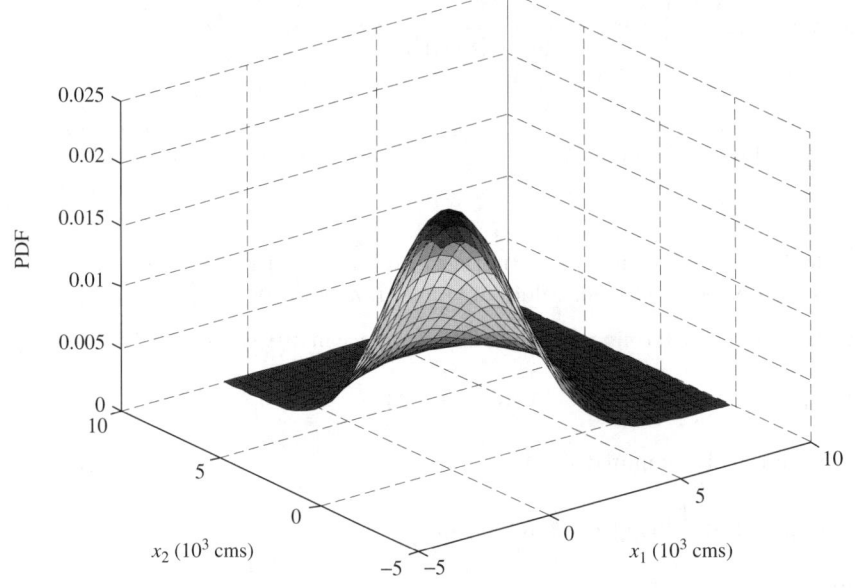

Figure 5.1 $f(x_1, x_2)$ for various values of σ_X^2 and $\gamma(1)$.

Figure 5.2 Joint probability density function $f(x_1, x_2)$ of monthly discharge of Brazos River at Waco, Texas.

Distribution entropy

The entropy of the joint distribution $f(x_1, x_2)$ is given by equation (5.6). Substituting equation (5.25) in equation (5.6) and writing x_1 and x_2 for $x_1 - \bar{x}$ or $x_1 - \mu$ and $x_2 - \bar{x}$ or $x_2 - \mu$ one obtains:

$$H(X_1, X_2) = -\int_{-\infty}^{\infty}\int_{-\infty}^{\infty} f(x_1, x_2)\{\ln \frac{1}{2\pi\sqrt{\sigma_X^4 - [\gamma(1)]^2}} - \frac{1}{2\{\sigma_X^4 - [\gamma(1)]^2\}}$$
$$\times [\sigma_X^2(x_1^2 + x_2^2) - 2\gamma(1)x_1x_2]\}dx_1 dx_2 \tag{5.37}$$

Equation (5.37) can be written as

$$H(X_1, X_2) = \left[\ln 2\pi + \frac{1}{2}\ln|\gamma_a|\right]\int_{-\infty}^{\infty}\int_{-\infty}^{\infty} f(x_1, x_2)dx_1 dx_2$$
$$+ \frac{1}{2\{\sigma_X^4 - [\gamma(1)]^2\}}\int_{-\infty}^{\infty}\int_{-\infty}^{\infty} f(x_1, x_2)[\sigma_X^2(x_1^2 + x_2^2) - 2\gamma(1)x_1x_2]dx_1 dx_2 \tag{5.38}$$

Equation (5.38) simplifies to:

$$H(x_1, x_2) = \ln 2\pi + \frac{1}{2}\ln|\gamma_a| + \frac{1}{2\{\sigma_X^4 - [\gamma(1)]^2\}}[\sigma_X^2\int_{-\infty}^{\infty}\int_{-\infty}^{\infty} x_1^2 f(x_1, x_2)dx_1 dx_2$$
$$+ \sigma_X^2\int_{-\infty}^{\infty}\int_{-\infty}^{\infty} x_2^2 f(x_1, x_2) - 2\gamma(1)\int_{-\infty}^{\infty}\int_{-\infty}^{\infty} f(x_1, x_2)x_1 x_2 dx_1 dx_2 \tag{5.39}$$

Equation (5.39) further simplifies to

$$H(x_1, x_2) = \ln 2\pi + \frac{1}{2}\ln|\gamma_a| + \frac{1}{2\{\sigma_X^4 - [\gamma(1)]^2\}}[\sigma_X^2\sigma_X^2 + \sigma_X^2\sigma_X^2 - 2\gamma(1)\gamma(1)] \tag{5.40}$$

or

$$H(x_1, x_2) = \ln(2\pi) + \frac{1}{2}\ln|\gamma_a| + 1 \tag{5.41}$$

Example 5.3: Compute the bivariate normal distribution entropy for the data of Example 5.2. Then, taking arbitrarily different values of γ_a, plot H as a function of γ_a.

Solution: First, the determinant of the autocovariance matrix is computed as:

$$|\gamma_a| = \sigma_X^4 - [\gamma(1)]^2 = 1.4910^2 - 0.9854^2 = 1.0426$$

Then, the entropy is computed as:

$$H = \ln(2\pi + 1) + \frac{1}{2}\ln|\gamma_a| = 2.8587 \text{ Napier}$$

For different values of $|\gamma_a|$, entropy is computed and plotted, as shown in Figure 5.3.

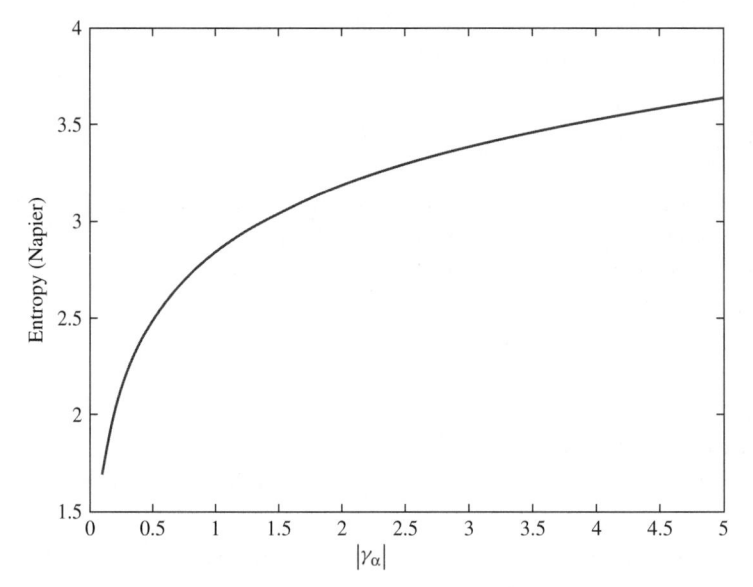

Figure 5.3 Entropy for different values of $|\gamma_a|$.

5.1.2 Two-lag serial dependence

Consider a univariate hydrologic process where observations are being made with two time lags apart, that is at a given time, one time lag apart and then another time lag apart. Thus there will be three sets of observations. Each set of observations represents a random variable. Essentially this case involves three random variables. Let X_1, X_2 and X_3 be then the three random variables of the same hydrologic process two time lags apart. The objective is to derive the joint distribution. Assume that X_1, X_2 and X_3 are normally distributed.

Specification of constraints

The constraints to be imposed on these variables are variance σ_X^2, autocovariance of the first lag $\gamma(1)$ and autocovariance of the second lag $\gamma(2)$:

$$\sigma_{X_i}^2 = \int_{-\infty}^{\infty} \int_{-\infty}^{\infty} \int_{-\infty}^{\infty} (x_i - \mu)^2 f(x_1, x_2, x_3) dx_1 dx_2 dx_3, \quad i = 1, 2, 3 \tag{5.42}$$

$$\gamma(1) = \int_{-\infty}^{\infty} \int_{-\infty}^{\infty} \int_{-\infty}^{\infty} (x_i - \mu)(x_{i+1} - \mu) f(x_1, x_2, x_3) dx_1 dx_2 dx_3, \quad i = 1, 2 \tag{5.43}$$

$$\gamma(2) = \int_{-\infty}^{\infty} \int_{-\infty}^{\infty} \int_{-\infty}^{\infty} (x_1 - \mu)(x_3 - \mu) f(x_1, x_2, x_3) dx_1 dx_2 dx_3 \tag{5.44}$$

where $f(x_1, x_2, x_3)$ is the joint PDF of three consecutive random variables $X : \{X_i, i = 1, 2, 3\}$ of a hydrologic process $X(t)$. $\sigma_X^2 i, \gamma(1)$, and $\gamma(2)$ are approximated by their sample values:

$$S_{X_i}^2 = \frac{1}{N} \sum_{t=0}^{N-1} (x_{t+i} - \bar{x})^2, i = 0, 1, 2 \tag{5.45}$$

$$c(1) = \frac{1}{N} \sum_{t=0}^{N+i-1} (x_{t+i} - \bar{x})(x_{t+i+1} - \bar{x}), i = 0, 1 \tag{5.46}$$

$$c(2) = \frac{1}{N} \sum_{t=0}^{N-2} (x_t - \bar{x})(x_{t+2} - \bar{x}) \tag{5.47}$$

where N denotes the number of observations and $t\varepsilon(0, T)$. Because the hydrologic process is the same, second order stationarity is assumed: $S_{X_1}^2 = S_{X_2}^2 = S_X^2$.

The total probability constraint can be expressed as

$$\int_{-\infty}^{\infty} \int_{-\infty}^{\infty} \int_{-\infty}^{\infty} f(x_1, x_2, x_3) dx_1 dx_2 dx_3 = 1 \tag{5.48}$$

Maximization of entropy

The three-variate entropy or the entropy of the joint distribution $f(x_1, x_2, x_3)$ can be written as:

$$H(X_1, X_2, X_3) = -\int_{-\infty}^{\infty} \int_{-\infty}^{\infty} \int_{-\infty}^{\infty} f(x_1, x_2, x_3) \ln[f(x_1, x_2, x_3)] dx_1 \, dx_2 dx_3 \tag{5.49}$$

Maximizing entropy $H(X_1, X_2, X_3)$, subject to equations (5.42) to (5.44) and equation (5.48), for deriving the trivariate distribution, one obtains

$$f(x_1, x_2, x_3) = \frac{1}{Z(\lambda_1, \lambda_2, \lambda_3, \lambda_4, \lambda_5, \lambda_6)} \exp\left[-\lambda_1(x_1 - \mu)^2 - \lambda_2(x_2 - \mu)^2 - \lambda_3(x_3 - \mu)^2\right.$$
$$\left. -\lambda_4(x_1 - \mu)(x_2 - \mu) - \lambda_5(x_2 - \mu)(x_3 - \mu) - \lambda_6(x_1 - \mu)(x_3 - \mu)\right] \tag{5.50a}$$

One can also write:

$$f(x_1, x_2, x_3) = \exp[-\lambda_0 - \lambda_1(x_1 - \mu)^2 - \lambda_2(x_2 - \mu)^2 - \lambda_3(x_3 - \mu)^2$$
$$- \lambda_4(x_1 - \mu)(x_2 - \mu) - \lambda_5(x_2 - \mu)(x_3 - \mu) - \lambda_6(x_1 - \mu)(x_3 - \mu)] \tag{5.50b}$$

where $\lambda_1, \lambda_2, \lambda_3, \lambda_4, \lambda_5$ and λ_6 are the Lagrange multipliers, and $Z(\lambda_1, \lambda_2, \lambda_3, \lambda_4, \lambda_5, \lambda_6)$ is the partition function.

Formulation of partition function

Using equation (5.48), the partition function can be written as

$$Z(\lambda) = \exp(\lambda_0) = \int_{-\infty}^{\infty} \int_{-\infty}^{\infty} \int_{-\infty}^{\infty} [-\lambda_1(x_1 - \mu)^2 - \lambda_2(x_2 - \mu)^2 - \lambda_3(x_3 - \mu)^2$$
$$- \lambda_4(x_1 - \mu)(x_2 - \mu) - \lambda_5(x_2 - \mu)(x_3 - \mu) - \lambda_6(x_1 - \mu)(x_3 - \mu)] dx_1 dx_2 dx_3 \tag{5.51}$$

For simplicity, one can write $x_1 - \bar{x} = x_1$ or $x_1 - \mu, x_2 - \bar{x} = x_2$ or $x_2 - \mu$, and $x_3 - \bar{x} = x_3$ or $x_3 - \mu$ keeping in mind that these values are mean-corrected. Also, $\lambda = \{\lambda_1, \lambda_2, \lambda_3, \lambda_4, \lambda_5, \lambda_6\}$. Separating the variables of integration, equation (5.51) can be written as

$$Z(\lambda) = \int_{-\infty}^{\infty} \exp[-\lambda_1 x_1^2] dx_1 \int_{-\infty}^{\infty} \exp(-\lambda_2 x_2^2 - \lambda_4 x_1 x_2) dx_2 \int_{-\infty}^{\infty} \exp[-\lambda_3 x_3^2 - x_3(\lambda_5 x_2 + \lambda_6 x_1)] dx_3$$

$$\tag{5.52}$$

The third integral can be solved using the formula:

$$\int_{-\infty}^{\infty} \exp[-a^2x^2 - bx]dx = \frac{\sqrt{\pi}}{a} \exp\left(\frac{b^2}{4a^2}\right), \text{ if } a = \sqrt{\lambda_3} > 0 \tag{5.53}$$

Thus,

$$Z(\lambda) = \int_{-\infty}^{\infty} \exp[-x_1^2\lambda_1]dx_1 \int_{-\infty}^{\infty} \exp(-\lambda_2 x_2^2 - \lambda_4 x_1 x_2)dx_2 \exp\left[\frac{(\lambda_5 x_2 + \lambda_6 x_1)^2}{4\lambda_3}\right]\sqrt{\frac{\pi}{\lambda_3}} \tag{5.54}$$

Rearranging the terms under the integral in equation (5.54), one obtains

$$Z(\lambda) = \sqrt{\frac{\pi}{\lambda_3}} \int_{-\infty}^{\infty} \exp\left[-x_1^2\lambda_1 + \frac{\lambda_6 x_1^2}{4\lambda_3}\right]dx_1 \int_{-\infty}^{\infty} \exp\left[-x_2^2\left(\lambda_2 - \frac{\lambda_5^2}{4\lambda_3}\right) - x_2\left(\lambda_4 - \frac{\lambda_5\lambda_6}{2\lambda_3}\right)\right]dx_2 \tag{5.55}$$

In a similar manner, the second integral in equation (5.55) can be solved as

$$Z(\lambda) = \sqrt{\frac{\pi}{\lambda_3}} \int_{-\infty}^{\infty} \exp\left[-x_1^2\left(\lambda_1 - \frac{\lambda_6^2}{4\lambda_3}\right)\right]dx_1 \, \exp\left[\frac{\left(\dfrac{2\lambda_3\lambda_4 - \lambda_5\lambda_6}{2\lambda_3}x_1\right)^2}{4\dfrac{4\lambda_2\lambda_3 - \lambda_5^2}{4\lambda_3}}\frac{\sqrt{\pi}}{\sqrt{\dfrac{4\lambda_2\lambda_3 - \lambda_5^2}{4\lambda_3}}}\right] \tag{5.56}$$

under the condition that

$$\frac{4\lambda_2\lambda_3 - \lambda_5^2}{4\lambda_3} > 0 \tag{5.57}$$

Rearranging the terms in equation (5.56), one gets

$$Z(\lambda) = \frac{2\pi}{\sqrt{4\lambda_2\lambda_3 - \lambda_5^2}} \int_{-\infty}^{\infty} \exp\left\{-x_1^2\left[\lambda_1 - \frac{\lambda_6^2}{4\lambda_3} - \frac{(2\lambda_3\lambda_4 - \lambda_5\lambda_6)^2}{4\lambda_3(4\lambda_2\lambda_3 - \lambda_5^2)}\right]\right\}dx_1 \tag{5.58}$$

Using equation (5.10), for $a > 0$ equation (5.58) simplifies to:

$$Z(\lambda) = \frac{2\pi}{\sqrt{4\lambda_2\lambda_3 - \lambda_5^2}} \frac{\sqrt{\pi}}{\sqrt{\dfrac{4\lambda_1\lambda_3(4\lambda_2\lambda_3 - \lambda_5^2) - \lambda_6^2(4\lambda_2\lambda_3 - \lambda_5^2) - (2\lambda_3\lambda_4 - \lambda_5\lambda_6)^2}{4\lambda_3(4\lambda_2\lambda_3 - \lambda_5^2)}}} \tag{5.59}$$

On simplification, equation (5.59) becomes

$$Z(\lambda) = \frac{2(\pi)^{3/2}}{\sqrt{4\lambda_1\lambda_2\lambda_3 - \lambda_1\lambda_5^2 - \lambda_2\lambda_6^2 - \lambda_3\lambda_4^2 + \lambda_4\lambda_5\lambda_6}} \tag{5.60}$$

To further simplify it, let

$$D = 4\lambda_1\lambda_2\lambda_3 - \lambda_1\lambda_5^2 - \lambda_2\lambda_6^2 - \lambda_3\lambda_4^2 + \lambda_4\lambda_5\lambda_6 \tag{5.61}$$

where $D > 0$.

Taking the logarithm of equation (5.60), one gets

$$\lambda_0 = \ln(2) + \frac{3}{2}\ln(\pi) - \frac{1}{2}\ln(D) \tag{5.62}$$

Relation between Lagrange multipliers and constraints

Differentiating the logarithm of equation (5.51) with respect to individual Lagrange multipliers separately, and making use of constraint equations (5.42) to (5.44), one obtains:

$$\frac{\partial\lambda_0}{\partial\lambda_1} = -\sigma_X^2 \tag{5.63}$$

$$\frac{\partial\lambda_0}{\partial\lambda_2} = -\sigma_X^2 \tag{5.64}$$

$$\frac{\partial\lambda_0}{\partial\lambda_3} = -\sigma_X^2 \tag{5.65}$$

$$\frac{\partial\lambda_0}{\partial\lambda_4} = -\gamma(1) \tag{5.66}$$

$$\frac{\partial\lambda_0}{\partial\lambda_5} = -\gamma(1) \tag{5.67}$$

$$\frac{\partial\lambda_0}{\partial\lambda_6} = -\gamma(2) \tag{5.68}$$

Differentiating λ_0 in equation (5.62) with respect to other Lagrange multipliers, $\lambda_1, \lambda_2, \lambda_3, \lambda_4, \lambda_5$ and λ_6, one gets

$$\frac{\partial\lambda_0}{\partial\lambda_1} = 4\lambda_2\lambda_3 - \lambda_5^2 \tag{5.69}$$

$$\frac{\partial\lambda_0}{\partial\lambda_2} = 4\lambda_1\lambda_3 - \lambda_6^2 \tag{5.70}$$

$$\frac{\partial\lambda_0}{\partial\lambda_3} = 4\lambda_1\lambda_2 - \lambda_4^2 + \lambda_5\lambda_6 \tag{5.71}$$

$$\frac{\partial\lambda_0}{\partial\lambda_4} = -2\lambda_3\lambda_4 + \lambda_5\lambda_6 \tag{5.72}$$

$$\frac{\partial\lambda_0}{\partial\lambda_5} = -2\lambda_1\lambda_5 + \lambda_4\lambda_6 \tag{5.73}$$

$$\frac{\partial\lambda_0}{\partial\lambda_6} = -2\lambda_2\lambda_6 + \lambda_4\lambda_5 \tag{5.74}$$

Equating equations (5.69) to (5.74) individually to corresponding equations (5.63) to (5.68), respectively, and then solving, one gets the relations between Lagrange multipliers and constraints as:

$$\lambda_1 = \frac{1}{4\lambda_2}(\lambda_4^2 + \sigma_X^2 2D) \tag{5.75}$$

$$\lambda_2 = \frac{1}{4\lambda_3}(\lambda_5^2 + \sigma_X^2 2D) \tag{5.76}$$

$$\lambda_3 = \frac{1}{4\lambda_1}(\lambda_6^2 + \sigma_X^2 2D) \tag{5.77}$$

$$\lambda_4 = -\frac{1}{2\lambda_3}(-\lambda_5\lambda_6 + \gamma(1)2D) \tag{5.78}$$

$$\lambda_5 = -\frac{1}{2\lambda_1}(\lambda_4\lambda_6 + \gamma(1)2D) \tag{5.79}$$

$$\lambda_6 = -\frac{1}{2\lambda_2}(-\lambda_4\lambda_5 + \gamma(2)2D) \tag{5.80}$$

where D is defined in equation (5.61). The solution of the system of equations (5.75) to (5.80) is found to be:

$$\lambda_1 = \lambda_3 = \frac{1}{2}\frac{\sigma_X^4 - [\gamma(1)]^2}{|\gamma_a|} \tag{5.81}$$

$$\lambda_2 = \frac{1}{2}\frac{\sigma_X^4 - [\gamma(2)]^2}{|\gamma_a|} \tag{5.82}$$

$$\lambda_4 = \lambda_5 = \frac{\gamma(1)\gamma(2) - \gamma(1)\sigma_X^2}{|\gamma_a|} \tag{5.83}$$

$$\lambda_6 = \frac{[\gamma(1)]^2 - \gamma(2)\sigma_X^2}{|\gamma_a|} \tag{5.84}$$

where $|\gamma_a|$ is the determinant of the autocovariance matrix given as:

$$\gamma_a = \begin{bmatrix} \sigma_X^2 & \gamma(1) & \gamma(2) \\ \gamma(1) & \sigma_X^2 & \gamma(1) \\ \gamma(2) & \gamma(1) & \sigma_X^2 \end{bmatrix} \tag{5.85}$$

The determinant is expressed as

$$|\gamma_a| = \sigma_X^6 - 2\sigma_X^2[\gamma(1)]^2 + 2[\gamma(1)]^2\gamma(2) - \sigma_X^2[\gamma(2)]^2 \tag{5.86}$$

Substituting equation (5.81) to (5.84) in equation (5.59), the partition function can now be expressed as

$$Z(\lambda) = 2(\pi)^{3/2}\Big[\frac{1}{2}\frac{\sigma_X^4 - [\gamma(1)]^2}{|\gamma_a|^3}\{\sigma_X^4 - [\gamma(2)]^2\} - \frac{1}{2}\frac{\sigma_X^4 - [\gamma(1)]^2}{|\gamma_a|}$$

$$\times \frac{[\gamma(1)\gamma(2) - \gamma(1)\sigma_X^2]^2}{|\gamma_a|} - \frac{1}{2}\frac{\{\sigma_X^4 - [\gamma(2)]^2\}}{|\gamma_a|}\frac{\{[\gamma(1)]^2 - \gamma(2)\sigma_X^2\}^2}{|\gamma_a|}\Big]^{-1/2} \tag{5.87}$$

Simplifying the denominator, collecting terms of the same power, and rearranging, equation (5.87) can be expressed as:

$$Z(\lambda) = (2\pi|\gamma_a|)^{3/2}\{\sigma_X^{12} + \sigma_X^4(2[\gamma(1)]^2 + [\gamma(2)]^2)^2 + 4[\gamma(1)]^4[\gamma(2)]^2 - 2\sigma_X^8\{2[\gamma(1)]^2$$
$$+ [\gamma(2)]^2\} + 4[\gamma(1)]^2\gamma(2)\sigma_X^6 + 4[\gamma(1)]^2\gamma(2)\{2\gamma(1)]^2 + [\gamma(2)]^2\}\sigma_X^2\}^{-1/2} \tag{5.88}$$

or

$$Z(\lambda) = \frac{(2\pi|\gamma_a|)^{3/2}}{\{\sigma_X^6 - \sigma_X^2[2[\gamma(1)]^2 + [\gamma(2)]^2] + 2[\gamma(1)]^2\gamma(2)\}^2} \tag{5.89}$$

Therefore,

$$\exp(\lambda_0) = (2\pi)^{3/2}|\gamma_a|^{1/2} \tag{5.90}$$

Substituting equations (5.81) to (5.84) and equation (5.90) in equation (5.50a), the PDF can be expressed as

$$f(x_1, x_2, x_3) = \frac{1}{(2\pi)^{3/2}|\gamma_a|^{1/2}} \exp\left\{-\frac{1}{2}[x_1 - \mu, x_2 - \mu, x_3 - \mu]\gamma_a^{-1}\begin{bmatrix} x_1 - \mu \\ x_2 - \mu \\ x_3 - \mu \end{bmatrix}\right\} \tag{5.91}$$

or

$$f(\vec{X}) = \frac{1}{(2\pi)^{3/2}|\gamma_a|^{1/2}} \exp\left\{-\frac{1}{2}(\vec{X} - \vec{\mu})\gamma_a^{-1}(\vec{X} - \vec{\mu})\right\} \tag{5.92}$$

where $\vec{X} = (x_1, x_2, x_3)$, $\vec{X} - \vec{\mu} = (x_1 - \mu, x_2 - \mu, x_3 - \mu)$ is the mean corrected vector $\vec{X}$, and γ_a^{-1} is the inverse of the autocovariance matrix γ_a given as

$$\gamma_a^{-1} = \begin{bmatrix} \dfrac{\sigma_X^4 - [\gamma(1)]^2}{|\gamma_a|} & \dfrac{\gamma(1)\gamma(2) - \sigma_X^2\gamma(1)}{|\gamma_a|} & \dfrac{[\gamma(1)]^2 - \sigma_X^2\gamma(2)}{|\gamma_a|} \\ \dfrac{\gamma(1)\gamma(2) - \sigma_X^2\gamma(1)}{|\gamma_a|} & \dfrac{\sigma_X^4 - [\gamma(2)]^2}{|\gamma_a|} & \dfrac{\gamma(1)\gamma(2) - \gamma(1)\sigma_X^2}{|\gamma_a|} \\ \dfrac{[\gamma(1)]^2 - \sigma_X^2\gamma(2)}{|\gamma_a|} & \dfrac{\gamma(1)\gamma(2) - \gamma(1)\sigma_X^2}{|\gamma_a|} & \dfrac{\sigma_X^4 - [\gamma(1)]^2}{|\gamma_a|} \end{bmatrix} \tag{5.93}$$

Now the connection between the Lagrange multipliers and the inverse matrix γ_a^{-1} given by equation (5.93) can be established. The inverse matrix can be rewritten as

$$\gamma_a^{-1} = \begin{bmatrix} 2\lambda_1 & \lambda_4 & \lambda_6 \\ \lambda_4 & 2\lambda_2 & \lambda_5 \\ \lambda_6 & \lambda_5 & 2\lambda_3 \end{bmatrix} \tag{5.94}$$

Since $\lambda_1 = \lambda_3$ and $\lambda_4 = \lambda_5$, equation (5.94) can be written as

$$\gamma_a^{-1} = \begin{bmatrix} 2\lambda_1 & \lambda_4 & \lambda_6 \\ \lambda_4 & 2\lambda_2 & \lambda_4 \\ \lambda_6 & \lambda_4 & 2\lambda_1 \end{bmatrix} \tag{5.95}$$

The solution is possible under the conditions:

$$\lambda_3 > 0, \quad 4\lambda_2\lambda_3 - \lambda_5^2 > 0 \tag{5.96}$$
$$|D| = |\gamma_a^{-1}| > 0 \tag{5.97}$$

For the inverse matrix given by equation (5.94), all principal minors must be positive.

It may be noted that there are four Lagrange multipliers that are responsible for carrying information in hydrologic processes. λ_1 and λ_2 depend only on the variance, λ_4 depends on the first lag autocovariance $\gamma(1)$, and λ_6 depends on the second lag autocovariance $\gamma(2)$. Define the Lagrange multiplier matrix as:

$$\Lambda = \frac{1}{2}\gamma_a^{-1} \tag{5.98}$$

For independent hydrologic processes, matrix Λ becomes diagonal:

$$\Lambda = \begin{bmatrix} \lambda_1 & 0 & 0 \\ 0 & \lambda_2 & 0 \\ 0 & 0 & \lambda_1 \end{bmatrix} \tag{5.99}$$

For the first lag-dependent process the Lagrange multiplier matrix is:

$$\Lambda = \begin{bmatrix} \lambda_1 & \dfrac{\lambda_4}{2} & 0 \\[2ex] \dfrac{\lambda_4}{2} & \lambda_2 & \dfrac{\lambda_4}{2} \\[2ex] 0 & \dfrac{\lambda_4}{2} & \lambda_1 \end{bmatrix} \tag{5.100}$$

and for the second lag dependency it is

$$\Lambda = \begin{bmatrix} \lambda_1 & \dfrac{\lambda_4}{2} & \dfrac{\lambda_6}{2} \\[2ex] \dfrac{\lambda_4}{2} & \lambda_2 & \dfrac{\lambda_4}{2} \\[2ex] \dfrac{\lambda_6}{2} & \dfrac{\lambda_4}{2} & \lambda_1 \end{bmatrix} \tag{5.101}$$

Example 5.4: Using the dataset of monthly discharges in Example 5.2, compute σ_X^2 and $\gamma(1), \gamma(2)$. Derive the joint probability distribution function $f(x_1, x_2, x_3)$ as a function of σ_X^2 and $\gamma(1), \gamma(2)$.

Solution: For the given dataset, the logarithm of discharge values is taken, first. Then, one gets: $\sigma_X^2 = 1.4190$ (cms)^2, $\gamma(1) = 0.9854$ (cms)^2 and $\gamma(2) = 0.7603$ (cms)^2. Then,

$$|\gamma_a| = \sigma_X^6 - 2\sigma_X^2[\gamma(1)]^2 + 2[\gamma(1)]^2\gamma(2) - \sigma_X^2[\gamma(2)]^2 = 0.7578$$

From equations (5.81)–(5.84), one gets:

$$\lambda_1 = \lambda_3 = \frac{1}{2}\frac{\sigma_X^4 - [\gamma(1)]^2}{|\gamma_a|} = 0.6879$$

$$\lambda_2 = \frac{1}{2}\frac{\sigma_X^2 - [\gamma(2)]^2}{|\gamma_a|} = 0.9472$$

$$\lambda_4 = \lambda_5 = \frac{\gamma(1)\gamma(2) - \gamma(1)\sigma_X^2}{|\gamma_a|} = -0.8566$$

$$\lambda_6 = \frac{\gamma(1)^2 - \gamma(2)\sigma_X^2}{|\gamma_a|} = -0.1423$$

Thus, one gets the matrix:

$$\gamma_a^{-1} = \begin{bmatrix} 2\lambda_1 & \lambda_4 & \lambda_6 \\ \lambda_4 & 2\lambda_2 & \lambda_4 \\ \lambda_6 & \lambda_4 & 2\lambda_1 \end{bmatrix} = \begin{bmatrix} 1.3758 & 0.8566 & -0.1423 \\ 0.8566 & 1.8944 & 0.8566 \\ -0.1423 & 0.8566 & 1.3758 \end{bmatrix}$$

Probability Density Function

Now, the PDF can be expressed as:

$$f(x_1, x_2, x_3) = \frac{1}{(2\pi)^{3/2}|\gamma_a|^{1/2}} \exp\left[-\frac{1}{2}(\vec{X} - \vec{\mu})\gamma_a^{-1}(\vec{X} - \vec{\mu})\right]$$

$$= 0.0729 \exp\left\{-\frac{1}{2}[x_1 - \mu, x_2 - \mu, x_3 - \mu] \begin{bmatrix} 1.3758 & 0.8566 & -0.1423 \\ 0.8566 & 1.8944 & 0.8566 \\ -0.1423 & 0.8566 & 1.3758 \end{bmatrix} \begin{bmatrix} x_1 - \mu \\ x_2 - \mu \\ x_3 - \mu \end{bmatrix}\right\}$$

Subtracting mean values from actual values as $x_1 - \mu$ or $x_1 - \bar{x}$ and $x_2 - \mu$ or $x_2 - \bar{x}$, and $x_3 - \mu$ or $x_3 - \bar{x}$ and then denoting the residuals simply as x_1, x_2 and x_2, the joint PDF can be written as:

$$f(x_1, x_2, x_3) = \frac{1}{(2\pi)^{3/2}|\gamma^a|^{1/2}} \exp\left\{-\frac{(\sigma_X^4 - [\gamma(1)]^2)x_1^2 + x_3^2}{2|\gamma_a|} - \frac{[\sigma_X^4 - [\gamma(2)]^2 x_2^2}{2|\gamma_a|}\right.$$
$$\left. - \frac{[\gamma(1)\gamma(2) - \gamma(1)\sigma_X^2](x_1 x_2 + x_2 x_3)}{|\gamma_a|} - \frac{[\{\gamma(1)\}^2 - \gamma(2)\sigma_X^2]x_1 x_3}{|\gamma_a|}\right\} \tag{5.102}$$

Distribution entropy

The entropy of the joint probability density function $f(x_1, x_2, x_3)$ can be expressed by substituting equation (5.102) in equation (5.49) as:

$$H(X_1, X_2, X_3) = -\int_{-\infty}^{\infty}\int_{-\infty}^{\infty}\int_{-\infty}^{\infty} f(x_1, x_2, x_3)\{\ln[(2\pi)^{-3/2}|\gamma_a|^{-1/2}] - \frac{1}{2(|\gamma_a|)}[\sigma_X^4 - [\gamma(1)]^2(x_1^2 + x_3^2)$$
$$+ [\sigma_X^4 - \gamma(2)]^2 x_2^2 + (2\gamma(1)\gamma(2) - \gamma(1)\sigma_X^2)(x_1 x_2 + x_2 x_3)$$
$$+ 2x_1 x_3[\{\gamma(1)\}^2 - \gamma(2)\sigma_x^2]]\}dx_1 dx_2 dx_3 \tag{5.103}$$

Equation (5.103) is essentially the sum of five integrals:

$$H(X_1, X_2, X_3) = \left(\frac{3}{2}\ln(2\pi) + \frac{1}{2}\ln|\gamma_a|\right)A_1 + \frac{1}{2|\gamma_a|}[A_2 + A_3 + A_4 + A_5] \tag{5.104}$$

where

$$A_1 = \int_{-\infty}^{\infty}\int_{-\infty}^{\infty}\int_{-\infty}^{\infty} f(x_1, x_2, x_3)dx_1 dx_2 dx_3 \tag{5.105}$$

$$A_2 = \{\sigma_X^4 - [\gamma(1)]^2\}\left[\int_{-\infty}^{\infty}\int_{-\infty}^{\infty}\int_{-\infty}^{\infty} f(x_1, x_2, x_3)x_1^2 dx_1 dx_2 dx_3 + \int_{-\infty}^{\infty}\int_{-\infty}^{\infty}\int_{-\infty}^{\infty} x_3^2 dx_1 dx_2 dx_3\right] \tag{5.106}$$

$$A_3 = \{\sigma_X^4 - [\gamma(2)]^2\}\left[\int_{-\infty}^{\infty}\int_{-\infty}^{\infty}\int_{-\infty}^{\infty} f(x_1, x_2, x_3)x_2^2 dx_1 dx_2 dx_3\right] \tag{5.107}$$

$$A_4 = 2[\gamma(1)\lambda(2) - \gamma(1)\sigma_X^2]\left[\int_{-\infty}^{\infty}\int_{-\infty}^{\infty}\int_{-\infty}^{\infty} f(x_1, x_2, x_3)x_1 x_2 dx_1 dx_2 dx_3\right.$$

$$+ \int_{-\infty}^{\infty} \int_{-\infty}^{\infty} \int_{-\infty}^{\infty} x_2 x_3 f(x_1, x_2, x_3) dx_1 dx_2 dx_3 \Bigg] \tag{5.108}$$

$$A_5 = 2\{[\gamma(1)]^2 - \gamma(2)\sigma_X^2\} \left[\int_{-\infty}^{\infty} \int_{-\infty}^{\infty} \int_{-\infty}^{\infty} f(x_1, x_2, x_3) x_1 x_3 dx_1 dx_2 dx_3 \right] \tag{5.109}$$

Using the constraint equations, equations (5.105) to (5.109) are solved as:

$$A_1 = 1 \tag{5.110}$$
$$A_2 = 2\sigma_X^2\{\sigma_X^4 - [\gamma(1)]^2\} \tag{5.111}$$
$$A_3 = \sigma_X^2\{\sigma_X^4 - [\gamma(2)]^2\} \tag{5.112}$$
$$A_4 = 4\gamma(1)[\gamma(1)\gamma(2) - \gamma(1)\sigma_X^2] \tag{5.113}$$
$$A_5 = 2\gamma(2)\{[\gamma(1)]^2 - \gamma(2)\sigma_X^2] \tag{5.114}$$

Inserting equation (5.110) to (5.114) in equation (5.104), and substituting (5.86), equation (5.49) becomes

$$H(X_1, X_2, X_3) = \frac{3}{2} \ln 2\pi + \frac{1}{2} \ln |\gamma_a|$$

$$+ \frac{\sigma_X^2\{\sigma_X^4 - [\gamma(1)]^2\} + \frac{\sigma_X^2}{2}\{\sigma_X^4 - [\gamma(2)]^2\} + 2\gamma(1)[\gamma(1)\gamma(2) - \gamma(1)\sigma_X^2] + \gamma(2)\{[\gamma(1)]^2 - \gamma(2)\sigma_X^2\}}{\sigma_X^6 - 2\sigma_X^2[\gamma(1)]^2 + 2[\gamma(1)]^2\gamma(2) - \sigma_X^2[\gamma(2)]^2} \tag{5.115}$$

On simplification, equation (5.115) reduces to:

$$H(X_1, X_2, X_3) = \frac{3}{2} \ln 2\pi + \frac{1}{2} \ln |\gamma_a| + \frac{\frac{3}{2}[\sigma_X^6 - 2\sigma_X^2[\gamma(1)]^2 - \sigma_X^2[\gamma(2)]^2 + 2[\gamma(1)]^2\gamma(2)]}{\sigma_X^6 - 2\sigma_X^2[\gamma(1)]^2 + 2[\gamma(1)]^2\gamma(2) - \sigma_X^2[\gamma(2)]^2} \tag{5.116}$$

Finally, equation (5.116) can be cast as

$$H(X_1, X_2, X_3) = \frac{3}{2} \ln(2\pi + 1) + \frac{1}{2} \ln |\gamma_a| \tag{5.117}$$

Example 5.5: Compute the distribution entropy for the data of Example 5.4.

Solution: $H(X_1, X_2, X_3) = \frac{3}{2}\ln(2\pi + 1) + \frac{1}{2}\ln|\gamma_a| = 4.1182$ Napier

5.1.3 Multi-lag serial dependence

The development of a three-time lag variate PDF can be extended to the case where the values of the hydrologic process $X(t) = \{x(t), t \in (0, T), m < T\}$ are dependent until an m-th lag.

Specification of constraints

From the previous discussion the resulting PDFs have the form of multivariate PDFs with the following changes:

a) The autocovariance matrix γ_a in the one variate PDF replaces the cross-covariance matrix γ_c in the multivariate counterpart.

b) The number of values $x \in X(t)$ serially correlated in the univariate PDF replaces the number of multivariables X in the multivariate counterpart.

The univariate distribution with multi-lag serial dependence can be used for any hydrologic process with strong serial dependency. An example may be a partial duration series with low cutoff level examining multiple flood events, hourly rainfall at one station, and so on.

To derive the multivariate PDF using POME, one must include all possible dependencies among values $X(t)$ until the m-th lag. For the development with no dependencies one constraint is sufficient, that is, variance – the statistic associated with the zeroth lag. For the case with one-lag serial dependence, two constraints – variance and autocovariance for the first lag – are sufficient. However, three Lagrange multipliers are needed for the specification of the autocovariance matrix. For two-lag serial dependence, three constraints and six Lagrange multipliers are necessary. For the m-th serial dependence, the autocorrelation matrix is of $(m+1) \times (m+1)$ dimensions. The number of constraints for POME is $m+1$; specifically these constraints are: $\sigma_X^2, \gamma(1), \gamma(2), \ldots, \gamma(m)$. The number of Lagrange multipliers is equal to the number of elements on and below the main diagonal of the autocovariance matrix γ_a, where

$$\gamma_a = \begin{bmatrix} \sigma_X^2 & \gamma(1) & . & . & . & \gamma(m) \\ \gamma(1) & . & & . & . & . \\ . & . & . & . & . & . \\ . & . & . & . & . & . \\ . & . & . & . & . & . \\ \gamma(m) & . & & . & . & \sigma_X^2 \end{bmatrix} \tag{5.118}$$

The inverse of the autocovariance matrix is

$$\gamma_a^{-1} = \begin{bmatrix} a_1 & . & b_1 & . & . & . & . \\ . & . & . & . & . & . \\ b_1 & . & . & . & . & . \\ . & . & . & . & . & b_j \\ . & . & . & . & . & . \\ . & . & . & b_j & . & a_{m+1} \end{bmatrix} \tag{5.119}$$

where matrix elements a's and b's are functions of $\sigma_X^2, \gamma(1), \gamma(2), \ldots, \gamma(m)$. The matrix of Lagrange multipliers is

$$\Lambda = \begin{bmatrix} a_1/2 & . & . & . & b_1/2 & . & . & . & . \\ . & . & . & . & . & . & . & . \\ . & . & . & . & . & . & . & . \\ . & . & . & . & . & . & b_j/2 \\ b_1/2 & . & . & . & . & . & . \\ . & . & . & . & . & . & . \\ . & . & . & . & . & . & . \\ . & . & b_j/2 & . & . & . & a_{m+1}/2 \end{bmatrix} \tag{5.120}$$

The matrix Λ determines the dependence structure of the hydrologic process up to the m-th lag. For example, in an independent process all off-diagonal elements in the Λ matrix are zero. For that case, equation (5.120) can be written as

$$
\Lambda = \begin{bmatrix}
\lambda_1 & \cdot & \cdot & \cdot & 0 & \cdot & \cdot & & \cdot \\
\cdot & \lambda_2 & \cdot & \cdot & \cdot & \cdot & \cdot & & \cdot \\
\cdot & \cdot & \cdot & \cdot & \cdot & \cdot & \cdot & & \cdot \\
\cdot & \cdot & \cdot & \cdot & \cdot & \cdot & \cdot & & 0 \\
\cdot & \cdot & \cdot & \cdot & \cdot & \cdot & \cdot & & \cdot \\
0 & \cdot & \cdot & \cdot & \cdot & \cdot & \cdot & & \cdot \\
\cdot & \cdot & \cdot & \cdot & \cdot & \cdot & \cdot & & \cdot \\
\cdot & \cdot & 0 & \cdot & \cdot & \cdot & \cdot & & \lambda_{m+1}
\end{bmatrix}
\tag{5.121}
$$

The existence of both partition function $Z(\lambda)$ and PDF is assured if all principle minors of γ_a^{-1} are positive.

Partition function

For the independent process this means that all Lagrange multipliers must be positive. The partition function is

$$
\exp(\lambda_0) = (2\pi)^{\frac{m+1}{2}} |\gamma_1|^{1/2}
\tag{5.122}
$$

POME-based PDF

The PDF is

$$
\begin{aligned}
f(X) &= \frac{1}{Z(\lambda)} \exp\left[-\frac{1}{2}(\vec{X} - \vec{\mu})\gamma_a^{-1}(\vec{X} - \vec{\mu})^* \right] \\
&= \frac{1}{Z(\lambda)} \exp\left[-(\vec{X} - \vec{\mu})\Lambda(\vec{X} - \vec{\mu})^* \right]
\end{aligned}
\tag{5.123}
$$

where $\vec{X} = (x_0, x_1, \dots, x_m)$ is the vector including serially dependent elements of the hydrological process, and $\vec{X} - \vec{\mu} = (x_0 - \mu, x_1 - \mu, \dots, x_m - \mu)$ is the mean corrected vector $\vec{X}$; γ_a^{-1} is the inverse of the autocovariance matrix γ_a; and $Z(\lambda)$ is the partition function.

Example 5.6: Take the data series of monthly discharges given in Example 5.4. Derive the joint probability distribution function $f(x_1, x_2, x_3, x_4)$ as a function of σ_X^2 and $\gamma(1), \gamma(2), \gamma(3)$.

Solution: Specification of Constraints: The constraints to be imposed on these variables are variance σ_X^2, autocovariances of the first, second and third lags: $\gamma(1), \gamma(2)$, and $\gamma(3)$:

$$
\sigma_{X_i}^2 = \int_{-\infty}^{\infty} \int_{-\infty}^{\infty} \int_{-\infty}^{\infty} \int_{-\infty}^{\infty} x_i^2 f(x_1, x_2, x_3, x_4) dx_1 dx_2 dx_3 dx_4, \quad i = 1, 2, 3, 4
\tag{5.124}
$$

$$
\gamma(1) = \int_{-\infty}^{\infty} \int_{-\infty}^{\infty} \int_{-\infty}^{\infty} \int_{-\infty}^{\infty} x_i x_{i+1} f(x_1, x_2, x_3, x_4) dx_1 dx_2 dx_3 dx_4 , i = 1, 2, 3
\tag{5.125}
$$

$$
\gamma(2) = \int_{-\infty}^{\infty} \int_{-\infty}^{\infty} \int_{-\infty}^{\infty} \int_{-\infty}^{\infty} x_i x_{i+2} f(x_1, x_2, x_3, x_4) dx_1 dx_2 dx_3 dx_4, i = 1, 2
\tag{5.126}
$$

$$
\gamma(3) = \int_{-\infty}^{\infty} \int_{-\infty}^{\infty} \int_{-\infty}^{\infty} \int_{-\infty}^{\infty} x_1 x_4 f(x_1, x_2, x_3, x_4) dx_1 dx_2 dx_3 dx_4
\tag{5.127}
$$

The total probability constraint can be expressed as

$$\int_{-\infty}^{\infty} \int_{-\infty}^{\infty} \int_{-\infty}^{\infty} \int_{-\infty}^{\infty} f(x_1, x_2, x_3, x_4) dx_1 dx_2 dx_3 dx_4 = 1 \tag{5.128}$$

Maximization of entropy

The entropy of the joint distribution $f(x_1, x_2, x_3, x_4)$ can be written as:

$$H(X_1, X_2, X_3, X_4) = -\int_{-\infty}^{\infty} \int_{-\infty}^{\infty} \int_{-\infty}^{\infty} \int_{-\infty}^{\infty} f(x_1, x_2, x_3, x_4) \ln f(x_1, x_2, x_3, x_4) dx_1 \, dx_2 dx_3 dx_4 \tag{5.129}$$

Maximizing entropy $H(X_1, X_2, X_3, x_4)$ subject to the constraints, one obtains

$$f(x_1, x_2, x_3, x_4) = \frac{1}{Z(\lambda)} \exp[-\lambda_1 x_1^2 - \lambda_2 x_2^2 - \lambda_3 x_3^2 - \lambda_4 x_4^2 \\ - \lambda_5 x_1 x_2 - \lambda_6 x_2 x_3 - \lambda_7 x_3 x_4 - \lambda_8 x_1 x_3 - \lambda_9 x_2 x_4 - \lambda_{10} x_1 x_4] \tag{5.130}$$

Partition function

The partition function can be expressed as:

$$Z(\lambda) = \exp(\lambda_0)$$

$$= \int_{-\infty}^{\infty} \int_{-\infty}^{\infty} \int_{-\infty}^{\infty} \int_{-\infty}^{\infty} \left(\begin{matrix} -\lambda_1 x_1^2 - \lambda_2 x_2^2 - \lambda_3 x_3^2 - \lambda_4 x_4^2 \\ -\lambda_5 x_1 x_2 - \lambda_6 x_2 x_3 - \lambda_7 x_3 x_4 - \lambda_8 x_1 x_3 - \lambda_9 x_2 x_4 - \lambda_{10} x_1 x_4 \end{matrix} \right) dx_1 dx_2 dx_3 dx_4$$

$$= \int_{-\infty}^{\infty} \exp[-\lambda_1 x_1^2] dx_1 \int_{-\infty}^{\infty} \exp(-\lambda_2 x_2^2 - \lambda_5 x_1 x_2) dx_2 \int_{-\infty}^{\infty} \exp[-\lambda_3 x_3^2 - x_3 (\lambda_6 x_2 + \lambda_8 x_1)] dx_3$$

$$\times \int_{-\infty}^{\infty} \exp[-\lambda_4 x_4^2 - x_4 (\lambda_7 x_3 + \lambda_9 x_2 + \lambda_{10} x_1)] dx_4$$

$$= \frac{2\pi^2}{\sqrt{D}} \tag{5.131}$$

where

$$D' = 4\lambda_1 \lambda_2 \lambda_3 \lambda_4 - \lambda_2 \lambda_4 \lambda_8^2 - \lambda_1 \lambda_3 \lambda_9^2 - \lambda_3 \lambda_4 \lambda_5^2 - \lambda_1 \lambda_2 \lambda_7^2 - \lambda_1 \lambda_4 \lambda_6^2 - \lambda_2 \lambda_3 \lambda_{10}^2 + \lambda_4 \lambda_5 \lambda_6 \lambda_8$$

$$+ \lambda_1 \lambda_6 \lambda_7 \lambda_9 + \lambda_3 \lambda_5 \lambda_9 \lambda_{10} + \lambda_2 \lambda_7 \lambda_8 \lambda_{10} - \frac{\lambda_5 \lambda_6 \lambda_7 \lambda_{10}}{2} - \frac{\lambda_6 \lambda_8 \lambda_9 \lambda_{10}}{2} - \frac{\lambda_5 \lambda_7 \lambda_8 \lambda_9}{2}$$

$$+ \frac{\lambda_6^2 \lambda_{10}^2}{4} + \frac{\lambda_8^2 \lambda_9^2}{4} + \frac{\lambda_5^2 \lambda_7^2}{4} \tag{5.132}$$

Relation between Lagrange multipliers and constraints

Differentiating the logarithm of $Z(\lambda)$ with respect to individual Lagrange multipliers separately, and making use of constraint equations, one obtains

$$\frac{\partial \lambda_0}{\partial \lambda_1} = -\sigma_X^2; \quad \frac{\partial \lambda_0}{\partial \lambda_2} = -\sigma_X^2; \quad \frac{\partial \lambda_0}{\partial \lambda_3} = -\sigma_X^2; \quad \frac{\partial \lambda_0}{\partial \lambda_4} = -\sigma_X^2; \quad \frac{\partial \lambda_0}{\partial \lambda_5} = -\gamma(1); \quad \frac{\partial \lambda_0}{\partial \lambda_6} = -\gamma(1);$$

$$\frac{\partial \lambda_0}{\partial \lambda_7} = -\gamma(1); \quad \frac{\partial \lambda_0}{\partial \lambda_8} = -\gamma(2); \quad \frac{\partial \lambda_0}{\partial \lambda_9} = -\gamma(2); \quad \frac{\partial \lambda_0}{\partial \lambda_{10}} = -\gamma(3) \tag{5.133}$$

The six equations used for solving the Lagrange multipliers can be expressed as:

$$4\lambda_2\lambda_3\lambda_4 - \lambda_3\lambda_9^2 - \lambda_2\lambda_7^2 - \lambda_4\lambda_6^2 + \lambda_6\lambda_7\lambda_9 \; = 2DS_x^2 \tag{5.134}$$

$$4\lambda_1\lambda_3\lambda_4 - \lambda_4\lambda_8^2 - \lambda_1\lambda_7^2 - \lambda_3\lambda_{10}^2 + \lambda_7\lambda_8\lambda_{10} \; = 2DS_x^2 \tag{5.135}$$

$$-2\lambda_3\lambda_4\lambda_5 + \lambda_4\lambda_6\lambda_8 + \lambda_3\lambda_9\lambda_{10} - \frac{\lambda_6\lambda_7\lambda_{10}}{2} - \frac{\lambda_7\lambda_8\lambda_9}{2} + \frac{\lambda_5\lambda_7^2}{2} \; = 2D\gamma(1) \tag{5.136}$$

$$-2\lambda_1\lambda_4\lambda_6 + \lambda_4\lambda_5\lambda_8 + \lambda_1\lambda_7\lambda_9 - \frac{\lambda_5\lambda_7\lambda_{10}}{2} - \frac{\lambda_{10}\lambda_8\lambda_9}{2} + \frac{\lambda_6\lambda_{10}^2}{2} \; = 2D\gamma(1) \tag{5.137}$$

$$-2\lambda_2\lambda_4\lambda_8 + \lambda_4\lambda_5\lambda_6 + \lambda_2\lambda_7\lambda_{10} - \frac{\lambda_6\lambda_9\lambda_{10}}{2} - \frac{\lambda_5\lambda_7\lambda_9}{2} + \frac{\lambda_8\lambda_9^2}{2} = 2D\gamma(2) \tag{5.138}$$

$$-2\lambda_2^2\lambda_{10} + 2\lambda_2\lambda_5\lambda_8 - \frac{\lambda_5^2\lambda_6}{2} - \frac{\lambda_6\lambda_8^2}{2} + \frac{\lambda_6^2\lambda_{10}}{2} = 2D\gamma(3) \tag{5.139}$$

where $\lambda_1 = \lambda_4$, $\lambda_2 = \lambda_3$, $\lambda_5 = \lambda_7$ and $\lambda_8 = \lambda_9$.
  Solving these equations above, one gets:

$$\lambda_1 = \lambda_4 = 0.6887, \; \lambda_2 = \lambda_3 = 0.9518; \; \lambda_5 = \lambda_7 = -0.8526;$$

$$\lambda_6 = -0.7866 \; \lambda_8 = \lambda_9 = -0.1130; \; \lambda_{10} = -0.0474$$

For the given data set, the autocovariance matrix can be expressed as:

$$\gamma_a = \begin{bmatrix} \sigma_X^2 & \gamma(1) & \gamma(2) & \gamma(3) \\ \gamma(1) & \sigma_X^2 & \gamma(1) & \gamma(2) \\ \gamma(2) & \gamma(1) & \sigma_X^2 & \gamma(1) \\ \gamma(3) & \gamma(2) & \gamma(1) & \sigma_X^2 \end{bmatrix} = \begin{bmatrix} 1.4190 & 0.9854 & 0.7603 & 0.6003 \\ 0.9854 & 1.4190 & 0.9854 & 0.7603 \\ 0.7603 & 0.9854 & 1.4190 & 0.9854 \\ 0.6003 & 0.7603 & 0.9854 & 1.4190 \end{bmatrix}$$

The matrix of Lagrange multipliers becomes

$$\Lambda = \begin{bmatrix} \lambda_1 & \frac{\lambda_5}{2} & \frac{\lambda_8}{2} & \frac{\lambda_{10}}{2} \\ \frac{\lambda_5}{2} & \lambda_2 & \frac{\lambda_6}{2} & \frac{\lambda_9}{2} \\ \frac{\lambda_8}{2} & \frac{\lambda_6}{2} & \lambda_3 & \frac{\lambda_7}{2} \\ \frac{\lambda_{10}}{2} & \frac{\lambda_9}{2} & \frac{\lambda_7}{2} & \lambda_4 \end{bmatrix} = \begin{bmatrix} 0.6887 & -0.4263 & -0.0565 & -0.0237 \\ -0.4263 & 0.9518 & -0.3933 & -0.0565 \\ -0.0565 & -0.3933 & 0.9518 & -0.4263 \\ -0.0237 & -0.0565 & -0.4263 & 0.6887 \end{bmatrix}$$

The partition function is:

$$Z(\lambda) = \exp(\lambda_0) = 2\pi^2 |D|^{-1/2} = 29.28$$

PDF

The probability distribution function can now be written as:

$$f(\vec{X}) = 0.0342 \exp\left\{ -(\vec{X} - \vec{\mu}) \begin{bmatrix} 0.6887 & -0.4263 & -0.0565 & -0.0237 \\ -0.4263 & 0.9518 & -0.3933 & -0.0565 \\ -0.0565 & -0.3933 & 0.9518 & -0.4263 \\ -0.0237 & -0.0565 & -0.4263 & 0.6887 \end{bmatrix} (\vec{X} - \vec{\mu})^* \right\}$$

Distribution entropy

Concurrent with this discussion, entropy of the univariate hydrologic process can be expressed as follows:

For the 0-th lag dependency:

$$H(X) = \frac{1}{2} \ln |\gamma_a| + \frac{1}{2} [\ln(2\pi) + 1] \tag{5.140}$$

where $\gamma_a = \sigma_X^2$.

For one-lag dependency:

$$H(X) = \frac{1}{2} \ln |\gamma_a| + \frac{2}{2} [\ln(2\pi) + 1] \tag{5.141}$$

where γ_a is given by equation (5.28)

For two-lag dependency:

$$H(X) = \frac{1}{2} \ln |\gamma_a| + \frac{3}{2} [\ln(2\pi) + 1] \tag{5.142}$$

where γ_a is given by equation (5.85).

For the m-lag dependency:

$$H(X) = \frac{1}{2} \ln |\gamma_a| + \frac{m+1}{2} [\ln(2\pi) + 1] \tag{5.143}$$

where γ_a is given by equation (5.118). Equation (5.143) is equivalent to the entropy of the multivariate normal distribution where γ_a is replaced by γ_c and m serially correlated values x_i $(i = 1, 2, \ldots, m)$ by m hydrologic multivariables.

Example 5.7: Compute the entropy using the data of Example 5.6.

$$H(X) = \frac{1}{2} \ln |\gamma_a| + \frac{m+1}{2} \left[\ln(2\pi) + 1 \right] = \frac{1}{2} \ln |\gamma_a| + 2 \left[\ln(2\pi) + 1 \right] = 5.97 \text{ (Napier)}$$

Note that the entropy is decreasing with increasing lag.

5.1.4 No serial dependence: bivariate case

Let X_1 and X_2 be two random variables of a stationary hydrologic process $X(t)$ Specifically the multivariables are: $X_1 = \{x_1(t), t \in (0, T)\}$ and $X_2 = \{x_2(t), t \in (0, T)\}$. To simplify notation, let X_1 and X_2, be designated respectively as X and Y, that is, $X \in X(t)$ and $Y \in Y(t)$. As an example, if $X(t)$ is a flood process, then X may represent the series of flood peaks and Y the series of flood volumes. From populations X and Y, representative sample measurements are obtained and used to compute relevant statistics.

For deriving the bivariate PDF by POME, one can use the results from one-lag serial dependent case, recognizing that multivariables are serially independent (0-lag dependence) and cross-correlated only at 0-th lag. Bivariate distribution with no serial dependence can be used for any two hydrologic processes strongly associated at the same lag, that is, rainfall depths and runoff peaks for the same storms, runoff and snowmelt for the same season, and so on. The objective is to derive the bivariate distribution with two variables dependent only at the 0-th lag and no serial dependency. The matrix that adequately expresses these characteristics is the cross covariance matrix γ_c of 2×2 dimensions:

$$\gamma_c = \begin{bmatrix} \sigma_X^2 & \gamma_{12}(0) \\ \gamma_{12}(0) & \sigma_Y^2 \end{bmatrix} \tag{5.144}$$

where $\gamma_{12}(0)$ is covariance of X and Y.

Specification of constraints

In a manner similar to what has been presented in the previous section on cases with serial dependency, the constraints necessary for deriving a bivariate PDF may be expressed as:

$$\sigma_1^2 = \sigma_X^2 = \int_{-\infty}^{\infty} \int_{-\infty}^{\infty} (x - \mu_x)^2 f(x, y) dx \, dy \tag{5.145}$$

$$\sigma_2^2 = \sigma_Y^2 = \int_{-\infty}^{\infty} \int_{-\infty}^{\infty} (y - \mu_y)^2 f(x, y) dx \, dy \tag{5.146}$$

$$\gamma_{12}(0) = \gamma_{XY}(0) = \int_{-\infty}^{\infty} \int_{-\infty}^{\infty} (x - \mu_x)(y - \mu_y) f(x, y) dx \, dy \tag{5.147}$$

where $f(x, y)$ is the bivariate joint PDF of random variables X and Y, and μ_x and μ_y are the means of X and Y, respectively. σ_X^2, σ_Y^2, and $\gamma_{12}(0)$ can be approximated by their sample values:

$$S_X^2 = \frac{1}{N} \sum_{t=0}^{N} (x_t - \bar{x})^2 \tag{5.148}$$

$$S_Y^2 = \frac{1}{N} \sum_{t=0}^{N} (y_t - \bar{y})^2 \tag{5.149}$$

$$c_{12} = \text{cov}(X, Y) = \frac{1}{N} \sum_{t=0}^{N} (x_t - \bar{x})(y_t - \bar{y}) \tag{5.150}$$

where $\bar{x}$ and $\bar{y}$ are sample means. If S_c is the sample approximation to γ_c then the number of Lagrange multipliers should be three, since S_c is symmetric and the number of elements on and below the main diagonal is 3.

POME-based PDF

Using POME, the PDF has the form:

$$f(x, y) = \frac{1}{Z(\lambda)} \exp[-\lambda_1 (x - \mu_x)^2 - \lambda_2 (y - \mu_y)^2 - \lambda_3 (x - \mu_x)(y - \mu_y)] \tag{5.151}$$

Equation (5.151) is equivalent to equation (5.7b).

Partition function

Writing x for $x - \mu_x$ and y for $y - \mu_y$ in equation (5.151) for simplicity, the partition function has the form:

$$Z(\lambda_1, \lambda_2, \lambda_3) = \int\limits_{-\infty}^{\infty} \int\limits_{-\infty}^{\infty} \exp[-\lambda_1 x^2 - \lambda_2 y^2 - \lambda_3 xy] dx \, dy \tag{5.152}$$

whose solution is

$$Z(\lambda_1, \lambda_2, \lambda_3) = Z(\lambda) = \frac{2\pi}{\sqrt{4\lambda_1 \lambda_2 - \lambda_3^2}} \tag{5.153}$$

Taking logarithm of equation (5.153),

$$\ln Z(\lambda_1, \lambda_2, \lambda_3) = \ln(2\pi) - \frac{1}{2} \ln(4\lambda_1 \lambda_2 - \lambda_3^2) \tag{5.154}$$

Relation between Lagrange multipliers and constraints

Differentiating the logarithm of equation (5.152) with respect to Lagrange multipliers separately and using constraint equations (5.145) to (5.147), and solving each derivative one obtains:

$$\frac{\partial \ln Z(\lambda)}{\partial \lambda_1} = -\sigma_X^2 \tag{5.155}$$

$$\frac{\partial \ln Z(\lambda)}{\partial \lambda_2} = -\sigma_Y^2 \tag{5.156}$$

$$\frac{\partial \ln Z(\lambda)}{\partial \lambda_3} = -\gamma_{12} \tag{5.157}$$

Differentiating equation (5.154) with respect to the Lagrange multipliers and substituting equations (5.155) to (5.157), one obtains:

$$\frac{2\lambda_2}{4\lambda_1 \lambda_2 - \lambda_3^2} = \sigma_X^2 \tag{5.158}$$

$$\frac{2\lambda_1}{4\lambda_1 \lambda_2 - \lambda_3^2} = \sigma_Y^2 \tag{5.159}$$

$$\frac{\lambda_3}{4\lambda_1 \lambda_2 - \lambda_3^2} = -\gamma_{12} \tag{5.160}$$

Solution of equation (5.158) to (5.160) is:

$$\lambda_1 = \frac{\sigma_Y^2}{2(\sigma_X^2 \sigma_Y^2 - \gamma_{12}^2)} \tag{5.161}$$

$$\lambda_2 = \frac{\sigma_X^2}{2(\sigma_X^2 \sigma_Y^2 - \gamma_{12}^2)} \tag{5.162}$$

$$\lambda_3 = -\frac{\gamma_{12}}{\sigma_X^2 \sigma_Y^2 - \gamma_{12}^2} \tag{5.163}$$

The common expression in the denominator of equations (5.161) to (5.163) is the determinant of the variance-covariance matrix γ_c given by equation (5.144). Thus,

$$\lambda_1 = \frac{\sigma_Y^2}{2|\gamma_c|} \tag{5.164}$$

$$\lambda_2 = \frac{\sigma_X^2}{2|\gamma_c|} \tag{5.165}$$

$$\lambda_3 = -\frac{\gamma_{12}}{|\gamma_c|} \tag{5.166}$$

and

$$Z(\lambda_1, \lambda_2, \lambda_3) = 2\pi |\gamma_c|^{1/2} \tag{5.167}$$

where the inverse of the γ_c matrix, γ_c^{-1}, is given as

$$\gamma_c^{-1} = \begin{bmatrix} \dfrac{\sigma_Y^2}{\sigma_X^2\sigma_Y^2 - \gamma_{12}^2} & -\dfrac{\gamma_{12}}{\sigma_X^2\sigma_Y^2 - \gamma_{12}^2} \\ -\dfrac{\gamma_{12}}{\sigma_X^2\sigma_Y^2 - \gamma_{12}^2} & \dfrac{\sigma_X^2}{\sigma_X^2\sigma_Y^2 - \gamma_{12}^2} \end{bmatrix} \tag{5.168}$$

The Lagrange multipliers continue to form the same pattern as in the one-lag serial dependence variate analysis. Thus, using equations (5.164) to (5.168), one obtains

$$\gamma_c^{-1} = \begin{bmatrix} 2\lambda_1 & \lambda_3 \\ \lambda_3 & 2\lambda_2 \end{bmatrix} \tag{5.169}$$

The necessary condition for the partition function $Z(\lambda)$ and PDF is the positiveness of the principal minors of the inverse matrix γ_c^{-1}:

$$\lambda_2 > 0; \quad 4\lambda_1\lambda_2 - \lambda_3^2 > 0 \tag{5.170}$$

Probability density function
Substituting the partition function and Lagrange multipliers [equation (5.164) to (5.167)] into equation (5.151), one obtains

$$f(x, y) = \frac{1}{2\pi |\gamma_c|^{1/2}} \exp\left\{ \frac{-[\sigma_Y^2(x - \mu_x)^2 + \sigma_X^2(y - \mu_y)^2 - 2\gamma_{12}(x - \mu_x)(y - \mu_y)]}{2|\gamma_c|} \right\} \tag{5.171}$$

This bivariate PDF is often expressed as a function of the correlation coefficient $\rho_{12} = \gamma_{12}/(\sigma_X\sigma_Y)$. Inserting this in equation (5.171) one obtains

$$f(x, y) = \frac{1}{2\pi |\gamma_c|^{1/2}} \exp\left\{ -\frac{1}{2(1 - \rho_{12}^2)} \left[\frac{(x - \mu_x)^2}{\sigma_X^2} + \frac{(y - \mu_y)^2}{\sigma_Y^2} - \frac{2\gamma_{12}(x - \mu_x)(y - \mu_y)}{\rho_{12}} \right] \right\} \tag{5.172}$$

The bivariate PDF can also be written in the matrix-vector form as:

$$f(\vec{X}) = \frac{1}{Z(\lambda)} \exp\left\{ -\frac{1}{2}(\vec{X} - \vec{\mu})\gamma_c^{-1}(\vec{X} - \vec{\mu})^* \right\} \tag{5.173}$$

or

$$f(\vec{X}) = \frac{1}{Z(\lambda)} \exp\left\{-(\vec{X} - \vec{\mu})\Lambda(\vec{X} - \vec{\mu})^*\right\} \tag{5.174}$$

where $\vec{X} = (x, y)$ is the row vector of the values of hydrologic random variables X and Y, $\vec{X} - \vec{\mu} = (x - \mu_x, y - \mu_y)$, the mean corrected vector $\vec{X}, \Lambda$ is the matrix of Lagrange multipliers, $Z(\lambda)$ is the partition function defined by equation (5.167).

For independent random variables X and Y, γ_c, γ_c^{-1} and Λ are diagonal matrices, and ρ_{12} in equation (5.172) is zero. For that case,

$$f(x, y) = f(x)f(y) \tag{5.175}$$
$$Z(\lambda) = 2\pi\sigma_X\sigma_Y \tag{5.176}$$

and

$$\lambda_i > 0, (i = 1, 2) \tag{5.177}$$

The entropy of two variables X and Y is equivalent to the entropy of equation (5.41) with γ_a replaced by γ_c, or to the entropy of equation (5.142).

Example 5.8: Take the flood peak (CMS = cubic meters per second) and volume data from a river and compute its bivariate probability density function.

Solution: Consider the flood peak and volume data are used by Yue et al. (1999). From the data, the statistics that are computed as: $u_x = 1.4265 \times 10^3 (\text{cms})$, $u_y = 5.2205 \times 10^4 (\text{cms})$, $\sigma_x^2 = 1.2944 \times 10^5 (\text{cms})^2$, $\sigma_y^2 = 1.5499 \times 10^8 (\text{cms})^2$, and $\gamma_{12}(0) = 2.6704 \times 10^6 (\text{cms})^2$. From equations (5.161) to (5.163), the Lagrange multipliers can be expressed as: $\lambda_1 = 5.9932$, $\lambda_2 = 0.50052$, and $\lambda_3 = -2.0652$. From equation (5.164), γ_c^{-1} can be expressed as:

$$\gamma_c^{-1} = \begin{bmatrix} 11.986 & -2.0652 \\ -2.0652 & 1.001 \end{bmatrix}$$

The partition function can be obtained as:

$$Z(\lambda) = 2\pi|\gamma_c|^{1/2} = 2.2593$$

The joint probability density function is plotted in Figure 5.4 and can be expressed as:

$$f(x, y) = 0.4426 \exp\left(-\frac{1}{2}[x - \mu_x, y - \mu_y]\begin{bmatrix} 11.986 & -2.0652 \\ -2.0652 & 1.001 \end{bmatrix}\begin{bmatrix} x - u_x \\ y - u_y \end{bmatrix}\right)$$

5.1.5 Cross-correlation and serial dependence: bivariate case

Now consider a bivariate case where each variable is serially dependent. First, consider the case when the dependency in each variable is one lag. The matrix that includes such dependencies is:

$$\sum = \begin{bmatrix} E_{11} & E_{12} \\ E_{21} & E_{22} \end{bmatrix} \tag{5.178}$$

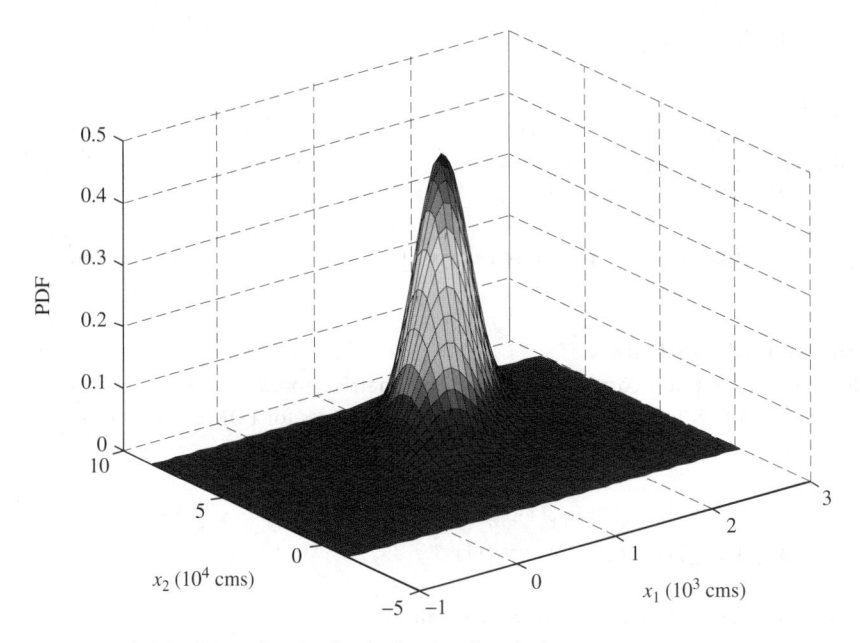

Figure 5.4 Joint probability density function for the flood peak and volume.

where E_{11} and E_{22} are autocovariance matrices of variables X and Y, while E_{12} and E_{21} are cross-variance matrices including the dependencies of zero order and first order:

$$E_{12} = \begin{bmatrix} \gamma_{12}(0) & \gamma_{12}(-1) \\ \gamma_{12}(1) & \gamma_{12}(0) \end{bmatrix}, \quad E_{21} = \begin{bmatrix} \gamma_{21}(0) & \gamma_{21}(-1) \\ \gamma_{21}(1) & \gamma_{21}(0) \end{bmatrix} \tag{5.179}$$

where γ_{ij} are cross-covariances of order $k = 0$ or 1. Sample approximations of these matrices are S_{12} and S_{21} matrices with elements $c_{ij}(k)$, Using the property $c_{12}(k) = c_{21}(-k)$, equation (5.179) can be written as

$$S_{12} = \begin{bmatrix} c_{12}(0) & c_{21}(1) \\ c_{12}(1) & c_{12}(0) \end{bmatrix}, \quad S_{21} = S_{12}^* \tag{5.180}$$

where * denotes the matrix transpose. Inserting equation (5.180) in equation (5.178), one obtains the symmetric Toeplitz matrix (A Toeplitz matrix is a matrix with constant diagonals):

$$S = \begin{bmatrix} S_{11} & S_{12} \\ S_{21} & S_{22} \end{bmatrix} = \begin{bmatrix} S_X^2 & c_1(1) & c_{12}(0) & c_{21}(1) \\ c_1(1) & S_X^2 & c_{12}(1) & c_{21}(0) \\ c_{12}(0) & c_{12}(1) & S_Y^2 & c_2(1) \\ c_{21}(1) & c_{12}(0) & c_2(1) & S_Y^2 \end{bmatrix} \tag{5.181}$$

The definition of $\sum$ and S matrices includes an expansion of submatrix elements according to random variables X or Y. For example, E_{11} is for X and E_{22} is for Y. However, submatrices may also be expanded according to lags, for example, E_{11} and E_{22} for the 0-th lag and E_{12} and E_{21}

for lag one. In that case,

$$
S = \begin{bmatrix}
S_X^2 & c_{12}(0) & c_1(1) & c_{12}(1) \\
c_{12}(0) & S_Y^2 & c_{21}(1) & c_2(1) \\
c_1(1) & c_{21}(1) & S_X^2 & c_{12}(0) \\
c_{12}(1) & c_2(1) & c_{12}(0) & S_Y^2
\end{bmatrix}
\tag{5.182}
$$

It may be noted that the determinant of matrix in equation (5.181) equals the determinant of matrix in equation (5.182).

Specification of constraints

The bivariate distribution corresponding to the S matrix would include dependencies among $(x_1, x_2) \in X(t)$ and $(y_1, y_2) \in Y(t)$ The number of constraints for POME is equal to the number of different elements in the S matrix:

$$
\sigma_X^2 = \sigma_1^2, \sigma_Y^2 = \sigma_2^2, \gamma_X(1) = \gamma_1(1), \gamma_Y(1) = \gamma_2(1), \mathrm{cov}(X, Y) = \gamma_{12}(0),
$$
$$
\gamma_{XY}(1) = \gamma_{12}(1), \gamma_{XY}(1) = \gamma_{YX}(1) = \gamma_{21}(1)
$$

The constraints are replaced by their sample values:

$$
S_X^2 = \frac{1}{N} \sum_{t=0}^{N} (x_t - \bar{x})
\tag{5.183}
$$

$$
S_Y^2 = \frac{1}{N} \sum_{t=0}^{N} (y_t - \bar{y})
\tag{5.184}
$$

$$
c_1(1) = \frac{1}{N} \sum_{t=0}^{N} (x_t - \bar{x})(x_{t+1} - \bar{x})
\tag{5.185}
$$

$$
c_2(1) = \frac{1}{N} \sum_{t=0}^{N} (y_t - \bar{y})(y_{t+1} - \bar{y})
\tag{5.186}
$$

$$
c_{12}(0) = \frac{1}{N} \sum_{t=0}^{N} (x_t - \bar{x})(y_t - \bar{y})
\tag{5.187}
$$

$$
c_{12}(1) = \frac{1}{N} \sum_{t=0}^{N} (x_t - \bar{x})(y_{t+1} - \bar{y})
\tag{5.188}
$$

$$
c_{21}(1) = \frac{1}{N} \sum_{t=0}^{N} (y_t - \bar{y})(x_{t+1} - \bar{x})
\tag{5.189}
$$

Maximization of entropy

Since the S matrix is symmetric, the number of Lagrange multipliers necessary for POME is equal to the number of elements under and on the main diagonal: ten. Therefore the POME-based PDF has the form:

$$
f(x_1, x_2; y_1, y_2) = \frac{1}{Z(\lambda)} \exp[-\lambda_1 x_1^2 - \lambda_2 x_2^2 - \lambda_3 y_1^2 - \lambda_4 y_2^2 - \lambda_5 x_1 x_2 - \lambda_6 x_2 y_1 - \lambda_7 y_1 y_2
$$
$$
- \lambda_8 x_1 y_1 - \lambda_9 x_2 y_2 - \lambda_{10} x_1 y_2]
\tag{5.190}
$$

where x_1, x_2, y_1, and y_2 are each corrected by respective means μ_x or $\bar{x}$ and μ_y or $\bar{y}$.

Partition function

Using the total probability constraint,

$$\int_{-\infty}^{\infty}\int_{-\infty}^{\infty}\int_{-\infty}^{\infty}\int_{-\infty}^{\infty} f(x_1,x_2,y_1,y_2)dx_1\,dx_2\,dy_1\,dy_2 = 1 \tag{5.191}$$

one solves for the partition function:

$$\exp(\lambda_0) = \int_{-\infty}^{\infty}\exp[-\lambda_1 x_1^2]dx_1 \int_{-\infty}^{\infty}\exp[-\lambda_2 x_2^2 - \lambda_5 x_1 x_2]dx_2 \int_{-\infty}^{\infty}\exp[-\lambda_3 y_1^2 - y_1(\lambda_6 x_2 + \lambda_8 x_1)]dy_1$$

$$\int_{-\infty}^{\infty}\exp[-\lambda_4 y_2^2 - y_2(\lambda_7 y_1 + \lambda_9 x_2 + \lambda_{10} x_1)]dy_2 \tag{5.192}$$

Relation between Lagrange multipliers and constraints

Equation (5.192) is solved by the successive use of the formula:

$$\int_{-\infty}^{+\infty}\exp[-p^2 x^2 \pm qx]dx = \frac{\sqrt{\pi}}{p}\exp\left[\frac{q^2}{4p^2}\right] \tag{5.193}$$

under the condition that $p > 0$. Final expressions for Lagrange multipliers are given as

$$\lambda_1 = \frac{S_Y^2 2D + \lambda_5^2 + \dfrac{\lambda_2\lambda_{10}^2}{\lambda_4} - \dfrac{\lambda_5\lambda_9\lambda_{10}}{\lambda_4}}{4\lambda_2 - \dfrac{\lambda_9^2}{\lambda_4}} \tag{5.194}$$

$$\lambda_2 = \frac{S_X^2 2D + \lambda_6^2 + \dfrac{\lambda_3\lambda_9^2}{\lambda_4} - \dfrac{\lambda_6\lambda_7\lambda_9}{\lambda_4}}{4\lambda_3 - \dfrac{\lambda_7^2}{\lambda_4}} \tag{5.195}$$

$$\lambda_3 = \frac{S_X^2 2D + \lambda_8^2 + \dfrac{\lambda_1\lambda_7^2}{\lambda_4} - \dfrac{\lambda_7\lambda_8\lambda_{10}}{\lambda_4}}{4\lambda_1 - \dfrac{\lambda_{10}^2}{\lambda_4}} \tag{5.196}$$

$$\lambda_4 = \frac{(4\lambda_1\lambda_2\lambda_3 - \lambda_2\lambda_8^2 - \lambda_1\lambda_6^2 - \lambda_3\lambda_5^2 + \lambda_5\lambda_6\lambda_8)}{2S_y^2\left(\begin{array}{c} 4\lambda_1\lambda_2\lambda_3 - \lambda_2\lambda_8^2 - \lambda_1\lambda_6^2 - \lambda_3\lambda_5^2 + \lambda_5\lambda_6\lambda_8 + \dfrac{\lambda_8^2\lambda_9^2}{4\lambda_4} + \dfrac{\lambda_6^2\lambda_{10}^2}{4\lambda_4} + \dfrac{\lambda_5^2\lambda_7^2}{4\lambda_4} - \dfrac{\lambda_2\lambda_3\lambda_{10}^2}{\lambda_4} \\[2mm] -\dfrac{\lambda_5\lambda_6\lambda_7\lambda_{10}}{2\lambda_4} - \dfrac{\lambda_1\lambda_3\lambda_9^2}{\lambda_4} - \dfrac{\lambda_1\lambda_2\lambda_7^2}{\lambda_4} - \dfrac{\lambda_6\lambda_8\lambda_9\lambda_{10}}{2\lambda_4} - \dfrac{\lambda_5\lambda_7\lambda_8\lambda_9}{2\lambda_4} + \dfrac{\lambda_2\lambda_7\lambda_8\lambda_{10}}{\lambda_4} \\[2mm] +\dfrac{\lambda_1\lambda_6\lambda_7\lambda_9}{\lambda_4} + \dfrac{\lambda_3\lambda_5\lambda_9\lambda_{10}}{\lambda_4} \end{array}\right)} \tag{5.197}$$

$$\lambda_5 = \frac{c_1(1)2D - \lambda_6\lambda_8 + \dfrac{\lambda_7\lambda_8\lambda_9}{2\lambda_4} - \dfrac{\lambda_3\lambda_9\lambda_{10}}{\lambda_4} + \dfrac{\lambda_6\lambda_7\lambda_{10}}{2\lambda_4}}{-2\lambda_3 + \dfrac{\lambda_7^2}{2\lambda_4}} \tag{5.198}$$

$$\lambda_6 = \frac{c_{12}(1)2D - \lambda_5\lambda_8 + \dfrac{\lambda_8\lambda_9\lambda_{10}}{2\lambda_4} - \dfrac{\lambda_1\lambda_7\lambda_9}{\lambda_4} + \dfrac{\lambda_5\lambda_7\lambda_{10}}{2\lambda_4}}{-2\lambda_1 + \dfrac{\lambda_{10}^2}{2\lambda_4}}$$

(5.199)

$$\lambda_7 = \frac{c_2(1)2D + \dfrac{\lambda_5\lambda_8\lambda_9}{2\lambda_4} - \dfrac{\lambda_2\lambda_8\lambda_{10}}{\lambda_4} - \dfrac{\lambda_1\lambda_6\lambda_9}{\lambda_4} + \dfrac{\lambda_5\lambda_6\lambda_{10}}{2\lambda_4}}{\dfrac{\lambda_5^2}{2\lambda_4} - 2\dfrac{\lambda_1\lambda_2}{\lambda_4}}$$

(5.200)

$$\lambda_8 = \frac{c_{12}(0)2D - \lambda_5\lambda_6 + \dfrac{\lambda_6\lambda_9\lambda_{10}}{2\lambda_4} + \dfrac{\lambda_5\lambda_7\lambda_9}{2\lambda_4} - \dfrac{\lambda_2\lambda_7\lambda_{10}}{\lambda_4}}{-2\lambda_2 + \dfrac{\lambda_9^2}{2\lambda_4}}$$

(5.201)

$$\lambda_9 = \frac{c_{12}(0)2D + \dfrac{\lambda_6\lambda_8\lambda_{10}}{2\lambda_4} + \dfrac{\lambda_5\lambda_7\lambda_8}{2\lambda_4} - \dfrac{\lambda_1\lambda_6\lambda_7}{\lambda_4} - \dfrac{\lambda_3\lambda_5\lambda_{10}}{\lambda_4}}{\dfrac{\lambda_8^2}{2\lambda_4} - 2\dfrac{\lambda_1\lambda_3}{\lambda_4}}$$

(5.202)

$$\lambda_{10} = \frac{c_{12}(1)2D + \dfrac{\lambda_6\lambda_8\lambda_9}{2\lambda_4} - \dfrac{\lambda_2\lambda_7\lambda_8}{\lambda_4} - \dfrac{\lambda_3\lambda_5\lambda_9}{\lambda_4} + \dfrac{\lambda_5\lambda_6\lambda_7}{2\lambda_4}}{\dfrac{\lambda_6^2}{2\lambda_4} - 2\dfrac{\lambda_2\lambda_3}{\lambda_4}}$$

(5.203)

where D is given as

$$D = 4\lambda_1\lambda_2\lambda_3\lambda_4 - \lambda_2\lambda_4\lambda_8^2 - \lambda_1\lambda_4\lambda_6^2 - \lambda_3\lambda_4\lambda_5^2 + \lambda_4\lambda_5\lambda_6\lambda_8 + \frac{\lambda_8^2\lambda_9^2}{4} + \frac{\lambda_6^2\lambda_{10}^2}{4} + \frac{\lambda_5^2\lambda_7^2}{4}$$
$$- \lambda_2\lambda_3\lambda_{10}^2 - \frac{\lambda_5\lambda_6\lambda_7\lambda_{10}}{2} - \lambda_1\lambda_3\lambda_9^2 - \lambda_1\lambda_2\lambda_7^2 - \frac{\lambda_6\lambda_8\lambda_9\lambda_{10}}{2} - \frac{\lambda_5\lambda_7\lambda_8\lambda_9}{2}$$
$$+ \lambda_2\lambda_7\lambda_8\lambda_{10} + \lambda_1\lambda_6\lambda_7\lambda_9 + \lambda_3\lambda_5\lambda_9\lambda_{10}$$

(5.204)

The system of equations (5.194) to (5.203) can be solved using the Newton-Raphson method or any other appropriate numerical method. The result can be presented as a matrix of Lagrange multipliers:

$$\Lambda = \begin{bmatrix} \lambda_1 & \dfrac{\lambda_5}{2} & \dfrac{\lambda_8}{2} & \dfrac{\lambda_{10}}{2} \\ \dfrac{\lambda_5}{2} & \lambda_2 & \dfrac{\lambda_6}{2} & \dfrac{\lambda_9}{2} \\ \dfrac{\lambda_8}{2} & \dfrac{\lambda_6}{2} & \lambda_3 & \dfrac{\lambda_7}{2} \\ \dfrac{\lambda_{10}}{2} & \dfrac{\lambda_9}{2} & \dfrac{\lambda_7}{2} & \lambda_4 \end{bmatrix} = \begin{bmatrix} \Lambda_{11} & \Lambda_{12} \\ \Lambda_{21} & \Lambda_{22} \end{bmatrix}$$

(5.205)

where $\Lambda_{11}, \Lambda_{12}, \Lambda_{21},$ and Λ_{22} are sub-matrices satisfying:

$$\Lambda_{11} = \frac{1}{2}(S_{11} - S_{12}\, S_{22}^{-1}\, S_{21})^{-1}$$

(5.206)

$$\Lambda_{12} = -\frac{1}{2}S_{11}^{-1}S_{12}(S_{22} - S_{21}\, S_{11}^{-1}\, S_{12})^{-1}$$

(5.207)

$$\Lambda_{21} = -\frac{1}{2}S_{22}^{-1}S_{21}(S_{11} - S_{12}\, S_{22}^{-1}\, S_{21})^{-1}$$

(5.208)

$$\Lambda_{22} = \frac{1}{2}(S_{22} - S_{21}\, S_{11}^{-1}\, S_{12})^{-1}$$

(5.209)

where S_{ii} and $S_{ij}(i, j = 1, 2)$ are autocovariance and cross-covariance defined by equations (5.179) and (5.181). The partition function can be written as

$$Z(\lambda) = 2\pi^2 D^{-1/2} \tag{5.210}$$

which can be shown to be equivalent to:

$$Z(\lambda) = (2\pi)^2 |S|^{1/2} \tag{5.211}$$

Probability Density Function

The PDF of equation (5.190) can be written as

$$f(\vec{X}, \vec{Y}) = \frac{1}{Z(\lambda)} \exp[-(\vec{X}\Lambda_{11}\vec{X}^* + \vec{Y}\Lambda_{21}\vec{X}^* + \vec{X}\Lambda_{12}\vec{Y}^* + \vec{Y}\Lambda_{22}\vec{X}^*)] \tag{5.212}$$

where $\vec{X} = (x_1 - \mu, x_2 - \mu) = (x_1 - \bar{x}, x_2 - \bar{x})$ and $\vec{Y} = (y_1 - \mu, y_2 - \mu) = (y_1 - \bar{y}, y_2 - \bar{y})$. The PDF can also be written more generally as

$$f(\vec{X}, \vec{Y}) = \frac{1}{Z(\lambda)} \exp[-(\vec{X}, \vec{Y})\Lambda(\vec{X}, \vec{Y})^*] \tag{5.213}$$

where Λ is defined by equation (5.205). In deriving equations (5.211) to (5.213), the necessary condition for the existence of the solution is the positiveness of all principal minors of the S matrix. The inverse of S, S^{-1}, can be expressed in terms of the Lagrange multipliers as a symmetric matrix:

$$S^{-1} = \begin{bmatrix} 2\lambda_1 & \lambda_5 & \lambda_8 & \lambda_{10} \\ \lambda_5 & 2\lambda_2 & \lambda_6 & \lambda_9 \\ \lambda_8 & \lambda_6 & 2\lambda_3 & \lambda_7 \\ \lambda_{10} & \lambda_9 & \lambda_7 & 2\lambda_4 \end{bmatrix} \tag{5.214}$$

Note that the number of variables analyzed here is $2 (M = 2)$ and the order of serial dependency is one $(m = 1)$. Equations (5.211) and (5.213) can be written more generally as

$$Z(\lambda) = (2\pi)^{M(m+1)/2} |S|^{1/2} \tag{5.215}$$

$$f(\vec{X}, \vec{Y}) = \frac{1}{Z(\lambda)} \exp\left[-\frac{1}{2}(\vec{X} - \vec{\mu})S^{-1}(\vec{X} - \vec{Y})^*\right] \tag{5.216}$$

where $\vec{X} - \vec{\mu} = (x_1 - \bar{x}, x_2 - \bar{x}, y_1 - \bar{y}, y_2 - \bar{y})$. If we use the covariance matrix S in equation (5.182), the same PDF as equation (5.216) is obtained but with a different arrangement of terms inside the mean corrected vector. Specifically, $\vec{X} - \vec{\mu} = (x_1 - \bar{x}, y_1 - \bar{y}, x_2 - \bar{x}, y_2 - \bar{y})$, while the determinants of S matrices remain the same.

Entropy

The entropy of the bivariate distribution with two variables with first order serial dependency is equal to the entropy of $2(1 + 1)$ variables. Using equation (5.142) and letting $M = 2$, and $m = 1$, the entropy of the bivariate distribution becomes

$$H(X) = \frac{1}{2} \ln |S| + 2[\ln(2\pi) + 1] \tag{5.217}$$

5.1.6 Multivariate case: no serial dependence

Univariate and bivariate normal distributions can be considered as building blocks for deriving the multivariate normal distribution. Let $X_1, X_2, X_3, \ldots, X_M$ be random multivariables of a stationary hydrologic process $X(t)$. If $X(t)$ is a flood process, then $X_1, X_2, X_3, \ldots, X_M$ are multivariables describing that process. Two cases are discussed: 1) no serial dependency, and (2) multi-serial dependency.

When there is no serial dependency, the process is strongly independent. For example, Peak, volume, duration, and time to peak of an extreme flood depend on one another but are independent of other floods. The development of a multivariate normal distribution in this case is analogous to the univariate case when:

i The number of lags $(m + 1)$ is replaced by the number of multivariables (M).

ii The autocovariance matrix S_a (or γ_a) is replaced by the cross-covariance matrix S_c (or γ_c) of zero lag cross-covariances or covariances between variables.

iii The number of Lagrange multipliers for POME is equal to the number of elements on and below the main diagonal of the cross-covariance matrix:

$$
S_c = \begin{bmatrix}
S_1^2 & c_{12} & \cdot & \cdot & \cdot & c_{1M} \\
c_{21} & S_2^2 & \cdot & \cdot & \cdot & \cdot \\
\cdot & \cdot & \cdot & \cdot & \cdot & \cdot \\
\cdot & \cdot & \cdot & \cdot & \cdot & \cdot \\
\cdot & \cdot & \cdot & \cdot & \cdot & \cdot \\
c_{M1} & \cdot & \cdot & \cdot & & S_M^2
\end{bmatrix}
\tag{5.218}
$$

where $c_{ij} = \text{cov}(X_i, X_j), (i, j = 1, 2, \ldots, M)$.

iv The existence of the partition matrix $Z(\lambda)$ and PDF $f(\vec{X})$ requires all principal minors of S_c to be positive.

The partition function for POME is given as:

$$
Z(\lambda) = (2\pi)^{M/2} |S_c|^{1/2}
$$

and the PDF as

$$
f(\vec{X}) = f(x_1, x_2, \ldots, x_M) = \frac{1}{Z(\lambda)} \exp\left[-\frac{1}{2}(\vec{X} - \vec{\mu}) S_c^{-1} (\vec{X} - \vec{\mu})^* \right]
\tag{5.219}
$$

where S_c^{-1} is the inverse of the S_c matrix given by equation (5.218) and $\vec{X} - \vec{\mu}$ is the mean corrected vector $\vec{X} = (x_1, x_2, \ldots, x_M)$. Each Lagrange multiplier in the Λ matrix corresponds to the appropriate element in the S_c^{-1} matrix. Specifically,

$$
\lambda_{ij}(\in \Lambda) = a_{ij} \ (\in S_c^{-1}) \text{ if } i = j \text{ (off diagonal elements)}
\tag{5.220a}
$$

$$
\lambda_{ij}(\in \Lambda) = \frac{1}{2} S_i^2 \ (\in S_c^{-1}) \text{ (diagonal elements)}
\tag{5.220b}
$$

The number of Lagrange multipliers determines the order of dependency between multivariables of the hydrologic process $X(t)$. The entropy of the multivariate process $\vec{X} = (X_1, X_2, \ldots, X_M)$ is:

$$
H(X) = \frac{1}{2} \ln |S_c| + \frac{M}{2} [\ln(2\pi) + 1]
\tag{5.221}
$$

5.1.7 Multi-lag serial dependence

This case applies where there is serial dependency in the multivariables of the process. For example, when one measures flood characteristics, such as peak, volume, and duration at very short intervals of time, then these characteristics are dependent serially as well on each other. The analysis of the bivariate case can be extended to the multivariate case. Let the total number of hydrologic variables be M and the order of serial dependency be m. The sample matrix that includes all possible dependencies is the matrix of all sample autocovariance and cross-covariance matrices:

$$S = \begin{bmatrix} S_{11} & \cdot & \cdot & \cdot & S_{1M} \\ S_{21} & \cdot & \cdot & \cdot & S_{2M} \\ \cdot & \cdot & \cdot & \cdot & \cdot \\ \cdot & \cdot & \cdot & \cdot & \cdot \\ S_{M1} & \cdot & \cdot & \cdot & S_{MM} \end{bmatrix} \tag{5.222}$$

where $S_{ii}(i = 1, 2, \ldots, M)$ are autocovariance matrices of $(m + 1) \times (m + 1)$ dimensions, and $S_{ij}(i, j = 1, 2, \ldots, M; \ i \neq j)$ are cross-covariance matrices of $(m + 1) \times (m + 1)$ dimensions:

$$S_{ij} = \begin{bmatrix} c_{ij}(0) & \cdot & \cdot & \cdot & c_{ij}(-m) \\ \cdot & \cdot & \cdot & \cdot & \cdot \\ \cdot & \cdot & \cdot & \cdot & \cdot \\ \cdot & \cdot & \cdot & \cdot & \cdot \\ c_{ij}(m) & \cdot & \cdot & \cdot & c_{ij}(0) \end{bmatrix} \tag{5.223}$$

Thus, the S matrix is of $[M \times (m + 1)][M \times (m + 1)]$ dimensions. It is known that the matrix of Lagrange multipliers will be proportional to S^{-1}. Expressions for the partition function and the PDF $f(X)$ are the same as equations (5.213) and (5.216) with $\vec{X} - \vec{\mu} = (X_1 - \overline{X_1}, X_2 - \overline{X_2}, \ldots, X_M - \overline{X_M})$ as the mean corrected vector. The entropy of the multivariate normal distribution with m-th order serial dependency is expressed as:

$$H(X) = \frac{1}{2} \ln |S| + \frac{M(m + 1)}{2} [\ln(2\pi) + 1] \tag{5.224}$$

5.2 Multivariate exponential distributions

Multivariate exponential distributions are useful for network design, flood frequency analysis, drought frequency analysis, reliability analysis, and so on. Here we derive these distributions of Marshall-Olkin form. First, the bivariate case is discussed.

5.2.1 Bivariate exponential distribution

Let X and Y be two hydrologic random variables each exponentially distributed as:

$$F_X(x) = 1 - \exp[-(a_1 + a_{12})x] \tag{5.225}$$

$$F_Y(y) = 1 - \exp[-(a_2 + a_{12})y] \tag{5.226}$$

where a_1, a_2, a_{12} are parameters: the first parameter is associated with X, the second with Y, and third with both X and Y. The mean and variance of the distributions are:

$$E[X] = \frac{1}{a_1 + a_{12}}, \quad \text{var}[X] = \frac{1}{(a_1 + a_{12})^2} \tag{5.227}$$

$$E[Y] = \frac{1}{a_2 + a_{12}}, \quad \text{var}[Y] = \frac{1}{(a_2 + a_{12})^2} \tag{5.228}$$

Since X and Y are dependent, the dependency is expressed by the coefficient of correlation:

$$\rho_{12} = \frac{a_{12}}{a_1 + a_2 + a_{12}} \tag{5.229}$$

The bivariate exponential distribution of Marshall-Olkin form can be expressed as

$$F(x, y) = 1 - \exp[-a_1 x - a_2 y - a_{12} \max(x, y)] \tag{5.230}$$

where $a_1 > 0, a_2 > 0, a_{12} > 0$ are distribution parameters, and $x, y > 0$. An alternative form of the distribution was introduced by Galambos (1978) as

$$F(x, y) = 1 - \exp[-(a_1 + a_{12})x] - \exp[-(a_2 + a_{12})y] - \exp[-a_1 x - a_2 y + a_{12} \max(x, y)] \tag{5.231}$$

This alternative form is employed in this chapter. In the derivation three cases are distinguished: 1) means are normalized, that is, $\mu_X = \mu_Y = 1$; 2) means are equal, that is, $\mu_X = \mu_Y = a$; and 3) means are unequal.

Derivation with normalized means

Here $E[X] = E[Y] = 1$. From equations (5.227) and (5.228), one gets

$$a_1 + a_{12} = 1, \quad a_2 + a_{12} = 1 \tag{5.232}$$

Let $a_{12} = b$. Equation (5.229) for the coefficient of correlation simplifies to:

$$\rho_{12} = \frac{b}{2 - b} \tag{5.233}$$

Equation (5.231) becomes

$$F(x, y) = 1 - \exp(-x) - \exp(-y) + \exp[-(1 - b)x - (1 - b)y - b \max(x, y)] \tag{5.234}$$

The PDF consists of two parts: continuous and discrete. The discrete part is associated with the singularity point and is not treated here. The continuous part is obtained after successive differentiation with respect to x and y. Two cases are distinguished: $x = \max(x, y)$, and $y = \max(x, y)$.

Specification of constraints: Case $X > Y > 0$

Equation (5.234) simplifies to:

$$F(x, y) = 1 - \exp(-x) - \exp(-y) + \exp[-x - (1 - b)y] \tag{5.235}$$

The PDF is:

$$f_1(x, y) = (1 - b)\exp[-x - (1 - b)y] \tag{5.236}$$

Note that

$$\int_0^\infty \int_0^\infty f_1(x, y)\,dxdy = 1 \tag{5.237}$$

The first moments are determined as

$$E[X] = \int_0^\infty \int_0^\infty xf_1(x, y)\,dxdy = (1 - b)\int_0^\infty x\exp(-x)\,dx \int_0^\infty \exp[-(1 - b)y]\,dy = 1 \tag{5.238}$$

$$E[Y] = \int_0^\infty \int_0^\infty yf_1(x, y)\,dxdy = (1 - b)\int_0^\infty \exp(-x)\,dx \int_0^\infty y\exp[-(1 - b)y]\,dy = \frac{1}{1 - b} \tag{5.239}$$

The joint entropy can be written as

$$H(X, Y) = -\log(1 - b)\int_0^\infty \int_0^\infty f_1(x, y)\,dxdy + \int_0^\infty \int_0^\infty xf_1(x, y)\,dxdy + (1 - b)\int_0^\infty \int_0^\infty yf_1(x, y)\,dxdy \tag{5.240}$$

Thus, the constraint equations are: equation (5.237) to (5.239). The theoretical means in equation (5.238) and (5.239) can be written as

$$\bar{x} = \frac{1}{n + 1}\sum_{i=0}^n x_i \tag{5.241}$$

$$\bar{y} = \frac{1}{n + 1}\sum_{i=0}^n y_i \tag{5.242}$$

Derivation of PDF: Case X > Y > 0
The POME-based PDF consistent with equations (5.237) to (5.239) can be expressed as:

$$f(x, y) = \frac{1}{Z(\lambda_1, \lambda_2)}\exp[-\lambda_1 x - \lambda_2 y] \tag{5.243}$$

Partition function
Using equation (5.237), the partition function is

$$Z(\lambda_1, \lambda_2) = \int_0^\infty \exp(-\lambda_1 x)\,dx \int_0^\infty \exp(-\lambda_2 y)\,dy = \frac{1}{\lambda_1 \lambda_2} \tag{5.244}$$

Relation between Lagrange multipliers and constraints
Differentiating the logarithm of equation (5.244) with respect to Lagrange multipliers and taking advantage of equations (5.243) and (5.244), one obtains

$$\frac{\partial[\log Z(\lambda_1, \lambda_2)]}{\partial \lambda_1} = -\frac{1}{\lambda_1} = -\mu_X = -1 \tag{5.245}$$

$$\frac{\partial[\log Z(\lambda_1, \lambda_2)]}{\partial \lambda_2} = -\frac{1}{\lambda_2} = -\mu_Y = -\frac{1}{1-b} \tag{5.246}$$

Equations (5.245) and (5.246) lead to

$$\lambda_1 = 1 \tag{5.247}$$

and

$$\lambda_2 = 1 - b \tag{5.248}$$

Substituting these values in equation (5.244), the partition function becomes

$$Z(\lambda_1, \lambda_2) = \int_0^\infty \exp(-\lambda_1 x)dx \int_0^\infty \exp(-\lambda_2 y)dy = \frac{1}{\lambda_1 \lambda_2} = \frac{1}{1-b} \tag{5.249}$$

Likewise, substituting equations (5.247) to (5.249) in equation (5.243), the PDF becomes equation (5.236) when $X > Y > 0$.

Specification of constraints: Case $\mathbf{Y > X > 0}$
Equation (5.234) simplifies to

$$F_2(x, y) = 1 - \exp(-x) - \exp(-y) + \exp[-(1-b)x - y] \tag{5.250}$$

The PDF is

$$f_2(x, y) = (1-b)\exp[-(1-b)x - y] \tag{5.251}$$

Note that

$$\int_0^\infty \int_0^\infty f_2(x, y)dxdy = 1 \tag{5.252}$$

The first moments are determined as

$$E[X] = \int_0^\infty \int_0^\infty xf_2(x, y)dxdy = (1-b)\int_0^\infty x\exp[(1-b)x]dx \int_0^\infty \exp(-y)dy = \frac{1}{1-b} \tag{5.253}$$

$$E[Y] = \int_0^\infty \int_0^\infty yf_2(x, y)dxdy = (1-b)\int_0^\infty \exp[(1-b)x]dx \int_0^\infty y\exp(-y)dy = 1 \tag{5.254}$$

The joint entropy can be written as

$$H(X, Y) = -\int_0^\infty \int_0^\infty [\log(1-b) - (1-b)x - y]f_2(x, y)dxdy$$
$$= -\log(1-b) + (1-b)\mu_X + \mu_Y \tag{5.255}$$

The constraints are equations (5.252)–(5.254).

Derivation of PDF: Case **X > Y > 0**

The POME-based PDF consistent with equations (5.252) to (5.254) can be expressed as:

$$f_2(x, y) = (1 - b) \exp[-(1 - b)x - y] \tag{5.256}$$

Combining equations (5.236) and (5.256) the complete PDF is:

$$f(x, y) = \begin{cases} (1 - b) \exp[-x - (1 - b)y], & x > y > 0 \\ (1 - b) \exp[-(1 - b)x - y], & y > x > 0 \end{cases} \tag{5.257}$$

The entropy for both cases is

$$H(X, Y) = 2 - \log(1 - b) - b \tag{5.258}$$

Derivation with equal means

Here $E[X] = E[Y] = a$. From equations (5.227) and (5.228), one obtains

$$a_1 + a_{12} = \frac{1}{a} \tag{5.259}$$

$$a_2 + a_{12} = \frac{1}{a} \tag{5.260}$$

Then, equation (5.229) becomes

$$\rho_{12} = \frac{a_{12}}{\dfrac{2}{a} - a_{12}} \tag{5.261}$$

With these changes, the bivariate exponential distribution transforms into ($b = a_{12}$):

$$F(x, y) = 1 - \exp\left(-\frac{x}{a}\right) - \exp\left(-\frac{y}{a}\right) + \exp\left[-\left(\frac{1}{a} - b\right)x - \left(\frac{1}{a} - b\right)y - b \max(x, y)\right] \tag{5.262}$$

Now two cases, as before, are discussed: Case 1: $X > Y > 0$ and Case 2: $Y > X > 0$.

Specification of constraints: Case **X > Y > 0**

Equation (5.262) simplifies to

$$F_1(x, y) = 1 - \exp\left(-\frac{x}{a}\right) - \exp\left(-\frac{y}{a}\right) + \exp\left[-\frac{x}{a} - \left(\frac{1}{a} - b\right)y\right] \tag{5.263}$$

The PDF is

$$f_1(x, y) = \frac{1}{a}\left(\frac{1}{a} - b\right) \exp\left(-\frac{x}{a} - \left(\frac{1}{a} - b\right)y\right) \tag{5.264}$$

The first moments of the distribution are:

$$E[X] = \int_0^\infty \int_0^\infty x f_1(x, y)\,dx\,dy = \int_0^\infty x \frac{1}{a}\left(\frac{1}{a} - b\right) \exp\left(-\frac{x}{a}\right) dx \int_0^\infty \exp\left[-\left(\frac{1}{a} - b\right)y\right] dy = a \tag{5.265}$$

$$E[Y] = \int_0^\infty \int_0^\infty y f_1(x,y)dxdy = \frac{1}{a}\left(\frac{1}{a}-b\right)\int_0^\infty \exp\left(-\frac{x}{a}\right)dx\int_0^\infty y\exp\left(-\left(\frac{1}{a}-b\right)\right)dy = \frac{1}{\frac{1}{a}-b}$$

(5.266)

Also,

$$\int_0^\infty \int_0^\infty f_1(x,y)dxdy = \frac{1}{a}\left(\frac{1}{a}-b\right)\int_0^\infty \exp\left(-\frac{x}{a}\right)dx\int_0^\infty \exp\left[-\left(\frac{1}{a}-b\right)\right]dy = 1$$

(5.267)

The joint entropy can be written as

$$H(X,Y) = -\int_0^\infty \int_0^\infty \left[\log\left[\frac{1}{a}\left(\frac{1}{a}-b\right)\right] - \frac{x}{a} - \left(\frac{1}{a}-b\right)y\right] f_1(x,y)\,dxdy$$

(5.268)

$$= -\log\left[\frac{1}{a}\left(\frac{1}{a}-b\right)\right] + \frac{1}{a}\mu_X + \left(\frac{1}{a}-b\right)\mu_Y$$

The constraints are equations (5.265) to (5.267).

Derivation of PDF: Case **X > Y > 0**

Using equation (5.267), the entropy-based PDF is:

$$f_1(x,y) = \frac{1}{Z(\lambda_1,\lambda_2)}\exp[-\lambda_1 x - \lambda_2 y]$$

(5.269)

This leads to

$$Z(\lambda_1,\lambda_2) = \int_0^\infty \exp(-\lambda_1 x)dx \int_0^\infty \exp(-\lambda_2 y)dy = \frac{1}{\lambda_1\lambda_2}$$

(5.270)

where $\lambda_1 = \frac{1}{a}, \lambda_2 = \frac{1}{a} - b$.
The partition function becomes

$$Z(\lambda_1,\lambda_2) = \frac{a}{\frac{1}{a}-b}$$

(5.271)

and the PDF becomes

$$f_1(x,y) = \frac{1}{a}\left(\frac{1}{a}-b\right)\exp\left[-\left(\frac{1}{a}-b\right)x - \frac{1}{a}y\right]$$

(5.272)

Derivation for the case **Y > X > 0**

The procedure is equivalent to the case $x > y > 0$, except that variables X and Y are interchanged. Thus,

$$f_2(x,y) = \frac{1}{a}\left(\frac{1}{a}-b\right)\exp\left[-\frac{1}{a}x - \left(\frac{1}{a}-b\right)y\right]$$

(5.273)

Combining equations (5.272) and (5.273), one gets

$$
f(x, y) = \begin{cases} \dfrac{1}{a}\left(\dfrac{1}{a} - b\right) \exp\left[-\left(\dfrac{1}{a} - b\right)x - \dfrac{1}{a}y\right], & x > y > 0 \\[3mm] \dfrac{1}{a}\left(\dfrac{1}{a} - b\right) \exp\left[-\dfrac{1}{a}x - \left(\dfrac{1}{a} - b\right)y\right], & y > x > 0 \end{cases}
$$

(5.274)

The entropy for both cases is:

$$
H(X, Y) = 2 - \log\left[\frac{1}{a}\left(\frac{1}{a} - b\right)\right] - b
$$

(5.275)

Derivation with unequal means
Here $\mu_X = E[X]$ and $\mu_Y = E[Y]$.

Case: **X > Y > 0**
Equation (5.231) simplifies to:

$$
F_1(x, y) = 1 - \exp[-(a_1 + a_{12})x)] - \exp[-(a_2 + a_{12})y)] + \exp[-(a_1 + a_{12})x - a_2 y] \quad (5.276)
$$

Then,

$$
f_1(x, y) = \frac{\partial^2 F_1(x, y)}{\partial x \partial y} = a_2(a_1 + a_{12}) \exp[-(a_1 + a_{12})x - a_2 y]
$$

(5.277)

The first moments of the distribution are:

$$
\begin{aligned}
E[X] &= \int_0^\infty \int_0^\infty x f_1(x, y) dx dy = a_2(a_1 + a_{12}) \int_0^\infty x \exp[-(a_1 + a_{12})x] dx \int_0^\infty \exp(-a_2 y) dy \\
&= \frac{1}{a_1 + a_{12}}
\end{aligned}
$$

(5.278)

$$
E[Y] = \int_0^\infty \int_0^\infty y f_1(x, y) dx dy = a_2(a_1 + a_{12}) \int_0^\infty \exp[-(a_1 + a_{12})x] dx \int_0^\infty y \exp(-a_2 y) dy = \frac{1}{a_2}
$$

(5.279)

Also,

$$
\int_0^\infty \int_0^\infty f_1(x, y) dx dy = a_2(a_1 + a_{12}) \int_0^\infty \exp[-(a_1 + a_{12})] dx \int_0^\infty \exp(-a_2 y)] dy = 1
$$

(5.280)

The joint entropy can be written as

$$
\begin{aligned}
H(X, Y) &= -\int_0^\infty \int_0^\infty \left[\log[a_2(a_1 + a_{12})] + [-(a_1 + a_{12})x - a_2 y] f_1(x, y) dx dy\right] \\
&= -\log[a_2(a_1 + a_{12})] + (a_1 + a_{12})\mu_x + a_2 \mu_Y
\end{aligned}
$$

(5.281)

The constraint equations are equations (5.278) to (5.280). The PDF is

$$f_1(x, y) = \frac{1}{Z(\lambda_1, \lambda_2)} \exp[-\lambda_1 x - \lambda_2 y]$$

The partition function is obtained as

$$Z(\lambda_1, \lambda_2) = \int_0^\infty \exp(-\lambda_1 x)dx \int_0^\infty \exp(-\lambda_2 y)dy = \frac{1}{\lambda_1 \lambda_2} \tag{5.282}$$

Differentiating the logarithm of equation (5.282) with respect to Lagrange multipliers and taking advantage of equations (5.278) through (5.280), one obtains

$$\frac{\partial[\log Z(\lambda_1, \lambda_2)]}{\partial \lambda_1} = -\frac{1}{\lambda_1} = -\mu_X = -\frac{1}{a_1 + a_{12}} \tag{5.283}$$

$$\frac{\partial[\log Z(\lambda_1, \lambda_2)]}{\partial \lambda_2} = -\frac{1}{\lambda_2} = -\mu_Y = -\frac{1}{a_2} \tag{5.284}$$

Equations (5.283) and (5.284) lead to

$$\lambda_1 = a_1 + a_{12} \tag{5.285}$$

and

$$\lambda_2 = a_2 \tag{5.286}$$

Substituting these values in equation (5.282), the partition function becomes

$$Z(\lambda_1, \lambda_2) = \int_0^\infty \exp(-\lambda_1 x)dx \int_0^\infty \exp(-\lambda_2 y)dy = \frac{1}{a_2(a_1 + a_{12})} \tag{5.287}$$

Likewise, substituting equations (5.285) to (5.287) in equation (5.243), the PDF becomes when $X > Y > 0$:

$$f_1(x, y) = (a_1 + a_{12})a_2 \exp[-(a_1 + a_{12})x - a_2 y] \tag{5.288}$$

Derivation with Y > X > 0
Equation (5.231) simplifies to:

$$F_2(x, y) = 1 - \exp[-(a_1 + a_{12})x)] - \exp[-(a_2 + a_{12})y)] + \exp[-a_1 x - (a_2 + a_{12})y] \tag{5.289}$$

Then, the continuous part of the PDF is

$$f_2(x, y) = \frac{\partial F^2(x, y)}{\partial x \partial y} = a_1(a_2 + a_{12}) \exp[-a_1 x - (a_2 + a_{12})y] \tag{5.290}$$

The first moments of the distribution are:

$$E[X] = \int_0^\infty \int_0^\infty x f_2(x, y)dxdy = a_1(a_2 + a_{12}) \int_0^\infty x \exp(-a_1 x)dx \int_0^\infty \exp[-(a_2 + a_{12})y]dy = \frac{1}{a_1}$$

$$\tag{5.291}$$

$$E[Y] = \int_0^\infty \int_0^\infty y f_2(x,y) \, dx \, dy = a_1(a_2 + a_{12}) \int_0^\infty \exp(-a_1 x) \, dx \int_0^\infty y \exp[-(a_2 + a_{12})y] \, dy$$

$$= \frac{1}{a_1 + a_{12}} \tag{5.292}$$

Also,

$$\int_0^\infty \int_0^\infty f_2(x,y) \, dx \, dy = 1 \tag{5.293}$$

The constraint equations are equations (5.291) to (5.293). Following the same procedure as before,

$$\lambda_1 = a_1, \lambda_2 = a_2 + a_{12} \tag{5.294}$$

The partition function becomes

$$Z(\lambda_1, \lambda_2) = \frac{1}{a_1(a_2 + a_{12})} \tag{5.295}$$

The PDF is

$$f_2(x,y) = a_1(a_2 + a_{12}) \exp[-a_1 x - (a_2 + a_{12})y] \tag{5.296}$$

Combining equations (5.288) and (5.296),

$$f(x,y) = \begin{cases} (a_1 + a_{12})a_2 \exp[-(a_1 + a_{12})x - a_2 y], & x > y > 0 \\ (a_2 + a_{12})a_1 \exp[-a_1 x - (a_2 + a_{12})y], & y > x > 0 \end{cases} \tag{5.297}$$

The joint entropy for the complete domain is:

$$H(x,y) = \begin{cases} -\log[a_2(a_1 + a_{12})] + (a_1 + a_{12})u_X - a_2 u_Y, & x > y > 0 \\ -\log[a_1(a_2 + a_{12})] - a_1 u_X - (a_2 + a_{12})u_Y, & y > x > 0 \end{cases} \tag{5.298}$$

The joint entropy can also be expressed as a function of Lagrange multipliers:

$$H(X,Y) = -\log(\lambda_1 \lambda_2) + \lambda_1 u_X + \lambda_2 u_Y \tag{5.299}$$

The number of domains where PDFs are defined is 2! for the bivariate exponential PDFs. Thus, POME must be based on each domain separately.

To determine parameters a_1, a_2, and a_{12} of the bivariate exponential distributions, the sample approximations of the means of X and Y and the sample correlation coefficient between X and Y are used:

$$\bar{x} = \frac{1}{a_1 + a_{12}} \tag{5.300}$$

$$\bar{y} = \frac{1}{a_2 + a_{12}} \tag{5.301}$$

$$r_{12} = \frac{a_{12}}{a_1 + a_2 + a_{12}} \tag{5.302}$$

5.2.2 Trivariate exponential distribution

Let X_1, X_2, and X_3 be three random variables of a hydrologic process $X(t): X_1 = \{x_1 \in (0, T)\}$, $X_2 = \{x_2 \in (0, T)\}$, and $X_3 = \{x_3 \in (0, T)\}$. The original Marshall-Olkin form of the trivariate exponential distribution is:

$$F(x_1, x_2, x_3) = P(X_1 < x_1, X_2 < x_2, X_3 < x_3)$$
$$= 1 - \exp(-a_1 x_1 - a_2 x_2 - a_3 x_3 - a_{12} m_{12} - a_{13} m_{13} - a_{23} m_{23} - a_{123} m_{123}) \quad (5.303)$$

where $m_{ij} = \max(x_i, x_j)$. This includes six regions of definitions, depending on the relations between x_1, x_2, and x_3.

$$F(x_1, x_2, x_3) = \begin{cases} 1 - \exp\left[-\sum_{i=1}^{3} a_i x_i - a_{12} x_1 - a_{13} x_1 - a_{23} x_2 - a_{123} x_1\right], & x_1 > x_2 > x_3 \\[2ex] 1 - \exp\left[-\sum_{i=1}^{3} a_i x_i - a_{12} x_1 - a_{13} x_1 - a_{23} x_3 - a_{123} x_1\right], & x_1 > x_3 > x_2 \\[2ex] 1 - \exp\left[-\sum_{i=1}^{3} a_i x_i - a_{12} x_2 - a_{13} x_1 - a_{23} x_2 - a_{123} x_2\right], & x_2 > x_1 > x_3 \\[2ex] 1 - \exp\left[-\sum_{i=1}^{3} a_i x_i - a_{12} x_2 - a_{13} x_3 - a_{23} x_2 - a_{123} x_2\right], & x_2 > x_3 > x_1 \\[2ex] 1 - \exp\left[-\sum_{i=1}^{3} a_i x_i - a_{12} x_1 - a_{13} x_3 - a_{23} x_3 - a_{123} x_3\right], & x_3 > x_1 > x_2 \\[2ex] 1 - \exp\left[-\sum_{i=1}^{3} a_i x_i - a_{12} x_2 - a_{13} x_3 - a_{23} x_3 - a_{123} x_3\right], & x_3 > x_2 > x_1 \end{cases}$$

$$(5.304)$$

Collecting the same variables, one can write the distribution as

$$F(x_1, x_2, x_3) = \begin{cases} 1 - \exp\left[-\sum_{i=1}^{3} a_i x_i - (a_{12} + a_{13} + a_{123})x_1 - a_{23} x_2\right], & x_1 > x_2 > x_3 \\[2ex] 1 - \exp\left[-\sum_{i=1}^{3} a_i x_i - (a_{12} + a_{13} + a_{123})x_1 - a_{23} x_3\right], & x_1 > x_3 > x_2 \\[2ex] 1 - \exp\left[-\sum_{i=1}^{3} a_i x_i - a_{13} x_1 - (a_{12} + a_{23} + a_{123})x_2\right], & x_2 > x_1 > x_3 \\[2ex] 1 - \exp\left[-\sum_{i=1}^{3} a_i x_i - (a_{12} + a_{23} + a_{123})x_2 - a_{13} x_3\right], & x_2 > x_3 > x_1 \\[2ex] 1 - \exp\left[-\sum_{i=1}^{3} a_i x_i - a_{12} x_1 - a_{13} x_3 - a_{23} x_3 - a_{123} x_3\right], & x_3 > x_1 > x_2 \\[2ex] 1 - \exp\left[-\sum_{i=1}^{3} a_i x_i - a_{12} x_2 - a_{13} x_3 - a_{23} x_3 - a_{123} x_3\right], & x_3 > x_2 > x_1 \end{cases}$$

$$(5.305)$$

The PDF is obtained by successive differentiation as

$$f(x_1, x_2, x_3) = \frac{\partial^3 F(x_1, x_2, x_3)}{\partial x_1 \partial x_2 \partial x_3} \tag{5.306}$$

which results in

$f(x_1, x_2, x_3)$

$$= \begin{cases} (a_1 + a_{12} + a_{13} + a_{123})(a_2 + a_{23})a_3 \exp\left[-\sum_{i=1}^{3} a_i x_i - (a_{12} + a_{13} + a_{123})x_1 - a_{23}x_2\right], \\ \quad x_1 > x_2 > x_3 \\[4pt] (a_1 + a_{12} + a_{13} + a_{123})a_2(a_3 + a_{23}) \exp\left[-\sum_{i=1}^{3} a_i x_i - (a_{12} + a_{13} + a_{123})x_1 - a_{23}x_3\right], \\ \quad x_1 > x_3 > x_2 \\[4pt] (a_1 + a_{13})a_3(a_2 + a_{12} + a_{23} + a_{123}) \exp\left[-\sum_{i=1}^{3} a_i x_i - a_{13}x_1 - (a_{12} + a_{23} + a_{123})x_2\right], \\ \quad x_2 > x_1 > x_3 \\[4pt] a_1(a_2 + a_{12} + a_{23} + a_{123})(a_{13} + a_3) \exp\left[-\sum_{i=1}^{3} a_i x_i - (a_{12} + a_{23} + a_{123})x_2 - a_{13}x_3\right], \\ \quad x_2 > x_3 > x_1 \\[4pt] (a_1 + a_{12})a_2(a_3 + a_{13} + a_{23} + a_{123}) \exp\left[-\sum_{i=1}^{3} a_i x_i - a_{12}x_1 - a_{13}x_3 - a_{23}x_3 - a_{123}x_3\right], \\ \quad x_3 > x_1 > x_2 \\[4pt] a_1(a_2 + a_{12})(a_3 + a_{13} + a_{23} + a_{123}) \exp\left[-\sum_{i=1}^{3} a_i x_i - a_{12}x_2 - a_{13}x_3 - a_{23}x_3 - a_{123}x_3\right], \\ \quad x_3 > x_2 > x_1 \end{cases} \tag{5.307}$$

Using POME, PDFs for each domain are derived separately.

Domain 1: $X_1 > X_2 > X_3$

$$f(x_1, x_2, x_3) = (a_1 + a_{12} + a_{13} + a_{123})(a_2 + a_{23})a_3$$
$$\times \exp\left[-\sum_{i=1}^{3} a_i x_i - (a_{12} + a_{13} + a_{123})x_1 - a_{23}x_2\right], x_1 > x_2 > x_3 \tag{5.308}$$

Note that

$$\int_0^\infty \int_0^\infty \int_0^\infty f(x_1, x_2, x_3)dx_1\,dx_2\,dx_3 = 1 \tag{5.309}$$

Specification of constraints

The first moments of the PDF in equation (5.308) are:

$$E[x_1] = (a_1 + a_{12} + a_{13} + a_{123}) \int_0^\infty x_1 \exp[-(a_1 + a_{12} + a_{13} + a_{123})x_1] dx_1$$

$$= \frac{1}{a_1 + a_{12} + a_{13} + a_{123}} \tag{5.310}$$

$$E[x_2] = (a_1 + a_{12} + a_{13} + a_{123})(a_2 + a_{23}) \int_0^\infty \exp[-(a_1 + a_{12} + a_{13} + a_{123})x_1] dx_1$$

$$\int_0^\infty x_2 \exp[-(a_2 + a_{23})x_2] dx_2 = \frac{1}{a_2 + a_{23}} \tag{5.311}$$

$$E[X_3] = \frac{1}{a_3} \tag{5.312}$$

Entropy of distribution

The entropy of equation (5.308) is

$$H(X_1, X_2, X_3) = -\log(const_1) + const_2 \mu_{X_1} + const_3 \mu_{X_2} + const_4 \mu_{X_3} \tag{5.313}$$

where the constraints in POME are the information constraint given by equation (5.309) and μ_{x_i}, $i = 1, 2, 3$ that are approximated, respectively, by equations (5.310), (5.311), and (5.312).

Derivation of probability density function

Maximizing entropy $H(X_1, X_2, X_3)$, subject to equations (5.309) to (5.312), one obtains

$$f_1(x_1, x_2, x_3) = \frac{1}{Z(\lambda_1, \lambda_2, \lambda_3)} \exp\left[-\sum_{i=1}^3 \lambda_i x_i \right] \tag{5.314}$$

Using equation (5.309), one obtains the partition function as

$$Z(\lambda_1, \lambda_2, \lambda_3) = \int_0^\infty \exp\left[-\lambda_1 x_1\right] dx_1 \int_0^\infty \exp[-\lambda_2 x_2] dx_2 \int_0^\infty \exp[-\lambda_3 x_3] dx_3 = \frac{1}{\lambda_1 \lambda_2 \lambda_3} \tag{5.315}$$

Taking the logarithm of equation (5.315), one gets

$$\log Z(\lambda_1, \lambda_2, \lambda_3) = -\log \lambda_1 - \log \lambda_2 - \log \lambda_3 \tag{5.316}$$

Differentiating equation (5.316) with respect to λ_1 and using equation (5.310), one gets

$$\frac{\partial \log Z(\lambda_1, \lambda_2, \lambda_3)}{\partial \lambda_1} = -\frac{1}{\lambda_1} = -\mu_{X_1} = -\frac{1}{a_1 + a_{12} + a_{13} + a_{123}}$$

or

$$\lambda_1 = a_1 + a_{12} + a_{13} + a_{123} \tag{5.317}$$

Similarly,

$$\lambda_2 = a_2 + a_{23} \tag{5.318a}$$
$$\lambda_3 = a_3 \tag{5.318b}$$

Therefore, the partition function is

$$Z(\lambda_1, \lambda_2, \lambda_3) = \frac{1}{(a_1 + a_{12} + a_{13} + a_{123})(a_2 + a_{23})a_3} \tag{5.319}$$

The PDF therefore is

$$f_1(x_1, x_2, x_3) = (a_1 + a_{12} + a_{13} + a_{123})(a_2 + a_{23})a_3 \exp[-(a_1 + a_{12} + a_{13} + a_{123})x_1 \\ -(a_2 + a_{23})x_2 - a_3 x_3 \tag{5.320}$$

which is equivalent to equation (5.308).

Determination of parameters

Parameters of the trivariate exponential distribution are: $a_1, a_2,$ and a_3 associated with random variables X_1, X_2 and X_3; $a_{12}, a_{13},$ and a_{23} are associated with maxima $m_{12}, m_{13},$ and m_{23}; and a_{123} is associated with m_{123}. These are seven parameters so seven equations are needed. The first three equations are obtained by the first three moment expressions given by equations (5.310) to (5.312). For the second three equations it is convenient to use statistics that best express dependencies among variables – partial correlation coefficients. Specifically,

$$\rho_{12} = \frac{a_{12}}{a_1 + a_2 + a_{12}} \tag{5.321}$$
$$\rho_{13} = \frac{a_{13}}{a_1 + a_3 + a_{13}} \tag{5.322}$$
$$\rho_{23} = \frac{a_{23}}{a_2 + a_3 + a_{23}} \tag{5.323}$$

In application sample approximations of correlation coefficients are obtained as

$$r_{ij} = \frac{\sum (x_i - \bar{x}_i)(x_j - \bar{x}_j)}{\sqrt{\sum (x_i - \bar{x}_i)}\sqrt{\sum (x_j - \bar{x}_j)}} \tag{5.324}$$

where $i, j = 1, 2$ and $3, i \neq j$, and $\bar{x}_i$ and $\bar{x}_j$ are sample means of X_i and X_j. To get the seventh parameter, one must compute a higher moment.

5.2.3 Extension to Weibull distribution

The Marshall-Olkin bivariate exponential distribution becomes the bivariate Weibull distribution when the following substitutions are made: Substitute X with $X^{1/b}$ and substitute Y with $Y^{1/c}$. The general form of the bivariate Weibull distribution is obtained as

$$F(x, y) = 1 - \exp[-a_1 x^b - a_2 y^c - a_{12} \max(x^b, y^c)] \tag{5.325}$$

or in alternative form:

$$F(x, y) = 1 - \exp[-(a_1 + a_{12})x^b] - \exp[-(a_2 + a_{12})y^c + \exp[-a_1 x^b - a_2 y^c - a_{12} \max(x^b, y^c)] \tag{5.326}$$

5.3 Multivariate distributions using the entropy-copula method

The preceding discussion shows that analytical derivation of multivariate probability distributions using the entropy theory becomes complicated for more than two variables, even when marginal distributions are the same. This complication can be overcome by employing the copula concept which was formulated by Sklar (1959) and which has been receiving a lot of attention in recent years. Consider random variables $X_1, X_2, \ldots, X_n$ with CDFs, respectively, as $F_1(x_1), F_2(x_2), \ldots, F_n(x_n)$. It should be emphasized that these CDFs can be obtained using entropy. Sklar's theorem enables a connection between the joint distribution function of $X_1, X_2, \ldots, X_n, F(x_1, x_2, \ldots, x_n)$, and marginal CDFs of these variables to be developed, that is, copulas are functions that provide that connection. Accordingly, there exists a copula C such that, for all $x \in R$, where $R \in (-\infty, \infty)$, the relationship between the joint distribution function $F(x_1, x_2, \ldots, x_n)$ and copula $C(x_1, x_2, \ldots, x_n)$ can be expressed as

$$
\begin{aligned}
F(x_1, x_2, \ldots, x_n) &= P(X_1 \leq x_1, X_2 \leq x_2, \ldots, X_n \leq x_n) \\
&= C[F_1(x_1), F_2(x_2), \ldots, F_n(x_d)] = C(u_1, u_2, \ldots, u_n)
\end{aligned}
\tag{5.327a}
$$

where $F_i(x_i) = u_i, i = 1, 2, \ldots, n$, with $U_i \sim U(0, 1)$ if F_i is continuous. Another way to think about the copula is as follows:

$$
C(u_1, u_2, \ldots, u_n) = F[F_1^{-1}(u_1), F_2^{-1}(u_2), \ldots, F_n^{-1}(u_n)]; (u_1, u_2, \ldots, u_n) \in [0, 1]^n
\tag{5.327b}
$$

Thus, an n dimensional copula is a mapping $[0, 1]^n \rightarrow [0, 1]$, and is a multivariate cumulative distribution function defined in the unit cube $[0, 1]^n$ with standard uniform univariate margins. It is implied that there is some dependence amongst random variables, and the copula captures the essential features of this dependence.

If $c(u_1, u_2, \ldots, u_n)$ denotes the probability density function (PDF) of copula $C(u_1, u_2, \ldots, u_n)$, then the mathematical relation between $c(u_1, u_2, \ldots, u_n)$ and $C(u_1, u_2, \ldots, u_n)$ can be expressed as

$$
\begin{aligned}
f(x_1, \ldots, x_n) &= \frac{\partial F(x_1, \ldots, x_n)}{\partial x_1 \ldots \partial x_n} = \frac{\partial C(x_1, \ldots, x_n)}{\partial x_1 \ldots \partial x_n} \\
&= \frac{\partial C(u_1, \ldots, u_n)}{\partial u_1 \ldots \partial u_n} \frac{\partial u_1}{\partial x_1} \ldots \frac{\partial u_n}{\partial x_n} = \frac{\partial C(u_1, \ldots, u_n)}{\partial u_1 \ldots \partial u_n} \frac{\partial F_1(x_1)}{\partial x_1} \ldots \frac{\partial F_n(x_n)}{\partial x_n} \\
&= c(u_1, \ldots, u_n) \prod_{i=1}^{n} \frac{\partial F_i(x_i)}{\partial x_i} = c(u_1, \ldots, u_n) \prod_{i=1}^{n} f_i(x_i)
\end{aligned}
\tag{5.328a}
$$

where $c(u_1, \ldots, u_n) = \frac{\partial C(u_1, \ldots, u_n)}{\partial u_1 \ldots \partial u_n}; f_i(x_i) = \frac{\partial F_i(x_i)}{\partial x_i}, i = 1, \ldots, n$. The PDF $c(u_1, \ldots, u_n)$ can be also expressed as

$$
c(u_1, \ldots, u_n) = \frac{f(x_1, \ldots, x_n)}{\prod_{i=1}^{n} f_i(x_i)}
\tag{5.328b}
$$

For illustration, consider the case of a bivariate copula $C(u, v)$ which is a mapping from $[0, 1] \times [0, 1]$ to $[0, 1]$. Then, the CDF and PDF can be written as

$$C(u, v) = F[F_1^{-1}(u), F_2^{-1}(v)]; u = F_1(x_1), v = F_2(x_2) \tag{5.329a}$$

$$c(u, v) = \frac{\partial C(u, v)}{\partial u \partial v} = \frac{f(x_1, x_2)}{f(x_1)f(x_2)} \tag{5.329b}$$

Likewise, for a trivariate copula $C(u, v, w)$ which is a mapping from $[0, 1] \times [0, 1] \times [0, 1]$ to $[0, 1]$, the CDF and PDF can be written as

$$C(u, v, w) = F[F_1^{-1}(u), F_2^{-1}(v), F_3^{-1}(w)]; u = F_1(x_1), v = F_2(x_2), w = F_3(x_3) \tag{5.330a}$$

$$c(u, v) = \frac{\partial C(u, v, w)}{\partial u \partial v \partial w} = \frac{f(x_1, x_2, x_3)}{f(x_1)f(x_2)f_3(x_3)} \tag{5.330b}$$

Now the conditional joint distribution based on the copula can be expressed. As an example, the conditional distribution function of U given $V = v$ can be expressed as:

$$C_{U|V=v}(u) = F[U \le u|V = v] = \lim_{\Delta v = 0} \frac{C(u, v + \Delta v) - C(u, v)}{\Delta v} = \frac{\partial}{\partial v} C(u, v)|V = v \tag{5.331}$$

Similarly, an equivalent formula for the conditional distribution function for variable V given $U = u$ can be obtained. Furthermore, the conditional distribution function of U given $V \le v$ can be expressed as:

$$C(u|V \le v) = F(U \le u|V \le v) = \frac{C(u, v)}{v} \tag{5.332}$$

Likewise, an equivalent formula for the conditional distribution function for V given $U \le u$ can be obtained.

5.3.1 Families of copula

The copula concept has led to defining a multitude of copulas and has led to a method that is capable of exhibiting the dependence between two or more random variables and has recently emerged as a practical and efficient method for modeling general dependence in multivariate data (Joe, 1997; Nelsen, 2006). The advantages in using copulas to model joint distributions are threefold. They provide 1) flexibility in choosing arbitrary marginals and the structure of dependence; 2) the capability for extension to more than two variables; and 3) a separate analysis of marginal distributions and dependence structure. As a result, hydrological applications of copulas have surged in recent years. For example, they have been used for rainfall frequency analysis, drought frequency analysis, rainfall and flood analysis, spillway and dam design, sea storm analysis, and some other theoretical analyses of multivariate extremes. Detailed theoretical background and description for the use of copulas can be found in Nelsen (2006) and Salvadori et al.(2007).

Copulas may be grouped into the Archimedean copulas, metaelliptical copulas, and quadratic form, copulas with cubic form. According to their exchangeable properties, copulas may also be classified as symmetric copulas and asymmetric copulas. For example, one parameter Archimedean copulas are symmetric copulas; and periodic copulas (Alfonsi and Brigo, 2005) and mixed copulas (Hu, 2006) belong to asymmetric copulas. The copulas which are more common in water resources engineering applications include: Archimedean, Plackett,

Table 5.1 Bivariate copulas and their generating functions: $\tilde{u} = -\ln u$ and $\tilde{v} = -\ln v$.

Copula	Generator $\phi(t)$	$C_\theta(u, v)$	Parameter Space
AMH	$\ln \dfrac{1 - \theta(1 - t)}{t}$	$\dfrac{uv}{1 - \theta(1 - u)(1 - v)}$	$[-1, 1)$
Clayton	$\dfrac{1}{\theta}(t^{-\theta} - 1)$	$[\max(u^{-\theta} + v^{-\theta} - 1, 0)]^{-1/\theta}$	$(0, \infty)$
Frank	$-\ln \dfrac{e^{-\theta t} - 1}{e^{-\theta} - 1}$	$-\dfrac{1}{\theta}\ln\left[1 - \dfrac{(1 - e^{-\theta u})(1 - e^{-\theta v})}{(1 - e^{-\theta})}\right]$	$(-\infty, \infty)\backslash\{0\}$
GH	$(-\ln t)^\theta$	$\exp[-(\tilde{u}^\theta + \tilde{v}^\theta)^{1/\theta}]$	$[1, \infty)$

metaelliptical, mixed, and empirical. The Archimedean copulas are widely applied due to their simple form, dependence structure, and other desirable properties. The Plackett copula has also been used in recent years. Metaelliptical copulas are a flexible tool for modeling multivariate data in hydrology. When data are analyzed with an unknown underlying distribution, the empirical data distribution can be transformed into what is called an "empirical copula" such that the marginal distributions become uniform. Parametric copulas place restrictions on the dependence parameter. When the process generating the data is heterogeneous, it is desirable to have additional flexibility in modeling the dependence. A mixture model, proposed by Hu (2006), is able to measure dependence structures that do not belong to the above copula families. By choosing component copulas in the mixture, a model can be constructed which is simple and flexible enough to generate most dependence patterns and provides flexibility in practical data. This also facilitates the separation of the degree of dependence and the structure of dependence. Considering three bivariate copulas $C_I(u_1, u_2)$, $C_{II}(u_1, u_2)$ and $C_{III}(u_1, u_2)$, the mixed copula can be defined as:

$$C_{mix}(u_1, u_2; \theta_1, \theta_2, \theta_3; w_1, w_2, w_3) = w_1 C_I(u_1, u_2; \theta_1) + w_2 C_{II}(u_1, u_2; \theta_2) + w_3 C_{III}(u_1, u_2; \theta_3)$$

(5.333)

where $C_{mix}(u_1, u_2; \theta_1, \theta_2, \theta_3; w_1, w_2, w_3)$ is the mixed copula; $C_I(u_1, u_2; \theta_1)$, $C_{II}(u_1, u_2; \theta_2)$ and $C_{III}(u_1, u_2; \theta_3)$ are three bivariate copulas, each with different dependence properties; and θ_1, θ_2, and θ_3 are the corresponding copula parameters; w_1, w_2, and w_3 may be interpreted as weights of the component copulas, $0 < w_j < 1, j = 1, 2, 3, \sum_{j=1}^{3} w_j = 1$.

From a hydrologic perspective, the Archimedean family of copulas has been most popular and the popular members of this family are the Gumbel-Hougaard (GH), Frank, Clayton, Ali-Mikhail-Haq (AMH), and Cook-Johnson (CJ). The Normal and t copulas have also been used for the bivariate case. The equations of these copulas are given in Table 5.1.

5.3.2 Application

First, the marginal distributions of random variables under consideration should be obtained and this can be done using the entropy theory. Then, the application of a copula entails the following components: 1) construction of copula, 2) dependence measures and properties, 3) copula parameter estimation, 4) copula model selection, and 5) goodness of fit tests. Each of these components is now briefly discussed.

Construction of copulas

Copulas may be constructed by different methods, that is, the inversion method, the geometric method, and the algebraic method. In the inversion method, the copula is obtained through the joint distribution function F and the continuous marginals. Nelson (1999) introduced the algebraic method by constructing the Plackett and Ali-Mikhail-Haq copulas through "odd" ratio in which the Plackett copula is constructed by measuring the dependence of 2×2 contingency tables, and the Ali-Mikhail-Haq copula is constructed by using the survival odds ratio. Nelson (2008) has discussed how to use these methods to construct copulas.

An n-dimensional Archimedean symmetric copula can be defined as (Salvadori et al., 2007; Savu and Trede, 2008):

$$C(u_1, \ldots, u_n) = \phi^{-1} \left[\sum_{k=1}^{n} \phi(u_k) \right] = \phi^{-1}[\phi(u_1) + \phi(u_2) + \ldots \phi(u_n)]; u_k \in [0, 1], k = 1, \ldots, n$$

(5.334)

where $\phi(.)$ is called the generating function of the Archimedean copula, which is a convex decreasing function satisfying $\phi(1) = 0$ and $\phi^{-1}(.)$ is equal to 0 when $u_k \geq \phi(0)$ Taking an example of a two-dimensional copula, the one-parameter Archimedean copula, $C_\theta(u, v)$, can be expressed from equation (5.334) as

$$C_\theta(u, v) = \phi^{-1}\{\phi(u), \phi(v)\}, 0 < u, v < 1$$

(5.335)

where subscript θ of copula C is a parameter hidden in the generating function ϕ; $u = F^{-1}(x)$ and $v = F^{-1}(y)$ are uniformly distributed random variables.

It may be noted that

$$C(u_1, u_2, u_3) = \phi^{-1}[[\phi(u_1) + \phi(u_2) + \phi(u_3)] = \phi^{-1}\{\phi(u_1) + \phi[\phi^{-1}(\phi(u_2) + \phi(u_3))]\}$$
$$= C[u_1, C(u_2, u_3)]$$

(5.336a)

and

$$C(u_1, u_2, u_3) = \phi^{-1}[\phi(u_1) + \phi(u_2) + \phi(u_3)] = \phi^{-1}\{\phi[\phi^{-1}(\phi(u_1) + \phi(u_2))] + \phi(u_3)\}$$
$$= C[C(u_1, u_2), u_3]$$

(5.336b)

that is

$$C(u_1, u_2, u_3) = C[u_1, C(u_2, u_3)] = C[C(u_1, u_2), u_3]$$

(5.336c)

Equation (5.336c) implies that given three random variables u_1, u_2, u_3, the dependence between the first two random variables taken together and the third one alone is the same as the dependence between the first random variable taken alone and the two last ones together. This implies a strong symmetry between different variables in that they are exchangeable but the associative property of the Archimedean copula is not satisfied by other copulas in general.

For the Gumbel-Hougaard copula, the generating function is defined as

$$\phi(u) = (-\ln u)^\theta, \phi^{-1}(t) = \exp(-t^{\frac{1}{\theta}}), t = u \text{ or } v$$

(5.337)

where copula parameter can be expressed in terms of Kendall's coefficient of correlation τ between X and Y as $\tau = 1 - \theta^{-1}$. Note that $(-\ln u)^{\theta} = \phi(u)$ and $(-\ln v)^{\theta} = \phi(v)$. Then we have

$$\phi(u_1) + \phi(u_2) = (-\ln u_1)^{\theta} + (-\ln u_2)^{\theta} \tag{5.338a}$$

and hence the copula can be expressed as

$$C(u_1, u_2) = \phi^{-1}[\phi(u_1) + \phi(u_2)] = \exp(-[(-\ln u_1)^{\theta} + (-\ln u_2)^{\theta}]^{\frac{1}{\theta}}) \tag{5.338b}$$

Likewise, for the three-dimensional copula, one can write

$$\phi(u_1) + \phi(u_2) + \phi(u_3) = (-\ln u_1)^{\theta} + (-\ln u_2)^{\theta} + (-\ln u_3)^{\theta} \tag{5.339a}$$

$$C(u_1, u_2, u_3) = \phi^{-1}\left[\phi(u_1) + \phi(u_2) + \phi(u_3)\right] = \exp\left(-[(-\ln u_1)^{\theta} + (-\ln u_2)^{\theta} + (-\ln u_3)^{\theta}]^{\frac{1}{\theta}}\right) \tag{5.339b}$$

For the Ali-Mikhail-Haq copula, the generating function is written as

$$\phi(t) = \ln\frac{1 - \theta(1 - t)}{t}, \tau = \left(\frac{3\theta - 2}{\theta}\right) - \frac{2}{3}\left(1 - \frac{1}{\theta}\right)^2 \ln(1 - \theta) \tag{5.340a}$$

Then the copula becomes

$$C(u, v) = C(F^{-1}(x), F^{-1}(y)) = F(x, y) = \frac{uv}{1 - \theta(1 - u)(1 - v)}, \theta \in [-1, 1) \tag{5.340b}$$

For the Frank copula, the generating function can be expressed as

$$\phi(t) = \ln\left[\frac{\exp(\theta t) - 1}{\exp(\theta) - 1}\right], \tau = 1 - \frac{4}{\theta}[D_1(-\theta) - 1] \tag{5.341}$$

where D_1 is the first order Debye function D_k which is defined as

$$D_k(\theta) = \frac{k}{x^k}\int_0^\theta \frac{t^k}{\exp(t) - 1} dt, \theta > 0 \tag{5.342a}$$

and the Debye function D_k with negative argument can be expressed as:

$$D_k(-\theta) = D_k(\theta) + \frac{k\theta}{k + 1} \tag{5.342b}$$

Thus, the Frank copula can be written as

$$C(u, v) = C(F^{-1}(x), F^{-1}(y)) = F(x, y) = \frac{1}{\theta}\ln\left[1 + \frac{(\exp(\theta u) - 1)(\exp(\theta v) - 1)}{\exp(\theta) - 1}\right], \theta \neq 0 \tag{5.343}$$

For the Cook-Johnson copula, the generating function can be written as

$$\phi(t) = t^{-\theta} - 1, \tau = \frac{\theta}{\theta + 2} \tag{5.344a}$$

and the copula becomes

$$C(u, v) = [u^{-\theta} + v^{-\theta} - 1]^{-1/\theta}, \theta \geq 0 \tag{5.344b}$$

Example 5.9: Consider two random variables, X and Y, with their marginal distributions as the extreme value type (EVI) or Gumbel distributions: $F(x) = \exp[-\exp(-x)]$ and $F(y) = \exp[-\exp(-y)]$. Construct the joint distribution using the Gumbel-Hougaard copula.

Solution: $u = \exp[-\exp(-x)]$ and $v = \exp[-\exp(-y)]$. Therefore, using equation (5.338b), the joint distribution of two random variables, $F(x, y)$, can be expressed as:

$$\begin{aligned} F(x, y) = C(u, v) &= \exp[-[(-\ln(\exp(-\exp(-x))))^{\theta} + (-\ln(\exp(-\exp(-y))))^{\theta}]^{1/\theta}] \\ &= \exp[-(\exp(-\theta x) + \exp(-\theta y))^{1/\theta}] \end{aligned} \tag{5.345}$$

Equation (5.345) is the Gumbel bivariate logistic distribution.

Example 5.10: Consider a random variable X which is EVI (Gumbel) distributed: $F(x) = \exp[-\exp(-x)]$, and another variable Y which is EVI (Gumbel) distributed as well: $F(y) = \exp[-\exp(-y)]$. Construct the joint distribution using the Ali-Mikhail-Haq (AMH) copula.

Solution: Here $u = \exp[-\exp(-x)]$ and $v = \exp[-\exp(-y)]$. The joint distribution of two random variables, $F(x, y)$ can be expressed as:

$$F(x, y) = C(u, v) = \frac{\exp[-\exp(-x) - \exp(-y)]}{1 - \theta[1 - \exp(-\exp(x)][1 - \exp(-\exp(-y)]}, \theta \in [-1, 1) \tag{5.346}$$

Example 5.11: Consider a random variable X which is uniformly distributed (i.e., $X \sim$ uniform $[-1, 1]$ and another variable Y which is exponentially distributed [i.e., $Y \sim \exp(1)$]). Construct the joint distribution using the Frank copula.

Solution: Here, $u = F(x)$ and $v = F(y)$. Then, using equation (5.343) the joint distribution of two random variables, $F(x, y)$, can be expressed as:

$$F(x, y) = C(u, v) = (x + 1)(e^y - 1)/[x + 2e^y - 1], (x, y) \in [-1, 1] \times [0, \infty) \tag{5.347}$$

Example 5.12: Consider a random variable X which is log-Pearson III distributed as $F(x)$ and another variable Y which is also log-Pearson III distributed as $F(y)$. Construct the joint distribution using the Cook-Johnson copula.

Solution: Here, $u = F(x)$ and $v = F(y)$. Then, using equation (5.344a), the joint distribution of the two random variables $F(x, y)$ is given as:

$$H(x, y) = C(u, v) = [F(x)^{-\theta} + F(y)^{-\theta} - 1]^{-1/\theta} \tag{5.348}$$

Dependence measures and properties

The dependence between random variables is important for multivariate analysis. The dependence properties include: positive quadrant and orthant dependence, stochastically increasing positive dependence, right-tail increasing and left-tail decreasing dependence, positive function dependence, and tail dependence. There are several measures of dependency or association amongst variables. Five popular measures of association are Pearson's classical correlation coefficient r_n, Spearman's ρ_n, Kendall's τ, Chi-Plots and K-Plots which were originally developed in the field of nonparametric statistics. Genest and Rivest (1993) described a procedure to identify the copula function. It is assumed that a random sample of bivariate observations $(x_1, y_1), (x_2, y_2), \ldots, (x_n, y_n)$ is available and that its underlying distribution function $F(x, y)$ has an associated Archimedean copula C_θ which also can be regarded as an alternative expression of F. Then the following steps are used to identify the appropriate copula:

1 Determine Kendall's τ (the dependence structure of bivariate random variables) from observations as:

$$\tau_N = \frac{2}{N(N-1)} \sum_{i=1}^{N} \sum_{j=1}^{N-1} sign[(x_{1i} - x_{1j})(x_{2i} - x_{2j})] \tag{5.349}$$

where N is the number of observations; $sign(x) = \begin{cases} 1; & x > 0 \\ 0; & x = 0 \\ -1; & x < 0 \end{cases}$; and τ_N is the estimate of τ

from the N observations.

2 Determine the copula parameter θ from the above value of τ_N for the copulas under consideration.

3 Obtain the generating function of each copula, Φ.

4 Obtain the copula function by inserting Φ.

5 Based on each generating function Φ and the parameter θ obtained from step 2, the copula can be identified.

Copula parameter estimation

If entropy is not used to derive parameters of marginal distributions, then their parameters also need to be estimated. Further, for certain copulas, the relation between Kendall's τ and the copula parameter(s) is not straightforward and other methods of parameter estimation may be needed. Two of the methods of parameter estimation are the exact maximum likelihood method, also called Full Maximum Likelihood (Full ML), which is a one stage method; and the inference function for marginal method (IFM) which is a two stage method. IMF entails estimating parameters of marginal distributions and dividing the copula into two steps. First, a semiparametric approach, which is a more flexible method, is used for estimating parameters of marginal distributions (empirical distribution functions). It consists of using a nonparametric estimator. Second, copula parameters are estimated using the maximum likelihood.

Copulas model selection

How to select a copula model that best fits the data is a difficult problem. In practice, the following methods are used: Root Mean Square Error (RMSE), Akaike Information Criterion (AIC), Bayesian Information Criterion (BIC), and goodness-of-fit tests for copulas. Goodness-of-fit tests are: 1) two tests based on the empirical copula; 2) two tests based on Kendall's transform; and 3) a test based on Rosenblatt's transform. A bootstrap version based on Rosenblatt's transformation is a popular method used to evaluate the goodness-of-fit.

5.4 Copula entropy

Let $x \in \mathbf{R}_N$ be random variables with marginal functions $u_i = F_i(x_i)$, with $U_i \sim U(0, 1)$, $i = 1, 2, \ldots, n,$. The copula entropy H_C can be defined as:

$$H_C(U_1, U_2, \ldots, U_n) = -\int_0^1 \int_0^1 \cdots \int_0^1 c(u_1, u_2, \ldots, u_n) \log(c(u_1, u_2, \ldots, u_n)) du_1 du_2 du_n$$

(5.350)

where $c(u_1, u_2, \ldots, u_n)$ is the probability density function of copula. The relation between joint entropy and copula entropy is now discussed. The joint probability density function of vector X can be defined as:

$$f(x_1, x_2, \ldots, x_n) = c(u_1, \ldots, u_n) \prod_{i=1}^n f(x_i)$$

(5.351)

The joint entropy can be expressed as:

$$
\begin{aligned}
H(X_1, X_2, \ldots, X_n) &= -\int_0^\infty \cdots \int_0^\infty f(x_1, x_2, \ldots, x_n) \log[f(x_1, x_2, \ldots, x_n)] dx_1 dx_2 \cdots dx_n \\
&= -\int_0^\infty \cdots \int_0^\infty c(u_1, \ldots, u_n) \prod_{i=1}^n f(x_i) \log[c(u_1, \ldots, u_n) \prod_{i=1}^n f(x_i)] dx_1 dx_2 \cdots dx_n \\
&= -\int_0^\infty \cdots \int_0^\infty c(u_1, \ldots, u_n) \prod_{i=1}^n f(x_i) \{\log[c(u_1, \ldots, u_n)] + \sum_{i=1}^n \log[f(x_i)]\} dx_1 dx_2 \cdots dx_n \\
&= -\int_0^\infty \cdots \int_0^\infty c(u_1, \ldots, u_n) \prod_{i=1}^n f(x_i) \cdot \log[c(u_1, \ldots, u_n)] dx_1 dx_2 \cdots dx_n \\
&\quad -\int_0^\infty \cdots \int_0^\infty c(u_1, \ldots, u_n) \prod_{i=1}^n f(x_i) \cdot \sum_{i=1}^n \log[f(x_i)] dx_1 dx_2 \cdots dx_n
\end{aligned}
$$

(5.352)

From equation (5.352),

$$
\begin{aligned}
&-\int_0^\infty \cdots \int_0^\infty c(u_1, \ldots, u_n) \prod_{i=1}^n f(x_i) \cdot \sum_{i=1}^n \log[f(x_i)] dx_1 dx_2 \cdots dx_n \\
&= -\int_0^\infty \cdots \int_0^\infty f(x_1, x_2, \ldots, x_n) \cdot \sum_{i=1}^n \log[f(x_i)] dx_1 dx_2 \cdots dx_n \\
&= -\sum_{i=1}^n \int_0^\infty f(x_i) \log[f(x_i)] dx_i = -\sum_{i=1}^n H(X_i)
\end{aligned}
$$

(5.353)

Recalling that $du = f(x_i)dx$,

$$-\int_0^\infty \cdots \int_0^\infty c(u_1, \ldots, u_n) \prod_{i=1}^n f(x_i) \cdot \log[c(u_1, \ldots, u_n)]dx_1 dx_2 \cdots dx_n$$

$$= -\int_0^\infty \cdots \int_0^\infty c(u_1, \ldots, u_n) \cdot \log[c(u_1, \ldots, u_n)]du_1 du_2 \cdots du_n = H_c(\mathbf{u}) \tag{5.354}$$

Therefore, the joint entropy can be expressed as the sum of the n univariate marginal entropies and the copula entropy as:

$$H(X_1, X_2, \ldots, X_n) = \sum_{i=1}^n H(X_i) + H_C(U_1, U_2, \ldots, U_n) \tag{5.355}$$

Equation (5.355) indicates that the joint entropy $H(X_1, X_2, \ldots, X_n)$ is divided into parts: the sum of the n marginal entropies $H(X_i)$ and the copula entropy $H_C(U_1, U_2, \ldots, U_n)$.

According to equation (5.350), the copula entropy can be derived using the multiple integration method. First, the parameters of the copula function need to be estimated, and then the copula probability density function can be determined. The multiple integration method, proposed by Berntsen et al. (1991), can be applied to calculate multiple integration. For more variables, maybe it is difficult to calculate multiple integration. The Monte Carlo method can then be used to calculate the copula entropy. For a multivariate vector with support in [0, 1], the copula entropy can be obtained by:

$$H_C(U_1, U_2, \ldots, U_n) = -\int_{[0,1]^n} c(U) \ln[c(U)]dU = E[\ln c(U)] \tag{5.356}$$

The copula entropy equals the expected value of $[\ln c(U)]$, which can be derived by the Monte Carlo method. Similar to the multiple integration method, first the dependence structure and parameters of the copula function need to be determined. M pairs of u are generated from the determined copula function, and then average values of the $\ln(c(u))$ were calculated.

Questions

Q.5.1 Obtain yearly discharge data for a number of years from a gaging station (say A) at a river and plot its histogram and check if it follows a normal distribution. Likewise, obtain yearly discharge data from another gaging station (say B) upstream at the same river and check if it follows a normal distribution. If either of the two data sets does not, then transform it using a power or log transformation so that it follows the normal distribution. Then, compute the joint probability distribution of yearly discharge. Compute the joint entropy.

Q.5.2 Compute the distribution of yearly discharge of station A conditioned on the yearly discharge at station B. Compute the conditional entropy.

Q.5.3 Compute 5, 10, 25, 50, and 100-year return periods of yearly discharge at Station A given a discharge value at station B given in Q.5.2.

Q.5.4 Obtain yearly peak rainfall depth and duration data for a rain gage in a basin. Assume the depth and duration each follow an exponential distribution. Compute the bivariate exponential distribution of rainfall depth and duration. Compute its entropy.

Q.5.5 Compute the distribution of rainfall depth conditioned on the duration. Compute the conditional entropy.

Q.5.6 Compute 5, 10, 25, 50, and 100-year return periods of rainfall depth given a duration.

Q.5.7 Obtain the flood peak and volume data from a gaging station at a river. Transform the data so they are normally distributed. Then, compute the bivariate probability density function.

Q.5.8 Compute the distribution of peak discharge conditioned on the volume. Compute the conditional entropy.

Q.5.9 Compute 5, 10, 25, 50, and 100-year return periods of peak discharge given a volume.

Q.5.10 Consider a random variable which is gamma distributed and another variable which is exponentially distributed. Construct the joint distribution using the Ali-Mikhail-Haq (AMH) copula.

Q.5.11 Consider a random variable which is uniformly distributed and another variable which is normally. Construct the joint distribution using the Gumbel-Hougaard copula.

Q.5.12 Consider a random variable which is log-Pearson III distributed and another variable which is log-normal distributed. Construct the joint distribution using the Cook-Johnson copula.

References

Alfonsi, A.E. and Brigo, D. (2005). New families of Copulas based on periodic functions. *Communications in Statistics: Theory and Methods*, Vol. 34, No. 7, pp. 1437–47.

Ali M.M., Mikhail, N.N. and Haq, M.S. (1978). A class of bivariate distributions including the bivariate logistic. *Journal of Multivariate Analysis*, Vol. 8, pp. 405–12.

Berntson, J., Espelid, T.O., and Genz, A. (1991). An adaptive algorithm for the approximate calculation of multiple integrals. *ACM Transactions of Mathematical Software*, Vol. 17, pp. 437–451.

Embrechts, P., Lindskog F. and McNeil, A. (2001). Modelling dependence with copulas and applications to risk management. http://www.risklab.ch/ftp/papers/DependenceWithCopulas.pdf

Galambos, J. (1978). *The Asymptotic Theory of Extreme Order Statistics*. John Wiley & Sons, New York.

Genest, C. and L.-P. Rivest (1993). Statistical inference procedures for bivariate Archimedean copulas. *Journal of the American Statistical Association* Vol. 88, No. 423, pp. 1034–43.

Hu, L. (2006). Dependence patterns across financial markets: a mixed copula approach. *Applied Financial Economics*, Vol. 16, pp. 717–29.

Joe H. (1997). *Multivariate Models and Dependence Concepts*, Chapman and Hall, London.

Krstanovic, P.F. and Singh, V.P. (1988). Application of entropy theory to multivariate hydrologic analysis, Vol. 1. Technical Report WRR 8, Water Resources Program, Department of Civil Engineering, Louisiana State University, Baton Rouge, Louisiana.

Krstanovic, P.F. and Singh, V.P. (1988). Application of entropy theory to multivariate hydrologic analysis, Vol. 2. Technical Report WRR 9, Water Resources Program, Department of Civil Engineering, Louisiana State University, Baton Rouge, Louisiana.

Nelson, R.B. (1999). *An Introduction to Copulas*. Springer, New York.

Nelsen, R.B. (2006) *An Introduction to Copulas*, second edition. Springer-Verlag, New York.

Salvadori, G., De Michele, C., Kottegoda, N.T. and Rosso, R. (2007) *Extremes in Nature: An Approach Using Copulas*. Springer, Dordrecht, Netherlands.

Savu, C. and Trede, M. (2008). Goodness-of-fit tests for parametric families of Archimedean copulas. *Quantitative Finance*, Vol. 8, No. 2, pp. 109–16.

Sklar, A. (1959) Fonctions de répartition à n dimensions et leurs marges. *Publ. Inst. Stat. Univ. Paris*, Vol. 8, pp. 229–231.

Wang, X., Gebremichael, M., Yan, J. (2010). Weighted likelihood copula modeling of extreme rainfall events in Connecticut. *Journal of Hydrology*, Vol. 390, No. 1-2, pp. 108–15.

Yue, S., Ouarda, T.B.M.J., Bobee, B., Legendre, P., Bruneau, P. (1999). The Gumbel mixed model for flood frequency analysis. *Journal of Hydrology*, Vol. 226, No. 1-2, pp. 88–100.

Additional Reading

Bacchi, B., Becciu, G. and Kottegoda, N.T. (1994). Bivariate exponential model applied to intensities and durations of extreme rainfall. *Journal of Hydrology*, Vol. 155, pp. 225–36.

Calsaverini, R. S. and Vicente, R. (2009). An information-theoretic approach to statistical dependence: Copula information. *European Physics Letters*, Vol. 88, doi: 10.1209/0295-5075/88/68003.

De Michele, C. and Salvadori, G. (2003). A generalized Pareto intensity duration model of storm rainfall exploiting 2-copulas. *Journal of Geophysical Research*, Vol. 108, No. D2, pp. 4067; doi:10.1029/2002JD002534.

De Michele, C., Salvadori, C. G., Passoni, G. and Vezzoli, R. (2007) A multivariate model of sea storms using copulas. *Coastal Engineering*, Vol. 54, No. 10, pp. 734–51.

Favre, A-C., Adlouni, S., Perreault, L., Thiémonge, N. and Bobée, B. (2004) Multivariate hydrological frequency analysis using copulas. *Water Resour. Res.*, Volume 40:W01101, 12.

Genest, C., Favre, A-C., Béliveau, J., Jacques, C. (2007) Metaelliptical copulas and their use in frequency analysis of multivariate hydrological data. *Water Resources Research*, Vol. 43:W09401, doi:10.1029/2006WR005275.

Genest, C. and Verret, F. (2005). Locally most powerful rank tests of independence for copula models, *Journal of Nonparametric Statistics*, Vol.17, pp. 521–39.

Grimaldi, S. and Serinaldi, F. (2006). Design hyetographs analysis with 3-copula function. *Hydrological Sciences Journal*, Vol. 51, No. 2, pp. 223–38.

Hu, L. (2006). Dependence patterns across financial markets: a mixed copula approach, *Applied Finance Economics*, Vol. 16, pp. 717–29.

Kao, S. C. and Govindaraju, R. S. (2007). A bivariate frequency analysis of extreme rainfall with implications for design. *Journal of Geophysical Research*, Vol. 112, D13119, doi:10.1029/2007JD008522.

Kao, S.C., and Govindaraju, R.S. (2010). A copula-based joint deficit index for droughts. *Journal of Hydrology*, Vol. 380, No. 1-2, pp. 121–34.

Kim, G., Silvapulle, M. and Silvapulle, P. (2007). Comparison of semiparametric and parametric methods for estimating copulas. *Computational Statistics and Data Analysis*, Vol. 51, No. 6, pp. 2836–50.

Krstanovic, P.F. and Singh, V.P. (1987). A multivariate stochastic flood analysis using entropy. in *Hydrologic Frequency Modeling* edited by V.P. Singh, D. Reidel Publishing Company, Boston, pp. 5l5–40.

Krstanovic, P.F. and Singh, V.P. (1988). A bivariate model for real time flood forecasting. Proceedings of the International Seminar on Hydrology of Extremes (Floods and Low Flows), pp. 235–45, Roorkee, India.

Krstanovic, P.F. and Singh, V.P. (1989a). Application of entropy theory to multivariate hydrologic analysis, Vol. 3, selected computer programs. Technical Report WRRl0, 234 p., Water Resources Program, Department of Civil Engineering, Louisiana State University, Baton Rouge, Louisiana.

Krstanovic, P.F. and Singh, V.P. (1989b). An entropy based method for flood forecasting. IAHS Publication No. l8l, Proceedings of the Baltimore Symposium (IAHS Third Scientific Assembly) on New Directions for Surface Water Modeling, pp. l05–ll3, Baltimore, Maryland.

Ma, J. and Sun, Z. (2011) Mutual information is copula entropy. *Tsinghua Science & Technology*, Vol. 16, No. 1, pp. 51–4.

Marshall, A.W. and Ingram, O. (1967). A multivariate exponential distribution. *Journal of American Statistical Association*, Vol. 62, No. 317, pp. 30–44.

Renard, B. and Lang, M. (2007). Use of a Gaussian copula for multivariate extreme value analysis: some case studies in hydrology. *Advances in Water Resources*, Vol. 30, No. 4, pp. 897–912.

Salvadori, G. and De Michele, C. (2010). Multivariate multiparameter extreme value models and return periods: a copula approach. *Water Resources Research*, Vol. 46, W10501, doi:10.1029/2009WR009040.

Schucany, W., Parr, W. and Boyer, J. (1978). Correlation structure in Falie-Gumbel-Morgenstern distributions. *Biometrika*, Vol. 65, pp. 650–3.

Shiau, J.T. (2006). Fitting drought duration and severity with two-dimensional copulas. *Water Resources Management*, Vol. 20, pp. 795–815.

Shiau, J.T., Wang, H.Y. and Chang, T.T. (2006). Bivariate frequency analysis of floods using copulas. *Journal of American Water Resources Association*, Vol. 42, No. 6, pp. 1549–64.

Singh, K. and Singh, V.P. (1991). Derivation of bivariate probability density functions with exponential marginals. *Journal of Stochastic Hydrology and Hydraulics*, Vol. 5, pp. 55–68.

Singh, V.P. and Zhang, L. (2007). IDF curves using the Frank Archimedean copula. *Journal of Hydrologic Engineering*, Vol. 12, No. 6, pp. 651–662, doi: 10.1061/(ASCE)1084-0699.

Song, S. and Singh, V. P. (2010). Meta-elliptical copulas for drought frequency analysis of periodic hydrologic data. *Stochastic Environmental Research and Risk Assessment*, Vol. 24, No. 3, pp. 425–44.

Wang, C., Chang, N.-B. and Yeh, G.-T. (2009). Copula-based flood frequency (COFF) analysis at the confluences of river systems. *Hydrological Processes*, E23 (10), 1471–86, doi:{10.1002/hyp.7273}.

Weiß, Gregor N. F. (2009). Copula Parameter Estimation by Maximum-Likelihood and Minimum-Distance Estimators - A Simulation Study. Computational Statistics, Forthcoming. Available at SSRN: http://ssrn.com/abstract=1334783

Zhang, L. and Singh, V.P. (2006). Bivariate flood frequency analysis using the copula method. *Journal of Hydrologic Engineering*, Vol. 11, No. 2, pp. 150–64.

Zhang, L. and Singh, V.P. (2007). Gumbel–Hougaard copula for trivariate rainfall frequency analysis. *Journal of Hydrologic Engineering*, Vol. 12, No. 4, pp. 409–19.

6 Principle of Minimum Cross-Entropy

The Principle of Minimum Cross-entropy (POMCE) is one of the building blocks of the entropy theory, and is a powerful principle. This chapter focuses on this principle.

6.1 Concept and formulation of POMCE

The principle of minimum cross-entropy (POMCE) was formulated by Kullback and Leibler (1951) and is detailed in Kullback (1959). Sometimes it is also referred to as the Kullback-Leibler (KL) principle. POMCE is also referred to as the principle of minimum discrimination information, principle of minimum directed divergence, principle of minimum distance, or principle of minimum relative entropy. Recalling that entropy of a random variable characterizes the uncertainty of the variable, it measures the amount of information required on average to describe the random variable. To explain POMCE, let one guess, based on intuition, experience or theory, a probability distribution $Q = \{q_1, q_2, \ldots, q_N\}$ for a random variable X which takes on N values. If Q is a uniform distribution then this would represent the maximum uncertainty. The guessed distribution constitutes the prior information in the form of a prior distribution. To verify one's guess, one takes some observations and computes some moments of the distribution using these observations and expresses them in the form of constraints. If the prior or guessed distribution satisfies the constraints then the guessed probability distribution $Q = \{q_1, q_2, \ldots, q_N\}$ is the desired distribution of X. If that is not the case which often happens then another probability distribution is to be sought. To derive the distribution $P = \{p_1, p_2, \ldots, p_N\}$ of X one takes all the given information and makes the distribution as near to Q (based on one's intuition and experience) as possible, that is, minimize the distance between P and Q. This means that the closer P is to Q, the greater will be its uncertainty. Thus, POMCE is expressed as

$$D(P, Q) = \sum_{i=1}^{N} P_i \ln \frac{p_i}{q_i} \tag{6.1}$$

where D is the cross-entropy or distance or discrimination information and the objective is to minimize D, the cross-entropy. In equation (6.1) we often use the convention that

Entropy Theory and its Application in Environmental and Water Engineering, First Edition. Vijay P. Singh.
© 2013 John Wiley & Sons, Ltd. Published 2013 by John Wiley & Sons, Ltd.

$0 \log(0/q) = 0$ and $p \log(p/0) = \infty$. If no a priori distribution is available in the form of constraints and Q is chosen to be a uniform distribution $Q = \{q_i = 1/N, i = 1, 2, \ldots, N\}$, then equation (6.1) takes the form:

$$D(P, Q) = \sum_{i=1}^{N} p_i \ln \left[\frac{p_i}{1/N} \right] = \ln N + \left(\sum_{i=1}^{N} p_i \ln p_i \right) = \ln N - H \tag{6.2}$$

where H is the Shannon entropy. Equation (6.2) shows that minimizing $D(P, Q)$ is equivalent to maximizing the Shannon entropy (Kapur, 1989; Kapur and Kesavan, 1987, 1992) defined as

$$H = - \sum_{i=1}^{N} p_i \ln p_i \tag{6.3}$$

Because D is a convex function, its local minimum is its global minimum.

Thus, a posterior distribution P is obtained by combining a prior Q with specified constraints. The distribution P minimizes the cross- (or relative) entropy with respect to Q, defined by equation (6.1), where the entropy of Q is defined or given. Minimization of cross- entropy results asymptotically from Bayes' theorem.

It is to be noted that POMCE involves two concepts: 1) a prior probability distribution and 2) a measure of distance. As stated earlier, $Q = \{q_1, q_2, \ldots, q_N\}$ is chosen based on some knowledge about X but this Q does not satisfy the prescribed constraints. Second, D is a measure of distance which has certain desirable properties, such as $D(P, Q) \geq 0$ and $D(P, Q) = 0$ if and only if $P = Q$. In essence, POMCE is a measure between two probability distributions one of which is related to the system to be characterized or the source to be estimated and assumed to be unknown and the other (called prior) to the model chosen to describe the system. The system is characterized by a set of moments or by the mean and any symmetric part of the covariance matrix of the system called constraints. The POMCE measure is obtained by minimizing the discrimination information with respect to the given prior distribution over all probabilistic descriptions of the system which concur with the given constraints.

One can express a measure of distance in many different ways. For example, one way to express it is in a root square sense as

$$D(P, Q) = \left[\sum_{i=1}^{N} (p_i - q_i)^2 \right]^{0.5} \tag{6.4}$$

However, one would want a measure D to be a convex function of $P = \{p_1, p_2, \ldots, p_N\}$. Such a measure is the Kullback-Leibler (KL) measure of cross-entropy given by equation (6.1).

6.2 Properties of POMCE

The KL measure has the following properties:

 1 The distance measure $D(P, Q)$ is non-negative: $D(P, Q) \geq 0$.
 2 The distance measure $D(P, Q)$ is asymmetric: $D(P, Q) \neq D(Q, P)$.

This is not a true distance between distributions, because it does not obey the triangular inequality and is not symmetric.

Example 6.1: Show this property of asymmetry assuming first Q as uniform and then P as uniform.

Solution: Let Q be a uniform distribution: $q_i = 1/N$. Then one obtains

$$D(P, Q) = \sum_{i=1}^{N} p_i \ln \frac{p_i}{1/N} = \ln N + \sum_{i=1}^{N} p_i \ln p_i = \ln N - H \tag{6.5}$$

where H is the Shannon entropy. On the other hand, if P is uniform $p_i = 1/N$, then

$$D(Q, P) = \sum_{i=1}^{N} q_i \ln \frac{q_i}{1/N} = \ln N + \sum_{i=1}^{N} q_i \ln q_i \tag{6.6}$$

Clearly, $D(P, Q)$ is not equal to $D(Q, P)$.

However, it turns out that the measure

$$W(P, Q) = D(P, Q) + D(Q, P) = \sum_{i=1}^{N} p_i \ln \frac{p_i}{q_i} + \sum_{i=1}^{N} q_i \ln \frac{q_i}{p_i} = \sum_{i=1}^{N} (p_i - q_i) \ln \frac{p_i}{q_i} \tag{6.7}$$

is symmetric, that is, $W(P, Q) = W(Q, P)$. This measure is sometimes referred to as a measure of symmetric cross-entropy or of symmetric divergence or W divergence (Lin, 1991). It may be noted that $D(P, Q)$ is undefined if $Q = 0$ and P is not equal to 0 for any x. This means that the distribution P must be absolutely continuous with respect to the distribution Q in order for $D(P, Q)$ to be defined. Likewise, P and Q should be continuous with respect to each other in order for $W(P, Q)$ to be defined.

Example 6.2: Let $P = \{p_1, p_2, \ldots, p_N\}$ and $Q = \{q_1, q_2, \ldots, q_N\}$, $p_i = ab^i$, and $q_i = 1/N$. Show that $D(P, Q)$ is not symmetric, but $W(P, Q)$ is.

Solution: First, from equation (6.1)

$$D(P, Q) = \sum_{i=1}^{N} ab^i \ln \left[\frac{ab^i}{1/N} \right] = a \sum_{i=1}^{N} b^i \ln[Nab^i] \tag{6.8}$$

$$D(Q, P) = \sum_{i=1}^{N} \frac{1}{N} \ln \left[\frac{1/N}{ab^i} \right] = \frac{1}{N} \sum_{i=1}^{N} \ln \left[\frac{1}{Nab^i} \right] \tag{6.9}$$

This shows that $D(P, Q)$ is not the same as $D(Q, P)$.

On the other hand,

$$W(P, Q) = \sum_{i=1}^{N} ab^i \ln \frac{ab^i}{1/N} + \sum_{i=1}^{N} \frac{1}{N} \ln \frac{1/N}{ab^i} \tag{6.10}$$

Likewise,

$$W(Q, P) = \sum_{i=1}^{N} \frac{1}{N} \ln \frac{1/N}{ab^i} + \sum_{i=1}^{N} ab^i \ln \frac{ab^i}{1/N} \tag{6.11}$$

It is seen that $W(P, Q) = W(Q, P)$.

3 $D(P, Q)$ is a continuous function of $P = \{p_1, p_2, \ldots, p_N\}$ and $Q = \{q_1, q_2, \ldots, q_N\}$.

4 $D(P, Q)$ does not change if $(p_1, q_1), (p_2, q_2), \ldots, (p_N, q_N)$ are permuted among themselves.

5 $D(P, Q) \geq 0$ and $D(P, Q) = 0$ if $P = Q$.

Example 6.3: Show that $D(P, Q) = 0$ if $P = Q$.

Solution: $P = \{p_1, p_2, \ldots, p_N\}$ and $Q = \{q_1, q_2, \ldots, q_N\}$. Substituting $Q = P$ in equation (6.1), one obtains

$$D(P, Q) = \sum_{i=1}^{N} p_i \ln \frac{p_i}{q_i} = \sum_{i=1}^{N} p_i \ln \frac{p_i}{p_i} = \sum_{i=1}^{N} p_i 0 = 0$$

6 $D(P, Q)$ is a convex function of P and Q. This can be shown by constructing the Hessian matrix of second order partial derivatives of D with respect to $p_1, p_2, \ldots, p_N$ and showing that it is positive definite. The Hessian matrix is defined in mathematics as a square matrix of second-order partial derivatives of a function. Similarly, the Hessian matrix of D with respect to $q_1, q_2, \ldots, q_N$ is positive definite. This will infer that $D(P, Q)$ is a convex function of P and Q.

Because $D(P, Q)$ is a convex function of $p_1, p_2, \ldots, p_N$, its maximum for a specified Q must coincide with one of the degenerate distributions. The maximum value must be:

$$\max(- \ln q_1 - \ln q_2 - \ldots - \ln q_N) = \ln \frac{1}{q_{\min}} \tag{6.12}$$

where

$$q_{\min} = \min(q_1, q_2, \ldots, q_N) \tag{6.13}$$

In a similar manner one can infer that the maximum value of $D(P, Q)$ can be made as large as desired by making some of q_i's sufficiently small. This is because $D(P, Q)$ is a convex function of Q.

7 For two independent variables X and Y with probability distributions $P = \{p_1, p_2, \ldots, p_N\}$ and $G = \{g_1, g_2, \ldots, g_M\}$, respectively, with two independent prior distributions $Q = \{q_1, q_2, \ldots, q_N\}$ and $R = \{r_1, r_2, \ldots, r_M\}$, $D(P, G; Q, R)$ satisfies the additive property:

$$D(P, G; Q, R) = \sum_{j=1}^{M} \sum_{i=1}^{N} p_i g_j \ln \frac{p_i g_j}{q_i r_j}$$

$$= \sum_{j=1}^{M} g_j \sum_{i=1}^{N} p_i \ln \frac{p_i}{q_i} + \sum_{i=1}^{N} p_i \sum_{j=1}^{M} g_j \ln \frac{g_j}{r_j}$$

$$= D(P, Q) + D(G, R) \tag{6.14}$$

8 For dependent variables X and Y with distribution as $P = \{p_{ij}, i = 1, 2, \ldots N; j = 1, 2, \ldots, M\}$ and prior as $Q = \{q_{ij}, i = 1, 2, \ldots N; j = 1, 2, \ldots, M\}$ one can write

$$D(P, Q) = \sum_{j=1}^{M} \sum_{i=1}^{N} p_{ij} \ln \frac{p_{ij}}{q_{ij}}$$

$$= \sum_{j=1}^{M} \sum_{i=1}^{N} p_{ij} \ln \frac{\displaystyle\sum_{j=1}^{M} p_{ij}}{\displaystyle\sum_{j=1}^{M} q_{ij}} + \sum_{i=1}^{N} \sum_{j=1}^{M} p_{ij} \sum_{j=1}^{M} \frac{p_{ij}}{\displaystyle\sum_{j=1}^{M} p_{ij}} \ln \frac{p_{ij} / \displaystyle\sum_{j=1}^{M} p_{ij}}{q_{ij} / \displaystyle\sum_{j=1}^{M} q_{ij}} \qquad (6.15)$$

Example 6.4: Show that the minimum cross-entropy $D(P, Q)$ is a convex function of both P and Q, where P and Q are vectors.

Solution: The Hessian matrix must be positive definite. In other words, all second partial derivatives of D with respect to P as well as Q must be positive definite. Equation (6.1) shows that $D(P, Q) = f(p_1, p_2, \ldots\ldots\ldots, p_N; q_1, q_2, \ldots\ldots\ldots, q_N)$. Therefore, the Hessian matrix is given as:

$$\begin{bmatrix} \dfrac{\partial^2 f}{\partial p_1^2} & \dfrac{\partial^2 f}{\partial p_1 \partial p_2} & .. & \dfrac{\partial^2 f}{\partial p_1 \partial p_N} \\[2ex] \dfrac{\partial^2 f}{\partial p_2 \partial p_1} & \dfrac{\partial^2 f}{\partial p_2^2} & .. & \dfrac{\partial^2 f}{\partial p_2 \partial p_N} \\[2ex] .. & .. & .. & .. \\[1ex] .. & .. & .. & .. \\[1ex] \dfrac{\partial^2 f}{\partial p_N \partial p_1} & \dfrac{\partial^2 f}{\partial p_N \partial p_2} & .. & \dfrac{\partial^2 f}{\partial p_N^2} \end{bmatrix} = \begin{bmatrix} \dfrac{1}{p_1} & 0 & .. & 0 \\[2ex] 0 & \dfrac{1}{p_2} & .. & 0 \\[2ex] .. & .. & .. & .. \\[1ex] .. & .. & .. & .. \\[1ex] 0 & 0 & .. & \dfrac{1}{p_N} \end{bmatrix} \qquad (6.16)$$

$$\Delta = \begin{vmatrix} \dfrac{1}{p_1} & 0 & .. & 0 \\[2ex] 0 & \dfrac{1}{p_2} & .. & 0 \\[2ex] .. & .. & .. & .. \\[1ex] .. & .. & .. & .. \\[1ex] 0 & 0 & .. & \dfrac{1}{p_N} \end{vmatrix} = \frac{1}{p_1 p_2 \cdots p_N} \qquad (6.17)$$

This is a positive definite quantity, as $p_i > 0$.

Similarly,

$$\begin{bmatrix} \dfrac{\partial^2 f}{\partial q_1^2} & \dfrac{\partial^2 f}{\partial q_1 \partial q_2} & .. & \dfrac{\partial^2 f}{\partial q_1 \partial q_N} \\[2ex] \dfrac{\partial^2 f}{\partial q_2 \partial q_1} & \dfrac{\partial^2 f}{\partial q_2^2} & .. & \dfrac{\partial^2 f}{\partial q_2 \partial q_N} \\[2ex] .. & .. & .. & .. \\[1ex] .. & .. & .. & .. \\[1ex] \dfrac{\partial^2 f}{\partial q_N \partial q_1} & \dfrac{\partial^2 f}{\partial q_N \partial q_2} & .. & \dfrac{\partial^2 f}{\partial q_N^2} \end{bmatrix} = \begin{bmatrix} \dfrac{p_1}{q_1^2} & 0 & .. & 0 \\[2ex] 0 & \dfrac{p_2}{q_2^2} & .. & 0 \\[2ex] .. & .. & .. & .. \\[1ex] .. & .. & .. & .. \\[1ex] 0 & 0 & .. & \dfrac{p_N}{q_N^2} \end{bmatrix} \qquad (6.18)$$

$$\Delta = \begin{vmatrix} \dfrac{p_1}{q_1^2} & 0 & \cdot\cdot & 0 \\[2mm] 0 & \dfrac{p_2}{q_2^2} & \cdot\cdot & 0 \\[2mm] \cdot\cdot & \cdot\cdot & \cdot\cdot & \cdot\cdot \\[1mm] \cdot\cdot & \cdot\cdot & \cdot\cdot & \cdot\cdot \\[1mm] 0 & 0 & \cdot\cdot & \dfrac{p_N}{q_N^2} \end{vmatrix} = \frac{p_1 p_2 \cdot\cdot p_N}{q_1^2 q_2^2 \cdot\cdot q_N^2} \tag{6.19}$$

This is a positive definite quantity, as $p_i, q_i > 0$. Thus, the minimum cross-entropy $D(P, Q)$ is a convex function of both P and Q.

6.3 POMCE formalism for discrete variables

$D(P, Q)$ in equation (6.1) is minimized subject to the moment constraints:

$$\sum_{i=1}^{N} p_i y_{ri}(x) = a_r, \ r = 1, 2, \ldots, m \tag{6.20}$$

where $y_r(x)$ is some function of x, and a_r is the average of $y_r(x)$. Equation (6.20) gives m constraints. Of course, the total probability theorem holds:

$$\sum_{i=1}^{N} p_i = 1 \tag{6.21}$$

In order to obtain the least-biased $P, D(P, Q)$ is minimized, subject to equations (6.20) and (6.21). This can be done using the method of Lagrange multipliers. The Lagrangean L is constructed as

$$L = \sum_{i=1}^{N} p_i \ln \frac{p_i}{q_i} + (\lambda_0 - 1)\left(\sum_{i=1}^{N} p_i - 1\right) + \sum_{j=1}^{m} \lambda_j \left(\sum_{i=1}^{N} p_i y_{rj}(x_i) - a_j\right) \tag{6.22}$$

where $\lambda_0, \lambda_1, \ldots, \lambda_m$ are the $(m+1)$ Lagrange multipliers corresponding to the $m+1$ constraints specified by equations (6.20) and (6.21). In equation (6.22), for convenience one uses $(\lambda_0 - 1)$ as the first Lagrange multiplier instead of λ_0.

Differentiating L in equation (6.22) with respect to p_i and equating it to zero, one obtains

$$\frac{\partial L}{\partial p_i} = 0 \Rightarrow \left(1 + \ln \frac{p_i}{q_i}\right) + (\lambda_0 - 1) + \sum_{r=1}^{m} \lambda_r y_{ri} = 0 \tag{6.23}$$

Equation (6.23) gives

$$p_i = q_i \exp[-\lambda_0 - \lambda_1 y_{1i} - \ldots - \lambda_m y_{mi}], \quad i = 1, 2, \ldots, N \tag{6.24}$$

Equation (6.24) is the POMCE-based probability distribution of X with the Lagrange multipliers $\lambda_0, \lambda_1, \ldots, \lambda_m$ which can be determined using equations (6.20) and (6.21) as follows. It may be noted that the probability distribution given by equation (6.24) is the product of the prior distribution and the probability distribution resulting from the maximization of the Shannon entropy.

Substituting equation (6.24) into equation (6.21), one gets

$$\exp(\lambda_0) = \sum_{i=1}^{N} q_i \, \exp(-\lambda_1 y_{1i} - \lambda_2 y_{2i} - \ldots - \lambda_m y_{mi}) \tag{6.25}$$

Substituting equation (6.24) in equation (6.20) one gets

$$a_r \, \exp(\lambda_0) = \sum_{i=1}^{N} q_i y_{ri} \, \exp(-\lambda_1 y_{1i} - \lambda_2 y_{2i} - \ldots - \lambda_m y_{mi}) \tag{6.26}$$

From equations (6.25) and (6.26), one can be write:

$$a_r = \frac{\displaystyle\sum_{i=1}^{N} q_i y_{ri} \, \exp(-\lambda_1 y_{1i} - \lambda_2 y_{2i} - \ldots - \lambda_m y_{mi})}{\displaystyle\sum_{i=1}^{N} q_i \, \exp(-\lambda_1 y_{1i} - \lambda_2 y_{2i} - \ldots - \lambda_m y_{mi})} \tag{6.27}$$

Equation (6.27) shows that $a_r, r = 1, 2, \ldots, m$, is a function of Lagrange multipliers as well as $q_1, q_2, \ldots, q_N$. Equation (6.25) shows that λ_0 is a function of $\lambda_1, \ldots, \lambda_m$. Differentiating equation (6.25) with respect to λ_1, one gets

$$\frac{\partial \lambda_0}{\partial \lambda_1} = \frac{-\displaystyle\sum_{i=1}^{N} q_i y_{1i} \, \exp(-\lambda_1 y_{1i} - \lambda_2 y_{2i} - \ldots - \lambda_m y_{mi})}{\exp(\lambda_0)} \tag{6.28}$$

Substituting equation (6.26) in equation (6.28) and then using equation (6.27) one obtains

$$\frac{\partial \lambda_0}{\partial \lambda_1} = -a_1 \tag{6.29}$$

Generalizing it, one gets

$$\frac{\partial \lambda_0}{\partial \lambda_r} = -a_r, r = 1, 2, \ldots, m \tag{6.30}$$

Similarly, taking the second derivative of λ_0 one obtains:

$$\frac{\partial^2 \lambda_0}{\partial \lambda_r^2} = E[y_r^2(x)] - \{E[y_r(x)]\}^2 = Var[y_r(x)] \tag{6.31}$$

and

$$\frac{\partial^2 \lambda_0}{\partial \lambda_r \partial \lambda_s} = E[y_r(x)y_s(x)] - E[y_r(x)]E[y_s(x)] = Cov[y_r(x)y_s(x)] \tag{6.32}$$

λ_0 is a convex function of $\lambda_1, \lambda_2, \ldots, \lambda_m$. This can be shown by constructing the Hessian matrix of the second order partial derivatives of λ_0 with respect to $\lambda_1, \lambda_2, \ldots, \lambda_m$ and then showing the matrix to be positive definite:

$$
M = \begin{bmatrix}
\partial^2 \lambda_0/\partial \lambda_1^2 & \partial^2 \lambda_0/\partial \lambda_1 \partial \lambda_2 & \cdots\cdots\cdots & \partial^2 \lambda_0/\partial \lambda_1 \partial \lambda_m \\
\partial^2 \lambda_0/\partial \lambda_2 \partial \lambda_1 & \partial^2 \lambda_0/\partial \lambda_2^2 & \cdots\cdots\cdots & \partial^2 \lambda_0/\partial \lambda_2 \partial \lambda_m \\
\cdot & & \cdot & \cdot \\
& & & \cdot \\
\partial^2 \lambda_0/\partial \lambda_m \partial \lambda_1 & \partial^2 \lambda_0/\partial \lambda_m \partial \lambda_2 & \cdots\cdots\cdots & \partial^2 \lambda_0/\partial \lambda_m^2
\end{bmatrix}
\tag{6.33}
$$

With the use of equations (6.31) and (6.32), the matrix in equation (6.33) can actually be expressed in terms of variances and co-variances as

$$
M = \begin{bmatrix}
Var[y_1(x)] & Cov[y_1(x), y_2(x)] & \cdots\cdots\cdots & Cov[y_1(x), y_2(x)] \\
Cov[y_2(x), y_1(x)] & Var[y_2(x)] & \cdots\cdots\cdots & Cov[y_2(x), y_m(x)] \\
\cdot & \cdot & & \cdot \\
& & \cdot & \\
Cov[y_m(x), y_1(x)] & Cov[y_m(x), y_2(x)] & \cdots\cdots\cdots & Var[y_m(x)]
\end{bmatrix}
\tag{6.34}
$$

Since all the principal minors are positive, M is therefore positive definite. (A minor of a matrix is called principal if it is obtained by deleting certain rows and the same numbered columns. This results in the diagonal elements of a principal minor being the diagonal elements of the matrix.)

The minimum value of $D(P, Q)$ can now be expressed by substituting equation (6.24) in equation (6.1) as:

$$
\begin{aligned}
D_{\min} &= \sum_{i=1}^{N} q_i \, \exp[-\lambda_0 - \lambda_1 y_{1i} - \lambda_2 y_{2i} - \ldots - \lambda_m y_{mi})][-\lambda_0 - \lambda_1 y_{1i} - \lambda_2 y_{2i} - \ldots - \lambda_m y_{mi}] \\
&= \sum_{i=1}^{N} p_i[-\lambda_0 - \lambda_1 y_{1i} - \lambda_2 y_{2i} - \ldots - \lambda_m y_{mi}] \\
&= -[\lambda_0 + \lambda_1 a_1 + \lambda_2 a_2 + \ldots + \lambda_m a_m]
\end{aligned}
\tag{6.35}
$$

Taking the derivative of equation (6.35) with respect to $a_r, r = 1, 2, \ldots, m$, one obtains

$$
\frac{\partial D_{\min}}{\partial a_r} = -\sum_{s=1}^{m} \frac{\partial \lambda_0}{\partial \lambda_s} \frac{\partial \lambda_s}{\partial a_r} - \lambda_r - \sum_{s=1}^{m} a_s \frac{\partial \lambda_s}{\partial a_r} = -\lambda_r, r = 1, 2, 3, \ldots, m
\tag{6.36}
$$

Taking the second derivative of $D_{\min}$ with respect to a_r, one gets

$$
\frac{\partial^2 D_{\min}}{\partial a_r^2} = -\frac{\partial \lambda_r}{\partial a_r}
\tag{6.37a}
$$

In a similar manner,

$$
\frac{\partial^2 D_{\min}}{\partial a_r \partial a_s} = -\frac{\partial \lambda_r}{\partial a_s}
\tag{6.37b}
$$

Now one can construct the Hessian matrix of D_{min} with respect to $a_1, a_2, \ldots, a_m$:

$$
N = \begin{bmatrix}
-\partial\lambda_1/\partial a_1 & -\partial\lambda_1/\partial a_2 \cdots & -\partial\lambda_1/\partial a_m \\
-\partial\lambda_2/\partial a_2 & -\partial\lambda_2/\partial a_2 \cdots & -\partial\lambda_2/\partial a_m \\
\cdot & \cdot & \cdot \\
& & \cdot \\
-\partial\lambda_m/\partial a_1 & -\partial\lambda_m/\partial a_2 \cdots & -\partial\lambda_m/\partial a_m
\end{bmatrix}
\tag{6.38}
$$

From equation (6.30),

$$
\frac{\partial a_s}{\partial \lambda_r} = -\frac{\partial^2 \lambda_0}{\partial \lambda_r \partial \lambda_s}
\tag{6.39a}
$$

Since

$$
\left[\frac{\partial \lambda_r}{\partial a_s}\right]\left[\frac{\partial a_s}{\partial \lambda_r}\right] = I
\tag{6.39b}
$$

it can be shown that the negative of M^{-1} is the variance-covariance matrix and is positive definite. Hence N is also positive definite. Thus, D_{min} is a convex function of $a_1, a_2, \ldots, a_m$.

Example 6.5: Determine the POMCE-based distribution of a random variable I which takes on values of 1, 2, 3, 4, 5, or 6, with probabilities of p_1, p_2, p_3, p_4, p_5, or p_6, respectively. It is assumed that the mean of I is known as:

$$
\sum_{i=1}^{6} i p_i = \bar{x} = 3.5
\tag{6.40}
$$

Also known is the a priori probability distribution $Q = \{q_i = 1/6, i = 1, 2, \ldots, 6\}$.

Solution: One needs to minimize $D(P, Q)$ given by equation (6.1) subject to equation (6.21) and the mean constraint given by equation (6.40). Equation (6.24) yields

$$
p_i = \frac{1}{6}\exp[-\lambda_0 - \lambda_1 i] = \frac{1}{6}ab^i, a = \exp(-\lambda_0), b = \exp(-\lambda_1)
\tag{6.41}
$$

This is the POMCE-based distribution whose parameters λ_0 and λ_1 or a and b can be determined using equations (6.24) and (6.40). Substituting equation (6.41) in equation (6.21) one obtains

$$
\sum_{i=1}^{6} \frac{1}{6}\exp[-\lambda_0 - \lambda_1 i] = \frac{1}{6}\sum_{i=1}^{6} ab^i = 1
\tag{6.42a}
$$

or

$$
\frac{a}{6}\sum_{i=1}^{6} b^i = 1
\tag{6.42b}
$$

Similarly, inserting equation (6.41) in equation (6.40), one gets

$$
\frac{a}{6}\sum_{i=1}^{6} ib^i = 3.5
\tag{6.43}
$$

Equations (6.42b) and (6.43) can be solved to get a and b and in turn λ_0 and λ_1:

$$\frac{\sum_{i=1}^{6} ib^i}{\sum_{i=1}^{6} b^i} = 3.5 \tag{6.44}$$

Solving the equation above, one gets:

$a = b = 1, \lambda_0 = \lambda_1 = 0.$

$p_i = \dfrac{1}{6}, i = 1, 2, \ldots, 6$

Example 6.6: A six-faced dice is given. It is observed that the mean of number appearing on the six-faced dice upon throw is found to be 4.5. Let X be the random variable expressing this number. Determine the POMCE-based distribution of X when an a priori probability distribution q is given by (0.05, 0.10, 0.15, 0.20, 0.22, 0.28).

Solution: Minimizing $\sum_{i=1}^{6} p_i \ln \frac{p_i}{q_i}$ subject to $\sum_{i=1}^{6} p_i = 1$ and $\sum_{i=1}^{6} ip_i = 4.5$, one gets $p_i = q_i ab^i$ where $\sum_{i=1}^{6} q_i ab^i = 1, \sum_{i=1}^{6} iq_i ab^i = 4.5$.

Substituting for $q_1, q_2, \ldots, q_6$ and solving for a, b one gets the distribution:

$P = \{0.035, 0.078, 0.131, 0.192, 0.234, 0.330\}$

Its entropy $H(P) = 1.606$ bits.

6.4 POMCE formulation for continuous variables

For a continuous variable X, equation (6.1) becomes

$$D(f, q) = D(x) = \int_a^b f(x) \ln \frac{f(x)}{q(x)} dx \tag{6.45}$$

where $f(x)$ is the probability density function of X, and $q(x)$ is the a priori PDF of X. For applying POMCE the following constraints can be specified:

$$\int_a^b f(x) dx = 1 \tag{6.46}$$

and

$$\int_a^b y_r(x) f(x) dx = \overline{y_r} = a_r, r = 1, 2, \ldots, m \tag{6.47}$$

where a and b are upper and lower limits of the integral or the range of variable X.

For minimization of $D(f, q)$ given by equation (6.45), subject to equations (6.46) and (6.47), one employs the method of Lagrange multipliers and obtains:

$$f(x) = q(x) \exp[-\lambda_0 - \lambda_1 y_1(x) - \ldots - \lambda_m y_m(x)] \tag{6.48}$$

where $q(x)$ is the density function of the a priori probability distribution. In this case, $D_{\min}$ is a convex function of $a_1, a_2, \ldots, a_m$, whereas $H_{\max}$ is concave function of $a_1, a_2, \ldots, a_m$.

Example 6.7: Determine the POMCE-based distribution of a random variable X which takes on values from $-\infty$ to $+\infty$ with probability density $f(x)$. It is assumed that the mean and the variance of X are known as: $E(x) = m = 0, E[(x - m)^2] = \sigma^2 = 1$. Also known is the a priori probability distribution $q(x) = \frac{1}{\sqrt{2\pi}} \exp[-\frac{1}{2}x^2]$.

Solution: One needs to minimize $D(f, q)$ given by equation (6.45), subject to equation (6.46), the mean and the variance constraints given above. Substituting equation (6.48) in equation (6.46) and the two constraints one obtains

$$f(x) = q(x) \exp[-\lambda_0 - \lambda_1 x - \lambda_2 x^2] \tag{6.49}$$

$$\int_{-\infty}^{+\infty} f(x) dx = \int_{-\infty}^{+\infty} \frac{1}{\sqrt{2\pi}} \exp\left[-\frac{1}{2}x^2\right] \exp[-\lambda_0 - \lambda_1 x - \lambda_2 x^2] dx = 1 \tag{6.50}$$

$$\int_{-\infty}^{+\infty} xf(x) dx = \int_{-\infty}^{+\infty} x \frac{1}{\sqrt{2\pi}} \exp\left[-\frac{1}{2}x^2\right] \exp[-\lambda_0 - \lambda_1 x - \lambda_2 x^2] dx = 0 \tag{6.51}$$

$$\int_{-\infty}^{+\infty} x^2 f(x) dx = \int_{-\infty}^{+\infty} x^2 \frac{1}{\sqrt{2\pi}} \exp\left[-\frac{1}{2}x^2\right] \exp[-\lambda_0 - \lambda_1 x - \lambda_2 x^2] dx = 1 \tag{6.52}$$

Solving these equations one gets $\lambda_0 = 0, \lambda_1 = 0, \lambda_2 = 0$, and the probability distribution is

$$f(x) = \frac{1}{\sqrt{2\pi}} \exp\left[-\frac{1}{2}x^2\right] \tag{6.53}$$

which is the standard normal distribution.

6.5 Relation to POME

In the case of POME, one maximizes the Shannon entropy subject to given constraints and determines the least-biased or maximum-uncertainty distribution. Thus, the emphasis is on the maximization of uncertainty. In the case of POMCE, one minimizes the distance between the distribution one seeks and the prior distribution and determines the least-biased probability distribution, subject to given constraints. Here the emphasis is on minimizing the distance between the prior and the posterior distributions. If the prior is uniform and no constraints are given, then POME becomes a special case of POMCE as seen from equation (6.2). For continuous variables, POMCE is invariant with coordinate transformation. POME yields the least-biased distribution as

$$p_i = \exp(-\lambda_0 - \lambda_1 y_{1i} - \ldots - \lambda_m y_{mi}) \tag{6.54}$$

whereas POMCE yields

$$p_i = q_i \exp(-\lambda_0 - \lambda_1 y_{1i} - \lambda_2 y_{2i} - \ldots - \lambda_m y_{mi}) \tag{6.55}$$

In the case of POME as well as POMCE, λ_0 is a convex function of $\lambda_1, \lambda_2, \ldots, \lambda_m$. D_{min} is a convex function of $a_1, a_2, \ldots, a_m$; where as H_{max} is a concave function of $a_1, a_2, \ldots, a_m$. Likewise $\partial D_{min}/\partial a_r = -\lambda_r$, where as $\partial H_{max}/\partial a_r = \lambda_r$.

Example 6.8: Referring to Example 6.5, compare the distributions resulting under POME and POMCE.

Solution: For POME, one does not need a prior q and hence it is not given. One would have used Laplace's principle of insufficient reason to use the uniform distribution instead of q. Thus, one would either minimize $\sum_{i=1}^{6} p_i \ln \frac{p_i}{1/6}$ or maximize $\sum_{i=1}^{6} p_i \ln p_i$, subject to the same constraints and would get the distribution:

$$P = \{0.0543, 0.0788, 0.1142, 0.1654, 0.2378, 0.3475\}$$

Its entropy $H(P) = 1.613$. Out of all distributions with mean 4.5, P has the maximum entropy.

6.6 Relation to mutual information

Let the joint probability distribution of random variables X and Y be $g(x, y)$. The marginal distributions of X and Y are: $f(x)$ and $q(y)$. Then, the mutual information or transinformation, $I(X, Y)$, is the relative entropy between the joint distribution and the product of distributions:

$$
\begin{aligned}
I(X, Y) &= \sum_{i=1}^{N} \sum_{j=1}^{M} g(x_i, y_i) \log \left[\frac{g(x_i, y_i)}{p(x_i)q(y_j)} \right] \\
&= D[g(x, y); p(x)q(y)] = E\left[\log \left(\frac{g}{pq} \right) \right] \tag{6.56}
\end{aligned}
$$

6.7 Relation to variational distance

The variational distance V between two probability distributions, P and Q, can be defined as

$$V(P, Q) = \sum_{x \in X} |P(x) - Q(x)| \tag{6.57}$$

There are several lower bounds for $D(P, Q)$ in terms of $V(P, Q)$ that have been derived, with the sharpest being

$$D(P, Q) \geq \max\{L_1[V(P, Q)], L_2[V(P, Q)]\} \tag{6.58a}$$

where

$$L_1[V(P, Q)] = \log \frac{2 + V(P, Q)}{2 - V(P, Q)} - \frac{2V(P, Q)}{2 + V(P, Q)}, 0 \leq V(P, Q) \leq 2 \quad (6.58b)$$

derived by Vajda (1970) and

$$L_2[V(P, Q)] = \frac{V^2(P, Q)}{2} + \frac{V^4(P, Q)}{36} + \frac{V^6(P, Q)}{288}, 0 \leq V(P, Q) \leq 2 \quad (6.59)$$

derived by Toussaint (1975). There is, however, no general upper bound either for $D(P, Q)$ or $W(P, Q)$.

6.8 Lin's directed divergence measure

Lin (1991) derived a new class of directed divergence measures based on the Shannon entropy and compared them with the above measures. One of the measures is defined as

$$K(P, Q) = \sum_{x \in X} p(x) \log \frac{p(x)}{\frac{1}{2}p(x) + \frac{1}{2}q(x)} \quad (6.60)$$

This measure possesses numerous desirable properties. $K(P, Q) = 0$ if and only if $P = Q$, and $K(P, Q) \geq 0$. The relation between $K(P, Q)$ and $D(P, Q)$ can be shown to be

$$K(P, Q) = D\left(P, \frac{1}{2}P + \frac{1}{2}Q\right) \quad (6.61)$$

The K-directed divergence is bounded by the D-divergence as

$$K(P, Q) \leq \frac{1}{2}D(P, Q) \quad (6.62)$$

The lower bound for the K-directed divergence can be expressed as

$$K(P, Q) \geq \max\left\{L_1\left(\frac{V(P, Q)}{2}\right), L_2\left(\frac{V(P, Q)}{2}\right)\right\} \quad (6.63)$$

where L_1 and L_2 are defined by equations (6.58a) and (6.59). It can be shown that

$$V\left(P, \frac{1}{2}P + \frac{1}{2}Q\right) = \frac{1}{2}V(P, Q) \quad (6.64)$$

A symmetric measure, $L(P, Q)$, can be defined based on $K(P, Q)$ which is not symmetric as

$$L(P, Q) = K(P, Q) + K(Q, P) \quad (6.65)$$

The L-divergence is related to the W-divergence in the same way as the K-divergence is related to the D-divergence:

$$L(P, Q) \leq \frac{1}{2}W(P, Q) \quad (6.66)$$

One can compare D, W, K and L divergences as shown in Figure 6.1, where $P = (x, 1 - x)$ and $Q = (1 - x, x), 0 \leq x \leq 1$. Divergences D and W have steeper slopes than do divergences K and L. When x approaches 0 or 1, divergences D and W approach infinity, whereas divergences K and L are well defined in the entire range $0 \leq x \leq 1$.

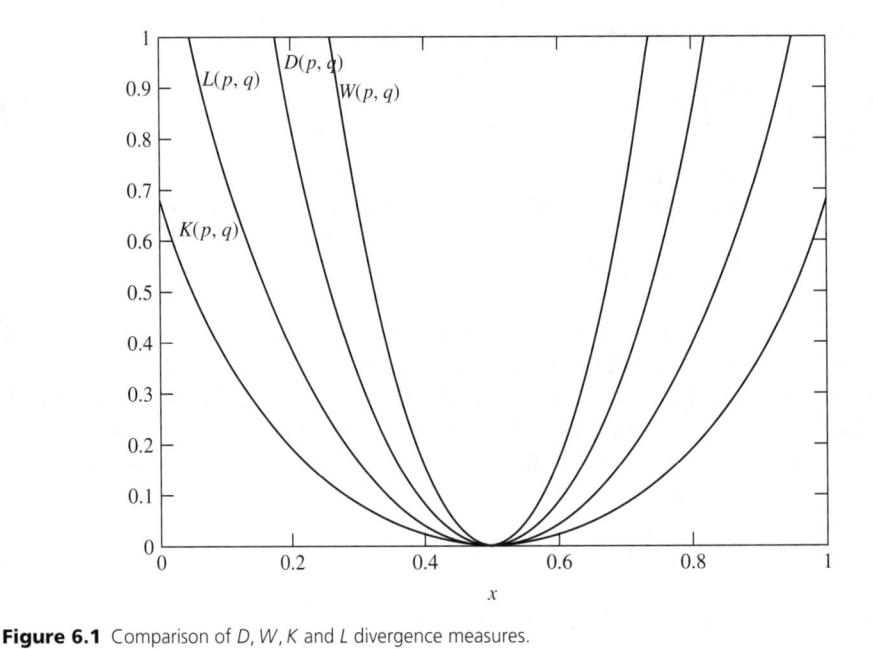

Figure 6.1 Comparison of D, W, K and L divergence measures.

The L divergence and the variational distance are related as

$$L(P, Q) \leq V(P, Q) \tag{6.67}$$

It may be noted that

$$H(w, 1 - w) \geq 2\min(w, 1 - w), 0 \leq w \leq 1 \tag{6.68}$$

and

$$1 - H(w, 1 - w) \leq |w - (1 - w)| \tag{6.69}$$

The bound for the K-divergence is also written as

$$K(P, Q) \leq V(P, Q) \tag{6.70}$$

This shows that the variational distance is an upper bound to both the K and L divergences. Both K and L are non-negative, finite, and semi-bounded:

$$K(P, Q) < +\infty, \quad K(P, Q) \geq K(P, P) \tag{6.71}$$
$$L(P, Q) < +\infty, \quad L(P, Q) \geq L(P, P) \tag{6.72}$$

for all probability distributions P and Q.

Another important property of the K and L divergences is their boundedness:

$$K(P, Q) \leq 1, \quad L(P, Q) \leq 2 \tag{6.73}$$

The L-divergence can be expressed as

$$L(P, Q) = 2H\left(\frac{P+Q}{2}\right) - H(P) - H(Q) \tag{6.74}$$

where H is the Shannon entropy.

The L-divergence can be generalized as the Jensen-Shannon divergence denoted as JS:

$$JS(P, Q) = H(w_1 P + w_2 Q) - w_1 H(P) - w_2 H(Q) \tag{6.75}$$

$JS(P, Q)$ is non-negative and is equal to 0 for $P = Q$. It also gives both upper and lower bounds to the Bayes probability of error. The advantage of JS is that one can assign weights to probability distributions, depending on their significance. Consider two classes $C = \{c_1, c_2\}$ with a priori probabilities $p(c_1) = w_1$ and $p(c_2) = w_2$. Let the corresponding conditional distributions be $P(X|c_1) = P(X)$, and $Q(X|c_2) = Q(X)$. The Bayes probability of error P_e is expressed (Hellman and Raviv, 1970) as

$$P_e(P, Q) = \sum \min[w_1 p(x), w_2 q(x)] \tag{6.76}$$

The upper bound for P_e can be written as

$$P_e(P, Q) \leq \frac{1}{2}[H(w_1, w_2) - JS(P, Q)] \tag{6.77}$$

where

$$H(w_1, w_2) = -w_1 \log w_1 - w_2 \log w_2 \tag{6.78}$$

The lower bound for P_e can be written as

$$P_e(P, Q) \geq \frac{1}{4}[H(w_1, w_2) - JS(P, Q)]^2 \tag{6.79}$$

It may be noted that

$$\frac{1}{2}H(s, 1-s) \leq \sqrt{s(1-s)}, \quad 0 \leq s \leq 1 \tag{6.80}$$

This is shown in Figure 6.2.

If $Q = \{q_i, i = 1, 2, \ldots, n\}, 0 \leq q_i \leq 1, 0 \leq i \leq 1$, and $\sum_{i=1}^{n} q_i = 1$, then

$$\frac{1}{2}H(Q) \leq \sum_{i=1}^{n-1} \sqrt{q_i(1 - q_i)} \tag{6.81}$$

Also,

$$JS(P, Q) \leq H(w_1, w_2) - 2P_e(P, Q) \leq H(w_1, w_2) \leq 1 \tag{6.82}$$

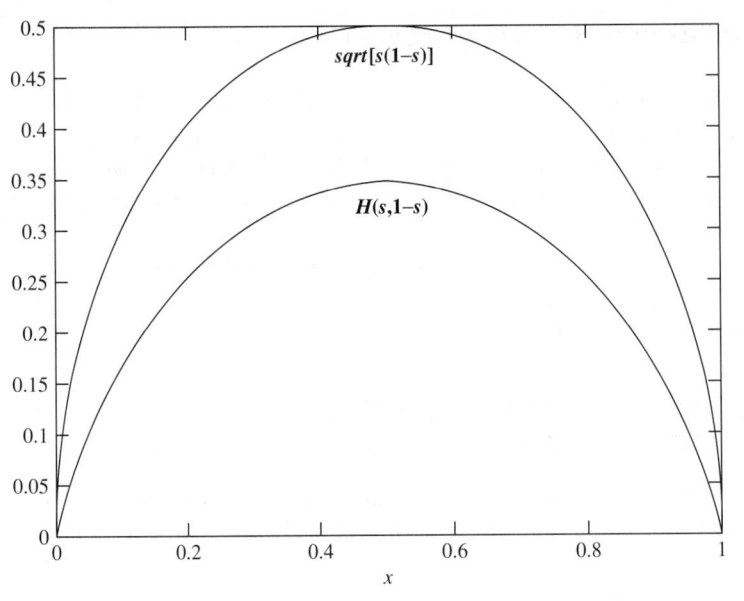

Figure 6.2 Comparison of $\frac{H(s,1-s)}{2}$ and $\sqrt{s(1-s)}$.

If there are n probability distributions, P_i, $i = 1, 2, \ldots, n$, with weights w_i, $i = 1, 2, \ldots, n$, then the generalized Jensen-Shannon divergence can be written as

$$JS(P_1, P_2, \ldots, P_n) = H\left(\sum_{i=1}^{n} w_i P_i\right) - \sum_{i=1}^{n} w_i H(P_i) \tag{6.83}$$

If there are n classes, $C = \{c_i, i = 1, 2, \ldots, n\}$, with prior probabilities w_i, $i = 1, 2, \ldots, n$, then the Bayes error for n classes can be expressed as

$$P(e) = \sum_{x \in X} p(x)\{1 - \max[p(w_1|x), p(w_2|x), \ldots, p(c_n|x)]\} \tag{6.84}$$

Now, the relationship between the generalized JS and the Bayes probability of error can be expressed as

$$P(e) \leq \frac{1}{2}[H(w) - JS(P_1, P_2, \ldots, P_n)] \tag{6.85}$$

where

$$H(w) = -\sum_{i=1}^{N} w_i \log w_i \tag{6.86}$$

and

$$P_i(X) = P(X|c_i), \quad i = 1, 2, 3, \ldots, n \tag{6.87}$$

The upper bound for $P(e)$ can be written as

$$P(e) \geq \frac{1}{4(n-1)}[H(w) - JS(P_1, P_2, \ldots, P_n)]^2 \tag{6.88}$$

6.9 Upper bounds for cross-entropy

Consider two distributions $P(x)$ and $Q(x)$ of random variable X. In statistics the distance between the two distributions is measured by the expectation of the logarithm of the likelihood ratio, whereas it is given by the relative or cross-entropy $D(p|q)$. In other words, $D(p|q)$. measures the inefficiency of the assumption that the distribution is Q when in fact the true distribution is P.

For the two distributions, Dragomir et al. (2000) have shown that

$$D(p,q) \leq \sum_{i=1}^{N} \frac{p^2(x_i)}{q(x_i)} - 1$$

$$= \frac{1}{2} \sum_{i=1}^{N} p(x_i)p(y_i) \left[\frac{p(x_i)}{q(x_i)} - \frac{p(y_i)}{q(y_i)} \right] \left[\frac{q(y_i)}{p(y_i)} - \frac{q(x_i)}{p(x_i)} \right]$$

with equality if and only if $p(x_i) = q(x_i)$ for all i's. For two random variables X and Y, two corollaries can be stated. The first is about the mutual information or transinformation:

$$I(X, Y) \leq \sum_{i=1}^{N} \frac{w^2(x_i, y_i)}{p(x_i)q(y_i)} - 1$$

The equality will hold only for X and Y being independent.

If $r(x) = p(x)/q(x)$, and $R = \max_{x \in X} r(x)$, and $r = \min_{x \in X} r(x)$, and the quotient $S = \dfrac{R}{r} \geq 1$, then

$$S \leq 1 + \varepsilon + \sqrt{\varepsilon(\varepsilon + 2)}$$

and

$$D(p,q) \leq \varepsilon$$

Likewise, let $M = \max\limits_{(x,y) \in X \times Y} \dfrac{g(x,y)}{p(x)q(y)}$, $m = \min\limits_{(x,y) \in X \times Y} \dfrac{g(x,y)}{p(x)q(y)}$ and $u = \dfrac{M}{m} \geq 1$. If $u \leq 1 + \varepsilon + \sqrt{\varepsilon(\varepsilon + 2)}$, $\varepsilon > 0$. Then one gets $I(X, Y) \leq \varepsilon$. Similarly, for two probability distributions which satisfy the condition:

$$0 < r \leq \frac{p(x)}{q(x)} \leq R$$

then the bound is

$$D(p,q) \leq \frac{(R-r)^2}{4rR}$$

The equality holds only if $p(x) = q(x)$. For mutual information,

$$0 \leq I(X, Y) \leq \frac{(M-m)^2}{4Mm}$$

Similarly, another upper bound is obtained using Diaz-Metcalf inequality:

$$D(p,q) \leq (1-r)(R-1) \leq \frac{1}{4}(R-r)^2$$

and for mutual information

$$0 \le I(X, Y) \le (1 - m)(M - 1) \le \frac{1}{4}(M - m)^2$$

Questions

Q.6.1 Obtain data on the number of rainy days (n) for a number of years (say, 30 or more years) for College Station in Texas. Use the data from Chapter 3. Using the mean number of rainy days as a constraint, determine the discrete distribution that n follows. Use the uniform distribution as a prior distribution. Fit this distribution to the histogram and discuss how well it fits. Compute the entropy of the distribution.

Q.6.2 Consider the number of rainy days as a continuous random variable. Then solve Q.6.1 for the continuous case.

Q.6.3 Using the mean number of rainy days as constraint and exponential distribution as a prior, determine the continuous distribution that n follows. Fit this distribution to the histogram and discuss how well it fits. Compute the cross-entropy of the distribution.

Q.6.4 Obtain the values of time interval between two successive rain events in College Station, Texas, for a number of years (say 30 or more years) and select the maximum value for each year. The maximum time interval between rainy days is considered here as a random variable. Using the mean time interval in days as a constraint and uniform distribution as prior, determine the discrete distribution that the time interval follows. Fit this distribution to the histogram and discuss how well it fits. Compute the cross-entropy of this distribution.

Q.6.5 Consider the time interval between two successive rain events and select the maximum value for each year for the data in Q.3.9. The maximum time interval is considered here as a continuous random variable. Using the mean time interval (maximum values) as a constraint, and exponential distribution as a prior, determine the continuous distribution that the time interval follows. Fit this distribution to the histogram and discuss how well it fits. Compute the cross-entropy of the distribution.

Q.6.6 Obtain the values of time interval between two successive rain events in College Station, Texas, for a number of years (say 30 or more years) and select the minimum value for each year. The minimum time interval between rainy days is considered here as a random variable. Using the mean time interval in days as a constraint and the uniform distribution as a prior, determine the discrete distribution that the time interval follows. Fit this distribution to the histogram and discuss how well it fits. Compute the cross-entropy of this distribution.

Q.6.7 Consider the time interval between two successive rain events and select the minimum value for each year. The minimum time interval is considered here as a continuous random variable. Using the mean time interval (minimum values) as a constraint, and the exponential distribution as a prior, determine the continuous distribution that the minimum time interval follows. Fit this distribution to the histogram and discuss how well it fits. Compute the cross-entropy of the distribution.

Q.6.8 Consider yearly rainfall for a number of years for College Station, Texas. Consider yearly rainfall as a discrete random variable. Using the mean yearly rainfall as a constraint and uniform distribution as prior, determine the discrete distribution that yearly rainfall follows. Fit this distribution to the histogram and discuss how well it fits. Compute the cross-entropy of the distribution.

Q.6.9 Consider yearly rainfall as a continuous random variable and obtain yearly rainfall values from a raing age in a watershed. Using the mean yearly rainfall as a constraint and exponential distribution as a prior, determine the continuous distribution that yearly rainfall follows. Fit this distribution to the histogram and discuss how well it fits. Compute the cross-entropy of the distribution.

Q.6.10 Obtain data on the number of days (n) having temperature above $36\,^{\circ}\text{C}$ ($100\,^{\circ}\text{F}$) for a number of years (say, 30 or more years) for College Station in Texas. Using the mean number of days as a constraint and uniform distribution as a prior, determine the discrete distribution that n follows. Fit this distribution to the histogram and discuss how well it fits. Compute the cross-entropy of the distribution.

Q.6.11 Consider the number of days (n) having temperature above $36\,^{\circ}\text{C}$ ($100\,^{\circ}\text{F}$) as a continuous random variable and obtain the needed data. Using the mean number of days as a constraint and exponential distribution as a prior, determine the continuous distribution that n follows. Fit this distribution to the histogram and discuss how well it fits. Compute the cross-entropy of the distribution.

Q.6.12 Obtain data on the number of days (n) having temperature equal to or below $0\,^{\circ}\text{C}$ ($32\,^{\circ}\text{F}$) for a number of years (say, 30 or more years) for College Station in Texas. Using the mean number of days as a constraint and uniform distribution as a prior, determine the discrete distribution that n follows. Fit this distribution to the histogram and discuss how well it fits. Compute the cross-entropy of the distribution.

Q.6.13 Consider the number of days (n) having temperature equal to or less than $0\,^{\circ}\text{C}$ ($36\,^{\circ}\text{F}$) as a continuous random variable and obtain the needed data. Using the mean number of days as a constraint and exponential distribution as a prior, determine the continuous distribution that n follows. Fit this distribution to the histogram and discuss how well it fits. Compute the cross-entropy of the distribution.

Q.6.14 Obtain the values of number of days without rainfall each year in College Station, Texas. The number of rainless days each year is considered here as a random variable. Plot a histogram of the number of rainless days and discuss what it looks like. Using the mean number of rainless days and uniform as a prior, determine the discrete distribution that the number of rainless days follows. Fit this distribution to the histogram and discuss how well it fits. Compute the cross-entropy of this distribution.

Q.6.15 Consider the number of rainless days as a continuous random variable and obtain the needed data. Using the mean number of rainless days as a constraint and exponential distribution as a prior, determine the continuous distribution that the number of rainless days follows. Fit this distribution to the histogram and discuss how well it fits. Compute the cross-entropy of the distribution.

References

Dragomir, S.S., Scholz, M.L. and Sunde, J. (2000). Some upper bounds for relative entropy and applications. *Computers and Mathematics with Applications*, Vol. 39, pp. 91–100.

Hellman, M.E. and Raviv, J., (1970). Probability of error, equivocation, and the Chernoff bound. *IEEE Transactions on Information Theory*, Vol. IT-16, No. 4, pp. 368–72.

Kullback, S., (1959). *Information Theory and Statistics*. John Wiley, New York.

Kullback, S. and Leibler, R.A., (1951). On information and sufficiency. *Annals of Mathematical Statistics*, Vol. 22, pp. 79–86.

Kapur, J.N. (1989). *Maximum Entropy Models in Science and Engineering*. Wiley Eastern, New Delhi, India.

Kapur, J.N. and Kesavan, H.K. (1987). *Generalized Maximum Entropy Principle (with Applications)*. Sandford Educational Press, University of Waterloo, Waterloo, Canada.

Kapur, J.N. and Kesavan, H.K. (1992). *Entropy Optimization Principles with Applications*. 408 p., Academic Press, New York.

Lin, J. (1991). Divergence measures based on the Shannon entropy. *IEEE Transactions on Information Theory*, Vol. 37, No. 1, pp. 145–51.

Toussaint, G.T. (1975). Sharper lower bounds for discrimination information in terms of variation. *IEEE Transactions on Information Theory*, Vol. IT-21, No. 1, pp. 99–100.

Vajda, I. (1970). Note on discrimination information and variation. *IEEE Transactions on Information Theory*, Vol. IT-16, pp. 771–6.

Additional Reading

Campbell, L.L. (1972). Characterization of entropy of probability distributions on the real line. *Information and Control*, Vol. 21, pp. 329–38.

Ephraim, Y., Lev-Ari, H. and Gray, R.M. (1988). Asymptotic minimum discrimination information measure for asymptotically weakly stationary processes. *IEEE Transactions on Information Theory*, Vol. 34, No. 5, pp. 1033–340.

Ephraim, Y., Dembo, A. and Rabiner, L.R. (1989). A minimum discrimination information approach for hidden Markov modeling. *IEEE Transactions on Information Theory*, Vol. 35, No. 5, pp. 1001–13.

Friedman, C., Huang, J. and Sandow, S. (2007). A utility approach to some information measures. *Entropy*, Vol. 9, pp. 1–26.

Hiai, F. and Petz, D. (1991). The proper formula for relative entropy and its asymptotics in quantum probability. *Communications in Mathematical Physics*, Vol. 143, pp. 99–114.

Rajski, C. (1961). A metric space of discrete probability distributions. *Information and Control*, Vol. 4, pp. 371–7.

Ziv, J. and Merhav, N. (1993). A measure of relative entropy between individual sequences with application to universal classification. *IEEE Transactions on Information Theory*, Vol. 39, No. 4, pp. 1270–9.

7 Derivation of POME-Based Distributions

In the preceding chapter on POMCE, it has been discussed that POMCE leads to a least-biased probability distribution corresponding to a given set of constraints and a prior distribution (Kapur, 1989; Kapur and Kesavan, 1992). In this chapter this discussion is extended further and a number of well-known distributions are derived. The constraints are usually expressed in terms of moments (or averages of some kind), although it is not a necessary condition. Examples of these constraints are $E[x], E[|x|], E[x^2], E[\ln x], E[\ln(1-x)]$, $E[\ln(1+x)], E[\{\ln(x)\}^2], E[\ln(1+x^2)]$, and so on. The prior distribution can be of any kind but is often specified as uniform, arithmetic, geometric, binomial, or normal. The prior distribution can be selected based on the hydrology of the variable under consideration. It must be noted that in the case of a continuous variable the limits of integration for entropy and specification of constraints must be compatible, or else POMCE would not lead to a probability distribution or POMCE-based distribution would not exist.

7.1 Discrete variable and mean E[x] as a constraint

Let a random variable X take on N values $x_1, x_2, x_3, \ldots, x_N$ with probabilities $p_1, p_2, p_3, \ldots, p_N$. such that

$$\sum_{i=1}^{N} p_i = 1 \tag{7.1}$$

Then the least biased POMCE-based distribution would depend on the specification of the prior distribution and the minimization of cross-entropy (Kullback and Leibler, 1951; Kullback, 1959):

$$D(P, Q) = \sum_{i=1}^{N} p_i \ln \frac{p_i}{q_i} \tag{7.2}$$

where $P = \{p_1, p_2, \ldots, p_N\}$ is the distribution to be determined, and $Q = \{q_1, q_2, \ldots, q_N\}$ is the prior distribution.

Entropy Theory and its Application in Environmental and Water Engineering, First Edition. Vijay P. Singh.
© 2013 John Wiley & Sons, Ltd. Published 2013 by John Wiley & Sons, Ltd.

7.1.1 Uniform prior distribution

The prior distribution Q is:

$$q_1 = q_2 = q_3, \ldots = q_N = \frac{1}{N} \tag{7.3}$$

The constraint is expressed as:

$$p_1 1 + p_2 2 + \ldots + p_N N = \sum_{i=1}^{N} i p_i = \bar{x} \tag{7.4}$$

From the previous chapter, the POMCE-based distribution can be written as

$$p_i = q_i \exp[-\lambda_0 - \lambda_i i] \tag{7.5a}$$

Equation (7.5a) can be recast as

$$p_i = \frac{1}{N} \exp[-\lambda_0 - \lambda_1 i] = \frac{\exp(-\lambda_0)}{N} \exp(-\lambda_1 i) = cd^i \tag{7.5b}$$

where $c = \exp(-\lambda_0)/N$, and $d = \exp(-\lambda_1)$.

Substitution of equation (7.5b) in equation (7.1) yields

$$\sum_{i=1}^{N} cd^1 = 1 \tag{7.6}$$

Inserting equation (7.5b) in equation (7.4), one obtains

$$\sum_{i=1}^{N} icd^i = \bar{x} \tag{7.7}$$

Parameter c can be eliminated by combining equations (7.6) and (7.7):

$$\frac{\sum_{i=1}^{N} id^i}{\sum_{i=1}^{N} d^i} = \bar{x} \tag{7.8}$$

Parameter d can be determined from equation (7.8). Equation (7.5b) shows that the POMCE-based distribution is a geometric distribution, that is, if the prior is a uniform distribution and the constraint is defined by the mean, then the posterior distribution is a geometric distribution.

Example 7.1: Determine the POMCE-based distribution of a random variable X expressing the number that occurs when a six-faced dice is thrown. Assume that the mean of the six-faced dice upon throw is given as 3.5, and the prior is given as a uniform distribution.

Solution: Let X take on values $x_1, x_2, x_3, \ldots, x_6$ with probabilities $p_1, p_2, p_3, \ldots, p_6$ such that

$$\sum_{i=1}^{6} p_i = 1 \tag{7.9}$$

Then the least biased POMCE-based distribution would depend on the specification of the prior distribution and minimization of cross-entropy given by equation (7.2). The prior distribution

Q is uniform, that is,

$$Q: q_1 = q_2 = q_3, \ldots = q_6 = \frac{1}{6} \tag{7.10}$$

The constraint is expressed as:

$$p_1.1 + p_2.2 + \ldots + p_6.6 = \sum_{i=1}^{6} ip_i = \bar{x} = 3.5 \tag{7.11}$$

The POMCE-based distribution can be written as

$$p_i = q_i \exp[-\lambda_0 - \lambda_1 i] \quad \text{or} \quad p_i = \frac{1}{6} \exp[-\lambda_0 - \lambda_1 i] = \frac{\exp(-\lambda_0)}{6} \exp(-\lambda_1 i) = cd^i \tag{7.12}$$

where $c = \exp(-\lambda_0)/6$, and $d = \exp(-\lambda_1)$. Substituting equation (7.12) in equation (7.9), one gets

$$\sum_{i=1}^{6} cd^i = 1 \tag{7.13}$$

Substituting equation (7.12) in equation (7.11), we get

$$\sum_{i=1}^{6} icd^i = \bar{x} = 3.5 \tag{7.14}$$

Dividing equation (7.14) by equation (7.13).

$$\frac{\displaystyle\sum_{i=1}^{6} id^i}{\displaystyle\sum_{i=1}^{6} d^i} = \bar{x} = 3.5 \tag{7.15}$$

Solving equation (7.15) for parameter d, we get $d = 1$. Inserting $d = 1$ in equation (7.13), we get $c = 1/6$. Hence, $c = \exp(-\lambda_0)/6$. This yields: $\lambda_0 = 0$; and $d = \exp(-\lambda_1)$, $\lambda_1 = 0$. Therefore, the POMCE-based distribution can be written as

$$p_i = \frac{1}{6} \exp[-\lambda_0 - \lambda_1 i] = \frac{1}{6} \exp(-0 - 0.i) = \frac{1}{6} \tag{7.16a}$$

or

$$P = \{1/6, 1/6, 1/6, 1/6, 1/6, 1/6\} \tag{7.16b}$$

The cross-entropy is:

$$D(P, Q) = \sum_{i=1}^{6} p_i \ln \frac{p_i}{q_i} = 6\left[\frac{1}{6}\ln\left(\frac{1/6}{1/6}\right)\right] = 0 \tag{7.17}$$

Thus, the posterior distribution P is the same as the prior distribution Q when Q has a uniform distribution. Hence, the Lagrangean multipliers λ_0, λ_1 as well as the cross-entropy are equal to zero. Therefore, plotting D as a function of Lagrange parameters or constraints is not meaningful.

7.1.2 Arithmetic prior distribution

The prior distribution is the arithmetic distribution expressed as:

$$Q = \{q_0, q_1, \ldots, q_N\} = \{Ka, K(a+b), K(a+2b), \ldots, K(a+Nb)\} = \sum_{i=0}^{N} K(a+ib) \tag{7.18}$$

where K, a, and b are constants. The constraint is given by equation (7.4). The POMCE-based least-biased probability distribution can be written as

$$p_i = q_i \exp[-\lambda_0 - \lambda_1 i] = K(a+bi) \exp[-\lambda_0 - \lambda_1 i] \tag{7.19}$$

Equation (7.19) can be recast as

$$p_i = k(a+ib)d^i \tag{7.20}$$

where

$$k = K \exp(-\lambda_0) \tag{7.21}$$

and

$$d = \exp(-\lambda_1) \tag{7.22}$$

The probability distribution given by equation (7.20) is the arithmetic-geometric distribution. Equation (7.20) contains two parameters k and d which can be determined with the aid of equations (7.1) and (7.4). Substitution of equation (7.20) in equations (7.1) and (7.4) yields, respectively,

$$k \sum_{i=0}^{N} (a+ib)d^i = 1 \tag{7.23}$$

and

$$k \sum_{i=0}^{N} i(a+ib)d^i = \bar{x} \tag{7.24}$$

Parameters k and d can be determined from equations (7.23) and (7.24).

Example 7.2: Determine the POMCE-based distribution of a random variable X expressing the number that occurs when a six-faced dice is thrown. The mean of the six-faced dice is given as 4.5 and a priori probability distribution Q is assumed to be an arithmetic distribution.

$$Q = \{q_1, \ldots, q_N\} = \{Ka, K(a+b), K(a+2b), \ldots, K(a+[N-1])b\} = \sum_{i=1}^{N} K(a+(i-1)b) \tag{7.25}$$

where a and b are parameters.

Solution: From equations (7.20), (7.23) and (7.24), one gets

$$p_i = k(a+(i-1)b)d^{i-1} \tag{7.26}$$

$$\sum_{i=1}^{N} p_i = k \sum_{i=1}^{N} [a + (i-1)b]d^{i-1} = 1 \tag{7.27}$$

$$p_1 1 + p_2 2 + \ldots + p_N N = \sum_{i=0}^{N} i p_i = k \sum_{i=1}^{N} ik[a + (i-1)b]d^{i-1} = \bar{x} \tag{7.28}$$

where $k = K \exp(-\lambda_0)$, K, a, and b are constants.

The POMCE-based least-biased probability distribution can be written as

$$p_i = q_i \exp[-\lambda_0 - \lambda_1 i] = K(a + bi) \exp[-\lambda_0 - \lambda_1 i] \tag{7.29}$$

The mean is given $= 4.5$ for number a six-faced dice and a priori probability distribution Q is an arithmetic distribution. Numerical iteration yields: $a = 0.025$, $b = 0.057$, $d = 2.270$, $K = 1$, $k = 0.035$, $\lambda_0 = 3.35$, and $\lambda_1 = -0.82$. This gives: $Q = (0.02, 0.08, 0.14, 0.20, 0.25, 0.31)$; $P = (0.001, 0.01, 0.03, 0.08, 0.23, 0.65)$. Changing the values of λ_0 and λ_1 one observes the changes in D (information gap):

Variation of λ_0 and λ_1 produces:

λ_0	λ_1	D
9.42	−2.78	3.11
7.56	−2.27	2.64
9.22	−2.75	3.09
4.87	−1.41	1.45
3.35	−0.82	0.34

7.1.3 Geometric prior distribution

The prior distribution is the geometric distribution expressed as

$$Q = \{q_0, q_1, \ldots, q_N\} = \{K, Ka, Ka^2, \ldots, Ka^N\}; q_i = Ka^i \tag{7.30}$$

The constraint is given by equation (7.4). Then the POMCE-based least-biased probability distribution can be written as

$$p_i = q_i \exp[-\lambda_0 - \lambda_1 i] = Ka^i \exp[-\lambda_0 - \lambda_1 i] \tag{7.31}$$

Equation (7.31) can be recast as

$$p_i = ka^i b^i \tag{7.32}$$

where

$$k = K \exp(-\lambda_0) \tag{7.33}$$

and

$$b = \exp(-\lambda_1) \tag{7.34}$$

Parameters b and k are determined by substituting equation (7.32) in equations (7.1) and (7.4), respectively,

$$k \sum_{i=0}^{N} a^i b^i = 1 \tag{7.35}$$

and

$$k \sum_{i=0}^{N} i a^i b^i = \bar{x} \tag{7.36}$$

Example 7.3: Solve Example 7.2 if the a priori probability distribution Q is assumed to be a geometric distribution.

$$Q = \{q_0, q_1, \ldots, q_N\} = \{K, Ka, Ka^2, \ldots, Ka^N\}, \text{ where } K = 0.1, a = 1 - K = 0.9, N = 7.$$

Solution: From equation (7.32), (7.35), and (7.36), one gets

$$p_i = ka^i b^i \tag{7.37}$$

$$k \sum_{i=0}^{N} a^i b^i = 1 \tag{7.38}$$

$$k \sum_{i=0}^{N} i a^i b^i = \bar{x} \tag{7.39}$$

Solving these equations one gets $b = 1.7081, k = 0.0279$. The distribution is

$$p_i = ka^i b^i = 0.0279 \times (1.5373)^i \tag{7.40}$$

7.1.4 Binomial prior distribution

The prior distribution is the binomial distribution expressed as

$$Q = \{q_0, q_1, q_2, \ldots, q_N\} = \{N_0 p^0 q^{N-0}, N_1 p^1 q^{N-1}, \ldots, N_N p^N q^0\} = N_i p^i q^{N-i}, N_i = \binom{N}{i} \tag{7.41}$$

where N_i is the i-th binomial coefficient, and $q = 1 - p$. The constraint is given by equation (7.4). Then the POMCE-based probability distribution is given by

$$p_i = N_i p^i q^{N-1} \exp(-\lambda_0 - \lambda_1 i) \tag{7.42}$$

Equation (7.42) can be recast as

$$p_i = a N_i p^i q^{N-1} b^i \tag{7.43}$$

where a and b are parameters and are defined as

$$a = \exp(-\lambda_0) \tag{7.44}$$

and

$$b = \exp(-\lambda_1) \tag{7.45}$$

Equation (7.43) is a binomial distribution. Parameters a and b in equations (7.43) and (7.44) can be determined, respectively, as

$$a \sum_{i=0}^{N} N_i p^i q^{N-i} b^i = 1 \tag{7.46}$$

and

$$a \sum_{i=0}^{N} i N_i p^i q^{N-i} b^i = \bar{x} \tag{7.47}$$

Equation (7.46) leads to

$$a(pb + q)^N = 1 \tag{7.48}$$

and equation (7.47) to

$$aNpb(pb + q)^N = \bar{x} = Np_* \tag{7.49}$$

for $p*$ such that

$$\frac{pb}{q} = \frac{m}{N-m} = \frac{p_*}{1 - p_*} \tag{7.50}$$

Substituting equation (7.50) in equation (7.48) one gets

$$aq^N \left(1 + \frac{m}{N-m}\right)^N = 1 \tag{7.51}$$

or

$$aq^N = q_*, \; q_* = 1 - p_* \tag{7.52}$$

Thus, POMCE-based distribution is:

$$p_i = N_i q_*^i \tag{7.53}$$

in which $p* = m/N$ is independent of the value of p in the prior distribution.

Example 7.4: Solve Example 7.2 if a priori probability distribution Q is assumed to a binomial distribution:

$$Q = \{q_0, q_1, q_2, \ldots, q_N\} = \{N_0 p^0 q^{N-0}, N_1 p^1 q^{N-1}, \ldots, N_N p^N q^0\} = N_i p^i q^{N-i}, N_i = \binom{N}{i} \tag{7.54}$$

where $N = 6, p = 1/6$.

Solution: From equation (7.53), one gets

$$p_i = N_i q_*^i = C_6^i (0.25)^i \tag{7.55}$$

in which C_6^i is the binomial coefficient, $p* = m/N = 4.5/6 = 0.75$ and $q_* = 1 - p_* = 0.25$.

7.1.5 General prior distribution

Consider a prior distribution $Q = \{q_1, q_2, ..., q_N\}$. One considers the case where there are N energy levels, $E_1, E_2, \ldots, E_N$, and the expected energy level is:

$$\sum_{i=1}^{N} E_i p_i = \overline{E} \tag{7.56}$$

Then the POMCE-based distribution is:

$$p_i = q_i \exp(-\lambda_0 - \lambda_1 E_i) \tag{7.57}$$

Substitution of equation (7.57) in equation (7.1) results in

$$\exp(\lambda_0) = \sum_{i=1}^{N} q_i \exp(-\lambda_1 E_1) \tag{7.58}$$

Inserting equation (7.58) in equation (7.57), one obtains

$$p_i = \frac{q_i \exp(-\lambda_1 E_i)}{\displaystyle\sum_{i=1}^{N} q_i \exp(-\lambda_1 E_i)} \tag{7.59}$$

Parameter λ_1 is determined by making use of equation (7.56).

$$\overline{E} = \frac{\displaystyle\sum_{i=1}^{N} E_i q_i \exp(-\lambda_1 E_i)}{\displaystyle\sum_{i=1}^{N} q_i \exp(-\lambda_1 E_i)} \tag{7.60}$$

Equation (7.59) looks like the Maxwell-Boltzmann distribution, but is, in general, different.
   Now let the a priori distribution be defined as

$$q_i = \frac{\exp(-\lambda_0 E_i)}{\displaystyle\sum_{i=1}^{N} \exp(-\lambda_0 E_i)} \tag{7.61}$$

The constant is still given by equation (7.56). Then the POMCE-based distribution becomes

$$p_i = q_i \exp(-\lambda_0 - \lambda_1 E_i) = \frac{\exp(-\lambda_0 E_i)}{\displaystyle\sum_{i=1}^{N} \exp(-\lambda_0 E_i)} \exp(-\lambda_0 - \lambda_1 E_i) \tag{7.62}$$

Inserting equation (7.62) in equation (7.1) one obtains

$$\frac{\displaystyle\sum_{i=1}^{N} \exp[-(\lambda_0 + \lambda_1)E_i]}{\displaystyle\sum_{i=1}^{N} \exp(-\lambda_0 E_i)} = \exp(\lambda_0) \tag{7.63}$$

Inserting equation (7.63) in equation (7.62), one gets

$$p_i = \frac{\exp[-(\lambda_0 + \lambda_1)E_i)]}{\sum_{i=1}^{N} \exp[-(\lambda_0 + \lambda_1)E_i]} \tag{7.64}$$

Also, inserting equation (7.64) in equation (7.56), one obtains

$$\overline{E} = \frac{\sum_{i=1}^{N} E_i \exp[-(\lambda_0 + \lambda_1)E_i)]}{\sum_{i=1}^{N} \exp[-(\lambda_0 + \lambda_1)E_i]} \tag{7.65}$$

Equation (7.64), in conjunction with equation (7.65), is the Maxwell-Boltzmann distribution. If $\lambda_0 = 0$ in equation (7.61), the resulting distribution is the uniform distribution.

7.2 Discrete variable taking on an infinite set of values

In this case, the variate takes on values 1, 2, 3, ... The constraint is given in terms of a mean value.

7.2.1 Improper prior probability distribution

The prior distribution is: $q_i = a/i^b$. The constraint is still the mean expressed by equation (7.4). Using POMCE, the least-biased probability distribution is obtained as

$$p_i = \frac{1}{i^b} \exp(-\lambda_0 - \lambda_1 i), i = 1, 2, 3, \ldots \tag{7.66}$$

Equation (7.66) can be cast as

$$p_i = \frac{c}{i^b d^i} \tag{7.67}$$

where

$$c = a \exp(-\lambda_0) \tag{7.68}$$
$$d = \exp(-\lambda_1) \tag{7.69}$$

Parameters c and d are obtained by substituting equation (7.67) in equations (7.1) and (7.4), which results in

$$c \sum_{i=1}^{N} \frac{d^i}{i^b} = 1 \tag{7.70}$$

and

$$c \sum_{i=1}^{N} \frac{i d^i}{i^b} = \overline{x} \tag{7.71}$$

In equation (7.66), b can be any value.

We now consider three special cases: $b = 0, 1,$ and -1. When $b = 0$, the a priori probability distribution becomes an improper uniform probability distribution: $p_i = cd^i$. Equations (7.70) and (7.71) become:

$$c \sum_{i=1}^{N} d^i = c \frac{d}{1-d} = 1 \tag{7.72}$$

and

$$c \sum_{i=1}^{N} i d^i = c \frac{d}{(1-d)^2} = \bar{x} \tag{7.73}$$

Equations (7.72) and (7.73) yield

$$\frac{1}{1-d} = \bar{x} \tag{7.74}$$

and

$$c = \frac{1}{\bar{x}-1} \tag{7.75}$$

Thus, equation (7.67), with $b = 0$, becomes

$$p_i = \frac{(\bar{x}-1)^{i-1}}{(\bar{x})^i} = \frac{1}{\bar{x}}\left(1 - \frac{1}{\bar{x}}\right)^{i-1}, i = 1, 2, \ldots \tag{7.76}$$

Equation (7.76) represents an infinite geometric progression with common ratio of $(\bar{x}-1)/\bar{x}$. Clearly, $\bar{x} > 1$ and the probability distribution is the infinite geometric distribution.

We now consider the case when $b = 1$. Then equation (7.67) becomes

$$p_i = \frac{c}{i} d^i \tag{7.77}$$

Therefore, equations (7.70) and (7.71) become

$$c \sum_{i=1}^{N} \frac{d^i}{i} = 1 \tag{7.78}$$

and

$$c \sum_{i=1}^{N} d^i = \bar{x} \tag{7.79}$$

Equations (7.78) and (7.79) produce

$$-c \ln(1 - d) = 1 \tag{7.80}$$

and

$$c \frac{d}{1-d} = \bar{x} \tag{7.81}$$

Equations (7.80) and (7.81) are used to determine parameters c and d. The POMCE-based probability distribution given by equation (7.77) becomes a log-series distribution.

Consider now $b = -1$. Then equation (7.67) becomes

$$p_i = cid^i \tag{7.82}$$

Equations (7.70) and (7.71) become

$$c \sum_{i=1}^{\infty} id^i = c \frac{d}{(1-d)^2} = 1 \tag{7.83}$$

and

$$c \sum_{i=1}^{\infty} i^2 d^i = c \frac{d+d^2}{(1-d)^3} = \bar{x} \tag{7.84}$$

Equations (7.83) and (7.84) can be solved for c and d:

$$d = \frac{\bar{x}-1}{\bar{x}+1} \tag{7.85}$$

and

$$c = \frac{4}{(\bar{x})^2 - 1} \tag{7.86}$$

Hence, equation (7.82) becomes

$$p_i = \frac{4i}{(\bar{x})^2 - 1} \left(\frac{\bar{x}-1}{\bar{x}-1} \right)^i \tag{7.87}$$

Note that the series

$$\sum_{i=1}^{\infty} \frac{d^i}{i^b} \tag{7.88}$$

always converges, as seen below:

$$\lim_{i \to \infty} \frac{d^{i+1}/(1+i)^b}{d^i/i^b} = d \lim_{i \to \infty} \left(\frac{i}{i+1} \right)^b = d \tag{7.89}$$

Example 7.5: Solve Example 7.2 but here random variable X which takes on the value of $1 \sim +\infty$ and a priori probability distribution Q is given by an improper distribution $q_i = a/i^b$.

Solution: When $b = 0$, one gets from equation (7.76).

$$p_i = \frac{(\bar{x}-1)^{i-1}}{(\bar{x})^i} = \frac{1}{\bar{x}} \left(1 - \frac{1}{\bar{x}} \right)^{i-1} = \frac{1}{4.5} \left(1 - \frac{1}{4.5} \right)^{i-1} = 0.2857 \times (0.7778)^i$$

When $b = 1$, solving equation (7.80).

$$-c \ln(1 - d) = 1$$

and equation (7.81)

$$c\frac{d}{1-d} = \bar{x}$$

one gets $d = 0.915$ and $c = 0.418$. From equation (7.77), $p_i = \frac{c}{i}d^i = \frac{0.418}{i}(0.915)^i$.
When $b = -1$, one gets from equation (7.87).

$$p_i = \frac{4i}{(\bar{x})^2 - 1}\left(\frac{\bar{x}-1}{\bar{x}+1}\right)^i = \frac{4i}{(4.5)^2 - 1}\left(\frac{4.5-1}{4.5+1}\right)^i = 0.2078 \times i \times (0.6364)^i$$

7.2.2 A priori Poisson probability distribution

The a priori probability distribution is given as

$$q_i = \frac{a^i}{i!}\exp(-a), i = 0, 1, 2, \ldots \tag{7.90}$$

The constraint is still given by equation (7.4). Then the POMCE-based distribution is obtained as

$$p_i = \frac{a^i}{i!}\exp(-a)\exp(-\lambda_0 - \lambda_1 i) \tag{7.91}$$

Equation (7.91) can be recast as

$$p_i = b\frac{a^i}{i!}\exp(-a)c^i \tag{7.92}$$

where

$$b = \exp(-\lambda_0), \quad c = \exp(-\lambda_1) \tag{7.93}$$

Parameters b and c can be determined by substituting equation (7.92) in equation (7.1) and (7.4) as

$$b\exp(-a)\sum_{i=1}^{\infty}\frac{a^i}{i!}c^i = 1 \tag{7.94}$$

and

$$b\exp(-a)\sum_{i=1}^{\infty}\frac{ia^i}{i!}c^i = \bar{x} \tag{7.95}$$

Equation (7.95) can be written as

$$b\exp(-a)ca\sum_{i=1}^{\infty}\frac{(ac)^{i-1}}{(i-1)!} = b\exp(-a)ca\sum_{i=1}^{\infty}\frac{(ac)^i}{i!} = \bar{x} \tag{7.96}$$

Equations (7.94) and (7.95) yield

$$b\exp(-a)\exp(ac) = 1 \tag{7.97}$$

and

$$b \exp(-a)ca \exp(ac) = \bar{x} \tag{7.98}$$

Solution of equations (7.97) and (7.98) gives

$$ac = \bar{x} \tag{7.99}$$

and

$$b = \exp(a - \bar{x}) \tag{7.100}$$

Inserting equations (7.99) and (7.100) in equation (7.92) yields

$$p_i = \exp(-\bar{x})\frac{(\bar{x})^i}{i!}, i = 0, 1, 2, \ldots \tag{7.101}$$

Equation (7.101) is a Poisson distribution with parameter given by $\bar{x}$. This shows that if the prior probability distribution is a Poisson distribution and the constraint is mean, then the POMCE-based distribution is also a Poisson distribution independent of the parameter of the a priori Poisson distribution.

Example 7.6: Solve Example 7.2 where the random variable X which takes on the value of $1 \sim +\infty$ with a mean of 4.5 and an a priori probability distribution Q is given by the Poisson distribution:

$$q_i = \frac{a^i}{i!} \exp(-a), i = 0, 1, 2, \ldots, \text{ where } a = 0.5. \tag{7.102}$$

Solution: From equation (7.101) one gets

$$p_i = \exp(-\bar{x})\frac{(\bar{x})^i}{i!} = \exp(-4.5)\frac{(4.5)^i}{i!} \tag{7.103}$$

The a priori probability distribution Q is given by the Poisson distribution:

$$q_i = \frac{a^i}{i!} \exp(-0.5), i = 0, 1, 2, \ldots \tag{7.104}$$

The constraints are given by

$$\sum_{i=0}^{\infty} i p_i = \bar{x}; \quad \sum_{i=1}^{\infty} p_i = 1 \tag{7.105}$$

Then the POMCE-based distribution is obtained as

$$p_i = q_i \exp[-\lambda_0 - \lambda_i i] = p_i = \frac{a^i}{i!} \exp(-a) \exp(-\lambda_0 - \lambda_1 i) \tag{7.106}$$

This equation can be recast as

$$p_i = b\frac{a^i}{i!} \exp(-a)c^i \tag{7.107}$$

where

$$b = \exp(-\lambda_0), c = \exp(-\lambda_1) \tag{7.108}$$

Parameters b and c can be determined by substituting these equations in the constraints, yielding

$$b\exp(-a) \sum_{i=1}^{\infty} \frac{a^i}{i!} c^i = 1 \tag{7.109}$$

and

$$b\exp(-a) \sum_{i=1}^{\infty} \frac{ia^i}{i!} c^i = \bar{x} \tag{7.110}$$

Simplification yields:

$$b\exp(-a)ca \sum_{i=1}^{\infty} \frac{(ac)^{i-1}}{(i-1)!} = b\exp(-a)ca \sum_{i=1}^{\infty} \frac{(ac)^i}{i!} = \bar{x} \tag{7.111}$$

$$b\exp(-a)\exp(ac) = 1 \tag{7.112}$$

$$b\exp(-a)ca\exp(ac) = \bar{x} \tag{7.113}$$

$$ac = \bar{x} \tag{7.114}$$

$$b = \exp(a - \bar{x}) \tag{7.115}$$

Therefore, the distribution is given by

$$p_i = \exp(-\bar{x}) \frac{(\bar{x})^i}{i!}, i = 0, 1, 2, \ldots \tag{7.116}$$

Substituting $\bar{x} = 4.5$

$$p_i = \exp(-\bar{x}) \frac{(\bar{x})^i}{i!} = \exp(-4.5) \frac{(4.5)^i}{i!}$$

and

$$b = e^{-4}; c = 9; \lambda_0 = 4; \lambda_1 = -\ln 9$$

Further,

$$D(P, Q) = \sum_{i=1}^{\infty} p_i \ln \frac{p_i}{q_i}$$

where

$$p_i = \exp(-\bar{x}) \frac{(\bar{x})^i}{i!} \tag{7.116}$$

$$\frac{p_i}{q_i} = \exp[-\lambda_0 - \lambda_i i] \tag{7.117}$$

Therefore,

$$D(P, Q) = e^{-\bar{x}} \sum_{i=1}^{\infty} \frac{\bar{x}^i}{i!} [-\lambda_0 - \lambda_1 i] \tag{7.118}$$

$$D(P, Q) = e^{-\bar{x}} \sum_{i=1}^{\infty} \frac{\bar{x}^i}{i!} [-\lambda_0] + e^{-\bar{x}} \sum_{i=1}^{\infty} \frac{\bar{x}^i}{i!} [-\lambda_1 i]$$

$$D(P, Q) = -\lambda_0 e^{-\bar{x}} \sum_{i=1}^{\infty} \frac{\bar{x}^i}{i!} - \lambda_1 e^{-\bar{x}} \sum_{i=1}^{\infty} \frac{\bar{x}^i}{i!} [i]$$

$$D(P, Q) = -\lambda_0 (e^{-2\bar{x}}) - \bar{x} \lambda_1 e^{-2\bar{x}}$$

$$D(P, Q) = -e^{-2\bar{x}} [\lambda_0 + \lambda_1 \bar{x}]$$

$$D(P, Q) = -e^{-9} [4 - 4.5 \ln 9] = 0.0013$$

7.2.3 A priori negative binomial distribution

The a priori distribution is a negative binomial distribution given as:

$$q_i = \frac{(N + i - 1)!}{(N - 1)! i!} a^N (1 - a)^i, i = 0, 1, 2, \dots \tag{7.119}$$

The constraint is given by equation (7.4). Then the POMCE-based distribution can be expressed as

$$p_i = \frac{(N + i - 1)!}{(N - 1)! i!} a^N (1 - a)^i \exp(-\lambda_0 - \lambda_1 i), i = 0, 1, 2, \dots \tag{7.120}$$

Equation (7.120) can be written as

$$p_i = b \frac{(N + i - 1)!}{(N - 1)! i!} a^N (1 - a)^i c^i \tag{7.121}$$

where

$$b = \exp(-\lambda_0) \tag{7.122}$$

and

$$c = \exp(-\lambda_1) \tag{7.123}$$

Parameters b and c can be determined by substituting equation (7.121) in equations (7.1) and (7.4) as

$$b \sum_{i=1}^{\infty} \frac{(N + i - 1)!}{(N - 1)! i!} a^N (1 - a)^i c^i = 1 \tag{7.124}$$

and

$$b \sum_{i=1}^{\infty} i \frac{(N+i-1)!}{(N-1)!i!} a^N (1-a)^i c^i = \bar{x} \tag{7.125}$$

Equation (7.124) can be simplified as

$$b a^N [1 - (1-a)c]^{-N} = 1 \tag{7.126}$$

and equation (7.125) as

$$b N a^N [1 - (1-a)c]^{-N-1} = \bar{x} \tag{7.127}$$

Equations (7.126) and (7.127) yield

$$b a^N = \left(\frac{N}{\bar{x}}\right)^N \tag{7.128}$$

and

$$(1-a)\bar{x} = \left(1 - \frac{N}{\bar{x}}\right) \tag{7.129}$$

Substituting equations (7.128) and (7.129) in equation (7.121), one gets

$$p_i = \frac{(N+i-1)!}{(N-1)!i!} \left(\frac{N}{\bar{x}}\right)^N \left(1 - \frac{N}{\bar{x}}\right)^i, i = 0, 1, 2, \ldots \tag{7.130}$$

Equation (7.130) is a negative binomial distribution with parameters N and $N/\bar{x}$, and is independent of parameter a of the a priori distribution.

Example 7.7: Solve Example 7.2 where the random variable X which takes on the value of $1^- + \infty$ with a mean of 4.5 and an a priori probability distribution q is given by a negative binomial distribution:

$$q_i = \frac{(N+i-1)!}{(N-1)!i!} a^N (1-a)^i, i = 0, 1, 2, \ldots$$

where $a = 0.1, N = 4$.

Solution: From equation (7.130) one gets

$$p_i = \frac{(N+i-1)!}{(N-1)!i!} \left(\frac{N}{\bar{x}}\right)^N \left(1 - \frac{N}{\bar{x}}\right)^i = \frac{(4+i-1)!}{(4-1)!i!} \left(\frac{4}{4.5}\right)^4 \left(1 - \frac{4}{4.5}\right)^i$$

7.3 Continuous variable: general formulation

In the previous chapter (Chapter 6) the principle of minimum cross-entropy (POMCE) for continuous variables has been formulated. That formulation is extended here. For a continuous random variable X in the range 0 and infinity, let $q(x)$ be the prior probability density function (PDF) and $p(x)$ be the posterior PDF. The objective is to determine $p(x)$ subject to specified constraints and given the prior PDF. To that end, the cross-entropy is expressed as

$$D(p, q) = D(x) = \int_0^\infty p(x) \ln \frac{p(x)}{q(x)} dx \tag{7.131}$$

where

$$\int_0^\infty q(x) dx = 1 \tag{7.132}$$

$$\int_0^\infty p(x) dx = 1 \tag{7.133}$$

In order to derive p(x) by applying POMCE, the following constraints can be specified:

$$\int_0^\infty g_r(x) p(x) dx = \overline{g_r} = C_r, \quad r = 1, 2, \ldots, m \tag{7.134}$$

The minimization of $D(p, q)$ subject to equations (7.133) and (7.134), can be achieved using the method of Lagrange multipliers where the Lagrangean function can be expressed as

$$L = \int_0^\infty p(x) \ln \frac{p(x)}{q(x)} dx + (\lambda_0 - 1) \left[\int_0^\infty p(x) dx - 1 \right] + \sum_{r=1}^m \lambda_r \left[\int_0^\infty g_r(x) p(x) dx - C_r \right] \tag{7.135}$$

Differentiating equation (7.135) with respect to $p(x)$ while recalling the calculus of variation and equating the derivative to zero, the following is obtained:

$$\ln \frac{p(x)}{q(x)} + 1 + (\lambda_0 - 1) + \sum_{r=1}^m \lambda_r g_r(x) = 0 \tag{7.136}$$

Equation (7.136) leads to the posterior PDF $p(x)$:

$$p(x) = q(x) \exp \left[-\lambda_0 - \sum_{r=1}^m \lambda_r g_r(x) \right] \tag{7.137}$$

Equation (7.136) is also conveniently written as

$$p(x) = \frac{1}{Z(\lambda_0)} q(x) \exp \left[-\sum_{r=1}^m \lambda_r g_r(x) \right] \tag{7.138}$$

where Z is called the partition function obtained by substituting equation (7.137) in equation (7.133).

$$Z = \exp(\lambda_0) = \int_0^\infty q(x) \exp \left[-\sum_{r=1}^m \lambda_r g_r(x) \right] dx \tag{7.139}$$

The Lagrange multipliers are determined using equation (7.134) with known values of constraints as

$$-\frac{1}{Z}\frac{\partial Z}{\partial \lambda_r} = -\frac{\partial \ln Z}{\partial \lambda_r} = \overline{g_r(x)} = C_r \tag{7.140}$$

Equation (7.140) does not lend itself to an analytical solution except for simple cases but numerical solution is not difficult. Equation (7.137) shows that specific forms of $p(x)$ depend on the specification of $q(x)$ and $g_r(x)$. Here two simple cases of the prior $q(x)$ are dealt with.

7.3.1 Uniform prior and mean constraint

It is assumed that the domain of X is bounded and therefore the assumption of a priori uniform density can be justified. Let

$$q(x) = \frac{1}{b-a} \tag{7.141}$$

where a and b are limits of X for q. The mean constraint is defined as

$$\int_0^\infty xp(x) = \overline{x} \tag{7.142}$$

The PDF $p(x)$ is determined, subject to equation (7.133) and (7.142) and given the prior as in equation (7.141). In this case equation (7.137) yields

$$p(x) = \frac{1}{b-a}\exp(-\lambda_0 - \lambda_1 x) \tag{7.143}$$

Substitution of equation (7.143) in equations (7.133) and (7.142) yields

$$\int_0^\infty \frac{1}{b-a}\exp(-\lambda_0 - \lambda_1 x)dx = 1 \tag{7.144}$$

$$\int_0^\infty \frac{1}{b-a}x\exp(-\lambda_0 - \lambda_1 x)dx = \overline{x} \tag{7.145}$$

Solution of equations yields

$$(b-a)\exp(\lambda_0) = \frac{1}{\lambda_1} \tag{7.146}$$

and

$$(b-a)\exp(\lambda_0)\overline{x} = \frac{1}{\lambda_1^2} \tag{7.147}$$

Equations (7.146) and (7.147) give

$$\frac{1}{\lambda_1} = \overline{x} \tag{7.148}$$

Substitution of equations (7.148) and (7.146) in equation (7.143) lead to

$$p(x) = \frac{1}{(b-a)\bar{x}} \exp(-x/\bar{x}) \tag{7.149}$$

which is of exponential type.

7.3.2 Exponential prior and mean and mean log constraints

Let the prior PDF be given as

$$q(x) = \frac{1}{k} \exp(-x/k) \tag{7.150}$$

The constraints are specified as equation (7.142) and

$$\int_0^\infty \ln x p(x) dx = \overline{\ln x} \tag{7.151}$$

Equation (7.137) in light of equation (7.150), (7.142), and (7.151) becomes

$$p(x) = \frac{1}{k} \exp(-x/k) \exp(-\lambda_0 - \lambda_1 x - \lambda_2 \ln x) \tag{7.152}$$

which can be expressed as

$$p(x) = \frac{x^{-\lambda_2}}{k} \exp(-x/k) \exp(-\lambda_0 - \lambda_1 x) \tag{7.153}$$

Substitution of equation (7.153) in equation (7.133), (7.142), and (7.151) yield, respectively,

$$\int_0^\infty \frac{x^{-\lambda_2}}{k} \exp(-x/k) \exp(-\lambda_0 - \lambda_1 x) dx = 1 \tag{7.154}$$

$$\int_0^\infty \frac{x^{1-\lambda_2}}{k} \exp(-x/k) \exp(-\lambda_0 - \lambda_1 x) dx = \bar{x} \tag{7.155}$$

$$\int_0^\infty \frac{x^{1-\lambda_2}}{k} \ln x \exp(-x/k) \exp(-\lambda_0 - \lambda_1 x) dx = \overline{\ln x} \tag{7.156}$$

Equations (7.154) to (7.156) can be solved numerically for the Lagrange multipliers.

Questions

Q.7.1 Show that the minimum cross-entropy $D(P, Q)$ is a convex function of both P and Q. P and Q are vectors here. [Hint: The Hessian matrix must be positive definite. In other words, all second partial derivatives of D with respect to P as well as Q must be positive definite].

Q.7.2 Determine the POMCF-based distribution of a random variable X expressing the number that occurs when a six-faced dice is thrown. Assume that the mean of the six-faced dice upon throw is given as 4.0, and the prior is given as a uniform distribution. Determine the value of D and plot D as a function of Lagrange multipliers as well as constraints.

Q.7.3 Determine the POMCF-based distribution of a random variable X expressing the number that occurs when a six-faced dice is thrown. Assume that the mean of the six-faced dice upon throw is given as 3.0, and the prior is a uniform distribution. Determine the value of D and plot D as a function of Lagrange multipliers as well as constraints. How does the POMCE-based distribution differ from that in Q.7.2?

Q.7.4 Solve Q.7.2 if the a priori distribution is arithmetic. How does the POMCE-based distribution differ from that in Q.7.2?

Q.7.5 Solve Q.7.3 if the priori distribution is arithmetic. How does the POMCE-based distribution differ from that in Q.7.3?

Q.7.6 Solve Q.7.2 if the a priori distribution is geometric. How does the POMCE-based distribution differ from that in Q.7.2?

Q.7.7 Solve Q.7.3 if the a priori distribution is geometric. How does the POMCE-based distribution differ from that in Q.7.3?

Q.7.8 Solve Q.7.2 if the a priori distribution is Poisson with the parameter a as 4.0. How does the POMCE-based distribution differ from that in Q.7.2?

Q.7.9 Solve Q.7.3 if the a priori distribution is Poisson with the parameter a as 3.0. How does the POMCE-based distribution differ from that in Q.7.2?

References

Kullback, S. (1959). *Information Theory and Statistics*. John Wiley, New York.

Kullback, S. and Leibler, R.A. (1951). On information and sufficiency. *Annals of Mathematical Statistics*, Vol. 22, pp. 79–87.

Kapur, J.N. (1989). *Maximum Entropy Models in Science and Engineering*. Wiley Eastern, New Delhi, India.

Kapur, J.N. and Kesavan, H.K. (1987). *Generalized Maximum Entropy Principle (with Applications)*. Sandford Educational Press, University of Waterloo, Waterloo, Canada.

Kapur, J.N. and Kesavan, H.K. (1992). *Entropy Optimization Principles with Applications*. 408 p., Academic Press, New York.

8   Parameter Estimation

Frequency distributions that satisfy the given information are often needed. The entropy theory is ideally suited to derive such distributions. Indeed POME and POMCE have been employed to derive a variety of distributions some of which have found wide applications in environmental and water engineering. Singh and Fiorentino (1992) and Singh (1998a) summarize many of these distributions. There are three entropy-based methods that have been employed for estimating parameters of frequency distributions in water and environmental engineering: 1) ordinary entropy method, 2) parameter space expansion method, and 3) numerical method. The objective of this chapter is to briefly discuss the first two methods and illustrate their application.

8.1 Ordinary entropy-based parameter estimation method

Recalling the definition of Shannon entropy:

$$H(P) = H(X) = -\sum_{i=1}^{N} p_i \ln p_i \tag{8.1a}$$

if X is a discrete variable and

$$H = -\int_{0}^{\infty} f(x) \ln f(x) dx \tag{8.1b}$$

if X is a continuous variable.

The general procedure for deriving an ordinary entropy-based parameter estimation method for a frequency distribution involves the following steps: 1) Define the given information in terms of constraints, 2) maximize the entropy subject to the given information, 3) construct the zeroth Lagrange multiplier, 4) relate the Lagrange multipliers to constraints, and 5) relate the parameters to the given information or constraints.

Entropy Theory and its Application in Environmental and Water Engineering, First Edition. Vijay P. Singh.
© 2013 John Wiley & Sons, Ltd. Published 2013 by John Wiley & Sons, Ltd.

8.1.1 Specification of constraints

For a continuous variable X, let the available information be given in terms of m linearly independent constraints $C_i(i = 1, 2, \ldots, m)$ in the form:

$$C_i = \int_a^b y_i(x)f(x)dx, \quad i = 1, 2, \ldots, m \tag{8.2a}$$

$$\int_a^b f(x)dx = 1 \tag{8.2b}$$

where $y_i(x)$ are some functions of X whose averages over the interval (a, b) specify the constraint values.

If X is a discrete variable, then the constraints can be defined as

$$\sum_{i=1}^N g_r(x_i)p_i = a_r \quad r = 1, 2, \ldots, m \tag{8.3a}$$

and

$$p_i \geq 0, \quad \sum_{i=1}^N p_i = 1 \tag{8.3b}$$

where a_r is the r-th constraint, g_r is the r-th function of X, N is the number of observations, and m is the number of constraints, $m + 1 \leq N$. When there are no constraints, then *POME* yields a uniform distribution. As more constraints are introduced, the distribution becomes more peaked and possibly skewed. In this way, the entropy reduces from a maximum for the uniform distribution to zero when the system is fully deterministic.

8.1.2 Derivation of entropy-based distribution

An increase in the number of constraints leads to less uncertainty about the information concerning the system. Considering the continuous random variable case, the maximum of H, subject to the conditions in equation (8.2a) and (8.2b), leads to:

$$f(x) = \exp\left[-\lambda_0 - \sum_{i=1}^m \lambda_i y_i(x)\right] \tag{8.4a}$$

where $\lambda_i, i = 0, 1, \ldots, m$, are the Lagrange multipliers. Equation (8.4a) is the entropy-based distribution. Similarly, for discrete X,

$$p_i = \exp[-\lambda_0 - \lambda_1\,g_1(x_i) - \lambda_2\,g_2(x_i) \cdots \cdots - \lambda_m\,g_m(x_i)], i = 1, 2, \ldots, N \tag{8.4b}$$

where $\lambda_i, i = 0, 1, \ldots, m$, are Lagrange multipliers.

8.1.3 Construction of zeroth Lagrange multiplier

For a continuous random variable X, POME specifies $f(x)$ by equation (8.4a). Inserting equation (8.4a) in equation (8.1b) yields

$$H[f] = \lambda_0 \int_a^b f(x)dx + \sum_{i=1}^m \lambda_i \int_a^b y_i(x)f(x)dx \tag{8.5a}$$

or

$$H[f] = \lambda_0 + \sum_{i=1}^{m} \lambda_i C_i \tag{8.5b}$$

where the Lagrange multipliers are determined using the information specified by equations (8.2a) and (8.2b). In addition, the zeroth Lagrange multiplier λ_0, also called the potential function, is obtained by inserting equation (8.4a) in equation (8.2b) as

$$\int_a^b \exp\left[-\lambda_0 - \sum_{i=1}^{m} \lambda_i y_i\right] dx = 1 \tag{8.6a}$$

resulting in

$$\lambda_0 = \ln \int_a^b \exp\left[-\sum_{i=1}^{m} \lambda_i y_i\right] dx \tag{8.6b}$$

For the discrete case, substitution of equation (8.4b) in equation (8.1a) yields

$$H = -\sum_{i=1}^{N} p_i \ln\left\{\exp\left[-\lambda_0 - \sum_{r=1}^{m} g(x_i)\lambda_r\right]\right\} \tag{8.7a}$$

or

$$H[p] = \lambda_0 + \sum_{r=1}^{m} \lambda_r a_r \tag{8.7b}$$

The zeroth Lagrange multiplier is obtained by inserting equation (8.4b) in equation (8.3b):

$$\lambda_0 = \ln\left\{\sum_{i=1}^{N} \exp\left[\sum_{r=1}^{m} \lambda_r g_r(x_i)\right]\right\} \tag{8.7c}$$

8.1.4 Determination of Lagrange multipliers

The Lagrange multipliers are related to the given information (or constraints) as

$$\frac{\partial \lambda_0}{\partial \lambda_i} = -C_i = E[y_i(x)], i = 1, 2, 3, \ldots\ldots, m \tag{8.8a}$$

$$\frac{\partial^2 \lambda_0}{\partial \lambda_i^2} = Var[y_i(x)] \tag{8.8b}$$

$$\frac{\partial^2 \lambda_0}{\partial \lambda_i \partial \lambda_j} = Cov[y_i(x), y_j(x)], i \neq j \tag{8.9}$$

$$\frac{\partial^3 \lambda_0}{\partial \lambda_i^3} = -\mu_3[y_i(x)] \tag{8.10}$$

where $E[.]$ is the expectation, $Var[.]$ is the variance, $Cov[.]$ is the covariance, and μ_3 is the third moment about the centroid, all for y_i.

For discrete X, similar relationships follow:

$$\frac{\partial \lambda_0}{\partial \lambda_i} = -a_i = E[g_i], \quad i = 1, 2, 3, \ldots \ldots, m \tag{8.11a}$$

$$\frac{\partial^2 \lambda_0}{\partial \lambda_i^2} = Var[g_i] \tag{8.11b}$$

$$\frac{\partial^2 \lambda_0}{\partial \lambda_i \partial \lambda_j} = Cov[g_i, g_j], i \neq j \tag{8.12a}$$

$$\frac{\partial^3 \lambda_0}{\partial \lambda_i^3} = -\mu_3[y_i(x)] \tag{8.12b}$$

The information available is in terms of constraints given by equation (8.3a). Then, the entropy-based distribution is given by equation (8.4b). Substitution of equation (8.4b) in equation (8.2b) yields

$$\exp(\lambda_0) = Z(\lambda_1, \lambda_2, \ldots, \lambda_N) = Z(\lambda) = \sum_{j=1}^{N} \exp\left[-\sum_{i=1}^{m} \lambda_i g_i(x_j)\right] \tag{8.13}$$

where $Z(\lambda)$ is called the partition function, and λ_0 is the zeroth Lagrange multiplier.

8.1.5 Determination of distribution parameters

With the Lagrange multipliers estimated from equations (8.7c) to (8.8a), the frequency distribution given by equation (8.4a) is uniquely defined. It is implied that the distribution parameters are uniquely related to the Lagrange multipliers. Clearly, this procedure states that a frequency distribution is uniquely defined by the specification of constraints and application of POME. This procedure is illustrated with a couple of examples.

Example 8.1: Derive the parameters of the Poisson distribution using the ordinary entropy method. The probability distribution function of the Poisson distribution is given as:

$$p_n = \frac{a^n e^{-a}}{n!}, n = 0, 1, 2, \ldots \tag{8.14}$$

where a is parameter and n is the n-th observation.

Solution: The Poisson distribution parameter is estimated as follows.

Specification of constraints Taking the natural logarithm of equation (8.14), one gets:

$$\ln(n!p_n) = -a + n \ln(a) \tag{8.15}$$

Applying the Shannon entropy formula for discrete case [i.e., equation (8.1a)], one gets:

$$H(P) = -\sum_{n=0}^{\infty} p_n \ln(n!p_n) = a \sum_{n=0}^{\infty} \frac{a^n e^{-a}}{n!} - \ln a \sum_{n=0}^{\infty} n \frac{a^n e^{-a}}{n!} \tag{8.16}$$

From equation (8.16), the constraints appropriate for equation (8.14) can be written as:

$$\sum_{n=0}^{\infty} p_n = 1 \tag{8.17a}$$

$$\sum_{n=0}^{\infty} n p_n = m, m : \text{sample mean} \tag{8.17b}$$

Construction of zeroth Lagrange multiplier The least-biased distribution p_n consistent with equations (8.17a) and (8.17b) and based on POME is:

$$p_n = \frac{1}{n!} \exp(-\lambda_0 - \lambda_1 n) \tag{8.18}$$

where λ_0, λ_1 are Lagrange multipliers. Substitution of equation (8.18) into equation (8.17a), one gets:

$$\sum_{n=0}^{\infty} \frac{1}{n!} \exp(-\lambda_0 - \lambda_1 n) = 1 \Rightarrow \exp(\lambda_0) = \sum_{n=0}^{\infty} \frac{\exp(-\lambda_1 n)}{n!} \tag{8.19a}$$

Recall the geometric series formula given as:

$$\sum_{n=0}^{\infty} \frac{x^n}{n!} = \exp(x) \tag{8.19b}$$

Equation (8.19a) can be simplified as:

$$\exp(\lambda_0) = \exp[\exp(-\lambda_1)] \tag{8.19c}$$

and the zeroth Lagrange multiplier λ_0 is given as:

$$\lambda_0 = \exp(-\lambda_1) \tag{8.19d}$$

Relation between Lagrange multiplier and constraints Differentiating equation (8.19a) with respect to λ_1 one gets:

$$\frac{\partial \lambda_0}{\partial \lambda_1} = -\frac{\sum_{n=0}^{\infty} n \dfrac{\exp(-\lambda_1 n)}{n!}}{\sum_{n=0}^{\infty} \dfrac{\exp(-\lambda_1 n)}{n!}} = -m \tag{8.20}$$

Similarly, differentiating equation (8.19d), one gets:

$$\frac{\partial \lambda_0}{\partial \lambda_1} = -\exp(-\lambda_1) \tag{8.21}$$

Equating equation (8.20) to equation (8.21), one gets:

$$\exp(-\lambda_1) = m \Rightarrow \lambda_1 = -\ln m \tag{8.22}$$

Substituting equation (8.22) back into equation (8.19d), one gets

$$\lambda_0 = m \tag{8.23}$$

Relation between Lagrange multipliers and parameters Substitution of equations (8.22) and (8.23) into equation (8.18), one gets:

$$p_n = \frac{1}{n!} \exp(-m + n \ln m) = \frac{e^{-m} m^n}{n!} \tag{8.24}$$

Comparison of equation (8.24) with equation (8.14), one gets:

$$a = m \tag{8.25}$$

Distribution entropy   According to equation (8.16), the Shannon entropy for the Poisson distribution can be rewritten as:

$$H(n) = a \sum_{n=0}^{\infty} p_n - \ln a \sum_{n=0}^{\infty} np_n + \sum_{n=0}^{\infty} \ln n! p_n$$

$$= m - m \ln m + e^{-m} \sum_{n=0}^{\infty} \frac{\ln(n!)m^n}{n!} \tag{8.26}$$

Example 8.2:   Derive the parameters of the geometric distribution using the ordinary entropy method.

Solution: The geometric distribution is given as:

$$p_n = (1 - p)^{n-1}p, n = 1, 2, \ldots \tag{8.27}$$

where p is the probability of success on the n-th try. The geometric distribution parameter is estimated as follows:

Specification of constraints   Taking the natural logarithm of equation (8.27), one gets:

$$\log p_n = (n - 1)\log(1 - p) + \log(p) = \log\left(\frac{p}{1 - p}\right) + n\log(1 - p) \tag{8.28}$$

Applying the Shannon entropy formula for discrete case [i.e., equation (8.1a)], one gets:

$$H(n) = -\sum_{n=1}^{\infty} p_n \log p_n = -\log\left(\frac{p}{1 - p}\right)\sum_{n=1}^{\infty} p_n - \log(1 - p)\sum_{n=1}^{\infty} np_n \tag{8.29}$$

From equation (8.29), the constraints appropriate for equation (8.27) can be written as:

$$\sum_{n=1}^{\infty} p_n = 1 \tag{8.30a}$$

$$\sum_{n=1}^{\infty} np_n = m, m : \text{sample mean} \tag{8.30b}$$

Construction of zeroth Lagrange multiplier   The least-biased probability distribution p_n consistent with equations (8.30a) and (8.30b) and based on POME is:

$$p_n = \exp(-\lambda_0 - \lambda_1 n) \tag{8.31}$$

where λ_0 and λ_1 are the Lagrange multipliers. Substitution equation (8.31) into equation (8.30a), one gets:

$$\sum_{n=1}^{\infty} \exp(-\lambda_0 - \lambda_1 n) = 1 \Rightarrow \exp(\lambda_0) = \sum_{n=1}^{\infty} \exp(-\lambda_1 n) \tag{8.32a}$$

Recall the geometric series formula given as:

$$\sum_{i=0}^{\infty} p^i = \frac{1}{1-p} \tag{8.32b}$$

Equation (8.32a) can be further simplified as:

$$\exp(\lambda_0) = \sum_{n=0}^{\infty} \exp(-\lambda_1 n) - 1 = \frac{\exp(-\lambda_1)}{1 - \exp(-\lambda_1)} \tag{8.32c}$$

and the zeroth Lagrange multiplier λ_0 is given as:

$$\lambda_0 = \ln \frac{\exp(-\lambda_1)}{1 - \exp(-\lambda_1)} \tag{8.32d}$$

Relation between Lagrange multiplier and constraints Differentiating equation (8.32c) with respect to λ_1, one gets:

$$\frac{\partial \lambda_0}{\partial \lambda_1} = -\frac{\displaystyle\sum_{n=1}^{\infty} n \exp(-\lambda_1 n)}{\displaystyle\sum_{n=1}^{\infty} \exp(-\lambda_1 n)} = -m \tag{8.33}$$

Differentiating equation (8.32d) with respect to λ_1, one gets:

$$\frac{\partial \lambda_0}{\partial \lambda_1} = -\frac{1}{1 - \exp(-\lambda_1)} \tag{8.34}$$

Equating equation (8.33) to equation (8.34), one gets:

$$\frac{1}{1 - \exp(-\lambda_1)} = m \Rightarrow \lambda_1 = -\ln \frac{m-1}{m} \tag{8.35}$$

Substituting equation (8.35) back into equation (8.32d), one gets:

$$\lambda_0 = \ln(m-1) \tag{8.36}$$

Relation between Lagrange multipliers and parameters Substitution of equations (8.35) and (8.36) into equation (8.31), one gets:

$$p_n = \exp\left[-\ln(m-1) + n\ln\left(\frac{m-1}{m}\right)\right] = \exp\left[-\ln \frac{(1-1/m)}{(1/m)} + n\ln\left(\frac{m-1}{m}\right)\right]$$

$$= \left(\frac{1}{m}\right)\left(1 - \frac{1}{m}\right)^{n-1} \tag{8.37}$$

Comparison of equation () with equation (8.27), one gets:

$$p = \frac{1}{m} \tag{8.38}$$

Distribution entropy Substitution equation (8.38) into equation (8.29), one gets:

$$H(n) = -\ln\left(\frac{1/m}{1-1/m}\right)\sum_{n=1}^{\infty}p_n - \ln\left(1-\frac{1}{m}\right)\sum_{n=1}^{\infty}np_n$$

$$= -\ln\left(\frac{1}{m-1}\right) - m\ln\left(\frac{m-1}{m}\right) = m\ln m - (m-1)\ln(m-1)$$

$$= \frac{-(1-p)\ln(1-p) - p\ln p}{p} \qquad (8.39)$$

Example 8.3: Derive the parameters of the normal distribution using the ordinary entropy method. The probability density function of the normal distribution is given as:

$$f(x) = \frac{1}{b\sqrt{2\pi}}\exp\left[-\frac{(x-a)^2}{2b^2}\right] \qquad (8.40)$$

where a and b are parameters.

Solution: The normal distribution parameters are estimated as follows.

Specification of constraints Taking the logarithm of equation (8.40) to the base e, one gets

$$\ln f(x) = -\ln\sqrt{2\pi} - \ln b - \frac{(x-a)^2}{2b^2} \quad or$$

$$\ln f(x) = -\ln\sqrt{2\pi} - \ln b - \frac{x^2}{2b^2} - \frac{a^2}{2b^2} + \frac{2ax}{2b^2} \qquad (8.41)$$

Multiplying equation (8.41) by $[-f(x)]$ and integrating between $-\infty$ to ∞, one gets

$$H(x) = -\int_{-\infty}^{\infty} f(x)\ln f(x)dx$$

$$= \left[\ln\sqrt{2\pi} + \ln b + \frac{a^2}{2b^2}\right]\int_{-\infty}^{\infty} f(x)dx + \frac{1}{2b^2}\int_{-\infty}^{\infty} x^2 f(x)dx - \frac{a}{b^2}\int_{-\infty}^{\infty} xf(x)dx \qquad (8.42)$$

From equation (8.42), the constraints appropriate for equation (8.40) can be written as

$$\int_{-\infty}^{\infty} f(x)dx = 1 \qquad (8.43a)$$

$$\int_{-\infty}^{\infty} xf(x)dx = E[x] = \bar{x} \qquad (8.43b)$$

$$\int_{-\infty}^{\infty} x^2 f(x)dx = E[x^2] = s_x^2 + \bar{x}^2 \qquad (8.43c)$$

where $\bar{x}$ is the sample mean, and s_x^2 is the sample variance of X.

Construction of zeroth Lagrange multiplier The least-biased probability density function $f(x)$ consistent with equations (8.43a) to (8.43c) and based on POME takes the form:

$$f(x) = \exp(-\lambda_0 - \lambda_1 x - \lambda_2 x^2) \tag{8.44}$$

where λ_0, λ_1, and λ_2 are the Lagrange multipliers. Substitution of equation (8.44) in the normality condition in equation (8.43a) gives

$$\int_{-\infty}^{\infty} f(x)dx = \int_{-\infty}^{\infty} \exp(-\lambda_0 - \lambda_1 x - \lambda_2 x^2)dx = 1 \tag{8.45}$$

Equation (8.45) can be simplified as

$$\exp(\lambda_0) = \int_{-\infty}^{\infty} \exp(-\lambda_1 x - \lambda_2 x^2)dx \tag{8.46}$$

Equation (8.46) defines the partition function. Making the argument of the exponential as a square in equation (8.46), one gets

$$\exp(\lambda_0) = \int_{-\infty}^{\infty} \exp\left(-\lambda_1 x - \lambda_2 x^2 + \frac{\lambda_1^2}{4\lambda_2} - \frac{\lambda_1^2}{4\lambda_2}\right) dx$$

$$= \exp\left(\frac{\lambda_1^2}{4\lambda_2}\right) \int_{-\infty}^{\infty} \exp-\left(x\sqrt{\lambda_2} + \frac{\lambda_1}{2\sqrt{\lambda_2}}\right)^2 dx \tag{8.47}$$

Let

$$t = x\sqrt{\lambda_2} + \frac{\lambda_1}{2\sqrt{\lambda_2}} \tag{8.48}$$

Then

$$\frac{dt}{dx} = \sqrt{\lambda_2} \tag{8.49}$$

Making use of equations (8.48) and (8.49) in equation (8.47), one gets

$$\exp(\lambda_0) = \frac{\exp\left(\frac{\lambda_1^2}{4\lambda_2}\right)}{\sqrt{\lambda_2}} \int_{-\infty}^{\infty} \exp\left(-t^2\right) dt = \frac{2\exp\left(\frac{\lambda_1^2}{4\lambda_2}\right)}{\sqrt{\lambda_2}} \int_{0}^{\infty} \exp(-t^2)dt \tag{8.50}$$

Consider the expression $\int_0^{\infty} \exp(-t^2)dt$. Let $k = t^2$. Then $[dk/dt] = 2t$ and $t = k^{0.5}$. Hence, this expression can be simplified by making substitution for t to yield

$$\int_0^{\infty} \exp(-t^2)dt = \int_0^{\infty} \exp(-k)\frac{dk}{2k^{0.5}} = \frac{1}{2}\int_0^{\infty} k^{-0.5} \exp(-k)dk$$

$$= \frac{1}{2}\int_0^{\infty} k^{[0.5-1]} \exp(-k)dk = \frac{\Gamma(0.5)}{2} = \frac{\sqrt{\pi}}{2} \tag{8.51}$$

Substituting equation (8.51) in equation (8.50), one gets

$$\exp(\lambda_0) = \frac{2 \exp\left(\dfrac{\lambda_1^2}{4\lambda_2}\right)}{\sqrt{\lambda_2}} \frac{\sqrt{\pi}}{2} = \exp\left(\frac{\lambda_1^2}{4\lambda_2}\right) \sqrt{\frac{\pi}{\lambda_2}}$$

(8.52)

Equation (8.52) is another definition of the partition function. The zeroth Lagrange multiplier λ_0 is given by taking the logarithm of equation (8.52) as

$$\lambda_0 = \frac{1}{2} \ln \pi - \frac{1}{2} \ln \lambda_2 + \frac{\lambda_1^2}{4\lambda_2}$$

(8.53)

One also obtains the zeroth Lagrange multiplier from equation (8.46) as

$$\lambda_0 = \ln \int\limits_{-\infty}^{\infty} \exp(-\lambda_1 x - \lambda_2 x^2) dx$$

(8.54)

Relation between Lagrange multipliers and constraints Differentiating equation (8.54) with respect to λ_1 and λ_2 respectively, one obtains

$$\frac{\partial \lambda_0}{\partial \lambda_1} = - \frac{\int\limits_{-\infty}^{\infty} x \exp(-\lambda_1 x - \lambda_2 x^2) dx}{\int\limits_{-\infty}^{\infty} \exp(-\lambda_1 x - \lambda_2 x^2) dx} = - \int\limits_{-\infty}^{\infty} x \exp(-\lambda_0 - \lambda_1 x - \lambda_2 x^2) dx$$

$$= - \int\limits_{-\infty}^{\infty} x f(x) dx = -\bar{x}$$

(8.55)

$$\frac{\partial \lambda_0}{\partial \lambda_2} = - \frac{\int\limits_{-\infty}^{\infty} x^2 \exp(-\lambda_1 x - \lambda_2 x^2) dx}{\int\limits_{-\infty}^{\infty} \exp(-\lambda_1 x - \lambda_2 x^2) dx} = - \int\limits_{-\infty}^{\infty} x^2 \exp(-\lambda_0 - \lambda_1 x - \lambda_2 x^2) dx$$

$$= - \int\limits_{-\infty}^{\infty} x^2 f(x) dx = -(s_x^2 + \bar{x}^2)$$

(8.56)

Differentiating equation (8.53) with respect to λ_1 and λ_2 respectively, one obtains

$$\frac{\partial \lambda_0}{\partial \lambda_1} = \frac{2}{4} \frac{\lambda_1}{\lambda_2} = \frac{\lambda_1}{2\lambda_2}$$

(8.57)

$$\frac{\partial \lambda_0}{\partial \lambda_2} = -\frac{1}{2\lambda_2} - \frac{\lambda_1^2}{4\lambda_2^2}$$

(8.58)

Equating equation (8.55) to equation (8.57) and equation (8.56) to equation (8.58), one gets

$$\frac{\lambda_1}{2\lambda_2} = -\bar{x}$$

(8.59)

$$\frac{1}{2\lambda_2} + \frac{1}{4}\left(\frac{\lambda_1}{\lambda_2}\right)^2 = s_x^2 + \bar{x}^2 \tag{8.60}$$

From equation (8.59), one gets

$$\lambda_1 = -2\lambda_2 \bar{x} \tag{8.61}$$

Substituting equation (8.61) in equation (8.60), one obtains

$$\frac{1}{2\lambda_2} + \frac{1}{4}\frac{4\lambda_2^2 \bar{x}}{\lambda_2^2} = s_x^2 + \bar{x}^2 \Rightarrow \frac{1}{2\lambda_2} = s_x^2 \Rightarrow \lambda_2 = \frac{1}{2s_x^2} \tag{8.62}$$

Eliminating λ_2 in equation (8.59) yields

$$\lambda_1 = -2\frac{1}{2s_x^2}\bar{x} = -\frac{\bar{x}}{s_x^2} \tag{8.63}$$

Relation between Lagrange multipliers and parameters Substitution of equation (8.53) in equation (8.44) yields

$$
\begin{aligned}
f(x) &= \left[-\frac{1}{2}\ln \pi + \frac{1}{2}\ln \lambda_2 - \frac{\lambda_1^2}{4\lambda_2} - \lambda_1 x - \lambda_2 x^2\right] \\
&= \exp\left[\ln(\pi)^{-0.5} + \ln(\lambda_2)^{0.5} - \frac{\lambda_1^2}{4\lambda_2} - \lambda_1 x - \lambda_2 x^2\right] \\
&= (\pi)^{-0.5}(\lambda_2)^{0.5}\exp\left[-\frac{\lambda_1^2}{4\lambda_2} - \lambda_1 x - \lambda_2 x^2\right]
\end{aligned}
\tag{8.64}
$$

Comparison of equation (8.64) with equation (8.40) shows that

$$\lambda_1 = -a/b^2 \tag{8.65}$$

$$\lambda_2 = 1/(2b^2) \tag{8.66}$$

Relation between parameters and constraints The normal distribution has two parameters a and b which are related to the Lagrange multipliers by equations (8.65) and (8.66), which themselves are related to the constraints through equations (8.62) and (8.63) [and in turn through equations (8.43b) and (8.43c)]. Eliminating the Lagrange multipliers between these two sets of equations, we obtain

$$a = \bar{x} \tag{8.67}$$

$$b = s_x \tag{8.68}$$

Distribution entropy Substitution of equations (8.67) and (8.68) in equation (8.42) yields

$$
\begin{aligned}
H(x) &= \left[\ln\sqrt{2\pi} + \ln s_x + \frac{\bar{x}^2}{2s_x^2}\right]\int_{-\infty}^{\infty} f(x)dx + \frac{1}{2s_x^2}\int_{-\infty}^{\infty} x^2 f(x)dx - \frac{\bar{x}}{s_x^2}\int_{-\infty}^{\infty} xf(x)dx \\
&= \left(\ln\sqrt{2\pi} + \ln s_x + \frac{\bar{x}^2}{2s_x^2}\right) + \frac{1}{2s_x^2}(\bar{x}^2 + s_x^2) - \frac{\bar{x}^2}{s_x^2} \\
&= \ln[s_x(2\pi e)^{0.5}]
\end{aligned}
\tag{8.69}
$$

Example 8.4: The gamma distribution is commonly employed for synthesis of instantaneous or finite-period unit hydrographs. If X has a gamma distribution then its probability density function (PDF) is given by

$$f(x) = \frac{1}{a\Gamma(b)} \left(\frac{x}{a}\right)^{b-1} e^{-x/a} \tag{8.70a}$$

where $a > 0$ and $b > 0$ are parameters. The gamma distribution is a two-parameter distribution. Its cumulative distribution function (CDF) can be expressed as:

$$f(x) = \int_0^\infty \frac{1}{a\Gamma(b)} \left(\frac{x}{a}\right)^{b-1} e^{-x/a} dx \tag{8.70b}$$

If $y = x/a$ then equation (8.70b) can be written as

$$F(y) = \frac{1}{\Gamma(b)} \int_0^y y^{b-1} \exp(-y) dy \tag{8.70c}$$

Abramowitz and Stegun (1958) expressed $F(y)$ as

$$F(y) = F(\chi^2 | v) \tag{8.71a}$$

where $F(\chi^2 | v)$ is the chi-square distribution with $v = 2b$ degrees of freedom and $\chi^2 = 2y$. According to Kendall and Stuart (1963), for v greater than 30, the following variable follows a normal distribution with zero mean and variance equal to one:

$$u = \left[\left(\frac{\chi^2}{v}\right)^{1/3} + \frac{2}{9v} - 1 \right] \left(\frac{9v}{2}\right)^{1/2} \tag{8.71b}$$

This helps compute $f(x)$ for a given x by first computing $y = x/a$ and $\chi^2 = 2y$ and then inserting these values into equation (8.71b) to obtain u. Given a value of u, $F(x)$ can be obtained from the use of normal distribution tables. Derive parameters of gamma distribution using the ordinary entropy method and express its entropy.

It may be useful to recall the definition of the gamma function as well as some of its properties in this chapter. The gamma function is defined as

$$\Gamma(b) = \int_0^\infty x^{b-1} \exp(-x) dx \tag{8.72}$$

which is convergent for $b > 0$. A recursive formula for the gamma function is:

$$\Gamma(b+1) = b\Gamma(b), \quad \Gamma(b=1) = 1, \Gamma\left(\frac{1}{2}\right) = \sqrt{\pi} \tag{8.73a}$$

If b is a positive integer then

$$\Gamma(b+1) = b!, \quad b = 1, 2, 3, \ldots \tag{8.73b}$$

This explains the reason that $\Gamma(b)$ is sometimes called the factorial function. If b is a large integer value then the value of the gamma function can be approximated using Stirling's factorial approximation or asymptotic formula for $b!$ as

$$b! \sim \sqrt{2\pi b}\, b^b \exp(-b) \tag{8.73c}$$

Also, Euler's constant γ is obtained from the gamma function as:

$$\Gamma'(1) = \int_0^\infty \exp(-x) \ln x\, dx = -\gamma \tag{8.74a}$$

where γ is defined as

$$\lim_{M \to \infty} \left(1 + \frac{1}{2} + \frac{1}{3} + \ldots + \frac{1}{M} - \ln M \right) = 0.577215 \tag{8.74b}$$

Solution: The gamma distribution parameters are estimated as follows.

Specification of constraints Taking the logarithm of equation (8.70a) to the base e, one gets

$$
\begin{aligned}
\ln f(x) &= -\ln a\Gamma(b) + (b-1)\ln x - (b-1)\ln a - \frac{x}{a} \\
&= -[\ln a\Gamma(b) + (b-1)\ln a] + (b-1)\ln x - \frac{x}{a}
\end{aligned} \tag{8.75}
$$

Multiplying equation (8.75) by $[-f(x)]$ and integrating between 0 and ∞, one obtains the function:

$$
\begin{aligned}
H(f) &= -\int_0^\infty f(x)\ln f(x)\,dx = [\ln a\Gamma(b) + (b-1)\ln a] \int_0^\infty f(x)\,dx \\
&\quad -(b-1)\int_0^\infty [\ln x]f(x)\,dx + \frac{1}{a}\int_0^\infty xf(x)\,dx
\end{aligned} \tag{8.76}
$$

From equation (8.76) the constraints appropriate for equation (8.70a) can be written (Singh et al., 1985, 1986) as

$$\int_0^\infty f(x) = 1 \tag{8.77}$$

$$\int_0^\infty xf(x)\,dx = \overline{x} \tag{8.78}$$

$$\int_0^\infty [\ln x]f(x)\,dx = E[\ln x] \tag{8.79}$$

Construction of zeroth Lagrange multiplier The least-biased PDF, based on the principle of maximum entropy (POME) and consistent with equations (8.77) to (8.79), takes the form:

$$f(x) = \exp[-\lambda_0 - \lambda_1 x - \lambda_2 \ln x] \tag{8.80}$$

where λ_0, λ_1, and λ_2 are the Lagrange multipliers. Substitution of equation (8.80) in equation (8.77) yields

$$\int_0^\infty f(x)dx = \int_0^\infty \exp[-\lambda_0 - \lambda_1 x - \lambda_2 \ln x]dx = 1 \tag{8.81}$$

This leads to the partition function as

$$\exp(\lambda_0) = \int_0^\infty \exp[-\lambda_1 x - \lambda_2 \ln x]dx = \int_0^\infty \exp[-\lambda_1 x]\exp[-\lambda_2 \ln x]dx$$

$$= \int_0^\infty \exp[-\lambda_1 x]\exp[\ln x^{-\lambda_2}]dx \tag{8.82}$$

Let $\lambda_1 x = y$. Then $dx = dy/\lambda_1$. Therefore, equation (8.82) becomes

$$\exp(\lambda_0) = \int_0^\infty \left(\frac{y}{\lambda_1}\right)^{-\lambda_2} \exp(-y)\frac{dy}{\lambda_1} = \frac{1}{\lambda_1^{1-\lambda_2}}\int_0^\infty y^{-\lambda_2}e^{-y}dy = \frac{1}{\lambda_1^{1-\lambda_2}}\Gamma(1-\lambda_2) \tag{8.83}$$

Thus, the zeroth Lagrange multiplier λ_0 is given by equation (8.83) as

$$\lambda_0 = (\lambda_2 - 1)\ln \lambda_1 + \ln \Gamma(1 - \lambda_2) \tag{8.84}$$

The zeroth Lagrange multiplier is also obtained from equation (8.82) as

$$\lambda_0 = \ln \int_0^\infty \exp[-\lambda_1 x - \lambda_2 \ln x]dx \tag{8.85}$$

Relation between Lagrange multipliers and constraints Differentiating equation (8.85) with respect to λ_1 and λ_2, respectively, produces

$$\frac{\partial \lambda_0}{\partial \lambda_1} = -\frac{\int_0^\infty x\exp[-\lambda_1 x - \lambda_2 \ln x]dx}{\int_0^\infty \exp[-\lambda_1 x - \lambda_2 \ln x]dx}$$

$$= -\int_0^\infty x\exp[-\lambda_0 - \lambda_1 x - \lambda_2 \ln x]dx = -\int_0^\infty xf(x)dx = -\overline{x} \tag{8.86}$$

$$\frac{\partial \lambda_0}{\partial \lambda_2} = -\frac{\int_0^\infty \ln x \exp[-\lambda_1 x - \lambda_2 \ln x] dx}{\int_0^\infty \exp[-\lambda_1 x - \lambda_2 \ln x] dx}$$

$$= -\int_0^\infty \ln x \exp[-\lambda_0 - \lambda_1 x - \lambda_2 \ln x] dx = -\int_0^\infty \ln x f(x) dx = -E[\ln x] \tag{8.87}$$

Also, differentiating equation (8.84) with respect to λ_1 and λ_2, respectively, gives

$$\frac{\partial \lambda_0}{\partial \lambda_1} = \frac{\lambda_2 - 1}{\lambda_1} \tag{8.88}$$

$$\frac{\partial \lambda_0}{\partial \lambda_2} = \ln \lambda_1 + \frac{\partial}{\partial \lambda_2} \ln \Gamma(1 - \lambda_2) \tag{8.89}$$

Let $1 - \lambda_2 = k$. Then

$$\frac{\partial k}{\partial \lambda_2} = -1 \tag{8.90}$$

and equation (8.89) can be written as

$$\frac{\partial \lambda_0}{\partial \lambda_2} = \ln \lambda_1 + \frac{\partial}{\partial k} \ln \Gamma(k) \frac{\partial k}{\partial \lambda_2} = \ln \lambda_1 - \psi(k) \tag{8.91a}$$

where $\psi(k)$ is the psi(ψ)-function defined as the derivative of the log-gamma function:

$$\psi(x) = \frac{\partial \ln \Gamma(x)}{\partial x} \tag{8.91b}$$

From equations (8.86) and (8.88) as well as equations (8.87) and (8.89) and (8.91a), one gets

$$\frac{\lambda_2 - 1}{\lambda_1} = -\bar{x}; \bar{x} = \frac{k}{\lambda_1} \tag{8.92}$$

$$\psi(k) - E[\ln x] = \ln \lambda_1 \tag{8.93}$$

From equation (8.92), $\lambda_1 = k/\bar{x}$, and substituting λ_1 in equation (8.93), one gets

$$E[\ln x] - \ln \bar{x} = \psi(k) - \ln k \tag{8.94}$$

One can find the value of 'k' $(= 1 - \lambda_2)$ from equation (8.94) and substitute it in equation (8.92) to get λ_1.

Relation between Lagrange multipliers and parameters Substituting equation (8.83) in equation (8.80) gives the entropy-based PDF as

$$
f(x) = \exp\left[(1 - \lambda_2)\ln\lambda_1 - \ln\Gamma(1 - \lambda_2) - \lambda_1 x - \lambda_2 \ln x\right]\exp\left[\ln\lambda_1^{1-\lambda_2}\right]
$$
$$
\times \exp\left[\ln(\frac{1}{\Gamma(1 - \lambda_2)})\right]\exp\left[-\lambda_1 x\right]\exp\left[\ln x^{-\lambda_2}\right]
$$
$$
= \lambda_1^{1-\lambda_2}\frac{1}{\Gamma(1 - \lambda_2)}\exp[-\lambda_1 x]x^{-\lambda_2} \tag{8.95}
$$

If $\lambda_2 = 1 - k$ then

$$
f(x) = \frac{\lambda_1^k}{\Gamma(k)}\exp[-\lambda_1 x]x^{k-1} \tag{8.96}
$$

Comparison of equation (8.96) with equation (8.70a) produces

$$
\lambda_1 = \frac{1}{a} \tag{8.97}
$$

and

$$
\lambda_2 = 1 - b \tag{8.98}
$$

Relation between parameters and constraints The gamma distribution has two parameters a and b which are related to the Lagrange multipliers by equations (8.97) and (8.98), which themselves are related to the known constraints by equations (8.92) and (8.93). Eliminating the Lagrange multipliers between these two sets of equations, we get parameters directly in terms of the constraints as

$$
ba = \bar{x} \tag{8.99}
$$

$$
\psi(b) - \ln b = E[\ln x] - \ln\bar{x} \tag{8.100}
$$

Distribution entropy Equation (8.76) gives the distribution entropy. Rewriting it, one gets

$$
H(x) = -\int_0^\infty f(x)\ln f(x)dx
$$
$$
= [\ln a\Gamma(b) + (b - 1)\ln a]\int_0^\infty f(x)dx - (b - 1)\int_0^\infty \ln x f(x)dx + \frac{1}{a}\int_0^\infty x f(x)dx
$$
$$
= [\ln a\Gamma(b) + \ln a^{b-1}] - (b - 1)E[\ln x] + \frac{\bar{x}}{a}
$$
$$
= \ln\{a\Gamma(b)a^{b-1}\} + \frac{\bar{x}}{a} - (b - 1)E[\ln x]
$$
$$
= \ln\{\Gamma(b)a^b\} + \frac{\bar{x}}{a} - (b - 1)E[\ln x] \tag{8.101}
$$

8.2 Parameter-space expansion method

This method, developed by Singh and Rajagopal (1986), employs an enlarged parameter space and maximizes entropy subject to both the parameters and the Lagrange multipliers.

An important implication of the enlarged parameter space is that the method is applicable to virtually any distribution, expressed in direct form, having any number of parameters. For a continuous random variable X having a probability density function $f(x, \theta)$ with parameters θ, the entropy can be expressed as

$$H(X) = H(f) = \int_{-\infty}^{\infty} f(x, \theta) \ln f(x, \theta) dx \tag{8.102}$$

Parameters of this distribution, θ, can be estimated by achieving the maximum of $H(f)$. To apply the method, the constraints are defined first. Next, the POME formulation of the distribution is obtained in terms of the parameters using the method of Lagrange multipliers. This formulation is used to define H whose maximum is sought. If the probability distribution has n parameters, $\theta_i, i = 1, 2, \ldots n$, and the $(n + 1)$ Lagrange multipliers are $\lambda_i, i = 1, 2, \ldots (n + 1)$, then the point where $H[f]$ is a maximum is a solution of $(2n + 1)$ equations:

$$\frac{\partial H[f]}{\partial \lambda_i} = 0 \quad i = 0, 1, 2, \ldots, n \tag{8.103}$$

and

$$\frac{\partial H[f]}{\partial \theta_i} = 0 \quad i = 1, 2, \ldots, n \tag{8.104}$$

Solution of equations (8.103) and (8.104) yields the estimates of distribution parameters. The method is illustrated using examples.

Example 8.5: Determine parameters of the normal distribution using the parameter expansion method.

Solution: Following Singh and Rajagopal (1986), the constraints for this method are given by equation (8.43a) and

$$\int_{-\infty}^{\infty} \left(\frac{xa}{b^2}\right) f(x) dx = E\left(\frac{xa}{b^2}\right) = \frac{a}{b^2} E(x) = \frac{\overline{ax}}{b^2} \tag{8.105}$$

$$\int_{-\infty}^{\infty} \left(\frac{x^2}{2b^2}\right) f(x) dx = E\left(\frac{x^2}{2b^2}\right) = \frac{s_x^2 + \overline{x}^2}{2b^2} \tag{8.106}$$

Derivation of entropy function The PDF corresponding to POME and consistent with equations (8.43a), (8.105), and (8.106) takes the form:

$$f(x) = \exp\left[-\lambda_0 - \lambda_1 \frac{xa}{b^2} - \lambda_2 \frac{x^2}{2b^2}\right] \tag{8.107}$$

where $\lambda_0, \lambda_1, \lambda_2$ are the Lagrange multipliers. Insertion of equation (8.107) into equation (8.43a) yields

$$\exp(\lambda_0) = \int_{-\infty}^{\infty} \exp\left(-\lambda_1 \frac{xa}{b^2} - \lambda_2 \frac{x^2}{2b^2}\right) dx = \frac{b\sqrt{2\pi}}{\sqrt{\lambda_2}} \exp\left(\frac{a^2 \lambda_1^2}{2\lambda_2 b^2}\right) \tag{8.108}$$

Equation (8.108) is the partition function. Taking the logarithm of equation (8.108) leads to the zeroth Lagrange multiplier which can be expressed as

$$\lambda_0 = \ln b + 0.5 \ln(2\pi) - 0.5 \ln \lambda_2 + \frac{a^2 \lambda_1^2}{2\lambda_2 b^2} \tag{8.109}$$

The zeroth Lagrange multiplier is also obtained from equation (8.108) as

$$\lambda_0 = \ln \int_{-\infty}^{\infty} \exp\left[-\lambda_1 \frac{xa}{b^2} - \lambda_2 \left(\frac{x^2}{2b^2}\right)\right] dx \tag{8.110}$$

Introduction of equation (8.108) in equation (8.107) gives

$$f(x) = \frac{\sqrt{\lambda_2}}{b\sqrt{2\pi}} \exp\left[-\left(\frac{a^2 \lambda_1^2}{2\lambda_2 b^2} + \frac{\lambda_1 xa}{b^2} + \frac{\lambda_2 x^2}{2b^2}\right)\right] \tag{8.111}$$

Comparison of equation (8.111) with equation (8.40) shows that $\lambda_2 = 1$ and $\lambda_1 = -1$. Taking the logarithm of equation (8.111) and multiplying by $[-1]$, one gets

$$-\ln f(x) = -\frac{1}{2} \ln \lambda_2 + \ln b + \frac{1}{2} \ln(2\pi) + \frac{a^2 \lambda_1^2}{2\lambda_2 b^2} + \frac{\lambda_1 xa}{b^2} + \frac{\lambda_2 x^2}{2b^2} \tag{8.112}$$

Multiplying equation (8.112) by $f(x)$ and integrating from minus infinity to positive infinity, we get the entropy function which takes the form:

$$H(f) = -\frac{1}{2} \ln \lambda_2 + \ln b + \frac{1}{2} \ln(2\pi) + \frac{a^2 \lambda_1^2}{2\lambda_2 b^2} + \frac{\lambda_1 a}{b^2} E[x] + \frac{\lambda_2}{2b^2} E[x^2] \tag{8.113}$$

Relation between distribution parameters and constraints   Taking partial derivatives of equation (8.113) with respect to λ_1, λ_2, a, and b individually, and then equating each derivative to zero, one obtains

$$\frac{\partial H}{\partial \lambda_1} = 0 = \frac{2a^2 \lambda_1}{2\lambda_2 b^2} + \frac{a}{b^2} E[x] \tag{8.114}$$

$$\frac{\partial H}{\partial \lambda_2} = 0 = -\frac{1}{2\lambda_2} - \frac{a^2 \lambda_1^2}{2\lambda_2^2 b^2} + \frac{1}{2b^2} E[x^2] \tag{8.115}$$

$$\frac{\partial H}{\partial a} = 0 = \frac{2a\lambda_1^2}{2b^2\lambda_2} + \frac{\lambda_1}{b^2} E[x] \tag{8.116}$$

$$\frac{\partial H}{\partial b} = 0 = \frac{1}{b} - \frac{2a^2 \lambda_1^2}{2\lambda_2 b^3} - \frac{2a\lambda_1}{b^3} E[x] - \frac{2\lambda_2}{2b^3} E[x^2] \tag{8.117}$$

Simplification of equation (8.114) through equation (8.117) results in

$$E[x] = a \tag{8.118}$$

$$E[x^2] = a^2 + b^2 \tag{8.119}$$

$$E[x] = a \tag{8.120}$$

$$E[x^2] = b^2 + a^2 \tag{8.121}$$

Equations (8.118) and (8.120) are the same, and so are equations (8.119) and (8.121). Thus the parameter estimation equations are equations (8.118) and (8.119).

Example 8.6: Determine parameters of the gamma distribution using the parameter expansion method.

Solution: The gamma distribution is given by equation (8.70a).

Specification of constraints Following Singh and Rajagopal (1986), the constraints for this method are equation (8.77) and

$$\int_0^\infty \frac{x}{a} f(x) dx = E\left[\frac{x}{a}\right] \tag{8.122}$$

$$\int_0^\infty \ln\left(\frac{x}{a}\right)^{b-1} f(x) dx = E\left[\ln\left(\frac{x}{a}\right)^{b-1}\right] \tag{8.123}$$

Derivation of entropy function The least-biased PDF corresponding to POME and consistent with equations (8.77), (8.122) and (8.123) takes the form

$$f(x) = \exp\left[-\lambda_0 - \lambda_1\left(\frac{x}{a}\right) - \lambda_2 \ln\left(\frac{x}{a}\right)^{b-1}\right] \tag{8.124}$$

where $\lambda_0, \lambda_1,$ and λ_2 are Lagrange multipliers. Insertion of equation (8.124) into equation (8.77) yields the partition function:

$$\exp(\lambda_0) = \int_0^\infty \exp\left[-\lambda_1\left(\frac{x}{a}\right) - \lambda_2 \ln\left(\frac{x}{a}\right)^{b-1}\right] dx = a\left(\lambda_1\right)^{\lambda_2(b-1)-1} \Gamma[1 - \lambda_2(b-1)] \tag{8.125}$$

The zeroth Lagrange multiplier is given by equation (8.125) as

$$\lambda_0 = \ln a - [1 - \lambda_2(b-1)] \ln \lambda_1 + \ln \Gamma[1 - \lambda_2(b-1)] \tag{8.126}$$

Also, from equation (8.126) one gets the zeroth Lagrange multiplier:

$$\lambda_0 = \ln \int_0^\infty \exp\left[-\lambda_1\left(\frac{x}{a}\right) - \lambda_2 \ln\left(\frac{x}{a}\right)^{b-1}\right] dx \tag{8.127}$$

Introduction of equation (8.126) in equation (8.124) produces

$$f(x) = \frac{1}{a}\left(\lambda_1\right)^{1-\lambda_2(b-1)} \frac{1}{\Gamma[1 - \lambda_2(b-1)]} \exp\left[-\lambda_1\frac{x}{a} - \lambda_2 \ln\left(\frac{x}{a}\right)^{b-1}\right] \tag{8.128}$$

Comparison of equation (8.128) with equation (8.70a) shows that $\lambda_1 = 1$ and $\lambda_2 = -1$. Taking logarithm of equation (8.128) yields

$$\ln f(x) = -\ln a + [1 - \lambda_2(b-1)] \ln \lambda_1 - \ln \Gamma[1 - \lambda_2(b-1)] - \lambda_1 \frac{x}{a} - \lambda_2 \ln \left(\frac{x}{a}\right)^{b-1}$$

(8.129)

Multiplying equation (8.129) by $[-f(x)]$ and integrating from 0 to ∞ yields the entropy function of the gamma distribution. This can be written as

$$H(f) = \ln a - [1 - \lambda_2(b-1)] \ln \lambda_1 + \ln \Gamma[1 - \lambda_2(b-1)] + \lambda_1 E\left[\frac{x}{a}\right] + \lambda_2 E\left[\ln \left(\frac{x}{a}\right)^{b-1}\right]$$

(8.130)

Relation between parameters and constraints Taking partial derivatives of equation (8.130) with respect to $\lambda_1, \lambda_2, a,$ and b separately and equating each derivative to zero, respectively, yield

$$\frac{\partial H}{\partial \lambda_1} = 0 = -[1 - \lambda_2(b-1)] \frac{1}{\lambda_1} + E\left(\frac{x}{a}\right)$$

(8.131)

$$\frac{\partial H}{\partial \lambda_2} = 0 = +(b-1) \ln \lambda_1 - (b-1)\psi(K) + E\left[\ln \left(\frac{x}{a}\right)^{b-1}\right], K = 1 - \lambda_2(b-1)$$

(8.132)

$$\frac{\partial H}{\partial a} = 0 = +\frac{1}{a} - \frac{\lambda_1}{a} E\left[\frac{x}{a}\right] + \frac{(1-b)}{a}\lambda_2$$

(8.133)

$$\frac{\partial H}{\partial b} = 0 = \lambda_2 \ln \lambda_1 - \lambda_2 \psi(K) + \frac{\lambda_2}{b-1} E\left[\ln \left(\frac{x}{a}\right)^{b-1}\right]$$

(8.134)

Simplification of equations (8.131) to (8.134), respectively, gives

$$E\left(\frac{x}{a}\right) = b$$

(8.135a)

$$E\left[\ln \left(\frac{x}{a}\right)\right] = \psi(K)$$

(8.135b)

$$E\left(\frac{x}{a}\right) = b$$

(8.135c)

Equation (8.135a) is the same as equation (8.135c). Therefore, equations (8.135a) and (8.135b) are the parameter estimation equations.

8.3 Contrast with method of maximum likelihood estimation (MLE)

Consider a discrete random variable X with known values $x_i, i = 1, 2, \ldots, N$. The probability mass function for the random variable X is known: $P(x, \theta)$; $\{p_i, i = 1, 2, \ldots, N\}$, where θ is a parameter vector. To estimate parameters θ, one maximizes the likelihood function L:

$$L = \prod_{i=1}^{N} p(x_i, \theta)$$

(8.136)

Quite often, one maximizes $\ln L$:

$$\ln L = \sum_{i=1}^{N} \ln p(x_i, \theta) \tag{8.137}$$

such that

$$\frac{\partial \ln L}{\partial \theta} = 0, \frac{\partial^2 \ln L}{\partial \theta^2} < 0 \tag{8.138}$$

where

$$\sum_{i=1}^{N} p(x_i, \theta) = 1 \tag{8.139}$$

Equation (8.138) can be written explicitly as

$$\sum_{i=1}^{N} \frac{1}{p(x_i, \theta)} \frac{\partial p(x_i, \theta)}{\partial \theta} = 0 \tag{8.140}$$

and

$$\sum_{i=1}^{N} \left[\frac{1}{p(x_i, \theta)} \frac{\partial^2 p(x_i, \theta)}{\partial \theta^2} - \frac{1}{p^2(x_i, \theta)} \left(\frac{\partial p(x_i, \theta)}{\partial \theta} \right)^2 \right] < 0 \tag{8.141}$$

When the Shannon entropy is maximized with respect to θ, one obtains

$$\sum_{i=1}^{N} \ln p(x_i, \theta) \frac{\partial p(x_i, \theta)}{\partial \theta} + \sum_{i=1}^{N} \frac{\partial p(x_i, \theta)}{\partial \theta} = 0 \tag{8.142}$$

$$\sum_{i=1}^{N} \left[\frac{1}{p(x_i, \theta)} \left[\frac{\partial p(x_i, \theta)}{\partial \theta} \right]^2 + \ln p(x_i, \theta) \frac{\partial^2 p(x_i, \theta)}{\partial \theta^2} \right] + \sum_{i=1}^{N} \frac{\partial^2 p(x_i, \theta)}{\partial \theta^2} = 0 \tag{8.143}$$

Note that from equation (8.139),

$$\sum_{i=1}^{N} \frac{\partial p(x_i, \theta)}{\partial \theta} = 0, \sum_{i=1}^{N} \frac{\partial^2 p(x_i, \theta)}{\partial \theta^2} = 0 \tag{8.144}$$

Using Jensen's inequality,

$$p(x) - 1 \geq \ln p(x) \geq 1 - \frac{1}{p(x)} \tag{8.145}$$

one gets

$$\sum_{i=1}^{N} p(x_i, \theta) \frac{\partial p(x_i, \theta)}{\partial \theta} \geq \sum_{i=1}^{N} \ln p(x_i, \theta) \frac{\partial p(x_i, \theta)}{\partial \theta} \geq -\frac{1}{p(x_i, \theta)} \frac{\partial p(x_i, \theta)}{\partial \theta} \tag{8.146}$$

Comparing equation (8.140) with equation (8.142) in view of equation (8.146),

$$\sum_{i=1}^{N} \ln p(x_i, \theta) \frac{\partial p(x_i, \theta)}{\partial \theta} \geq -\sum_{i=1}^{N} \frac{1}{p(x_i, \theta)} \frac{\partial p(x_i, \theta)}{\partial \theta} \tag{8.147}$$

If $\partial p(x_i, \theta)/\partial \theta = 0$, the two methods will yield the same parameter estimates. The uniform and exponential distributions are examples for which the two methods are the same. For other distributions they may not lead to the same estimates. For example, for gamma and Weibull distributions parameter estimates by the two methods are not the same.

Furthermore, consider cross-entropy $H(p|q)$ of two density functions p and q:

$$H(p|q) = \sum p(x_i, \theta) \ln \frac{p(x_i, \theta)}{q(x_i, \theta)} \tag{8.148}$$

where p is deduced from maximizing $H(f)$ and q is obtained from MLE. If $p = q$, then

$$H(p|q) = 0 \tag{8.149}$$

Otherwise,

$$H(p|q) \geq 0 \tag{8.150}$$

Using Jensen's inequality $x - x^2 \geq \ln x$, one can show that

$$\sum_{i=1}^{N} q(x_i, \theta) \left[\frac{p(x_i, \theta)}{q(x_i, \theta)} \right]^2 - 1 \geq H(p|q) \geq 0 \tag{8.151}$$

Equation (8.151) gives a measure of the relative information contained in the two methods. Indeed one can use cross-entropy and equation (8.151) to compare POME with any other parameter estimation scheme.

8.4 Parameter estimation by numerical methods

It is difficult to analytically determine the Lagrange multipliers contained in equation (8.4a) unless $y_i(x), i = 1, 2, \ldots, m$, are quite simple. In general, an analytical solution for obtaining the Lagrange multipliers (for $m > 2$) does not exist and numerical solution is the only resort. It has been shown that the problem of solving the set of nonlinear equations is equivalent to finding the minimum of a function Γ expressed as (Agmon et al., 1979; Mead and Papanicolaou, 1984):

$$\Gamma = \ln Z + \sum_{r=1}^{m} \lambda_r \bar{y}_r, r = 1, 2, \ldots, m \tag{8.152}$$

where

$$Z = \exp(\lambda_0) = \int_{0}^{\infty} \exp\left[-\sum_{r=0}^{m} \lambda_r y_r(x) \right] dx \tag{8.153}$$

The minimization can be achieved by employing Newton's method. Starting from some initial value $\lambda_{(0)}$, one can solve for Lagrange parameters by updating $\lambda_{(1)}$ through the equation given as:

$$\lambda_{(1)} = \lambda_{(0)} - H^{-1}\frac{\partial \Gamma}{\partial \lambda_i}, i = 0, 1, 2, \ldots, m \tag{8.154}$$

where the gradient Γ is expressed as:

$$\frac{\partial \Gamma}{\partial \lambda_i} = \bar{y}_i - \int_0^\infty \exp\left[-\sum_{r=0}^m \lambda_r y_r(x)\right] y_i(x) dx, \quad i = 0, 1, 2, \ldots, m \tag{8.155}$$

and H is the Hessian matrix whose elements are expressed as:

$$H_{i,j} = \int_0^\infty \exp\left(-\sum_{r=0}^m \lambda_r y_r(x)\right) y_i(x) y_j(x) dx, i, j = 0, 1, 2, \ldots, m \tag{8.156}$$

and H^{-1} is the inverse of Hessian matrix H. Clearly, the minimization of equation (8.152) is done numerically. One can employ the MATLAB function *fminsearch* to obtain the minimum of equation (8.155) and hence the Lagrange multipliers.

Questions

Q.8.1 Take a sample data of annual peak discharge from a gaging station on a river near your town. Fit the gamma distribution to the discharge data. Then using this distribution, determine the effect of sample size on the value of the Shannon entropy.

Q.8.2 Use the same gamma distribution as in Q.8.1. Changing the parameter values determine the Shannon entropy and discuss the effect of parameter variation.

Q.8.3 Determine the effect of discretization on the Shannon entropy.

Q.8.4 Determine the constraints for the log-gamma distribution needed for estimation of its parameters using entropy. Then determine its parameters in terms of the constraints.

Q.8.5 Determine the constraints for the Pearson type III distribution needed for estimation of its parameters using entropy. Then determine its parameters in terms of the constraints.

Q.8.6 Consider the log-Pearson type III distribution and determine its parameters using both the ordinary entropy method and the parameter space expansion method. Plot the distribution entropy as a function of its parameters.

Q.8.7 Determine the constraints for the Weibull distribution needed for estimation of its parameters using entropy. Then determine its parameters in terms of the constraints.

Q.8.8 Determine the constraints for the Gumbel distribution needed for estimation of its parameters using entropy. Then determine its parameters in terms of the constraints.

Q.8.9 Consider the log-Gumbel distribution and determine its parameters using both the ordinary entropy method and the parameter space expansion method. Plot the distribution entropy as a function of its parameters.

Q.8.10 Determine the constraints for three-parameter log-normal distribution needed for estimation of its parameters using entropy. Then determine its parameters in terms of the constraints.

Q.8.11 Determine the constraints for the logistic distribution needed for estimation of its parameters using entropy. Then determine its parameters in terms of the constraints.

Q.8.12 Consider the log-logistic distribution and determine its parameters using both the ordinary entropy method and the parameter space expansion method. Plot the distribution entropy as a function of its parameters.

Q.8.13 Determine the constraints for the Pareto distribution needed for estimation of its parameters using entropy. Then determine its parameters in terms of the constraints.

Q.8.14 Determine the constraints for the log-Pearson type III distribution needed for estimation of its parameters using entropy. Then determine its parameters in terms of the constraints.

References

Abramowitz, M. and Stegun, I.S. (1965). *Handbook of Mathematical Functions*. U.S. Department of Commerce, National Bureau of Standards, Washington, D.C.

Agmon, N., Alhassid, Y. and Levine, R.D. (1979). An algorithm for finding the distribution of maximal entropy. *Journal of Computational Physics*, Vol. 30, pp. 250–8.

Kapur, J.N. (1989). *Maximum Entropy Models in Science and Engineering*. Wiley Eastern, New Delhi, India.

Kapur, J.N. and Kesavan, H.K. (1988). *Generalized Maximum Entropy Principle (with Applications)*. Sandford Educational Press, University of Waterloo, Waterloo, Canada.

Kapur, J.N. and Kesavan, H.K. (1992). *Entropy Optimization Principles with Applications*. 408 p., Academic Press, New York.

Kendall, M. G. and Stuart, A. (1973). *The Advanced Theory of Statistics*, vol. 2. Inference and relationship. Charles Griffin, London, UK.

Mead, L.R. and Papanicolaou, N. (1984). Maximum entropy in the problem of moments. *Journal of Mathematical Physics*, Vol. 25, No. 8, pp. 2404–17.

Singh, V.P. (1997). The use of entropy in hydrology and water resources. *Hydrological Processes*, Vol. 11, pp. 587–626.

Singh, V.P. (1998a). *Entropy-based Parameter Estimation in Hydrology*. Kluwer Academic Publishers, Boston.

Singh, V.P. (1998b). Entropy as a decision tool in environmental and water resources. *Hydrology Journal*, Vol. XXI, No. 1-4, pp. 1–12.

Singh, V.P. (2000). The entropy theory as tool for modeling and decision making in environmental and water resources. *Water SA*, Vol. 26, No. 1, pp. 1–11.

Singh, V.P. and Fiorentino, M. (1992). A historical perspective of entropy applications in water resources. In: *Entropy and Energy Dissipation in Water Resources*, edited by V.P. Singh and M. Fiorentino, pp. 21–62, Kluwer Academic Publishers, Dordrecht, The Netherlands.

Singh, V.P. and Rajagopal, A.K. (1986). A new method of parameter estimation for hydrologic frequency analysis. *Hydrological Science and Technology*, Vol. 2, No. 3, pp. 33–40.

Singh, V.P., Rajagopal, A. K. and Singh, K. (1986). Derivation of some frequency distributions using the principle of maximum entropy (POME). *Advances in Water Resources*, Vol. 9, No. 2, pp. 91–106.

Singh, V.P., Singh, K. and Rajagopal, A. K. (1985). Application of the principle of maximum entropy (POME) to hydrologic frequency analysis, Completion Report 06, 144 p., Louisiana Water Resources Research Institute, Louisiana State University, Baton Rouge, Louisiana.

Additional Reading

Akaike, H. (1973). Information theory and an extension of the maximum likelihood principle. Proceedings, 2nd International Symposium on Information Theory, Akademia Kiado, Budapest, Hungary.

Basu, P.C. and Templeman, A.B. (1984). An efficient algorithm to generate maximum entropy distributions. *International Journal of Numerical Methods in Engineering*, Vol. 20, pp. 1039–55.

Brown, C.B. (1980). Entropy constructed probabilities. *Journal of Engineering Mechanics*, Vol. 106, No. 4, pp. 633–40.

Ciulli, S., Mounsif, M., Gorman, N. and Spearman, T.D. (1991). On the application of maximum entropy to the moments problem. *Journal of Mathematical Physics*, Vol. 32, No. 7, pp. 1717–19.

Collins, R. and Wragg, A. (1977). Maximum entropy histograms. *Journal of Physics, A: Math. Gen.*, Vol. 10. No. 9, pp. 1441–64.

Csiszar, I. (1991). Why least squares and maximum entropy? An axiomatic approach to inference for linear inverse problems. *Annals of Statistics*, Vol. 19, No. 4, pp. 2032–66.

Dempster, A. P., Laird, N.M. and Rubin, D.B. (1977). Maximum likelihood from incomplete data via the EM algorithm. *Journal of the Royal Statistical Society*, Series B, Vol. 39, No. 1, pp. 1–38.

Dowson, D.C. and Wragg, A. (1973). Maximum-entropy distributions having prescribed first and second moments. *IEEE Transactions on Information Theory*, Vol. IT-19, pp. 689–93.

Dutta, M. (1966). On the maximum (information-theoretic) entropy estimation. *Sankhya*, Vol. 28, Series A, pp. 319–28.

Frontini, M. and Tagliani, A. (1994). Maximum entropy in the finite Stieltjes and Hamburger moment problem. *Journal of Mathematical Physics*, Vol. 35, No. 12, pp. 6748–56.

Gabardo, J.P. (1992). A maximum entropy approach to the classical moment problem. *Journal of Functional Analysis*, Vol. 106, pp. 80–94.

Griffeath, D.S. (1972). Computer solution of the discrete maximum entropy problem. *Technometrics*, Vol. 14, No. 4, pp. 891–7.

Kam, T.Y. and Brown, C.B. (1983). Updating parameters using fuzzy entropies. *Journal of Engineering Mechanics*, Vol. 109, No. 6, pp. 1334–43.

Landau, H.J. (1987). Maximum entropy and the moment problem. Bulletin, *American Mathematical Society*, Vol. 16, No. 1, pp. 47–77.

Li, X. (1992). An entropy-based aggregate method for minimax optimization. *Engineering Optimization*, Vol. 18, pp. 277–85.

Liu, I.S. (1972). Method of Lagrange multipliers for exploration of the entropy principle. *Archive for Rational Mechanics and Analysis*, Vol. 46, No. 2, pp. 131–48.

Louis, T.S. (1982). Finding the observed information matrix when using the EM algorithm. *Journal of Royal Statistical Society*, Series B, Vol. 44, No. 2, pp. 226–33.

Ormoneit, D. and White, H. (1999). An efficient algorithm to compute maximum entropy densities. *Econometric Reviews*, Vol. 18, No. 2, pp. 127–40.

Taglani, A. (1993). On the application of maximum entropy to the moments problem. *Journal of Mathematical Physics*, Vol. 34, No. 1, pp. 326–37.

Tagliani, A. (2003). Entropy estimate of probability densities having assigned moments: Hausdorff case. *Applied Mathematics Letters*, Vol. 15, pp. 309–14.

Weidemann, H.L. and Stear, E.B. (1969). Entropy analysis of parameter estimation. *Information and Control*, Vol. 14, pp. 493–506.

Wragg, A. and Dowson, D.C. (1970). Fitting continuous probability density functions over $(0, \infty)$ using information theory ideas. *IEEE Transactions on Information Theory*, Vol. 16, p. 226–30.

Zellner, A. and Highfield, R.A. (1988). Calculation of maximum entropy distributions and approximation of marginal posterior distributions. *Journal of Econometrics*, Vol. 37, pp. 195–209.

9 Spatial Entropy

In water and environmental engineering, many phenomena need to be spatially analyzed. For example, consider land use and land cover (LULC) and their changes in a large urban area. From the standpoint of determining the hydrologic and water quality consequences, it may be desirable to partition the area into a number of zones. Within each zone also, there can be more than one LULC and each LULC has its own area. In this case the variable of interest is the area of each zone. Another dimension of great interest is the location of each zone from a reference point, such as watershed outlet. One can then analyze the spatial distribution of LULCs, as well as the distribution of their locations. The area can be divided into more zones of smaller areas or less zones of larger areas. This means that the number of zones determines the level of detail and the areas will determine the geometry or architecture of LULCs. For urban development and planning, one may ask the question: What is the best way to partition the area into zones and how can these zones be configured in order to minimize the environmental impact? In this case, spatial analysis of the environmental consequences would involve two dimensions of area and location, and can be undertaken using entropy. In a similar vein, best management practices and their consequences can be analyzed.

Consider another case where one needs to construct the time-area (TA) diagram of a watershed for purposes of constructing the unit hydrograph. The TA diagram requires computing times of travel from the outlet, drawing isochrones or contours of equal travel time, and computing areas between two consecutive isochrones. Sometimes equal travel time is equated with equal distance. From the perspective of spatial analysis, the watershed outlet is the point of reference. The watershed is partitioned into a number of sub-areas based on the number of isochrones selected. Each sub-area is enclosed by two isochrones. This subarea is represented by a mean distance or mean travel time. Within each sub-area there is a large number of sub-sub-areas that have somewhat different travel times or distances. Thus, there are two variables here: area between two consecutive isochrones and travel time.

In an urban area, population is the primary determinant for the planning and development of water supply, land use, roads, shopping centers, schools, hospitals, energy supply, waste disposal facilities, pipe lines, and a host of other infrastructure facilities. People decide where to live based on a number of factors, including such amenities as shops, schools, hospitals, roads, parks, job location, transportation facilities, socio-economic level, types of houses available, lawns and garden requirements, and so on. The net result of these factors working in concert is that the distribution of residential locations will be quite heterogeneous such that many zones will have widely varying population densities. For urban planning and development, an urban area can be divided into a number of zones or spatial units called sub-areas. Each unit has an area and has a number of people living in it, and can be considered as a residential area. Considering the city center as a point of reference, the distances from these sub-areas to the city center can be computed. The distances roughly translate into times of travel from residential areas to the city center and back. The variables of interest are: location of each zone and area of the zone. Entropy theory can be applied to study the population distribution in different zones or distribution of distances. In a similar manner, entropy can be employed to analyze spatial distributions of various geographical phenomena, such as economic level, crime rate, drug use, economic dependence on government, emergency preparedness, athletic facilities, and so on. Analysis of spatial phenomena is not straightforward and requires special attention.

There has been a considerable body of literature on spatial analysis using both discrete and continuous Shannon entropies that have been discussed in Chapter 2. Since the work of Leopold and Langbein (1962) who were amongst the first to apply entropy to derive the most probable profile of a river, spatial applications of entropy in water and environmental engineering have received comparatively limited attention. However, in geography numerous spatial applications of entropy have been reported. Chapman (1970), Semple and Gollege (1970), and Gurevich (1969) applied entropy to measure the amount of information in spatial probability distributions. Anderson (1969), Medvedkov (1970), Marchand (1972), Mogridge (1972), amongst others, provided informative reviews of spatial applications. Curry (1964) suggested the use of entropy in investigating spatial series. Later, Curry (1971, 1972a, b) applied entropy to derive rank-size frequency distributions of human settlements. Wilson (1970) laid the foundation for geographical model building. Batty (1972, 1974, 1976, 1983, 2010) and Batty and Sammons (1978) presented entropy concepts and techniques for spatial analyses. The discussion in this chapter on application of entropy for spatial analysis draws significantly from the work of Batty and his associates.

9.1 Organization of spatial data

Spatial data are gathered for analysis of spatial phenomena. Often the interest is in discerning the pattern that may underlie the phenomenon under consideration. This depends on the manner in which the data are organized. Spatial data can be organized in terms of the number and configuration of spatial units (or classes or zones). The number and configuration of zones influence the level of detail and information imparted. Therefore, the spatial data should be organized so that the maximum information is imparted.

The number of zones relates to the level of detail characterizing the phenomenon and configuration relates to the geometric structure used to divide the two-dimensional spatial phenomenon. The number of zones also determines the spatial scale of analysis and hence the level of aggregation and in turn configuration. Configuration also affects the level of detail, but its central concern is with the homogeneity and regularity of the phenomenon. When spatial units are differently aggregated, different patterns may result; in other words, as the number and configuration of spatial units change, different spatial patterns can emerge. These patterns will exhibit different values of variance and co-variance of the phenomenon. However, there is uncertainty as to the pattern for two reasons. First, the process of observation is subjective. Second, the pattern is an abstraction. This means that the set of observations is of crucial importance. The question is: Is the set of observations organized in the best possible way? Further, different configurations based on the same number of units or different numbers based on the same configuration can give rise to the same or different attributes. This suggests that prior to any spatial analysis data should be organized efficiently and optimally.

Any set of observations exhibits a measurable amount of information and this information can be increased or decreased by changing the level of detail characterizing the observations. This level of detail depends on the number and configuration of zones. This suggests that one should define an information measure so that the maximum information is imparted by the pattern to be achieved. Once the optimal pattern is achieved, it will lead to the most informative spatial pattern.

9.1.1 Distribution, density, and aggregation

The number and configuration of zones characterize a two-dimensional spatial distribution. They approximate the distribution and determine the level of detail thereof. The number of zones and the ensuing configuration may not be independent of each other. It may be noted that equally good and informative approximations to the true distribution may be achieved with large or small number of zones and different configurations.

The probability density function (PDF) underlying the spatial phenomenon under consideration is seldom known a priori. A PDF can be enumerated by an infinite number of classes or points defining the density and these points imply a regular configuration. In a one dimensional case, the sets of points can be approximated by equal class intervals. In a two-dimensional case, the spatial density function can be employed to define an information statistic in which both the number and configuration are included. As an example, consider the configuration in which all spatial units are of the same value, that is, they are of equal area. Equal area means the same weight is assigned to each spatial unit and the phenomenon within it. Here, space will act in a neutral way as regards spatial analysis. In this case of equal area configuration, the density rather than the distribution is the main concern.

Now consider a region encompassing the phenomenon of interest and let the region be defined by a set S. Let this set be partitioned into a series of N mutually exclusive subsets s_i, $i = 1, 2, \ldots, N$, referred to as zones:

$$S = \left\{ s_i, i = 1, 2, ..., N \right\} = \left\{ s_1, s_2, ...s_N \right\} \tag{9.1}$$

in which N is the number of subsets characterizing the distribution of phenomenon. For zones s_i, $i = 1, 2, \ldots, N$, two characteristics must be noted. First, the distribution of phenomenon across zones can be defined as

$$\varphi = \left\{ \varphi_i, i = 1, 2, ..., N \right\} = \left\{ \varphi_1, \varphi_2, ..., \varphi_N \right\} \tag{9.2}$$

which, for example, may be the distribution of LULCs or population. Second, the areal size of zones s_i can be defined as

$$x = \left\{ x_i, i = 1, 2, ..., N \right\} = \left\{ x_1, x_2, ..., x_N \right\} \tag{9.3}$$

Now the main point is to define an information measure relating the distribution $\{\varphi_i\}$ to the distribution $\{x_i\}$ such that it is maximized when these distributions approach the limit or N tends to be large. Consider an aggregation of the original N zones into M zones or sub-regions:

$$S = \left\{ s_1, s_2, ...s_M \right\}, \quad for \ \ M \le N \tag{9.4}$$

The partition of S is such that the information measure is maximized.

Note that if the phenomenon is one of population then the total population denoted by Φ and the total area denoted by X can be expressed as:

$$\Phi = \sum_{i=1}^{N} \varphi_i, \qquad X = \sum_{i=1}^{N} x_i \tag{9.5}$$

By normalizing $\{\varphi_i\}$ and $\{x_i\}$, one can obtain probabilities of populations in different areas:

$$p_i = \frac{\varphi_i}{\Phi}, \qquad \sum_{i=1}^{N} p_i = 1 \tag{9.6}$$

Likewise, for the land area

$$q_i = \frac{x_i}{X}, \qquad \sum_{i=1}^{N} q_i = 1 \tag{9.7}$$

Equation (9.6) yields the probability (mass) distribution of the phenomenon (i.e., LULC or population) as:

$$\{p_i\} = \{p_1, p_2, ..., p_N\} \tag{9.8}$$

and equation (9.7) defines the probability (mass) distribution of land area as:

$$\{q_i\} = \{q_1, q_2, ..., q_N\} \tag{9.9}$$

One can employ the probability mass function $P = \{p_i\}$ for approximating the probability density function $p(x_i)$ of the phenomenon associated with zones s_i. The probability density denotes the probability of the phenomenon occurring in a unit of space of the relevant zone and this is the zonal average, that is, the mean estimate of the probability density across a zone. Since this is the average probability of occurrence in a unit of space, the total probability of occurrence or probability mass, p_i, is determined by multiplying the density $p(x_i)$ by the area associated with s_i, that is, x_i:

$$p_i = p(x_i) x_i \tag{9.10}$$

This shows that

$$p(x_i) = \frac{p_i}{x_i} \tag{9.11}$$

is normalization of p_i with respect to x_i. Thus, $p(x_i)$ is the probability over unit of space and is probability density.

Equation (9.11) shows the relation between probability density function and probability distribution or probability mass function. Two points may be noted. First, the approximation to the probability density function $p(x_i)$ may likely become more accurate with the decreasing size of area x_i. This means that as the number of zones increases the probability density underlying the whole spatial system is more closely approximated. Second, if all the zones were of the same size, $x_1 = x_2 = ... = x_N$, then the probability density $p(x_i)$ and zonal area x_i would be irrelevant, and therefore p_i would be a measure of $p(x_i)$. The actual density of population in each zone is defined by $\varphi(x_i)$ and the total (system wide) density of population by D. Then

$$\varphi(x_i) = \frac{\varphi_i}{x_i}, \qquad D = \frac{\Phi}{X} \qquad (9.12)$$

Here $\varphi(x_i)$ denotes the number of attributes (e.g. persons, land uses, trees, etc.) located in a unit of space, x_i, in zone s_i. The total population in s_i, that is, φ_i, is obtained by multiplying the actual population density $\varphi(x_i)$ by the associated area x_i. Equations (9.6), (9.11), and (9.12) show that $p(x_i)$ is proportional to $\varphi(x_i)$, that is,

$$p(x_i) = \frac{\varphi(x_i)}{\Phi} \quad \text{or} \quad \varphi(x_i) = \Phi p(x_i) \qquad (9.13)$$

Extending the probability densities to the continuous case when $x_i \to 0$, $\forall i$, for $p(x)$ and $\varphi(x)$,

$$\lim_{\max(x_i) \to 0} \sum_{i=1}^{N} p(x_i)x_i = \lim_{N \to \infty} \sum_{i=1}^{N} p(x_i)x_i = \int_x p(x)dx \qquad (9.14)$$

and

$$\lim_{\max(x_i) \to 0} \sum_{i=1}^{N} \varphi(x_i)x_i = \lim_{N \to \infty} \sum_{i=1}^{N} \varphi(x_i)x_i = \int_x \varphi(x)dx \qquad (9.15)$$

Now the question of most informative spatial pattern can be addressed. Corresponding to the original partition of S into N zones, $\{s_i, i = 1, 2, \ldots, N\}$ or $\rho_N(S)$, a measure of information $I(P:Q|N)$, is derived from the probability distributions $\{q_i\}$ and $\{p_i\}$. The objective is to determine the optimal value of $I(P:Q|N)$ by aggregating sub-sets $S_N = \{s_i, i = 1, 2, \ldots, N\}$ consistent with partition $\rho_N(S)$ into sets $S_M = \{s'_i, i = 1, 2, \ldots, M\}$ consistent with the partition $\rho_M(S)$. The optimum value of $I(P:Q|N)$ is attained by aggregating $\{q_i\}$ to $\{p_i\}$ so as to achieve the idealized number of zones and configurations. Here $\{q_i\}$ and $\{p_i\}$ can be regarded as prior and posterior probability distributions, since $\{q_i\}$ can be considered as a hypothetical (a priori) probability distribution based on the assumption that the zone or area is divided into N subzones and $\{p_i\}$ as a (posterior) probability distribution of the phenomenon after aggregation. The prior probability distribution of the phenomenon $\{q_i\}$ can be deemed as proportional to the area size x_i or more generally as lineally related to the area size. Sets S_M can be defined as $S_M = \{s'_i, i = 1, 2, \ldots, M\}$, where s'_i is the union of subsets of the partition $S_N = \{s_i, i = 1, 2, \ldots, N\}$. Clearly $s_i \subseteq S_M$, that is, zone s_i is an element of a sub-region of S_M.

9.2 Spatial entropy statistics

Assuming there is a random variable X divided into N intervals Δx_i, $i = 1, 2, \ldots, N$ with probability for $X \in [x_i - 0.5\Delta x_i, x_i + 0.5\Delta x_i]$ approximated as $p_i = p(x_i)\Delta x_i$, then the discrete Shannon entropy, denoted as H_d, is defined as

$$H_d = -\sum_{i=1}^{N} p_i \ln(p_i) \qquad (9.16)$$

whereas the continuous counterpart of the discrete Shannon entropy, that is, the continuous Shannon entropy, denoted as H_c, is defined as

$$H_c = -\int p(x) \ln [p(x)]dx \tag{9.17}$$

The spatial entropy for random variable X, denoted by H_s, is defined as

$$H_s = -\lim_{\Delta x_i \to 0} \sum_{i=1}^{N} p_i \ln \left(\frac{p_i}{\Delta x_i} \right) \tag{9.18}$$

Equation (9.18) shows that the spatial entropy explicitly takes into account the effects of the way in which the spatial system is partitioned. Also, it can actually be shown that equation (9.18) is equivalent to the continuous Shannon entropy:

$$H_s = H_c = -\lim_{\Delta x_i \to 0} \sum_{i=1}^{N} p_i \log \left(\frac{p_i}{\Delta x_i} \right) \tag{9.19}$$

In real application, for a given interval size Δx_i, the discrete approximation of the spatial entropy (i.e., continuous Shannon entropy) is usually computed as

$$\hat{H}_s = -\sum_{i=1}^{N} p_i \ln \left(\frac{p_i}{\Delta x_i} \right) \tag{9.20}$$

Here the subscripts aim to avoid confusion. In the following all the subscripts will be omitted, so that no misunderstanding is introduced. For simplicity, subscripts in equations (9.16) will be omitted throughout this chapter. It should be obvious where such omissions are made.

Example 9.1: Consider a 100 square kilometer urban area. Divide the area into a number of equal zones. Take the number as 10, 20, 30, 40 and 50. Compute the discrete entropy and plot it as a function of number of zones. Now divide the area into unequal zones as shown in Tables 9.1 and 9.2 and then compute entropy and plot it. Finally compute the entropy reflecting the spatial interval size and plot it.

Table 9.1 Division of area into unequal zones.

Number of zones/ partitions	Randomly assigning number of unequal subzones such that the total number of subzones satisfies each partition				
10	0	2	3	5	0
20	1	2	2	5	10
30	1	2	2	5	20
40	1	6	3	10	20
50	1	2	2	20	25
60	1	3	6	20	30
70	2	3	5	25	35
80	2	8	15	25	30
90	0	15	20	25	30
100	10	10	20	25	35

Table 9.2 Areas of different numbers of unequal zones for different partitions.

Number of zones	Areas (km²) for corresponding number of zones in Table 9.1					Total area (km²)
10	0 (each 0)	20 (each 10)	35 (each 11.7)	45 (each 9)	0 (each 0)	100
20	10 (each 10)	25 (each 12.5)	35 (each17.5)	10 (each 2)	20 (each 2)	100
30	10 (each 10)	25 (each 12.5)	35 (each 17.5)	20 (each4)	10 (each 0.5)	100
40	10 (each 10)	20 (each 3.3)	25 (each 8.3)	30 (each 3)	15 (each 0.75)	100
50	15 (each 15)	15 (each 7.5)	25 (each 12.5)	25 (each1.25)	20 (each 0.8)	100
60	10 (each 10)	20 (each 6.67)	25 (each 4.17)	30 (each 1.5)	15 (each 0.5)	100
70	10 (each 5)	20 (each 6.67)	30 (each 6)	30 (each 1.2)	10 (each 0.29)	100
80	10 (each 5)	20 (each 2.5)	25 (each 1.67)	30 (each 1.2)	15 (each 0.5)	100
90	0 (each 0)	10 (each 0.67)	15 (each 0.75)	40 (each 1.6)	35 (each 1.17)	100
100	10 (each 1)	15 (each 1.5)	25 (each 1.25)	30 (each 1.2)	20 (each 0.57)	100

Solution: First, compute the discrete entropy when partitioning the system into equal zones. Consider the number of zones as N and the total area as Φ. Then the discrete probability mass can be computed as

$$p_i = \frac{\varphi_i}{\Phi} = \frac{\Phi/N}{\Phi} = \frac{1}{N}$$

According to equation (9.16), the discrete entropy can be calculated as

$$H_{equal}(N) = -\sum_{i=1}^{N} p_i \ln(p_i) = -\sum_{i=1}^{N} \frac{1}{N} \ln\left(\frac{1}{N}\right) = -\ln\left(\frac{1}{N}\right) = \ln(N) \text{ nats}$$

Take $N = 10$ as an example, the discrete entropy is calculated as

$$H_{equal}(10) = \ln(10) = 2.3026 \text{ nats}$$

For each partition of the given urban area, the discrete entropy is shown in the left panel of Figure 9.1. As expected, with increase in the number of zones, the discrete entropy increases

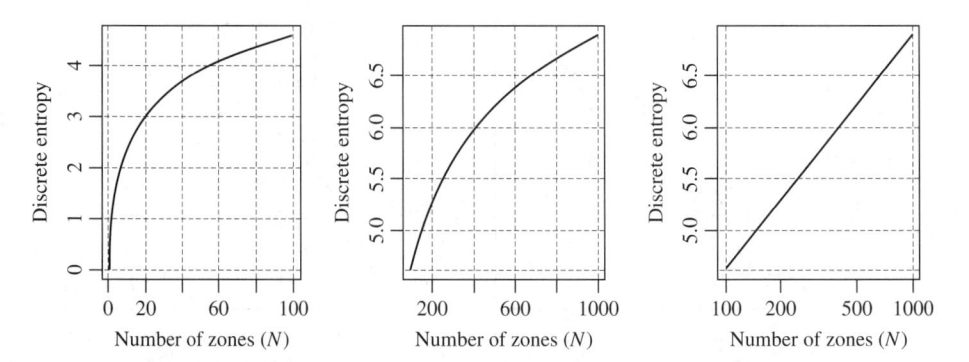

Figure 9.1 Relationship between number of zones and discrete entropy. (Left: N increases from 1 to 100; Middle: N increases from 100 to 1000; Right: the same as figure in the middle panel but using a semi-logarithmic scale).

without bound. This can be explicitly seen from Figure 9.1. In addition, when the system is equally partitioned, the discrete entropy tends to increase linearly with the logarithm of the number of zones, indicated by the right panel of Figure 9.1.

Second, compute the discrete entropy when partitioning the system into unequal zones. For entropies of unequally divided zones, the assumed number of zones for each partition is shown in Table 9.1. Areas of different numbers of unequal zones for different partitions are shown in Table 9.2. For instance, if the total area is divided into 10 zones, from Table 9.1 it is known that there are three types of zones for this partition. The numbers of different types are 2, 3, and 5, respectively. From Table 9.2, it is found that for type 1, it consists of two zones with an area of $10\,km^2$, respectively. For type 2, it consists of three zones of $11.7\,km^2$ respectively. For type 3, it consists of five zones of $9\,km^2$, respectively. Likewise, the distribution of unequal zones for other partitions (i.e., 20, 30, ..., 100) can be obtained from Tables 9.1 and 9.2. Here one should be careful that the system is not partitioned into zones all with different sizes but some of them with the same area, as explained above. Also using equation (9.16) the discrete entropy can be calculated as follows:

$$H_{unequal}(10) = -\sum_{i=1}^{10} p_i \ln(p_i) = -\frac{10}{100}\ln\left(\frac{10}{100}\right) - \frac{10}{100}\ln\left(\frac{10}{100}\right)$$

$$-\frac{11.7}{100}\ln\left(\frac{11.7}{100}\right) - \frac{11.7}{100}\ln\left(\frac{11.7}{100}\right) - \frac{11.7}{100}\ln\left(\frac{11.7}{100}\right)$$

$$-\frac{9}{100}\ln\left(\frac{9}{100}\right) - \frac{9}{100}\ln\left(\frac{9}{100}\right) - \frac{9}{100}\ln\left(\frac{9}{100}\right) - \frac{9}{100}\ln\left(\frac{9}{100}\right) - \frac{9}{100}\ln\left(\frac{9}{100}\right)$$

$$= 2.297 \text{ nats}$$

Discrete entropies for other numbers of intervals are computed following the same way and plotted as shown in the left panel of Figure 9.2. For unequal zones, the entropy is not only determined by the number of zones, but also by the specific partition of the system. In general, the entropy for unequal zones is less than that for equal zones (see the right panel of Figure 9.2).

Third, to compute the entropy reflecting the effect of spatial interval size, the cases regarding equal and unequal intervals should be considered separately. In the case of equal size interval,

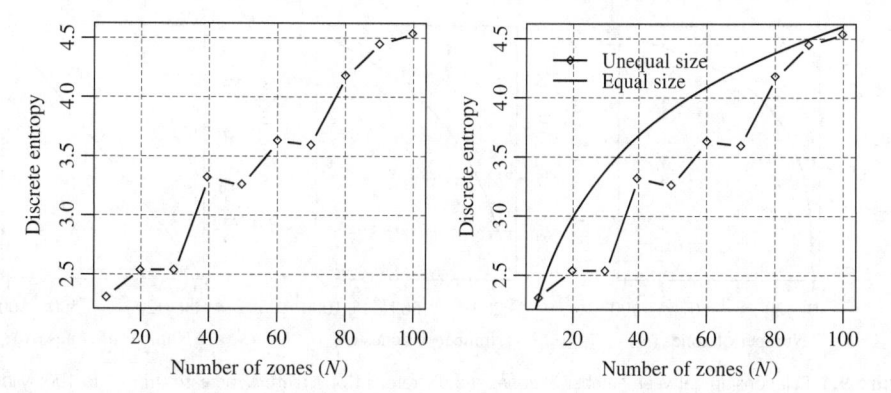

Figure 9.2 Relationship between number of zones with unequal areas and discrete entropy (right panel) and comparison of relationships between number of zones for equal and unequal partitions of zones (left panel).

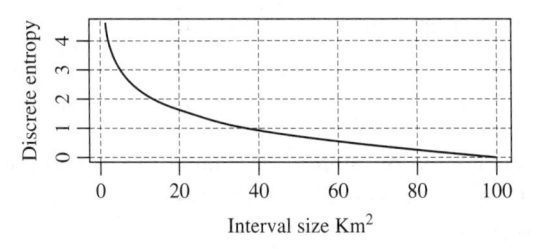

Figure 9.3 Relationship between spatial interval size and discrete entropies for equal zones.

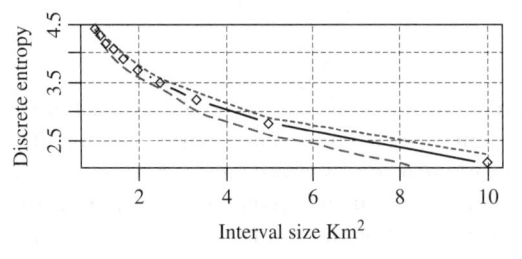

Figure 9.4 Relationship between averaged spatial interval size and discrete entropies for unequal zones.

the interval number determines the interval size. Therefore, plotting the computed discrete entropies versus the corresponding interval sizes, one can obtain a general picture about the relationship between spatial interval size and discrete entropy for equal zones. The results are shown in Figure 9.3. As the interval size increases, the discrete entropy decreases. In the case of unequal size interval, for the same number of intervals there are many different possible partitions. In turn, different partitions lead to different discrete entropies. In order to show the effect of interval size on the computed discrete entropy, one can figure out the relationship between discrete entropy and the averaged interval size, which is determined only by the interval numbers, given a spatial system. For each given interval number, partition the system into 100 patterns, and for each pattern there is one discrete entropy. Therefore, corresponding to each interval number, there are 100 discrete entropies, the maximum, the mean and the minimum of which can be computed. Assuming the interval numbers to be $10, 20, 30, \ldots, 100$, the mean of the computed discrete entropies as a function of averaged interval size is shown in Figure 9.4, in which the maximum and the minimum values are also presented by dashed lines. Similar to the case of equal size intervals, it is apparent that the discrete entropy also decreases as the averaged interval size increases.

9.2.1 Redundancy

Consider a measure of redundancy Z defined as one minus the ratio of the actual entropy to the maximum entropy:

$$Z = 1 - \frac{H}{H_{\text{max}}} \tag{9.21}$$

For a discrete case, H is given by the Shannon discrete entropy:

$$H = -\sum_{i=1}^{N} p_i \ln p_i \tag{9.22}$$

and the maximum entropy from equation (9.22) results as

$$H_{\max} = \ln N \qquad (9.23)$$

if and only if the phenomenon of interest (e.g., population) is uniformly distributed over different zones, that is, $p_i = 1/N$, $i = 1, 2, \ldots, N$. Substituting equations (9.22) and (9.23) in equation (9.21), the redundancy Z is obtained as:

$$Z = 1 + \frac{\displaystyle\sum_{i=1}^{N} p_i \ln p_i}{\ln N} \qquad (9.24)$$

which is information redundancy measured by the discrete Shannon entropy. One defect of the information redundancy Z defined in equation (9.24) is that it changes as the number of intervals N changes. Comparison of redundancy between different systems is therefore difficult. Since spatial entropy as defined in equation (9.20) explicitly takes into account the effect of spatial interval size and/or the number of intervals, it is intuitive to replace the discrete entropy in equation (9.24) by the spatial entropy.

Now consider a spatial system (e.g., a watershed) having an area X divided into N zones. Let Δx_i be defined as X/N, which means the spatial system is divided into N equal size intervals. Assuming the spatial phenomenon of interest is distributed over different zones with probability mass function $\{p_i, i = 1, 2, \ldots, N\}$, the spatial entropy of the phenomenon is computed as

$$H = -\sum_{i=1}^{N} p_i \ln \left(\frac{p_i}{\Delta x_i} \right) = -\sum_{i=1}^{N} p_i \ln \left(\frac{N p_i}{X} \right)$$

$$= -\sum_{i=1}^{N} p_i \ln (N) - \sum_{i=1}^{N} p_i \ln (p_i) + \sum_{i=1}^{N} p_i \ln(x)$$

$$= -\ln (N) + \ln(x) - \sum_{i=1}^{N} p_i \ln (p_i) \qquad (9.25)$$

Apparently, if and only if the spatial phenomenon is uniformly distributed over different zones, that is, $p_i = 1/N$, $i = 1, 2, \ldots, N$, the spatial entropy is maximized. That is

$$H_{\max} = -\ln (N) + \ln(x) - \sum_{i=1}^{N} \frac{1}{N} \ln \left(\frac{1}{N} \right)$$

$$= \ln(x) \qquad (9.26)$$

Then, applying the spatial entropy [equation (9.20)], the redundancy measure can be defined as

$$Z = 1 + \frac{\displaystyle\sum_{i=1}^{N} p_i \ln \left(\frac{p_i}{\Delta x_i} \right)}{\ln(x)} \qquad (9.27)$$

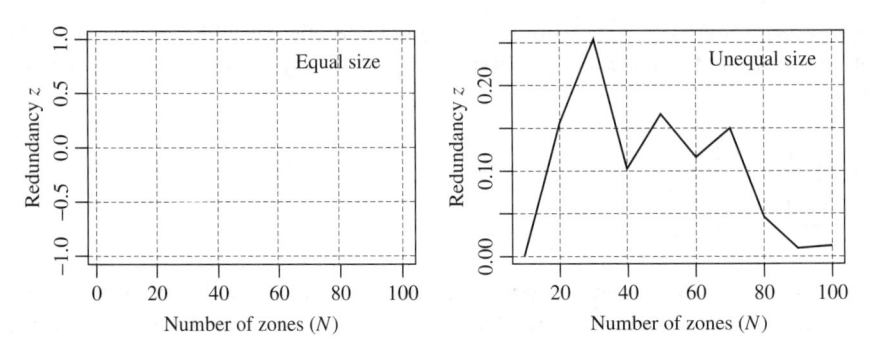

Figure 9.5 Relationship between number of zones and redundancy for equal (left) and unequal partitions (right).

which is the information redundancy measured by spatial entropy or more precisely by discrete approximation of spatial entropy. Substituting equation (9.25) into equation (9.27) yields

$$Z = \frac{\ln(N) + \sum_{i=1}^{N} p_i \ln(p_i)}{\ln(x)} \tag{9.28}$$

Equations (9.27) and (9.28) show that as the number of zones increases, Z converges to a limiting value.

Example 9.2: Compute the value of information redundancy Z using discrete entropy for the data in Example 9.1.

Solution: Since the spatial phenomenon of interest is the area of zones, p_i is proportional to the area of the zone i. In the case of even division, $p_i = 1/N$. From equation (9.24) one can see that the redundancy Z is a constant of 0 no matter how many intervals are used to divide the system. The relationship between the number of zones and the redundancy for equal partitions is presented in the left panel of Figure 9.5.

In the case of uneven division, the probability associated with each zone is calculated using the assumed numbers in Tables 9.1 and 9.2. First, the discrete entropy is computed for each interval number ($N = 10, 20, \ldots, 100$). The detailed calculation procedure and the final results can be found in Example 9.1. Then, applying equation (9.24), in which N takes on values of $10, 20, \ldots, 100$, redundancy Z can be computed. The relationship between the number of zones and redundancy for unequal zones is shown in the right panel of Figure 9.5.

9.2.2 Information gain

Statistical information gain is defined as the difference between the posterior distribution $\{p_i\}$ and the prior distribution $\{q_i\}$. Following Kullback (1959),

$$I = \sum_{i=1}^{N} p_i \ln\left(\frac{p_i}{q_i}\right), \quad \sum_{i=1}^{N} p_i = \sum_{i=1}^{N} q_i = 1 \tag{9.29}$$

where I varies between 0 and ∞. It is often referred to as cross-entropy of one distribution $P = \{p_1, p_2, \ldots, p_N\}$ with respect to another distribution $Q = \{q_1, q_2, \ldots, q_N\}$. The information gain can be considered as a measure of divergence, distance or difference between the two distributions $\{q_i\}$ and $\{p_i\}$. If $p_i = q_i$, $I = 0$, meaning no information is gained by moving from $\{q_i\}$ to $\{p_i\}$. If

$$q_i = \frac{\Delta x_i}{\sum\limits_{i=1}^{N} \Delta x_i} = \frac{\Delta x_i}{X} \tag{9.30}$$

where X is the area of the entire system. In this case, the spatial phenomenon of interest is assumed a priori as uniformly distributed over the entire spatial system or distributed proportionally or more generally, linearly, related to the zonal area over the system. Then,

$$I = \sum_{i=1}^{N} p_i \ln\left(\frac{p_i}{\Delta x_i/X}\right) = \ln(X) + \sum_{i=1}^{N} p_i \ln\left(\frac{p_i}{\Delta x_i}\right) = \ln(X) - H \tag{9.31}$$

where H represents the discrete approximation of spatial entropy as defined in equation (9.20). If $\Delta x_i = \Delta x$, $\forall i$, that is, the system X is partitioned into equal size subsystems, then $q_i = 1/N, i = 1, 2, \ldots, N$. In turn the information gain in equation (9.31) becomes

$$I = \ln(N) + \sum_{i=1}^{N} p_i \ln(p_i) = \ln N - H \tag{9.32}$$

where H represents the discrete Shannon entropy defined in equation (9.16). In this case, $\{q_i\}$ is noninformative.

Another useful measure is given by the difference between the maximum entropy and the actual entropy:

$$I = H_{\max} - H \tag{9.33}$$

where I denotes the information gain defined by Theil (1967). It can be shown that equation (9.33) would converge to the same value for both discrete and continuous entropies. Substituting equations (9.22) and (9.23) in equation (9.33), one obtains

$$I = \ln N + \sum_{i=1}^{N} p_i \ln(p_i) \tag{9.34}$$

which is the information gain measured by the discrete Shannon entropy.

Now, using spatial entropy or more precisely the discrete approximation of spatial entropy as defined in equation (9.20) to substitute for the discrete Shannon entropy in equation (9.34), the information gain, with system X being divided into N zones, can be expressed as

$$I = \ln X + \sum_{i=1}^{N} p_i \ln\left(\frac{p_i}{\Delta x_i}\right) \tag{9.35}$$

which is the information gain measured by spatial entropy or more precisely by discrete approximation of spatial entropy. The derivation from equation (9.33) to equation (9.35) is clear, considering the maximum value of the discrete approximation of spatial entropy is $\ln(X)$

as shown in Section 9.2.1. It can be seen that equation (9.35) will reduce to equation (9.34) when system X is equally divided, that is, $\Delta x_i = X/N$.

It seems helpful to make clear the relationship between statistical information gain defined in equation (9.29) and information gain defined in equation (9.33) or more specifically in equations (9.34) and (9.35). Statistical information gain in equation (9.29) is a general form of information gain in equation (9.33). Statistical information gain defines a measure of information change when moving from a prior distribution $\{q_i\}$ to a posterior distribution $\{p_i\}$. It may assume any real values. Assuming the prior distribution as $\{q_i\} := \{1/N\}$ for all i, that is, the phenomenon is uniformly distributed over the system and the system is equally partitioned, equation (9.29) will reduce to equation (9.34). On the other hand, assuming the prior distribution as $\{q_i\} := \{\Delta x_i/X\}$, that is, the phenomenon is uniformly distributed but the system may not be equally partitioned, equation (9.29) will reduce to equation (9.35). Following the framework of statistical information gain, that is, it is a measure of information change when moving from the prior distribution to the posterior distribution, equation (9.33) can be explained as the information gained when moving from the noninformative distribution (uniform distribution) to another distribution. $\{q_i\} := \{1/N\}$ and $\{q_i\} := \{\Delta x_i/X\}$ can be understood as two types of uniform distribution. The relation between statistics for redundancy and information gain using equations (9.21) and (9.33) becomes

$$Z = 1 - \frac{H}{H_{\max}} = \frac{H_{\max} - H}{H_{\max}} = \frac{I}{H_{\max}} \tag{9.36}$$

With the concept of information gain, the effect of interval size Δx_i can be evaluated. A system may be continuous, as for example, a watershed is. When its area is divided into zones, then a discrete representation becomes an approximation of the continuous system. In other words an integral is approximated by a finite sum, as for example, partitioning the watershed into zones or discretizing river flow. If the discrete entropy approximation is compared with its continuous form, the difference between the two would yield the loss of information and this loss would depend on the choice of Δx_i. Then one can set a criterion for the loss and in turn Δx_i:

$$\left| -\int_a^b f(x) \ln f(x)\, dx + \sum_{i=1}^N p_i \ln\left(\frac{p_i}{\Delta x_i}\right) \right| < \varepsilon \tag{9.37}$$

where a and b are the end points of the support of X, and ε is tolerance limit.

Example 9.3: Compute the value of information gain I for the data in Example 9.1.

Solution: Prior to solution, it is better to make clear that the information gain I can be measured either by discrete Shannon entropy as in equation (9.34) or by (discrete approximation of) spatial entropy as in equation (9.35).

First, compute the information gain by equation (9.34) for equal and unequal partitions, respectively. Let us assume the system is equally partitioned into $N = 10$ zones. According to equation (9.34), we need to calculate the probabilities p_i and then the discrete Shannon entropy. Since the spatial phenomenon of interest is the zonal area, p_i can be computed as $p_i = 1/N = 0.1$, and the discrete entropy is calculated as

$$H_{equal}(10) = -\sum_{i=1}^{10} \frac{1}{10} \ln\left(\frac{1}{10}\right) = \ln(10)$$

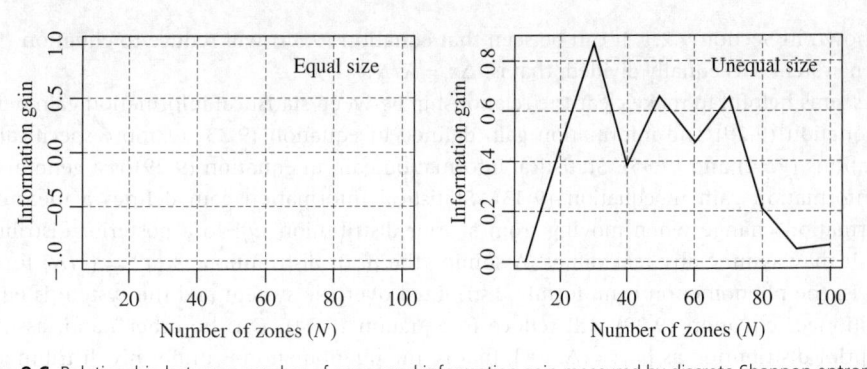

Figure 9.6 Relationship between number of zones and information gain measured by discrete Shannon entropy for equal (left) and unequal (right) partitions.

Substituting the discrete entropy in equation (9.33) yields

$$I_{equal}(10) = \ln(10) - H_{equal}(10) = \ln(10) - \ln(10) = 0$$

In the same way, the information gain for other interval numbers can be computed and plotted in the left panel of Figure 9.6. Apparently, the information gain is 0 in the case of equal size intervals no matter how many interval numbers are used.

When the system is divided into unequal size intervals, the probability associated with each sub-zone can be computed using the assumed numbers in Tables 9.1 and 9.2. The calculation can be found in Example 9.1. For ease of understanding, we repeat the calculation procedure here. Again take $N = 10$ as an example, Tables 9.1 and 9.2 show that the system is divided into three types of sub-zones. For type 1 whose the interval size is $10\,km^2$, there are in total two sub-zones with an area of 10. For type 2, the interval size is $11.7\,km^2$ and there are in total three sub-zones with a size of $11.7\,km^2$. Similarly for type 3, the interval size is $9\,km^2$ and the total number of sub-zones with this interval size is 5. Knowing this the discrete probability for each sub-zone can be computed by dividing the corresponding interval size by the total area of the system, which is 100 here. Then, the discrete entropy is computed as:

$$H_{unequal}(10) = -\sum_{i=1}^{10} p_i \ln(p_i) = \underbrace{-\frac{10}{100}\ln\left(\frac{10}{100}\right) - \frac{10}{100}\ln\left(\frac{10}{100}\right)}_{n_1=2}$$

$$\underbrace{-\frac{11.7}{100}\ln\left(\frac{11.7}{100}\right) - \frac{11.7}{100}\ln\left(\frac{11.7}{100}\right) - \frac{11.7}{100}\ln\left(\frac{11.7}{100}\right)}_{n_2=3}$$

$$\underbrace{-\frac{9}{100}\ln\left(\frac{9}{100}\right) - \frac{9}{100}\ln\left(\frac{9}{100}\right) - \frac{9}{100}\ln\left(\frac{9}{100}\right) - \frac{9}{100}\ln\left(\frac{9}{100}\right) - \frac{9}{100}\ln\left(\frac{9}{100}\right)}_{n_3=5}$$

$$= 2.297 \text{ nats}$$

where n_i denotes the number of type i sub-zones. Similarly for the case of $N = 20$, the discrete entropy is computed as:

$$H_{unequal}(20) = -\sum_{i=1}^{20} p_i \ln(p_i) = \underbrace{-\frac{10}{100} \ln\left(\frac{10}{100}\right)}_{n_1=1}$$

$$\underbrace{-\frac{12.5}{100} \ln\left(\frac{12.5}{100}\right) - \frac{12.5}{100} \ln\left(\frac{12.5}{100}\right)}_{n_2=2}$$

$$\underbrace{-\frac{17.5}{100} \ln\left(\frac{17.5}{100}\right) - \frac{17.5}{100} \ln\left(\frac{17.5}{100}\right)}_{n_3=2}$$

$$\underbrace{-\frac{2}{100} \ln\left(\frac{2}{100}\right) - \frac{2}{100} \ln\left(\frac{2}{100}\right) - \frac{2}{100} \ln\left(\frac{2}{100}\right) - \frac{2}{100} \ln\left(\frac{2}{100}\right) - \frac{2}{100} \ln\left(\frac{2}{100}\right)}_{n_4=5}$$

$$\underbrace{-\frac{2}{100} \ln\left(\frac{2}{100}\right) - \frac{2}{100} \ln\left(\frac{2}{100}\right) - \ldots\ldots - \frac{2}{100} \ln\left(\frac{2}{100}\right)}_{n_5=10}$$

$$= 2.5338 \text{ nats.}$$

Using equation (9.35), the corresponding information gains can be computed as

$$I_{unequal}(10) = \ln(10) - H_{unequal}(10) = 2.3026 - 2.2972 = 0.0054 \text{ nats.}$$

$$I_{unequal}(20) = \ln(20) - H_{unequal}(20) = 2.9957 - 2.5338 = 0.4619 \text{ nats.}$$

We caution that for a given number of intervals, there are many different ways to partition the system. The partition given in Tables 9.1 and 9.2 represents only one of the possible partitions. For other interval numbers the information gains can be computed similarly and are plotted in the right panel of Figure 9.6.

Second, compute the information gain by equation (9.35) for equal and unequal partitions, respectively. When the system is equally partitioned, equation (9.26) will reduce to equation (9.34), as shown below:

$$I = \ln X + \sum_{i=1}^{N} p_i \ln\left(\frac{p_i}{\Delta x_i}\right) = \ln X + \sum_{i=1}^{N} p_i \ln\left(\frac{p_i}{X/N}\right) = \ln(N) + \sum_{i=1}^{N} p_i \ln(p_i)$$

Therefore the information gain measured by spatial entropy will be the same as that measured by discrete Shannon entropy, that is, the information gain [computed by equation (9.35)] is 0 when the system is equally partitioned no matter how many interval numbers are used, as illustrated in the left panel of Figure 9.7.

When the system is unequally divided, first using numbers given in Tables 9.1 and 9.2, the probabilities p_i can be computed. Then using equation (9.35) the information gain can be

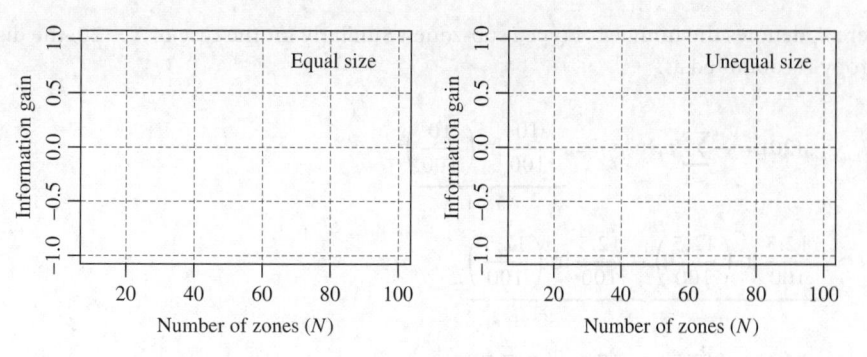

Figure 9.7 Relationship between number of zones and information gain measured by (discrete approximation of) spatial entropy for equal (left) and unequal (right) partitions.

computed. Here, again taking $N = 10$ as an example, the information gain measured by spatial entropy is

$$I(10) = \ln(X) + \sum_{i=1}^{10} p_i \ln \left(\frac{p_i}{\Delta x_i} \right)$$

$$= \ln(100) + \underbrace{\frac{10}{100} \ln \left(\frac{10/100}{10} \right) + \frac{10}{100} \ln \left(\frac{10/100}{10} \right)}_{n_1 = 2}$$

$$+ \underbrace{\frac{11.7}{100} \ln \frac{11.7/100}{11.7} + \frac{11.7}{100} \ln \frac{11.7/100}{11.7} + \frac{11.7}{100} \ln \frac{11.7/100}{11.7}}_{n_2 = 3}$$

$$+ \underbrace{\frac{9}{100} \ln \frac{9/100}{9} + \frac{9}{100} \ln \frac{9/100}{9} + \frac{9}{100} \ln \frac{9/100}{9} + \frac{9}{100} \ln \frac{9/100}{9} + \frac{9}{100} \ln \frac{9/100}{9}}_{n_3 = 5}$$

$$= \ln(100) + \ln \left(\frac{1}{100} \right) = 0 \text{ nats}$$

The above calculation signifies that in the present example the information gain measured by spatial entropy (or more precisely discrete approximation of spatial entropy) is always null, no matter how the system is partitioned, as indicated in the right panel of Figure 9.7. This result is expected since the spatial phenomenon of interest of the present example is the zonal area which is definitely uniformly distributed over the spatial system, which means that probability distribution of the zonal area can be fully represented by $\{\Delta x_i / X\}$. Considering the fact that the information gain in equation (9.35) assumes a prior distribution of $\{q_i\} := \{\Delta x_i / X\}$, which means in this example the posterior distribution is the same as the prior distribution.

Example 9.4: How much loss in information will there be when a continuous form of entropy is represented by a discrete form employing different class interval sizes?

Solution: Prior to the solution, one point worth noting is that the discrete Shannon entropy as defined in equation (9.16) is not an approximation of its analogous continuous Shannon entropy defined in equation (9.17). In other words, the continuous Shannon entropy cannot be derived from the discrete Shannon entropy by letting the number of intervals in equation (9.16) tend to infinity and passing to the limit. The continuous Shannon entropy is exactly equal to the spatial entropy as defined in equation (9.18). In practice, the spatial entropy is commonly approximated by its discrete form as in equation (9.20). In this question, therefore, the discrete form of entropy is referred to the discrete approximation of the spatial entropy, that is, equation (9.20).

Referring to equation (9.33), with the increase in the number of zones, the difference between continuous form and discrete form of entropy should be reduced. For purposes of illustration, it is supposed that the continuous distribution is a truncated exponential distribution at 10 from the right, whose probability density function is given as

$$f(x) = A \exp(-x), \quad A = \frac{1}{1 - \exp(-10)} \approx 1, 0 \le x \le 10$$

First, compute the continuous Shannon entropy (or the spatial entropy) as

$$H = -\int_0^{10} f(x) \ln f(x) dx = \exp(-x)(Ax - A - A \ln A)|_0^{10} = 1.0004 \text{ nats}$$

The simplest case is considering the entire support as a single interval, then the discrete probability is $p = 1$ and the interval size is $\Delta x = 10$. According to equation (9.20), the discrete approximation of spatial entropy is

$$H_s(1) = -1 \times \ln\left(\frac{1}{10}\right) = 2.3026 \text{ nats}$$

Similarly, when dividing the support into two equal sized intervals, that is, $\Delta x_i = 5$, the probability mass p can be computed as

$$p_1 = \int_0^5 A \exp(-x) dx = 0.9933$$

$$p_2 = \int_5^{10} A \exp(-x) dx = 1 - \int_0^5 A \exp(-x) dx = 0.0067$$

Substituting p_1 and p_2 into equation (9.20), one obtains

$$H_s(2) = -\sum_{i=1}^2 p_i \ln\left(\frac{p_i}{\Delta x_i}\right) = -0.9933 \times \ln\left(\frac{0.9933}{5}\right) - 0.0067 \times \ln\left(\frac{0.0067}{5}\right) = 1.6496 \text{ nats}$$

Following the same procedure, one can compute $H_s(3), H_s(4)$ .... The difference between the continuous entropy and its discrete approximations with respect to different numbers of intervals is illustrated in Figure 9.8. This figure shows that as the number of intervals increases, the divergence between the continuous entropy and its discrete approximation becomes smaller and smaller, and the discrete approximation will converge to the continuous entropy as the number of intervals approaches a sufficiently large value.

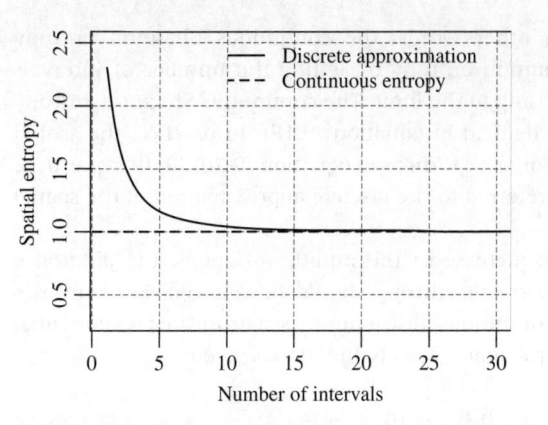

Figure 9.8 Spatial entropy as a function of number of intervals.

9.2.3 Disutility entropy

In the case of spatial distribution and location, the Shannon entropy

$$H = -k\sum_{i=1}^{N} p_i \ln p_i \tag{9.38}$$

can be considered as a measure of accessibility or interactivity (Erlander and Stewart, 1978), wherein p_i is the probability of locating in zone i of a city with N zones, and k is a constant of proportionality. If the cost of location or accessibility is regarded as a random variable then the probability distribution of location or accessibility entropy can be determined by maximizing entropy. Because location, accessibility, and cost of location are related, the probability of location can be regarded as a function of the cost of location. H can be maximized subject to

$$\sum_{i=1}^{N} p_i = 1 \tag{9.39}$$

$$\sum_{i=1}^{N} r_i p_i = \bar{r} = r_m \tag{9.40}$$

where r_i is the cost of locating in zone i ($r_i \geq 0$) and r_m is the mean cost of locating in the city. One can also interpret r_i as a measure of distance from some point in the city, such as central area, to zone i. Thus, H corresponds to accessibility and r_m to cost, and these can be used to classify spatial distributions of different areas or cities.

A weighted entropy measure (Belis and Guiasu, 1968) can be defined by combining the two measures as

$$I = -k\sum_{i=1}^{N} r_i p_i \ln p_i \tag{9.41}$$

With k disregarded, equation (9.41) is the Shannon entropy with $p_i \ln p_i$ weighted by r_i, and is referred to as disutility entropy (Theil, 1980). The disutility of any location i is a function of the cost r_i of locating there (Beckmann, 1974). The weight r_i indicates the utility of event i.

If $r_i = r$, for all i, that is, the distribution of weights is immaterial, then H and I measure the same variation. In general I and H are the same. If $H = 0$, all activity is concentrated in one zone, accessibility is at a minimum. When the probability distribution is uniform, $H = \ln N$, the accessibility is maximum. On the other hand, disutility entropy reaches its maximum if the probabilities are distributed as a negative exponential function of the inverse of the disutility (Guiasu, 1977) but has a minimum equivalent to H.

It can be shown that $H(N) \geq H(N-1)$ and $H(N) \to \infty$ as $N \to \infty$. This means that the greater the number of zones characterizing the system, the greater the information imparted to an observer of the system. One can make the same argument for the disutility entropy $I(N)$, but only if the aggregation is nested and aggregate weights are weighted averages. As an example, consider two zones i and j which are combined to form zone k in moving from the N-th to the $(N-1)$-th level. If

$$r_k = \frac{r_i p_i + r_j p_j}{p_i + p_j} \tag{9.42}$$

it can be shown that $I(N) \geq I(N-1)$ (Guiasu, 1977). This means that as activity grows the city is described by more and more zones both accessibility and disutility entropies will increase. This would correspond to finer and finer partition of a fixed space.

9.3 One dimensional aggregation

In hydrology, geography, climatology, and watershed sciences, frequently one aggregates spatially varying units into a finite number of relatively homogeneous units such that the system heterogeneity is more or less preserved. For example, an urban watershed may have a large number of land uses and it may be desirable to group them into a smaller number of relatively uniform land use zones. Likewise, a watershed may be divided into a number of zones each having its own population density. A state may be divided into a number of meteorologically homogenous regions. Consider a watershed whose time-area diagram is constructed. To that end, it is divided into different zones based on isochrones or contours of equal travel time from the outlet. Each zone or sub-area enclosed by two isochrones has an area. The rain falling on this area will produce water. Thus, the water from this area will take a certain amount of time to reach the watershed outlet. Let r define the distance in terms of time of travel of water from that area to the outlet. The farthest sub-area or zone is bounded by R, the boundary of the upstream watershed, the time of concentration or the longest time of travel. The question is: How many isochrones or subdivisions should one use?

It may be useful to discuss aggregation using a simple truncated exponential probability density function $f(r)$:

$$f(r) = K \exp(-ar) \tag{9.43}$$

where a is parameter of the probability density function and is related to the mean travel distance to the outlet, r_m, and K is a normalizing constant so that the density function integrates to 1 over its support, that is, from 0 to its boundary R, and defined as

$$K = \frac{a}{1 - \exp(-aR)} \tag{9.44}$$

Equation (9.43) is bounded from $0 \leq r \leq R$, where R is the boundary. The mean travel distance can be obtained from the first moment of $f(r)$ about the origin located at the outlet as:

$$r_m = \int_0^R rK \exp(-ar) dr = \frac{1}{a} - \frac{\exp(-aR)R}{1 - \exp(-aR)} \tag{9.45}$$

Solving equation (9.45), the value of parameter a can be obtained for known r_m. It is seen from equation (9.45) that if R is very large, that is, $R \to \infty$, the second term on the right side vanishes which can be easily obtained from the L'Hospital's rule, and one simply gets $a = 1/r_m$.

In urban areas where aggregating land uses is a main concern, the choice of the regional boundary becomes important. This can be addressed by computing the entropy for $R \to \infty$ and the entropy for finite R. For any R, the continuous entropy of equation (9.43) can be expressed as

$$H = -\int_0^R f(r) \ln[f(r)] dr = -\int_0^R K \exp(-ar) \ln[K \exp(-ar)] dr$$

$$= -\frac{a}{1 - \exp(-aR)} \int_0^R \exp(-ar) \{\ln a - ar - \ln[1 - \exp(-aR)]\} dr \tag{9.46}$$

Integrating the right hand side by parts, equation (9.46) simplifies to

$$H = -\ln a + 1 - \frac{a \exp(-aR)R}{1 - \exp(-aR)} + \ln[1 - \exp(-aR)] \tag{9.47}$$

For $R \to \infty$, equation (9.47) yields

$$H = -\ln a + 1 \tag{9.48}$$

Subtracting equation (9.47) from equation (9.48), the difference in entropy $H(e)$ (between finite R and R tending to ∞) is:

$$H(e) = -\int_R^{\infty} f(r) \ln f(r) dr = \frac{a \exp(-aR)R}{1 - \exp(-aR)} - \ln[1 - \exp(-aR)] \tag{9.49}$$

Taking the ratio of equation (9.49) to equation (9.47) and setting the tolerance limit as 0.05 or any other value, one gets:

$$\frac{a \exp(-aR)R - [1 - \exp(-aR)] \ln[1 - \exp(-aR)]}{(-\ln a + 1)[1 - \exp(-aR)] - a \exp(-aR)R + [1 - \exp(-aR)] \ln[1 - \exp(-aR)]} = 0.05 \tag{9.50}$$

For different ratios, different values of boundaries would be obtained. In this manner boundaries of any monocentric region can be determined. Also if the boundary is given according to prior information, solving equation (9.50) numerically yields estimate for the

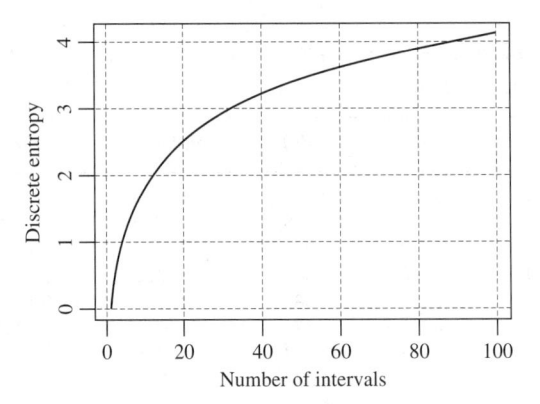

Figure 9.9 The relationship between discrete entropy and the number of intervals, assuming the distribution given in equation (9.43) with parameter $R = 10$, $a = 0.3979$.

parameter a. Now consider $H(e)/H = 0.05$. If R is set to equal 10, a is calculated as 0.3979 from equation (9.50). The discrete entropy is calculated using the following equations:

$$H = -\sum_i p_i \ln p_i = -\sum_i p(x_i)\Delta x_i \ln\left[p(x_i)\Delta x_i\right] \tag{9.51}$$

$$p(x_i) = \frac{a}{1 - \exp(-aR)} \exp(-ax_i) \tag{9.52}$$

The relationship between the number of intervals and the discrete entropy is shown in Figure 9.9. Obviously, the discrete entropy increases without bound as the number of interval increases. The continuous entropy computed from equation (9.47) is 1.8646 nats. This confirms that the continuous Shannon entropy cannot be derived from the discrete Shannon entropy by letting the number of intervals tend to infinity and passing to the limit. The continuous Shannon entropy is equivalent to the spatial entropy, which, in practice, can be approximated by the discrete approximation of spatial entropy, as defined in equation (9.20).

Example 9.5: Compute the value of entropy for different values of R: 10, 20, 30, 40, 50, 60, 70, and 80 minutes and plot $H(R)$. Notice that time is taken here as a measure of distance. Compute the value of a for different values of R. Also compute parameter a when $R \to \infty$. Then compute the difference in entropy $H(e)$ between $H(R)$ and $H(R \to \infty)$ and plot it.

Solution: Different values of R will correspond to different values of a. The relationship of R and a is described in equation (9.44). Given a value of R, numerically solving for the zero of equation (9.50) yields a. For example, assuming the ratio $H(e)/H$ be 0.05 and R be 10, a will be 0.3979. After knowing R and a, the continuous entropy $H(R)$ can be computed via equation (9.47). In the case of $R = 10$ minutes, the continuous entropy $H(R)$ is 1.8646 nats. When $a = 0.3979$, the limiting entropy $H(R \to \infty)$ is computed by equation (9.48), that is, $H(R \to \infty) = -\ln(0.3979) + 1 = 1.9216$ nats. Thus, the difference between $H(R)$ and $H(R \to \infty)$ is computed as

$$H(R \to \infty) - H(R) = 1.9216 - 1.8646 = 0.057 \text{ nats}$$

Following the same way, the values of a, $H(R)$ and $H(e)$ can be computed and are tabulated in Table 9.3a. The relationships between R and $H(R)$, and between R and $H(e)$ is plotted in the left and right panels of Figure 9.9a, respectively.

Table 9.3a Parameter a, $H(R)$, $H(R \to \infty)$ and $H(e)$ for different R values with $H(e)/H = 0.05$.

R	a	$H(R)$	$H(R \to \infty)$	$H(e)$
10	0.3979	1.8268	1.9216	0.0947
20	0.1773	2.5953	2.7299	0.1346
30	0.1118	3.0340	3.1910	0.1570
40	0.0808	3.3427	3.5158	0.1731
50	0.0629	3.5806	3.7662	0.1856
60	0.0513	3.7743	3.9701	0.1957
70	0.0432	3.9376	4.1419	0.2043
80	0.0373	4.0778	4.2888	0.2109

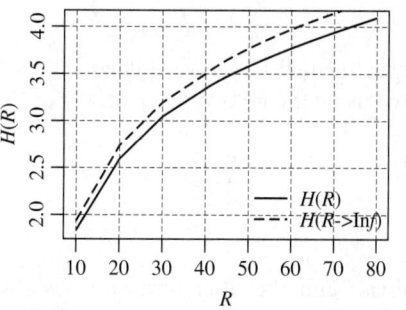 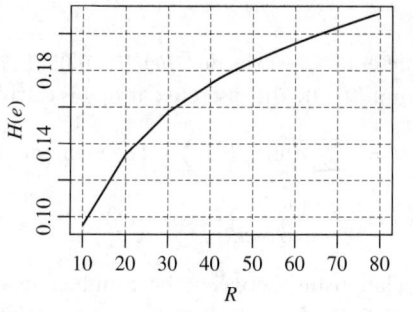

Figure 9.9a Relationship between R represented in minutes and continuous entropy, the continuous entropy evaluated when R approaches infinity and the difference entropy $H(e)$ with the ratio of $H(e)/H = 0.05$.

Table 9.3b Parameter a, $H(R)$, $H(R \to \infty)$ and $H(e)$ for different R values with $H(e)/H = 0.10$.

R	a	$H(R)$	$H(R \to \infty)$	$H(e)$
10	0.2961	2.0023	2.2171	0.2148
20	0.1279	2.7611	3.0565	0.2954
30	0.0791	3.1953	3.5370	0.3417
40	0.0565	3.5000	3.8735	0.3835
50	0.0436	3.7349	4.1327	0.3978
60	0.0352	3.9270	4.4367	0.4197
70	0.0295	4.0878	4.5234	0.4355
80	0.0253	4.2271	4.6769	0.4498

For different values of ratio $H(e)/H$, results of calculations are shown in Tables 9.3b and 9.3c, and plotted in Figures 9.9b and 9.9c.

So far, the above solutions illustrate a way to determine the distribution parameter a from a given reasonable boundary approximation. Now assuming parameter a is already known, say to be a value of 0.033, which implies that the expected travel distance r_m is 30 minutes. This implication can be easily seen from equation (9.45). We intend to explore the difference between the real continuous entropy (as R approaches infinity) and

Table 9.3c Parameter a, $H(R)$, $H(R \rightarrow \infty)$ and $H(e)$ for different R values with $H(e)/H = 0.15$.

R	a	$H(R)$	$H(R \rightarrow \infty)$	$H(e)$
10	0.2348	2.1005	2.4490	0.3485
20	0.0984	2.8485	3.3187	0.4702
30	0.0598	3.2770	3.8168	0.5397
40	0.0421	3.5785	4.1677	0.5892
50	0.0322	3.8106	4.4358	0.6252
60	0.0258	4.0001	4.6574	0.6573
70	0.0215	4.1592	4.8397	0.6805
80	0.0183	4.2973	5.0008	0.7036

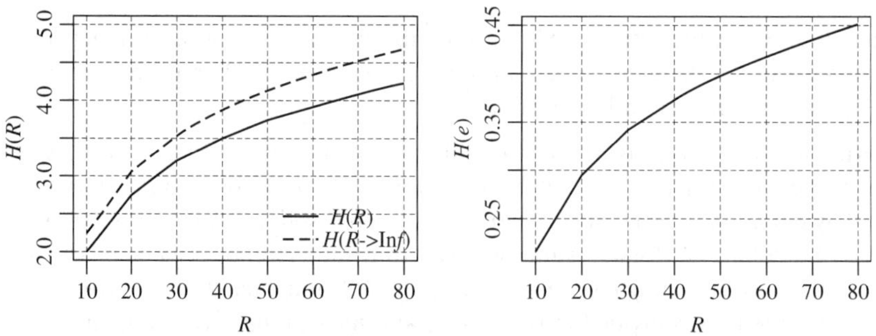

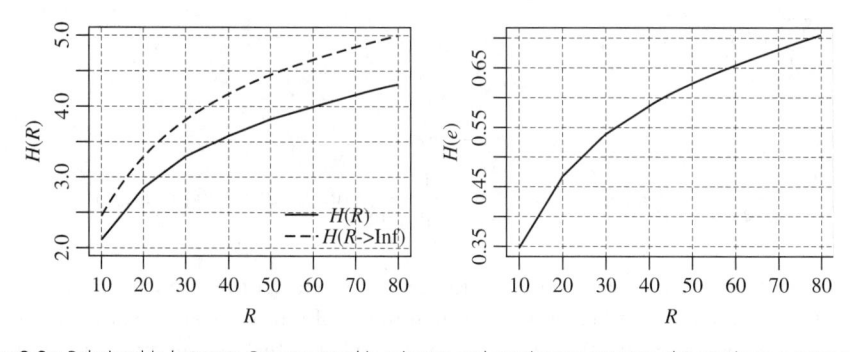

Figure 9.9b Relationship between R represented in minutes and continuous entropy, the continuous entropy evaluated when R approaches infinity and the difference entropy $H(e)$ with the ratio of $H(e)/H = 0.10$.

Figure 9.9c Relationship between R represented in minutes and continuous entropy, the continuous entropy evaluated when R approaches infinity and the difference entropy $H(e)$ with the ratio of $H(e)/H = 0.15$.

its approximation, assuming a finite boundary R. The real value of continuous entropy is computed by equation (9.48), that is, $H = - \ln(0.033) + 1 = 4.4112$ nats. Assuming boundary R be a finite value, the approximation of continuous entropy is evaluated by equation (9.47). Finally, the difference $H(e)$ can be computed. Or it can also be computed from equation (9.49) directly. The relationship between R represented in minutes and $H(e)$ with $a = 0.033$ is computed as shown in Figure 9.10. From the figure it is clear that when R approaches infinity, $H(e)$ approaches zero.

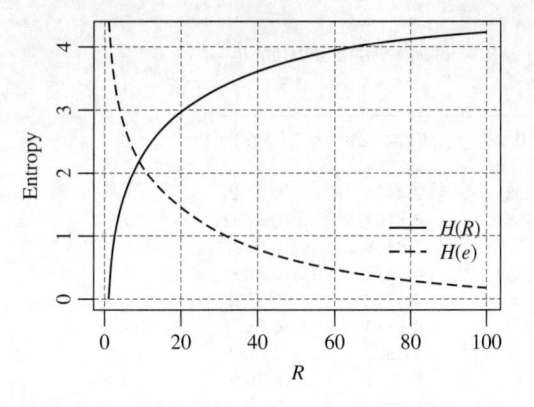

Figure 9.10 Relationship between R represented in minutes and $H(e)$ with $a = 0.033$ and $r_m = 30$.

Example 9.6: Compute discrete (Shannon) entropy as a function of number of zones for different values of R. Take the number of zones up to 80. Plot the discrete entropy as a function of number of zones on a semi-logarithm coordinates.

Solution: Three cases, that is, $R = 20$, $R = 40$ and $R = 60$, are considered for illustrating the relationship between discrete entropy and the number of zones. For each R, parameter a can be estimated by equation (9.50). Take $R = 20$ as an example, solving equation (9.50) numerically yields the value of 0.1773 for a, assuming the entropy ratio $H(e)$ be 0.05. Similarly, for $R = 40$ and $R = 60$ the corresponding a estimates will be 0.0808 and 0.0512, respectively.

A specific value of R is divided into N zones, and their probability p_i and discrete entropy H are calculated, respectively. Suppose points $r_1, r_2, \ldots, r_{N+1}$ divide the spatial system into N equally sized intervals. The discrete entropy can be calculated as follows:

$$H = -\sum_{i=1}^{N} p_i \ln p_i$$

where

$$p_i = \int_{r_i}^{r_{i+1}} \frac{a}{1 - \exp(-aR)} \exp(-ar)\,dr = \frac{1}{a - \exp(-aR)}\left[\exp(-ar_i) - \exp(-ar_{i+1})\right]$$

For each boundary R and each interval number, discrete entropy can be computed. The relationship between discrete entropy and the number of intervals can be shown by plotting the discrete entropy as a function of the interval number, which is presented in Figure 9.11. Obviously, the discrete entropy increases without bound as the number of zones increases. In the semi-logarithm coordinates, the discrete entropy approximately linearly increases with the logarithm of the number of intervals.

Example 9.7: Compute the discrete approximation of continuous entropy or spatial entropy as a function of number of zones for different values of R. Take N up to 80 on a log scale. Also compute the continuous entropy corresponding to each boundary value R and see how close does the discrete approximation of continuous entropy or spatial entropy get to become the continuous entropy?

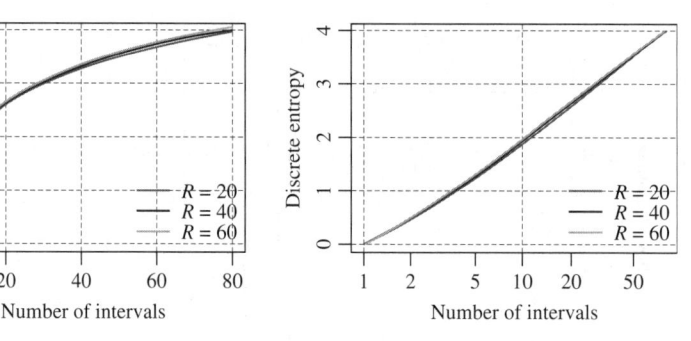

Figure 9.11 Relationship between the discrete entropy and the number of zones in natural coordinate (left) and semi-logarithm coordinates (right).

Solution: Similarly as in Example 9.6, three cases are considered, that is, $R = 20$, $R = 40$, and $R = 60$. Also for each R parameter a can be estimated by equation (9.44), assuming the entropy ratio $H(e)$ to be 0.05. The estimated a values corresponding to $R = 20$, $R = 40$, and $R = 60$ are 0.1733, 0.0808, and 0.0512, respectively. For a given R, the spatial system is divided into $1, 2, \ldots, 80$ zones. Their probability p_i and discrete entropy H are calculated. For generalization, suppose the spatial system is divided into N equally sized intervals by $r_1, r_2, \ldots, r_{N+1}$. The discrete approximation of continuous entropy (spatial entropy) is expressed as follows:

$$H = -\sum_{i=1}^{N} p_i \ln \frac{p_i}{\Delta x_i}$$

where

$$p_i = \int_{r_i}^{r_{i+1}} \frac{a}{1 - \exp(-aR)} \exp(-ar)dr = \frac{1}{a - \exp(-aR)} \left[\exp(-ar_i) - \exp(-ar_{i+1}) \right]$$

Then the discrete approximation of continuous entropy is computed and is plotted, as shown in Figure 9.12 which shows that with the increase in the number of intervals the discrete approximation of continuous entropy decreases but approaches a bound.

Actually the boundary value of the discrete approximation of continuous entropy is the continuous entropy or spatial entropy, which can be directly calculated from equation (9.47) for a given R and thus a given a. For example, when $R = 20$, the parameter a can be obtained from the solution of equation (9.50) as 0.1733, assuming an entropy ratio $H(e)$ of 0.05. Substituting $R = 20$ and $a = 0.1733$ into equation (9.47), the continuous entropy is computed as

$$H = -\ln a + 1 - \frac{a\exp(-aR)R}{1 - \exp(-aR)} + \ln\left[1 - \exp(-aR)\right]$$

$$= -\ln(0.1733) + 1 - \frac{0.1733 \times \exp(-0.1733 \times 20) \times 20}{1 - \exp(-0.1733 \times 20)} + \ln\left[1 - \exp(-0.1733 \times 20)\right]$$

$$= 2.5953 \text{ nats.}$$

Investigating the behavior of the discrete approximation of continuous entropy with respect to the continuous entropy, one can get a picture about how closely the discrete approximation of

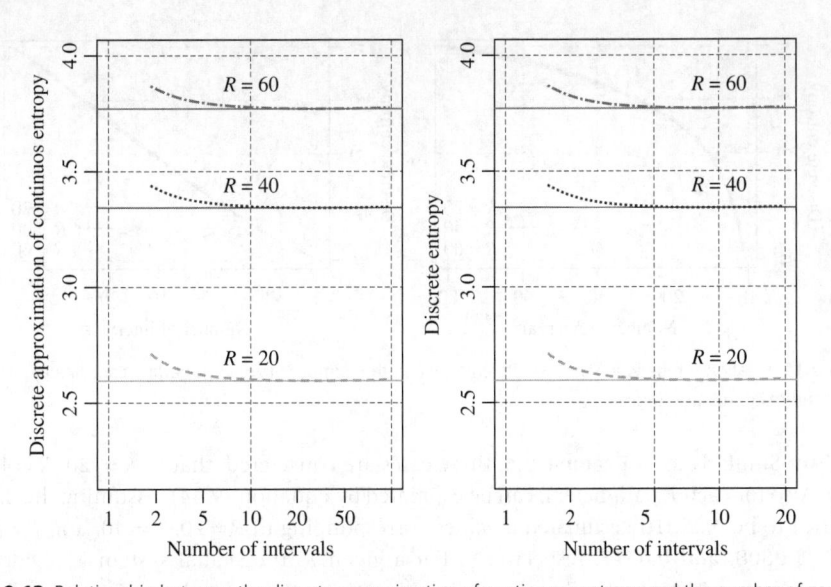

Figure 9.12 Relationship between the discrete approximation of continuous entropy and the number of zones in semi-logarithm coordinates. Left: the interval numbers range from 1 to 80. Right: the interval numbers range from 1 to 20 for the convenience to show the decreasing trend. The corresponding continuous entropies are depicted as horizontal lines.

continuous entropy approaches the continuous entropy as the number of intervals increases. As expected, Figure 9.12 clearly shows that as the number of intervals increases the discrete approximation of continuous entropy convergence to the corresponding continuous entropy, which is shown by the horizontal line.

9.4 Another approach to spatial representation

Consider the cost of location C as a random variable with a probability density function $f(c)$, minimum value as c_0 and a maximum value as c_{max}. Let it be assumed that the average cost of location is empirically known. The Shannon entropy for the cost can be written as

$$H(C) = - \int_{c_0}^{c_{max}} f(c) \ln f(c) dc \tag{9.53}$$

The objective is to determine $f(c)$ subject to

$$\int_{c_0}^{c_{max}} f(c) dc = 1 \tag{9.54}$$

$$\int_{c_0}^{c_{max}} c f(c) dc = \bar{c} = c_m \tag{9.55}$$

Applying POME, the PDF of C using the method of Lagrange multipliers is

$$f(c) = \exp(-\lambda_0 - \lambda_1 c) \tag{9.56}$$

where λ_0 and λ_1 are the Lagrange multipliers which can be determined by using equations (9.54) and (9.55). Substituting equation (9.56) in equation (9.54), one gets

$$\exp(-\lambda_0) \int_{c_0}^{c_{max}} \exp(-\lambda_1 c) dc = 1 \tag{9.57}$$

or

$$\exp(\lambda_0) = \frac{1}{\lambda_1} \left[\exp(-\lambda_1 c_0) - \exp(-\lambda_1 c_{max}) \right] \tag{9.58}$$

Therefore,

$$\lambda_0 = \ln \left[\exp(-\lambda_1 c_0) - \exp(-\lambda_1 c_{max}) \right] - \ln \lambda_1 \tag{9.59}$$

Now substituting equation (9.56) in equation (9.55), one gets

$$\int_{c_0}^{c_{max}} c \exp(-\lambda_0 - \lambda_1 c) dc = \bar{c} = c_m \tag{9.60}$$

Equation (9.60), with the use of equation (9.59), leads to

$$c_m \frac{1}{\lambda_1} \left[\exp(-\lambda_1 c_0) - \exp(-\lambda_1 c_{max}) \right] = \frac{1}{\lambda_1} \left[c_0 \exp(-\lambda_1 c_0) - c_{max} \exp(-\lambda_1 c_{max}) \right] +$$

$$+ \frac{1}{\lambda_1^2} \left[\exp(-\lambda_1 c_0) - \exp(-\lambda_1 c_{max}) \right] \tag{9.61}$$

Equation (9.61) contains one unknown λ_1 and can therefore be solved in terms of c_0, c_m, and c_{max}.

Now it is hypothesized that

$$F(c) = P(C \le c) = \frac{x}{L}, \quad x_0 \le x \le L \tag{9.62}$$

Here x is travel time. Differentiation of equation (9.62) yields

$$dF(c) dc = f(c) dc = \exp(-\lambda_0 - \lambda_1 c) dc = \frac{dx}{L} \tag{9.63}$$

Integrating equation (9.63), the result is

$$\int_{c_0}^{c} \exp(-\lambda_0 - \lambda_1 c) dc = \frac{1}{L} \int_{x_0}^{x} dx \tag{9.64}$$

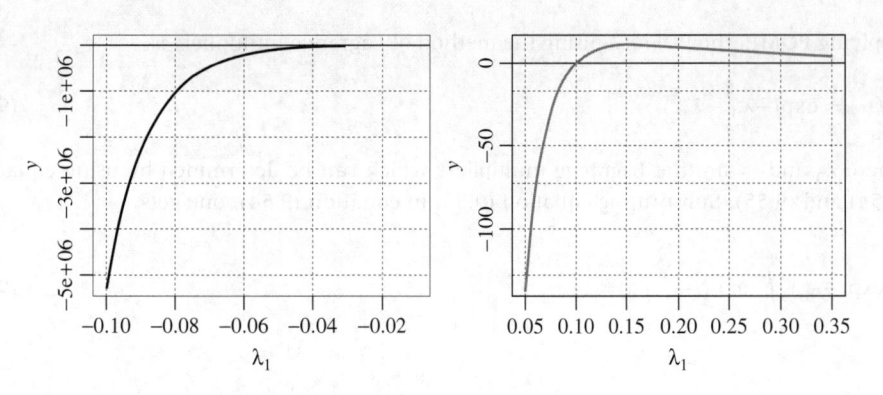

Figure 9.13 Behavior of equation (9.61) as the Lagrange multiplier λ_1 varies.

Equation (9.64) results in

$$c = -\frac{1}{\lambda_1} \ln\left[\exp(-\lambda_1 c_0) - \frac{\lambda_1}{L}(x - x_0)\exp(\lambda_0)\right] \tag{9.65}$$

The entropy in this case is

$$H(C) = \lambda_0 + \lambda_1 \bar{c} \tag{9.66}$$

Example 9.8: Considering the average time of travel as 15 minutes, the minimum time of travel as 5 minutes and the maximum as 90 minutes, compute the values of the Lagrange multipliers λ_0 and λ_1. Plot the probability density function of time of travel.

Solution: Substituting $c_m = 15$ minutes, $c_0 = 5$ minutes, and $c_{max} = 90$ minutes into equation (9.61) and solving it with the use of the bisection method or some other more advanced numerical techniques, one can get $\lambda_1 = 0.1$. To obtain ideas about the behavior of equation (9.61) with respect to the Lagrange multiplier λ_1, Figure 9.13 presents the plots of equation (9.61) versus λ_1, from which it is clear where to locate the zero of the equation. One may find that there are some other intervals of λ_1, for which equation (9.61) is also pretty close to 0. It is necessary to caution that more efforts are required when solving for the Lagrange multiplier.

After obtaining λ_1, λ_0 is calculated by equation (9.59) as follows:

$$\lambda_0 = \ln\left[\exp(-\lambda_1 c_0) - \exp(-\lambda_1 c_{max})\right] - \ln\lambda_1$$
$$= \ln\left[\exp(-0.1 \times 5) - \exp(-0.1 \times 90)\right] - \ln(-0.1) = 1.8$$

Finally, substituting λ_0 and λ_1 into equation (9.56), the probability density function of time of travel can be achieved and plotted as shown in Figure 9.14.

Example 9.9: Let the maximum distance of travel, L, be 20 kilometers and the minimum distance of travel be 0.5 kilometer. Using the values of the Lagrange multipliers obtained in Example 9.8, determine and plot the time of travel as a function of distance.

Solution: Substituting $\lambda_0 = 1.8$, $\lambda_1 = 0.1$, $x_0 = 0.5$ and $L = 90$ into equation (9.65), the function between the travel time and distance can be obtained and plotted as shown in Figure 9.15.

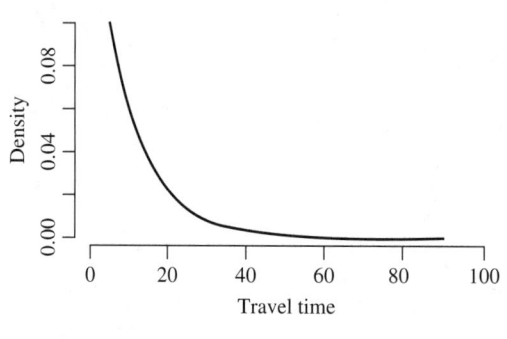

Figure 9.14 Probability density of the travel time.

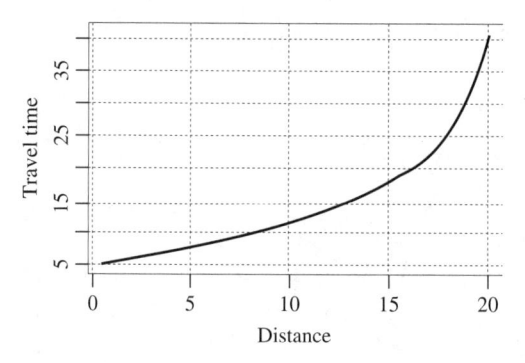

Figure 9.15 Relationship between the time of travel and the distance.

9.5 Two-dimensional aggregation

For a two-dimensional case it is more convenient to use polar coordinates in place of rectangular coordinates. Thus, a location in a field is defined by coordinates r and θ, where r is the distance from the pole, and θ is the angle of variation. Now consider a two-dimensional probability density function which describes the probability of locating at a given distance r from the origin O in a radially symmetric density field. The underlying implication of this radially symmetric density is that the distance is isotropically distributed. Let the density function be defined as

$$f(r,\theta) = K\exp(-ar) \tag{9.67}$$

where a is a parameter, and K is a normalizing constant so that

$$\int_0^{2\pi}\int_0^R rf(r,\theta)\,dr\,d\theta = 1 \tag{9.68}$$

One may wonder why r comes into equation (9.68). Actually it involves a small trick when integrating in a polar coordinate system. Since such two-dimensional integration plays an important role in the following two-dimensional spatial entropy analysis, it is necessary to explain it a bit more clearly. Suppose one wants to integrate $f(r,\theta)$ over the region as shaded

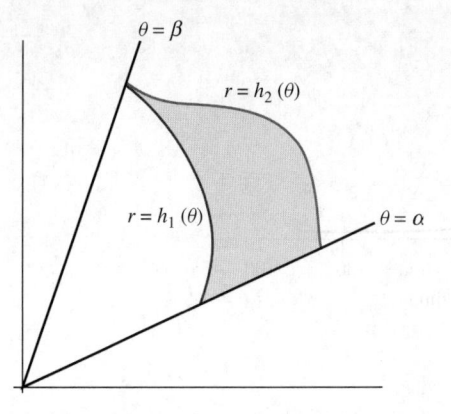

Figure 9.16 Integral regions in a polar coordinate system.

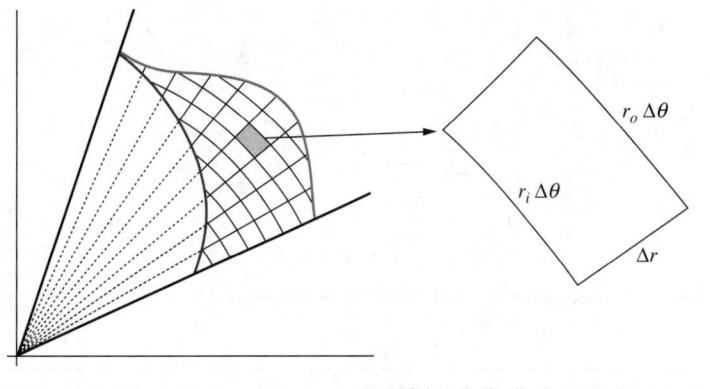

Figure 9.17 Partition of the integral region and an illustration of the small cell.

in Figure 9.16. Mathematically, the integral region can be expressed as $h_1(\theta) \leq r \leq h_2(\theta)$, $\alpha \leq \theta \leq \beta$. Suppose the integral region is partitioned into small cells by radial lines and arcs as shown in Figure 9.17, and then the integration, denoted as I for example, can be approximated by $I \approx \sum_k f(r_k, \theta_k) \Delta A_k$, where the summation means sum over all small cells and ΔA_k represents the area of the k-th cell. Now we need to find the expression for ΔA_k. Picking one cell out for illustration, which is almost a rectangle but is not, basic geometry tells that the length of inner and outer arcs are, respectively, $r_i \Delta \theta$ and $r_o \Delta \theta$, where r_i and r_o are radii of the inner and outer arcs and $\Delta \theta$ is the angle between the two radial lines that form the small cell. Now assuming the cell is small enough such that $r_i \approx r_o$, denoting this quantity as r, therefore we have $r_i \Delta \theta \approx r_o \Delta \theta = r \Delta \theta$. When the cell is small enough, it can be approximately considered as a rectangle and thus its area can be approximated by $\Delta A_k = r \Delta \theta \Delta r$. Substituting ΔA_k back yields $I \approx \sum_k f(r_k, \theta_k) r \Delta r \Delta \theta$. Finally, the integral is

computed as $I = \lim\limits_{N \to \infty} \sum\limits_k f(r_k, \theta_k) r \Delta r \Delta \theta = \int\limits_{\alpha}^{\beta} \int\limits_{h_1(\theta)}^{h_2(\theta)} f(r, \theta) \, r \, dr \, d\theta$.

Now let us go back to the two-dimensional spatial aggregation problem. Substitution of equation (9.67) in equation (9.68) yields the value of K:

$$K = \frac{a}{2\pi \left\{ \frac{1}{a} \left[1 - \exp(-aR) \right] - \exp(-aR) R \right\}} \tag{9.69}$$

Parameter a can be related to the mean travel time from pole r_m as

$$r_m = \int_0^{2\pi} \int_0^R r^2 K \exp(-ar) dr d\theta = 2\pi K \left[\frac{2}{a^3} - \frac{\exp(-aR)}{a} \left(R^2 + \frac{2R}{a^2} + \frac{2}{a} \right) \right] \tag{9.70}$$

Equation (9.70) can be solved iteratively to get parameter a for known r_m. A first approximation is obtained by taking R as ∞:

$$r_m = \int_0^{2\pi} \int_0^\infty r^2 K \exp(-ar) dr d\theta = \frac{2}{a} \tag{9.71}$$

It is interesting to note that for a one-dimensional case $a = 1/r_m$ from equation (9.45), whereas for a two-dimensional case $a = 2/r_m$.

For determining the value of R, analogous to the one-dimensional case, one may relate the entropy for finite R to the entropy where $R \rightarrow \infty$. The entropy H can be expressed for the two-dimensional case as

$$H = -\int_0^{2\pi} \int_0^R f(r,\theta) \ln(f(r,\theta)) dr d\theta = -2\pi K \int_0^R \exp(-ar)(\ln K - ar) r dr$$

$$= -2\pi K \ln K \int_0^R \exp(-ar) r dr + 2\pi a K \int_0^R \exp(-ar) r^2 dr \tag{9.72}$$

The first integral in equation (9.72) is integrated by parts as follows:

$$\int \exp(-ar) r dr = -\frac{1}{a} \int r d(\exp(-ar)) = -\frac{1}{a} \left[r \exp(-ar) - \int \exp(-ar) dr \right]$$

$$= -\frac{1}{a} \left[r \exp(-ar) + \frac{1}{a} \exp(-ar) \right] \tag{9.73}$$

Then we have

$$\int_0^R \exp(-ar) r dr = -\frac{1}{a} \left[\exp(-aR) R + \frac{1}{a} \exp(-aR) \right] + \frac{1}{a} \left[0 \times \exp(0) + \frac{1}{a} \exp(0) \right]$$

$$= -\frac{1}{a} \left[\exp(-aR) R + \frac{1}{a} \exp(-aR) \right] + \frac{1}{a^2} \tag{9.74}$$

Similarly, the second integral in equation (9.72) can be integrated as follows:

$$\int \exp(-ar) r^2 \, dr = -\frac{1}{a} \int r^2 \, d(\exp(-ar)) = -\frac{1}{a} \left[r^2 \exp(-ar) - 2 \int \exp(-ar) \, r \, dr \right]$$

$$= -\frac{1}{a} \left[r^2 \exp(-ar) + \frac{2}{a} \left(r \exp(-ar) + \frac{1}{a} \exp(-ar) \right) \right] \tag{9.75}$$

Then we obtain

$$\int_0^R \exp(-ar) r^2 \, dr = -\frac{1}{a} \left[\exp(-aR) R^2 + \frac{2}{a} \left(\exp(-aR) R + \frac{1}{a} \exp(-aR) \right) \right]$$

$$+ \frac{1}{a} \left[0 \times \exp(0) + \frac{2}{a} \left(0 \times \exp(0) + \frac{1}{a} \exp(0) \right) \right]$$

$$= -\frac{1}{a} \left[R^2 \exp(-aR) + \frac{2}{a} \left(\exp(-aR) R + \frac{1}{a} \exp(-aR) \right) \right] + \frac{2}{a^3} \tag{9.76}$$

Substituting equation (9.74) and equation (9.76) into equation (9.72), one can get

$$H = -2\pi K \ln(K) \left\{ -\frac{1}{a} \left[\exp(-aR) R + \frac{1}{a} \exp(-aR) \right] + \frac{1}{a^2} \right\}$$

$$+ 2\pi Ka \left\{ -\frac{1}{a} \left[R^2 \exp(-aR) + \frac{2}{a} \left(\exp(-aR) R + \frac{1}{a} \exp(-aR) \right) \right] + \frac{2}{a^3} \right\}$$

$$= \frac{2\pi K \ln(K)}{a} \left[\exp(-aR) R + \frac{1}{a} \exp(-aR) \right] - \frac{2\pi K \ln(K)}{a^2}$$

$$- 2\pi K \left[R^2 \exp(-aR) + \frac{2}{a} \left(\exp(-aR) R + \frac{1}{a} \exp(-aR) \right) \right] + \frac{4\pi K}{a^2}$$

$$= \frac{\frac{2\pi K \ln(K)}{a} \left[R + \frac{1}{a} \right] - 2\pi K \left[R^2 + \frac{2R}{a} + \frac{2}{a^2} \right]}{\exp(aR)} + \frac{2\pi K}{a^2} [2 - \ln(K)] \tag{9.77}$$

Simplifying equation (9.77) by replacing K with equation (9.74) and after performing a series of algebraic operations, it would reduce to

$$H = -\ln(a) + \ln(2\pi) + \ln \left[\frac{1}{a} (1 - \exp(-aR)) - \exp(-aR) R \right]$$

$$+ \frac{\frac{2}{a} - a \exp(-aR) \left(R^2 + \frac{2R}{a^2} + \frac{2}{a} \right)}{\frac{1}{a} [1 - \exp(-aR)] - \exp(-aR) R} \tag{9.78}$$

In the limiting case as $R \to \infty$, equation (9.78) is

$$H = -2 \ln(a) + 2 + \ln(2\pi) \tag{9.79}$$

The error entropy $H(e)$ can be obtained by subtracting equation (9.79) from equation (9.77) or equation (9.78):

$$H(e) = -2\ln(a) + 2 + \ln(2\pi)$$

$$-\frac{\frac{2\pi K \ln(K)}{a}\left[R + \frac{1}{a}\right] - 2\pi K\left[R^2 + \frac{2R}{a} + \frac{2}{a^2}\right]}{\exp(aR)} - \frac{2\pi K}{a^2}[2 - \ln K] \tag{9.80}$$

or

$$H(e) = -\ln(a) + 2 - \ln\left[\frac{1}{a}(1 - \exp(-aR)) - \exp(-aR)R\right]$$

$$-\frac{\frac{2}{a} - a\exp(-aR)\left(R^2 + \frac{2R}{a^2} + \frac{2}{a}\right)}{\frac{1}{a}[1 - \exp(-aR)] - \exp(-aR)R} \tag{9.81}$$

Taking the ratio of equation (9.80) or equation (9.81) to equation (9.79) and setting the tolerance limit as a threshold, say 0.05, one gets:

$$\frac{H(e)}{H} = 0.05 \tag{9.82}$$

where H is given by equation (9.78). For a boundary value R, solving equation (9.69) numerically yields estimate for parameter a. Determining an initial value for a is important for any numerical solver. As previously mentioned, $a = 2/r_m$ is a reasonable initial guess of a for known r_m, which can be determined based on the knowledge about the problem under study. After knowing all the distribution parameters, the discrete entropy, the discrete approximation of continuous entropy or spatial entropy can be calculated, respectively, as

$$H = -\sum_i p_i \ln p_i \tag{9.83}$$

$$H = -\sum_i p_i \ln \frac{p_i}{\Delta x_i} \tag{9.84}$$

where

$$p_i = \int_0^{2\pi}\int_{r_i}^{r_{i+1}} K\exp(-ar)\,r\,dr\,d\theta = -2\pi\frac{K}{a}\left[r\exp(-ar) + \frac{1}{a}\exp(-ar)\right]\Big|_{r_i}^{r_{i+1}} \tag{9.85}$$

$$\Delta x_i = r_{i+1} - r_i \tag{9.86}$$

Example 9.10: Compute and plot two-dimensional (a) continuous entropy versus distance, and (b) error entropy versus distance, for different values of R: 10, 20, 30, 40, 50, 60, 70, and 80 minutes.

Solution: Prior to the computation of continuous entropy and error entropy, first we need to determine parameter a for a given spatial boundary R. Parameter a can be calculated for different values of R according to the ratio of $H(e)$ to the entropy at infinity, that is,

Table 9.4a Parameter a, $H(R)$, $H(R \to \infty)$ and $H(e)$ for different R values with $H(e)/H = 0.05$.

R	a	$H(R)$	$H(R \to \infty)$	$H(e)$
10	0.5163	4.9144	5.1604	0.2460
20	0.2672	6.1688	6.4774	0.3086
30	0.1843	6.8766	7.2203	0.3437
40	0.1421	7.3721	7.7403	0.3682
50	0.1163	7.7539	8.1410	0.3871
60	0.0988	8.0646	8.4672	0.4026
70	0.0861	8.3266	8.7424	0.4158
80	0.0764	8.5529	8.9814	0.4286

equation (9.82). For illustrative purposes, take $R = 10$ as an example. The continuous entropy and error entropy $H(e)$ are calculated, respectively, by equation (9.78) and equation (9.80) or equation (9.81). Assuming the criterion ratio as 0.05, the value of a can be achieved by solving equation (9.82) using a numerical method. When $R = 10$, a is estimated to be 0.5163. Then, the continuous entropy is calculated by substituting the corresponding R and a into equation (9.78). Here $R = 10$, $a = 0.5163$, from equation (9.78) the total entropy can be computed as

$$H = -\ln(0.5163) + \ln(2\pi) + \ln\left[\frac{1}{0.5163}(1 - \exp(-0.5163 \times 10)) - \exp(-0.5163 \times 10) \times 10\right]$$

$$+ \frac{\dfrac{2}{0.5163} - 0.5163 \times \exp(-0.5163 \times 10)\left(10^2 + \dfrac{2 \times 10}{0.5163^2} + \dfrac{2}{0.5163}\right)}{\dfrac{1}{0.5163}\left[1 - \exp(-0.5163 \times 10)\right] - \exp(-0.5163 \times 10) \times 10}$$

$$= 4.9144 \text{ nats}$$

Following the same procedure, a and H corresponding to other boundary values of R can be calculated as shown in Table 9.4a. Then, the error entropies can be easily computed as

$$H(R \to \infty) = -2\ln(0.5163) + 2 + \ln(2\pi) = 5.1604 \text{ nats}$$
$$H(e) = H(R \to \infty) - H(R) = 5.1604 - 4.9144 = 0.2460 \text{ nats}$$

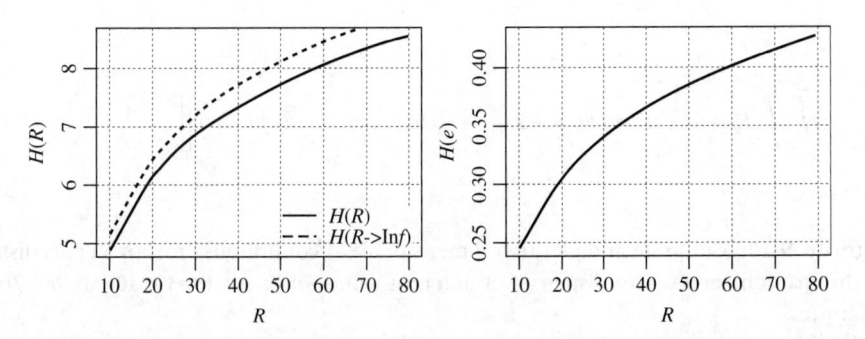

Figure 9.18a Relationship between R represented in minutes and continuous entropy, the continuous entropy evaluated when R approaches infinity and the difference entropy $H(e)$ with the ratio of $H(e)/H = 0.05$.

Table 9.4b Parameter a, $H(R)$, $H(R \rightarrow \infty)$ and $H(e)$ for different R values with $H(e)/H = 0.10$.

R	a	$H(R)$	$H(R \rightarrow \infty)$	$H(e)$
10	0.4280	5.0319	5.5351	0.5032
20	0.2279	6.1777	6.7956	0.6179
30	0.1593	6.8296	7.5118	0.6823
40	0.1238	7.2877	8.0161	0.7283
50	0.1019	7.6419	8.4504	0.7735
60	0.0869	7.9298	8.7239	0.7941
70	0.0760	8.1738	8.9919	0.8181
80	0.0677	8.3861	9.2232	0.8371

Table 9.4c Parameter a, $H(R)$, $H(R \rightarrow \infty)$ and $H(e)$ for different R values with $H(e)/H = 0.15$.

R	a	$H(R)$	$H(R \rightarrow \infty)$	$H(e)$
10	0.3804	5.1084	5.7709	0.7526
20	0.2066	6.0799	6.9918	0.9120
30	0.1456	6.6876	7.6917	1.0041
40	0.1138	7.1177	8.1845	1.0668
50	0.0940	7.4410	8.5668	1.1168
60	0.0804	7.7206	8.8794	1.1587
70	0.0705	7.9522	9.1422	1.1810
80	0.0629	8.1523	9.3703	1.2180

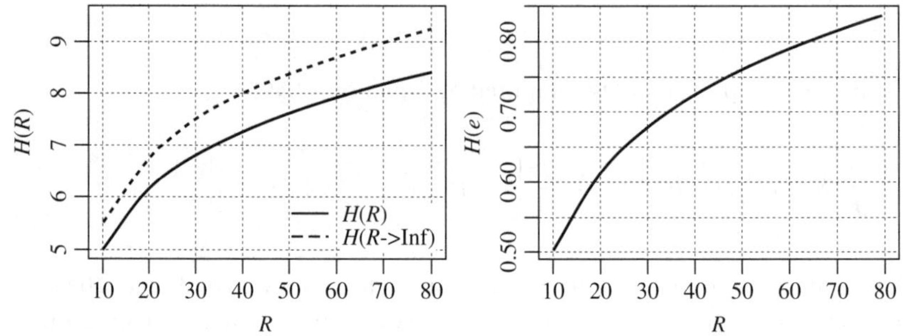

Figure 9.18b Relationship between R represented in minutes and continuous entropy, the continuous entropy evaluated when R approaches infinity and the difference entropy $H(e)$ with the ratio of $H(e)/H = 0.10$.

In the same way, the values of a, $H(R)$ and $H(e)$ can be computed and are tabulated in Table 9.4a. The relationships between R and $H(R)$ and between R and $H(e)$ are plotted in the left and right panels of Figure 9.18a, respectively. Similarly, for different threshold ratios, one can perform the same calculations to obtain the corresponding parameter a, continuous entropy at R, continuous entropy when R approaches infinity, and error entropy $H(e)$. Results for other threshold values are tabulated in Tables 9.4b and 9.4c and are plotted in Figures 9.19b and 9.18c.

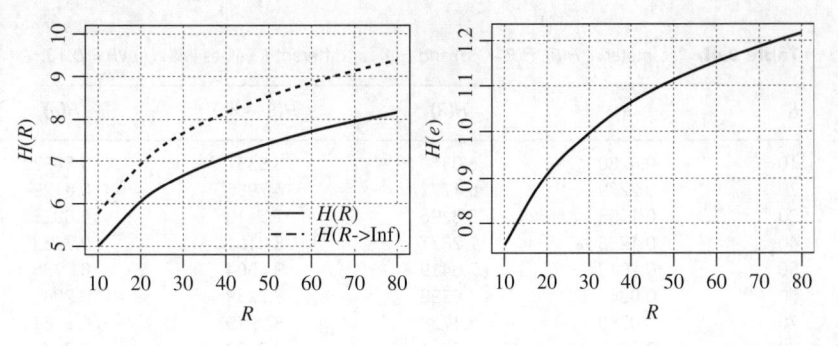

Figure 9.18c Relationship between R represented in minutes and continuous entropy, the continuous entropy evaluated when R approaches infinity and the difference entropy $H(e)$ with the ratio of $H(e)/H = 0.15$.

Example 9.11: Compute discrete (Shannon) entropy given by equation (9.83) as a function of number of zones for different values of R.

Solution: Similar to Example 9.6, three cases, that is, $R = 20$, $R = 40$ and $R = 60$, are considered for illustrating the relationship between discrete Shannon entropy and the number of zones. For each R, parameter a can be estimated by numerically solving equation (9.82). Also take $R = 20$ as an example, solving equation (9.82) yields the value of 0.2279 for a. In this example, the entropy ratio threshold is assumed be 0.05. Solving equation (9.82) given $R = 40$ and $R = 60$ yields 0.1238 and 0.0869 for the parameter a, respectively.

A specific value of R is divided into N zones, and their probability p_i and discrete entropy H are calculated, respectively. Suppose circles with radius of $r_1, r_2, \ldots, r_{N+1}$ partition the spatial system into N equally sized intervals. The discrete entropy can be calculated by equation (9.83):

$$H = -\sum_{i=1}^{N} p_i \ln p_i$$

in which the discrete probability is computed by equation (9.50)

$$p_i = \int_{0}^{2\pi} \int_{r_i}^{r_{i+1}} K \exp(-ar)\, r dr d\theta = -2\pi \frac{k}{a} \left[r \exp(-ar) + \frac{1}{a} \exp(-ar) \right]\Big|_{r_i}^{r_{i+1}}$$

For each boundary R and each interval number, discrete entropy is computed. The relationship between discrete entropy and the number of intervals can be shown by plotting the discrete entropy as a function of interval numbers, which is presented in Figure 9.19. Obviously, the discrete entropy increases without bound as the number of zones increases. In the semi-logarithm coordinates, the discrete entropy approximately linearly increases with the logarithm of the number of intervals.

Example 9.12: Compute and plot two-dimensional spatial entropy versus number of zones. Also determine the optimum number of zones for describing the density [N on the log scale: take $N = 1000$, and $R = 20, 40, 60$].

Solution: Here we take $R = 20$, $R = 40$, and $R = 60$ as examples. For each R parameter a can be estimated by equation (9.82), assuming the threshold for the entropy ratio $H(e)$ be 0.05. The estimated a values corresponding to $R = 20$, $R = 40$, and $R = 60$ are, respectively, 0.2672,

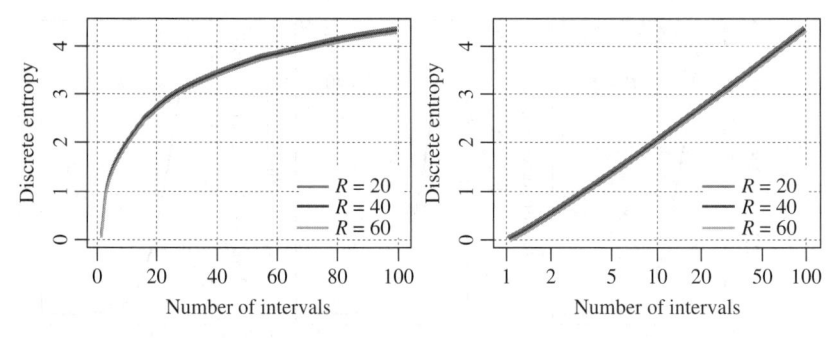

Figure 9.19 Relationship between the discrete Shannon entropy and the number of zones in natural coordinate (left) and semi-logarithm coordinates (right).

0.1421, and 0.0988. Suppose the spatial system is divided into N equally sized intervals by $r_1, r_2, \ldots, r_{N+1}$. The discrete approximation of continuous entropy (spatial entropy) is calculated by equation (9.84):

$$H = -\sum_{i=1}^{N} p_i \ln \frac{p_i}{\Delta x_i}$$

where p_i and Δx_i are computed by equation (9.85) and equation (9.86), respectively, that is,

$$p_i = \int_{0}^{2\pi} \int_{r_i}^{r_{i+1}} K \exp(-ar) \, r dr d\theta = -2\pi \frac{K}{a} \left[r \exp(-ar) + \frac{1}{a} \exp(-ar) \right] \Bigg|_{r_i}^{r_{i+1}}$$

$\Delta x_i = \pi \left(r_{i+1}^2 - r_i^2 \right)$. Then the discrete approximation of continuous entropy is computed and is plotted, as shown in Figure 9.20 which shows that with the increase in the number of intervals the discrete approximation of continuous entropy decreases but approaches a bound. From Figure 9.20 one can detect that the discrete approximation of spatial entropy reaches a flat plateau, after that the region is partitioned into about 100 regions. Therefore, the optimal number of zones is 100 in the sense that the density can be well described.

Example 9.13: Compute and plot two-dimensional information gain I and information redundancy Z measured by spatial entropy as a function of number of intervals N for different values of boundary R. Take $R = 20, 40, 60$ as examples to illustrate the behavior of I and Z.

Solution: First, let us compute the information gain. As mentioned previously, information gain can be measured by discrete Shannon entropy and spatial entropy (continuous entropy) or more precisely the discrete approximation of spatial entropy (continuous entropy). In this example, we adopted the second one, that is, by the spatial entropy. According to equation (9.35)

$$I = \ln X + H$$

in which X is the size of the spatial system and H is the spatial entropy. Here X assumes 20, 40, and 60, for the cases of $R = 20, 40$, and 60, respectively. Substituting the spatial entropy, which is computed in Example 9.12, into equation (9.35), the information can be obtained.

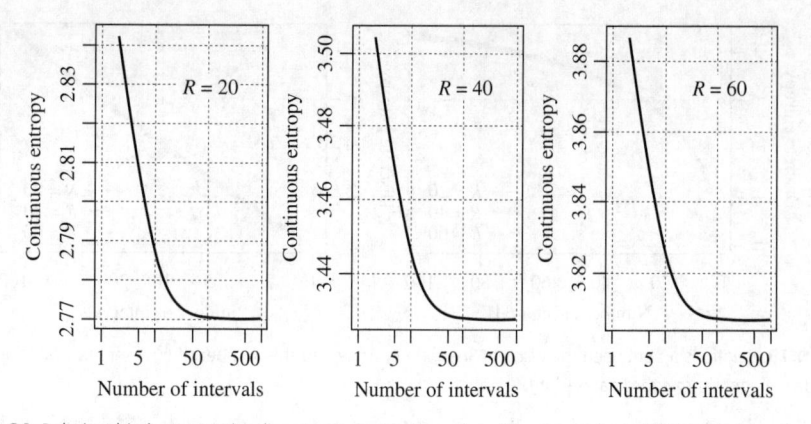

Figure 9.20 Relationship between the discrete approximation of continuous entropy and the number of zones in semi-logarithm coordinates.

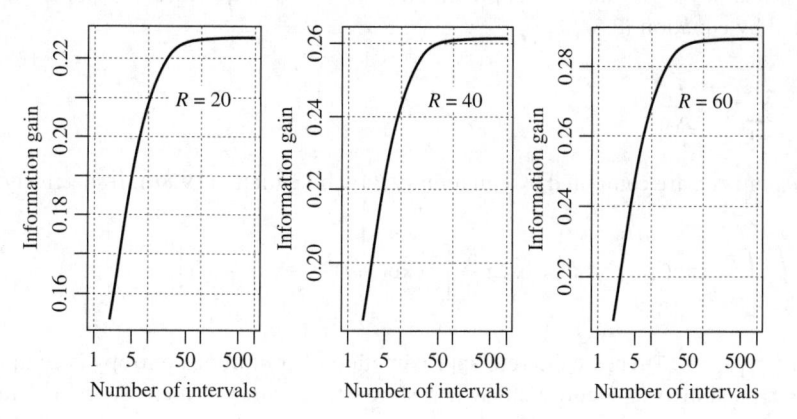

Figure 9.21 Relationship between information gain and the number intervals for $H(e)/H = 0.05$.

The behavior of information gain with respect to the number of intervals is presented in Figure 9.21 for each boundary value R. Remember that parameter a is estimated assuming an entropy ratio threshold of 0.05. As the number of intervals increases, the information gain increases and then approaches a plateau.

Then, we compute the information redundancy. According to equation (9.36), the information redundancy can be obtained directly via dividing the information gain by the maximum entropy. In Section 9.2.2, we have already shown that for a given spatial system with size X the maximum spatial entropy is $\ln(X)$, that is, $Z = 1/\ln(X)$. The behavior of information redundancy with respect to the number of intervals is presented in Figure 9.22 for each boundary value R. Also one should remember that parameter a is estimated, assuming an entropy ratio threshold of 0.05. As the number of intervals increases, the information redundancy also increases and then approaches a plateau.

9.5.1 Probability density function and its resolution

It is assumed that the probability of location is some function of location cost, and the location cost is measured as a distance or time of travel from some point in the city. Probabilities

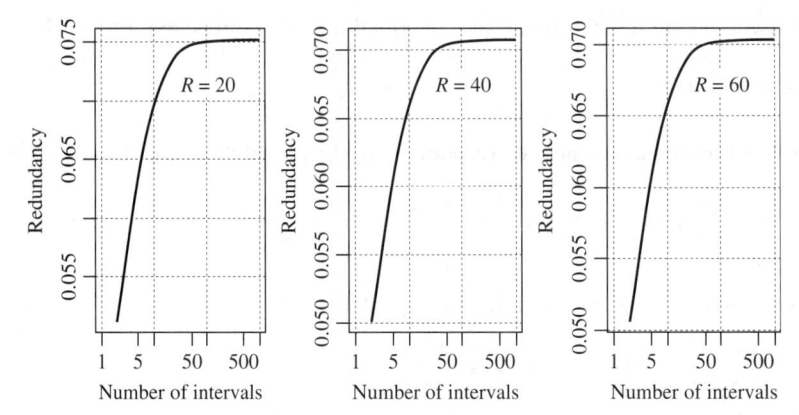

Figure 9.22 Relationship between information redundancy and the number intervals for $H(e)/H = 0.05$.

are defined over two dimensions. For simplicity, consider M radial sectors of the city, and n zones within each sector. Thus, $N = Mn$ zones in total. Each sector can be considered as a segment of a circle centered on the central point, and each sector j subtends an angle $\Delta\theta_j$ about the center, as shown by the small cell in Figure 9.17. Within each sector, the boundaries of each zone i define the average distance as $(r_i - r_{i-1})$. The area of a typical zone, Δr_{ij}, can be approximated as

$$\Delta r_{ij} \approx r_i \Delta\theta_j \left(r_i - r_{i-1}\right) \tag{9.87}$$

It is assumed that each sector has the same regular variation $\Delta\theta_j = \Delta\theta$, $\forall j$ and each distance difference $(r_i - r_{i-1}) = \Delta r$, $\forall i$. The area of the system, city, R can be written as

$$R = \sum_i \sum_j \Delta r_{ij} \tag{9.88}$$

From equations (9.87) and (9.88), the probability of locating in zone i in sector j, denoted by p_{ij} is consistent with a two-dimensional density function defined by polar coordinates. It is noted that

$$\sum_i \sum_j p_{ij} = 1 \tag{9.89}$$

The probability density function $p(r, \theta)$ can be approximated as

$$p(r_{ij}) = \frac{p_{ij}}{\Delta r_{ij}} \tag{9.90}$$

This yields

$$p_{ij} = p(r_{ij}) \Delta r_{ij} \tag{9.91}$$

The Shannon entropy, $H(N)$, for the two-dimensional case can be written as

$$H(N) = -\sum_i \sum_j p_{ij} \ln p_{ij} \tag{9.92}$$

With the use of equations (9.90) and (9.91), equation (9.92) can be expressed as

$$H(N) = S(N) + L(N) \tag{9.93}$$

where $S(N)$ is the spatial entropy, corresponding to the probability density, defined as

$$S(N) = -\sum_i \sum_j p(r_{ij}) \ln[p(r_{ij})] \Delta r_{ij} = -\sum_i \sum_j p_{ij} \ln \frac{p_{ij}}{\Delta r_{ij}} \tag{9.94}$$

and $L(N)$ denotes the level of resolution entropy defined as

$$L(N) = -\sum_i \sum_j p(r_{ij}) \ln[\Delta r_{ij}] \Delta r_{ij} = -\sum_i \sum_j p_{ij} \ln \Delta r_{ij} \tag{9.95}$$

$H(N)$ is unbounded (Goldman, 1953). Taking $p(r_{ij})$ as an approximation to a continuous single-valued function $p(r, \theta)$, one can write equations (9.93), (9.94) and (9.95) as

$$\lim_{N \to \infty} H(N) = \lim_{N \to \infty} S(N) + \lim_{N \to \infty} L(N) = S + L \tag{9.96}$$

in which

$$S = -\int \int_R p(r, \theta) \ln \left[p(r, \theta) \right] r d\theta dr \tag{9.97}$$

$$L = -\lim_{N \to \infty} \sum_i \sum_j p(r_{ij}) \ln \left[\Delta r_{ij} \right] \Delta r_{ij} \tag{9.98}$$

Now the disutility entropy $I(N)$ gain in equation (9.41) can be recast as

$$I(N) = -\sum_i \sum_j r_i p_{ij} \ln p_{ij} \tag{9.99}$$

Equation (9.99) can be decomposed using equation (9.94) and (9.95) as

$$I(N) = D(N) + T(N) \tag{9.100}$$

in which $D(N)$ denotes the disutility entropy, spatial disutility, corresponding to the density, defined as

$$D(N) = -\sum_i \sum_j r_i p(r_{ij}) \ln[p(r_{ij})] \Delta r_{ij} = -\sum_i \sum_j r_i p_{ij} \ln \frac{p_{ij}}{\Delta r_{ij}} \tag{9.101}$$

and $T(N)$ denotes the level of resolution disutility defined as

$$T(N) = -\sum_i \sum_j r_i p(r_{ij}) \ln[\Delta_{ij}] \Delta r_{ij} = -\sum_i \sum_j r_i p_{ij} \ln \Delta r_{ij} \tag{9.102}$$

For $N \to \infty$, equation (9.100) can be written as

$$\lim_{N \to \infty} I(N) = \lim_{N \to \infty} D(N) + \lim_{N \to \infty} T(N) = D + T \tag{9.103}$$

where

$$D = - \int_R \int p(r,\theta) \ln\left[p(r,\theta)\right] r^2 d\theta\, dr \tag{9.104}$$

and

$$T = - \lim_{N\to\infty} \sum_i \sum_j p(r_{ij}) \ln\left[\Delta r_{ij}\right] r_i \Delta r_{ij} \tag{9.105}$$

For a nested aggregation, $S(N) \leq S(N-1)$ and $S(1) = \log R$. Over a wide range of aggregations $S(N)$ and $D(N)$ remain unchanged and the variation in $H(N)$ and $I(N)$ is due to $L(N)$ and $T(N)$.

9.5.2 Relation between spatial entropy and spatial disutility

For illustrative purposes, consider a gamma density function in polar coordinates. This density function has been used as a spatial distribution function:

$$p(r,\theta) = \frac{b^{a+2}}{\Gamma(a+2)\, 2\pi} r^a \exp(-br) \tag{9.106}$$

where a and b are parameters of the gamma density function. Of course,

$$\int_0^{2\pi} \int_0^\infty p(r,\theta)\, r\, d\theta\, dr = 1 \tag{9.107}$$

Parameters a and b can be estimated using the maximum likelihood function or the principle of maximum entropy:

$$\int_0^{2\pi} \int_0^\infty p(r,\theta)\, r^2\, d\theta\, dr = r_m = \frac{a+2}{b} \tag{9.108}$$

$$\int_0^{2\pi} \int_0^\infty p(r,\theta)\, r \ln r\, d\theta\, dr = \overline{\ln r} = \frac{d}{da}\ln(a+2) - \ln b = \psi(a+2) - \ln b \tag{9.109}$$

$\Psi(a+2)$ is called the psi or digamma function. Equations (9.107) to (9.109) are constraints for equation (9.106). The spatial entropies S and D in equation (9.97) and (9.104) can be obtained, subject to equations (9.107) to (9.109), as

$$S = - \int_0^{2\pi} \int_0^\infty p(r,\theta) \ln[p(r,\theta)\, r d\theta\, dr = -2\ln b + \ln\Gamma(a+2) + \ln 2\pi - a\psi(a+2) + a = 2 \tag{9.110}$$

$$D = - \int_0^{2\pi} \int_0^\infty p(r,\theta) \ln\left[p(r,\theta)\right] r^2\, dr\, d\theta$$
$$= -r_m[-2\ln b + \ln\Gamma(a+2) + \ln 2\pi - a\psi(a+3) + a + 3] \tag{9.111}$$

From the definition of the gamma function

$$\Gamma\left(a+3\right) = \left(a+2\right)\Gamma\left(a+2\right) \tag{9.112a}$$

and the psi function:

$$\psi\left(a+3\right) = \frac{d\ln\Gamma\left(a+3\right)}{da}$$

or

$$\psi\left(a+3\right) = \psi\left(a+2\right) + \frac{1}{a+2} \tag{9.112b}$$

Inserting equation (9.112b) in equation (9.111) and then substituting for S in equation (9.110), one obtains a relation between D and S as

$$D = r_m \left(S + 1 - \frac{a}{a+2}\right) \tag{9.113}$$

Example 9.14: Consider a one-dimensional probability density function $p(r)$. Determine S and D.

Solution: $D = r_m \left(S + 1 - \frac{a}{a+1}\right)$, where S is the approximate entropy (Tribus, 1968). If $a = 0$, the density becomes negative exponential, and $D = r_m(S + 1)$. The spatial disutility is directly related to the product of average cost and accessibility measured by spatial entropy.

9.6 Entropy maximization for modeling spatial phenomena

Consider two constraints:

$$\sum_{i=1}^{N} p_i = 1 \tag{9.114}$$

$$\sum_{i=1}^{N} r_i p_i = r_m \tag{9.115}$$

where r_i is a measure of location cost i or travel cost between 0 and location i, and r_m is the mean location cost in the system.

Maximizing equation (9.16), subject to equations (9.114) and (9.115), the probability of locating any i is:

$$p_i = \frac{\exp\left(-ar_i\right)}{\sum_{i=1}^{N} \exp\left(-ar_i\right)} \tag{9.116}$$

Equation (9.116) does not explicitly show the effect of zone size or shape. To explicitly incorporate the zone size, the spatial entropy given by

$$H = -\sum_{i=1}^{N} p_i \ln\left(\frac{p_i}{\Delta x_i}\right) \tag{9.117}$$

can be maximized, subject to equations (9.114) and (9.115) using the method of Lagrange multipliers:

$$L = -\sum_{i=1}^{N} p_i \ln\left(\frac{p_i}{\Delta x_i}\right) - \lambda_0\left(\sum_{i=1}^{N} p_i - 1\right) - \lambda_1\left(\sum_{i=1}^{N} r_i\, p_i - r_m\right) \tag{9.118}$$

Differentiating equation (9.118) with respect to p_i and equating the derivative to zero, one obtains

$$\frac{\partial L}{\partial p_i} = -\ln p_i + \ln \Delta x_i - \lambda_0 - \lambda_1 r_i \tag{9.119}$$

Equation (9.119) yields

$$p_i = \Delta x_i \exp\left(-\lambda_0 - \lambda_1 r_i\right) \tag{9.120}$$

Substituting equation (9.120) in equation (9.114), one gets

$$\sum_{i=1}^{N} p_i = \exp\left(-\lambda_0\right) \sum_{i=1}^{N} \Delta x_i \exp(-\lambda_1 r_i) \tag{9.121}$$

From equation (9.121),

$$\exp\left(-\lambda_0\right) = \frac{1}{\displaystyle\sum_{i=1}^{N} \Delta x_i \exp\left(-\lambda_1 r_i\right)} \tag{9.122}$$

The location probability mass function then becomes

$$p_i = \frac{\Delta x_i \exp\left(-\lambda_1 r_i\right)}{\displaystyle\sum_{i=1}^{N} \Delta x_i \exp\left(-\lambda_1 r_i\right)} \tag{9.123}$$

If $\Delta x_i = \Delta x$, then equation (9.123) becomes

$$p_i = \frac{\exp\left(-\lambda_1 r_i\right)}{\displaystyle\sum_{i=1}^{N} \exp\left(-\lambda_1 r_i\right)} \tag{9.124}$$

To further explore the effect of zone size and shape, consider the probabilities of location as equal in each zone. Then

$$\frac{1}{N} = \frac{\Delta x_i \exp\left(-\lambda_1 r_i\right)}{\displaystyle\sum_{i=1}^{N} \Delta x_i \exp\left(-\lambda_1 r_i\right)} \tag{9.125}$$

which yields

$$\Delta x_i = \frac{1}{N \exp\left(-\lambda_1 r_i\right)} \sum_{i=1}^{N} \Delta x_i \exp(-\lambda_1 r_i) \tag{9.126}$$

Equation (9.126) can be employed to fix zone sizes which yield location probabilities. It shows that such zones vary directly with the exponential distribution of location costs.

It may be interesting to dwell a little bit further on equation (9.124) which would directly result from the specification of constraint equation (9.115) without consideration of the spatial size or interval. Let c_i be the cost of locating in zone i, and $\bar{c}$ be the average cost of locating in any zone:

$$\bar{c} = \sum_{i=1}^{N} p_i c_i \tag{9.127}$$

In that case

$$p_i = \exp\left(-\lambda_0 - \lambda_1 c_i\right) \tag{9.128}$$

Here

$$\lambda_0 = \ln\left[\sum_{i=1}^{N} \exp\left(-\lambda_1 c_i\right)\right] \tag{9.129}$$

If the Lagrange multiplier for the average cost of locating is redundant, that is, $\lambda_1 = 0$, the exponential distribution reduces to a uniform distribution $p_i = 1/N$. The maximum entropy would be

$$H_{\max} = \lambda_0 + \lambda_1 \bar{x} \tag{9.130}$$

Thus, the maximum entropy is a function of the Lagrange multipliers and constraints. It depends on the spread of the distribution which in turn depends on the constraints. Thus, the entropy can be regarded as a system wide accessibility function where the partition and the cost of locating relate to the spread of probabilities across the system.

Going from the discrete form to the continuous form, $p_i = p(x_i)\Delta x_i$ and $c_i = c(x_i)$, where $p(x_i)$ can be an approximation of the size of the population x at the point location i to the probability density over the interval or area defined by Δx_i and $c(x_i)$ is an equivalent approximation to the cost density in zone i. If $\Delta x_i \to 0$, $p(x_i) \to p(x)$ and $c(x_i) \to c(x)$. Then,

$$\lim_{\Delta x_i \to 0} \sum_{i=1}^{N} p(x_i)\Delta x_i = \int_0^\infty p(x)dx = \int_0^\infty \exp\left(-\lambda_0\right) \exp\left[-\lambda_1 c(x)\right]dx$$

$$= \frac{\exp\left(-\lambda_0\right)}{\lambda_1} = 1 \tag{9.131}$$

which simplifies to

$$\exp\left(-\lambda_0\right) = \lambda_1 \quad \text{or} \quad \lambda_0 = -\ln\lambda_1 \tag{9.132}$$

$$\lim_{\Delta x_i \to 0} \sum_{i=1}^{N} p(x_i) c(x_i) \Delta x_i = \int_0^\infty p(x) c(x) dx$$

$$= \int_0^\infty \lambda_1 \exp\left[-\lambda_1 c(x)\right] c(x) dx = \frac{1}{\lambda_1} = \bar{c} \tag{9.133}$$

Thus,

$$p_i = \frac{1}{\bar{c}} \exp\left(-\frac{c(x)}{\bar{c}}\right) \tag{9.134}$$

Quantity $c(x)$ in thermodynamics is the energy (thermal) at location x and $\bar{c}$ is related to the average temperature, because $\bar{c} = kT$, where k is the Boltzmann constant.

In continuous form,

$$H = -\int_0^\infty p(x) \log p(x) dx$$

$$= \int p(x) \log\left\{\frac{1}{\bar{c}} \exp\left[-\frac{c(x)}{\bar{c}}\right]\right\} dx = \ln \bar{c} + 1 = -\ln \lambda_1 + 1 = 1 - \lambda_0 \tag{9.135}$$

which is not the same as in the case of discrete form but is of similar form. S varies with the log of the average cost or temperature, and the Lagrange multipliers λ_0 and λ_1 can be related to this average cost which can be an indicator of accessibility, and entropy can be regarded as a measure.

The continuous form of the spatial entropy then is written as

$$\lim_{\Delta x \to 0} H = -\int_0^\infty p(x) \log p(x) dx - \int_0^\infty p(x) \log p(x) dx \tag{9.136}$$

$H \to \infty$ as in the limit $\Delta x_i \to 0$. If $\Delta x_i = x/n$, $\forall i$, then $H \approx \log N$ and tends to infinity in an equivalent manner. Thus,

$$H_s = -\sum_{i=1}^{N} p_i \log\left[\frac{p_i}{\Delta x_i}\right] \tag{9.137}$$

which is the spatial entropy. Maximization of entropy leads to the augmented Boltzmann-Gibbs exponential model:

$$p_i = \frac{\Delta x_i \exp\left(-\lambda_1 c_i\right)}{\sum_{i=1}^{N} \Delta x_i \exp\left(-\lambda_1 c_i\right)} \tag{9.138}$$

The interval size Δx_i is introduced as a weight on the probability. The entropy can then be written as:

$$H_s = -\sum_{i=1}^{N} p_i \log p_i + \sum_{i=1}^{N} p_i \log \Delta x_i$$

$$= H + \sum_{i=1}^{N} p_i \log \Delta x_i \tag{9.139}$$

The second term $\sum\limits_{i=1}^{N} p_i \log \Delta x_i$ is the expected value of the logarithm of interval sizes and it can be considered as a constraint on the discrete entropy:

$$\sum_{i=1}^{N} p_i \log \Delta x_i = \overline{\log x} \tag{9.140}$$

Then,

$$L = -\sum_{i=1}^{N} p_i \log p_i - (\lambda_0 - 1)\left(\sum_{i=1}^{N} p_i - 1\right) - \lambda_1 \left(\sum_{i=1}^{N} p_i c_i - \bar{c}\right) - \lambda_2 \left(\sum_{i=1}^{N} p_i \log \Delta x_i - \overline{\log x}\right) \tag{9.141}$$

$$\frac{\partial L}{\partial p_i} = 0 \Rightarrow p_i = \exp(-\lambda_0 - \lambda_1 c_i - \lambda_2 \log \Delta x_i) \tag{9.142}$$

which can be cast as

$$p_i = \frac{(\Delta x_i)^{-\lambda_2} \exp(-\lambda_1 c_i)}{\sum\limits_{i=1}^{N} (\Delta x_i)^{-\lambda_2} \exp(-\lambda_1 c_i)} \tag{9.143}$$

Thus, the interval or zone size appears in the distribution as a scaling factor – a kind of benefit rather than cost. If $\lambda_2 = 1$ then this reduces to what is obtained from the maximization of spatial entropy.

9.7 Cluster analysis by entropy maximization

In spatial aggregation and regionalization the objective is to preserve the heterogeneity of the system as far as possible. Following Theil (1967), an aggregation procedure is based on between-set entropy and within-set entropy. Consider S_j is a larger set.

$$H = -\sum_{j=1}^{N} P_j \ln P_j - \sum_{j} P_j \left(\sum_{i \in S_j} \frac{p_i}{P_j} \ln \frac{p_i}{P_j}\right) \tag{9.144}$$

$$P_j = \sum_{i \in S_j} p_i \tag{9.145}$$

$$\sum_{j} P_j = \sum_{j} \sum_{i \in S_j} p_i = 1 \tag{9.146}$$

Equation (9.144) is a sum of between-set entropy denoted by the first term on the right side and within-set entropy denoted by the second term. Equation (9.144) is constant for any aggregation of spatial units noted by i in larger set S_j. When spatial units are aggregated in S_j sets, it is observed that the between-set entropy monotonically decreases as the size of sets

S_j increases, whereas the within-set entropy monotonically increases. This has been proved by Ya Nutenko (1970).

Consider an initial aggregation of spatial units $N+1$ sets where the set S_{N+1} contains a single unit. For the next level of aggregation, set S_N and set S_{N+1} are combined to form set S_M. The difference E between the entropy of the new M sets and the old $N+1$ sets is written as

$$E = \sum_{j=1}^{N+1} P_j \ln P_j - \sum_{j=1}^{M} P_j \ln P_j \tag{9.147}$$

Equation (9.147) can be simplified as

$$E = P_N \ln P_N + P_{N+1} \ln P_{N+1} - P_M \ln P_M \tag{9.148}$$

Equation (9.148) can be expressed solely in terms of the first $N+1$ sets as

$$E = P_N \ln P_N + P_{N+1} \ln P_{N+1} - (P_N + P_{N+1}) \ln (P_N + \ln P_{N+1}) \tag{9.149}$$

or

$$E = P_N \left[\ln P_N - \ln(P_N + P_{N+1}) \right] + P_{N+1} \left[\ln P_{N+1} - \ln(P_N + P_{N+1}) \right] \tag{9.150}$$

E is negative and can exhibit the increasing value of within-set entropy at higher levels of aggregation.

Now consider the discrete equivalent of the continuous entropy function. This means that equation (9.146) can be replaced by its continuous equivalent, including the zonal size in the aggregation problem. To that end, equation (9.20) can be written as

$$H = -\sum_i p_i \ln p_i + \sum_i p_i \ln \Delta x_i \tag{9.151}$$

The second term on the right side of equation (9.151) can be written as

$$\sum_{i=1}^{N} p_i \ln \Delta x_i = \sum_j \sum_{i \in S_j} p_i \ln \Delta x_i = \sum_j P_j \sum_{i \in S_j} \frac{p_i}{P_j} \ln \Delta x_i \tag{9.152}$$

Term $\ln \Delta x_i$ can be written as:

$$\ln \Delta x_i = \ln \frac{\Delta x_i}{\sum\limits_{i \in S_j} \Delta x_i} + \ln \sum_{i \in S_j} \Delta x_i \tag{9.153}$$

Let $\Delta X_j = \sum\limits_{i \in S_j} \Delta x_i$. Equation (9.152) can be cast as

$$\sum_i p_i \ln \Delta x_i = \sum_j P_j \left[\sum_{i \in S_j} \frac{p_i}{P_j} \left(\ln \frac{\Delta x_i}{\Delta X_j} + \ln \Delta X_j \right) \right]$$

$$= \sum_j P_j \left(\sum_{i \in S_j} \frac{p_i}{P_j} \ln \frac{\Delta x_i}{\Delta X_j} \right) + \sum_j P_j \ln \Delta X_j \tag{9.154}$$

The first term on the right side of equation (9.154) can be expanded as equation (9.144). Adding equations (9.144) and (9.154) yields the spatial form of Theil's aggregation formula:

$$
H = -\sum_j P_j \ln P_j + \sum_j P_j \ln \Delta X_j - \sum_j P_j \left(\sum_{i \varepsilon S_j} \frac{p_i}{P_j} \ln \frac{p_i}{P_j} \right) + \sum_j P_j \left(\sum_{i \varepsilon S_j} \frac{p_i}{P_j} \ln \frac{\Delta x_i}{\Delta X_j} \right)
$$

$$
= -\sum_j P_j \ln \left(\frac{P_j}{\Delta X_j} \right) - \sum_j P_j \left[\sum_{i \varepsilon S_j} \frac{p_i}{P_j} \ln \left(\frac{p_i \Delta X_j}{\Delta x_i P_j} \right) \right] \tag{9.155}
$$

For maximizing the first term using the probability constraint given by equation (9.146), the Lagrangean function can be written as

$$
L = -\sum_j P_j \ln \left(\frac{P_j}{\Delta X_j} \right) - \lambda_0 \left(\sum_j P_j - 1 \right) \tag{9.156}
$$

Differentiating equation (9.156), one gets

$$
\frac{\partial L}{\partial P_j} = -\ln P_j + \ln \Delta X_j - \lambda_0 \tag{9.157}
$$

Equation (9.157) yields

$$
P_j = \frac{\Delta X_j}{\sum_j \Delta X_j} \tag{9.158}
$$

Equation (9.158) shows that probabilities of location P_j depend on the geometry of the system. Equation (9.155) can be employed for aggregating zones where the population can be used to determine the probabilities of location.

Example 9.15: Considering that a spatial system is partitioned into three regions, each region is then partitioned into different sub-regions. The probability associated with each sub-region is given in Figure 9.23. Compute the between-set and within-set entropies.

Solution: Following equation (9.144), between-set and within-set entropies of the given system can be computed. First let us compute the between-set entropy as:

$$
H = -\sum_{j=1}^{3} P_j \ln P_j
$$

where

$$
P_1 = 0.3 + 0.25 + 0.15 = 0.7
$$
$$
P_2 = 0.05 + 0.02 + 0.03 = 0.1
$$
$$
P_3 = 0.1 + 0.05 + 0.03 + 0.02 = 0.2
$$

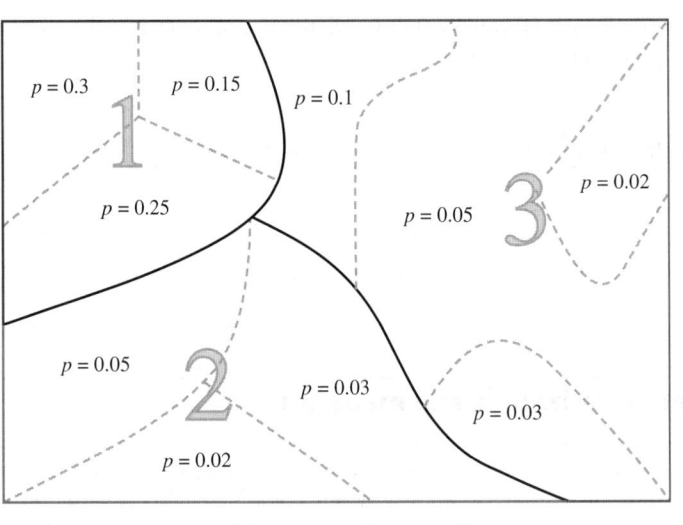

Figure 9.23 The spatial system partition and the corresponding probabilities.

Therefore

$$H = -0.7 \log (0.7) - 0.1 \log (0.1) - 0.2 \log (0.2) = 0.8018 \text{ nats}$$

Then we compute the within-set entropy region by region. For region 1, the within set entropy is

$$H_1 = -\sum_{i \in S_1} \frac{p_i}{P_1} \ln \frac{p_i}{P_1}$$

$$= -\frac{0.3}{0.7} \log \left(\frac{0.3}{0.7}\right) - \frac{0.25}{0.7} \log \left(\frac{0.25}{0.7}\right) - \frac{0.15}{0.7} \log \left(\frac{0.15}{0.7}\right)$$

$$= 1.0694 \text{ nats}$$

For region 2

$$H_2 = -\sum_{i \in S_2} \frac{p_i}{P_2} \ln \frac{p_i}{P_2}$$

$$= -\frac{0.05}{0.1} \log \left(\frac{0.05}{0.1}\right) - \frac{0.02}{0.1} \log \left(\frac{0.02}{0.1}\right) - \frac{0.03}{0.1} \log \left(\frac{0.03}{0.1}\right)$$

$$= 1.0297 \text{ nats}$$

For region 3

$$H_3 = -\sum_{i \in S_3} \frac{p_i}{P_3} \ln \frac{p_i}{P_3}$$

$$= -\frac{0.1}{0.2} \log \left(\frac{0.1}{0.1}\right) - \frac{0.05}{0.2} \log \left(\frac{0.05}{0.2}\right) - \frac{0.02}{0.2} \log \left(\frac{0.02}{0.2}\right) - \frac{0.03}{0.2} \log \left(\frac{0.03}{0.2}\right)$$

$$= 1.2080 \text{ nats}$$

Also we compute the total entropy of the system by summing over between-set and within-set entropies in a weighted manner as

$$H = -\sum_{j=1}^{3} P_j \ln P_j + \sum_{j=1}^{3} P_j H_j$$

$$= 0.8018 + 0.7 \times 1.0694 + 0.1 \times 1.0297 + 0.2 \times 1.0280$$

$$= 1.8890 \text{ nats}$$

9.8 Spatial visualization and mapping

Kriging is an interpolator which performs linear averaging using the observed data, and is employed to compute spatial variations. When combined with entropy, a method can be developed to determine the optimum number and spatial distribution of data. Entropy yields information at each location of observation and data at locations can be re-constructed. By calculating entropy and transmitted information, areas for measurement can be prioritized. In this manner, the optimum spatial distribution and the minimum number of observation stations in the network can be determined.

The outcoming kriging estimators of the spatial phenomena depend on the number of observations. The question then is: What is the number of observations that yield no more useful information than the outcoming spatial modeling? The outcoming spatial model of the phenomenon yields the next amount of information as a function of the number of modeled points. The spatial function is a convex function. This means that the next added interval to the visualization of the spatial phenomenon yields much less information than the previous one. Different organizations of intervals (the same number of intervals) yield different amounts of information.

Quite often, we want to know the status of land use/land cover (LULC) within specific regions, river basins, watersheds, geopolitical units, or agro-ecological zones. The data available on LULC may be from satellite observations, biophysical crop suitability assessments, periodic surveys, crop production surveys, forest management and characterization, and population density. However, there are often gaps in data with respect to geographical and geospatial data, time period, and vegetative coverage. Then there are complexities in the conversion of local measurement units into standard units of area and quantity.

Often data on land use in terms of area and type is lacking or insufficient at best. A spatial allocation approach proposed by You and Wood (2003, 2005) can be employed, which permits allocations of land use type (say, crop production) at the scale of individual pixels (arbitrary scale). Accordingly, the boundary of a geographical area/watershed within which crops are grown is divided into pixels whose actual sizes depend on the spatial resolution of land use/land cover data and crop suitability surfaces. In the absence of detailed complete data, the spatial allocation approach allocates these land use/land cover areas in those pixels within the watershed/geo-political boundary where individual land uses are most likely to be found. In this manner, some pixels may be allocated no land use, some one type of land use/cover, and some may be allocated multiple covers/uses.

The spatial mapping of land cover/use, crop productivity, ecological diversity, and so on is needed. Such mapping can be constructed with the use of cross-entropy. The advantage of the entropy-based approach is that it can utilize information even if incomplete, partially correct, and sometimes conflicting. Recalling the cross-entropy,

$$D(P, Q) = \sum_{i=1}^{N} p_i \log\left(\frac{p_i}{q_i}\right) = \sum_{i=1}^{N} p_i \log p_i - \sum_{i=1}^{N} p_i \log q_i$$

$$= -H + E[\log q] \tag{9.159}$$

Cross-entropy assumes that there is some information available about the probability distribution P which is derived by the minimization of D, subject to specified constraints. This information can be regarded as an initial estimate of the unknown distribution P, and this estimate is called prior Q which may not satisfy all the specified constraints. The principle of minimum cross-entropy (POMCE) then states that minimization of D results in the minimization of difference between P and Q. You and Wood (2003, 2005, 2006) and You et al. (2009) employed cross entropy to do spatial mapping of crop distributions. The cross-entropy-based approach can be described using an example of land use mapping as follows:

1 The area under consideration is a river basin or a watershed which is divided by major land use systems, that is, urban, agricultural, forestry, and so on. Let k represent the major land use division, $k = 1, 2, \ldots, K$.

2 Each land use system is further divided into different land uses. For example, agricultural land area may be divided based on crops – wheat, rice, maize, soybean, sugarcane, and so on; urban land use into residential areas, commercial buildings, schools, hospitals, parks, gardens, paved areas, and so on; and forest areas into different types of trees – pines, oaks, post-oaks, water-oaks, and so on. Let j represent the land use category, $j = 1, 2, \ldots, J$.

3 Divide the area into a grid of pixels, where the coarseness of grid depends on the objective of study. Let i represent the i-th pixel for land use category j at the k-th system within the basin. Here $i = 1, 2, \ldots, Ijk$.

4 Convert all the real-value parameters into probabilities.

5 Generate the initial land use distribution using the empirical data.

Let q_{ijk} define the area of pixel i and land use type j at the division or watershed level k within a certain watershed; u_{jk} define the total physical area of land use type j at system level k; and A_{ijk} define the area allocated to pixel i for land use type j at system level k in the watershed. Then,

$$p_{ijk} = \frac{A_{ijk}}{u_{jk}} \tag{9.160}$$

$$D\left(p_{ijk}; q_{ijk}\right) = \sum_{k}\sum_{j}\sum_{i} p_{ijk} \log p_{ijk} - \sum_{k}\sum_{j}\sum_{i} p_{ijk} \log q_{ijk} \tag{9.161}$$

Equation (9.161) is minimized, subject to the following constraints:

$$\sum_{i} p_{ijk} = 1 \quad \forall j, k \tag{9.162}$$

As pointed out by Zellner (1988), the advantage of the cross-entropy approach is that it satisfies the "information conservation principle," meaning that the estimation procedure should neither ignore any input nor it should use any false information (Robinson et al., 2001). It permits the use of prior knowledge about land cover/use. It also enables the employment of equality or inequality constraints reflecting conditions under which spatial allocation of land cover/use should be made.

The prior LULC distribution $\{q_{ijk}\}$ should be specified. One way is to normalize the potential LULC for land cover/use j at input level k for pixel i as:

$$q_{ijk} = \frac{A}{N} \quad \forall i, j, k \tag{9.163}$$

where A is the area of the watershed and N is the total number of pixels. Equation (9.163) is based on biophysical conditions which may not reflect realistic conditions in the field. For example, some areas which are quite suitable for forestry have not yet been afforested. To make the prior more realistic, population density may be superimposed, such that higher density may correspond to urban, zero density to forest, and low density to rural. Similarly, zero density with little cover may represent desert.

9.9 Scale and entropy

Consider a set Z of n objects (the spatial zones of a geographical system) divided into K sets, Z_k, $k = 1, 3, \ldots, K$, each with n_k objects such that $\sum_{k=1}^{K} n_k = n$. The sets are mutually exclusive and collectively exhaustive. $Z = \overset{K}{\underset{k=1}{\cup}} Z_k$ and $\phi = \overset{K}{\underset{k=1}{\cap}} Z_k$, ϕ is the empty set. Each probability $p_i \in Z_k$ is defined as

$$p_k = \sum_{i \in Z_k} p_i, \; \sum_{k=1}^{K} p_k = \sum_{k=1}^{K} \sum_{i \in Z_k} p_i = 1 \tag{9.164}$$

Then

$$H = -\sum_{k=1}^{K} p_k \log p_k - \sum_{k=1}^{K} p_k \sum_{i \in Z_k} \frac{p_i}{p_k} \log \frac{p_i}{p_k}$$

$$= H_B + \sum_{k=1}^{K} p_k H_k, \quad H_k = -\sum_{i \in Z_k} \frac{p_i}{p_k} \log \frac{p_i}{p_k} \tag{9.165}$$

where H_B is the between set entropy at the higher system level, and the second term is the sum of within-set entropies H_k weighted by their probability of occurrence p_k at the higher level. If K decreases, meaning fewer Z_k sets and each set becoming larger then between-set entropy decreases but within-set entropy increases. In the limit $K = 1$, $H_B \to 0$ and $\sum_{k=1}^{K} p_k H_k \to H$. On the other hand, if K increases then H_B increases and within-set entropy decreases in sum. In the limit, $K \to n$, that is, $n_k = 1$, then the within set entropy vanishes and $H_B \to H$.

Now note that

$$X_k = \sum_{i \in Z_k} \Delta x_i \tag{9.166}$$

where X_k is the sum of the intervals (areas) in each aggregate set Z_k, the spatial entropy can be decomposed as

$$H_s = -\sum_{k=1}^{K} p_k \log\left[\frac{p_k}{X_k}\right] - \sum_{k=1}^{K} p_k \sum_{i \in Z_k} \frac{p_i}{p_k} \log\left[\frac{(p_i/p_k)}{(\Delta x_i/X_k)}\right]$$

$$= H_{sB} + \sum_{k=1}^{K} p_k H_{sk} \tag{9.167}$$

where H_{sB} is the between-set spatial entropy, and $\sum_{k=1}^{K} p_k H_{sk}$ is the sum of the weighted within-set entropies. Entropies can be nested into a hierarchy of levels, that is, between-set entropies can be further subdivided into sets that are smaller than Z_k but larger than basic sets for each object or zone Z_i. The concepts are employed to redistrict zones so that they have equal populations in the case of discrete entropy and equal population densities in the case of continuous entropies.

The decomposed entropy measures can be employed in accordance with POME to derive probability distributions subject to constraints at different system levels. Suppose the mean cost constraint applies to the entire system, but the entropy is maximized so that aggregate probabilities sum to those that are set by the level of decomposition or aggregation selected. The Lagrangean L function can be written as

$$L = -\sum_{k=1}^{K} p_k \log p_k - \sum_{k=1}^{K} p_k \sum_{i \in Z_k} \frac{p_i}{p_k} \log \frac{p_i}{p_k} - \sum_{k=1}^{K} \left(\lambda_0^k - 1\right)\left(\sum_{i \in Z_k}(p_i - p_k)\right) - \lambda_1\left(\sum_{i \in Z_k} p_i c_i - \bar{c}\right)$$

$$\tag{9.168}$$

where λ_0^k is the zeroth Lagrange multiplier associated with total probability law and λ_1 is the Lagrange multiplier associated with the mean cost constraint.

Minimizing the L function,

$$\frac{\partial L}{\partial p_i} = -\log p_i - \lambda_0^k - \lambda_1 c_i = 0, \quad i \in Z_k \tag{9.169}$$

which yields

$$p_i = \exp\left(-\lambda_0^k - \lambda_1 c_i\right), \quad i \in Z_k \tag{9.170}$$

The partition function can be expressed as

$$\exp\left(-\lambda_0^k\right) = \frac{p_k}{\displaystyle\sum_{i \in Z_k} \exp\left(-\lambda_1 c_i\right)} \tag{9.171}$$

or

$$\lambda_0^k = \log \frac{\displaystyle\sum_{i \in Z_k} \exp\left(-\lambda_1 c_i\right)}{p_k} \tag{9.172}$$

Thus,

$$p_i = p_k \frac{\exp\left(-\lambda_1 c_i\right)}{\sum\limits_{i \in Z_k} \exp\left(-\lambda_1 c_i\right)}, \quad i \in Z_k \quad \text{and} \quad \sum_{i \in Z_k} p_i = p_k \tag{9.173}$$

Note the cost of constraint is for the entire system and couples the distributions for each sub-set.

If there is no system-wide constraint then entropy maximizing is separable in K sub-problems. To illustrate, the cost constraints or subsets are:

$$\sum \frac{p_i}{p_k} c_i = \overline{c_k}, \quad \forall k \tag{9.174}$$

This permits satisfying the system wide constraint as

$$\sum_{k=1}^{K} p_k \sum_{i \in Z_k} \frac{p_i}{p_k} c_i = \sum_{k=1}^{K} p_k \overline{c_k} = \sum_{k=1}^{K} \sum_{i \in Z_k} p_i c_i = \overline{c}, \forall k \tag{9.175}$$

Using the method of Lagrange multipliers with k multipliers,

$$p_i = p_k \frac{\exp\left(-\lambda_1^k c_i\right)}{\sum\limits_{i \in Z_k} \exp\left(-\lambda_1^k c_i\right)}, \quad i \in Z_k \tag{9.176}$$

This is not separable for each sub-set.

By the maximization of spatial entropy,

$$p_i = p_k \frac{\Delta x_i \exp\left(-\lambda_1^k c_i\right)}{\sum\limits_{i \in Z_k} \Delta x_i \exp\left(-\lambda_1^k c_i\right)}, i \in Z_k \tag{9.177}$$

9.10 Spatial probability distributions

If the population size X is the random variable in place of cost of location, then maximization of entropy leads to the city size distributions. Let the probability p_i of each event be denoted with its frequency $f(.)$, the size of the event be x_i, and the discrete probability frequency be $f(x_i)$. Then

$$H(X) = -\sum_{i=1}^{N} f(x_i) \log f(x_i) \tag{9.178}$$

The usual constraints are defined as

$$\sum_{i=1}^{N} f(x_i) = 1, \quad \sum_{i=1}^{N} f(x_i) x_i = \overline{x}, \quad \sum_{i=1}^{N} f(x_i) \left[x_i^2 - (\overline{x})^2\right] = \sigma^2 \tag{9.179}$$

where σ^2 is the variance of the distribution.

Using the first constraints, one gets

$$\log f(x_i) = -\lambda_0 - \lambda_1 x_i \tag{9.180}$$

Therefore,

$$f(x_i) = \exp\left(-\lambda_0 - \lambda_1 x_i\right) \tag{9.181}$$

and

$$f(x_i) = \frac{\exp\left(-\lambda_1 x_i\right)}{\displaystyle\sum_{i=1}^{N} \exp\left(-\lambda_1 x_i\right)} \tag{9.182}$$

This shows that the larger size leads to lower probability. From the above one can write

$$x_i = -\frac{\lambda_0}{\lambda_1} - \frac{1}{\lambda_1} \log f(x_i) \tag{9.183}$$

The size can be in terms of the number of people living in the area. However, in this the size cannot be directly equated to cost, because more people will prefer to live in low cost areas. However, the size distributions can be developed by maximization of entropy.

Consider the mean log constraint as

$$\sum_{i=1}^{N} f(x_i) \log x_i = \overline{\log x} \tag{9.184}$$

Then the maximization of entropy leads to

$$\log f(x_i) = -\lambda_0 - \lambda_1 \log x_i \tag{9.185}$$

or

$$f(x_i) = \exp\left(-\lambda_0\right) x_i^{-\lambda_1} \tag{9.186}$$

This is a power function and can be written as

$$f(x_i) = \frac{x_i^{-\lambda_1}}{\displaystyle\sum_{i=1}^{N} x_i^{-\lambda}} \tag{9.187}$$

From equation (9.187) one can also write

$$x_i = \exp\left(-\frac{\lambda_0}{\lambda_1}\right) f(x_i)^{-1/\lambda_1} \tag{9.188}$$

Equation (9.188) shows that x_i varies inversely with the power of frequency. This relation can be used to generate rank-size rule, first explored by Pareto (1906) for income size and by Zipf (1949) for city sizes.

Consider another constraint expressed by variance as

$$\sum_{i=1}^{N} f(x_i)(x_i - \bar{x})^2 = \sum_{i=1}^{N} f(x_i)x_i^2 - (\bar{x})^2 = \sigma^2 \tag{9.189}$$

Then, maximization of entropy leads to

$$f(x_i) = \exp\left(-\lambda_0 - \lambda_1 x_i - \lambda_2 x_i^2\right) \tag{9.190}$$

Using the total probability,

$$\exp(\lambda_0) = \sum_{i=1}^{N} \exp\left(-\lambda_1 x_i - \lambda_2 x_i^2\right) \tag{9.191}$$

Equation (9.190), with the use of equation (9.191), can be written as

$$f(x_i) = \frac{\exp\left(-\lambda_1 x_i - \lambda_2 x_i^2\right)}{\displaystyle\sum_{i=1}^{N} \exp\left(-\lambda_1 x_i - \lambda_2 x_i^2\right)} \tag{9.192}$$

This is a form of normal distribution with the contribution of the mean and variance associated with the Lagrange multipliers λ_1 and λ_2. Parameter λ_1 is negative making the exponential positive, and parameter λ_2 is positive making the exponential negative. For $\lambda_1 \ll \lambda_2$, the variance of the distribution becomes increasingly smaller, but the skewness becomes increasingly peaked.

If the size is defined as its logarithm, then constraints can be defined as

$$\sum_{i=1}^{N} f(x_i) = 1 \tag{9.193}$$

$$\sum_{i=1}^{N} f(x_i)\log x_i = \overline{\log x} \tag{9.194}$$

$$f(x_i)\left[(\log x_i)^2 - \left(\overline{\log x}\right)^2\right] = \sigma^2 \tag{9.195}$$

On maximizing entropy, one obtains

$$\begin{aligned}
f(x_i) &= \frac{\exp\left(-\lambda_1 \log x_i - \lambda_2 [\log x_i]^2\right)}{\displaystyle\sum_{i=1}^{N} \exp\left(-\lambda_1 \log x_i - \lambda_2 [\log x_i]^2\right)} \\
&= \frac{v_i^{-\lambda_1}\left(v_i^2\right)^{-\lambda_2}}{\displaystyle\sum_{i=1}^{N} x_i^{-\lambda_1}\left(x_i^2\right)^{-\lambda_2}}
\end{aligned} \tag{9.196}$$

For $\lambda_1 \ll \lambda_2$, the log-normal distribution reduces to the inverse of the power law form but only for the largest values of v_i. This shows that power laws tend to dominate in the upper or heavy tail of the logarithmic-normal distribution.

9.11 Scaling: rank size rule and Zipf's law

The cumulative probability distribution function of size (continuous case) can be interpreted in terms of size rank (Batty 2010). To that end, all size values are arranged in descending order from the largest value to the smallest value. Consider rank one r_1 for the largest value of x_i and r_n for the smallest value. Then, when ranking the values from the smallest rank to the largest rank, $i = m < n$. The ranking is the accumulation of frequencies (Adamic, 2002). To illustrate, consider the exponential distribution where one can write $f(x_i) = f(x)$, when $\Delta x_i \to 0$.

9.11.1 Exponential law

For the continuous exponential distribution, $f(x) \sim \exp(-\lambda_1 x)$. The cumulative probability distribution becomes

$$F(x) = \int_x^\infty f(x) dx \sim \int_x^\infty \exp\left(-\lambda_1 x\right) dx = \frac{1}{\lambda_1} \exp\left(-\lambda_1 x\right) \Big|_x^\infty \tag{9.197}$$

where for rank $F(x) \sim x_{n-m} = r_k$, $i = m$, $k = n - i$. Thus, one can express

$$r_k \sim \exp\left(-\lambda_1 x_k\right) \tag{9.198}$$

Taking the logarithm of equation (9.198),

$$\log r_k \sim -\lambda_1 x_k \tag{9.199}$$

or

$$x_k \sim \frac{1}{\lambda_1} \log\left(\frac{1}{r_k}\right) \tag{9.200}$$

Equation (9.200) defines rank as a function of population or population as a function of rank, manifesting the log-linear structure of the exponential rank-size relationship.

9.11.2 Log-normal law

There are myriad cases where processes follow log-normal rather than power laws. Expressing the first Lagrange multiplier in equation (9.185) as α and the second as β and taking the limit $f(x) \sim x^\alpha x^{-2\beta}$, one expresses the counter cumulative frequency as

$$F(x) = \int_x^\infty f(x) dx \sim \int_x^\infty x^\alpha x^{-2\beta} dx = \frac{1}{\alpha - 2\beta + 1} x^{\alpha - 2\beta} \Big|_x^\infty \tag{9.201}$$

Equation (9.201) shows that the shape of the log-normal distribution depends on parameters α and β. The logarithmic equation yields

$$\log r_k \sim \left(-\lambda_1 + 1\right) \log x_k \tag{9.202}$$

or

$$x_k \sim r_k^{1/(-\lambda_1 + 1)} \tag{9.203}$$

Equation (9.203) shows size as a function of rank or rank as a function of size.

The rank and size relationship can be expressed as

$$\log r_k \sim (\alpha + 1) \log v_k - 2\beta \log v_k \tag{9.204}$$

or

$$v_k \sim r_k^{\frac{1}{\alpha+1-2\beta}} \tag{9.205}$$

If $\alpha+1 \ll 2\beta$, then for the largest values of v_k the second term becomes dominant. This suggests that the rank-size relation is more like a power law in its upper or heavy tail.

9.11.3 Power law

The rank-size relationship is commonly derived using a power law for the relationship between size and frequency. Power laws emerge from two sources: 1) the constraint for maximization of entropy is a geometric mean, and 2) the constraints are those leading to log-normal distribution but for very large values of the variable where the variance of the distribution is also very large. This means that heavy tails occur over several orders of magnitude. The continuous form of power equation (9.205) can be expressed as $f(x) \sim x^{-\lambda_1}$. Then the counter cumulative frequency $F(x)$ can be defined as

$$F(x) = \int_x^\infty f(x)\, dx \sim \int_x^\infty x^{-\lambda_1}\, dx = \frac{1}{-\lambda_1 + 1}\left(x^{-\lambda_1+1}\right)\Big|_x^\infty \tag{9.206}$$

in which $F(x)$ corresponds to the rank r_k which can be expressed as

$$r_k \sim x_k^{-\lambda_1+1} \tag{9.207}$$

Equations (9.207) and (9.205) show that these power laws are scale invariant. To show this property, consider a scale parameter α. Then size is scaled as αx_k but the rank r_k clearly does not change. Only the power laws have this property.

9.11.4 Law of proportionate effect

Consider Gibrat's (1931) law of proportionate effect. An object is of size x_{it}. It grows or declines to size x_{it+1} by a random amount ζ_{it} whose value is proportional to the size of the object already attained. This can be written as

$$x_{it+1} = x_{it} + \zeta_{it}x_{it} = x_{it}\left(1 + \zeta_{it}\right) \tag{9.208}$$

where t indicates time. Operating continually for any periods this process results in log normal. If the process is constrained such that objects do not decrease in size below a certain minimum (+ve) then the resulting distribution is scaling in the form of an inverse power function.

Solomon (2000) discussed the generalized Lotka-Volterra (GLV) model where in the steady state power laws arise from processes involving random proportionate growth. The GV type models can lead to log-normal and power laws.

Questions

Q.9.1 Consider an urban area of 500 km², with seven types of average land use areas as follows: 10 km² industrial development, 40 km² roads, and other pavements, 120 km² agricultural-horticultural, 30 km² dairy farms, and 50 km² forest. Define the probability of each land use as the area of land use divided by the total area. For each land use the value of the spatial interval is different and can be taken as equal to the area of land use. Compute the spatial entropy of the urban area.

Q.9.2 Consider that the urban area in Q.9.1 has equal area land uses. Then compute the spatial entropy and then compare it with the entropy value computed in Q.9.1.

Q.9.3 Compute the redundancy measure for the data in Q.9.1.

Q.9.4 Compute the information gain or loss considering the land use distribution in Q.9.1 and in Q.9.2.

Q.9.5 Consider an urban area having 10 subdivisions (Subdiv.) located at different distances from the city center. Residents travel to the city center for meeting their daily needs, such as work, shopping, eating, and so on. The average distances (Av. Dist.) from these subdivisions and the average times of travel (Trav. Time) are given. It is surmised that the cost of travel is a direct function of the distance or time of travel. The probability (Prob.) of locating is almost inversely proportional to the average time of travel. The weighting factor associated with each average distance is computed.

Subdiv.	Av. Dist. (km)	Weigh. Factor	Av. Trav. Time (h)	Prob. of Locating
1	1	0.0250	0.10	0.04
2	1.5	0.0750	0.15	0.06
3	2.5	0.0625	0.20	0.08
4	4.5	0.1125	0.30	0.12
5	5.0	0.1250	0.30	0.12
6	3.0	0.0750	0.20	0.08
7	3.5	0.0875	0.25	0.10
8	2.0	0.0500	0.20	0.08
9	8.0	0.2000	0.35	0.14
10	10.0	0.2500	0.45	0.18

Compute the disutility considering the distance of travel.

Q.9.6 Compute the disutility considering the time of travel given in Q.9.5. Compare this value with the value computed in Q.9.5.

Q.9.7 For a residential area located 10 km away from the city center, the minimum time of travel is 40 minutes and the maximum time of travel is 80 minutes. Compute the probability density function of time of travel and the value of entropy.

Q.9.8 For a residential area the minimum distance to the city park is 5 km and the maximum distance is 8 km. The average distance is 6 km. Compute the probability density function of the distance and the value of entropy.

Q.9.9 Consider an urban area that is partitioned into five regions and then each region is portioned into different sub-regions. The probability associated with each sub-region is given below. Compute the between-set and within-set entropies.

Probability values of regions and sub-regions																		
Region 1			Region 2						Region 3				Region 4			Region 5		
0.25			0.30						0.15				0.20			0.10		
0.05	0.10	0.10	0.05	0.10	0.08	0.03	0.04	0.03	0.04	0.06	0.02	0.1	0.05	0.05	0.02	0.03	0.04	0.01

Q.9.10 Consider an urban area that is partitioned onto four regions where each region is partitioned into different sub-regions. The probability associated with each sub-region is given. Compute the between-set entropy and within-set entropy.

Probability values of regions and sub-regions															
Region 1				Region 2					Region 3				Region 4		
0.35				0.30					0.15				0.20		
0.07	0.13	0.14	0.01	0.05	0.10	0.08	0.03	0.04	0.03	0.04	0.06	0.02	0.1	0.05	0.05

References

Adamic, L.A. (2002). Zpf, power-laws, and Pareto-ranking tutorial. Information Dynamics Lab, HP Labs, Palo Alto, California. Available at http://www.hpl.hp.com/research/idl/papers/ranking/ranking.html.

Anderson, J. (1969). On general system theory and the concept of entropy in urban geography. Research Paper No. 31, London School of Economics, Department of Geography, London, U.K.

Beckmann, M.J. (1974). Entropy, gravity and utility in transportation modeling. In: *Information, Inference and Decision*, edited by G. Menges, pp. 155–63, D. Reidel Publishing Company, Dordrecht, Holland.

Batty, M. (1972). Entropy and spatial geometry. *Area*, Vol. 4(4), pp. 230–6.

Batty, M. (1974). Spatial entropy. *Geographical Analysis*, Vol. 6, pp. 1–31.

Batty, M. (1976). Entropy in spatial aggregation. *Geographical Analysis*, Vol. 8, pp. 1–21.

Batty, M. (1983). Cost, accessibility, and weighted entropy. *Geographical Analysis*, Vol. 15, pp. 256–67.

Batty, M. (2010). Space, scale, and scaling in entropy maximizing. *Geographical Analysis*, Vol. 42, pp. 395–421.

Batty, M. and Sammons, R. (1978). On searching for the most informative spatial pattern. *Environment and planning A*, Vol. 10, pp. 747–79.

Batty, M. and Sammons, R. (1979). A conjecture on the use of Shannon's formula for measuring spatial information. *Geographical Analysis*, Vol. 11, No. 3, pp. 304–10.

Belis, M. and Guiasu, S. (1968). A quantitative-qualitative measure of information in cybernetic systems. *IEEE Transactions on Information Theory*, Vol. 14, pp. 593–4.

Chapman, G.P. (1970). The application of information theory to the analysis of population distributions in space. *Economic Geography, Supplement*, Vol. 46, pp. 317–31.

Curry, L. (1964). The random spatial economy: An exploration in settlement theory. *Annals of the Association of American Geographers*, Vol. 54, pp. 138–46.

Curry, L. (1971). Application of space-time moving-average forecasting. In: *Regional Forecasting*, edited by M. Chisholm, A. Frey and P. Haggett, Colston Papers No. 22, Butterworths, London, U.K.

Curry, L. (1972a). A spatial analysis of gravity flows. *Regional Studies*, Vol. 6, pp. 131–47.

Curry, L. (1972b). Spatial entropy. In: *International Geography 2*, P1204, edited by W.P. Adams and F.M. Helleiner, University of Toronto Press, Toronto, Canada.

Erlander, S. and Stewart, N.F. (1978). Interactivity, accessibility and cost in trip distribution. *Transportation Research*, Vol. 12, pp. 291–3.

Goldman, S. (1953). *Information Theory*. Prentice Hall, Englewood Cliffs, New Jersey.

Gibrat, R. (1931). *Les Inegalites Economiques*. Paris: Librarie du Recueil, Sirey.

Guiasu, S. (1977). *Information Theory with Applications*. McGraw-Hill, New York.

Gurevich, B.L. (1969). Geographical differentiation and its measures in a discrete system. *Soviet Geography: Reviews and Translations*, Vol. 16, pp. 387–413.

Leopold, L.B. and Langbein, W.B. (1962). The concept of entropy in landscape evolution. U.S. Geological Survey Professional Paper 500-A, Government Printing Office, Washington, D.C.

Marchand, B. (1972). Information theory and geography. *Geographical Analysis*, Vol. 4, pp. 234–57.

Medvedkov, Y. (1970). Entropy: An assessment of potentialities in geography. *Economic Geography*, Supplement, Vol. 46, pp. 306–16.

Mogridge, M.J.H. (1972). The use and misuse of entropy in urban and regional modeling of economic and spatial systems. Working Note No. 320, Centre for Environment Studies, London, U.K.

Pareto, V. (1906). *Manual of Political Economy*. (English Translation by Ann S. Schwier, 1971), Augustus M. Kelley Publishers, New York.

Robinson, S., Cattaneo, A. and El-Said, M. (2001). Updating and estimating in a social accounting matrix using cross entropy methods. *Economic Systems Research*, Vol. 13, No. 1, pp. 47–64.

Semple, R.K. and Golledge, R.G. (1970). An analysis of entropy changes in a settlement pattern over time. *Economic Geography*, Vol. 46, pp. 157–60.

Theil, H. (1967). *Economics and Information Theory*. North Holland Publishing Company, Amsterdam, The Netherlands.

Theil, H. (1980). Disutility as a probability: An interpretation of weighted information measures. *Management Science*, Vol. 26, pp. 1281–4.

Tribus, M. (1969). *Rational Descriptions, Decisions and Designs*. Pergamon Press, New York.

Wilson, A.G. (1970). *Entropy in Urban and Regional Modeling*. Pion, London, England.

Ya Nutenko, L. (1970). An information theory approach to the partitioning of an area. *Soviet Geography: Reviews and Translations*, Vol. 11, pp. 540–4.

You, L. and Wood, S. (2003). Patial allocation of agricultural production using a cross-entropy approach. Environment and Production Technology Division Discussion Paper No. 126, International Food Policy Research Institute, Washington, D.C.

You, L. and Wood, S. (2005). Assessing the spatial distribution of crop areas using a cross-entropy method. *International Journal of Applied Earth Observation and Geoinformatics*, Vol. 7, pp. 310–23.

You, L. and Wood, S. (2006). An entropy approach to spatial disaggregation of agricultural production. *Agricultural Systems*, Vol. 90, No. 1–3, pp. 329–47.

You, L., Wood, S. and Wood-Sichra U. (2009). Generating plausible crop distribution maps for sub-Saharan Africa using a spatially disaggregated data fusion and optimization approach. *Agricultural Systems*, Vol. 99, pp. 126–40.

Zellner, A. (1988). Optimal information processing and Bayes theorem. *American Statistician*, Vol. 42, pp. 278–84.

Zipf, G.K. (1949). *Human Behavior and the Principle of Least Effort*. Addison-Wesley, Cambridge, Massachusetts.

Further Reading

Angel, S. and Hyman, G. (1971). Urban spatial interaction. Working Paper No. 69, Centre for Environmental Studies, London, U.K.

Batty, M. (1974). Urban density and entropy functions. *Journal of Cybernetics*, Vol. 4, pp. 41–55.

Batty, M. (1977). Speculations on an information theoretic approach to spatial representation. In: *Spatial Representation and Spatial Interaction*, edited by I. Masser and P. Brown, Martinus Nijhoff, Leiden, Holland, pp. 115–47.

Batty, M. (2008). Cities as complex systems: scaling, interactions, networks, dynamics and urban morphologies. Working Papers Series 131, UCL Centre for Advanced Spatial Analysis, University College London, U.K.

Batty, M. and Sikdar, P.K. (1982). Spatial aggregation in gravity models: 1. An information theoretic framework. *Environment and Planning A*, Vol. 14, pp. 377–405.

Batty, M. and Sikdar, P.K. (1982). Spatial aggregation in gravity models: 2. One-dimensional population density models. *Environment and Planning A*, Vol. 14, pp. 525–53.

Batty, M. and Sikdar, P.K. (1982). Spatial aggregation in gravity models: 3. Two-dimensional trip distribution and location models. *Environment and Planning A*, Vol. 14, pp. 629–58.

Batty, M. and Sikdar, P.K. (1982). Spatial aggregation in gravity models: 4. Generalizations and large-scale applications. *Environment and Planning A*, Vol. 14, pp. 795–822.

Batty, M. and Sikdar, P.K. (1984). Proximate aggregation-estimation of spatial interaction models. *Environment and Planning A*, Vol. 16, pp. 467–6.

Berry, B.J.L. and Schwind, P.J. (1969). Information and entropy in migrant flows. *Geographical Analysis*, Vol. 1, pp. 5–14.

Broadbent, T.A. (91969). Zone size and singly-constrained interaction models. Working Note No. 132, Centre for Environmental Studies, London, U.K.

Bussiere, R. and Snickers, F. (1970). Derivation of the negative exponential model by an entropy maximizing method. *Environment and Planning*, Vol. 2, pp. 295–301.

Curry, L. (1971). Applicability of space-time moving average forecasting. Regional Forecasting, *Colston Papers*, Vol. 22, pp. 11–24.

Eeckhout, J. (2004). Gibrat's law for (all) cities. *American Economic Review*, Vol. 94, pp. 1429–51.

Foley, D.K. (1994). A statistical equilibrium theory of markets. *Journal of Economic Theory*, Vol. 62, No. 2, pp. 321–45.

Hobson, A. and Cheng, B.K. (1973). A comparison of the Shannon and Kullback information measure. *Journal of Statistical Physics*, Vol. 7, pp. 301–10.

Kalkstein, L.S., Tan, G. and Skindlov, J.A. (1987). An evaluation of three clustering procedures for use in synoptic climatological classifications. *Journal of Climate and Applied Meteorology*, Vol. 26, pp. 717–30.

Levy, M. and Solomon, S. (1986). Power laws are logarithmic Boltzmann laws. *International Journal of Modern Physics, C*, Vol. 7, No. 4, pp. 595–601.

Luce, R.D. (1960). The theory of selective information and some of its behavioral applications. In: *Developments in Mathematical Psychology*, edited by R.D. Luce, The Free Press, Glencoe.

Martin, M.J.R., De Pablo, C.L., De Agar, P.M. (2006). Landscape changes over time: Comparison of land uses, boundaries and mosaics. *Landscape Ecology, Vol.* 21, p. 1075–88.

Montroll, E.W. and Schlesinger, M.F. (1982). On 1/f noise and other distributions with long tails. *Proceedings of the National Academy of Sciences*, Vol. 79, pp. 3380–3.

Morphet, R. (2010). Thermodynamic potentials and phase change for transportation systems. UCL Working Papers Series, Paper 156, 45 p., University College London, England.

Neff, D.S. (1966). Statistical analysis for areal distributions. Monograph Series No. 2, Regional Science Research Institute, Philadelphia.

Newman, M.E.J. (2005). Power laws, Pareto distributions and Zipf's law. *Contemporary Physics*, Vol. 46, pp. 323–51.

Paes, A. and Scott, D.M. (2004). Spatial statistics for urban analysis: A review of techniques with examples. *GeoJournal*, Vol. 61, pp. 53–67.

Perline, R. (2005). Strong, weak and false inverse power laws. *Statistical Science*, Vol. 20, No. 1, pp. 68–88.

Richmond, R. and Solomon, S. (2001). Power laws are disguised Boltzmann laws. *International Journal of Modern Physics, C*, Vol. 12, No. 3, pp. 333–43.

Rojdestvenski, I. and Cottam, M.G. (2000). Mapping of statistical physics to information thery with application to biological systems. *Journal of Theoretical Biology*, Vol. 202, pp. 43–54.

Roy, J.R. and Thill, J.C. (2004). Spatial interaction modeling. *Papers in Regional Science*, Vol. 83, pp. 339–61.

Sammons, R. (1976). Zoning systems for spatial models. *Geographical Papers*, No. 52, 53 p., Department of Geography, University of Reading, England.

Sammons, R. (1977). A simplistic approach to the redistricting problem. In: *Spatial Representation and Spatial Interaction*, edited by I. Masser and P. Brown, Martinus Nijhoff, Leiden, Holland, pp. 71–94.

Siradeghyan, Y., Zakarian, A. and Mohanty, P. (2008). Entropy-based associative classification algorithm for mining manufacturing data. *International Journal of Computer Integrated Manufacturing*, Vol. 21, No. 7, pp. 825–38.

Sornette, D. (2006). *Critical Phenomena in Natural Sciences*. 2nd Edition, Springer, Heidelberg, Germany.

Sy, B.K. (2001). Information-statistical pattern based approach for data mining. *Journal of Statistics and Computer Simulation*, Vol. 69, pp. 171–201.

Turner, M.G. (1987). Spatial simulation of landscape changes in Georgia: A comparison of 3 transition models. *Landscape Ecology*, Vol. 1, No. 1, pp. 29–36.

Turner, M.G. (1990). Spatial and temporal analysis of landscape patterns. *Landscape Ecology*, Vol. 4, No. 1, pp. 21–30.

Verburg, P.H., Schot, P.P., Dijst, M.J. and Veldkamp, A. (2004). Land use change modeling: Current practice and research priorities. *GeoJournal*, Vol. 61, pp. 309–24.

von Foerster, H. (1960). On self-organizing systems and their environments. In: *Self Organizing Systems*, edited by M.C. Yovits and S. Cameron, Pergamon Press, Oxford, pp. 31–50.

Wang, L., Ji, H. and Gao, X. (2004). Clustering based on possibilistic entropy. ICSP'04, pp. 1467–70.

Ward, J.H. (1963). Hierarchical grouping to optimize an objective function. *Journal of American Statistical Association*, Vol. 58, p. 236–44.

Zhou, P., Fan, L.W. and Zhou, D.Q. (2010). Data aggregation in constructing composite indicators: A perspective of information loss. *Expert Systems with Applications*, Vol. 37, p. 360–5.

10 Inverse Spatial Entropy

The concept of spatial entropy was discussed in Chapter 9. This chapter extends the discussion of spatial entropy to inverse spatial entropy. There are many situations where a watershed or a component thereof is divided into a number of parts called zones and calculations are then based on these zones. For example, to compute the time-area diagram of a watershed, isochrones (contours of equal travel time) are constructed. If there are M isochrones then they divide the watershed into $M+1$ or N zones. Each zone is bounded by two isochrones and has an area. Thus, these N zones have N areas, with the sum of these areas being equal to the watershed area. Another example is one of computing evapotranspiration or evaporation from soil, soil moisture accounting, or soil moisture routing. In general, soil is heterogeneous both in the vertical direction and the horizontal direction. At a given location in the vertical plane, the soil is divided into a number of layers, such that each layer is more or less uniform in its texture. For example, one layer may be sand, the other clay, another silt, and so on. Frequently, two to five layers are found sufficient to represent the soil horizon. Thus, the vertical soil column is represented by these layers whose thicknesses are different. In a similar manner, a watershed is divided into a number of areas each represented by a linear reservoir. Then the watershed unit hydrograph is derived. A channel is divided into a number of segments such that the sum of segments is equal to the length of the channel reach. These segments may be of different lengths. In a similar vein, when computing the curve number, the watershed is divided into a number of land use areas such that each area has more or less the same land use.

In the above examples, there are two parameters: number of zones and configuration of zones characterized by size or length or some geometric parameter. The question arises: What information can be gleaned from the division into different zones and how does the configuration affect the information? The discussion in this chapter focuses on this and related questions.

10.1 Definition

Consider a random variable X divided into N unequal intervals (or classes) Δx_i, $i = 1, 2, \ldots, N$, with probability of $X = x_i$, $p_i = p(x_i)$. Here p_i is the probability of occurrence in each class

Entropy Theory and its Application in Environmental and Water Engineering, First Edition. Vijay P. Singh.
© 2013 John Wiley & Sons, Ltd. Published 2013 by John Wiley & Sons, Ltd.

interval. The information content of event $X = x_i$ is

$$I(p_i) = \log \frac{1}{p_i} \tag{10.1}$$

The expected information or the discrete Shannon entropy H of X can be written as

$$H = -\sum_{i=1}^{N} p_i \log p_i; \quad \sum_{i=1}^{N} p_i = 1 \tag{10.2}$$

Equation (10.2) can be employed to measure the variance of spatial probability distributions. Entropy in equation (10.2) is dimensionless, because it depends only on the number of classes or events but not on any system spatial characteristic. Equation (10.2) is useful if the number of events or class intervals is the central focus. If systems have different numbers comprising the values of X, comparison of these systems using entropy then becomes difficult. Consider an example where a watershed is divided into two different zoning systems such that the total number of land use zones in each system is different. For example, one zoning system has five types of land uses and another has eight types of land uses. In this case, the entropy characteristic of each system (set of zones) may be different and to compare these two systems will then be difficult. It is therefore desirable to explicitly incorporate the effect of zone class size (measured by interval size or coordinate system) in the entropy expression. Such incorporation leads to the definition of spatial entropy:

$$H = -\sum_{i=1}^{N} p_i \log \left(\frac{p_i}{\Delta x_i} \right) \tag{10.3}$$

The spatial entropy can also be derived in another way. Consider a spatial system with the total number of land uses (or population) being M. Let m_i be the number of land uses assigned to zone i whose capacity (or area) is Δx_i. Then $X = \sum_{i=1}^{N} x_i$. Thus these land uses can be divided into a number of zones N. Clearly each zone will have more than one land use and each zone will have its own area. There are two variables here: number of land uses per zone and zone area. There are many ways in which the M land uses can be assigned to N zones. Let W be the number of ways in which m_i land uses can be assigned to zone i with area Δx_i. The function yielding W can be obtained from the multinomial distribution:

$$W = M! \prod_i \left(\frac{\Delta x_i^{m_i}}{m_i!} \right) \tag{10.4}$$

Equation (10.4) is analogous to the likelihood function whose maximization can be achieved by first using logarithmic transformation and then using Stirling's formula for simplifying the log-factorial. Stirling's formula can be expressed as (Blundell and Blundell, 2006):

$$\ln N! \approx N \ln N - N \tag{10.5a}$$

when N is very large. Thus, one obtains from equation (10.4):

$$\log W = -\sum_i m_i \log \left(\frac{m_i}{M \Delta x_i} \right) = M \sum_i p_i \log \left(\frac{p_i}{\Delta x_i} \right) \tag{10.5b}$$

where $m_i/M = p_i$. Equation (10.5b) is similar to equation (10.3) and states that spatial entropy is proportional to the average number of ways each land use can be assigned to each zone. Dividing equation (10.5b) by M, one gets the spatial entropy function described in Chapter 9:

$$H = \frac{\log W}{M} = -\sum_i p_i \log\left(\frac{p_i}{\Delta x_i}\right) \tag{10.6}$$

which is the same as equation (10.3). There are, however, two problems with equation (10.6). First, it can give rise to negative values which make physical interpretation of entropy difficult. Second, if spatial systems are of different sizes which is not unusual then it is difficult to compare relative distributions of spatial phenomena. Thus, it is suggested to normalize spatial entropy by its maximum value.

Considering Δx_i as probability distributed or as a weighting factor, that is, $w_i = \Delta x_i / X$, then the maximum spatial entropy from equation (10.3) can be written as $\log X$. Therefore, the relative spatial entropy (or information gain) can be expressed as

$$I = H_{\max} - H = \log X + \sum_i p_i \log\left(\frac{p_i}{\Delta x_i}\right) \tag{10.7}$$

Equation (10.7) can be written as

$$I = \sum_{i=1}^{N} p_i \log\left(\frac{X p_i}{\Delta x_i}\right) = \sum_{i=1}^{N} p_i \log p_i - \sum_{i=1}^{N} p_i \log\left(\frac{\Delta x_i}{X}\right) \tag{10.8}$$

Equation (10.8) denotes what is referred to by Batty (1976, 2010) as inverse spatial entropy. Equations (10.7) and (10.8) show that inverse spatial entropy is virtual relative spatial entropy. The sum of equations (10.6) and (10.8) is always equal to the logarithm of the system size $\log X$.

Equation (10.8) can be considered as a special case of a general case. To that end, let

$$q_i = \frac{\Delta x_i}{X} \tag{10.9}$$

which can be interpreted as the probability of occurrence of Δx_i. Here the probability q_i is like a weighting factor or q_i is equivalent to the weight assigned to the ith zone. Inserting equation (10.9) in equation (10.8) one obtains

$$I = \sum_i p_i \log\left(\frac{p_i}{q_i}\right) = \sum_i p_i \log p_i - \sum_i p_i \log q_i \tag{10.10}$$

which is the same as cross or relative entropy due to Kullback and Leibler (1951) discussed in Chapter 6. Here the probability distribution $Q = \{q_i, i = 1, 2, \ldots, N\}$ can be considered as a prior distribution and the probability distribution $P = \{p_i, i = 1, 2, \ldots, N\}$ as the posterior distribution. Equation (10.10) is the sum of two components and defines the information gain by $P = \{p_i\}$ over $Q = \{q_i\}$. The quantity $\log(p_i/q_i)$ may be positive, zero or negative, depending on whether p_i is greater than, equal to, or less than q_i. If the a priori probability distribution leads to the maximum entropy $\log N$, say when the spatial system is divided into N equal zones, then

$$I(P; Q) = \log N + \sum_i p_i \log p_i = \log N - H(p) \tag{10.11}$$

Equation (10.10) measures the difference between prior distribution $\{q_i\}$ and posterior distribution $\{p_i\}$. It also denotes the information gain due to the posterior distribution over the prior distribution. In other words, the posterior distribution reflects the additional knowledge over the prior distribution which reflects a state of prior knowledge or ignorance. In the spatial context equation (10.10) determines the difference between the observed spatial probability function $\{q_i\}$ and a hypothetical probability distribution function $\{p_i\}$ based on the premise that probability is proportional to the zone size. Following Gibbs' inequality, equation (10.10) is always positive for any distributions $\{q_i\}$ and $\{p_i\}$. It is zero when $\{p_i\}=\{q_i\}$.

The entropy given by equation (10.10) approximates a χ^2 statistic and can be interpreted as the logarithm of a weighted geometric measure as follows. Taking the exponential of equation (10.8)

$$\exp(I) = \prod_i \left(\frac{Xp_i}{\Delta x_i}\right) p_i \tag{10.12}$$

In equation (10.12), term $Xp_i/\Delta x_i$ is the ratio of the density of population of, say, land uses in zone i, $m_i/\Delta x_i$, to the average density of land uses M/X in the system. Equation (10.12) is the geometric mean of all land use density ratios in the system weighted by location probabilities of land uses. Equations (10.7), (10.8), (10.10), and (10.11) express the decomposition of various entropies into simple additive terms.

Example 10.1: Consider a 1000 acre watershed having five land use types as follows:

Land use	Crops	Forest	Pasture	Garden	Urban
Area (acres)	300	350	200	50	100

Compute spatial entropy, inverse spatial entropy, maximum spatial entropy, and relative spatial entropy.

Solution: Here $N=5$, and $\Delta x_1 = 300$, $\Delta x_2 = 350$, $\Delta x_3 = 200$, $\Delta x_4 = 50$, and $\Delta x_5 = 100$ acres. Assume that different land uses are equally likely, that is, $p_i = 1/5 = 0.2$, $i = 1, 2, \ldots, 5$. Following equation (10.3) or equation (10.6), the spatial entropy is computed as:

$$H = -\sum_{i=1}^{5} p_i \log\left(\frac{p_i}{\Delta x_i}\right)$$

$$= -0.2\log\left(\frac{0.2}{300}\right) - 0.2\log\left(\frac{0.2}{350}\right) - 0.2\log\left(\frac{0.2}{200}\right) - 0.2\log\left(\frac{0.2}{50}\right) - 0.2\log\left(\frac{0.2}{100}\right)$$

$$= 6.6849 \text{ nats}$$

Following equation (10.8), the inverse spatial entropy is calculated as:

$$I = \sum_{i=1}^{N} p_i \log\left(\frac{Xp_i}{\Delta x_i}\right)$$

$$= 0.2\log\left(\frac{1000 \times 0.2}{300}\right) + 0.2\log\left(\frac{1000 \times 0.2}{350}\right) + 0.2\log\left(\frac{1000 \times 0.2}{200}\right)$$

$$+ 0.2\log\left(\frac{1000 \times 0.2}{50}\right) + 0.2\log\left(\frac{1000 \times 0.2}{100}\right)$$

$$= 0.2229 \text{ nats}$$

The system (watershed) size $X = 1000$ acres. The maximum spatial entropy thereof is:

$$H_{max} = \log(1000) = 6.9078 \text{ nats}$$

Following equation (10.7), the relative spatial entropy can be computed as

$$I = H_{max} - H = 6.9078 - 6.6849 = 0.2229 \text{ nats}.$$

The above computations show that relative spatial entropy (or information gain measured by spatial entropy) is indeed equal to the inverse spatial entropy.

10.2 Principle of entropy decomposition

Theil (1972) defined the principle of decomposition (POD) consisting of two terms, the first measuring variation at the aggregate level and the second measuring variation at the disaggregation level. This principle has been applied extensively to measure locational concentration in geographical analysis (Chapman, 1970; Semple and Golledge, 1970). It can form the basis for hierarchical analysis (Batty, 1976).

Let a spatial system Z be decomposed into K mutually exclusive subsystems or regions denoted by z_k, $k = 1, 2, \ldots, K$, where each subsystem is divided into M_k parts. The principle of decomposition states that the entropy of system Z is the sum of two entropies: 1) between-set entropy defined over the set of z_k regions, and 2) the weighted within-set entropy defined within each of the z_k regions. With the inverse spatial entropy defined by equation (10.8), the principle of decomposition can be expressed as

$$I = I_K + \sum_k w_k I_k, \quad k = 1, 2, \ldots, K \tag{10.13}$$

where I_K is the between-set entropy, I_k is the within-set entropy for each z_k, and w_k is the weight assigned to I_k. Equation (10.13) represents I at two levels of aggregation.

Before applying equation (10.13), the following concerning the aggregation of probabilities and zones should be defined:

$$P_k = \sum_{i \in z_k} p_i \tag{10.14}$$

$$\sum_k P_k = \sum_k \sum_{i \in z_k} p_i = 1 \tag{10.15}$$

$$\Delta X_k = \sum_{i \in z_k} \Delta x_i \tag{10.16}$$

$$\sum_k \Delta X_k = \sum_k \sum_{i \in z_k} \Delta x_i = X \tag{10.17}$$

where P_k is the regional probability for aggregated zone z_k and ΔX_k is the land area or size of z_k. On the basis of equation (10.13), equation (10.8) can be expressed as

$$I = \sum_k P_k \log\left(\frac{P_k X}{\Delta X_k}\right) + \sum_k P_k \sum_{i \varepsilon z_k} \frac{p_i}{P_k} \ln\left(\frac{p_i}{P_k} \frac{\Delta X_k}{\Delta x_i}\right) \tag{10.18}$$

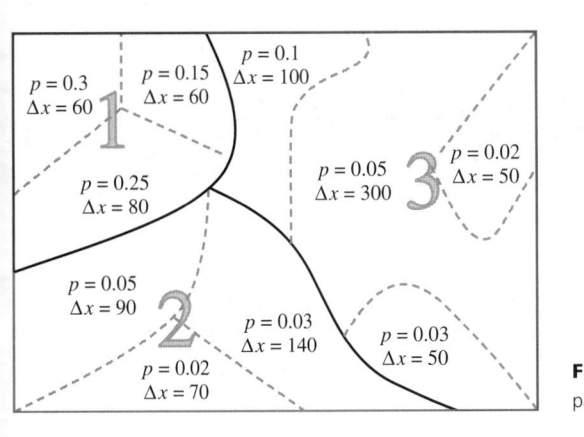

Figure 10.1 Sketch demonstrating the partition of the spatial system.

Equation (10.18) shows that the first term, designated as between-set entropy, is the inverse of spatial entropy defined over sets z_k; whereas the second term, designated as within-set entropy, is an inverse spatial entropy for zone i within each set z_k but weighted by the regional probability P_k. The between-set inverse entropy is positive but within-set inverse entropy is negative. Equations (10.7), (10.8), (10.10), (10.11), and (10.13) express decomposition of various entropies into simple additive terms.

Example 10.2: Suppose a spatial system is partitioned as shown in Figure 10.1. Compute the between-set and within-set inverse spatial entropies.

Solution: From Figure 10.1 and equation (10.14) to equation (10.17), we can compute the probability associated with each region as

$$P_1 = 0.3 + 0.15 + 0.25 = 0.7$$
$$P_2 = 0.05 + 0.03 + 0.02 = 0.1$$
$$P_3 = 0.1 + 0.05 + 0.02 + 0.03 = 0.2$$

Also we can compute the area of each region as

$$\Delta X_1 = 60 + 60 + 80 = 200 \text{ acres}$$
$$\Delta X_2 = 90 + 70 + 140 = 300 \text{ acres}$$
$$\Delta X_3 = 100 + 300 + 50 + 50 = 500 \text{ acres}$$

The total area of the system is $X = \Delta X_1 + \Delta X_2 + \Delta X_3 = 1000$ acres.
Then, following equation (10.18), the between-set inverse spatial entropy is calculated as

$$I_K = \sum_{k=1}^{3} P_k \log\left(\frac{P_k X}{\Delta X_k}\right)$$

$$= 0.7 \log\left(\frac{0.7 \times 1000}{200}\right) + 0.1 \log\left(\frac{0.1 \times 1000}{300}\right) + 0.2 \log\left(\frac{0.2 \times 1000}{500}\right)$$

$$= 0.5838 \text{ nats}$$

The within-set inverse spatial entropy is computed as:

For region 1:

$$I_1 = \frac{0.3}{0.7} \log \left(\frac{0.3}{0.7} \frac{200}{60} \right) + \frac{0.15}{0.7} \log \left(\frac{0.15}{0.7} \frac{200}{60} \right) + \frac{0.25}{0.7} \log \left(\frac{0.25}{0.7} \frac{200}{80} \right)$$
$$= 0.0403 \text{ nats}$$

For region 2:

$$I_2 = \frac{0.05}{0.1} \log \left(\frac{0.05}{0.1} \frac{300}{90} \right) + \frac{0.03}{0.1} \log \left(\frac{0.03}{0.1} \frac{300}{70} \right) + \frac{0.02}{0.1} \log \left(\frac{0.02}{0.1} \frac{300}{140} \right)$$
$$= 0.1613 \text{ nats}$$

For region 3:

$$I_2 = \frac{0.1}{0.2} \log \left(\frac{0.1}{0.2} \frac{500}{100} \right) + \frac{0.05}{0.2} \log \left(\frac{0.05}{0.2} \frac{500}{300} \right) + \frac{0.02}{0.2} \log \left(\frac{0.02}{0.2} \frac{500}{50} \right)$$
$$+ \frac{0.03}{0.2} \log \left(\frac{0.03}{0.2} \frac{500}{50} \right)$$
$$= 0.3001 \text{ nats}$$

Then, the weighted within-set inverse spatial entropy is

$$\sum_k P_k I_k = 0.7 \times 0.0403 + 0.1 \times 0.1613 + 0.2 \times 0.3001 = 0.1044 \text{ nats}$$

Finally, from equation (10.13) the total inverse spatial entropy of the given spatial system is

$$I = 0.5838 + 0.1044 = 0.6882 \text{ nats}$$

On the other hand, according to the definition as in equation (10.8), the system inverse spatial entropy also can be computed as

$$I = \sum_{i=1}^{10} p_i \log \left(\frac{X p_i}{\Delta x_i} \right)$$
$$= 0.3 \log \left(\frac{1000 \times 0.3}{60} \right) + 0.15 \log \left(\frac{1000 \times 0.15}{60} \right) + 0.25 \log \left(\frac{1000 \times 0.25}{80} \right)$$
$$+ 0.05 \log \left(\frac{1000 \times 0.05}{90} \right) + 0.03 \log \left(\frac{1000 \times 0.03}{70} \right) + 0.02 \log \left(\frac{1000 \times 0.02}{140} \right)$$
$$+ 0.1 \log \left(\frac{1000 \times 0.1}{100} \right) + 0.05 \log \left(\frac{1000 \times 0.05}{300} \right) + 0.02 \log \left(\frac{1000 \times 0.02}{50} \right)$$
$$+ 0.03 \log \left(\frac{1000 \times 0.03}{50} \right)$$
$$= 0.6882 \text{ nats}$$

From the computations one can see that the total inverse spatial entropies computed in two different ways are the same.

10.3 Measures of information gain

In a given area, usually water supply, energy supply, waste disposal facilities are directly related to the population in the area. Thus, the migration of people from one area to the other becomes important. Migration of people will determine the planning and design of such facilities, and spatial analysis of migrant flows can be analyzed using entropy. In a similar vein, consider a watershed receiving rainfall. Different areas of the watershed receive different rainfall amounts. Part of the rainfall received in any area is converted to surface water which flows as runoff to the nearest channel or stream which then flows into a higher order stream and so on, until the water reaches the watershed outlet. In a similar manner, all areas of the watershed will contribute flows to the flow that reaches the outlet. Thus, the movement of water can be considered as a two dimensional problem and can be analyzed using entropy.

10.3.1 Bivariate measures

One can characterize the movement of a body of water as an event E. This event originates in an area denoted by a_i and the probability that this event occurs in area a_i is p_i ($i = 1, 2, \ldots, N$). The probability that the water (or event) reaches the channel j, c_j, is p_j ($j = 1, 2, \ldots, M$). The probability of the movement between a_i and c_j is p_{ij}. If the information is received to the effect that the flow did occur between i and j, then each of these probabilities is raised to unity. The smaller the value of probability, the message about the actual movement is more informative. Let the a priori probabilities be q_{ij} and a posteriori probabilities of the movement of water or flow be p_{ij}. Then one can write

$$\sum_{i=1}^{N} \sum_{j=1}^{M} q_{ij} = 1 \tag{10.19}$$

$$\sum_{j} q_{ij} = q_{i\cdot} \tag{10.20}$$

$$\sum_{i} q_{ij} = q_{\cdot j} \tag{10.21}$$

$$\sum_{i} q_{i\cdot} = \sum_{j} q_{\cdot j} = 1.0 \tag{10.22}$$

Putting in matrix form,

$$\begin{bmatrix} q_{11} & q_{12} & \cdot & \cdot & \cdot & q_{1M} \\ q_{21} & q_{22} & \cdot & \cdot & \cdot & q_{2M} \\ \cdot & \cdot & \cdot & \cdot & \cdot & \cdot \\ \cdot & \cdot & \cdot & \cdot & \cdot & \cdot \\ \cdot & \cdot & \cdot & \cdot & \cdot & \cdot \\ q_{N1} & q_{N2} & \cdot & \cdot & \cdot & q_{NM} \end{bmatrix}$$

Analogous relations for p_{ij} can be written as

$$\sum_{i} \sum_{j} p_{ij} = 1.0 \tag{10.23}$$

$$\sum_{j} p_{ij} = p_{i\cdot} \tag{10.24}$$

$$\sum_i p_{ij} = p_{\cdot j} \tag{10.25}$$

$$\sum_i p_{i\cdot} = \sum_j p_{\cdot j} = 1.0 \tag{10.26}$$

$$\begin{bmatrix} p_{11} & p_{12} & \cdot & \cdot & \cdot & p_{1M} \\ p_{21} & p_{22} & \cdot & \cdot & \cdot & p_{2M} \\ \cdot & & \cdot & & \cdot & \cdot \\ \cdot & & & \cdot & & \cdot \\ \cdot & & \cdot & & \cdot & \cdot \\ p_{N1} & p_{N2} & \cdot & \cdot & \cdot & p_{NM} \end{bmatrix}$$

These constitute matrices of probabilities. Note that rows represent flow origins and columns flow destinations.

Comparison of a priori and a posteriori probability matrices results in three issues. First, row and column entropies may or may not be maximized. If rows and columns are at the maximum of log N then there is complete uncertainty about the origins and destinations. If the entropies are less than log N, then origins and destinations have some degree of concentration or organization reflecting systematic regularity. Second, if joint entropy and expected joint entropy or transinformation are the same, there is interchange reflecting movement within the bounds specified by marginal probability distributions. The equality of joint and expected joint entropies reveals pure random movement and in that case marginal entropies are maximized. If the joint and expected joint entropies are not equal, flows depart from order or organization along margins. The transmitted information ultimately gets degraded to noise and thus the information entropy can be viewed in the same way as thermal (thermodynamics) entropy. Of the most value is the non-noisy or significant information, and this may be revealed by the differences between a priori expectations and the a posteriori expectations.

Matrix P implies the joint, row, and column entropies:

$$H\left(p_{ij}\right) = -\sum_{i=1}^{N}\sum_{j=1}^{M} p_{ij} \log p_{ij} \qquad \text{(Joint entropy)} \tag{10.27}$$

$$H(p_{i\cdot}) = -\sum_{i=1}^{N} p_{i\cdot} \log p_{i\cdot} \qquad \text{(Row entropy)} \tag{10.28}$$

$$H\left(p_{\cdot j}\right) = -\sum_{j=1}^{M} p_{\cdot j} \log p_{\cdot j} \qquad \text{(Column entropy)} \tag{10.29}$$

Analogous expressions can be written for matrix Q as:

$$H\left(q_{ij}\right) = -\sum_{i=1}^{N}\sum_{j=1}^{M} q_{ij} \log q_{ij} \qquad \text{(Joint entropy)} \tag{10.30}$$

$$H(q_{i\cdot}) = -\sum_{i=1}^{N} q_{i\cdot} \log q_{i\cdot} \qquad \text{(Row entropy)} \tag{10.31}$$

$$H\left(q_{\cdot j}\right) = -\sum_{j=1}^{M} q_{\cdot j} \log q_{\cdot j} \qquad \text{(Column entropy)} \tag{10.32}$$

If flows at origins and destinations are independent the following measures of expected joint entropy (transinformation) can be expressed, respectively, as:

$$I\left(q_{ij}\right) = \log\left(\frac{q_{ij}}{q_{i\cdot}q_{\cdot j}}\right) \tag{10.33}$$

$$T\left(q_{ij}\right) = \sum_{i=1}^{N}\sum_{j=1}^{M} q_{ij}\log\left(\frac{q_{ij}}{q_{i\cdot}q_{\cdot j}}\right) \tag{10.34}$$

$$T\left(p_{ij}\right) = \sum_{i=1}^{N}\sum_{j=1}^{M} p_{ij}\log\left(\frac{p_{ij}}{p_{i\cdot}p_{\cdot j}}\right) \tag{10.35}$$

One may also write

$$T\left(q_{ij}\right) = -\sum_{i=1}^{N} q_{i\cdot}\log q_{i\cdot} - \sum_{j=1}^{M} q_{\cdot j}\log q_{\cdot j} + \sum_{i=1}^{N}\sum_{j=1}^{M} q_{ij}\log q_{ij} \tag{10.36}$$

Therefore,

$$T\left(q_{ij}\right) = H\left(q_{i\cdot}\right) + H\left(q_{\cdot j}\right) - H\left(q_{ij}\right) \tag{10.37}$$

This means that

$$H\left(q_{ij}\right) < H\left(q_{i\cdot}\right) + H\left(q_{\cdot j}\right) \tag{10.38}$$

Likewise,

$$T\left(p_{ij}\right) = H\left(p_{i\cdot}\right) + H\left(p_{\cdot j}\right) - H\left(p_{ij}\right) \tag{10.39}$$

The transinformation is zero if the difference between the sum of marginal entropies and the joint entropy (based on the margins) is zero. The transinformation increases with increasing deviations of q_{ij} and p_{ij} from their expectations $\langle q_{i\cdot}q_{\cdot j}\rangle$ and $\langle p_{i\cdot}p_{\cdot j}\rangle$ (usually $<\cdot>$ denotes expectation). Furthermore, $\log(q_{ij}/q_{i\cdot}q_{\cdot j})$ is positive, zero or negative, depending on whether the probabilities of movement from origin i to destination j are greater than, equal to, or less than the independence level.

Comparing flows further, one can write the a priori information of Q and the a posteriori information of P, then the gain of information by p over q can be stated as $I(p; q) = \sum p\log\dfrac{p}{q}$ in which $\log(p/q)$ may take on a positive, zero, or negative value, based on whether p is greater than, equal to, or less than q. If the a priori probabilities correspond to the maximum entropy then equation (10.10) becomes

$$I(p; q) = \log N + \sum p\log p = \log N - H(p)$$

Now one can express specific aspects of comparison as:

$$I\left(p_{i\cdot} : q_{i\cdot}\right) = \sum_{i=1}^{N} p_{i\cdot}\log\left(\frac{p_{i\cdot}}{q_{i\cdot}}\right) \tag{10.40}$$

$$I\left(p_{.j} : q_{.j}\right) = \sum_{j=1}^{M} p_{.j} \log \left(\frac{p_{.j}}{q_{.j}}\right) \tag{10.41}$$

$$I\left(p_{i.}p_{.j} : q_{i.}q_{.j}\right) = \sum_{i=1}^{N} \sum_{j=1}^{M} p_{i.}p_{.j} \log \left(\frac{p_{i.}p_{.j}}{q_{i.}q_{.j}}\right) \tag{10.42}$$

$$I\left(p_{ij} : q_{ij}\right) = \sum_{i=1}^{N} \sum_{j=1}^{M} p_{ij} \log \left(\frac{p_{ij}}{q_{ij}}\right) \tag{10.43}$$

Several interesting aspects of flows can be investigated using these comparative information gains. For example, if the marginals of Q and P are constant, or if the a priori and a posteriori distributions of flows among origins and destinations change, one can focus on the interchange of flows among areas. The direction of interchange is of interest in any case. The question of a change in steady state on the margins, if generated probabilistically, occurs or there is a differing degree and pattern of organization exhibiting systematic regularity.

Example 10.3: Consider five interconnected reservoirs in a large urban area. Flows can occur from one reservoir to another, depending on the requirements. In other words, one reservoir can be a supplier at one time but can be a receiver at another time.

Reservoir		Flows ($\times 1000\,\text{m}^3$)			
		1	2	3	4
1	Expected	5000	1000	1500	1200
	Observed	5500	1800	750	1400
	Residual	500	800	−750	200
2	Expected	5000	1300	2800	1700
	Observed	4500	2400	4600	2500
	Residual	−500	1100	1800	800
3	Expected	1000	1400	950	1300
	Observed	2200	3600	700	2250
	Residual	1200	2200	−250	950
4	Expected	1500	2850	950	1220
	Observed	670	3700	410	1450
	Residual	−830	850	−540	230
5	Expected	1200	1810	2420	1240
	Observed	580	2910	3710	1525
	Residual	−620	1100	1290	285

Compute q_{ij}, p_{ij}, $p_{i.}$, $p_{.j}$, $q_{i.}$, $q_{.j}$. Compute entropies for rows and columns for the observed flows as well as expected flows. Also, compute entropies for the flows. Determine if there is some systematic ordering present in the flows, if there is concentration both in origin and destination patterns, and if inflows are exhibiting more concentration than outflows. What do residuals reveal?

Solution: From the given observed flow data, we can obtain the expected flow matrix as:

5000	1000	1500	1200
5000	1300	2800	1700
1000	1400	950	1300
1500	2850	950	1220
1200	1810	2420	1240

Dividing each element by the summation of the above matrix yields the expected flow probabilities matrix $Q = [q_{ij}]$ as

0.1339	0.0268	0.0402	0.0321
0.1339	0.0348	0.0750	0.0455
0.0268	0.0375	0.0254	0.0348
0.0402	0.0762	0.0254	0.0327
0.0321	0.0485	0.0648	0.0332

In a similar manner, the observed flow matrix is:

5500	1800	750	1400
4500	2400	4600	2500
2200	3600	700	2250
670	3700	410	1450
580	2910	3710	1525

Also dividing each element by the summation of the above matrix yields the expected flow probabilities matrix $P = [p_{ij}]$ as

0.1166	0.0382	0.0159	0.0297
0.0954	0.0509	0.0976	0.0530
0.0467	0.0763	0.0148	0.0477
0.0142	0.0785	0.0087	0.0307
0.0123	0.0617	0.0787	0.0323

Obviously, the sums of Q and P are 1, respectively.

$p_{i.}$ is equivalent to the probabilities by summing up all columns (j) of P for each row (i):

$$\sum_j p_{ij} = p_{i.}$$

Similarly, $p_{.j}$ is equivalent to the probabilities by summing up all rows (j) of P for each column (i):

$$\sum_i p_{ij} = p_{.j}$$

Likewise, q_i and q_j can be calculated by the above procedure by manipulating matrix Q. The resulting $p_{i.}$, $p_{.j}$, q_i and q_j are shown as below:

$p_{i.}$	$p_{.j}$	$q_{i.}$	$q_{.j}$
0.2004	0.2852	0.2330	0.3669
0.2969	0.3056	0.2892	0.2239
0.1856	0.2157	0.1245	0.2309
0.1321	0.1935	0.1746	0.1784
0.1850		0.1786	
1	1	1	1

Following equations (10.27) to (10.32), the associated entropies can be calculated as follows:

Joint entropy: $H\left(p_{ij}\right) = -\sum_{i=1}^{N}\sum_{j=1}^{M} p_{ij} \log p_{ij} = 2.7966$ nats

Row entropy: $H\left(p_{i.}\right) = -\sum_{i=1}^{N} p_{i.} \log p_{i.} = 1.5748$ nats

Column entropy: $H\left(p_{.j}\right) = -\sum_{j=1}^{M} p_{.j} \log p_{.j} = 1.3688$ nats

Joint entropy: $H\left(q_{ij}\right) = -\sum_{i=1}^{N}\sum_{j=1}^{M} q_{ij} \log q_{ij} = 2.8355$ nats

Row entropy: $H\left(q_{i.}\right) = -\sum_{i=1}^{N} q_{i.} \log q_{i.} = 2.5700$ nats

Column entropy: $H\left(q_{.j}\right) = -\sum_{j=1}^{M} q_{.j} \log q_{.j} = 1.3489$ nats

Then from equations (10.37) and (10.39) one can obtain the transinformation as

$$T\left(p_{ij}\right) = H(p_{i.}) + H\left(p_{.j}\right) - H\left(p_{ij}\right) = 1.5748 + 1.3688 - 2.9766 = 0.147 \text{ nats}$$
$$T\left(q_{ij}\right) = H\left(q_{i.}\right) + H\left(q_{.j}\right) - H\left(q_{ij}\right) = 1.5700 + 1.3489 - 2.8355 = 0.0834 \text{ nats}$$

Results indicate that there is some systematic ordering present in the flows because $T(p_{ij}) = 0.147$ nats. Residuals reveal that the a posteriori flows are generally larger than the a priori flows.

10.3.2 Map representation
A map can be considered as a carrier of information. The distribution of its contents can be designed using entropy. The quantity of information of a map can be expressed as

$$I(Y, X) = H(X) - \phi(Z) \tag{10.44}$$

where $I(Y, X)$ or $I_{y \to x}$ denotes the quantity of information of base map (system) X received by plotting map (system) Y in the process of development and transformation of the base map X (expressed in binary units); and $\phi(z)$ denotes the loss of information of the map incurred due to its transformation and development in the process of plotting.

The loss of information $\phi(z)$ can be measured by the relative entropy in binary units. The loss is due to the generalization of the base map and external and internal noise. For estimating the loading of information one can write

$$I(Y, X) = H(X|Y) \tag{10.45}$$

where $I(Y,X)$ is the loading information in binary units, say, for 1 cm^2 of the map; $H(X)$and $H(X|Y)$ denote entropies corresponding to the marginal and conditional probability density functions expressed in bits for a unit area (say 1 cm^2) of the map. For estimating spatial information,

$$I\left(y_j, x_i\right) = \log_2 \frac{p\left(x_i|y_j\right)}{p\left(x_i\right)} \tag{10.46}$$

or

$$I\left(y_j, x_i\right) = \log_2 \frac{p\left(x_i, y_j\right)}{p\left(x_i\right) p\left(y_j\right)} \tag{10.47}$$

where x_i denotes that X is in a state of x_i and y_j denotes that Y is in a state of y_j; $p(x_i)$, $p(y_j)$, and $p(x_i, y_j)$ correspond to their probabilities. The spatial information $I(y_j, x_i)$ from equation (10.46) corresponds to the information quantity per unit area (e.g., 1 cm^2) of the map.

Information loss by map plotting is determined as the difference in entropy in a base map and that in a plotted map. Let $A_1, A_2, \ldots, A_m$ shown on the map (land use, settlements, for example) be associated with probabilities $p(x_1), p(x_2), \ldots, p(x_m)$. The system (map) entropy can be written as

$$H(X) = -\sum_{i=1}^{m} p\left(x_i\right) \log_2 \left[p\left(x_i\right)\right] \tag{10.48}$$

where $H(X)$ denotes the entropy of contents in bits for an elemental cell, a symbol, a sign, a letter, and so on. $H(X)$ denotes the entropy loading (in bits) per unit area (1 cm^2) of a map. The map information capacity or informativity can be regarded as a special case of entropy when the states are equiprobable:

$$p\left(x_1\right) = p\left(x_2\right) = \ldots = p\left(x_m\right) \tag{10.49a}$$

In this case,

$$H_0 = \log_2 m \tag{10.49b}$$

The relative entropy H_{rel} is

$$H_{rel} = \frac{H}{H_0} \tag{10.50}$$

The deviation of relative entropy from unity means excessivity and indicates the degree of information under loading or "reserve":

$$R = 1 - \frac{H}{H_0} \tag{10.51}$$

Some elements of map contents (for example, relief shown by contours) are continuous and can be expressed by entropy as

$$H(X) = -\int p(x) \log_2 \left[p(x) \right] dx \tag{10.52}$$

where $p(x)$ is the probability density function.

Consider the case of loading estimation for a small scale map. The loading estimation can be made using map graticule, hydrography, relief, settlements, roads, administrative, and territorial division, and so on.

10.3.3 Construction of spatial measures

Three desirable properties of a suitable information measure include the following: 1) The measure should increase with the increase in the number of alternatives (e.g., number of zones). 2) The measure should increase as the configuration of alternatives (e.g., the sizes of zones) becomes closer to the equal area configuration. 3) Different forms of the underlying distribution should exhibit different amounts of information, the uniform density leading to the maximum uncertainty or lack of information. One way is to disaggregate the measure into two components corresponding to the first two properties and then combine them to form the composite function.

To explicitly reflect the effect of number of zones, the Shannon entropy function $H(P|N)$ is suitable and is expressed as:

$$H(P|N) = -\sum_{i=1}^{N} p_i \log p_i, \ \ 0 \le H(P|N) \le \log N \tag{10.53}$$

Clearly, $H(P|N) \to \infty$, as $N \to \infty$, and $H(P|N+1) > H(P|N)$, regardless of $\{p_i\}$. For a fixed number of zones, the Shannon entropy attains its maximum $\log N$ as $p_i = 1/N, \forall i$.

However, equation (10.52) does not consider the configuration of zones, and it is a measure of the number rather than the density of the phenomenon under consideration. What this means is that for very different configurations for the same number of partitions of the system or a given area and possibly different densities, identical entropy values can be obtained. Thus, it is necessary for the measure of information to relate to configuration, that is, relate $\{p_i\}$ to $\{q_i\}$. Kerridge (1961) defined an inaccuracy function as

$$K(P:Q|N) = -\sum_{i=1}^{N} p_i \log q_i = -\sum_{i=1}^{N} p_i \log \left(\frac{q_i p_i}{p_i} \right)$$

$$= -\sum_{i=1}^{N} p_i \log p_i + \sum_{i=1}^{N} p_i \log \left(\frac{p_i}{q_i} \right), \ \ 0 \le K(P:Q|N) \le \infty \tag{10.54}$$

which is a kind of cross-entropy and measures the difference between $P = \{p_i\}$ and $Q = \{q_i\}$. If $\{q_i\} = \{p_i\}$, then $K(P:Q|N)$ reduces to $H(P|N)$. If the two distributions are extreme-peaked such that $p_j = 1$ and $p_i = 0, \forall i, i \ne j$, and $q_j = 1$, and $q_i = 0, \forall i, i \ne j$, then $K(P:Q|N) = 0$. If only the distribution $\{q_i\}$ is extreme-peaked, then $K(P:Q|N) = \infty$. If $\{q_i\}$ is uniform, that is, the zones are of equal area, then

$$K(P:Q|N) = \log N = H(P|N) \tag{10.55}$$

The inaccuracy function has a positive value but less than $\log N$, if the densities are uniform, that is, $p_i = q_i, \forall i$, and if $\{q_i\}$ is not uniform.

The inaccuracy measure $K(P : Q|N)$ depends on N. Normalizing $K(P : Q|N)$ with respect to the number of zones N, one can define another measure, called accuracy measure, $A(P : Q|N)$ as

$$A(P : Q|N) = \log N - K(P : Q|N)$$
$$= \log N + \sum_{i=1}^{N} p_i \log q_i, \quad -\infty \leq A(P : Q|N) \leq \log N \tag{10.56}$$

$A(P : Q|N)$ does not directly depend on N and can be construed as a measure for comparing the two distributions $\{p_i\}$ and $\{q_i\}$ in terms of their relative forms. If $K(P : Q|N) = 0$ (i.e., $\{p_i\}$ and $\{q_i\}$ are extreme-peaked), then $A(P : Q|N) = \log N$. If the system configuration is an equal area, the accuracy measure is zero, and when the configuration is extreme-peaked, the accuracy measure is $-\infty$.

A measure of information relating to the zone number, configuration, and density, denoted as $\hat{I}(P : Q|N)$, can be obtained by summing $H(P|N)$ and $A(P : Q|N)$ as:

$$\hat{I}(P : Q|N) = H(P|N) + A(P : Q|N)$$
$$= -\sum_{i=1}^{N} p_i \log p_i + \log N + \sum_{i=1}^{N} p_i \log q_i \tag{10.57}$$

If $q_i = 1/N, \forall i$, that is, $\{q_i\}$ is equal area, $A(P : Q|N) = 0$ and $H(P|N)$ dominates. This implies that if configuration is irrelevant, the Shannon entropy is the measure to be used. If the elements of the distribution $\{p_i\}$ are weighted unequally then the Shannon entropy should be modified.

The maximum value of $\hat{I}(P : Q|N)$ can occur in two ways. If $\{p_i\}$ and $\{q_i\}$ are both uniform then $H(P|N)$ dominates and takes its maximum value of $\log N$. If $p_i = q_i, \forall i$ then $\hat{I}(P : Q|N) = \log N$. In this case the distribution $\{q_i\}$ is not uniform (or equal area). Although the weighting of zones is not equal, it is consistent across both $\{q_i\}$ and $\{p_i\}$. If $\{p_i\}$ is an extreme valued distribution, $A(P : Q|N)$ dominates. If $\{q_i\}$ is extreme-peaked but not $\{p_i\}$, the minimum value of accuracy $-\infty$ occurs. If both these distributions are extreme-peaked, that is, $p_j = q_j = 1$, and $p_i = q_i = 0, \forall i, i \neq j$, this is equivalent to uniform density simplification, $\hat{I}(P : Q|N) = \log N$. Furthermore, $\hat{I}(P : Q|N) = 0$ if $\{p_i\}$ is extreme-peaked and $\{q_i\}$ is uniform.

Equation (10.57) shows that $H(P|N)$ and $K(P : Q|N)$ can be combined to constitute a measure of relative information, $I(P : Q|N)$. Thus, $\hat{I}(P : Q|N)$ can be expressed as

$$\hat{I}(P : Q|N) = \log N - I(P : Q|N) = \log N - \sum_{i=1}^{N} p_i \log \frac{p_i}{q_i} \tag{10.58}$$

The quantity $I(P : Q|N)$ is the well-known cross-entropy (Kullback and Leibler, 1951). Theil (1972) termed $I(P : Q|N)$ as information gain which is independent of N in terms of its magnitude. From equation (10.58) the effect of zone number and configuration can be re-interpreted. It can be argued that $\log N$ refers to the zone number and $I(P : Q|N)$ relates to the configuration. There is a relationship between $H(P|N)$ and $K(P : Q|N)$ and $I(P : Q|N)$. One can also define $H(P|N)$ in terms of $K(P : Q|N)$ and $I(P : Q|N)$. According to Aczel and Daroczy (1975), $K(P : Q|N)$ is inaccuracy and $I(P : Q|N)$ is information error, and $K(P : Q|N) - I(P : Q|N)$ is inaccuracy minus error, and $H(P|N)$ is uncertainty. If $\{p_i\} = \{q_i\}$, $I(P : Q|N) = 0$; this reflects the tendency to uniform density.

The other interpretation is via improvement revision or improvement (Theil, 1972). Consider a pre-prior $\{q_i^*\}$ and let it be an equal area distribution, $q_i^* = 1/N, \forall i$. The posterior distribution is the same. The accuracy can be expressed as

$$A(P:Q|N) = \sum_{i=1}^{N} p_i \log \frac{q_i}{q_i^*} = \sum_{i=1}^{N} p_i \log \frac{p_i}{q_i^*} - \sum_{i=1}^{N} p_i \log \frac{p_i}{q_i} \tag{10.59}$$

Thus, accuracy is the difference between the equal area information gain and the actual gain. If the prior and the posterior distributions are the same, implying uniform density, the accuracy is the greatest. If the pre-prior, prior, and posterior distributions are the same, $\{q_i^*\} = \{q_i\} = \{p_i\}$, $A(P:Q|N) = 0$. The information measure $\hat{I}(P:Q|N)$ can be expressed as

$$\hat{I}(P:Q|N) = -\sum_{i=1}^{N} p_i \log p_i + \sum_{i=1}^{N} p_i \log \frac{q_i}{q_i^*} \tag{10.60}$$

This shows that accuracy is a modification of the Shannon entropy because of the deviation from the equal area concept. This deviation can lead to an additional amount of information over and above $H(P|N)$, due to the tendency toward uniform density, or a loss in information from $H(P|N)$ due to the deviation from the equal area notion.

Example 10.4: Consider five zones dividing a 1000-acre watershed. Compute the inaccuracy function $K(P; Q|N)$. The zones are given as follows:

	Zones				
Configuration	1	2	3	4	5
1	300	350	200	50	100
2	250	300	150	175	125
3	200	200	200	200	200
4	500	50	150	120	180
5	400	225	175	80	120

Suppose $p_1 = 0.15$, $p_2 = 0.30$, $p_3 = 0.25$, $p_4 = 0.2$, and $p_5 = 0.1$. Compute the Shannon entropy of the zones. Compute the inaccuracy accounting for the configuration types. Also compute the accuracy measure $N = 5$, and $p_i = 1/5 = 0.2$, $i = 1, 2, \ldots, 5$, for maximum entropy.

Solution: Following the definition as shown in equation (10.52), the discrete Shannon entropy for each configuration can be computed as:

$$H(P|N) = -\sum_{i=1}^{N} p_i \log p_i$$

$$= -0.15 \log_2 (0.15) - 0.30 \log_2 (0.30) - 0.25 \log_2 (0.25) - 0.2 \log_2 (0.2) - 0.1 \log_2 (0.1)$$

$$= 2.2282 \text{ bits}$$

To take into account zonal configuration, we compute the inaccuracy measure as given in equation (10.54). In this case, p_is are given without due consideration of the zonal configuration and q_is are based on the zonal configuration. Therefore, for configuration 1,

$$K_1\,(P;\,Q|N = 5) = -\sum_{i=1}^{N} p_i \log_2 (q_i)$$

$$= -0.15 \log_2 \left(\frac{300}{1000}\right) - 0.30 \log_2 \left(\frac{350}{1000}\right)$$

$$- 0.25 \log_2 \left(\frac{200}{1000}\right) - 0.2 \log_2 \left(\frac{50}{1000}\right) - 0.1 \log_2 \left(\frac{100}{1000}\right)$$

$$= 2.4920 \text{ bits}$$

For configuration 2

$$K_2\,(P;\,Q|N = 5) = -\sum_{i=1}^{N} p_i \log_2 (q_i)$$

$$= -0.15 \log_2 \left(\frac{250}{1000}\right) - 0.30 \log_2 \left(\frac{300}{1000}\right)$$

$$- 0.25 \log_2 \left(\frac{150}{1000}\right) - 0.2 \log_2 \left(\frac{175}{1000}\right) - 0.1 \log_2 \left(\frac{125}{1000}\right)$$

$$= 2.3082 \text{ bits}$$

For configuration 3

$$K_3\,(P;\,Q|N = 5) = -\sum_{i=1}^{N} p_i \log_2 (q_i)$$

$$= -0.15 \log_2 \left(\frac{200}{1000}\right) - 0.30 \log_2 \left(\frac{200}{1000}\right)$$

$$- 0.25 \log_2 \left(\frac{200}{1000}\right) - 0.2 \log_2 \left(\frac{200}{1000}\right) - 0.1 \log_2 \left(\frac{200}{1000}\right)$$

$$= 2.2319 \text{ bits}$$

For configuration 4

$$K_4\,(P;\,Q|N = 5) = -\sum_{i=1}^{N} p_i \log_2 (q_i)$$

$$= -0.15 \log_2 \left(\frac{500}{1000}\right) - 0.30 \log_2 \left(\frac{50}{1000}\right)$$

$$- 0.25 \log_2 \left(\frac{150}{1000}\right) - 0.2 \log_2 \left(\frac{120}{1000}\right) - 0.1 \log_2 \left(\frac{180}{1000}\right)$$

$$= 2.9900 \text{ bits}$$

For configuration 5

$$K_5\,(P;\,Q|N = 5) = -\sum_{i=1}^{N} p_i \log_2 (q_i)$$

$$= -0.15 \log_2 \left(\frac{400}{1000}\right) - 0.30 \log_2 \left(\frac{225}{1000}\right)$$

$$- 0.25 \log_2 \left(\frac{175}{1000}\right)$$

$$- 0.2 \log_2 \left(\frac{80}{1000} \right) - 0.1 \log_2 \left(\frac{120}{1000} \right)$$
$$= 2.5072 \, \text{bits}$$

The accuracy measure can be computed using equation (10.56). Take configuration 1 as an example. The accuracy measure can be computed as:

$$A_1 \, (P : Q|N) = \log N - K_1 \, (P : Q|N) = \log_2 5 - 2.4920 = -0.1701 \, \text{bits}$$

In a similar manner, the accuracy measures for other configurations can be obtained as:

$$A_2 \, (P : Q|N) = \log N - K_2 \, (P : Q|N) = \log_2 5 - 2.3082 = \quad 0.0137 \, \text{bits}$$
$$A_3 \, (P : Q|N) = \log N - K_3 \, (P : Q|N) = \log_2 5 - 2.2319 = \quad 0.0900 \, \text{bits}$$
$$A_4 \, (P : Q|N) = \log N - K_4 \, (P : Q|N) = \log_2 5 - 2.9900 = -0.6681 \, \text{bits}$$
$$A_5 \, (P : Q|N) = \log N - K_5 \, (P : Q|N) = \log_2 5 - 2.5072 = -0.1853 \, \text{bits}$$

Example 10.5: Compute the information measure $\hat{I}$ relating zone number, configuration, and density using the data given in Example 10.4. Use $N = 5$, $p_1 = 0.15$, $p_2 = 0.3$, $p_3 = 0.25$, $p_4 = 0.2$, and $p_5 = 0.1$. Then, consider $p_i = q_i = 1/N$. Compute $\hat{I}$. Compute information error and uncertainty.

Solution: The measure of information relating to zone number, configuration, and density, denoted as $\hat{I} (P : Q|N)$, is defined in equation (10.57). Then for the configuration 1, such information measure is:

$$\hat{I}_1 \, (P : Q|N) = H(P|N) + A_1 \, (P : Q|N)$$

From Example 10.5, we have already obtained:

$$H(P|N) = 0.2282 \, \text{bits}$$
$$A_1 \, (P : Q|N) = -0.1701$$

Therefore, of $\hat{I}_1 \, (P : Q|N) = 0.2282 - 0.1701 = 0.0581 \, \text{bits}$
Similarly, we have

$$\hat{I}_2 \, (P : Q|N) = H(P|N) + A_2 \, (P : Q|N) = 0.2282 + 0.0137 = 0.2419 \, \text{bits}$$
$$\hat{I}_3 \, (P : Q|N) = H(P|N) + A_3 \, (P : Q|N) = 0.2282 + 0.0900 = 0.3182 \, \text{bits}$$
$$\hat{I}_4 \, (P : Q|N) = H(P|N) + A_4 \, (P : Q|N) = 0.2282 - 0.6681 = -0.4399 \, \text{bits}$$
$$\hat{I}_5 \, (P : Q|N) = H(P|N) + A_5 \, (P : Q|N) = 0.2282 - 0.1853 = 0.0429 \, \text{bits}$$

When $p_i = q_i = 1/5$, also following equation (10.57) the information measure is computed as:

$$I = -\sum_{i=1}^{N} p_i \log p_i + \log N + \sum_{i=1}^{N} p_i \log q_i$$
$$= -\sum_{i=1}^{5} 0.2 \log_2 (0.2) + \log_2 (5) + \sum_{i=1}^{5} 0.2 \log_2 (0.2) = \log_2 (5)$$
$$= 2.2319 \, \text{bits}$$

Information error, also known as cross-entropy, as presented in equation (10.58), for configuration 1 can be computed as:

$$I_1\,(P:Q|N) = \sum_{i=1}^{5} p_i \log \frac{p_i}{q_i}$$

where

$$p_1 = 0.15,\ p_2 = 0.3,\ p_3 = 0.25,\ p_4 = 0.2,\ p_5 = 0.1$$
$$q_1 = \frac{300}{1000} = 0.3,\ q_2 = \frac{350}{1000} = 0.35,\ q_3 = \frac{200}{1000} = 0.2,\ q_4 = \frac{50}{1000} = 0.05,$$
$$q_5 = \frac{100}{1000} = 0.1$$

Therefore,

$$I_1\,(P:Q|N) = 0.15 \log_2 \left(\frac{0.15}{0.3}\right) + 0.3 \log_2 \left(\frac{0.3}{0.35}\right) + 0.25 \log_2 \left(\frac{0.25}{0.2}\right) + 0.2 \log_2 \left(\frac{0.2}{0.05}\right)$$
$$+ 0.1 \log_2 \left(\frac{0.1}{0.1}\right) = 0.2638\ \text{bits}$$

Similarly, information errors for other configurations are

$$I_2\,(P:Q|N) = 0.0800\ \text{bits}$$
$$I_3\,(P:Q|N) = 0.0937\ \text{bits}$$
$$I_4\,(P:Q|N) = 0.7618\ \text{bits}$$
$$I_5\,(P:Q|N) = 0.2790\ \text{bits}$$

$H(P|N)$ also known as uncertainty. From Example 10.4, it is 2.2282 bits.

10.4 Aggregation properties

The information of the proposed measure ensues from two sources: from the number of zones N describing the phenomenon and from zonal configuration based on the equal area or uniform density assumption. The information owing to the number of zones can be traded off with information imparted by the zonal configuration. The objective of aggregation is to find a configuration of M zones from N, $M < N$, where the change in information is positive. This change in information can be expressed as

$$\Delta \hat{I}\,(P:Q|N \rightarrow M) = \hat{I}\,(P:Q|M) - I\big(P:\hat{Q}|N\big) \tag{10.61}$$

For the aggregation to be meaningful, $\Delta \hat{I} > 0$.

Consider the aggregation from N to M zones. For the zonal aggregation of $S_i,\ i = 1, 2, \ldots, N$, into sub-regions $S_j,\ j = 1, 2, \ldots, M$, the associated aggregation of probabilities can be written as

$$P_j = \sum_{i \in S_j} p_i \quad \text{and} \quad Q_j = \sum_{i \in S_j} Q_i \tag{10.62}$$

Equation (10.61) can now be recast as

$$\Delta \hat{I}\,(P:Q|N \to M) = -\sum_j P_j \log P_j + \log M + \sum_j P_j \log Q_j + \sum_i p_i \log p_i - \log N - \sum_i p_i \log q_i$$

$$(10.63)$$

From the theorem of entropy decomposition (Theil, 1972),

$$\sum_i p_i \ln p_i = \sum_j P_j \log P_j + \sum_j P_j \sum_{i \in S_j} \frac{p_i}{P_j} \ln \frac{p_i}{P_j}$$

$$(10.64)$$

and

$$\sum_i p_i \ln q_i = \sum_j P_j \log Q_j + \sum_j P_j \sum_{i \in S_j} \frac{p_i}{P_j} \ln \frac{q_i}{Q_j}$$

$$(10.65)$$

Introducing equation (10.65) in equation (10.63), the information change can be cast as

$$\Delta \hat{I}\,(P:Q|N \to M) = \Delta H(P|N \to M) + \Delta A(P:Q|N \to M)$$

$$(10.66)$$

which can be written as

$$\Delta I(P:Q|\hat{N} \to M) = \sum_j P_j \sum_{i \in S_j} \frac{p_i}{P_j} \log \frac{p_i}{P_j} + \log \frac{M}{N} - \sum_j P_j \sum_{i \in S_j} \frac{p_i}{P_j} \log \frac{q_i}{Q_j}$$

$$(10.67)$$

The first term on the right side of equation (10.67) represents $\Delta H(P|N \to M)$, and the second and third terms represent $\Delta A(P:Q|N \to M)$. Clearly, $\Delta H(P|N \to M)$ is negative. In order for $\hat{I}\,(P:Q|N \to M)$ to be positive, $\Delta A(P:Q|N \to M)$ must be positive and greater than $\Delta H(P|N \to M)$, that is,

$$\Delta A(P:Q|N \to M) > -\Delta H(P|N \to M)$$

$$(10.68)$$

The term $\log\,(M/N) < 0$. Therefore, for $I\big(P:Q|\hat{N} \to M\big)$ to be positive,

$$-\sum_j P_j \sum_{i \in S_j} \frac{p_i}{P_j} \log \left(\frac{q_i}{Q_j}\right) > \log \left(\frac{N}{M}\right) - \sum_j P_j \sum_{i \in S_j} \frac{p_i}{P_j} \log \left(\frac{q_i}{P_j}\right)$$

$$(10.69)$$

The loss in zonal information $\log\,(M/N)$ is owing to the change in the way accuracy is normalized at the level of M and N zones.

From equations (10.58) and (10.61) one can write

$$\Delta \hat{I}\,(P:Q|N \to M) = \log M - \sum_j P_j \log \left(\frac{P_j}{Q_j}\right) - \log N + \sum_i p_i \log \left(\frac{p_i}{q_i}\right)$$

$$(10.70)$$

Thus, the change in information is a function of change in the number of zones and change in the information gain:

$$\Delta I(P:Q|N \to M) = I(P:Q|M) - I(P:Q|N)$$

$$(10.71)$$

Using the entropy decomposition theorem,

$$\sum_i p_i \log \frac{p_i}{q_i} = \sum_j P_j \log \frac{P_j}{Q_j} + \sum_j P_j \sum_{i \in S_j} \frac{p_i}{P_j} \log \left(\frac{p_i/P_j}{q_i/Q_j} \right) \tag{10.72}$$

Substituting equation (10.72) in equation (10.70),

$$\Delta \hat{I} (P : Q | N \rightarrow M) = \log \frac{M}{N} + \sum_j P_j \sum_{i \in S_j} \frac{p_i}{P_j} \log \left(\frac{p_i/P_j}{q_i/Q_j} \right) \tag{10.73}$$

For $\hat{I} (P : Q | N \rightarrow M) > 0$, one gets

$$\sum_j P_j \sum_{i \in S_j} \frac{p_i}{P_j} \log \left(\frac{p_i/P_j}{q_i/Q_j} \right) > \log \frac{N}{M} \tag{10.74}$$

Equation (10.74) shows that the gain in information from moving toward the equal area-uniform density assumption must be larger than the loss owing to the adjustment of the zonal number information.

Example 10.6: Consider three unequal zones and the corresponding probability distribution. The distribution of phenomenon at the three-zone level is uneven. Aggregation of three zones to two equal zones is made, as shown in Figure 10.2.

Solution: The aggregation leads to a uniform distribution of the phenomenon. The accuracy measure will be zero and the Shannon entropy will change from $\Delta H(P|3) = 0.9433$ to $\Delta H(P|2) = \log 2 = 0.6931$. Therefore, $\Delta H(P|3 \rightarrow 2) = -0.2502$ bits.
  The Kerridge inaccuracy:

$$K(P : Q|3) = 1.3593, K(P : Q|2) = 0.6931, \Delta K(P : Q|3 \rightarrow 2) = -0.6661 \text{ bits.}$$

Pure zonal information:

$$\log 3 = 1.0986, \log 2 = 0.6931, \Delta \log N = -0.4055 \text{ bits.}$$

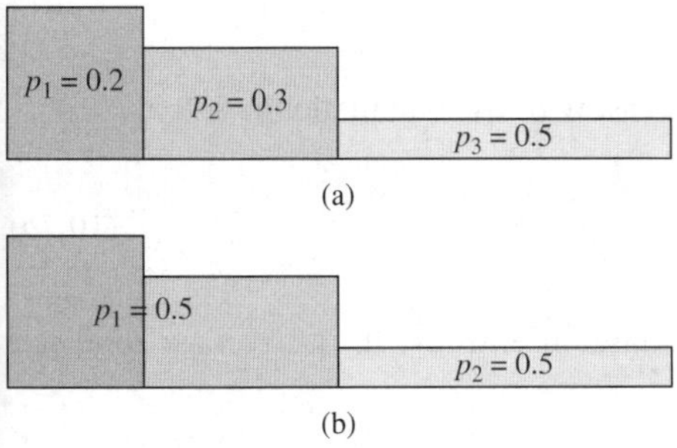

(a)

(b)

Figure 10.2 Aggregation from three zones to two zones.

Table 10.1 Information measures before and after aggregation.

	ΔH	K	A	I	$\hat{I}$
Zones 3	0.9433	1.3593	−0.2606	0.4159	0.6827
Zones 2	0.6931	0.6931	0.0000	0.0000	0.6931
Changes	−0.2502	−0.6661	0.2606	−0.4159	0.0104

Accuracy:

$$A(P:Q|3) = -0.2606, A(P:Q|2) = 0.00, \Delta A(P:Q|3 \rightarrow 2) = 0.2606 \, \text{bits.}$$

Information gain:

$$I(P:Q|3) = 0.4159, I(P:Q|2) = 0.00. \Delta I(P:Q|3 \rightarrow 2) = -0.4159 \, \text{bits.}$$

Information:

$$\hat{I}(P:Q|N=3) = 0.6827, \hat{I}(P:Q|N=2) = 0.6931, \Delta\hat{I}(P:Q|N \rightarrow 3\,to\,M \rightarrow 2) = 0.0104 \, \text{bits.}$$

Results are shown in Table 10.1. The change in information $\Delta\hat{I}(P:Q|3 \rightarrow 2)$ is positive. This means that the two-zone system conveys slightly more information than the three-zone system. In this case the loss of information $\Delta H(P|3 \rightarrow 2)$ is more than compensated for by the gain in inaccuracy $A(P:Q|3 \leftarrow 2)$. The loss in the Shannon entropy is 26% of $H(P|3)$, even though $\{p_i\}$ is aggregated to a uniform distribution, but the gain in information is also of this order and offsets the loss. The slight increase in information is achieved by quite a radical change in the configuration of zones. In practical applications, aggregation to equal area can never be achieved. As n increases, it becomes more difficult to achieve a positive value of $\Delta\hat{I}(P:Q|N \rightarrow M)$ when $M = N-1$.

Referring to Table 10.1, the loss in zonal information $\log(2/3)$ is about 36% and the gain through the change in information gain just offsets this amount. The change in accuracy, a positive quantity in the information, is on the order of 49% of the inaccuracy at the three-zone level, but the loss with inequality is $\log(3/2) + \Delta H(P|3 \rightarrow 2)$ is about 32% of $\log 3 + H(P|3)$. Improvements in information through aggregation entail radical changes in configuration.

10.5 Spatial interpretations

Information in equation (10.57) or (10.58) can be cast with use of equation (10.9) as

$$\hat{I}(P;Q|N) = -\sum_{i=1}^{N} p_i \log \frac{p_i}{x_i} - \log \frac{X}{N} \tag{10.75}$$

in which the first term on the right side is the Shannon entropy. If, the average zone size $X/N = \bar{x}$, is defined as

$$\bar{x} = \frac{X}{N} \tag{10.76}$$

Then equation (10.75) yields

$$I\left(P; \hat{Q}|N\right) = -\sum_{i=1}^{N} p_i \log \frac{p_i}{x_i} - \log \overline{x} \tag{10.77}$$

Equation (10.76) can be interpreted as the information proportional to the spatial entropy modified by the system information measured by the average zonal size.

Rearranging equation (10.76), and recalling $p_i = p(x_i)x_i$ (x_i is the area of the corresponding zone i) and taking the limits, one obtains

$$\lim_{N\to\infty}\left[\hat{I}\left(P; Q|N\right) + \log\left(\frac{X}{N}\right)\right] = -\int_X p(x) \log\left[p(x)\right] dx \tag{10.78}$$

Equation (10.78) shows that spatial entropy converges to continuous Shannon entropy as $x_i \to 0, \forall i$.

Using equation (10.58) in equation (10.78) and recalling definitions of $q(x)$ and $q(x_i)$ analogous to those of $p(x)$ and $p(x_i)$, the continuous information gain or divergence can be defined as

$$\lim_{N\to\infty}\left[\hat{I}\left(P; Q|N\right) + \log\left(\frac{1}{N}\right)\right] = -\int_X p(x) \log\left[\frac{p(x)}{q(x)}\right] dx \tag{10.79}$$

The difference between the above two measures entailing $\Delta\hat{I}\left(P; Q|N\right)$ is solely in terms of X, the system size. Thus, the zone size or configuration enters the modified Shannon entropy in an absolute way.

Using equation (10.58), equation (10.79), in terms of $p(x_i)$, can be cast as

$$\hat{I}\left(P; Q|N\right) = -\sum_{i=1}^{N} p\left(x_i\right) x_i \log p\left(x_i\right) - \log \overline{x} = -\sum_i p\left(x_i\right) x_i \log\left[p\left(x_i\right)\overline{x}\right] \tag{10.80}$$

$\Delta\hat{I}\left(P; Q|N\right)$ is the modified Shannon entropy. Equation (10.80) converges to $H(P|N)$. As $x_i \to \overline{x}$, with $p(x_i)$ redefined, $\hat{I}\left(P; Q|N\right) \to H(P|N)$.

Using equations (10.48), (10.52), and (10.80), accuracy $A(P; Q|N)$ can be re-interpreted:

$$A(P; Q|N) = \hat{I}\left(P; Q|N\right) - H(P|N) = -\sum_i p\left(x_i\right) x_i \log\left[p\left(x_i\right)\overline{x}\right] + \sum_i p\left(x_i\right) x_i \log\left[p\left(x_i\right) x_i\right]$$

$$\tag{10.81}$$

Equation (10.81) can be rearranged as

$$A(P; Q|N) = \sum_i p\left(x_i\right) x_i \log\left(\frac{x_i}{\overline{x}}\right) \tag{10.82}$$

Equation (10.82) is the explicit density equivalent of the information-improvement or revision interpretation of accuracy expressed by equation (10.59). Accuracy can be positive or negative within the range given by equation (10.58). It can also be inferred as a measure of $\Delta I(P; Q|N) \to H\left(P|\hat{N}\right)$ is $A(P; Q|N) \to 0$, which can occur only when $x_i \to \overline{x}, \forall i$, with $p(x_i)$ defined appropriately.

Now expressing $\Delta \hat{I}(P; Q|N)$ in terms of actual values of the phenomenon $\{\varphi_i\}$ using equations (10.75) and $p_i = \Delta x_i/X$, $\varphi_i = \varphi_i/\Phi$, where φ_i can be, say the number of land uses in zone I, and Φ is the total number of land uses,

$$\hat{I}(P; Q|N) = -\sum_{i=1}^{N} p_i \log\left(\frac{\varphi_i}{x_i}\right) + \log\left(\frac{N\Phi}{X}\right) \tag{10.83}$$

Noting $\varphi_i = \Phi p(x_i)$, $D = \Phi/X$, (D = density of population system wise) and using equation (10.49b), equation (10.83) can be written as

$$\hat{I}(P; Q|N) = \log N - \sum_{i=1}^{N} p_i \log \varphi_i + \log D \tag{10.84}$$

Comparing equation (10.84) with equation (10.58) one obtains

$$I(P; Q|N) == \sum_{i} p_i \log\left[\varphi(x_i)\right] - \log D \tag{10.85}$$

Equation (10.85) is the information gain as the average of logarithms of population density minus the logarithm of average density. When $\varphi(x_i) = D, \forall i$, the information gain is zero. This implies a uniform distribution and a maximum value for $\Delta \hat{I}(P; Q|N)$ of $\log N$. The zonal component is isolated as $\log N$.

Thus, the information gain measure is a function of the difference between zonal and system wide densities. Using equation (10.84), one gets

$$\Delta \hat{I}(P; Q|N \to M) = \log M - \sum_{j} P_j \log\left[\varphi\left(X_j\right)\right] + \log D - \log N + \sum_{i} p_i \log\left[\varphi(x_i)\right] - \log D \tag{10.86}$$

Using the principle of entropy decomposition and by analogy with equation (10.74), the condition for positive change can be stated as

$$\sum_{j} P_j \sum_{i \in S_j} \frac{p_i}{P_j} \log\left(\frac{\varphi(x_i)}{\varphi\left(X_j\right)}\right) > \log \frac{N}{M} \tag{10.87}$$

The term on the left side of the inequality can be seen as a density difference or divergence, and the right side is the absolute gain in information from moving from N to M zones.

Let

$$\mu(x) = \frac{1}{X} \tag{10.88}$$

where X is partitioned into N equal areas of size $\bar{x}$,

$$\mu(x) = \frac{1}{N\bar{x}} \tag{10.89}$$

Using equation (10.48), (10.52), and (10.89), the Shannon entropy can be expressed in terms of $p(x_i)$ and $\bar{x}$:

$$H(P|N) = -\sum_{i} \frac{p(x_i)}{N\mu(x)} \log\left[\frac{p(x_i)}{\mu(x)}\right] \bar{x} \tag{10.90}$$

This can be simplified as

$$H(P|N) = \log N - \sum_i p(x_i) \log \left[\frac{p(x_i)}{\mu(x)} \right] \bar{x} \tag{10.91}$$

Therefore,

$$\lim_{N \to \infty} H(P|N) = \lim_{N \to \infty} \log N - \int_X p(x) \log \left[\frac{p(x)}{\mu(x)} \right] dx \tag{10.92}$$

Equation (10.92) is equivalent to equation (10.78).

Let the total area X be partitioned into equal areas x_i. The measure $\mu(x)$ is a density estimate:

$$\mu(x_i) = \frac{1}{\sum_i x_i} = \frac{1}{X}, \ \forall i \tag{10.93}$$

It is uniform and equivalent to $\mu(x)$. Then

$$\Delta \hat{I}(P; Q|N \to M) = \log N - \sum_j p(x_i) \log \left[\frac{p(x_i)}{\mu(x_i)} \right] x_i = \log N - \sum_i p_i \log \left[\frac{p_i}{q_i} \right] \tag{10.94}$$

This explains the approximation of the Shannon entropy by the measure $\Delta \hat{I}(P; Q|N)$ which allows for unequal zone size or area.

Example 10.7: Consider aggregating zones $N = 3$ unequal zones into $M = 2$ equal zones, as shown in Figure 10.1. It is assumed that the aggregation of three zones in two zones results in uniform configuration. Compute the Shannon entropy for the two cases and the resulting change in the Shannon entropy. Also compute the change in information, loss of information, and the gain in accuracy, inaccuracy, zonal information $\log N$, and information $\hat{I}(P; Q|N)$. Discuss the results.

Figure 10.1a shows the density and distribution over three zones, and Figure 10.1b shows the aggregation to two equal area zones.

Solution: See Example 10.6

Example 10.8: Consider two hypothetical urban areas named A and B. Their major characteristics are shown in Table 10.2. Compute different information measures: $\hat{I}(P; Q|N)$, $H[P(N)]$, $A(P; Q|N)$, $\hat{I}(P; Q|N) / \log N$, $\hat{I}(P; Q|N) / H[P(N)]$, and $R[P(N)]$.

Solution: For urban area A

$$\begin{aligned}
\hat{I}_A(P; Q|N) &= -\sum_{i=1}^{5} p_i \log p_i + \log N + \sum_{i=1}^{5} p_i \log q_i \\
&= -0.2 \log 0.2 - 0.3 \log 0.3 - 0.15 \log 0.15 - 0.25 \log 0.25 - 0.1 \log 0.1 + \log 5 \\
&\quad + 0.2 \log 0.5 + 0.3 \log 0.1 + 0.15 \log 0.05 + 0.25 \log 0.15 + 0.1 \log 0.2 \\
&= 1.2399 \text{ bits}
\end{aligned}$$

Table 10.2 Major characteristics of two hypothetical areas.

Urban area	No. of zones N	Land use X (mi^2)	Total no. of trees (Φ)	Av. Area/zone $\bar{x} = X/N$	Av. no. of trees $\phi = \Phi/N$
A	5	15	100,000	3	20,000
B	10	20	150,000	2	15,000

Urban A Zone	1	2	3	4	5	Total
x_i	7.5	1.5	0.75	2.25	3.0	15
q_i	0.5	0.1	0.05	0.15	0.2	1
p_i	0.2	0.3	0.15	0.25	0.1	1

Urban B Zone	1	2	3	4	5	6	7	8	9	10	Total
x_i	2.4	3.6	3.0	1.0	0.5	1.5	2.7	1.3	1.7	2.3	20
q_i	0.12	0.18	0.15	0.05	0.025	0.075	0.135	0.065	0.085	0.115	1
p_i	0.1	0.05	0.06	0.08	0.25	0.21	0.05	0.04	0.06	0.10	1

$$H_A[P(N)] = -\sum_{i=1}^{5} p_i \log p_i$$
$$= -0.2 \log 0.2 - 0.3 \log 0.3 - 0.15 \log 0.15 - 0.25 \log 0.25 - 0.1 \log 0.1$$
$$= 1.5445 \text{ bits}$$

$$A_A(P; Q|N) = \log N + \sum_{i=1}^{5} p_i \log q_i$$
$$= \log 5 + 0.2 \log 0.5 + 0.3 \log 0.1 + 0.15 \log 0.05 + 0.25 \log 0.15 + 0.1 \log 0.2$$
$$= -0.3046 \text{ bits}$$

$$\hat{I}_A(P; Q|N) / \log N = \frac{1.2399}{\log 5} = 0.7706 \text{ bits}$$

$$\hat{I}_A(P; Q|N) / H_A[P(N)] = \frac{1.2399}{1.5445} = 0.8028 \text{ bits}$$

$$R_A[P(N)] = 1 - \frac{H_A[P(N)]}{\log N} = 1 - \frac{1.5445}{\log 5} = 0.04$$

Similarly, for urban area B, we have

$$\hat{I}_B(P; Q|N) = 1.7143 \text{ bits}$$

$$H_B[P(N)] = 2.1028 \text{ bits}$$

$$A_B(P; Q|N) = -0.3885 \text{ bits}$$

$$\hat{I}_B(P; Q|N) / \log N = \frac{1.7143}{\log 5} = 0.7444 \text{ bits}$$

$$\hat{I}_B(P; Q|N) / H[P(N)] = \frac{1.7143}{2.1028} = 0.8152 \text{ bits}$$

$$R_B[P(N)] = 1 - \frac{H_B[P(N)]}{\log N} = 1 - \frac{2.1028}{\log 5} = 0.0869 \text{ bits}$$

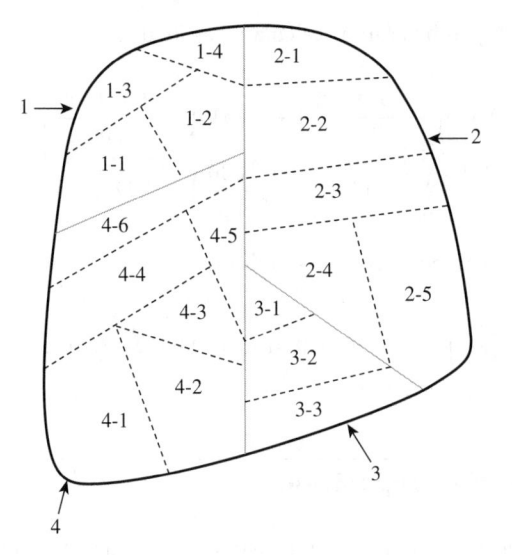

Figure 10.3 Decomposition of a spatial system into four zones.

Example 10.9: Consider a hypothetical 1000-acre urban area decomposed into four zones as shown in Figure 10.3.

Area	1	2	3	4		
Acres	200	300	150	450		
p_i	0.30	0.25	0.35	0.10		
Subdivision of areas						
Area 1	1-1	1-2	1-3	1-4		
Acres	35	45	50	70		
p_i	0.105	0.075	0.075	0.045		
Area 2	2-1	2-2	2-3	2-4	2-5	
Acres	60	40	55	65	80	
p_i	0.055	0.0625	0.05	0.045	0.0375	
Area 3	3-1	3-2	3-3			
Acres	35	55	60			
p_i	0.1575	0.105	0.0875			
Area 4	4-1	4-2	4-3	4-4	4-5	4-6
Acres	55	90	20	95	65	75
p_i	0.020	0.010	0.030	0.010	0.017	0.013

Compute the aggregation probabilities, between-set entropy, and within-set entropy

Solution: Aggregation probabilities can be shown below:

$P_1 = 0.3, P_2 = 0.25, P_3 = 0.35,$ and $P_1 = 0.3,$ and
$Q_1 = 0.1818, Q_2 = 0.2727, Q_3 = 0.1364,$ and $Q_4 = 0.4091$

The between-set entropy in bits can be calculated as follows:

$$\sum_{k=1}^{4} P_k \ln \left(\frac{P_k X}{\Delta X_k} \right) = 0.3 \ln \left(\frac{0.3 \times 1000}{200} \right) + 0.25 \ln \left(\frac{0.25 \times 1000}{300} \right)$$

$$+ 0.35 \ln \left(\frac{0.35 \times 1000}{150} \right) + 0.1 \ln \left(\frac{0.1 \times 1000}{450} \right) = 0.2222$$

The within-set entropy can be calculated as follows:

$$\sum_{k=1}^{4} P_k \sum_{i \subset z_k} \frac{p_i}{P_k} \ln \left(\frac{p_i}{P_k} \frac{\Delta X_k}{\Delta x_i} \right) = 0.0426 + 0.0190 + 0.0412 + 0.0523 = 0.1550 \text{ bits}$$

10.6 Hierarchical decomposition

Using POD one can measure the variation in activity at a series of related levels of aggregation forming a hierarchy. From equation (10.13), the within-set entropies $\sum_{k} w_k I_k$ attributable to each level of hierarchy can be obtained by subtracting the between-set entropy at level $l+1$ from the between-set entropy at level l. Noting that the between-set entropy monotonically decreases for increasing l, one can express

$$I = \left(I - I_k^1 \right) + \left(I_k^1 - I_k^2 \right) + \dots + \left(I_k^l - I_k^{l+1} \right) + \dots + \left(I_k^{L-1} - I_k^L \right) \tag{10.95}$$

where I_K^L is the between-set entropy at the highest possible level L, which would be equal to zero. Equation (10.95) can be cast as the sum of the within-set entropies:

$$I = \sum_{l=1}^{L} \sum_{k=1}^{K^l} w_k^l I_k^l, \quad 1 \leq K^{l+1} \leq K^l \leq N \tag{10.96}$$

in which K^l denotes the number of aggregated zones or sets at level l. With this notation, the entropy at level l can be written as

$$I_K^l = I_K^{l+1} + \sum_{k} w_k^{l+1} I_k^{l+1} \tag{10.97}$$

Equation (10.97) represents the partitioning of entropy at two levels of hierarchy. Thus the within-set entropy can be calculated from equation (10.97) as

$$I_K^l - I_K^{l+1} = \sum_{k} w_k^{l+1} I_k^{l+1} \tag{10.98}$$

This partitioning of entropy at each level of hierarchy helps assess the variation in the phenomenon under consideration at any geographic scale. It is important to emphasize the proper understanding of entropy. First, the total inverse spatial entropy at each level I_K^l indicates the significance of scale. A higher value of this entropy would indicate a higher difference between the actual distribution and the uniform distribution at this level. Second, the within-set entropy from equation (10.98) indicates the extent to which entropy I_K^l would be captured by the next level of aggregation up.

The Shannon entropy given by equation (10.2), the spatial entropy given by equation (10.3), and the inverse spatial entropy given by equation (10.8) can now be decomposed in a similar manner. Relating probabilities to particular hierarchical levels l and $l+1$, the Shannon discrete entropy can be expressed as

$$H = -\sum_l \sum_k P_k^{l+1} \sum_{i \in z_k^{l+1}} \frac{p_i^l}{P_k^{l+1}} \ln\left(\frac{p_i^l}{P_k^{l+1}}\right) \tag{10.99}$$

in which the aggregated probabilities are obtained as

$$P_k^{l+1} = \sum_{i \in z_k^{l+1}} p_i^l \tag{10.100}$$

The spatial entropy can now be decomposed as

$$H = \ln X - \sum_l \sum_k P_k^{l+1} \sum_{i \in z_k^{l+1}} \frac{p_i^l}{P_k^{l+1}} \ln\left(\frac{p_i^l \Delta X_k^{l+1}}{P_k^{l+1} \Delta x_i^l}\right) \tag{10.101}$$

in which zonal areas are aggregated

$$\Delta X_K^{l+1} = \sum_{i \in z_k^{l+1}} \Delta x_i^l \tag{10.102}$$

In equation (10.101), as each term in the hierarchy of within-set spatial entropies is negative, the value of H decreases with increasing zonal size. Equation (10.101) converges to the continuous form in the following sense:

$$H = \ln X - \lim_{\substack{\Delta x_i \to 0 \\ L \to \infty}} \sum_{l=1}^{L-1} \sum_{k=1}^{K^l} P_k^{l+1} \sum_{i \in z_k^{l+1}} \frac{p_i^l}{P_k^{l+1}} \ln\left(\frac{p_i^l \Delta X_k^{l+1}}{P_k^{l+1} \Delta x_i^l}\right) \tag{10.103}$$

This suggests that a continuous probability density function can be used to compute H. The second term in equation (10.103) would evaluate the difference between the continuous entropy of the PDF and the maximum value of log X. Each measure of spatial entropy at hierarchical level $l = 1, 2, \ldots, L-1$ is negative, and hence the hierarchical spatial entropy cannot be used for aggregation analysis. However, since the inverse spatial entropy does not take on negative values, it can be used. Therefore,

$$I = \sum_l \sum_k P_k^{l+1} \sum_{i \notin z_k^{l+1}} \frac{p_i^l}{P_k^{l+1}} \ln\left(\frac{p_i^l \Delta X_k^{l+1}}{P_k^{l+1} \Delta x_i^l}\right) \tag{10.104}$$

in which I_K^L is zero.

10.7 Comparative measures of spatial decomposition

Moellering and Tobler (1972) employed an analysis of variance scheme. Consider the distribution of population (M_i, $i = 1, 2, \ldots, N$). Let μ^l_{ik} denote a mean of population for each basic zone i over an aggregated set of zone z^l_k at each level of hierarchy l. This can be defined as

$$\mu^l_{ik} = \frac{\sum\limits_{i \in z^l_k} M_i}{n^l_k}, \quad k = n^L_k, \ldots, n^l_k, \ldots, n^1_k \tag{10.105}$$

where n^l_k is the number of zones in each set z^l_k at the hierarchical level l. Note the number of zones at each hierarchical level is related to its higher level as

$$n^{l-1}_k = \sum\limits_{i \in z^{l-1}_k} n^l_j \tag{10.106}$$

At the highest level of hierarchy $n^L_k = 1$, and at the base level $n^l_k = N$ zones. The analysis of variance principle in the context of geographical hierarchies says that the deviation of each zone from the overall mean at the base level of the hierarchy can be written as a sum of the deviations between successive means at each successive level of hierarchy, that is,

$$M_i - \mu^l_{ik} = \left(\mu^2_{ik} - \mu^1_{ik}\right) + \left(\mu^3_{ik} - \mu^2_{ik}\right) + \ldots + \left(\mu^{l+1}_{ik} - \mu^l_{ik}\right) + \ldots + \left(\mu^L_{ik} - \mu^{L-1}_{ik}\right)$$

$$= \sum\limits_{l=1}^{L-1} \left(\mu^{l+1}_{ik} - \mu^l_{ik}\right) \tag{10.107}$$

It may be noted that $n^L_k = 1$, $\mu^L_{ik} = M_i$. Squaring the terms in equation (10.107) and summing over all levels of hierarchy, the proportion of variance at each level of hierarchy is given by the sum of deviations at each level, that is,

$$V = \sum\limits_i \left(M_i - \mu^l_{ik}\right) = \sum\limits_{l=1}^{L-1} \sum\limits_{i=1}^{n^l_k} \left(\mu^{l+1}_{ik} - \mu^l_{ik}\right)^2 \tag{10.108}$$

Equation (10.108) includes a special case in which the number of zones in each set at any hierarchical level is constant and the ratio of zones in each set to zones in sets at a higher level is also constant. This happens if $n^l_1 = n^l_2 = \ldots$, and

$$\frac{n^{l-1}_k}{n^l_k} = M \tag{10.109}$$

For all k and l. This was suggested by Curry (1971) who computed entropy from probabilities computed for each set k at level $l + 1$. Thus,

$$P^{l+1}_k = Q \sum\limits_{i \in z^{l+1}_k} \left|\mu^{l+1}_{ik} - \mu^l_{ik}\right| \tag{10.110}$$

where Q is a normalizing factor making sure that

$$\sum_{l=1}^{L-1}\sum_{k=1}^{n_k^l} P_k^{l+1} = 1 \tag{10.111}$$

and is evaluated as

$$Q = \cfrac{1}{\sum_{l=1}^{L-1}\sum_{k=1}^{n_k^l}\sum_{i\in z_k^{l+1}} \left| \mu_{ik}^{l+1} - \mu_{ik}^{l} \right|} \tag{10.112}$$

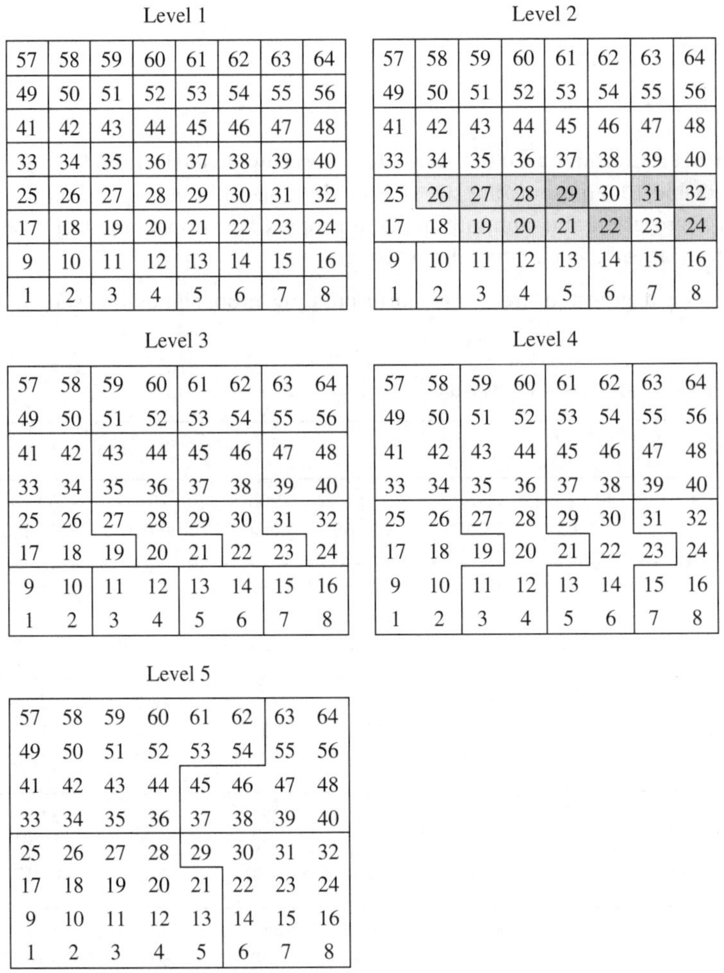

Level 1

57	58	59	60	61	62	63	64
49	50	51	52	53	54	55	56
41	42	43	44	45	46	47	48
33	34	35	36	37	38	39	40
25	26	27	28	29	30	31	32
17	18	19	20	21	22	23	24
9	10	11	12	13	14	15	16
1	2	3	4	5	6	7	8

Level 2

57	58	59	60	61	62	63	64
49	50	51	52	53	54	55	56
41	42	43	44	45	46	47	48
33	34	35	36	37	38	39	40
25	26	27	28	29	30	31	32
17	18	19	20	21	22	23	24
9	10	11	12	13	14	15	16
1	2	3	4	5	6	7	8

Level 3

57	58	59	60	61	62	63	64
49	50	51	52	53	54	55	56
41	42	43	44	45	46	47	48
33	34	35	36	37	38	39	40
25	26	27	28	29	30	31	32
17	18	19	20	21	22	23	24
9	10	11	12	13	14	15	16
1	2	3	4	5	6	7	8

Level 4

57	58	59	60	61	62	63	64
49	50	51	52	53	54	55	56
41	42	43	44	45	46	47	48
33	34	35	36	37	38	39	40
25	26	27	28	29	30	31	32
17	18	19	20	21	22	23	24
9	10	11	12	13	14	15	16
1	2	3	4	5	6	7	8

Level 5

57	58	59	60	61	62	63	64
49	50	51	52	53	54	55	56
41	42	43	44	45	46	47	48
33	34	35	36	37	38	39	40
25	26	27	28	29	30	31	32
17	18	19	20	21	22	23	24
9	10	11	12	13	14	15	16
1	2	3	4	5	6	7	8

Figure 10.4 Hierarchy based on two-fold cascading.

Table 10.3 Zones and their probabilities.

Zones	1	2	3	4	5	6	7	8
Area (%)	10	18	20	15	12	10	8	7
(acres)	1000	1800	2000	1500	1200	1000	800	700
p_i	0.10	0.11	0.08	0.17	0.13	0.14	0.12	0.15

The marginal entropies H^{l+1} for each level l are defined by Curry (1971) as

$$H^{l+1} = -\sum_{k=1}^{n_k^l} P_k^{l+1} \ln P_k^{l+1} \tag{10.113}$$

showing the proportion of entropy at each level. The total entropy is calculated as

$$H = -\sum_{l=1}^{L-1}\sum_{k=1}^{n_k^l} P_k^{l+1} \ln P_k^{l+1} \tag{10.114}$$

Example 10.10: Consider a 10,000-acre area originally divided into 64 zones as shown in Figure 10.4. Table 10.3 presents the size of each division and tree population therein. Do the hierarchical aggregation and compute spatial entropy, inverse spatial entropy, variance, and Curry's entropy at each level. Then discuss the results.

Solution: The aggregated areas and corresponding probabilities are shown Tables 10.4 to 10.14.

Table 10.4 Hierarchy based on $l = 5$ and $i = 64$.

Zones	1	2	3	4	5	6	7	8
a_i (acres)	100	140	250	150	200	75	60	25
p_i	0.02	0.0085	0.0025	0.0065	0.0075	0.0100	0.0300	0.0150
Zones	9	10	11	12	13	14	15	16
a_i (acres)	140	180	340	290	270	200	220	160
p_i	0.030	0.0085	0.0025	0.0065	0.0055	0.0045	0.0150	0.0375
Zones	17	18	19	20	21	22	23	24
a_i (acres)	100	260	320	280	315	310	215	200
p_i	0.020	0.006	0.015	0.015	0.005	0.004	0.008	0.007
Zones	25	26	27	28	29	30	31	32
a_i (acres)	160	220	210	190	230	170	180	140
p_i	0.040	0.012	0.030	0.034	0.008	0.014	0.017	0.015
Zones	33	34	35	36	37	38	39	40
a_i (acres)	120	260	140	200	210	80	90	100
p_i	0.013	0.003	0.022	0.030	0.010	0.021	0.020	0.011

Table 10.4 (*continued*)

Zones	41	42	43	44	45	46	47	48
a_i (acres)	25	250	150	140	200	75	60	100
p_i	0.015	0.005	0.022	0.035	0.011	0.021	0.020	0.011

Zones	49	50	51	52	53	54	55	56
a_i (acres)	80	120	125	115	105	95	85	75
p_i	0.013	0.003	0.012	0.030	0.010	0.021	0.020	0.011

Zones	57	58	59	60	61	62	63	64
a_i (acres)	60	80	115	90	110	95	85	65
p_i	0.018	0.007	0.025	0.036	0.012	0.021	0.020	0.011

For $l = 1$

Table 10.5 Areas of level 1.

60	80	115	90	110	95	85	65
80	120	125	115	105	95	85	75
25	250	150	140	200	75	60	100
120	260	140	200	210	80	90	100
160	220	210	190	230	170	180	140
100	260	320	280	315	310	215	200
140	180	340	290	270	200	220	160
100	140	250	150	200	75	60	25

Table 10.6 Probabilities of level 1.

0.018	0.007	0.025	0.036	0.012	0.021	0.02	0.011
0.013	0.003	0.012	0.03	0.01	0.021	0.02	0.011
0.015	0.005	0.022	0.035	0.011	0.021	0.02	0.011
0.013	0.003	0.022	0.03	0.01	0.021	0.02	0.011
0.04	0.012	0.03	0.034	0.008	0.014	0.017	0.015
0.02	0.006	0.015	0.015	0.005	0.004	0.008	0.007
0.03	0.0085	0.0025	0.0065	0.0055	0.0045	0.015	0.0375
0.02	0.0085	0.0025	0.0065	0.0075	0.01	0.03	0.015

For $l = 2$

Table 10.7 Areas of level 2.

140	200	240	205	215	190	170	140
145	510	290	340	410	155	150	200
260	480	530	470	545	480	395	340
240	320	590	440	470	275	280	185

Table 10.8 Probabilities of level 2.

0.031	0.01	0.037	0.066	0.022	0.042	0.04	0.022
0.028	0.008	0.044	0.065	0.021	0.042	0.04	0.022
0.06	0.018	0.045	0.049	0.013	0.018	0.025	0.022
0.05	0.017	0.005	0.013	0.013	0.0145	0.045	0.0525

For $l = 3$

Table 10.9 Areas of level 3.

340	445	405	310
655	630	565	350
740	1000	1025	735
560	1030	745	465

Table 10.10 Probabilities of level 3.

0.041	0.103	0.064	0.062
0.036	0.109	0.063	0.062
0.078	0.094	0.031	0.047
0.067	0.018	0.0275	0.0975

For $l = 4$

Table 10.11 Areas of level 4.

995	1075	970	660
1300	2030	1770	1200

Table 10.12 Probabilities of level 4.

0.077	0.212	0.127	0.124
0.145	0.112	0.0585	0.1445

For $l = 5$

Table 10.13 Areas of level 5.

2070	1630
3330	2970

Table 10.14 Probabilities of level 5.

0.289	0.251
0.257	0.203

Therefore, the spatial entropy, inverse spatial entropy, variance, and Curry's entropy at each level are calculated as follows:
The spatial entropy:

$$H = \ln X - \lim_{\substack{\Delta x_i \to 0 \\ L \to \infty}} \sum_{l=1}^{L-1} \sum_{k=1}^{K^l} P_k^{l+1} \sum_{i \in z_k^{l+1}} \frac{p_i^l}{P_k^{l+1}} \ln \left(\frac{p_i^l \Delta X_k^{l+1}}{P_k^{l+1} \Delta x_i^l} \right)$$

$$= \ln 10000 - (0.0674 + 0.0805 + 0.0717 + 0.0420)$$

$$= 8.9487 \text{ bits}$$

where $l = 1, 2, 3$, and 4 in order to guarantee $l + 1 = 5$.
The inverse spatial entropy:

$$I = \sum_{l} \sum_{k} P_k^{l+1} \sum_{i \notin z_k^{l+1}} \frac{p_i^l}{P_k^{l+1}} \ln \left(\frac{p_i^l \Delta X_k^{l+1}}{P_k^{l+1} \Delta x_i^l} \right)$$

$$= 0.0674 + 0.0805 + 0.0717 + 0.0420$$

$$= 0.2616 \text{ bits}$$

The variance can be calculated as

$$V = \sum_{i} \left(M_i - \mu_{ik}^l \right) = \sum_{l=1}^{L-1} \sum_{i=1}^{n_k^l} \left(\mu_{ik}^{l+1} - \mu_{ik}^l \right)^2 = \left(\mu_{ik}^2 - \mu_{ik}^1 \right)^2 + \left(\mu_{ik}^3 - \mu_{ik}^2 \right)^2 + \left(\mu_{ik}^4 - \mu_{ik}^3 \right)^2$$

$$+ \left(\mu_{ik}^5 - \mu_{ik}^4 \right)^2 = 7812500$$

and Curry's entropy can be calculated as

$$H = -\sum_{l=1}^{L-1} \sum_{k=1}^{n_k^l} P_k^{l+1} \ln P_k^{l+1} = 0.3304 + 0.3585 + 0.3125 + 0.3531 = 5.4562 \text{ bits}$$

Questions

Q.10.1 Consider an urban area of 100 km², with seven types of average land use areas as follows: 4 km² industrial development, 16 km² roads and other pavements, 48 km² agricultural-horticultural, 12 km² dairy farms, and 20 km² forest. Define the probability of each land use as the area of land use divided by the total area. For each land use the value of the inverse spatial interval is different and can be expressed in terms of the area of land use. Compute the inverse spatial entropy of the urban area.

Q.10.2 Consider that the urban area in Q.10.1 has equal area land uses. Then compute the inverse spatial entropy and then compare it with the entropy value computed in Q.10.1.

Q.10.3 Suppose a spatial system is partitioned as shown below:

Area							
Land use 1		Land use 2		Land use 3		Land use 4	
Δx (km^2)	Probability	Δx (km^2)	Probability	Δx (km^2)	Probability	Δx (km^2)	Probability
20	0.04	25	0.04	60	0.25	40	0.07
40	0.06	40	0.01	40	0.15	80	0.015
60	0.10	35	0.06			70	0.015
80	0.15						

Compute the between-set and within-set inverse spatial entropies.

Q.10.4 Consider four zones dividing a 2000-acre watershed. Compute the inaccuracy function $K(P; Q|N)$. The zones are given as follows:

Configuration	Zones			
	1	2	3	4
1	600	700	400	300
2	500	600	350	550
3	500	500	500	500
4	800	100	650	450
5	700	500	400	400

Suppose $p_1 = 0.20$, $p_2 = 0.35$, $p_3 = 0.25$, and $p_4 = 0.2$. Compute the Shannon entropy of the zones. Compute the inaccuracy accounting for the configuration types. Also compute the accuracy measure. $N = 4$, and $p_i = 1/4 = 0.5$, $i = 1, 2, \ldots, 4$, for maximum entropy.

Q.10.5 Compute the information measure $\hat{I}$ relating zone number, configuration, and density using the data given in Q.10.4. Use $N = 4$, $p_1 = 0.20$, $p_2 = 0.35$, $p_3 = 0.25$, and $p_4 = 0.2$. Then, consider $p_i = q_i = 1/N$. Compute $\hat{I}$. Compute information error and uncertainty.

Q.10.6 Consider five unequal zones and their corresponding probabilities. The distribution of phenomenon at the five-zone level is uneven. Then, aggregate the five zones to four equal zones, then three zones and finally two zones. Compute the Kerridge inaccuracy, information and information gain or loss as the number of zones is reduced.

Q.10.7 Consider a hypothetical 2000-acre urban area and decompose it into five zones. Compute the aggregation probabilities, between-set entropy, and within-set entropy.

Q.10.8 Consider a 50,000-acre area and divide it into 50 zones. Each zone has its own land use configuration. Do the hierarchical aggregation and compute spatial entropy, inverse spatial entropy, variance, and Curry's entropy at each level. Then discuss the results.

References

Aczel, J. and Daroczy, Z. (1975). *On Measures of Information and Their Characterizations*. Academic Press, New York.

Batty, M. (1976). Entropy in spatial aggregation. *Geographical Analysis*, Vol. 8, pp. 1–21.

Batty, M. (2010). Space, scale, and scaling in entropy maximizing. *Geographical Analysis*, Vol. 42, pp. 395–421.

Blundell, S.J. and Blundell, K.M. (2006). *Concepts in Thermal Physics*, Oxford University Press, Oxford.

Chapman, G.P. (1970). The application of information theory to the analysis of population distributions in space. *Economic Geography, Supplement*, Vol. 46, pp. 317–31.

Curry, L. (1971). Applicability of space-time moving average forecasting. In M. Chisolm, A, Frey and P. Haggett, editors, *Regional Forecasting*, Colston Papers, No. 22, Butterworths, London, U.K.

Curry, L. (1972a). A spatial analysis of gravity flows. *Regional Studies*, Vol. 6, pp. 131–47.

Curry, L. (1972b). Spatial entropy. P1204 in W.P. Adams and F.M. Helleiner, editors, *International Geography*, Vol. 2, Toronto and Buffalo, University of Toronto Press.

Kerridge, D.F. (1961). Inaccuracy and inference. *Journal of the Royal Statistical Society, Series B*, Vol. 23, pp. 184–94.

Kulback, S. (1959). *Statistics and Information Theory*. John Wiley, New York.

Moellering, H. and Tobler, W. (1972). Geographical variances. *Geographical Analysis*, Vol. 4, pp. 34–50.

Semple, R.K. and Golledge, R.G. (1970). An analysis of entropy changes in a settlement pattern over time. *Economic Geography*, Vol. 46, pp. 157–60.

Theil, H. (1967). *Economics and Information*. North Holland, Amsterdam.

Theil, H. (1972). *Statistical Decomposition Analysis*. North Holland, Amsterdam.

Tobler, W. (1963). Geographic area and map projections. *Geographical Review*, Vol. 53, pp. 59–78.

11 Entropy Spectral Analyses

In environmental and water engineering, time series of many variables are available at locations of measurement. For example, stream flow is continuously measured at several gauging stations along a river and thus time series of stage and discharge are available at each of these gauging stations. Similarly, rainfall is measured at a number of points (rain gage stations) in a watershed and hence the time series of rainfall at each of the stations is available. Similarly, time series of pollutant concentration in air as well as in water, sediment yield, temperature, wind velocity, air pressure, radiation, and so on are available. Time series is an empirical series and contains a wealth of information about the system under study. It can be discrete or continuous and its analysis permits estimation of system transfer function, identification of system parameters, forecasting, detection of trends and periodicities, determination of power spectrum, and system design. Autocovariance (or autocorrelation) function is employed to characterize the evolution of a process (using time series) through time. Frequency properties of the time series are considered using spectral analysis. This chapter discusses spectral analyses based on maximum entropy and minimum cross entropy.

11.1 Characteristics of time series

A time series is a record of variation of an observed quantity in time. For example, river flow at a specified location is an example of a time series. Let the time series under consideration be a stochastic process $X(t)$, which can be defined as random variable X at any given time t. At this given time, the random variable can take on a number of possible values. For example, if time is the month of January, then stream flow for January in one year will be different from that in another year and the January stream flow is a random variable. In a similar manner, a stochastic process can be thought of as the collection of all possible records of variation (or realizations) of the observed quantity with time. This collection is referred to as ensemble. For example, if air quality is observed at a number of places (or stations) in an area and time series of air quality is available at each of the stations, then those time series together constitute the ensemble. Likewise, soil moisture is observed at a number of locations in a watershed as a function of time. Considering soil moisture as a random variable, its time series is a stochastic process, and the collection of these time series of soil moisture constitutes the ensemble.

Entropy Theory and its Application in Environmental and Water Engineering, First Edition. Vijay P. Singh.
© 2013 John Wiley & Sons, Ltd. Published 2013 by John Wiley & Sons, Ltd.

It is assumed that a time series $X(t)$ consists of a set of observations $x_0, x_1, x_2, \ldots, x_N$:

$$X(t) = \left[x_0 \; x_1 \; x_2 \; \ldots \; x_N \right]^T \tag{11.1}$$

where continuous time t is broken down into the N number of discrete intervals for observations and N corresponds to the time when observation ends, x_i is the i-th ordered variate or the value of $X(t)$ at the i-th time, and T is transpose. It is further assumed that the time interval Δt between successive observations (or measurements) is constant; in other words, observations are equi-spaced in time; that is, the time interval (Δt) between x_0 and x_1 is the same as x_2 and x_3; this is often the case with stream flow observations. For example, in daily stream flow series $\Delta t = 1$ day.

It may often be advantageous to transform the observations so that the mean of transformed values is zero. To that end, the mean of observations $\bar{x}$ is first computed:

$$\bar{x} = \frac{1}{N} \sum_{i=1}^{N} x_i \tag{11.2}$$

where N is the number of observations and x_i is the i-th observation. Then the mean is subtracted from each individual observation:

$$x_{*i} = x_i - \bar{x} \tag{11.3}$$

where x_{*i} is the i-th reduced observation, and the reduced observations, x_{*i}, $i = 1, \ldots, N$, will now have zero mean. It may now be worthwhile recalling the definitions of mean, variance, covariance, and correlation of the stochastic process here before describing other characteristics of the time series.

11.1.1 Mean

Let a random variable time at t be designated as $X(t)$. It is assumed that the probability distribution of $X(t)$ is given explicitly in the form of probability mass function (*PMF*), probability density function (*PDF*), $f(x, t)$, or cumulative distribution function (*CDF*), $F(x, t)$. The mean is one of the most important parameters of a distribution and is defined as the first moment of the probability distribution about the origin, and is usually designated by the Greek letter μ. For a continuous probability distribution, at time t, $\mu(t)$ is defined as

$$\mu(t) = E\left[X(t)\right] = \int_{-\infty}^{+\infty} x f(x, t)\, dx = \int_{0}^{1} x dF(x, t) \tag{11.4}$$

where x is the value of random variable $X(t)$ at time t, E is the expectation operator, $f(x, t)$ is the PDF of $X(t)$, $F(x, t)$ is the CDF of $X(t)$, and $dF(x, t) = f(x, t)dx$.

For a discrete probability distribution $p(x, t)$, $\mu(t)$ is defined as

$$\mu(t) = \sum_{i=1}^{N} x_i \, p\left(x_i, t\right) \tag{11.5}$$

where N is the number of observations, x_i is the ith observation of X, $p(x_i, t)$ is the probability (or probability mass) if $x = x_i$, and $i = 1, 2, 3, \ldots, N$. The mean gives the distance of the center of gravity of the probability mass from the origin and describes the location of the probability mass, and is calculated from sample data as:

$$m(t) = \overline{x(t)} = \frac{1}{N} \sum_{i=1}^{N} x_i(t) \tag{11.6}$$

where $m(t) = \overline{x(t)}$ is the sample mean approximating $\mu(t)$.

11.1.2 Variance

Variance measures the variability of a random variable and is the second most important descriptor of its probability distribution. A small variance of a variable indicates that its values are likely to stay near the mean value, while a large variance implies that the values have large dispersion around the mean. If the stage of a river at a gauging station is independently measured in a quick succession a number of times in a survey, then there will likely be a variability in the stage measurements. The magnitude of variability is a measure of both the natural variation and the measurement error.

Variance, designated by the Greek letter σ^2, measures the deviation from the mean and is universally accepted as given by the second moment of the probability distribution about the mean. Sometimes, the notation $VAR(x)$ or $Var(x)$ is also used. For a continuous variable $X(t)$ it is expressed as

$$\sigma^2(t) = \int_{-\infty}^{+\infty} [x - \mu(t)]^2 f(x, t) \, dx \tag{11.7}$$

and for a discrete variable $X(t)$,

$$\sigma^2(t) = \sum_{i=1}^{N} [x_i - \mu(t)]^2 \, p(x_i, t) \tag{11.8}$$

The sample variance, denoted as $s^2(t)$, is calculated as:

$$s^2(t) = \frac{1}{N} \sum_{i=1}^{N} [x_i - m(t)]^2 \tag{11.9}$$

When the number of values in the samples $N \leq 30$, an unbiased estimate of variance is obtained as

$$s^2(t) = \frac{1}{N-1} \sum_{i=1}^{N} (x_i - m(t))^2 \tag{11.10}$$

The variance can also be computed as

$$s^2(t) = \frac{1}{N} \sum_{i=1}^{N} x_i^2 - m^2(t) \tag{11.11}$$

In structural engineering, the variance is the moment of inertia of the probability mass about the center of gravity.

The variance has a dimension equal to the square of the dimension of the random variable. If X is discharge which is measured in m^3/s, then its variance σ^2 is in $(m^3/s)^2$. This makes it quite difficult to visualize the degree of variability associated with a given value of the variance. For this reason, the positive square root of the variance, called standard deviation, denoted by σ, is often used. Its mechanical analogy is the radius of gyration. Variance has four important properties: 1) The variance of a constant, a, is zero: $Var\ [a] = 0$, $a =$ constant. 2) The variance of X multiplied by a constant a is equal to the variance of X multiplied by the square of a: $Var\ [aX] = a^2 \times Var\ [X]$. 3) The variance of X is the difference between the second moment of X about the origin, μ_2', and the square of first moment of X about the origin μ^2: $Var\ [X] = \mu_2' - \mu^2$. 4) The variance of the sum of n independent random variables is equal to the sum of variances of the n individual variables: $Var[X_1 + X_2 + X_3 + \ldots + X_n] = Var[X_1] + Var[X_2] + Var[X_3] + \ldots + Var[X_n]$.

A dimensionless measure of dispersion is the coefficient of variation, c_v, which is computed as the ratio of standard deviation and mean:

$$C_v = \frac{\sigma}{\mu} = \frac{s}{m} \tag{11.12}$$

where s is the sample standard deviation estimate of the population standard deviation σ, and m is the sample mean estimate of the population mean μ. When the mean of the data is zero, C_v is undefined. The coefficient of variation is useful in comparing different populations or their distributions. For example, if the two samples of aggregates of water quality are analyzed, the one with larger C_v will have more variation. If each value of a variable is multiplied by a constant α, the mean, variance and the standard deviation are obtained by multiplying the original mean, variance and standard deviation by α, α^2 and α, respectively; the coefficient of variation remains unchanged. If a constant α is added to each value of the variable, the new mean is equal to the old mean $+ \alpha$; variance and the standard deviation remain unchanged; but the coefficient of variation changes because the unchanged standard deviation is divided by the new (changed) mean.

Example 11.1: A time series of annual stream flow for a USGS gaging station 08109000 on the Brazos River near Bryan, Texas, is given in Table 11.1. Plot the time series. Compute its mean and variance. Compute its reduced time series and show that its mean is zero.

Table 11.1 Annual stream flow (cfs) for USGS 08109000.

Year	Flow	Year	Flow	Year	Flow	Year	Flow
1941	15110	1954	1965	1967	1584	1980	2320
1942	9369	1955	1835	1968	9196	1981	3575
1943	3705	1956	2053	1969	4770	1982	6161
1944	7396	1957	12680	1970	4889	1983	2143
1945	9061	1958	8469	1971	1078	1984	627
1946	6137	1959	1923	1972	3100	1985	3504
1947	5561	1960	6674	1973	5321	1986	5273
1948	1836	1961	10190	1974	3697	1987	9135
1949	3185	1962	3538	1975	9012	1988	1390
1950	2912	1963	1896	1976	3930	1989	4204
1951	1258	1964	1334	1977	7131	1990	7081
1952	1155	1965	7707	1978	930	1991	4394
1953	2319	1966	6479	1979	6716	1992	21720

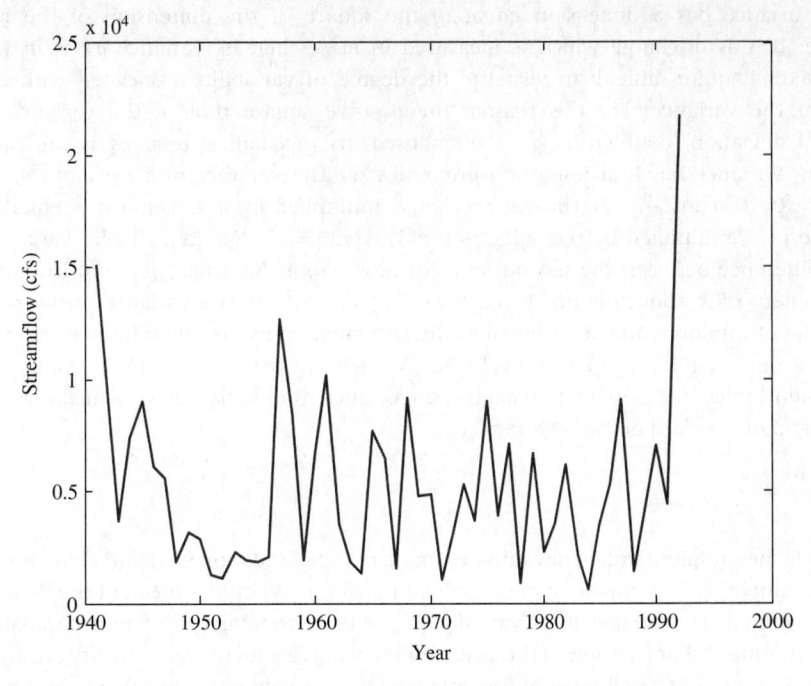

Figure 11.1 Plot of annual stream flow for the Brazos River near Bryan from 1941–92.

Solution: The annual stream flow data for the Brazos River near Bryan (USGS station 08109000) from 1941–92, given in Table 11.1, is plotted in Figure 11.1. The mean and variance of annual flow data are computed as

$$\bar{x} = \frac{1}{N}\sum_{i=1}^{N} x_i = 5166 \text{ cfs}$$

$$s^2(t) = \frac{1}{N-1}\sum_{i=1}^{N}(x_i - \bar{x})^2 = 1.574 \times 10^7 (\text{cfs})^2$$

The mean of the reduced series is computed as:

$$\bar{x}_* = \frac{1}{N}\sum_{i=1}^{N}(x_{*i} - \bar{x}) = \frac{1}{N}\sum_{i=1}^{N}(x_i - \bar{x}) = \frac{1}{N}\sum_{i=1}^{N}x_i - \bar{x} = 0$$

11.1.3 Covariance

Another basic characteristic of a time series is its covariance at a certain lag. Consider x_i and x_{i+k}, where k is the time lag or simply lag. If $k=1$, then x_{i+1} is one time lag away from x_i; similarly, x_{i+4} is four time lags away from x_i. Since x_i and x_{i+k} belong to the same $X(t)$, the covariance in this case is called the autocovariance, defined at lag k as:

$$r_k = Cov(x_i, x_{i+k}) = E[(x_i - \mu)(x_{i+k} - \mu)] = \int_{-\infty}^{\infty}\int_{-\infty}^{\infty}(x_i - \mu)(x_{i+k} - \mu)f(x_i, x_{i+k})\,dx_i dx_{i+k}$$

$$(11.13)$$

where $f(x_i, x_{i+k})$ is the joint PDF of x_i and x_{i+k}, $k = 0, 1, 2, \ldots$; μ is the mean of $X(t)$, approximated by m or $\bar{x}$, r_k is the autocovariance at lag k, and $Cov(x_i, x_{i+k})$ is the covariance of (x_i, x_{i+k}). The stationarity assumption implies that the joint PDF is the same at all times separated by a constant time interval k. In practice, the autocovariance is estimated as

$$r_k = Cov\left(x_i, x_{i+k}\right) = \frac{1}{N} \sum_{i=1}^{N-k} \left(x_i - \bar{x}\right)\left(x_{i+k} - \bar{x}\right) \tag{11.14}$$

where $\bar{x}$ is the mean of the observed series.

If the number of observations at equally spaced intervals is large, say infinite, then one can determine autocovariances r_k at all lags and since the autocovariance is a function of the lag and can be plotted against the lag, it is called the autocovariance function. When $k = 0$, covariance becomes the variance of $X(t)$, $Var[X]$ or r_0:

$$Var(X) = r_0 = Cov[x_i, x_{i+0}] = E[(x_i - \mu)(x_{i+0} - \mu)] = E[(x_i - \mu)^2]$$
$$= \frac{1}{n} \sum_{i=1}^{n} (x_i - \mu)^2 \cong \frac{1}{n-1} \sum_{i=1}^{n} (x_i - \mu)^2 \tag{11.15}$$

Variance can be regarded as a measure of power, and covariance can be regarded as a measure of cross-power.

For computational purposes, equation (11.14) can be expressed as

$$r_k = Cov\left(x_i, x_{i+k}\right) = \frac{1}{N} \sum_{i=1}^{N-k} \left[x_i, x_{i+k} - (\bar{x})^2\right], k = 0, 1, 2, \ldots \tag{11.16}$$

If the covariance is independent of t but depends only on k, which is true for a stochastic process stationary in mean and covariance, then

$$Cov\left(x_i, x_{i+k}\right) = Cov(k) = r_k = r_0 \sigma_x^2 \tag{11.17}$$

Since r_k varies with k, it is called the autocovariance function and is an even function for a stationary stochastic process, $r_k = r_{-k}$ or $Cov(k) = Cov(-k)$.

Mathematically, the negative exponential function has often been used to represent the covariance function as:

$$r(k) = \sigma^2 \exp\left[-|k|/\alpha\right]$$

where σ^2 is the variance, and α is the integral scale expressed as

$$\alpha = \int_0^{\infty} \rho(k)\, dk, \quad \rho(k) = r(k)/\sigma^2$$

11.1.4 Correlation

Covariance is a dimensional quantity and it is often useful to nondimensionalize it by normalization which gives rise to autocorrelation. Then, it is easier to compare different time series having different scales of measurement, because autocorrelation is independent of the scale of measurement. Thus, the autocorrelation at lag k can be defined as

$$\rho_k = \rho(k) = \frac{E\left[(x_i - \mu)(x_{i+k} - \mu)\right]}{\sqrt{E\left[(x_i - \mu)^2\right]E\left[(x_{i+k} - \mu)^2\right]}} = \frac{E\left[(x_i - \mu)(x_{i+k} - \mu)\right]}{\sigma_x^2} = \frac{Cov(x, k)}{\sigma_x^2} \quad (11.18)$$

Since autocorrelation varies with k, $\rho_k = \rho(k)$ is called the autocorrelation function. The plot of autocorrelation function is sometimes called correlogram. The autocorrelation function of data reflects the dependence of data at one time on the data at another time. Thus, the autocorrelation at lag k can simply be expressed as

$$\rho_k = \rho(k) = \frac{r_k}{r_0} \quad (11.19)$$

Equation (11.19) implies that when $k = 0$, $\rho_0 = \rho(0) = 1$. Since the autocorrelation coefficient is a measure of linear dependence, its decease with time or lag suggests a decrease in the memory of the process. In hydrology, it is frequently observed that processes are more linearly dependent at short intervals of time than at long intervals. Thus, the correlogram provides information about the dependence structure of the time series. In many hydrologic processes, the correlogram does not decrease continuously with lag but exhibits a periodic increase and decrease, suggesting a deterministic component in the processes such as seasonal variability.

Further, knowing the autocorrelation function and the variance is equivalent to knowing the autocovariance function. For observations made for a stationary time series at n successive times, one can construct a covariance matrix as

$$\Gamma_n = r_{ij} = \begin{bmatrix}
r_0 & r_1 & r_2 & r_3 & \cdot & \cdot & \cdot & \cdot & \cdot & r_{n-1} \\
r_1 & r_0 & r_1 & r_2 & r_3 & \cdot & \cdot & \cdot & \cdot & r_{n-2} \\
r_2 & r_1 & r_0 & r_1 & r_2 & r_3 & \cdot & \cdot & \cdot & r_{n-3} \\
r_3 & r_2 & r_1 & r_0 & r_1 & r_2 & r_3 & \cdot & \cdot & \cdot \\
\cdot & r_3 & r_2 & r_1 & r_0 & r_1 & r_2 & \cdot & \cdot & \cdot \\
\cdot & \cdot & r_3 & r_2 & r_1 & r_0 & r_1 & \cdot & \cdot & \cdot \\
\cdot & \cdot & \cdot & r_3 & r_2 & r_1 & r_0 & \cdot & \cdot & \cdot \\
\cdot & \cdot & \cdot & \cdot & \cdot & \cdot & r_1 & r_0 & r_1 & r_2 \\
\cdot & \cdot & \cdot & \cdot & \cdot & \cdot & \cdot & r_1 & r_0 & r_1 \\
r_{n-1} & r_{n-2} & r_{n-3} & \cdot & \cdot & \cdot & \cdot & r_2 & r_1 & r_0
\end{bmatrix}$$

$$= \sigma_x^2 \begin{bmatrix}
1 & \rho_1 & \rho_2 & \rho_3 & \cdot & \cdot & \cdot & \cdot & \cdot & \rho_{n-1} \\
\rho_1 & 1 & \rho_1 & \rho_2 & \rho_3 & \cdot & \cdot & \cdot & \cdot & \rho_{n-2} \\
\rho_2 & \rho_1 & 1 & \rho_1 & \rho_2 & \rho_3 & \cdot & \cdot & \cdot & \rho_{n-3} \\
\rho_3 & \rho_2 & \rho_1 & 1 & \rho_1 & \rho_2 & \rho_3 & \cdot & \cdot & \cdot \\
\cdot & \rho_3 & \rho_2 & \rho_1 & 1 & \rho_1 & \rho_2 & \rho_3 & \cdot & \cdot \\
\cdot & \cdot & \rho_3 & \rho_2 & \rho_1 & 1 & \rho_1 & \rho_2 & \cdot & \cdot \\
\cdot & \cdot & \cdot & \rho_3 & \rho_2 & \rho_1 & 1 & \rho_1 & \rho_2 & \cdot \\
\cdot & \cdot & \cdot & \cdot & \cdot & \cdot & 1 & \rho_1 & \rho_2 \\
\cdot & \cdot & \cdot & \cdot & \cdot & \cdot & \rho_1 & 1 & \rho_1 \\
\rho_{n-1} & \rho_{n-2} & \rho_{n-3} & \cdot & \cdot & \cdot & \cdot & \rho_2 & \rho_1 & 1
\end{bmatrix} = \sigma_x^2 P_n \quad (11.20)$$

Here Γ_n is the autocovariance matrix, and P_n is the autocorrelation matrix. Both of these matrices are symmetric and have constant elements on any diagonal. These matrices are positive definite. This characteristic of the autocorrelation matrix means that its determinant and all principal minors are greater than zero. For example, when $n = 2$,

$$\begin{vmatrix} 1 & \rho_1 \\ \rho_1 & 1 \end{vmatrix} > 0 \quad (11.21)$$

implying that

$$1 - \rho_1^2 > 0 \tag{11.22}$$

That is,

$$-1 < \rho_1 < 1 \tag{11.23}$$

In a similar manner, for $n = 3$, it can be shown that

$$\begin{vmatrix} 1 & \rho_1 & \rho_2 \\ \rho_1 & 1 & \rho_2 \\ \rho_2 & \rho_1 & 1 \end{vmatrix} > 0, \quad \begin{vmatrix} 1 & \rho_1 \\ \rho_1 & 1 \end{vmatrix} > 0, \quad \begin{vmatrix} 1 & \rho_2 \\ \rho_2 & 1 \end{vmatrix} > 0 \tag{11.24}$$

Equation (11.24) shows that

$$-1 < \rho_1 < 1, \ -1 < \rho_2 < 1, \ -1 < \frac{\rho_2 - \rho_1^2}{1 - \rho_1^2} < 1 \tag{11.25}$$

and so on.

The autocorrelation function is symmetric about zero, that is,

$$\rho_k = \rho_{-k} \tag{11.26}$$

11.1.5 Stationarity

A commonly made assumption in practice is that the stochastic process $X(t)$ is stationary; in other words, properties of its time series, such as mean, variance, skewness, spectrum, probability distribution, and so on, do not change with shift in origin, that is, two processes $X(t)$ and $X(t + \tau)$ have the same statistics regardless of what the value of τ is, where $\tau = t_1 - t_2$ (time difference). In other words, the stationary process has a similar structure with respect to the variability in time, implying some kind of repetition in the process. This suggests that statistical interpretations can be based on a single realization. Then, the first order PDF of X will be independent of time, but the second order PDF will depend only on the time difference. Likewise, the mean will be constant in time but the autocorrelation, function, autocovariance function, and autocorrelation coefficient will depend on the time difference. If the time series is nonstationary then it is rendered stationary by the use of a suitable transformation. One such transformation is data differencing:

$$x_i^* = x_{i+1} - x_i, \ i = 1, 2, 3, \ldots, N - 1 \tag{11.27}$$

This operation can be continued until data are rendered stationary. Another transformation can be done by taking the logarithm.

Example 11.2: Consider an ensemble of time series $X(t)$ from say 1 to m, and plot these on the same graph. Show on the graph $X_i(t)$, $X_i(t+\tau)$, $i = 1, 2, \ldots, m$; and τ is the lag. This will make clear which values to take when computing time series characteristics.

Solution: An ensemble of time series is shown in Figure 11.2.

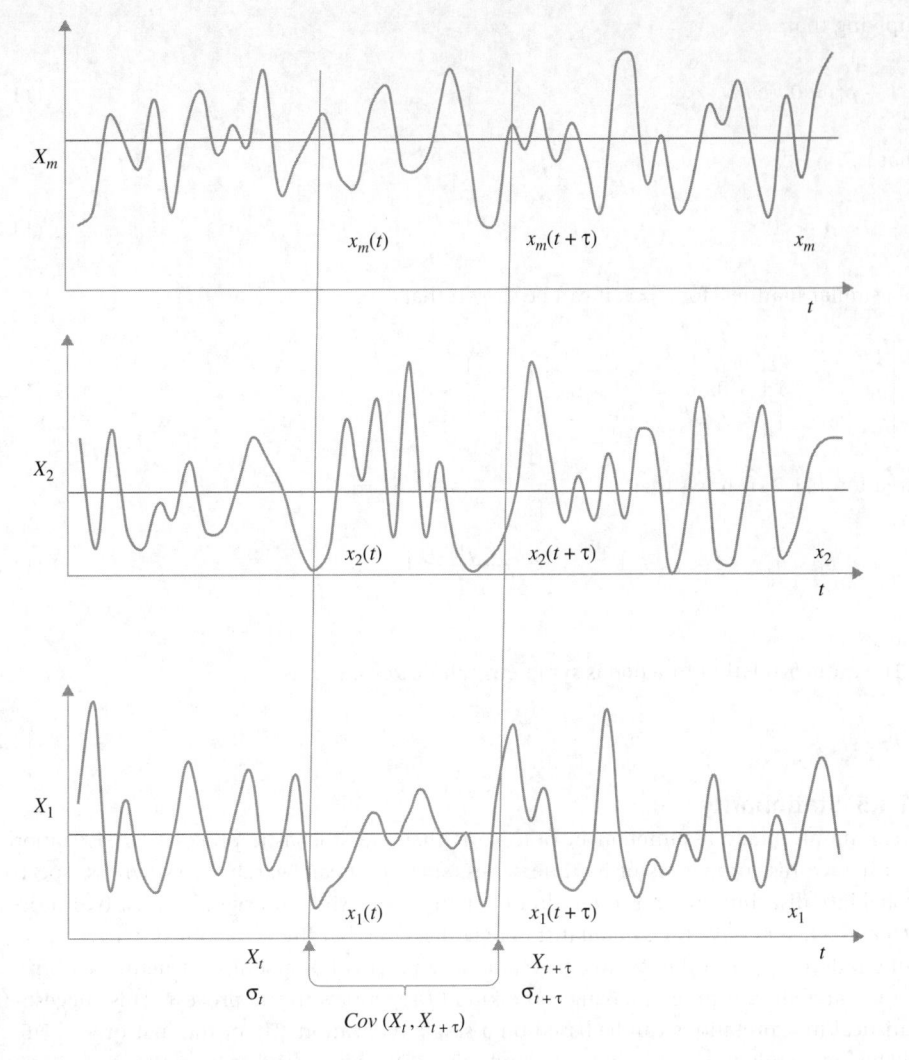

Figure 11.2 Ensemble of time series.

Example 11.3: Using the stream flow time series given in Table 11.1 for a gauging station at Brazos River, compute the autocovariance function by taking different values of lag. Note, in general, the maximum lag should not be more than one fourth of the length of the series. For example, if 60 years of data are available, then the maximum lag would be no more than 15 years. Then, compute the autocorrelation function (or correlogram) and plot it, and discuss the plot. Also, when doing the computation of autocovariance, use equation (11.16). Note for lags greater than 0, the number of values in the sum proportionately becomes less. Do not use the circular method where the number remains the same. Use the same method (non-circular) for other questions.

Solution: $r_0 = Cov(x_i, x_{i+0}) = 1.574 \times 10^7 \, (cfs)^2$. Similarly $r_1 = 1.307 \times 10^6 \, (cfs)^2$ and $r_2 = 9.390 \times 10^4 \, (cfs)^2$. Likewise, the coefficient of correlation is computed for other lags up to 12 years. Table 11.2 shows the autocovariance and autocorrelation values for six lags. The

Table 11.2 Autocovariance and autocorrelation values for annual stream flow at station USGS 0819000.

Lags (year)	Autocovariance (cfs)^2	Autocorrelation
0	1.57E + 07	1.00E + 00
1	1.31E + 06	8.31E − 02
2	9.39E + 04	5.97E − 03
3	8.61E + 05	5.47E − 02
4	8.93E + 05	5.68E − 02
5	1.20E + 06	7.61E − 02
6	−2.87E + 06	−1.82E − 01

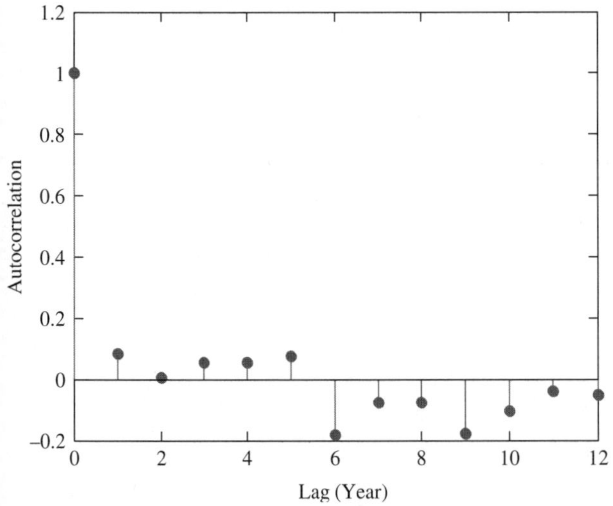

Figure 11.3 Autocorrelation of annual stream flow of Brazos River near Bryan from 1941–92.

autocorrelation is shown in Figure 11.3. It is seen that the autocorrelation function fluctuates and the memory seems to be long for stream flows of this river.

Example 11.4: Now consider monthly stream flow for the same gauging station at Brazos River as in Table 11.1. Plot the stream flow as a function of month. Now compute the autocorrelation function (ACF) and plot as a function of lag. Also, transform the series logarithmically and then compute the autocorrelation function and plot it. What is the effect of transformation?

Solution: Monthly stream flow for the USGS gauging station at Brazos River (USGS 08109000) near Bryan, Texas, from 1941–92 is plotted as shown in Figure 11.4. For monthly flow data, the autocorrelation function is shown in Figure 11.5, which shows that it fluctuates. Table 11.3 gives the autocorrelation values for monthly stream flow for the Brazos River near Bryan for 24 lags. Then monthly stream flow is transformed logarithmically and is then plotted as shown in Figure 11.6. For the transformed data, the autocorrelation function is computed, as shown in Figure 11.7. Again, the autocorrelation function is a fluctuating function.

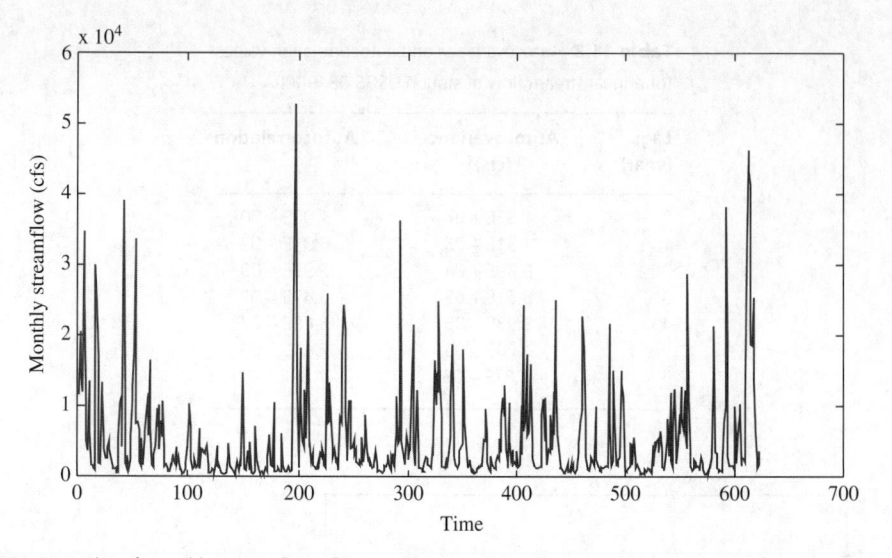

Figure 11.4 Plot of monthly stream flow of Brazos River near Bryan from 1941–92.

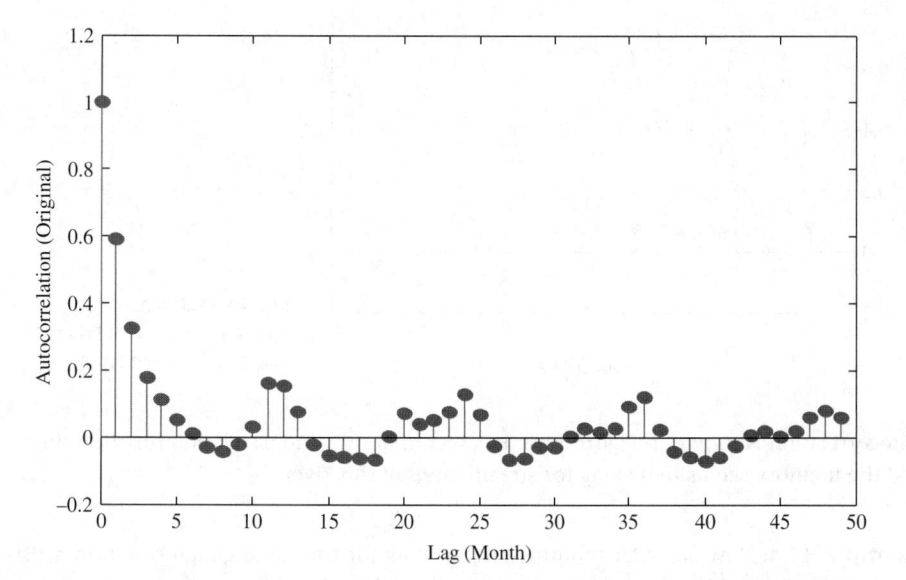

Figure 11.5 Autocorrelation functions of monthly stream flow of Brazos River near Bryan from 1941–92.

11.2 Spectral analysis

Environmental and water resources processes, such as rainfall, stream flow, drought, soil moisture, evapotranspiration, temperature, radiation, relative humidity, water quality, and so on, have inherent periodicities which are not easily observed in raw data. For example, consider the drought process which is stochastic. Then, Fourier analysis of drought data in certain parts of the world, say northern India, may reveal a significant period of five years. This means that, on average, a cycle of drought occurs once every five years. Extending the

Table 11.3 Autocorrelation values for monthly stream flow with 24 lags (lag 0–lag 23 months).

Lag	ACF	Lag	ACF	Lag	ACF	Lag	ACF
0	1.00	6	0.01	12	0.15	18	−0.07
1	0.59	7	−0.03	13	0.08	19	0.00
2	0.32	8	−0.04	14	−0.02	20	0.07
3	0.18	9	−0.02	15	−0.05	21	0.04
4	0.11	10	0.03	16	−0.06	22	0.05
5	0.05	11	0.16	17	−0.06	23	0.07

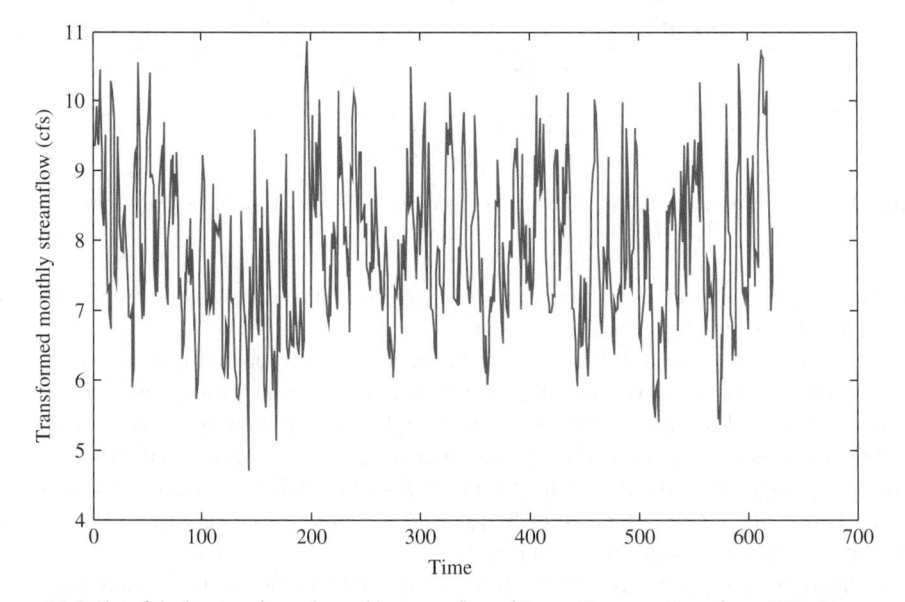

Figure 11.6 Plot of the log-transformed monthly stream flow of Brazos River near Bryan from 1941–92.

argument further, a sample may exhibit oscillations at all possible frequencies, suggesting the stochastic process may be composed of oscillations of all possible frequencies. Identifying these individual frequencies helps understand periodic patterns of the process. The term frequency is invoked whenever a function of a physical phenomenon fluctuates in time/or space. The fluctuating characteristic is then defined by the rate of fluctuation in terms of frequency or the wave number. Thus, the description of the phenomenon is now in the frequency domain in place of time or space domain. When dealing with data, the arrangement of data is now made in terms of frequency. This paves the way for spectral analysis.

To develop a physical appreciation of spectrum, consider a beam of white light passing through a prism. It is then observed that the beam is decomposed into an array of colored lights and the resulting colors are the primary colors of the spectrum. Then, it can be said that any color may be the result of a particular combination of primary colors. The deflection of each color is in proportion to its harmonic frequency or rate of energy oscillation, where the violet has the highest frequency and the red has the lowest frequency. In a similar vein, any sound can be partitioned into a combination of fundamental sounds. In hydraulics, there are

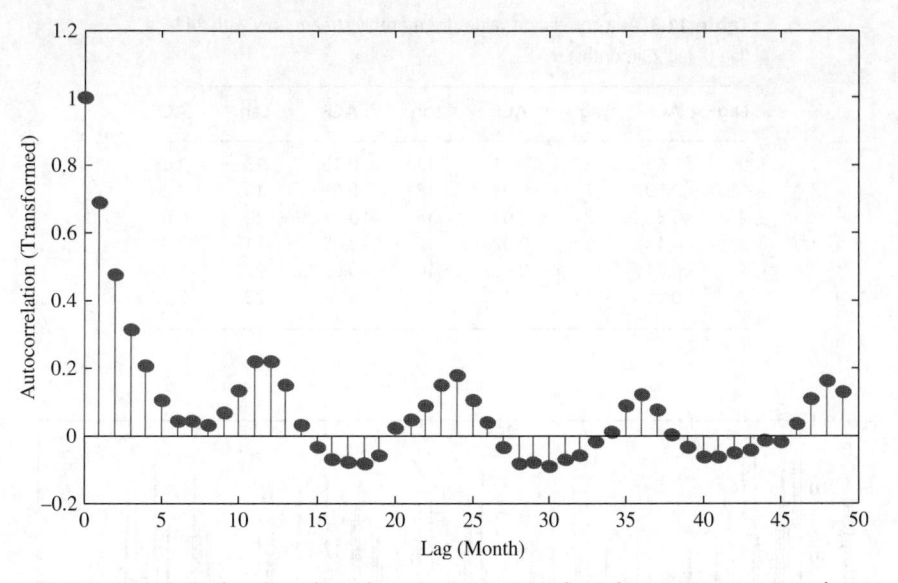

Figure 11.7 Autocorrelation functions of transformed monthly stream flow of Brazos River near Bryan from 1942–92.

three fundamental quantities, such as density, viscosity, and compressibility in terms of which most hydraulic parameters can be expressed.

The spectrum of a record is defined by the transform of the record into the frequency (or wave number) domain and is essentially a function defined by amplitude, power or any other property. It depends on an independent variable which is generally frequency or/and wave number. In statistics, the spectrum is generally a complex function defined either as a sum of real and imaginary parts or a product thereof. Spectral analysis is employed to determine the spectrum over the whole range of frequencies. It entails estimating the spectral density function or spectrum of a given time series.

Correlogram may experience periodic fluctuations and may not monotonically decrease as a function of lag, thus indicating a presence of a deterministic component in the stochastic process. From Fourier or harmonic analysis, it is known that any periodic function can be represented by the sum of a series of sine terms and/or cosine terms of increasing frequencies. This suggests that hidden periods of oscillations in samples of stochastic processes can be identified by Fourier or harmonic analysis. Spectral analysis is a modification of Fourier analysis. For analyzing a time series made up of sine and cosine waves of different frequencies the periodogram is often employed. Spectral analysis is employed to detect periodicities in time series data. The spectrum so determined should have sharp peaks only at frequencies corresponding to frequencies of actual periodicities in the data. The methods which determine such spectra are said to have a high degree of resolution. Entropy-based spectral analysis has this property.

11.2.1 Fourier representation

Recall a periodic deterministic function, $f(t)$, with a period of T. Here period T is just a number such that

$$f(t) = f(t + T) \tag{11.28}$$

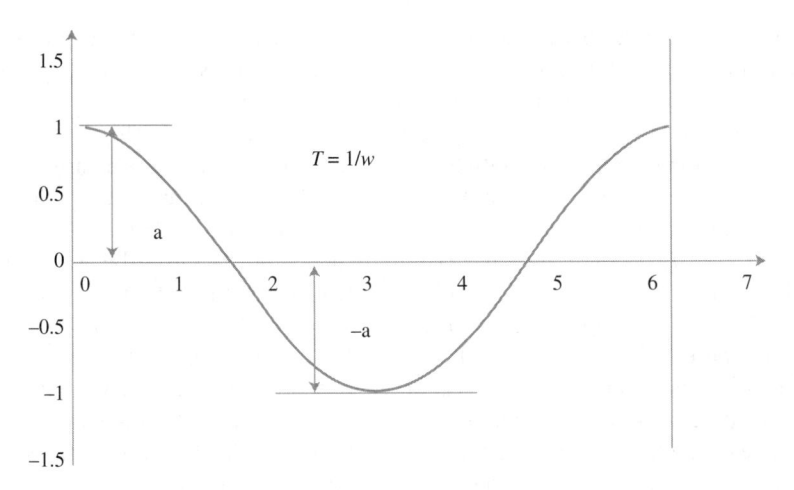

Figure 11.8 Illustration of a periodic function with period T and frequency w.

for all t. The function $f(t)$ between t and $t+T$ can be of any shape. Consider, for example, a simple, cosine function $f(t) = a \cos 2\pi wt$ with a period $w = 1/T$, then

$$a \cos 2\pi wt = a \cos 2\pi w (t + T) \tag{11.29}$$

where a is called amplitude. Term $2\pi w$ is called angular frequency in cycles per 2π and w is the cyclic frequency in cycles per unit, with $w = 1$/the period T of cycle, as shown in Figure 11.8.

Now let there be a periodic function in time, $f(t)$, with periodicity of T. If there is another function $g(t)$ such that $g(t) = f[tT/(2\pi)]$, then $g(t)$ is periodic with period 2π. According to the Fourier theorem, $g(t)$ can be expanded by an infinite Fourier series consisting of sine and cosine functions as

$$g(s) = a_0 + \sum_{n=1}^{\infty} \left(a_n \cos ns + b_n \sin ns \right) \tag{11.30}$$

Now, writing $s = 2\pi t/T$ and denoting $f(t) = g(tT/2\pi)$, one can write equation (11.30) as

$$f(t) = g\left(\frac{2\pi t}{T} \right) = a_0 + \sum_{n=1}^{\infty} \left(a_n \cos \frac{2\pi nt}{T} + b_n \sin \frac{2\pi nt}{T} \right) \tag{11.31a}$$

or simply with $2\pi = T$,

$$f(t) = a_0 + \sum_{n=1}^{\infty} \left(a_n \cos nt + b_n \sin nt \right) \tag{11.31b}$$

where a_0, a_n, b_n, $n = 1, 2, \ldots, \infty$, are called Fourier coefficients and they are evaluated by Fourier analysis or harmonic analysis. One can take $a_0 = 0$, as is sometimes done, since that is just the base for measurement for $f(t)$. The first term corresponding to $n = 0$, a_0, is often denoted by $(a_0/2)$ rather than by a_0; the next term corresponding to $n = 1$ is called the fundamental and represents a cosine/sine wave whose period matches exactly that of the given function $f(t)$. The next term corresponding to $n = 2$ is called the first harmonic and represents a cosine/sine wave whose period is exactly half of that of the given function $f(t)$. The terms corresponding to $n = 3, 4, 5, \ldots$, are called the second harmonic, the third harmonic, the fourth harmonic, the fifth harmonic, and so on.

In equation (11.31a), the function $f(t)$ has a periodicity of T. This is seen because $\sin(\pi t/T)$ and $\cos(\pi t/T)$ have periods $2\pi/(\pi T) = 2T$, or $4T$, $6T$, $\ldots$ Similarly, $\sin(2\pi t/T)$ and $\cos(2\pi t/T)$ have periods $2\pi/(2\pi T) = T$, or $2T$, $3T$, $\ldots$ In general, it should be noted that each of $\sin(2n\pi t/T)$ and $\cos(2n\pi t/T)$ has a period equal to $2\pi/(2n\pi T) = T/n$, $2T/n$, $3T/n$, $\ldots$ $2nT/2n = T$, $\ldots$ All terms have a common period of T – this is the least period of all terms. Thus, if the infinite series of equation (11.31a) or equation (11.31b) is equal to $f(t)$, where t lies in the interval of length T, then it will be true for any other interval provided $f(t)$ has a period equal to T. Often we limit ourselves to the interval of length $(-T/2, T/2)$ but results would be valid for any interval of length T. To illustrate it may be instructive to consider a continuous series, say stream flow, of duration T sampled at discrete times and the series is to be expanded into periodic functions. Let the sampling interval be Δt, as shown in Figure 11.9. Then in period T, this produces $N = T/\Delta t$ sample values f_r:

$$f_r = f(t = r\Delta t) \tag{11.32}$$

where r is a number.

Note that the highest frequency for fitting the data is $w = \pi$ $(0 < w < \pi)$, which is the Nyquist frequency, while the lowest frequency one can reasonably fit completes one cycle in the whole length of the time series. Equating the cycle length $2\pi/w$ to N, the lowest frequency is obtained as $2\pi/N$. If observations are taken at equal intervals of time Δt then the Nyquist frequency is $w_N = \pi/\Delta t$. The equivalent frequency stated in cycles per unit time is $f_N = w_N/2\pi = 1/2\Delta t$. The Nyquist frequency is the highest frequency about which meaningful information can be extracted from the data. As an example, consider evaporation readings taken at 1:00 P.M. each day. These observations will not tell anything about the variation in evaporation whether it is more at noon or less in the morning and even less during nights. With only one observation the Nyquist frequency is $w_N = \pi$ radians per day or $f_N = 1$ cycle per two days. This is lower than the frequencies that consider variation within a day. The variation with a wave length of one day has frequency (angular) $w = 2\pi$ radians per day or $f = $ one cycle per day. Clearly, more observations will be needed to obtain information about variation of evaporation within a day.

It may be instructive to label the terms in the Fourier series in terms of their frequencies. Terms $\cos nt$ and $\sin nt$ each have period $(2\pi/n)$, meaning each term will go through $(n/2\pi)$ complete cycles. If time t is measured in seconds then term $a_n \cos nt + b_n \sin nt$ has a frequency of $(n/2\pi)$ cycles per second. The quantity n can be interpreted as the angular frequency of the term $(a_n \cos nt + b_n \sin nt)$ and is measured in radians per second (if t is measured in seconds). Geometrically, the angle (nt) passes through n radians each time t moves through one unit of time. This leads to a relation between frequency (cycles per second) and angular frequency (in radians per second) as

$$Angular\,frequency\,(w) = (frequency, f) \times 2\pi = \frac{2\pi}{period} \tag{11.33}$$

where period is units of time.

It may be remarked that if a function is represented by a Fourier series for $0 \le x \le L$ instead of $0 \le x \le 2\pi$, then one can substitute $t = (2\pi x)/L$ and use the Fourier series in terms of t. If the interval is $a \le x \le b$, then one can take $y = x - a$ and use $t = (2\pi y)/(b - a)$. The Fourier

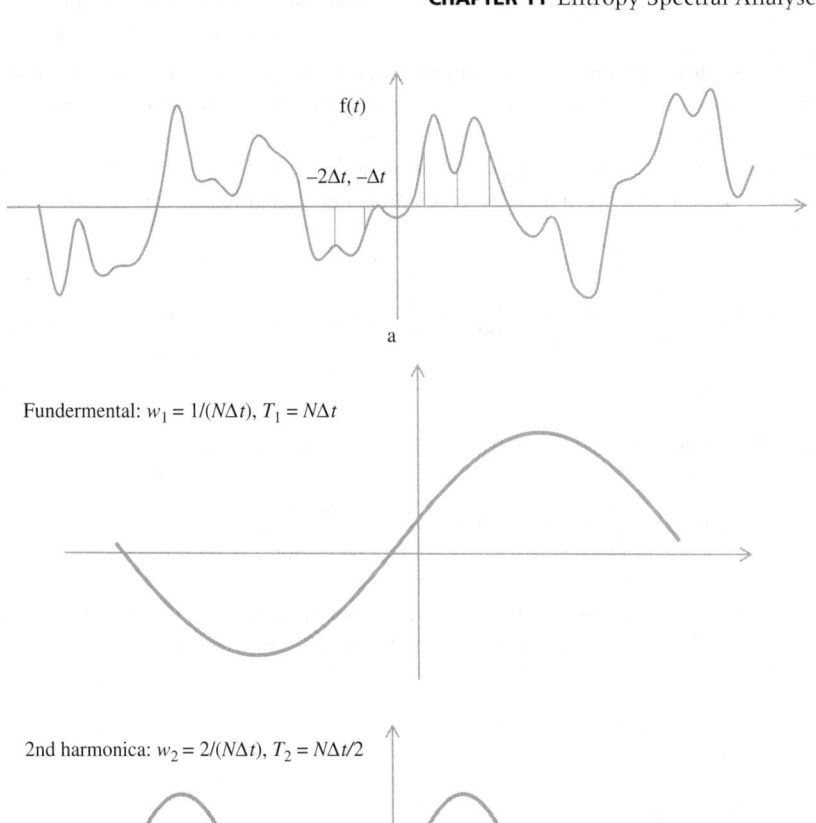

Fundermental: $w_1 = 1/(N\Delta t)$, $T_1 = N\Delta t$

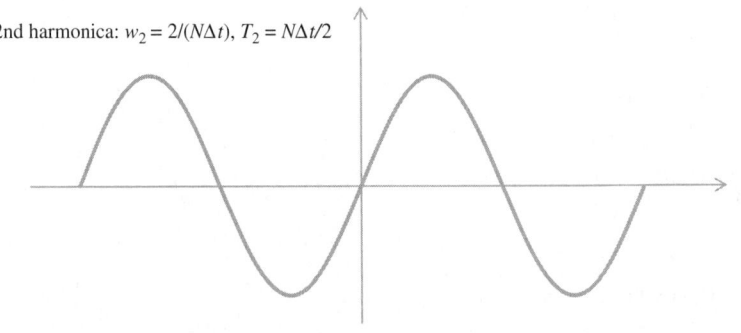

2nd harmonica: $w_2 = 2/(N\Delta t)$, $T_2 = N\Delta t/2$

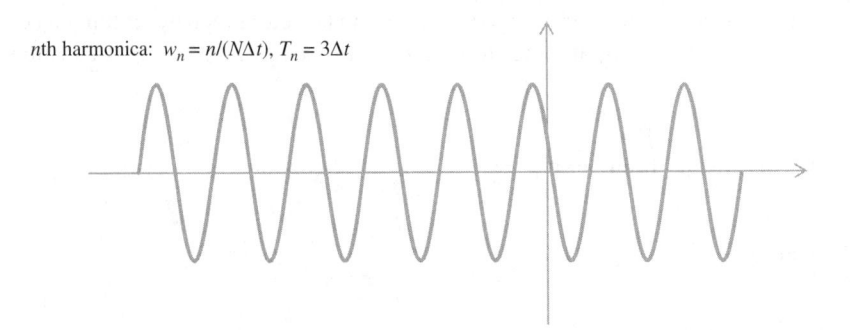

nth harmonica: $w_n = n/(N\Delta t)$, $T_n = 3\Delta t$

Figure 11.9 Waves and harmonics for a discrete signal. a: a discrete signal sampling from a continuous signal; b: fundamental $w_1 = 1/(N\Delta t)$, $T_1 = N\Delta t$; c: 2nd harmonic $w_2 = 2/(N\Delta t)$, $T_2 = N\Delta t/2$; d: n-th harmonics, $w_n = 1/(2\Delta t)$, $T = 2\Delta t$.

series can be employed to represent functions that cannot be represented by power series, for the former depends on the existence of certain integrals whereas the latter depends on the existence of derivatives.

The Fourier coefficients can be determined as follows. For each integer m, both sides of equation (11.31b) can be multiplied by $\cos mt$ and then integrated from $-\pi$ to π. Integrating term by term we get

$$\int_{-\pi}^{\pi} f(t) \cos mt \, dt = \frac{a_0}{2} \int_{-\pi}^{\pi} \cos mt \, dt + \sum_{n=1}^{\infty} \left\{ a_n \int_{-\pi}^{\pi} \cos nt \cos mt \, dt + b_n \int_{-\pi}^{\pi} \sin nt \cos mt \, dt \right\},$$

$$m = 0, 1, 2, \ldots \tag{11.34}$$

Now recall the property of orthogonality of sine and cosine functions:

$$\int_{-\pi}^{\pi} \cos(mt) \sin(nt) \, dt = \int_{-\pi}^{\pi} \sin(mt) \cos(nt) \, dt = \begin{cases} \pi \text{ for } m = n \\ 0 \text{ for } m \neq n \end{cases} \tag{11.35}$$

Taking advantage of equation (11.35), all terms on the right side of equation (11.34) will vanish except for the cosine integral term in which $n = m$. Thus equation (11.34) becomes

$$\int_{-\pi}^{\pi} f(t) \cos mt \, dt = a_m \int_{-\pi}^{\pi} \cos^2 mt \, dt = \pi a_m \tag{11.36}$$

Equation (11.36) yields

$$a_0 = \frac{1}{2\pi} \int_{-\pi}^{\pi} f(t) dt \tag{11.37}$$

$$a_m = \frac{1}{\pi} \int_{-\pi}^{\pi} f(t) \cos mt \, dt, \quad m = 1, 2, \ldots \tag{11.38}$$

Likewise, the Fourier coefficients b_n, $n = 0, 1, 2, \ldots$, can be determined by multiplying equation (11.31b) by $\sin mt$, and integrating term by term from $-\pi$ to π, and using the orthogonality property as

$$\int_{-\pi}^{\pi} f(t) \sin mt \, dt = b_m \int_{-\pi}^{\pi} \sin^2 mt \, dt = \pi m_m \tag{11.39}$$

which yields

$$b_m = \frac{1}{\pi} \int_{-\pi}^{\pi} f(t) \sin mt \, dt, \quad m = 1, 2, \ldots \tag{11.40}$$

Equations (11.38) and (11.40) are called the Euler-Fourier equations.

Note that the limits of integration can also be taken as 0 to 2π. If the fundamental period (periodicity) is T, instead of 2π, then the expressions for the Fourier series and the associated coefficients become:

$$f(t) = a_0 + \sum_{n=1}^{\infty} \left(a_n \cos \frac{2n\pi t}{T} + b_n \sin \frac{2n\pi t}{T} \right) \tag{11.41}$$

$$a_0 = \frac{1}{T} \int_{-T/2}^{T/2} f(t)\, dt \tag{11.42}$$

$$a_n = \frac{2}{T} \int_{-T/2}^{T/2} f(t) \cos \frac{2n\pi t}{T}\, dt \tag{11.43}$$

$$b_n = \frac{2}{T} \int_{-T/2}^{T/2} f(t) \sin \frac{2n\pi t}{T}\, dt \tag{11.44}$$

The above integral expressions for the coefficients [equations (11.38) and (11.40) or (11.43) and (11.44)] are called Cauchy integrals.

11.2.2 Fourier transform

Let w be the Fourier transform variable. Recalling $\exp(-it) = \cos(t) - i\sin(t)$ where $i = \sqrt{-1}$, the Fourier transform of function $f(t)$, denoted by $F(w)$, can be expressed as

$$F(w) = \int_{-\infty}^{\infty} f(t) \exp(-it)\, dt = \int_{-\infty}^{\infty} f(t) \left[\cos wt - i\sin wt \right] dt \tag{11.45}$$

Equation (11.45) transforms function f in the time domain to function F in the Fourier domain. The Fourier variable w, called angular frequency, is defined as

$$w = \frac{2\pi}{T} = 2\pi w_* = 2\pi f, \quad w_* = \frac{1}{T} = f \tag{11.46}$$

where T is the period in units of time, and w_* is the frequency in cycles per unit of time. Term w_i is the i-th harmonic of the fundamental frequency $1/N$. Thus, quantities frequency, period, and number of cycles per unit of time are related.

The Fourier transform (also called Fourier spectrum) of function $f(t)$, $F(\omega)$, can be written as

$$F(w) = a(w) - ib(w) = |F(w)| \exp\left[i\Phi(w)\right] \tag{11.47}$$

where

$$|F(w)| = \left[a^2(w) + b^2(w) \right]^{0.5} \tag{11.48}$$

is the amplitude spectrum, and

$$\Phi(w) = \tan^{-1}\left[-b(w)/a(w) \right] + 2n\pi \quad (n = 0, \pm 1, \pm 2, \ldots) \tag{11.49}$$

is the phase spectrum and its negative is called phase-lag spectrum. Determination of $F(w)$ is called the Fourier analysis of $f(t)$.

Thus, the Fourier spectrum is defined in terms of the amplitude spectrum and the phase spectrum. The real part, $a(w)$, the Fourier cosine transform and the imaginary part $b(w)$, the Fourier sine transform, of $F(w)$ are called, respectively, co-spectrum and quad-spectrum:

$$a(w) = \int_{-\infty}^{\infty} f(t) \cos wt \, dt \tag{11.50}$$

$$b(w) = \int_{-\infty}^{\infty} f(t) = \sin wt \, dt \tag{11.51}$$

These spectra correspond to:

$$a(w) \to \frac{1}{2} a_n; \quad b(w) \to \frac{1}{2} b_n \tag{11.52}$$

The inverse Fourier transform is now defined as

$$f(t) = \frac{1}{2\pi} \int_{-\infty}^{\infty} F(w) \exp(iwt) \, dw \tag{11.53}$$

and is indeed the Fourier synthesis of $f(t)$. Functions $f(t)$ and $F(w)$ are said to constitute a Fourier pair and there is a one to one correspondence between them. However, this correspondence is not point to point in the two domains but is only curve by curve. In a physical sense, $F(w)$ represents an average of $f(t)\exp(-iwt)$ over the interval of integration. Term $\exp(-iwt)$ selects from $f(t)$ only those terms that have frequency w, that is, $F(w)$ is an average of those components of $f(t)$ that have frequency w. When the frequency interval is unity, $F(w)$ is called density or more specifically spectral density and $|F(w)|$ is called the amplitude density. Both $f(t)$ and $F(w)$ have the same dimensions. One can also refer the spectral density to unit time interval in place of unit frequency interval. Then, $F(w)$ should be multiplied by the number of cycles per second or any proportional quantity, such as w. Hence $wF(w)$ would be the spectral density for the unit time interval.

11.2.3 Periodogram

The periodogram is employed to detect and determine the amplitude of a sine component of known frequency in the random series and can be calculated as follows. Let there be $N = 2m + 1$ odd observations of series $X(t)$. The Fourier series is fitted to this series as:

$$x(t) = a_0 + \sum_{i=1}^{m} \left[a_i \cos \left(\frac{2\pi it}{N} \right) + b_i \sin \left(\frac{2\pi it}{N} \right) \right] + e(t) \tag{11.54}$$

where i/N is the i-th harmonic of the fundamental frequency $1/N$, and $e(t)$ represents the residual or error term. The least square estimate of coefficients a_0, a_i, and b_i can be written as

$$a_0 = \bar{x} \tag{11.55}$$

$$a_i = \frac{2}{N} \sum_{i=1}^{N} x(t) \cos \left(\frac{2\pi it}{N} \right) \quad i = 1, 2, \dots, m \tag{11.56}$$

$$b_i = \frac{2}{N} \sum_{i=1}^{N} x(t) \sin \left(\frac{2\pi it}{N} \right) \quad i = 1, 2, \dots, m \tag{11.57}$$

The periodogram, denoted as $P(i/N)$, then is constituted by $m = (N-1)/2$ values determined as

$$P(i/N) = \frac{N}{2}\left(a_i^2 + b_i^2\right), \quad i = 1, 2, \ldots, m \tag{11.58}$$

Term $P(i/N)$ represents the intensity at frequency i/N.

If the number of observations N is even, then $N = 2m$ and equations (11.56) and (11.57) will hold for $i = 1, 2, \ldots, m-1$ but for $i = m$ the following will apply:

$$a_m = \frac{1}{N}\sum_{i=1}^{N}(-1)^i x_i(t) \tag{11.59}$$

$$b_m = 0 \tag{11.60}$$

$$P(m/N) = P(0.5) = N a_m^2 \tag{11.61}$$

It may be noted that the smallest period is two intervals and hence the highest frequency is 0.5 cycles per interval.

Example 11.5: Represent the annual stream flow time series, given in Table 11.1 in Example 11.1, by a Fourier series and compute the Fourier coefficients. Specify the values of a_n and b_n. Also, plot the original series and the fitted Fourier series and discuss how they are determined.

Solution: There are 52 years of data, that is, 52 points for the annual stream flow for the dataset ($N = 52$). The corresponding discrete Fourier series can be given as:

$$y_n = a_0 + \sum_{t=1}^{N/2}\left(a_t \cos \frac{2n\pi t}{T} + b_t \sin \frac{2n\pi t}{T}\right), \quad n = 1, \ldots, N$$

The first coefficient a_0 is computed as:

$$a_0 = \frac{1}{N}\sum_{i=1}^{N} y_i = 5165.9$$

The coefficients from $n = 1$ to $N/2 - 1$ are computed as

$$a_t = \sum_{i=1}^{N} y_i \cos \frac{2n\pi i}{T} \quad t = 1, \ldots, N/2 - 1$$

$$b_t = \sum_{i=1}^{N} y_i \sin \frac{2n\pi i}{T} \quad t = 1, \ldots, N/2 - 1$$

For $n = N$,

$$a_n = \frac{1}{N}\sum_{i=1}^{N}(-1)^i x_i(t) = -180.1, \; b_n = 0$$

The first 26 coefficients are computed, as shown in Figure 11.10. The original series and the fitted Fourier series are shown in Figure 11.11 which shows that these two series are close to each other.

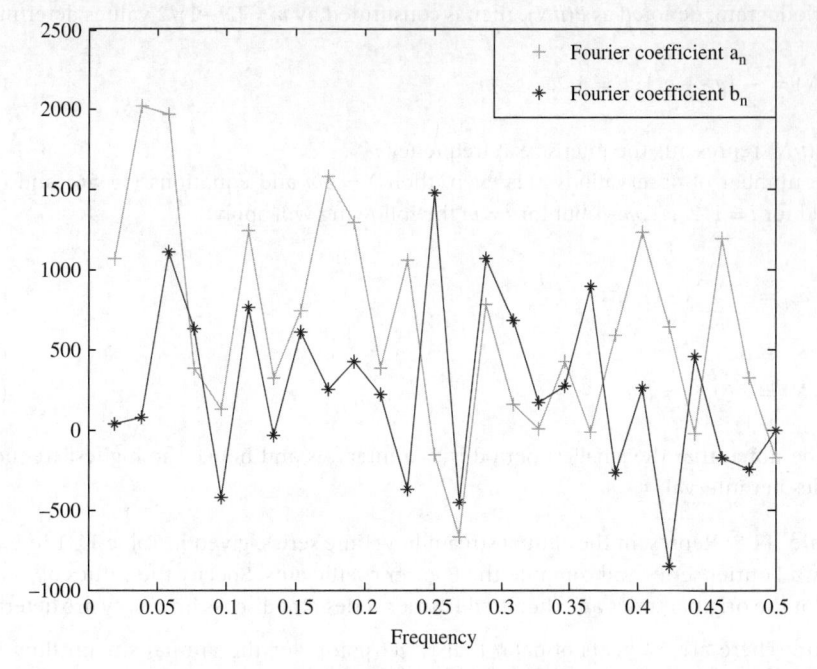

Figure 11.10 Fourier coefficients for annual stream flow of Brazos River near Bryan for the period 1941–92.

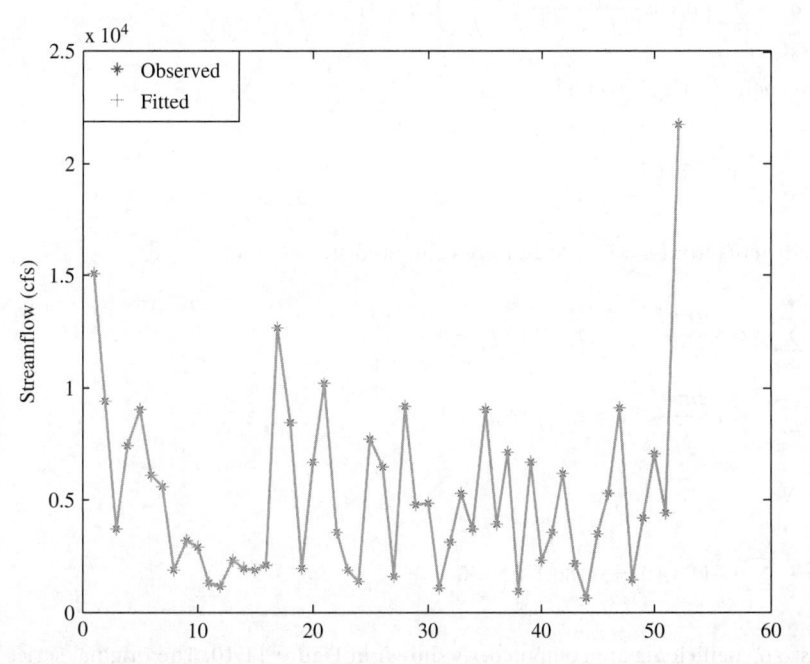

Figure 11.11 Plot of the original series and the fitted Fourier series for annual stream flow of Brazos River near Bryan for the period 1941–92.

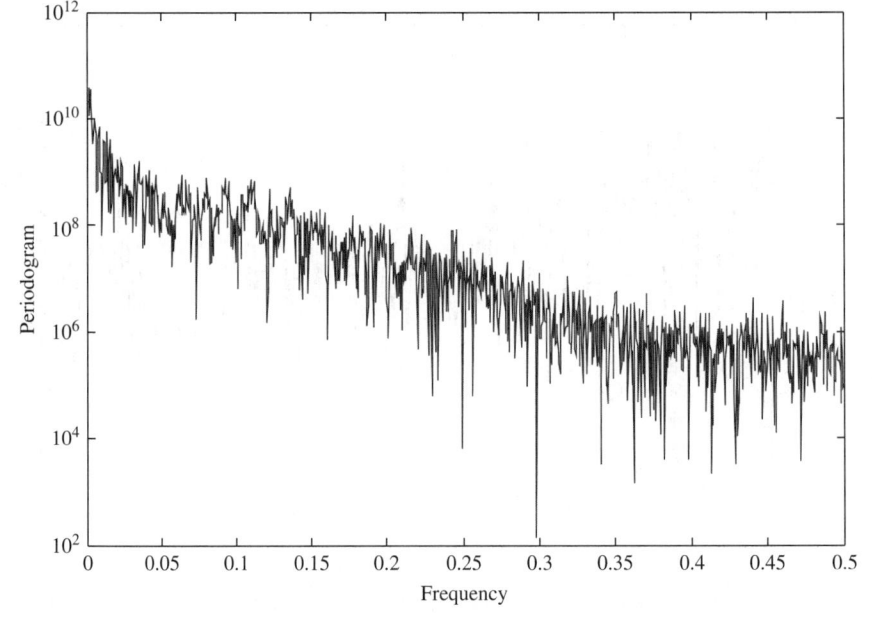

Figure 11.12 Periodogram of daily stream flow for Brazos River near Bryan, Texas, from 1987–92.

Example 11.6: Compute and plot the periodogram of the following data: (a) daily time series of stream flow of Brazos River near Bryan; (b) mean monthly stream flow of Brazos River near Bryan; and (c) mean monthly temperature of College Station.

Solution: For the daily stream flow of Brazos River from 1987–92, the number of observations is 2192 which is even. The coefficients can be obtained with the procedure in Example 11.5. The periodogram for the daily stream flow is then plotted as shown in Figure 11.12. Similarly, the periodogram for monthly stream flow and monthly temperature are plotted in Figure 11.13 and Figure 11.14.

11.2.4 Power

If a physical function, denoted as $f(t)$, is deterministic expressed as equation (11.31b) then its periodic representation can be interpreted in terms of energy/frequency. Let the period of the function be 2π. The total energy over the interval $(-\pi, \pi)$ is expressed as

$$\text{Total energy over interval } (-\pi,\pi) = \int_{-\infty}^{\infty} [f(t)]^2 \, dt \tag{11.62}$$

Squaring both sides of equation (11.31b), integrating from $-\pi$ to π, and using the orthogonality property, equation (11.62) can be written as

$$\int_{-\pi}^{\pi} [f(t)]^2 dt = \pi \left[a_0^2 + \sum_{n=1}^{\infty} \left(a_n^2 + b_n^2 \right) \right] \tag{11.63}$$

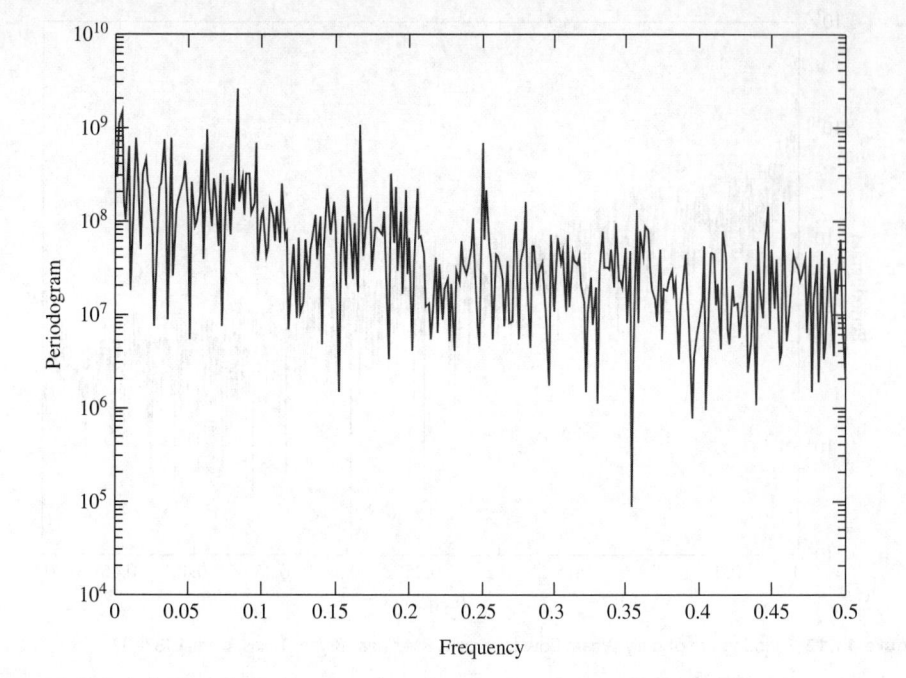

Figure 11.13 Periodogram of monthly stream flow of Brazos River near Bryan from 1941–92.

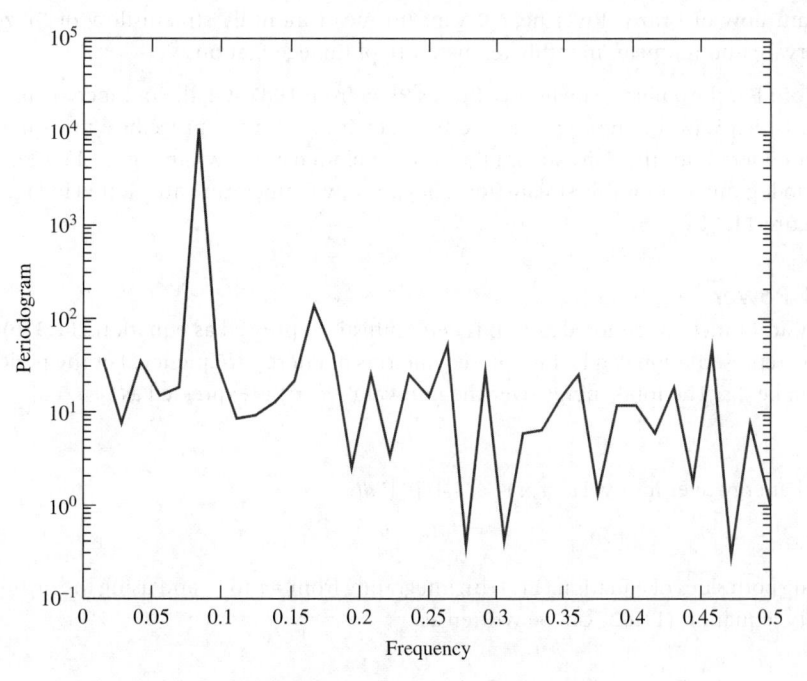

Figure 11.14 Periodogram of monthly temperature of College Station from 2003–8.

Denoting $c_0 = \frac{a_0}{\sqrt{2}}$, $c_n = \left[\left(a_n^2 + b_n^2 \right) / 2 \right]^{0.5}$, $n = 1, 2, \ldots$, one can write equation (11.62) as

$$\text{Total energy over interval} \, (-\pi, \pi) = 2\pi \left[\sum_{n=0}^{\infty} c_n^2 \right] \tag{11.64}$$

This is the amount of energy dissipated over a time period of 2π. The energy dissipated per unit of time, called power, can also be expressed as

$$\text{Total power} = \frac{\text{Total energy over interval} \, (-\pi, \pi)}{2\pi} = \sum_{n=0}^{\infty} c_n^2 \tag{11.65}$$

Consider a special case as

$$f(t) = a_n \cos nt + b_n \sin nt \tag{11.66}$$

In this case, all Fourier coefficients, excepting a_n and b_n vanish and consequently all cs, excepting c_n, vanish. Then, equation (11.65) yields

$$\text{Total power} = c_n^2 \tag{11.67}$$

This shows that c_n^2 represents the contribution of the term $a_n \cos nt + b_n \sin nt$ to the total power. Thus, the quantity c_n^2 describes the contribution to the total power from the term in the Fourier series having frequency $(n/2\pi)$ cycles per second or angular frequency n radians per second.

If function $f(t)$ is periodic with period T as equation (11.41), then the total energy dissipated in time interval $(-T/2, T/2)$ can be expressed as

$$\text{Total energy over interval} \, (-T/2, T/2) = \int_{-T/2}^{T/2} f^2(t) dt = T \left(\sum_{n=0}^{\infty} c_n^2 \right) \tag{11.68}$$

Similarly,

$$\text{Total power over} \, (-T/2, T/2) = \frac{\text{Total energy over} \, (-T/2, T/2)}{T} = \sum_{n=0}^{\infty} c_n^2 \tag{11.69}$$

Thus, term c_n^2 represents the contribution to the total power from the term in the Fourier series of $f(t)$ having a frequency of n/T (cycles per second) or angular frequency $2\pi n/T$ (radians per second). A plot of c_n^2 against $n/2\pi$, which is a discrete power spectrum, shows graphically the distribution of total power over the various frequency components of function $f(t)$.

Let us now discuss the concept of power. Consider the time function $f(t)$, say representing stream flow. Then the corresponding frequency function $F(w)$ related to $f(t)$ through equations (11.45) and (11.53) is the stream flow spectrum. Sometimes it is more meaningful to consider the power of the signal rather than the amplitude. In general, power is expressed as the square of the amplitude. That is the reason that the corresponding spectra are referred to as power spectra. For any time function (real), the average power (or variance) is defined in the time domain as

$$s_T^2 = \frac{1}{T} \int_{-T/2}^{T/2} |f(t)|^2 dt = \sum_{m=-\infty}^{\infty} |X_m|^2 \tag{11.70}$$

where X_m denotes the complex amplitude at the harmonic frequency $w_m = m/T$, measuring the amplitudes of sine and cosine terms at frequency w_m in series $f(t)$. The complex amplitude can be computed as

$$X_m = \frac{1}{T} \int_{-T/2}^{T/2} f(t) \exp\left(-\frac{i2\pi mt}{T}\right) dt \tag{11.71}$$

where

$$\exp\left(-\frac{2i\pi mt}{T}\right) = \cos\left(\frac{2\pi mt}{T}\right) - i\sin\left(\frac{2\pi mt}{T}\right) \tag{11.72}$$

Note the Fourier decomposition of $f(t)$ as

$$f(t) = \sum_{m=-\infty}^{\infty} X_m \exp\left(\frac{2\pi imt}{T}\right) \tag{11.73}$$

If function $f(t)$ is discrete observed at times $t = -n\Delta t, -(n-1)\Delta t, \ldots, (n-1)\Delta t$, then the average power can be decomposed into contributions at a finite number of harmonics of the fundamental frequency $w_1 = 1/N\Delta t$ $(N = 2n)$. The corresponding relations for the discrete case can be expressed as

$$s_T^2 = \frac{1}{N} \sum_{t=-n}^{n-1} f_t^2 = \sum_{m=-n}^{n-1} |X_m|^2 \tag{11.74}$$

$$X_m = \frac{1}{N} \sum_{t=-n}^{n-1} f_t \exp\left(-\frac{2\pi im\Delta t}{n\Delta t}\right) = \frac{1}{N} \sum_{t=-n}^{n} f_t \exp\left(-\frac{2\pi imt}{N}\right) \tag{11.75}$$

Equation (11.75) represents the decomposition of power at harmonics $w_m = m/T$ of the fundamental frequency $w_1 = 1/T$. Term $|f(t)|^2$ denotes the instantaneous power of $f(t)$ and the integral $\int_{-T/2}^{T/2} |f(t)|^2 dt$ represents the total energy of $f(t)$. Using Parseval's theorem, it can be shown that

$$\int_{-\infty}^{\infty} |f(t)|^2 \, dt = \frac{1}{\pi} \int_{0}^{\infty} |F(w)|^2 dw \tag{11.76}$$

where $F(w)$ is the Fourier transform of $f(t)$. Further, a simple shift in $f(t)$ will keep the power spectrum unchanged. That is,

$$\int_{-\infty}^{\infty} |f(t \pm a)|^2 \, dt = \frac{1}{2\pi} \int_{-\infty}^{\infty} |\exp(\pm iaw) F(w)|^2 = \frac{1}{2\pi} \int_{-\infty}^{\infty} |F(w)|^2 dw \tag{11.77}$$

11.2.5 Power spectrum

The power spectrum of a given time series, $S(w)$, is defined by the Fourier transform of its autocovariance function $r(\tau)$:

$$S(w) = \frac{1}{2\pi} \int_{-\infty}^{\infty} e^{-iw\tau} r(\tau) d\tau \tag{11.78}$$

or

$$S(w) = \frac{1}{2\pi} \int_{-\infty}^{\infty} r(\tau) \cos(w\tau) d\tau \tag{11.79}$$

where τ is the lag. The spectrum is an even function, $S(w) = S(-w)$. From the inverse Fourier transform, the covariance function $Cov(\tau)$ can be expressed as:

$$Cov(\tau) = r(\tau) = \int_{-\infty}^{\infty} e^{iwt} S(w) dw \tag{11.80}$$

or

$$r(\tau) = \int_{-\infty}^{\infty} S(w) \cos(w\tau) dw \tag{11.81}$$

Equation (11.80) shows that the covariance function can be expressed as the inverse Fourier transform of the spectrum. Fourier transform $S(w)$ provides information about the time series in the frequency domain, in contrast with the autocovariance function which gives information in the time domain. The advantage of $S(w)$ is that it helps uncover hidden periodicities in data and determines closely spaced frequency peaks.

For the discrete case the spectrum can likewise be defined as:

$$S(w) = \Delta t \sum_{j=-(N-1)}^{(N-1)} r_k \exp\left[-2\pi i w j \Delta t\right], \quad -\frac{1}{2\Delta t} \leq w \leq \frac{1}{2\Delta t} \tag{11.82}$$

where $i = \sqrt{-1}$, $\Delta t =$ the sampling time interval, and $w =$ frequency. Term $1/(2\Delta t)$ is called the Nyquist frequency. Equation (11.82) implies that r_k is known for finite values of lag, that is, observations are finite.

Taking the inverse Fourier transform of $S(w)$ in equation (11.82) yields the autocovariance function:

$$r_k = \int_{-1/2\Delta t}^{1/2\Delta t} S(w) \exp(2\pi k i w) dw, \quad -N\Delta t \leq k \leq N\Delta t \tag{11.83}$$

If the observation interval $\Delta t = 1$, that is, normalized to unit length, then equation (11.82) results in the normalized power spectrum:

$$S(w) = \sum_{j=-\infty}^{\infty} r_k \exp\left[-2j\pi i w\right], \quad |w| < \frac{1}{2}, i = \sqrt{-1} \tag{11.84}$$

Likewise, equation (11.83) becomes

$$r_k = \int_{-1/2}^{1/2} S(w) \exp(2\pi kiw) dw \tag{11.85}$$

Let $\theta = 2\pi w$. Then, equation (11.85) becomes

$$r_k = \frac{1}{2\pi} \int_0^{2\pi} S\left(\frac{\theta}{2\pi}\right) \exp(i\theta k) d\theta \tag{11.86}$$

The quantity $|F(w)|^2/T$ is called power spectral density. For comparing time series with different scales of measurement, one may normalize the spectrum by dividing it by the variance σ_X^2:

$$g(w) = \frac{S(w)}{\sigma_X^2} \tag{11.87}$$

Equation (11.87) defines the spectral density function $g(w)$. This means that the covariance function in equation (11.78) can be replaced by the autocorrelation function $\rho_*(\tau)$:

$$g(w) = \frac{1}{2\pi} \int_{-\infty}^{\infty} \rho_X(\tau) e^{-w\tau} d\tau = \frac{1}{2\pi} \int_{-\infty}^{\infty} \rho_X(\tau) \cos(w\tau) d\tau \tag{11.88}$$

where $g(w)$ is the normalized spectral density function. Likewise, the inverse of equation (11.88) can be cast as

$$\rho_X(\tau) = \int_{-\infty}^{\infty} g(w) e^{iw\tau} d\omega = \int_{-\infty}^{\infty} g(w) \cos(w\tau) dw \tag{11.89}$$

Equation (11.89) indicates that the area under the spectral density function is unity, since $\rho(0) = \cos(0) = 1$:

$$\int_0^{1/2} g(w) dw = 1 \tag{11.90}$$

If $\tau = 0$, $\text{Cov}(0) = \sigma^2 = \text{variance}$, then equation (11.89) reduces to

$$r(0) = \sigma^2 = \int_{-\infty}^{\infty} g(w) dw \tag{11.91}$$

Equation (11.91) shows that the variance of the process $X(t)$ is distributed over frequency in the same manner as variance of one particular sample of length T is distributed over frequency shown by equation (11.85). More specifically, the variance of $X(t)$ due to frequencies in the range w and $w + \Delta w$ is approximately $S(w)\Delta w$.

Interpreted physically, the spectrum represents a distribution of variance over frequency. When divided by the variance, the spectrum is analogous to a probability density function.

This explains the designation of spectral density function. When divided by the variance, the integrated spectrum is analogous to the cumulative probability distribution function. The spectral density function helps determine the frequencies that dominate the variance. A graph of the spectral density function shows predominant frequencies relative to less dominant frequencies. For example, a partial area between a band of two frequencies, f_1 and f_2, under the spectral density function indicates the contribution of the total normalized variance in the band. Thus, the spectral density is a function of frequency in cycles per unit of time (w_m), frequency in radians per unit of time (w_m), or period in units of time (T). Therefore,

$$S(w) = 2\pi S(w_m) \tag{11.92}$$

Equations (11.78) and (11.79) can also be expressed in terms of w_m. If x's are random variables then one can express the joint probability density $p(X)$, where $X = \{x_1, x_2, \ldots, x_N\}$ to describe the stochastic signal. Here $p(X)$ is the spectral power probability density function.

It may be worthwhile to recall several important aspects of covariance-spectrum relationships: 1) The spectrum must be non-negative at all frequencies, for it corresponds to a square amplitude. 2) The spectrum is a distribution of variance over frequency. When divided by variance, the spectrum is analogous to a probability density function. Likewise, the integrated spectrum, when divided by variance, is analogous to the cumulative probability distribution function. 3) The spectrum is an even function of frequency, because the covariance is an even function of lag.

Example 11.7: Consider the covariance function expressed by the negative exponential function. Determine the spectrum.

Solution: Substituting the negative exponential covariance function equation (11.78), one gets

$$S(w) = \frac{1}{2\pi} \int_{-\infty}^{\infty} \exp(-iwk) \frac{1}{\sigma^2} \exp(-|k|/\alpha] dk$$

$$= \frac{\sigma^2}{2\pi} \left\{ \int_{0}^{\infty} \exp\left[(-iwk) - \left(\frac{1}{\alpha}|k|\right)\right] dk + \int_{-\infty}^{0} \exp\left[(-iwk) - \left(|k|\frac{1}{\alpha}\right)\right] dk \right\}$$

$$= \frac{\sigma^2 \alpha}{\pi(1 + \alpha^2 w^2)}$$

Example 11.8: Consider an autocorrelation function which is periodic in lag measured in months. Assume the function has the form:

$$\rho(\tau) = \frac{1}{\pi} \cos\left(\frac{\pi\tau}{12}\right)$$

where τ is the lag in months. Compute the spectral density function and what does it tell?

Solution: Substitute given function into equation (11.79), one gets

$$S(w) = \frac{1}{2\pi} \int_{-\infty}^{\infty} \frac{1}{\pi} \cos\left(\frac{\pi\tau}{12}\right) \cos(w\tau) d\tau$$

$$= \left(\frac{1}{2\pi}\right)^2 \int\limits_{-\infty}^{\infty} \cos\left(\frac{\pi\tau}{12} + w\tau\right) + \cos\left(\frac{\pi\tau}{12} - w\tau\right) d\tau$$

$$= \left(\frac{1}{2\pi}\right)^2 \frac{2}{\left(\frac{\pi}{12}\right)^2 - w^2}$$

11.3 Spectral analysis using maximum entropy

In spectral analysis, the main goal is twofold: 1) determination of the spectrum and 2) extension of the spectrum beyond the length of data. The first goal can be accomplished in three ways: 1) application of the principle of maximum entropy (POME) to Burg entropy, 2) consideration of the spectrum as a PDF, and 3) use of linear filter. Before discussing these methods, we discuss the reasoning for treating spectral densities as PDFs. Consider (frequency) w as random variable. Because the power spectrum (non-negative) is normalized to have an area equal to one, it can be perceived as the PDF. It is known that the Fourier transform pair (autocovariance function and power spectrum), $r(\tau) \leftrightarrow S(w)$, has the same properties as those of the pair of characteristic function $\leftrightarrow$ PDF of a random variable differing only by a scaling factor. To illustrate this point, let us consider another random variable φ whose characteristic function satisfies $\varphi(1) = \varphi(0) = 0$. Random variable w is independent of φ with an event density function $f(w)$. Let a stochastic process be formed as

$$X(t) = a\cos(wt + \varphi)$$

where a is coefficient. If $X(t)$ is wide sense stationary, it has a power spectrum expressed as $S(w) = \pi a^2 f(w)$. This shows that the power spectrum is identical to an underlying PDF of the frequency of the process multiplied by a constant.

Now consider a more general case entailing two bounded signals: deterministic $x(t)$ and stochastic $\varphi(t)$. Both signals can be related to one another one on one through the transformation:

$$x(t) = A\sin\varphi(t), \quad |x(t)| \leq A, \quad |\varphi(t)| \leq \pi/2, \quad \forall(t)$$

where A is coefficient. Then, the stochastic process $X(t)$, composed of realizations as $x_1(t), x_2(t), \ldots, x_N(t)$, has a one on one relation with the stochastic process $\varphi(t)$ composed of realizations $\varphi_1(t), \varphi_2(t), \ldots, \varphi_N(t)$ as:

$$x_1(t) = A\sin(\varphi_1(t)), \quad x_2(t) = A\sin(\varphi_2(t)), \ldots, \quad x_N(t) = A\varphi_N(t)$$

Therefore, the two random processes, $X(t)$ and $\varphi(t)$, are connected one on one to each other as

$$X(t) = A\sin\varphi(t)$$

Tzannes et al. (1985) show that the normalized power spectrum of any wide sense stationary stochastic process can be considered as equivalent to the first-order probability density function (PDF) of the underlying stochastic process, that is, the instantaneous frequency process. For the stochastic process $X(t)$, its instantaneous frequency can be expressed as

$$w(t) = \frac{d\varphi(t)}{dt}$$

and the PDF of $w(t)$, denoted as $P(w)$, can be assumed as time invariant and an even function of w. Then the power spectral density of $X(t)$ can be shown to be

$$S(w) = \pi A^2 P(w)$$

Thus the normalized power spectrum of a wide stationary random process is equivalent to the first order PDF of an underlying process, namely the instantaneous frequency process of $X(t)$. This discussion establishes a basis for the interpretation of normalized spectral density as PDF and it can be so treated in entropy-based spectral analysis.

The power spectrum $S(w)$ of a band limited stationary stochastic process is related to its autocorrelation function $\rho(\tau)$ by Fourier transform. It is easy to determine $\rho(\tau)$ from data. Spectral analysis techniques are based on two concepts. In the first concept, the power spectrum is determined by taking the Fourier transform of the product of the autocorrelation function $\rho(\tau)$, $\tau \leq T$, and a window function $W(t)$, where $W(t) = 0$, $t > T$. The unknown power spectrum is then determined by making use of the convolution theorem which states that the Fourier transform of the product of two time functions equals the convolution of their Fourier transforms in the frequency domain. There are, however, two shortcomings of this concept. First, it assumes that $\rho(\tau) = 0$, $\tau > T$, which in general is not true. Second, it distorts the known values of $\rho(t)$. In the second concept, $\rho(\tau)$ is extended beyond T and then the power spectrum $S(w)$ is determined by taking the Fourier transform of the extended $\rho(\tau)$. However, it requires an extension of $\rho(\tau)$, $t > T$, and it is not clear how best to accomplish the extension.

For a given set of data, it is assumed that specific information is contained in the autocorrelation values. Further, power spectra must be non-negative. These two conditions can be satisfied by an infinite number of spectra. Thus, the goal is to find a single spectrum which will hopefully be representative of the class of all possible spectra. Burg (1967) proposed choosing such spectrum by maximizing entropy which will correspond to the most random or the most unpredictable time series whose autocorrelation function will concur with known values. The spectrum so derived will be most noncommittal to the unknown values of the autocorrelation function. Burg (1975) suggested extending $\rho(t)$ by maximizing entropy of the underlying stochastic process and thus he proposed maximum entropy spectral analysis (MESA). The principle of maximum entropy (POME) is applied in MESA somewhat indirectly. The Burg entropy expresses the entropy gain in a stochastic process that is transmitted through a linear filter having a characteristic function $Y(w)$, where $S(w) = |Y(w)|^2$. If the input is white noise then the output has spectral power density $s(w)$ or $g(w)$. The implication here is that the process entropy can be maximized by maximizing the entropy gain $H(w)$ of the filter that produces the process. Thus, the Burg entropy is maximized subject to constraints defined by autocorrelation functions.

Burg (1975) discussed desirable properties of entropy-based spectrum. First, the resolution of the maximum entropy spectral estimate is greater than that obtained by conventional methods. Second, since the most random time series has a white or flat spectrum, the maximum entropy spectrum is expected to be white, consistent with the autocorrelation values. The white spectrum usually has sharp spectral peaks of high resolution. Third, the maximum entropy spectra can be computed from known autocorrelation functions.

11.3.1 Burg method

Spectral analysis by the Burg method involves the following steps: 1) defining Burg entropy, 2) specification of constraints in terms of autocorrelation functions, 3) maximization of Burg entropy for obtaining the least-biased spectrum, 4) determining parameters, 5) determining the spectral density, 6) Levinson-Burg algorithm, and 7) extension of autocorrelation function.

Burg entropy

The Burg entropy, $H(w)$, discussed in Chapter 2, can be re-written as

$$H(w) = \int_{-W}^{W} \log[S(w)]\, dw \tag{11.93}$$

where w is the frequency, W is the band width, and $S(w)$ is the power spectrum at frequency w related to the autocovariance function ρ_n of the time series (n denoting the lag). It is assumed that the time series is sampled at a uniform period of Δt, $W = 1/(2\Delta t) =$ the Nyquist frequency and the power spectrum of the series is band limited to $\pm W$. The objective is to maximize the Burg entropy subject to specified constraints.

Specification of constraints

For maximizing the Burg entropy given by equation (11.93), the constraints are given as autocorrelations:

$$\rho_n = \int_{-W}^{W} S(w) \exp(i2\pi w n\Delta t)\, dw, \quad -N \leq n \leq N \tag{11.94}$$

where N is the specified maximum lag.

Maximization of Burg entropy

First, the power spectrum is expressed in terms of the Fourier series:

$$S(w) = \frac{1}{2W} \sum_{n=-\infty}^{\infty} \rho_n \exp(-i2\pi w n\Delta t) \tag{11.95}$$

Substituting equation (11.95) in the Burg entropy in equation (11.93), one obtains

$$H(w) = \int_{-W}^{W} \log\left[\frac{1}{2W} \sum_{n=-\infty}^{\infty} \rho_n \exp(-i2\pi n w\Delta t) \right] dw \tag{11.96}$$

Differentiating equation (11.96) partially with respect to ρ_n, and equating to zero, where $|n| > N$,

$$\int_{-W}^{W} \left[\frac{1}{2W} \sum_{n=-\infty}^{\infty} \rho_n \exp(-i2\pi n w\Delta t) \right]^{-1} \frac{1}{2W} \exp(-i2\pi n w\Delta t)\, dw$$

$$= \frac{1}{2W} \int_{-W}^{W} \frac{1}{S(w)} \exp(-i2\pi n w\Delta t)\, dw = 0, \quad |n| > N \tag{11.97}$$

Expanding $1/S(w)$ in a Fourier series, one gets

$$\frac{1}{S(w)} = \sum_{n=-\infty}^{\infty} C_n \exp(-i2\pi n w\Delta t) \tag{11.98}$$

where C_n are undetermined coefficients. Equation (11.98) shows that $C_n = 0$ for $|n| > N$. This means that the Fourier series of the function $1/S(w)$ truncates at $n = N$. Then, equation (11.98) yields the power spectrum as

$$S(w) = \frac{1}{\displaystyle\sum_{n=-N}^{N} C_n \exp(-i2\pi nw\Delta t)} \tag{11.99}$$

Equation (11.99) can also be derived using the method of Lagrange multipliers where C_n will be the Lagrange multipliers.

Example 11.9: Derive equation (11.99) using the method of Lagrange multipliers.

Solution: Denote the spectrum as $S(f)$, where $f = 2\pi w$ and w is the frequency. The autocorrelations are denoted as $r(k)$, $k = -n, \ldots, -1, 0, 1, \ldots, n$. Then the problem can be formalized by maximizing the entropy function H:

$$H = -\int_{-\pi}^{\pi} \log S(f)\,df$$

subject to the constraints:

$$r(k) = \int_{-\pi}^{\pi} S(f) \exp(ifk)df, \quad k = -n, \ldots, -1, 0, 1, \ldots, n$$

By introducing the Lagrange multipliers λ_k ($k = -n, \ldots, -1, 0, 1, \ldots, n$), the Lagrange function can be defined as:

$$L = \int_{-\pi}^{\pi} \log S(f)\,df + \sum_{k=-n}^{n} \lambda_k \left[r(k) - \int_{-\pi}^{\pi} S(f) \exp(ifk)df \right]$$

Taking derivative of L with respect to the spectra $S(f)$ and setting the derivatives to be zero, one obtains the spectrum as:

$$S(f) = \frac{1}{\displaystyle\sum_{k=-n}^{n} \lambda_k \exp(ifk)} = \frac{1}{\displaystyle\sum_{k=-n}^{n} \lambda_k \exp(-i2\pi wk)}, \quad k = -n, \ldots, -1, 0, 1, \ldots, n$$

Determination of parameters C_n

Burg (1975) determined parameters C_n in two ways: 1) direct integration and 2) z-transform. The direct integration method is followed here. Substitution of equation (11.99) in equation (11.94) yields

$$\int_{-W}^{W} \frac{\exp(i2\pi nw\Delta t)}{\displaystyle\sum_{n=-N}^{N} C_s \exp(-i2\pi wn\Delta t)}\,dw = \rho_n, \quad -N \le n \le N \tag{11.100}$$

Let $z = \exp(-i2\pi w\Delta t)$ (the z-transform notation). Then,

$$dz = -i2\pi \Delta t \exp(-i2\pi w\Delta t)\, dw = -i2\pi \Delta t z\, dw \tag{11.101a}$$

or

$$dw = -\frac{dz}{i2\pi \Delta t z} \tag{11.101b}$$

Substituting these quantities in equation (11.100), one gets

$$\int_{-W}^{W} \frac{z^{-n}}{\displaystyle\sum_{n=-N}^{N} C_s z^n}\, dw = \frac{1}{2\pi i \Delta t} \oint \frac{z^{-n-1} dz}{\displaystyle\sum_{n=-N}^{N} C_n z^n} = \rho_n \tag{11.102}$$

in which the contour integral is around the unit circle in the counter-clock wise direction. Since $S(w)$ must be real and positive for $|z| = 1$, it is possible to write from equation (11.99) using the z-transform notation ($z = \exp(-i2\pi w\Delta t)$):

$$\sum_{n=-N}^{N} C_n z^n = [P_N \Delta t]^{-1} [1 + a_1 z + a_2 z^2 + \ldots + a_N z^N][1 + a_1^* z^{-1} + a_2^* z^{-2} + \ldots + a_N^* z^{-N}]$$

$$= [P_N \Delta t]^{-1} \sum_{n=0}^{N} a_n z^n \sum_{n=0}^{N} a_n^* z^{-n} \tag{11.103}$$

in which $P_N > 0$ and $a_0 = 1$. All of the roots of the first polynomial in z in equation (11.103) can be chosen to lie outside of the unit circle; thus all of the roots of the second polynomial will be inside the unit circle. Inserting equation (11.103) in equation (11.102), one obtains:

$$\frac{P_N}{2\pi \Delta t} \oint \frac{z^{-n-1} dz}{\displaystyle\sum_{n=0}^{N} a_s z^n \sum_{n=0}^{N} a_s^* z^{-n}} = \rho_n, \quad -N \leq n \leq N \tag{11.104}$$

Forming the summations,

$$\sum_{n=0}^{N} a_n^* \rho_{n-j} = \frac{P_N}{2\pi \Delta t} \oint \frac{z^{j-1} \displaystyle\sum_{n=0}^{N} a_n^* z^{-n}}{\displaystyle\sum_{n=0}^{N} a_n z^n \sum_{n=0}^{N} a_n^* z^{-n}}\, dz = \frac{P_N}{2\pi \Delta t} \oint \frac{z^{j-1}}{\displaystyle\sum_{n=0}^{N} a_n z^n}\, dz, \quad j \geq 0 \tag{11.105}$$

Note one form of Cauchy's integral formula:

$$\frac{1}{2\pi \Delta t} \oint \frac{g(z)\, dz}{z} = g(0) \tag{11.106}$$

in which $g(z)$ is analytic on and inside the contour of integration. With the use of this formula and noting that $a_0 = 1$, it is observed that equation (11.105) equals P_N for $j = 0$. Using the complex conjugate of equation (11.105), the result is the prediction error filter equation:

$$\sum_{n=0}^{N} \rho_{j-n} a_n = P_N, \ j = 0 \tag{11.107a}$$

and

$$\sum_{n=0}^{N} \rho_{j-n} a_n = 0, \ j \geq 1 \tag{11.107b}$$

For $0 \leq j \leq N$, equations (11.107a) and (11.107b) can be written in matrix form as

$$\begin{bmatrix} \rho_0 & \rho_{-1} & \cdot & \cdot & \rho_{-N} \\ \rho_1 & \rho_0 & \cdot & \cdot & \rho_{1-N} \\ \cdot & & \cdot & & \cdot \\ \cdot & & \cdot & \rho_0 & \cdot \\ \rho_N & \rho_{N-1} & \cdot & \cdot & \rho_0 \end{bmatrix} \begin{Bmatrix} 1 \\ a_1 \\ \cdot \\ \cdot \\ a_N \end{Bmatrix} = \begin{Bmatrix} P_N \\ 0 \\ \cdot \\ \cdot \\ 0 \end{Bmatrix} \tag{11.108}$$

Equation (11.108) is the N-th order prediction error filter and can be solved for P_N and a_n $(n = 1, 2, \ldots, N)$ using the Levinson-Burg algorithm. Note that substitution of equations (11.95) and (11.103) into equation (11.99) yields

$$S(w) = \frac{P_N \Delta t}{\displaystyle\sum_{s=0}^{N} a_s z^s \sum_{n=0}^{N} a_n^* z^{-n}} = \frac{1}{2W} \sum_{r=-\infty}^{\infty} \rho_r z^r \tag{11.109}$$

Substitution of values obtained from the Levinson-Burg algorithm into equation (11.109) yields the maximum entropy spectrum.

Solution by the Levinson-Burg algorithm
It is assumed that the N by N Toeplitz matrix in equation (11.108) is positive definite and the full $N + 1$ by $N + 1$ Toeplitz matrix is at least nonnegative definite. This means that equation (11.108) will have a unique solution. The algorithm is recursive wherein the solution of the $N + 1$th set of equations is obtained from the solution of the Nth set of equations. Thus, from the solution of N equations:

$$\begin{bmatrix} \rho_0 & \rho_{-1} & \cdot & \cdot & \rho_{1-N} \\ \rho_1 & \rho_0 & \cdot & \cdot & \rho_{2-N} \\ \cdot & & \cdot & & \cdot \\ \cdot & & \cdot & \rho_0 & \cdot \\ \rho_{N-1} & \rho_{N-2} & \cdot & \cdot & \rho_0 \end{bmatrix} \begin{Bmatrix} 1 \\ b_1 \\ \cdot \\ \cdot \\ b_{N-1} \end{Bmatrix} = \begin{Bmatrix} P_{N-1} \\ 0 \\ \cdot \\ \cdot \\ 0 \end{Bmatrix} \tag{11.110a}$$

the solution of $N + 1^{\text{st}}$ equations is obtained by evaluating the matrix equation:

$$\begin{bmatrix} \rho_0 & \rho_{-1} & \cdot & \cdot & \rho_{-N} \\ \rho_1 & \rho_0 & \cdot & \cdot & \rho_{1-N} \\ \cdot & & \cdot & & \cdot \\ \rho_{N-1} & \rho_{N-2} & \cdot & \rho_0 & \cdot \\ \rho_N & \rho_{N-1} & \cdot & \cdot & \rho_0 \end{bmatrix} \left[\begin{Bmatrix} 1 \\ b_1 \\ \cdot \\ b_{N-1} \\ 1 \end{Bmatrix} + C_N \begin{Bmatrix} 0 \\ b_{N-1}^* \\ \cdot \\ b_1^* \\ 1 \end{Bmatrix} \right] = \left[\begin{Bmatrix} P_{N-1} \\ 0 \\ \cdot \\ 0 \\ \Delta_N \end{Bmatrix} + C_N \begin{Bmatrix} \Delta_n^* \\ 0 \\ 0 \\ 0 \\ P_{N-1} \end{Bmatrix} \right] \tag{11.110b}$$

where $b*_1, \ldots, b*_{N-1}$ and $\Delta*n$ denotes the conjugate reverse of $b_1, \ldots, b_{N-1}$ and Δn. In equation (11.110b), the second column vector on both sides is the simple complex conjugate reverse of the first column. The autocorrelation matrix has the last row as complex conjugate reverse of the first row, the second to the last row as reverse of the second row, and so on. From equation (11.111b),

$$\Delta_N = \sum_{n=0}^{N-1} \rho_{N-n} b_n \tag{11.111}$$

where $b_0 = 1$. The Nth order reflection coefficient C_N is specified as

$$C_N = -\frac{\Delta_N}{P_{N-1}} \tag{11.112}$$

Thus, Nth order prediction filter is obtained from C_N and the N-1 the order filter as:

$$\begin{Bmatrix} 1 \\ a_1 \\ \cdot \\ \cdot \\ \cdot \\ a_{N-1} \\ a_N \end{Bmatrix} = \begin{Bmatrix} 1 \\ b_1 \\ \cdot \\ \cdot \\ \cdot \\ b_{N-1} \\ 0 \end{Bmatrix} + C_N \begin{Bmatrix} 0 \\ b^*_{N-1} \\ \cdot \\ \cdot \\ \cdot \\ b^*_1 \\ 1 \end{Bmatrix} \tag{11.113}$$

The value of P_N is now obtained as

$$P_N = P_{N-1}\left(1 - C_N C^*_N\right) = P_{N-1}\left(1 - |C_N|^2\right) \tag{11.114}$$

The recursive algorithm is often started with $P_0 = \rho_0$ and the zeroth order prediction error filter which is one point filter with unit weight.

Extension of autocorrelation function

The autocorrelation function values for lag n are known, that is, $|n| \leq N$, and are to be determined for n greater than N. Entropy maximization leads to

$$\begin{bmatrix} \rho_0 & \rho_1 & \cdot & \cdot & \cdot & \rho_{N-1} & \rho_N \\ \rho_1 & \rho_0 & \rho_1 & \cdot & \cdot & \cdot & \rho_{N-1} \\ \cdot & \cdot & \cdot & \cdot & \cdot & & \cdot \\ \cdot & \cdot & \cdot & \cdot & \cdot & & \cdot \\ \cdot & \cdot & \cdot & \cdot & \cdot & & \cdot \\ \rho_{N-1} & \rho_{N-1} & \cdot & \cdot & \cdot & \rho_0 & \rho_1 \\ \rho_N & \rho_{N-1} & \cdot & \cdot & \cdot & \rho_1 & \rho_0 \end{bmatrix} \begin{Bmatrix} 1 \\ C_1 \\ \cdot \\ \cdot \\ \cdot \\ C_{N-1} \\ C_N \end{Bmatrix} = \begin{Bmatrix} P_N \\ 0 \\ \cdot \\ \cdot \\ \cdot \\ \cdot \\ \cdot \end{Bmatrix} \tag{11.115}$$

where P_N is one step ahead prediction error power, and C_N are the coefficients of filter. Both P_N and C_N are estimated recursively. Then the spectrum is estimated as

$$S(w) = \frac{P_N}{2W\left|1 + \sum_{j=1}^{N} C_j \exp(-j2\pi wj)\right|^2} \tag{11.116}$$

where W is the Nyquist frequency.

Example 11.10: Select the first eight years of annual stream flow data (1941–8) of the Brazos River near Bryan, Texas, used in Example 11.1 and compute the maximum entropy spectrum.

Solution: First, the time series of stream flow is logarithmically transformed is taken and the mean of the time series is than subtracted. Assume the autocorrelation is known up to lag-2 ($m = 2$), which are $\rho_0 = 1.0000$, $\rho_1 = 0.0899$; $\rho_2 = -0.2112$. Second, the Levinson-Burg algorithm is applied to solve for the Lagrange multipliers as illustrated below.

Three equations are used sequentially to obtain Δ_N, C_N, and P_N, which are expressed as:

$$\Delta_N = \sum_{n=0}^{N-1} \rho_{N-n} b_n \tag{11.117}$$

$$C_N = -\frac{\Delta_N}{P_{N-1}} \tag{11.118}$$

$$P_N = P_{N-1} \left(1 - C_N C_N^*\right) = P_{N-1}\left(1 - |C_N|^2\right) \tag{11.119}$$

Step 1: Equation (11.110a) can be written as:

$$[\rho_0][1] = [P_0]$$

From this equation, one obtains: $P_0 = 1$. Note that $b_0 = 1$.

Step 2: Equation (11.110b) can be written as:

$$\begin{bmatrix} \rho_0 & \rho_1 \\ \rho_1 & \rho_0 \end{bmatrix}\left[\begin{pmatrix} 1 \\ 0 \end{pmatrix} + C_1 \begin{pmatrix} 0 \\ 1 \end{pmatrix}\right] = \begin{bmatrix} P_0 \\ \Delta_1 \end{bmatrix} + C_1 \begin{bmatrix} \Delta_1 \\ P_0 \end{bmatrix}$$

Using equations (11.117), (11.118) and (11.119),

$$\Delta_1 = \rho_1 b_0 = \rho_1$$

$$C_1 = -\frac{\Delta_1}{P_0} = -\rho_1$$

$$P_1 = P_0 \left(1 - |C_1|^2\right) = 1 - \rho_1^2$$

Since

$$\begin{pmatrix} 1 \\ a_1 \end{pmatrix} = \begin{pmatrix} 1 \\ 0 \end{pmatrix} + C_1 \begin{pmatrix} 0 \\ 1 \end{pmatrix}$$

one obtains:

$$a_1 = C_1 = -\rho_1$$

Step 3:
From the results in Step 2, equation (11.110a) can be written as:

$$\begin{bmatrix} \rho_0 & \rho_1 \\ \rho_1 & \rho_0 \end{bmatrix}\begin{bmatrix} 1 \\ a_1 \end{bmatrix} = \begin{bmatrix} P_1 \\ 0 \end{bmatrix}$$

Then equation (11.110b) can be written as:

$$
\begin{bmatrix} \rho_0 & \rho_{-1} & \rho_{-2} \\ \rho_1 & \rho_0 & \rho_{-1} \\ \rho_2 & \rho_1 & \rho_0 \end{bmatrix} \left[\begin{pmatrix} 1 \\ b_1 \\ 0 \end{pmatrix} + C_2 \begin{pmatrix} 0 \\ b_1 \\ 1 \end{pmatrix} \right] = \begin{bmatrix} P_1 \\ 0 \\ \Delta_2 \end{bmatrix} + C_2 \begin{bmatrix} \Delta_2 \\ 0 \\ P_1 \end{bmatrix}
$$

where $b_1 = a_1 = -\rho_1$.

Using equations (11.117), (11.118), and (11.119), one obtains:

$$
\Delta_2 = \rho_2 b_0 + \rho_1 b_1 = \rho_2 + \rho_1 b_1
$$

$$
C_2 = -\frac{\Delta_2}{P_1} = -\frac{\rho_2 + \rho_1 b_1}{1 - \rho_1^2}
$$

$$
P_2 = P_1 \left(1 - |C_2|^2 \right) = (1 - \rho_1^2) \left[1 - \left(\frac{\rho_2 + \rho_1 b_1}{1 - \rho_1^2} \right)^2 \right]
$$

Since

$$
\begin{pmatrix} 1 \\ a_1 \\ a_2 \end{pmatrix} = \begin{pmatrix} 1 \\ b_1 \\ 0 \end{pmatrix} + C_2 \begin{pmatrix} 0 \\ b_1 \\ 1 \end{pmatrix}
$$

one obtains:

$$
a_1 = b_1 + C_2 b_1 = -\rho_1 \left(1 - \frac{\rho_2 + \rho_1 b_1}{1 - \rho_1^2} \right)
$$

$$
a_2 = C_2 = -\frac{\rho_2 + \rho_1 b_1}{1 - \rho_1^2}
$$

Substituting $\rho_0 = 1.0000$, $\rho_1 = 0.0899$; $\rho_2 = -0.2112$, one obtains:

$a_1 = -0.1098$, $a_2 = 0.2211$, and $P_2 = 0.9434$.

Step 4:

From the results in Step 3, equation (11.110a) can be written as:

$$
\begin{bmatrix} \rho_0 & \rho_{-1} & \rho_{-2} \\ \rho_1 & \rho_0 & \rho_{-1} \\ \rho_2 & \rho_1 & \rho_0 \end{bmatrix} \begin{bmatrix} 1 \\ a_1 \\ a_2 \end{bmatrix} = \begin{bmatrix} P_2 \\ 0 \\ 0 \end{bmatrix}
$$

or

$$
\begin{bmatrix} 1.0000 & 0.0899 & -0.2112 \\ 0.0899 & 1.0000 & 0.0899 \\ -0.2112 & 0.0899 & 1.0000 \end{bmatrix} \begin{bmatrix} 1 \\ -0.1098 \\ 0.2211 \end{bmatrix} = \begin{bmatrix} 0.9434 \\ 0 \\ 0 \end{bmatrix}
$$

Then the maximum entropy spectrum can be expressed as: (use the actual values in the equation below.)

$$
S(f) = \frac{P_m}{\sum\limits_{s=0}^{m} a_s z^s \sum\limits_{n=0}^{m} a_n^* z^{-n}} = \frac{P_2}{\left| 1 + \sum\limits_{k=1}^{2} a_k \exp\left(-2\pi ikf \right) \right|^2}
$$

11.3.2 Kapur-Kesavan method

Kapur and Kesavan (1992) presented a method for maximum entropy spectral analysis. The method entails the same steps as in the Burg method.

Specification of constraints

Let the variates be $x_{-m}, x_{-m+1}, \ldots, x_0, x_1, \ldots, x_m$ or $X = \left[x_{-m}, x_{-m+1}, \ldots, x_0 x_1, \ldots, x_m \right]^T$. These variates and their variances and covariances vary from $-\infty$ to $+\infty$. The means of these variates are zero, for the mean is subtracted from the actual values. Since observations of these variates are assumed to be available, their autocovariance functions $m+1$ can be determined from data and these form the constraints:

$$r_k = \frac{1}{n} \sum_{j=0}^{n-k} x_j x_{j-k} \tag{11.120}$$

For real values which are being considered here, $r_{-k} = r_k$.

11.3.3 Maximization of entropy

From the discussion in Chapter 4, it is clear that POME-based probability density function with zero mean and nonzero autocovariances will be a multivariate normal which can be written for a random vector $Y = [y_1, y_2, \ldots, y_M]^T$ as

$$p(y) = \frac{1}{(2\pi)^{M/2} \left| \sum \right|^{1/2}} \exp\left[-\frac{1}{2} y^T \sum^{-1} y \right] \tag{11.121}$$

where $y = [y_1, y_2, \ldots, y_M]$ has zero mean and covariance matrix $\sum$ which will be clear in what follows. For random vector X, $x^T = \left[x_0\, x_1, \ldots, x_n \right]^T$, the entropy-based probability density function can be derived, using POME subject to equation (11.121) and the method of Lagrange multipliers, as

$$p(x) = \frac{1}{Z(\lambda)} \exp\left\{ -\left[\sum_{k=-m}^{m} \lambda_k \sum_{i=0}^{n-k} x_i xi + k \right] \right\}$$
$$= \frac{1}{Z(\lambda)} \exp\left[-\frac{1}{2} x^T \Lambda x \right] \tag{11.122}$$

where $Z(\lambda)$ is a partition function, a function of Lagrange multipliers $\lambda_{-m}, \ldots, \lambda_{-2}, \lambda_{-1}, \lambda_0, \lambda_1, \lambda_2, \ldots, \lambda_m$; and Λ is an equidiagonal $(m+1) \times (m+1)$ Toeplitz matrix of the Lagrange multipliers represented as

$$c_{ij} = \begin{bmatrix} \lambda_0 & \lambda_1 & \cdot & \cdot & \lambda_m \\ \cdot & \cdot & \cdot & \cdot & \cdot \\ \cdot & \cdot & \cdot & \cdot & \cdot \\ \cdot & \cdot & \cdot & \cdot & \cdot \\ \lambda_{-m} & \cdot & \cdot & \cdot & \lambda_0 \end{bmatrix} \tag{11.123}$$

It is noted that each entry at (i, j) in the matrix is c_{ij} where $c_{ij} = \lambda_{j-i}$ if $|j - i| \leq m$ and $c_{ij} = 0$ if $|j - i| > m$. In order for the matrix to be related to the constraints via the covariance matrix, it must be positive definite. Hence, matrix Λ is also a Hessian matrix.

The PDF given by equation (11.122) has the same form as equation (11.121). Therefore, X can be regarded as having a multivariate normal distribution (with zero mean) and covariance matrix Λ^{-1}. The covariance matrix $\sum$ is expressed as

$$\sum = \Lambda^{-1} \tag{11.124}$$

Comparing equation (11.124) with equation (11.122), the partition function $Z(\lambda)$ is obtained as:

$$Z(\lambda) = (2\pi)^{\frac{n+1}{2}} \left|\Lambda^{-1}\right|^{1/2} \tag{11.125}$$

It should be noted that equation (11.122) is valid only if the autocovariances r_k's are given such that the determinant of the autocovariance matrix is not zero, thus permitting the eigenvalues of the matrix to be positive and real.

Determination of Lagrange multipliers

The values of Lagrange multipliers in the partition function $Z(\lambda) = Z(\lambda_{-m}, \ldots, \lambda_{-1}, \lambda_0, \lambda_1, \ldots, \lambda_m)$ can be obtained as follows. Taking the logarithm of equation (11.125) and denoting the real and positive eigenvalues of matrix Λ as $w_0, w_1, w_2, \ldots, w_n$, one obtains

$$\ln Z(\lambda) = \frac{n+1}{2} \ln(2\pi) - \frac{1}{2} \ln|\Lambda| \tag{11.126}$$

Equation (11.126) relates the eigenvalues to Lagrange multipliers or vice versa. Since Λ is a Toeplitz matrix, its eigenvalues can be expressed analytically in the limiting case $m \ll n$. One can write the determinant of the covariance matrix as

$$|\Lambda| = \prod_{j=0}^{n} w_j \tag{11.127}$$

Inserting equation (11.127) in equation (11.126), one gets

$$\ln Z(\lambda) = \frac{n+1}{2} \ln(2\pi) - \frac{1}{2} \sum_{j=0}^{n} \ln w_j \tag{11.128}$$

For Toeplitz matrices, the eigenvalues can be expressed, if $n \gg m$, as

$$w_j = g\left(z_j\right) = \sum_{k=-m}^{m} \lambda_k z_j^k, \quad j = 0, 1, 2, \ldots, n \tag{11.129}$$

where z's are the $(n+1)$ roots of $z^{n+1} = 1$ lying on the unit circle in the complex plane. Because Λ is a Hermitian matrix (A matrix is a Hermitian matrix if the element a_{ij} in the i-th row and j-th column is equal to the complex conjugate of the element in the j-th row and i-th column) and all the eigenvalues $w_0, w_1, w_2, \ldots, w_n$ are real, one can write the roots as

$$z_j = \exp\left(\frac{2\pi ij}{n+1}\right), \quad j = 0, 1, \ldots, n; \quad i = \sqrt{-1} \tag{11.130}$$

Substituting equation (11.130) into equation (11.129), the result is

$$w_j = \sum_{k=-m}^{m} \lambda_k \exp\left(\frac{2\pi ijk}{n+1}\right) \tag{11.131}$$

Substitution of equation (11.131) in equation (11.128) produces

$$\ln Z(\lambda) \cong \frac{n+1}{2}\left\{\ln(2\pi) - \frac{1}{n+1}\sum_{j=0}^{n}\ln g\left[\exp\left(\frac{2\pi ij}{n+1}\right)\right]\right\} \tag{11.132}$$

Recall Riemann's definition of a definite integral:

$$\frac{1}{2\pi}\int_{0}^{2\pi} f(\theta)d\theta = \lim_{n\to\infty}\frac{1}{n+1}\sum_{j=0}^{n}f\left(\frac{2\pi j}{n+1}\right) \tag{11.133}$$

For $n \to \infty$, equation (11.132), with the use of equation (11.133) and $f(\theta) = g[\exp(i\theta)]$, can be approximated as

$$\frac{2}{n+1}\ln Z(\lambda) \cong \ln(2\pi) - \frac{1}{2\pi}\int_{0}^{2\pi}\ln g\left[\exp(i\theta)\right]d\theta \tag{11.134}$$

Equation (11.134), with the use of equation (11.129), becomes

$$\frac{2}{n+1}\ln Z(\lambda) \cong \ln(2\pi) - \frac{1}{2\pi}\int_{0}^{2\pi}\ln\left[\sum_{k=-m}^{m}\lambda_k\exp(ik\theta)\right]d\theta \tag{11.135}$$

Equation (11.135) is now differentiated with respect to z_k:

$$\frac{2}{n+1}\frac{\partial \ln Z(\lambda)}{\partial \lambda_k} \cong -\frac{1}{2\pi}\int_{0}^{2\pi}\frac{\partial}{\partial \lambda_k}\ln\left[\sum_{k=-m}^{m}\lambda_k\exp(ik\theta)\right]d\theta$$

$$= -\frac{1}{2\pi}\int_{0}^{2\pi}\frac{\exp(i\theta k)}{g[\exp(i\theta)]}d\theta, \quad k = 0, \pm1, \pm2, \ldots, \pm m \tag{11.136}$$

In the multivariate maximum entropy formulation, λ_0 is expressed in terms of the Lagrange multipliers $\lambda_1, \lambda_2, \ldots, \lambda_m$ and derivatives of λ_0 with respect to other Lagrange multipliers are equated to the constraints:

$$\frac{\partial \lambda_0}{\partial \lambda_k} = -r_k = -E\left[\sum_{j=0}^{n-k}x_j^*x_{j+k}\right] \tag{11.137}$$

In equation (11.122), $Z(\lambda)$ is the partition function and serves the same role as λ_0 and there constraints r_k are the same as $(n+1)/2\overline{r}_k$. Therefore, one obtains

$$\frac{\partial}{\partial \lambda_k}\ln Z(\lambda) = -\frac{n+1}{2}\overline{r}_k = -E\left(\sum_{j=0}^{n-k}x_j^*x_{j+k}\right) \tag{11.138}$$

Equation (11.137) becomes

$$\overline{r}_k = r_k\left(\frac{2}{n+1}\right) = \frac{1}{2\pi}\int_{0}^{2\pi}\frac{\exp(i\theta k)}{g[\exp(i\theta)]}d\theta = E\left[\frac{2}{n+1}\sum_{j=0}^{n-k}x_j^*x_{j+k}\right] \tag{11.139}$$

Equation (11.139) permits $(2m+1)$ equations that express autocovariances in terms of the Lagrange multipliers to be generated. The left side of equation (11.136) is obtained from data for all values of k between $-m$ and m. On the right side, $g[\exp(i\theta)]$ is a function of $\lambda_0, \lambda_1, \lambda_2, \ldots, \lambda_{m-1}, \lambda_m$. Thus, equation (11.139) leads to $(2m+1)$ equations which can be employed to determine $(2m+1)$ unknown Lagrange multipliers. Then, $Z(\lambda)$ is obtained and the POME-based distribution is given by equation (11.122) in terms of autocovariances.

Example 11.11: Compute the right side of equation (11.139) for the time series used in Example 11.4.

Solution: For the given stream flow data, covariances are computed first. From the equation:

$$r_k = \frac{1}{8}\sum_{j=0}^{n-k} x_j x_{j-k}$$

One obtains r_k as: $r_k =$ [0.3621 0.0326 -0.0765 -0.0203 0.0219], $k = 0, 1, \ldots, 4$. The right side of equation (11.139) can then be computed as:

$$\overline{r_k} = r_k\left(\frac{2}{8}\right)$$

The result for $k = 0, 1, \ldots 4$ is then obtained as:

$$\overline{r_k} = \quad [0.0905 \quad 0.0081 \quad -0.0191 \quad -0.0051 \quad 0.0055], \quad k = 0, 1, \ldots, 4$$

11.3.4 Determination of Lagrange multipliers λ_k

To solve for $\lambda_k s$, equation (11.139) is transformed as a contour integral over the unit circle in an anticlockwise direction in the complex plane. Let

$$z = \exp(i\theta), dz = i\exp(i\theta)\, d\theta \tag{11.140}$$

Substituting in equation (11.139), one obtains

$$\overline{r_k} = \frac{1}{2\pi i}\oint \frac{z^{k-1}}{g(z)}\,dz \tag{11.141}$$

where the closed contour C is the unit circle.

The term $g(z)$ in equation (11.141) can be factored as

$$g(z) = \sum_{k=-m}^{m}\lambda_k z^k = G_m(z)\, G_m^*(z^*), \quad zz^* = 1 \tag{11.142}$$

where

$$G_m(z) = g_0 + g_1 z^{-1} + g_2 z^{-2} + \ldots + g_m z^{-m}$$

$$G_m^*(z^*) = g_0^* + g_1^* z + g_2^* z^2 + \ldots + g_m^* z^m$$

since on a unit circle, $zz^* = 1$, where z^* is the conjugate of z. Terms g's are selected such that $G_m(z)$ comprises all its zeros inside the unit circle (minimum phase polynomial) and $G^*_m(z^*)$ comprises all its zeros outside the unit circle (maximum phase polynomial).

When coefficients in equation (11.140) are equated to those in equation (11.142), one obtains

$$\lambda_k = \sum_{j=0}^{m-k} g_j g_{j+k}^* \tag{11.143}$$

and $\lambda_{-k} = \lambda_k^*$. To solve for g_is so that λ_k can be determined, substitute equation (11.143) in equation (11.141),

$$\overline{r_k} = \frac{1}{2\pi i} \oint z^{k-1} \left[G_m(z) G_m^*(z^*) \right]^{-1} dz \tag{11.144}$$

This can also be written as

$$\sum_{j=0}^{m} g_j \overline{r_{k-j}} = \frac{1}{2\pi i} \oint \frac{\left(\sum_{j=0}^{m} g_j z^{-j} \right) z^{k-1}}{G_m(z) G_m^*(z^*)} dz$$

$$= \frac{1}{2\pi i} \oint \frac{z^{k-1}}{g_o^* + g_1^* z + \ldots + g_m^* z^m} dz \tag{11.145}$$

Since $G_m^*(z^*)$ comprises zeros only outside the unit circle, the integral in equation (11.145) can be evaluated as:

$$\frac{1}{2\pi i} \oint \frac{z^{k-1}}{G_m^*(z^*)} dz = \begin{cases} 0 & k > 0 \\ \frac{1}{g_0^*} & k = 0 \end{cases} \tag{11.146a}$$

Hence,

$$\sum_{j=0}^{m} g_j \overline{r_{k-j}} = \begin{cases} \frac{1}{g_o^*} & k > 0 \\ 0 & k = 0 \end{cases} \tag{11.146b}$$

Equation (11.146b) leads to $(m+1)$ equations which can be solved for g_js, since autocovariances are known from data.

Let

$$a_j = \frac{g_j}{g_0} \tag{11.147}$$

Then equation (11.146b) can be expressed as

$$\begin{bmatrix} \overline{r_0} & \overline{r_{-1}} & \cdots & \overline{r_{-m}} \\ \overline{r_1} & \overline{r_0} & \cdots & \overline{r_{-m+1}} \\ \cdot & & & \cdot \\ \cdot & & & \cdot \\ \cdot & & & \cdot \\ \overline{r_m} & \overline{r_{m-1}} & \cdots & \overline{r_0} \end{bmatrix}_{(m+1)\times(m+1)} \begin{bmatrix} 1 \\ a_1 \\ \cdot \\ \cdot \\ \cdot \\ a_m \end{bmatrix}_{(m+1)\times 1} = \begin{bmatrix} 1/g_0 g_0^* \\ 0 \\ \cdot \\ \cdot \\ \cdot \\ 0 \end{bmatrix}_{(m+1)\times 1} \tag{11.148a}$$

or in matrix form:

$$R_m a = s_m \tag{11.148b}$$

The autocovariance matrix in equation (11.148a) is Hermitian and Toeplitz.

Recall that equation (11.129) can be recast as

$$\sum_{k=-m}^{m} \lambda_k z^k = g_0 g_0^* \left(1 + a_1 z^{-1} + \ldots + a_m z^{-m}\right)\left(1 + a_1^* z + \ldots + a_m^* z^m\right) \tag{11.149}$$

Comparing the coefficients of various powers of z,

$$\frac{\lambda_0}{g_0 g_0^*} = 1 + a_1 a_1^* + a_2 a_2^* + \ldots + a_m a_m^*$$

$$\frac{\lambda_1}{g_0 g_0^*} = a_1^* + a_1 a_2^* + a_2 a_3^* + \ldots + a_{m-1} a_m^* \tag{11.150a}$$

$$\vdots$$

$$\frac{\lambda_m}{g_0 g_0^*} = a_m^*$$

More compactly,

$$\lambda_k = g_0 g_0^* \sum_{j=0}^{m-k} a_j a_{j+k}^*, \quad k = 0, 1, 2, \ldots, m \tag{11.150b}$$

Equation (11.150b) permits a determination of $\lambda_{-m}, \lambda_{-m+1}, \ldots, \lambda_1, \lambda_2, \ldots, \lambda_m$ in terms of $g_0, g_0^*, a_1, a_1^*, \ldots, a_m, a_m^*$ which themselves are determined in terms of autocovarainces from equation (11.148).

Example 11.12: Compute the Lagrange multipliers for the time series used in Example 11.4.

Solution: For $m = 2$, the matrix in equation (11.148a) can be solved to obtain the parameters $g_0, g_0^*, a_1, \ldots, a_m$:

$$\begin{bmatrix} 0.0905 & 0.0081 & -0.0191 \\ 0.0081 & 0.0905 & 0.0081 \\ -0.0191 & 0.0081 & 0.0905 \end{bmatrix} \begin{Bmatrix} 1 \\ a_1 \\ a_2 \end{Bmatrix} = \begin{Bmatrix} 1/g_0 g_0^* \\ 0 \\ 0 \end{Bmatrix}$$

The solution is:

$$a_1 = -0.1098 \quad a_2 = 0.2211 \quad g_0 g_0^* = 11.7079$$

Then, the Lagrange multipliers are computed as:

$$\lambda_0 = g_0 g_0^* \left(1 + a_1 a_1^* + a_2 a_2^*\right) = 12.4214$$
$$\lambda_1 = g_0 g_0^* \left(a_1 + a_1 a_2^*\right) = -1.5701$$
$$\lambda_2 = g_0 g_0^* \left(a_2^*\right) = 2.5883$$

Example 11.13: If X is not a complex number then represent equation (11.122).

Solution: From the Lagrange multipliers above, one can write the Lagrange multiplier matrix as:

$$\Lambda = \begin{bmatrix} \lambda_0 & \lambda_1 & \lambda_2 & 0 & 0 & 0 \\ \lambda_{-1} & \lambda_0 & \lambda_1 & \lambda_2 & 0 & 0 \\ \lambda_{-2} & \lambda_{-1} & \lambda_0 & \lambda_1 & \lambda_2 & 0 \\ \cdots & & & \cdots & & \cdots \\ 0 & 0 & 0 & \lambda_{-1} & \lambda_0 & \lambda_1 \\ 0 & 0 & 0 & \lambda_{-2} & \lambda_{-1} & \lambda_0 \end{bmatrix}_{(12\times12)}$$

where $\lambda_{-2} = 2.5883, \lambda_{-1} = -1.5701$, $\lambda_0 = 12.4214$, $\lambda_1 = -1.5701$, and $\lambda_2 = 2.5883$.

The eigenvalues can be obtained from equation (11.131) and then $|\Lambda|$ can be computed as:

$$|\Lambda| = \prod_{j=1}^{8} w_j = 3.5106 \times 10^8$$

Then the joint distribution is given by:

$$p(y) = 12.0218 \exp\left[-\frac{1}{2} y^T \Lambda y\right]$$

11.3.5 Spectral density

From the limited data available in practice, the autocovariances that can be calculated are $\overline{r_k}$, $k = -m$ to m. The power spectrum $S(w)$ requires r_k, $k = -\infty$ to ∞, and hence cannot be calculated exactly. Thus, it is desired to estimate $S(w)$ with the least assumptions regarding the unknown autocovaraince functions. To that end, POME is utilized to obtain the maximum entropy spectral analysis (MESA). Recall that the spectral density at frequency w is the Fourier transform of the autocovariance functions,

$$S(w) = \sum_{n=-\infty}^{\infty} \overline{r_k} \exp(-i2\pi wn) \tag{11.151}$$

The autocovariance can be recovered by:

$$\overline{r_k} = \int_{-W}^{W} S(w) \exp(i2\pi wn)\,dw \tag{11.152}$$

Substituting $\theta = 2\pi w$ into equation (11.152), one obtains:

$$\overline{r_k} = \int_{-W}^{W} S\left(\frac{\theta}{2\pi}\right) \exp(i2\pi wn)\,dw \tag{11.153}$$

Comparison of equations (11.153) and (11.141) points to an estimate of spectral density function:

$$\hat{S}\left(\frac{\theta}{2\pi}\right) = \frac{1}{g\left[\exp(i\theta)\right]}, \quad i = \sqrt{-1} \tag{11.154a}$$

or

$$\hat{S}(w) = \frac{1}{g\left[\exp\left(i2\pi w\right)\right]} = \frac{1}{\displaystyle\sum_{k=-m}^{m} \lambda_k \exp\left(i2\pi kw\right)} \tag{11.154b}$$

If the Lagrange multipliers λ_k are known, which can be determined from autocovariances of data, then the spectral density function $\hat{S}(w)$ can be obtained from equation (11.154b). Note that $\lambda_k = \lambda_{-k}$, and $\lambda_{-k} = \lambda_k^*$, and $S(w)$ is real. Therefore,

$$\hat{S}(w) = \frac{1}{\displaystyle\sum_{k=-m}^{m} \lambda_k \exp\left(i2\pi kw\right)} = \frac{1}{\displaystyle\sum_{k=-m}^{m} \lambda_{-k} \exp\left(-i2\pi kw\right)} = \frac{1}{\displaystyle\sum_{k=-m}^{m} \lambda_k^* \exp\left(-i2\pi kw\right)} \tag{11.155}$$

Using equation (11.142), equation (11.155) can be written as

$$\hat{S}(w) = \frac{1}{G_m\left[\exp\left(i2\pi wk\right)\right] G_m^*\left[\exp\left(-i2\pi wk\right)\right]} \tag{11.156}$$

or

$$\hat{S}(w) = \frac{1}{\left|G_m\left[\exp\left(i2\pi wk\right)\right]\right|^2} \tag{11.157}$$

Equation (11.157) can be expressed as

$$\hat{S}(w) = \frac{1}{\left|g_0\right|^2 \left|1 + a_1 \exp\left(-i2\pi w\right) + a_2 \exp\left(-i4\pi w\right) + \ldots + a_m \exp\left(-i2m\pi w\right)\right|} \tag{11.158}$$

where a_i's are the same as in equation (11.147). Equation (11.158) can be further simplified as

$$\hat{S}(w) = \frac{1}{\left|g_0\right|^2} \left[1 + \sum_{j=1}^{m} a_j^2 + 2\sum_{j=1}^{m} a_j \cos\left(2\pi jw\right) + 2\sum_{j=1}^{m}\sum_{k>j}^{m} a_j a_k \cos\left(2\pi w \left|k-j\right|\right)\right]^{-1} \tag{11.159}$$

The power spectrum given by equation (11.159) is said to be most unbiased, smoothest, most random, uniform, and consistent with the given information. Also, an explicit connection between the power spectrum and Lagrange multipliers is found. If additional constraints are specified then it turns out that their Lagrange multipliers tend to be zero, and they contribute little to the structure of the power spectrum as there will be less poles within the unit circle in equation (11.159).

The assumption underlying equation (11.159) is that $n \gg m$ and equation (11.126) is approximated by equation (11.135). If the approximation is not good then equation (11.159) will, in general, not be a maximum entropy estimate. Then an exact solution can be derived as follows:

$$\overline{r_k} = -\frac{2}{n+1}\left(-\frac{1}{2}\frac{\partial}{\partial \lambda_k}\ln|\Lambda|\right) = -\frac{1}{n+1}\frac{\partial}{\partial \lambda_k}\ln|\Lambda|, \quad k = 0, \pm 1, \pm 2, \ldots, \pm m \tag{11.160}$$

It involves estimating λ_k directly from equation (11.126) along with equations (11.122) and (11.138). From equation (11.160), $(2m+1)$ equations are generated for determining the

$(2m + 1)$ Lagrange multipliers. Assuming that $\lambda_k = 0$ for $(m + 1) \leq n$, equation (11.151), with the use of equation (11.160), becomes

$$S(w) = \frac{1}{n+1} \sum_{k=-n}^{n} \exp(-i2\pi wk) \frac{\partial}{\partial \lambda_k} \ln |\Lambda|, \quad |w| \leq \frac{1}{2} \tag{11.161}$$

A major drawback here is that the partial derivatives of a finite dimensional Toeplitz matrix must be calculated. This can be done using the identity:

$$\frac{\partial \ln |\Lambda|}{\partial \lambda_{ij}} = \frac{1}{|\Lambda|} (-1)^{i+j} \left| \Lambda_{ij} \right| \tag{11.162}$$

where λ_{ij} is the (i, j) entropy in matrix Λ and its associated co-factor is $(-1)^{i+j} |\Lambda_{ij}|$. Because $\lambda_{-k} = \lambda_k$ for real time series, Λ is symmetric and Toeplitz. This means that $m(m+1)/2$ out of $(m+1)^2$ partial derivatives need to be evaluated:

$$\frac{\partial \ln |\Lambda|}{\partial \lambda_{ij}} = \frac{\partial \ln |\Lambda|}{\partial \lambda_{ji}} \tag{11.163}$$

The partial derivatives with respect to λ_k appear along one diagonal above and one diagonal below the main diagonal of Λ for real time series, except for $k = 0$, as shown by equation (11.122).

Now the appropriate co-factors can be summed up to compute the needed partial derivatives:

$$\frac{\partial \ln |\Lambda|}{\partial \lambda_k} = \begin{cases} \dfrac{2}{|\Lambda|} \sum_{i=1}^{n+1-k} (-1)^{2i+k} \left| \Lambda_{i,i+k} \right|, & k > 0 \\[4mm] \dfrac{1}{|\Lambda|} \sum_{i=1}^{n+1-k} (-1)^{2i+k} \left| \Lambda_{i,i+k} \right|, & k = 0 \end{cases} \tag{11.164}$$

If equation (11.164) proves to be difficult, then equation (11.162) can be employed.

Example 11.14: Compute the spectral density of the stream flow time series used in Example 11.11.

Solution: The spectrum is given by:

$$\hat{S}(w) = \frac{1}{|g_0|^2 \left| 1 + a_1 \exp(-i2\pi w) + a_2 \exp(-i4\pi w) + \ldots + a_m \exp(-2m\pi w) \right|}$$

$$= \frac{1}{|g_0|^2 \left| 1 + a_1 \exp(-i2\pi w) + a_2 \exp(-i4\pi w) \right|^2}$$

From Example 11.10, the parameter a is obtained as $a_1 = -0.1098$ and $a_2 = 0.2211$. From the results given in the previous example, the spectral density can be expressed as:

$$\hat{S}(w) = \frac{1}{11.7079 \left| 1 - 0.1098 \exp(-i2\pi w) + 0.2211 \exp(-i4\pi w) \right|^2}$$

The spectral density is plotted as below:

11.3.6 Extrapolation of autocovariance functions

Equation (11.139) is valid for $m \ll n$ for all k, including $|k| > m$. Solution of this matrix equation (11.148a) or (11.148b) yields λ. For extrapolating for the expected value of the autocovariance function beyond lag m, $\overline{r_k}$, an extra row is added to matrix R_m for each lag beyond m:

$$R_{m+1}a = S_{m+j}, \quad j = 1, 2, \ldots, n-m \tag{11.165}$$

One can then write just the extra equations that need to be evaluated:

$$\overline{r_{m+1}}a_0 + \overline{r_m}a_1 + \ldots + \overline{r_1}a_m = 0$$

$$\overline{r_{m+2}}a_0 + \overline{r_{m+1}}a_1 + \ldots + \overline{r_2}a_m = 0$$

$$\ldots$$

$$\ldots$$

$$\ldots$$

$$\overline{r_n}a_0 + \overline{r_{N-1}}a_1 + \ldots + \overline{r_{n-m}}a_m = 0 \tag{11.166}$$

Equations (11.166) can be solved sequentially, since each would contain only one unknown. One can indeed write the recursive relation:

$$\overline{r_k} = -\left(a_1\overline{r_{k-1}} + a_2\overline{r_{k-2}} + \ldots + a_j\overline{r_{k-j}} + a_{j+1}\overline{r_{j+1}} + \ldots + a_m\overline{r_{k-m}}\right) \tag{11.167}$$

for $k > m$ and $(k-j) > m$. However, knowledge of these extra autocovaraince functions does not change $p(x)$.

Example 11.15: Extend the autocovariance functions.

Solution: The auto-correlation functions are extended through the equation below:

$$\overline{r_3} = -\left(a_1\overline{r_2} + a_2\overline{r_1}\right) = -0.0039$$

$$\overline{r_4} = -\left(a_1\overline{r_3} + a_2\overline{r_2}\right) = 0.0038$$

From observations, we get $r_3 = -0.0051$ and $r_4 = 0.0055$. It can be seen that the autocovariance function for lags 3 and 4 from the maximum entropy extrapolation is close to the values obtained from observations.

11.3.7 Entropy of power spectrum

Using equations (11.122), (11.126), and (11.135), entropy H can be written as

$$H = H_{\max} = \ln Z(\lambda) + \frac{1}{2}(n+1) = \frac{1}{2}(n+1)\ln(2\pi e) - \frac{n+1}{4\pi}\int_0^{2\pi} \ln g\left[\exp\left(i\theta\right)\right]d\theta$$

$$= \frac{1}{2}(n+1)[\ln(2\pi e) + \int_0^1 \ln S\left(w\right)dw] \tag{11.168}$$

Example 11.16: Compute entropy of the stream flow time series used in Example 11.8.

Solution: The entropy can be computed through the equation below:

$$H_{max} = \frac{1}{2}(n+1)\ln(2\pi e) - \frac{1}{2}\ln|\Lambda|$$

From Example 11.11,

$$|\Lambda| = \prod_{j=1}^{8} w_j = 3.5106 \times 10^8$$

Then the entropy can be obtained as:

$$H_{max} = 1.5133\,\text{Napier}$$

11.4 Spectral estimation using configurational entropy

For spectral analysis Frieden (1972) suggested using entropy defined as

$$H(w) = \int_{-W}^{W} S(w)\log S(w)\,dw \qquad (11.169)$$

where W can be equal to π. Equation (11.169) defines what is called configurational (Gull and Daniell, 1978) or simply spectral entropy (Tzannes and Avgeris, 1981; Wu, 1983) and can be used in place of the Burg entropy. Of course, $S(w)$ is the (normalized) spectral power or spectral density (PDF of w). The problem is one of determining $S(w)$ by maximizing H given by equation (11.169), subject to given constraints. The spectral density so obtained will be optimum. The constraints can be defined in terms of the autocorrelation function:

$$r_k = \int_{-\pi}^{\pi} S(w)\exp(jwk)\,dw, \quad |k| = 0, 1, 2, \ldots, m \qquad (11.170)$$

The autocorrelation function r_k is the inverse Fourier transform or the characteristic function of the PDF $S(w)$ of random variable w. Also,

$$\int_{-\pi}^{\pi} S(w)\,dw = 1 \qquad (11.171)$$

The constraints can also be specified differently. Note that $r_{-k} = r_k$ and $S(w) = S(-w)$. $S(w), -\pi \leq w \leq \pi$, where $w = 2\pi f$ or $f = w/(2\pi)$, is to be determined.

For maximization of entropy, the method of Lagrange multipliers is used. The Lagrangean function L is constructed as

$$L = -\int_{-\pi}^{\pi} S(w)\log S(w)\,dw - \sum_{k=-m}^{m} \lambda_k \left[\int_{-\pi}^{\pi} S(w)\exp(jwk)\,dw - r_k \right] \qquad (11.172)$$

where λ_k are the Lagrange multipliers where $\lambda_{-k} = \lambda_k$, because $r_{-k} = r_k$. Differentiating equation (11.172) with respect to S(w) and equating the derivative to zero, the result is

$$\frac{\partial L}{\partial S(w)} = 0 = \int_{-\pi}^{\pi} \left\{ -[\log S(w) + 1] - \sum_{k=-m}^{m} \lambda_k \exp(jwk) \right\} dw \tag{11.173}$$

Therefore,

$$S(w) = \exp\left[-1 - \sum_{k=-m}^{m} \lambda_k \exp(jwk) \right] \tag{11.174}$$

This is the entropy-based spectral density function.

Now the Lagrange multipliers must be determined from the known constraints in terms of the autocorrelation function. Substitution of equation (11.174) in equation (11.170) yields

$$\int_{-\pi}^{\pi} \frac{\exp(jkw)\, dw}{\exp\left[1 + \sum_{k=-m}^{m} \lambda_k \exp(jkw) \right]} = r_k \tag{11.175}$$

Let $x = \exp(jw)$. Then $jw = \ln x$ and $dw = dx/(jx)$. With the substitution of these quantities, equation (11.175) becomes

$$\int_{-x_L}^{x_u} \frac{x^{k-1}\, dx}{(je) \exp\left[\sum_{k=-m}^{m} \lambda_k x^k \right]} = r_k \tag{11.176}$$

where x_u and x_L are, respectively, upper and lower limits of integration. Equation (11.176) is a system of $(m+1)$ nonlinear equations wherein the integration is done over the frequency band of the expected spectral density. The solution can be obtained numerically.

Tzannes and Avgeris (1981) suggested that spectral analysis can be undertaken in a similar manner as above by specifying constraints in addition to the autocorrelation function given by equation (11.170). These can be defined as functions of w as $g_i(w)$ which are supposed to be known a priori:

$$\int_{-\pi}^{\pi} g_i(w) S(w)\, dw = \overline{g_i(w)}, i = 1, 2, \ldots, J \tag{11.177}$$

Then, using the method of Lagrange multipliers, the Lagrangean function becomes

$$L = -\int_{-\pi}^{\pi} S(w) \log S(w)\, dw - \sum_{k=-m}^{m} \lambda_k \left[\int_{-\pi}^{\pi} S(w) \exp(jwk)\, dw - r_k \right]$$
$$- \sum_{i=1}^{J} \rho_i \left(\int_{-\pi}^{\pi} g_i(w) S(w)\, dw - \overline{g_i(w)} \right) \tag{11.178}$$

Then, the entropy-based spectral density will become:

$$S(w) = \exp\left[-1 - \sum_{k=-m}^{m} \lambda_k \exp(jwk) - \sum_{i=1}^{J} \rho_i g_i(w)\right] \tag{11.179}$$

in which the Lagrange multipliers λ_k and ρ_i can be estimated using equations (11.170) and (11.177).

Example 11.17: Compute the spectral density of the stream flow monthly data used in Example 11.11.

Solution: The autocorrelation is known up to lag-2 $(m=2)$: $\rho_0 = 1.0000$, $\rho_1 = 0.0899$; $\rho_2 = -0.2112$ as used in Example 11.10. Using equation (11.176), one can form three nonlinear equations to numerically solve for λ_0, λ_1 and λ_2. Before solving equations formed from (11.176), according to the method by Wragg and Dowson (1970), let a_k denote the approximation of λ_k, so that the residual defined by the equation $\lambda_k = a_k + \varepsilon_k$ is small. Thus, the left side of equation (11.176) can be linearized as follows:

$$\int_{-x_L}^{x_u} \frac{x^{i-1} dx}{(je) \exp\left[\sum_{k=-m}^{m} \lambda_k x^k\right]} = \int_{-x_L}^{x_u} \frac{x^{i-1} dx}{(je) \exp\left[\sum_{k=-m}^{m} (a_k + \varepsilon_k) x^k\right]}$$

$$= \frac{1}{je} \int_{-x_L}^{x_u} x^{i-1} \exp\left[-\sum_{k=-m}^{m} (a_k + \varepsilon_k) x^k\right] dx$$

$$= \frac{1}{je} \int_{-x_L}^{x_u} x^{i-1} \exp\left[-\sum_{k=-m}^{m} a_k x^k\right] \exp\left[-\sum_{k=-m}^{m} \varepsilon_k x^k\right] dx$$

$$= \frac{1}{je} \int_{-x_L}^{x_u} x^{i-1} \exp\left[-\sum_{k=-m}^{m} a_k x^k\right]\left[1 - \sum_{k=-m}^{m} \varepsilon_k x^k\right] dx$$

Here, we used the property that for small ε, $\exp(\varepsilon) = 1 + \varepsilon$. Therefore,

$$= \frac{1}{je} \int_{-x_L}^{x_u} x^{i-1} \exp\left[-\sum_{k=-m}^{m} a_k x^k\right] dx - \sum_{k=-m}^{m} \varepsilon_k \int_{-x_L}^{x_u} x^{k+i-1} \exp\left[-\sum_{k=-m}^{m} a_k x^k\right] dx$$

Let $C_i = \int_{-x_L}^{x_u} x^{i-1} \exp\left[-\sum_{k=-m}^{m} a_k x^k\right] dx$, solving equation (11.176) is now equivalent to solving

$$r_i = C_i + \sum_{k=-m}^{m} \varepsilon_k C_{i+k}$$

With an initial approximation of a_k, one can obtain an acceptable ε_k and end the iteration.

For given data, applying the above iteration, solution is: $\lambda_0 = 1.36$, $\lambda_1 = -1.67$ and $\lambda_2 = 1.66$. The spectral density is computed by equation (11.174) and plotted in Figure 11.16, which is similar to Figure 11.15.

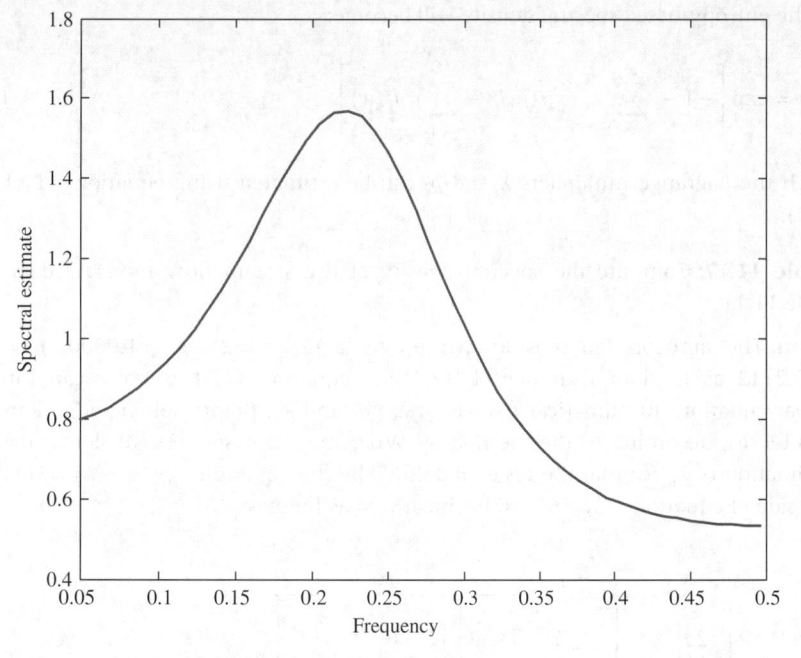

Figure 11.15 Spectral power density for the monthly data with lag 2.

11.5 Spectral estimation by mutual information principle

If a random process $x(t)$ is observed through a system, linear or nonlinear, which introduces noise (t), then the output process $y(t)$ can be considered as the noisy quantized form of the input process $x(t)$. Let the power spectrum of the input process, $S_x(w)$, be estimated, and the power spectrum of the output process, $S_y(\varphi)$, be observed, and let these spectra be PDFs. Here w and φ are input and output frequencies, respectively, and are random variables. Now the system spectral function can be represented as a two-dimensional spectral density denoted as $S(w, \varphi)$. It may be noted that each input frequency w leads to many output frequency components φ given by two-dimensional joint spectral density $S(w, \varphi)$. The joint spectrum of the input and output processes $x(t)$ and $y(t)$ is a measure of the power falling in each frequency φ for a given frequency w. The mutual information $I(w, \varphi)$ can be formulated as

$$I(w, \varphi) = \int \int S(w, \varphi) \log \frac{S(w, \varphi)}{S_x(w) \, S_y(\varphi)} dw d\varphi \qquad (11.180)$$

In order to ensure compatibility with the joint PDF, $S(w, \varphi)$ must satisfy

$$\int_{-\infty}^{\infty} S(w, \varphi) \, d\varphi = S_x(w) \qquad (11.181)$$

$$\int_{-\infty}^{\infty} S(w, \varphi) \, dw = S_y(\varphi) \qquad (11.182)$$

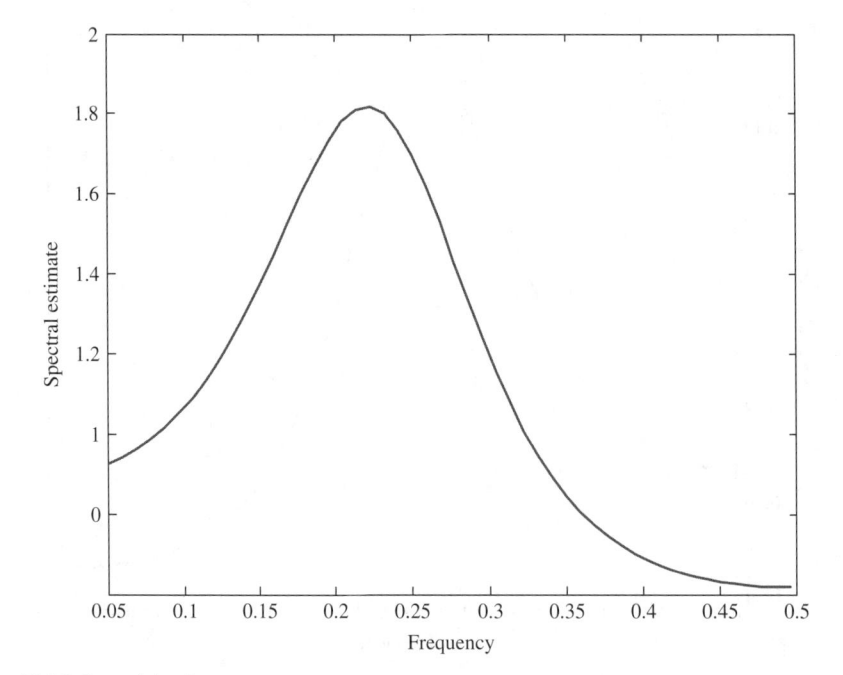

Figure 11.16 Spectral density.

Now the conditional spectral density can be defined as

$$S(w|\varphi) = \frac{S(w, \varphi)}{S_y(\varphi)} \qquad (11.183a)$$

$$S(\varphi|w) = \frac{S(w, \varphi)}{S_x(w)} \qquad (11.183b)$$

The optimum $S(w,\varphi)$ can be determined by minimizing the spectral mutual information given by equations (11.180).

The mutual spectral information can also be expressed in terms of conditional spectral density as

$$I(w, \varphi) = \int_{-\infty}^{\infty}\int_{-\infty}^{\infty} S(w, \varphi) S_y(\varphi) \log \frac{S(w|\varphi) S_y(\varphi)}{\left[\int_{-\infty}^{\infty} S(w|\varphi) S_y(\varphi)\, dw\right] S_y(\varphi)}\, dw d\varphi \qquad (11.184)$$

The output spectral density $S_y(\varphi)$ is assumed to be known under a constraint of the form:

$$E[\rho(w, \varphi)] \leq D \qquad (11.185)$$

where E is the expectation, D is some value, and ρ is an error function of random variables w and φ which depend on the errors introduced by the observation system.

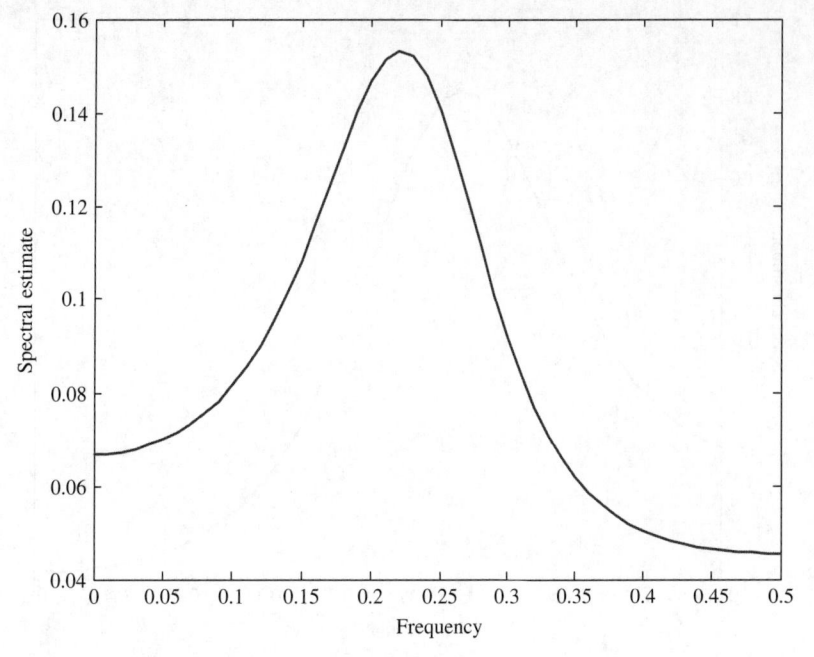

Figure 11.17 Spectral density.

Tzannes and Avgeris (1981) discuss that given a PDF, $S_x(w)$, minimization of mutual information between w and φ, subject to equation (11.185), leads to a unique conditional PDF $S(\varphi|w)$. For minimization, the constraint can be defined as

$$\sum_{k=-m}^{m} C_k \left| \int_{-\infty}^{\infty} S(w) \exp(-ikw)\,dw - r_k \right|^2 \leq \sigma^2 \tag{11.186}$$

where C_k are weights accounting for the differing degree of confidence placed in each measured autocorrelation value. The value of σ^2 is a measure of the mean squared noise introduced by the system. In accord with equation (11.186), the function to be minimized can be written as

$$L(t) = I(t) - \sum_j a_j \sum_k T(k|j) - b \sum_{n=-N}^{N} C_n \left| \sum_{j,k} S_x(w_j) T(k,j) \exp(-inw_j) - \rho_n \right|^2 \tag{11.187}$$

where $T = S(\varphi_k|w_j)$. Defining coefficients $b_n, n = 0, \pm 1, \pm 2, \ldots, \pm N$,

$$b_n = bC_n \left\{ \sum_{j,k} S_x(w_j) T(k|j) \exp(-inw_j) - \rho_n \right\} = b^*_{-n} \tag{11.188}$$

and differentiating equation (11.187) with respect to T and letting the derivative go to zero, the equations based on the spectral mutual information principle are obtained as

$$\sum_k S_y(\varphi_k) \exp[bg(j,k)] = \frac{1}{\lambda_j}, \quad j = 1, 2, \ldots, M-1 \tag{11.189}$$

$$\sum_j \lambda_j S_x(w_j) \exp\left[bg(j,k)\right] = 1, \quad j = 1, 2, \ldots, N - 1 \tag{11.190}$$

where

$$g(j,k) = \sum_{n=-N}^{N} b_n^* b_n \tag{11.191}$$

The conditional spectral density $T(k|j)$ is given as

$$T(k|j) = \lambda_j S_y(\varphi_k) \exp\left[bg(j,k)\right] \tag{11.192}$$

Considering the equality in equation (11.186) and combining with equation (11.188), the result is the Lagrange multiplier:

$$b = \frac{1}{\sigma}\left[\sum_{m=-N}^{N} b_m \frac{b_m^*}{C_m}\right]^{1/2} \tag{11.193}$$

Substituting equation (11.193) into equation (11.188), parameter b can be eliminated and one obtains

$$\rho_n + \sigma b_n/C_n \left[\sum_{m=-N}^{N} b_m \frac{b_m^*}{C_m}\right]^{-1/2} = \sum_{j,k} S_x\left(w_j\right) T(k|j) \exp\left(-inw_j\right) \tag{11.194}$$

Equations (11.189) to (11.194) can be solved following Newman (1977). To start the solution procedure, it can be assumed that there is no measurement error, that $\sigma = 0$ and b_n obtained from equation (11.194), $g(j, k)$ is obtained from equation (11.191). Then, with given $S_y(\varphi_k)$ equation (11.189) is solved for λ_js. With these values of λ_js and $S_y(\varphi_k)$, equation (11.190) is solved to obtain the estimated $S_x(w_j)$ and $T(k|j)$ is got from equation (11.192). Substituting $S_x(w_j)$ and $T(k|j)$ into equation (11.186), the value of σ is obtained, which then is used for the next iteration. The procedure is repeated until convergence is achieved.

Example 11.18: Compute the spectral density of the stream flow monthly data used in Example 11.11.

Solution: Consider the autocorrelation known up to lag 2 ($m = 2$), which are $\rho_0 = 1.0000$, $\rho_1 = 0.0899$; $\rho_2 = -0.2112$ as used in Example 11.10. Assuming equal weight for each degree of confidence in each measured autocorrelation value, the computation of the spectral density can be obtained through the following iteration, starting with $\sigma = 0$. As there is no measurement error, $S_x(w_j)$ is conserved to be equal to $S_y(\varphi_k)$, and the conditional spectral $T(k|j)$ can be considered as unity. Thus, b_n is computed from equation (11.194) and $g(j,k)$ can be computed from equation (11.192).

$$b_n = [1.15\ 0.67\ 0.28]$$

$$g(j,k) = [2.54\ 0.37\ 3.12;\ 0.58\ 3.67\ 1.11;\ 1.15\ 0.77\ 1.05]$$

With given $S_y(\varphi_k)$, solving for λ_j from equation (11.189), one obtains: $\lambda_1 = 2.576$, $\lambda_2 = -0.183$, $\lambda_3 = 0.105$. Now $S_x(w_j)$ can be estimated using equation (11.190) with the above parameters. Substituting $S_x(w_j)$ and $T(k|j)$ in equation (11.186), the error becomes $\sigma = 4.615$, which is used for new iteration. Repeating the above steps, the final result is $\lambda_1 = 1.754$, $\lambda_2 = -2.589$, $\lambda_3 = 0.201$ for $\sigma = 0.482$ and the spectral density is plotted in Figure 11.17.

References

Burg, J.P. (1967). Maximum entropy spectral analysis. Paper presented at 37th Annual Meeting of Society of Exploration Geophysicists, Oklahoma City, Oklahoma.

Burg, J.P. (1969). A new analysis technique for time series data. Paper presented at NATO Advanced Institute on Signal Processing with Emphasis on Underwater Acoustics.

Burg, J.P. (1972a). The relationship maximum entropy spectra and maximum likelihood spectra. *Geophysics*, Vol. 37, No. 2, pp. 375–6.

Burg, J.P. (1972b). The relationship between maximum entropy spectra and maximum likelihood spectra. in: *Modern Spectral Analysis*, edited by D.G. Childers, pp. 130–1, M.S.A.

Burg, J.P. (1975). Maximum entropy spectral analysis. Unpublished Ph.D. dissertation, Stanford University, Palo Alto, California.

Dalezios, N.R. and Tyraskis, P.A. (1999). Maximum entropy spectra for regional precipitation analysis and forecasting. *Journal of Hydrology*, Vol. 109, pp. 25–42.

Fougere, P.F., Zawalick, E.J. and Radoski, H.R. (1976). Spontaneous line splitting in maximum entropy power spectrum analysis. *Physics of the Earth and Planetary Interiors*, Vol. 12, pp. 201–7.

Frieden, B.R. (1972). Restoring with maximum likelihood and maximum entropy. *Journal of Optical Society of America*, Vol. 62, pp. 511–18.

Gull, S.F. and Daniell, G.J. (1978). Image reconstruction from incomplete and noisy data. *Nature*, Vol. 272, pp. 686–90.

Kapur, J.N. (1999). *Maximum Entropy Models in Science and Engineering*. Wiley Eastern, New Delhi, India.

Kapur, J.N. and Kesavan, H.K. (1997). *Generalized Maximum Entropy Principle (with Applications)*. Sandford Educational Press, University of Waterloo, Waterloo, Canada.

Kapur, J.N. and Kesavan, H.K. (1992). *Entropy Optimization Principles with Applications*. 409 p., Academic Press, New York.

Padmanabhan, G. and Rao, A.R. (1999). Maximum entropy spectral analysis of hydrologic data. *Water Resources Research*, Vol. 24, No. 9, pp. 1519–33.

Shore, J.E. (1979). Minimum cross-entropy spectral analysis. NRL Memorandum Report 3921, 32 p., Naval Research Laboratory, Washington, D.C.

Tzannes, N.S. and Avgeris, T.G. (1981). A new approach to the estimation of continuous spectra. *Kybernetes*, Vol. 10, pp. 123–33.

Tzannes, M.A., Politis, D. and Tzannes, N.S. (1985). A general method of minimum cross-entropy spectral estimation. *IEEE Transactions on Acoustics, Speech, and Signal Processing*, Vol. ASSP-33, No. 3, pp. 748–52.

Ulrych, T.J. and Clayton, R.W. (1976). Time series modeling and maximum entropy. *Physics of the Earth and Planetary Interiors*, Vol. 12, pp. 199–9.

Wu, N.-L. (1983). An explicit solution and data extension in the maximum entropy method. *IEEE Transactions on Acoustics, Speech, and Signal Processing*, Vol. ASSP-31, No. 2, pp. 486–91.

Additional Reading

Bath, M. (1974). Spectral Analysis in Geophysics. Vol. 7, *Developments in Solid Earth Geophysics*, Elsevier, Amsterdam.

Bendat, J.S. and Piersol, A.G. (1971). *Random Data: Analysis and Measurement Procedures*. Wiley-Interscience, New York.

Box, G.E.P. and Jenkins, G.M. (1976). *Time Series Analysis: Forecasting and Control*. Holden-Day, San Francisco, California.

Capon, J. (1969). High-resolution frequency-wave number spectrum analysis. *Proceedings of the IEEE*, Vol. 57, No. 8, pp. 1408–969.

Chatfield, C. (1975). *The Analysis of Time Series: Theory and Practice*. Chapman and Hall, London.

Edwards, J.A. and Fitelson, M.M. (1973). Notes on maximum-entropy processing. *IEEE transactions on Information Theory*, Vol. IT-19, pp. 232–4.

Harris, F.J. (1978). On the use of windows for harmonic analysis with the discrete Fourier transform. *IEEE Proceedings*, Vol. 66, No. 1, pp. 51–83.

Hedgge, B.J. and Masselink, G. (1996). Spectral analysis of geomorphic time series: auto-spectrum. *Earth Surface Processes and Landforms*, Vol. 21, pp. 1021–40.

Jenkins, G.M. and Watts, D.G. (1968). *Spectral Analysis and its Applications*. Holden-Day, San Francisco, California.

Kaveh, M. and Cooper, G.R. (1976). An empirical investigation of the properties of the autoregressive spectral estimator. *IEEE Transactions on Information Theory*, Vol. IT-22, No. 3, pp. 313–23.

Kerpez, K.J. (1989). The power spectral density of maximum entropy charge constrained sequences. *IEEE Transactions*, Vol. 35, No. 3, pp. 692–5.

Krastanovic, P.F. and Singh, V.P. (1991a). A univariate model for long-term streamflow forecasting: 1. Development. *Stochastic Hydrology and Hydraulics*, Vol. 5, pp. 173–88.

Krastanovic, P.F. and Singh, V.P. (1991b). A univariate model for long-term streamflow forecasting: 2. Application. *Stochastic Hydrology and Hydraulics*, Vol. 5, pp. 189–205.

Krastanovic, P.F. and Singh, V.P. (1993a). A real-time flood forecasting model based on maximum entropy spectral analysis: I. Development. *Water Resources Management*, Vol. 7, pp. 109–29.

Krstanovic, P.F. and Singh, V.P. (1993b). A real-time flood forecasting model based on maximum entropy spectral analysis: II. Application. *Water Resources Management*, Vol. 7, pp. 131–51.

Lacoss, R.T. (1971). Data adaptive spectral analysis methods. *Geophysics*, Vol. 36, No. 4, pp. 661–75.

Newman, W.I. (1977). Extension to the Maximum entropy method. *IEEE Transactions on Information Theory*, Vol. IT-23, No. 1, pp. 89–93.

Noonan, J.P., Tzannes, N.S. and Costfello, T. (1976). On the inverse problem of entropy maximizations. *IEEE Transactions on Information Theory*, Vol. IT-22, pp. 120–3.

Padmanabhan, G. and Rao, A.R. (1986). Maximum entropy spectra of some rainfall and river flow time series from southern and central India. *Theoretical and Applied Climatology*, Vol. 37, pp. 63–73.

Papoulis, A. (1973). Minimum-bias windows for high-resolution spectral estimates. *IEEE Transactions on Information Theory*, Vol. IT-19, No. 1, pp. 9–12.

Pardo-Iguzquiza, E. and Rodriguez-Tovar, F.J. (2005a). MAXENPER: A program for maximum spectral estimation with assessment of statistical significance by the permutation test. *Computers and Geosciences*, Vol. 31, pp. 555–67.

Pardo-Iguzquiza, E. and Rodriguez-Tovar, F.J. (2005b). Maximum entropy spectral analysis of climatic time series revisited: Assessing the statistical significance of estimated spectral peaks. *Journal of Geophysical Research*, Vol. 111, D010102, doi:10.1029/2005JD006293, pp. 1–8.

Tzannes, N.S. and Noonan, J.P. (1973). The mutual information principle and applications. *Information and Control*, Vol. 22, pp. 1–12.

Tzannes, M.A., Politis, D. and Tzannes, N.S. (1985). A general method of minimum cross-entropy spectral estimation. *IEEE Transactions on Acoustics, Speech, and Signal Processing*, Vol. ASSP-33, No. 3, pp. 748–752.

Van Den Bos, A. (1971). Alternative interpretation of maximum entropy spectral analysis. *IEEE Transactions on Information Theory*, Vol. IT-17, pp. 493–4.

Wragg, A. and Dowson, D.C. (1970). Fitting continuous probability density functions over $[0, \infty)$ using information theory ideas, *IEEE Transactions on Information Theory*, Vol. 16, pp. 226–30.

Yevjevich, V. (1972). *Stochastic Processes in Hydrology*. Water Resources Publications, Fort Collins, Colorado.

12 Minimum Cross Entropy Spectral Analysis

In the preceding chapter the maximum entropy spectral analysis (MESA) was presented for analyzing time series of environmental and water engineering variables. In environmental and water resources, processes, such as streamflow, rainfall, radiation, water quality constituents, sediment yield, drought, soil moisture, evapotranspiration, temperature, wind velocity, air pressure, radiation, and so on, have inherent periodicities. This chapter discusses the application of the principle of minimum cross entropy (POMCE) (minimum directed divergence, minimum discrimination information, or minimum relative entropy) for estimating power spectra. The estimation of power spectra is done in different ways wherein values of autocorrelation function and a prior estimate of the spectrum are given or the probability density function underlying the given process is determined or the cross-entropy between the input and output of linear filters is minimized.

12.1 Cross-entropy

Let X be a random variable. The cross-entropy between any two probability density functions $p(x)$ and $q(x)$ of the random variable X, $H(p, q)$, can be expressed as

$$H(p, q) = \int_D p(x) \log \left[\frac{p(x)}{q(x)} \right] dx \tag{12.1}$$

where D is domain of X. $H(p, q)$ is a measure of the information divergence or information dissimilarity between $p(x)$ and $q(x)$. It can be assumed that the probability density function $q(x)$ reflects the current estimate of the PDF of X and can be called a prior. Then, the probability density function $p(x)$ is to be determined. It is implied that there can be a set of probability densities on D to which $p(x)$ belongs but it is not known. Of course,

$$\int_D p(x) dx = 1 \tag{12.2}$$

$$\int_D q(x) dx = 1 \tag{12.3}$$

Entropy Theory and its Application in Environmental and Water Engineering, First Edition. Vijay P. Singh.
© 2013 John Wiley & Sons, Ltd. Published 2013 by John Wiley & Sons, Ltd.

If new information is obtained about $p(x)$, expressed in the form of constraints, as

$$\int_D g_j(x)p(x)dx = \bar{g}_j, \quad j = 1, 2, \ldots, m \tag{12.4}$$

where $g_j(x)$ is some function of X, "bar" indicates average, and m is the number of constraints. It may be noted that there may be other distributions satisfying these constraints or these constraints may not determine $p(x)$ completely. The question then arises: Which single PDF should be chosen from the subset to be the true $p(x)$ and how should the prior $q(x)$ and constraints be used to make that choice? POMCE states that of all the densities one should choose that $p(x)$ that has the minimum cross-entropy with respect to the prior $q(x)$. Thus, the PDF $p(x)$ can be determined, using the method of Lagrange multipliers, by minimizing cross-entropy, $H(p,q)$, defined by equation (12.1), subject to the specified constraints defined by equations (12.2) and (12.4), as:

$$p(x) = q(x)\exp\left[-\lambda_0 - \sum_{k=1}^{m} \lambda_k g_k(x)\right] \tag{12.5}$$

where $\lambda_k = 0, 1, 2, \ldots, m$, are the Lagrange multipliers which are determined using the known constraints. It may be noted that the above discussion can be extended if X is a vector. Cross-entropy is now applied to spectral analysis.

12.2 Minimum cross-entropy spectral analysis (MCESA)

The minimum cross-entropy spectral analysis (MCESA) can be undertaken in several ways. First, the power spectrum is estimated by the minimization of cross-entropy given values of autocorrelation function and a prior estimate of the power spectrum. Second, cross-entropy minimization is applied to the input and output of a linear filter. Third, the probability density function of the process is determined by minimizing the cross-entropy given a prior and constraints and then determining the power spectrum. The prior probability density function is usually taken as a uniform or a Gaussian distribution prior. The constraints can be specified as the total probability law, and the information in terms of moments, autocorrelation functions, or expected spectral powers. The estimation of power spectra using POMCE does not assume any specific PDF structure of the random process under consideration.

12.2.1 Power spectrum probability density function

Consider a stationary stochastic process $z(t)$. This process can be obtained as the limit of a sequence of processes with discrete spectra. Let the time-domain processes be expressed by Fourier series as

$$y(t) = \sum_{k=1}^{n} a_k \cos(w_k t) + b_k \sin(w_k t) \tag{12.6}$$

where a_k and b_k are coefficients, t is time, and $w_k = 2\pi f_k$ is frequency that need not to be equally spaced. It may be noted that w is the frequency in cycles per unit time, whereas f is angular frequency in radians per unit time, and $w = 2\pi/T$, $T = $ period in units of time. The discrepancy between $z(t)$ and $y(t)$, expressed by mean square error $E[|z(t) - y(t)|^2]$, can be minimized by an appropriate choice of frequencies and random variables. The power x_k at each frequency can be expressed as

$$x_k = a_k^2 + b_k^2 \tag{12.7}$$

If x_k are considered random variables, then the stochastic process described by equation (12.6) can be described in terms of a joint probability density function $p(\vec{x})$, $\vec{x} : (x_1, x_2, \ldots, x_n)$ is the vector. Here $p(\vec{x})$ is the spectral power probability density function of the stationary stochastic process which can simply be called process.

The objective is to derive $p(\vec{x})$ by minimizing cross-entropy. To that end,

$$\int_D p(\vec{x}) d\vec{x} = 1 \tag{12.8}$$

In order to apply POMCE, a prior estimate of the spectral power PDF, $q(\vec{x})$, of $p(\vec{x})$ is needed. Of course,

$$\int_D q(\vec{x}) d\vec{x} = 1 \tag{12.9}$$

Since the magnitude of x_k will be finite and the domain of $\vec{x}$ will be bounded, it is not entirely unreasonable to assume a uniform prior for the spectral power density function which can be specified as

$$q(\vec{x}) = \frac{1}{b-a} \tag{12.10}$$

or assume an exponential prior

$$q(\vec{x}) = \frac{1}{\alpha} \exp\left(\frac{\vec{x}}{\alpha}\right) \tag{12.11}$$

Then, a constraint on the spectral power can be specified in terms of the total expected power per discrete frequency expressed as:

$$P = \frac{1}{n} \int_D \left[\sum_{k=1}^{n} x_k\right] p(\vec{x}) d\vec{x} \tag{12.12}$$

where $d\vec{x} = dx_1 dx_2 \ldots dx_n$; and D is a possible set of system states.

Minimization of cross-entropy, subject to equations (12.8) and (12.12), can be done using the method of Lagrange multipliers. The Lagrangean function can be written as

$$L = \int_D p(\vec{x}) \log\left[\frac{p(\vec{x})}{q(\vec{x})}\right] d\vec{x} + (\lambda_0 - 1)\left[\int_D p(\vec{x}) d\vec{x} - 1\right] + \lambda\left\{\frac{1}{n} \int_D \left[\sum_{k=1}^{n} x_k\right] p(\vec{x}) d\vec{x} - P\right\} \tag{12.13}$$

where λ_0 and λ the Lagrange multipliers associated with equation (12.8) and (12.12), respectively. Differentiating the Lagrangean function L with respect to $p(\vec{x})$ and equating the derivative to zero, one obtains

$$\frac{\partial L}{\partial p(\vec{x})} = 0 = 1 + \log \frac{p(\vec{x})}{q(\vec{x})} + (\lambda_0 - 1) + \lambda \sum_{k=1}^{n} x_k \tag{12.14}$$

Equation (12.14) yields

$$p(\vec{x}) = q(\vec{x}) \exp \left[-\lambda_0 - \lambda \sum_{k=1}^{n} x_k \right] \tag{12.15}$$

Equation (12.15) has two Lagrange multipliers that need to be determined. Substituting equation (12.15) in equation (12.8) one obtains

$$\int_D q(\vec{x}) \exp \left[-\lambda_0 - \lambda \sum_{k=1}^{n} x_k \right] d\vec{x} = 1 \tag{12.16}$$

Equation (12.16) can be written as

$$\exp(\lambda_0) = \int_D q(\vec{x}) \exp \left[-\lambda \sum_{k=1}^{n} x_k \right] d\vec{x} \tag{12.17}$$

Substitution of equation (12.17) in equation (12.15) yields

$$p(\vec{x}) = \frac{q(\vec{x}) \exp \left[-\lambda \sum_{k=1}^{n} x_k \right]}{\int_D q(\vec{x}) \exp \left[-\lambda \sum_{k=1}^{n} x_k \right] d\vec{x}} \tag{12.18}$$

Integration in the denominator of equation (12.18) depends on the form of the prior distribution $q(\vec{x})$ and will lead to a constant quantity. Recognizing that the prior is assumed as a uniform distribution, equation (12.15) can be written as

$$p(\vec{x}) = A \exp \left[-\lambda \sum_{k=1}^{n} x_k \right] \tag{12.19}$$

in which

$$A^{-1} = \int dx_1 \int dx_2 \ldots \int dx_n \exp \left(-\lambda \sum_{k=1}^{n} x_k \right) \tag{12.20}$$

where λ is the Lagrange multiplier corresponding to equation (12.8); and A has absorbed the uniform prior and the Lagrange multiplier corresponding to equation (12.8).

If P is much less than the maximum value of x_k, the integration limits of $(0, \infty)$ can be used in equation (12.19) which then yields $A = \lambda^n$. Substitution of equation (12.15) in equation (12.12) yields

$$P = \frac{\lambda^n}{n} \sum_k \int dx_k x_k \exp(-\lambda x_k) \prod_{m \neq k} \exp(-\lambda x_m) dx_m = \frac{1}{\lambda} \tag{12.21}$$

The posterior, $p(\vec{x})$, in terms of the spectral power, therefore, is:

$$p(\vec{x}) = \prod_{k=1}^{n} \frac{1}{P} \exp\left(-\frac{x_k}{P}\right) \qquad (12.22)$$

Equation (12.22) for $p(\vec{x})$ is a multivariate exponential distribution where each spectral power x_k is exponentially distributed with mean P.

Example 12.1: Consider annual streamflow at a gaging station (USGS 08201000) on Brazos River, Texas, given in Table 12.1 Fit the Fourier series to the streamflow time series and specify the values of a_k and b_k and x_k. Tabulate the values of a_k and b_k at each frequency.

Solution: The Fourier series is fitted to the given annual stream flow as shown in Figure 12.1. The Fourier coefficients determined by fitting are tabulated in Table 12.2, and power at each frequency x_k is computed using sum of squares of a_k and a_k. Spectral powers are plotted against frequency, as shown in Figure 12.2.

Example 12.2: Determine and plot the PDF $p(\vec{x})$ using equation (12.19) and determine parameters A and λ using the values of x_k determined from observations.

Solution: For the given data, the expected value for x_k is $P = \frac{1}{n} \int_{D} \left[\sum_{k=1}^{n} x_k\right] p(\vec{x})d\vec{x} = 0.214$, and $\lambda = 46.645$ from equation (12.21). Then, $A = \lambda^n = 2.598 \times 10^8$. The plot of $p(x)$ is shown in Figure 12.3

Table 12.1 Annual stream flow (cfs) for station USGS 08201000 on Brazos River, Texas.

Year	Flow (cfs)	Year	Flow (cfs)	Year	Flow (cfs)
1940	96.3	1964	158.6	1988	125.3
1941	733.2	1965	60.2	1989	243.1
1942	364.7	1966	166.1	1990	224.6
1943	172	1967	48.8	1991	302.8
1944	175	1968	276.5	1992	388.4
1945	464.3	1969	251.1	1993	305.6
1946	368.3	1970	61.7	1994	134
1947	437.4	1971	22.8	1995	653
1948	69.7	1972	81.7	1996	89.4
1949	195.3	1973	491.9	1997	236.8
1950	436.6	1974	386.6	1998	378.5
1951	34.6	1975	530.7	1999	515.1
1952	86.1	1976	178.5	2000	47.3
1953	201.6	1977	239.2	2001	322.3
1954	50.3	1978	122.5	2002	205.1
1955	94.5	1979	498.9	2003	363.2
1956	32.2	1980	155	2004	434.8
1957	100.1	1981	74.4	2005	299.6
1958	314.2	1982	189.6	2006	45.8
1959	199.8	1983	451.1	2007	336.4
1960	200.4	1984	240.2	2008	128.1
1961	484.6	1985	244.8	2009	90.3
1962	81.6	1986	320.2	2010	125.1
1963	122.1	1987	265.1		

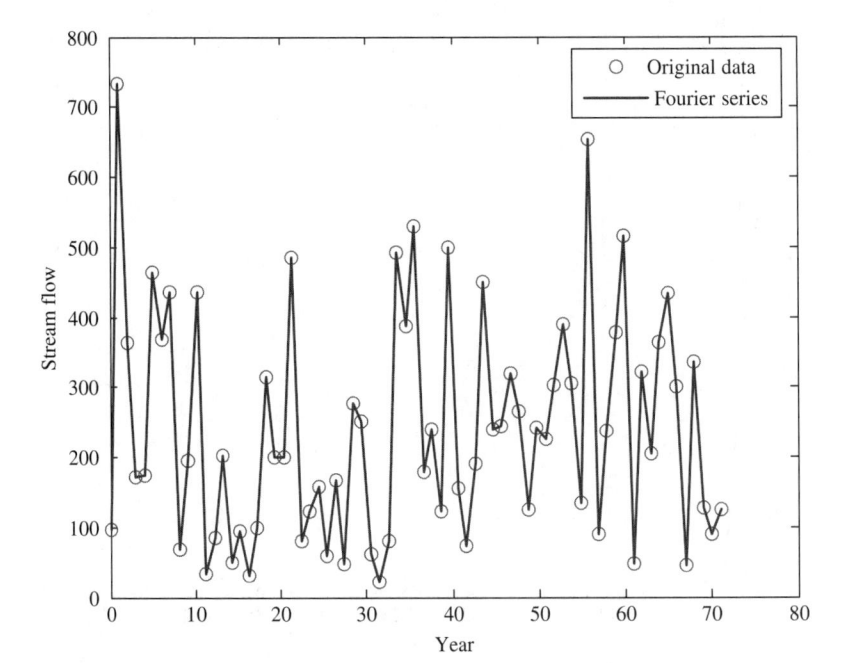

Figure 12.1 Fitted Fourier series.

Table 12.2 Fourier series coefficients and power.

Frequency	a_k	b_k	x_k
0.0000	239.7986	0.0000	57503.3645
0.0078	130.9574	15.0196	17149.8481
0.0156	45.0139	19.6673	2026.2484
0.0234	47.0654	48.2899	2215.1548
0.0313	37.4613	11.2689	1403.3515
0.0391	36.4513	−15.5981	1328.6972
0.0469	20.5041	18.3882	420.4220
0.0547	25.5351	38.5774	652.0429
0.0625	36.3263	−3.6214	1319.6069
0.0703	26.9672	20.4958	727.2335
0.0781	8.5758	43.9508	73.5505
0.0859	9.5198	−4.0530	90.6346
0.0938	19.9633	31.1986	398.5421
0.1016	5.4451	−18.0767	29.6590
		...	
0.4531	15.0716	4.0530	227.3574
0.4609	15.7400	−43.9508	247.9606
0.4688	16.5653	−20.4958	274.6304
0.4766	9.1356	3.6214	83.6864
0.4844	18.1669	−38.5774	330.2715
0.4922	24.8573	−18.3882	618.1275
0.5000	44.3310	15.5981	1965.4861

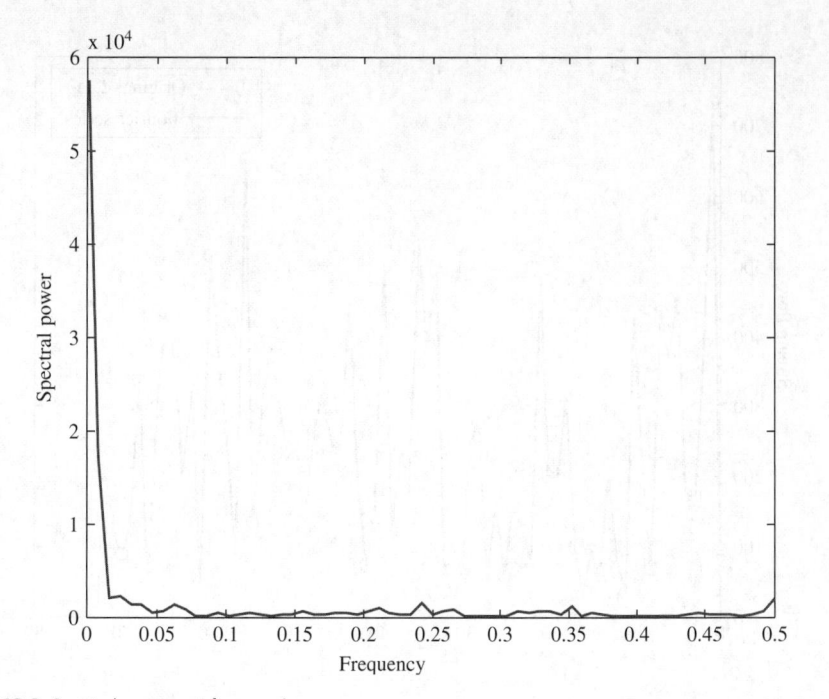

Figure 12.2 Spectral powers at frequencies.

It is benoted that the PDF $p(\vec{x})$, determined using equation (12.22) where parameters A and λ are determined using the values of x_k obtained from observations, would be the same as in Figure 12.3.

12.2.2 Minimum cross-entropy-based probability density functions given total expected spectral powers at each frequency

Now consider the case where the expected spectral power P_k at each frequency is known:

$$P_k = E[x_k] = \int x_k p(\vec{x}) d\vec{x} \tag{12.23}$$

Again, a uniform prior is chosen. Then, the maximum cross-entropy-based $p(\vec{x})$ becomes:

$$p(\vec{x}) = \prod_{k=1}^{n} \frac{1}{P_k} \exp\left(-\frac{x_k}{P_k}\right) \tag{12.24}$$

If equation (12.22) is used as a prior, then equation (12.24) is still obtained. This shows that using equation (12.22) or a uniform prior leads to the same result, that is, equation (12.24).

Example 12.3: Determine $p(\vec{x})$ using equation (12.24). Using streamflow values from Example 12.1, the expected spectral powers at each frequency P_k is obtained as in the solution of Example 12.1.

Solution: Take frequency $= 0.5$ as an example, $P_k = E[x_k] = \int x_k p(\vec{x}) d\vec{x} = 100.74$. Using equation (12.24), the posterior power PDF is determined as shown in Figure 12.4.

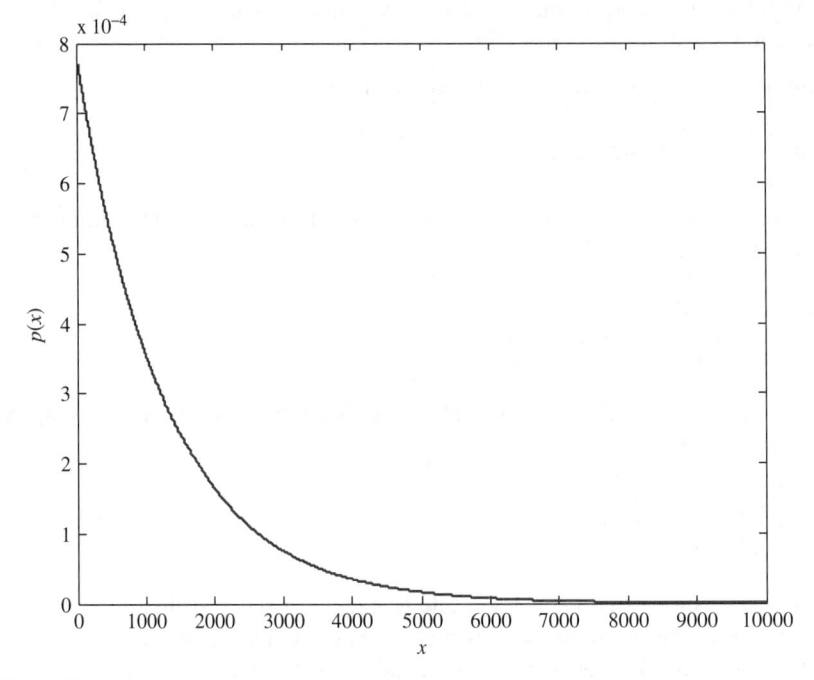

Figure 12.3 PDF of spectral power.

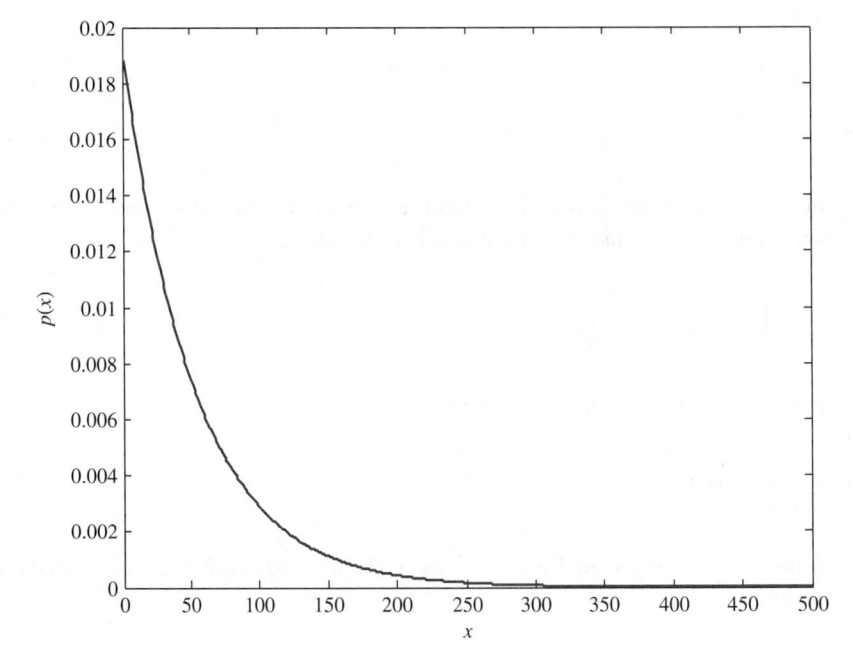

Figure 12.4 PDF of spectral power.

Example 12.4: Consider a uniform prior for variables a_k and b_k, rather than for variables x_k, in equation (12.7). Then, derive $p(\vec{x})$.

Solution: In this case the constraint takes on the form:

$$P_k = \overline{a_k^2 + b_k^2} = \int (a_k^2 + b_k^2) p(\vec{a}, \vec{b}) d\vec{a}\, d\vec{b} \tag{12.25}$$

In a manner similar to the previous discussion, cross-entropy minimization, using the method of Lagrange multipliers, yields

$$p(\vec{a}, \vec{b}) = A \exp\left[-\sum_k \lambda_k (a_k^2 + b_k^2) \right] \tag{12.26}$$

Solving for A using equation (12.8), and for the Lagrange multipliers λ_k using equation (12.25), one obtains

$$p(\vec{a}, \vec{b}) = \prod_{k=1}^{n} \frac{1}{\pi P_k} \exp\left[-\frac{a_k^2 + b_k^2}{P_k} \right] \tag{12.27}$$

Equation (12.27) shows that variables a_k and b_k have Gaussian distribution with zero means and variances $P_k/2$. Since variances are related to expectations of power a_k^2 or b_k^2, the expected power P_k is evenly divided between the two quadrature components.

In order to transform equation (12.27) to a probability density function in terms of spectral power variables x_k, and compare the result with equation (12.24), coordinates (a_k, b_k) are transformed to (r_k, θ_k) as:

$$r_k^2 = a_k^2 + b_k^2 \tag{12.28a}$$

$$\theta_k = \tan^{-1}\left(\frac{b_k}{a_k}\right) \tag{12.28b}$$

Then, there is a relation between volume elements in the two coordinate systems: $da_k db_k = r_k dr_k d\theta_k$ and $p(\vec{a}, \vec{b}) da\, db = p(\vec{r}, \vec{\theta}) d\vec{r}\, d\vec{\theta}$. Therefore,

$$p(\vec{r}, \vec{\theta}) = \prod_{k=1}^{N} \frac{r_k}{\pi P_k} \exp\left(-\frac{r_k^2}{P_k}\right) \tag{12.29}$$

and integrating over the θ_k coordinates, the result is

$$p(\vec{r}) = \prod_k \frac{2 r_k}{P_k} \exp\left(-\frac{r_k^2}{P_k}\right) \tag{12.30}$$

Recalling that r_k and x_k are related as $x_k = r_k^2$, $dx_k = 2 r_k dr_k$, and $p(\vec{x}) dx = p(\vec{r}) dr$, then equation (10.30) becomes

$$p(\vec{x}) = \prod_{k=1}^{n} \frac{1}{P_k} \exp\left(-\frac{x_k}{P_k}\right) \tag{12.31}$$

which is the same as equation (12.24).

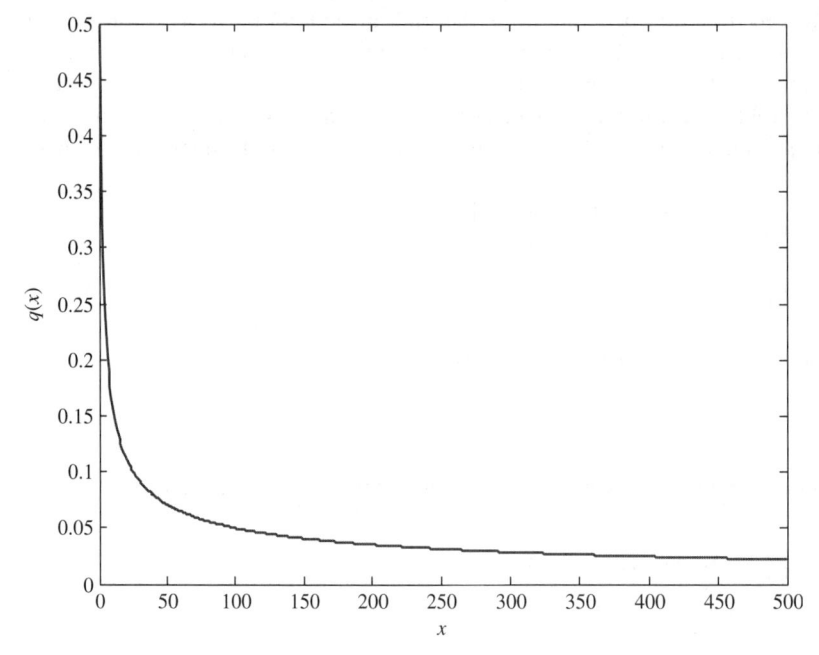

Figure 12.5 Prior density of spectral powers.

This shows that if the information is given in terms of expected spectral powers, it is immaterial if the prior probability density function is taken as uniform in the amplitude variables a_k, b_k or in the power variables x_k. The posterior probability distribution is a multivariate exponential distribution in the variables x_k or multivariate Gaussian distribution in the variables a_k and b_k.

It may be interesting to employ $(\vec{r}, \vec{\theta})$ coordinates and use a prior $q(r, \theta)$ that is uniform with respect to the volume element $drd\theta$. Then, integrating over θ and transforming to the x_k coordinates result in:

$$q(\vec{x}) = \frac{1}{2} \prod_{k=1}^{n} (x_k)^{-1/2} \tag{12.32}$$

Equation (12.32) is a nonuniform prior in contrast with a uniform prior $q(\vec{a}, \vec{b})$ which leads to a uniform prior $q(\vec{x})$. This explains rejecting the probability of a uniform prior $q(\vec{r}, \vec{\theta})$, since there is no reason to have a nonuniform prior $q(\vec{x})$.

Example 12.5: Determine $q(\vec{x})$ using equation (12.32). Use streamflow data from Example 12.1.

Solution: For computed spectral powers from Example 12.1, the prior is computed using equation (12.32) and is plotted in Figure 12.5.

12.2.3 Spectral probability density functions for white noise

The term "white noise" means that the expected spectral powers $\overline{x_k}$ are all equal. There are two possibilities for representing the probability density function of the white noise. The first is the uniform prior $q(\vec{x})$ for which $x_k = x_{\max}/2$, where $x_{\max}$ is the maximum value of x_k. This can be called the uniform white noise. Second, if the total power per discrete frequency is given then equation (12.24) would be an appropriate probability density function. This can be referred to as Gaussian white noise.

It is plausible that the total power per discrete frequency of the stochastic process is not known but its upper limit is. That also implies that a limit for the quantity is known:

$$P(x) = P(p) = \frac{1}{n} \sum_k \overline{x_k} = \frac{1}{n} \int_D \left[\sum_k^n x_k \right] p(\vec{x}) d\vec{x} \tag{12.33}$$

$$P(p) \leq P_{max} \tag{12.34}$$

The uniform prior will satisfy equation (12.34) if

$$\frac{x_{max}}{2} \leq P_{max} \tag{12.35}$$

If equation (12.35) is not satisfied then the appropriate PDF from equation (12.24) follows:

$$p(\vec{x}) = \prod_{k=1}^n \left(\frac{1}{P_{max}} \right) \exp \left(-\frac{x_k}{P_{max}} \right) \tag{12.36}$$

Example 12.6: Determine $p(\vec{x})$ using equation (12.36). Use data from Example 12.1.

Solution: From Example 12.1, $P_{max} = 404.95$. Substituting into equation (12.36), one gets the PDF of spectral powers which is plotted in Figure 12.6.

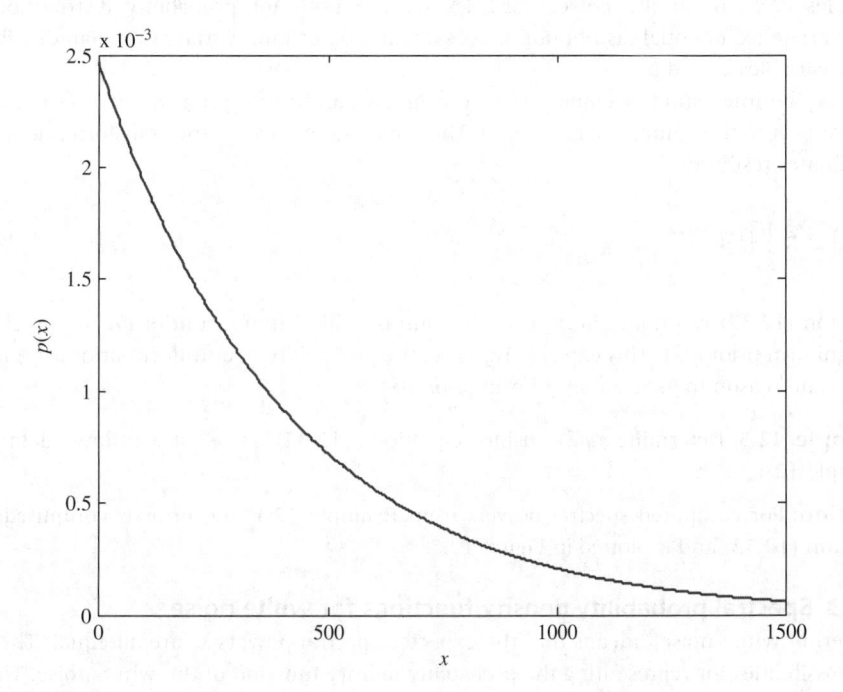

Figure 12.6 PDF of spectral power.

The quantity x_{max} reflects the knowledge about a physical limit, whereas P_{max} reflects the knowledge about a power limitation. It is likely that equation (12.35) would not be true; in that case the white noise density given by equation (12.36) should be employed.

12.3 Minimum cross-entropy power spectrum given auto-correlation

Let an unknown signal $p(x)$ have a power spectrum $G(f)$ and autocorrelation function $r(t)$. Now suppose that information about $G(f)$ is given in terms of a set of samples of autocorrelation function $r_k = r(t_k)$:

$$r_k = r(t_k) = \int_{-W}^{W} G(f) \exp(2\pi i t_k f) df, \quad k = 1, 2, \ldots, m \tag{12.37}$$

where t_k can be equally or unequally spaced. If the frequency spectrum $S(w)$ is discrete as in equation (12.6), the power spectrum $G(f)$ can be expressed as

$$G(f) = \sum_{k=-n}^{n} G_k \delta(w - f_k) \tag{12.38}$$

where $f_k = -f_{-k}$, $G_k = G_{-k} = G(f_k)$, and $G_0 = 0$. Equation (12.37) becomes

$$r_j = \sum_{k=-n}^{n} G_k \exp(2\pi i t_j f_k) \tag{12.39}$$

In the noncomplex form, equation (12.39) becomes

$$r_j = \sum_{k=1}^{n} G_k C_{jk} \tag{12.40}$$

where

$$C_{jk} = 2 \cos (2\pi t_j f_k) \tag{12.41}$$

Since G_k satisfies

$$G_k = \overline{x_k} = \int x_k p(\vec{x}) d\vec{x} \tag{12.42}$$

one can write

$$r_j = \int \left[\sum_k x_k C_{jk} \right] p(\vec{x}) d\vec{x} \tag{12.43}$$

Equation (12.43), however, involves unknown $p(\vec{x})$ which can be estimated using the principle of minimum cross-entropy (POMCE) for which constraints can be given as

$$\int_D g_j(\vec{x}) p(\vec{x}) d\vec{x} = E[g_j] = \overline{g_j} \tag{12.44}$$

where

$$g_j = \sum_k x_k C_{jk} \tag{12.45}$$

This minimum cross-entropy-based formulation differs from the preceding one where knowledge of the expected spectral powers in the form of equation (12.42) was assumed, whereas here all that is available is equation (12.43). Usually $m < n$, equation (12.43) is less informative than equation (12.42).

12.3.1 No prior power spectrum estimate is given

If there is no prior information on $p(\vec{x})$, then a uniform prior $q(\vec{x})$ can be used. Using POMCE subject to equation (12.43) and (12.19),

$$p(\vec{x}) = A \exp\left[-\sum_{j=1}^{m} \lambda_j \sum_{k=1}^{n} x_k C_{jk} \right] \tag{12.46}$$

where λ_j are the m Lagrange multipliers corresponding to the autocorrelation constraints given by equation (12.43). For simplicity, let

$$u_k = \sum_{j=1}^{m} \lambda_j C_{jk} \tag{12.47}$$

Equation (12.46) can be written as

$$p(\vec{x}) = A \exp\left[-\sum_{k=1}^{n} u_k x_k \right] \tag{12.48}$$

Solving for A yields

$$p(\vec{x}) = \prod_{k=1}^{n} u_k \exp(-u_k x_k) \tag{12.49}$$

which has the form of a multivariate exponential distribution with parameters u_k.

For estimating the power spectrum G_k, equation (12.49) is utilized to compute

$$G_k = \overline{x_k} = 1/u_k \tag{12.50}$$

or

$$G_k = \frac{1}{\displaystyle\sum_{j=1}^{m} \lambda_j C_{jk}} \tag{12.51}$$

in which the Lagrange multipliers λ_j are determined using equations (12.40),

$$r_j = \sum_{k=1}^{n} \left[\frac{C_{jk}}{\displaystyle\sum_{i=1}^{m} \lambda_i C_{ik}} \right] \tag{12.52}$$

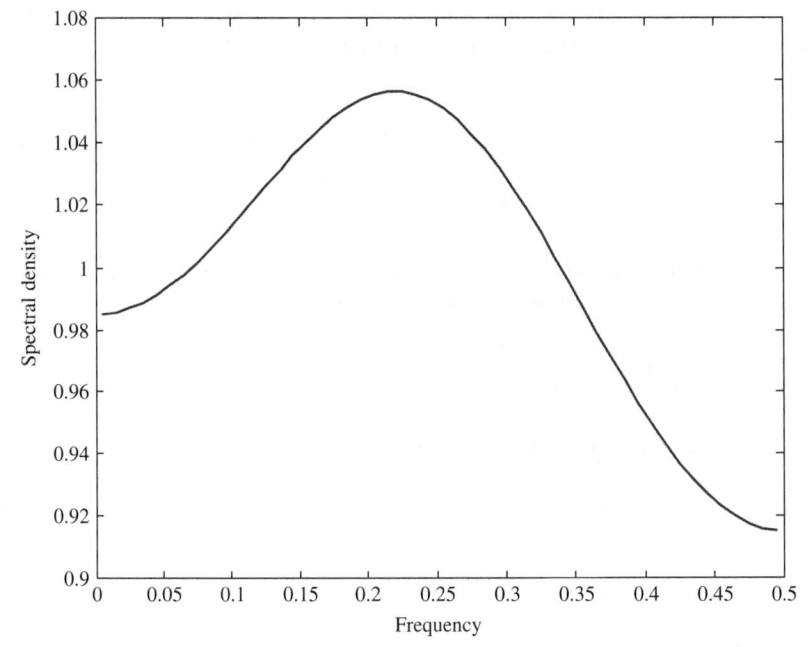

Figure 12.7 Spectral density.

The minimum cross-entropy power spectrum given by equation (10.51) is identical to that for MESA.

Example 12.7: Determine the spectral density from equation (12.49) using observations from Example 12.1.

Solution: From the given data, $r_0 = 1.0000, r_1 = 0.0184$, and $r_2 = -0.0256$. The Lagrange multipliers computed from equation (12.52) are: $\lambda_1 = 1, \lambda_2 = -0.675$, and $\lambda_3 = 0.483$. With u_k computed from equation (12.47), the spectral density computed from equation (12.49) is plotted in Figure 12.7.

12.3.2 A prior power spectrum estimate is given

In this case we obtain autocorrelation function r_k from equation (12.43) with the knowledge of an estimate of P_k of the power spectrum G_k (equation (12.42)). The prior probability density function (exponential) may be given by equation (12.24):

$$q(\vec{x}) = \prod_{k=1}^{n} \frac{1}{P_k} \exp\left(-\frac{x_k}{P_k}\right) \tag{12.53}$$

Equation (12.53) is the minimum cross-entropy-based probability density function given expected spectral powers P_k and uniform prior. If the prior estimate P_k is the estimate Q_k, then the appropriate prior density for obtaining a new estimate given the new autocorrelation function is the posterior given by equation (12.49). Equation (12.53) is equivalent to equation (12.48), since $u_k = 1/Q_k$.

In order to obtain G_k, given P_k and equation (12.43), we minimize cross-entropy with a prior given by equation (12.53) and equation (12.12). This yields

$$p(\vec{x}) = q(\vec{x}) \exp\left[-\lambda - \sum_k u_k x_k\right]$$

$$= \exp(-\lambda_0) \prod_{k=1}^{n} \frac{1}{P_k} \exp\left[-\left(u_k + \frac{1}{P_k}\right)x_k\right] \tag{12.54}$$

where u_k are defined by equation (12.54), and λ_0 is the zeroth Lagrange multiplier. Using the total probability theorem, equation (12.54) can be written as

$$p(\vec{x}) = \prod_{k=1}^{n} \left(u_k + \frac{1}{P_k}\right) \exp\left[-\left(u_k + \frac{1}{P_k}\right)x_k\right] \tag{12.55}$$

For an estimate of the power spectrum Q_k, equation (12.55) is utilized to compute:

$$Q_k = E[x_k] = \overline{x_k} = \frac{1}{u_k + P_k^{-1}} \tag{12.56a}$$

or

$$Q_k = \frac{1}{\frac{1}{P_k} + \sum_j \lambda_j C_{jk}} \tag{12.56b}$$

where the Lagrange multipliers are determined by the condition that Q_k satisfies equation (12.40).

Example 12.8: Determine the spectral density from equation (12.55). Use observations from Example 12.1.

Solution: From the given data, $r_0 = 1.0000, r_1 = 0.0184$, and $r_2 = -0.0256$ from Example 12.7. The Lagrange multipliers computed from equation (12.56b) are: $\lambda_1 = 1, \lambda_2 = -0.638$, and $\lambda_3 = 0.452$. With P_k computed in Example 12.3, the spectral density computed from equation (12.55) is shown in Figure 12.8.

12.3.3 Given spectral powers: $T_k = G_j, G_j = P_k$

Let the actual spectral powers of $p(\vec{x})$ be given as

$$T_k = \int x_k p(\vec{x}) d\vec{x} \tag{12.57}$$

and information about $p(\vec{x})$ is obtained in terms of $M + 1$ values of the autocorrelation function $r(t_k)$ as:

$$r_k = r(t_k) = \sum_{k=-N}^{N} T_k \exp(2\pi i t_k f_k) \tag{12.58}$$

where $k = 0, 1, \ldots, M$, and $t_0 = 0$. It is preferable to write in noncomplex form as

$$r_k = \sum_{j=1}^{N} T_k C_{kj} \tag{12.59}$$

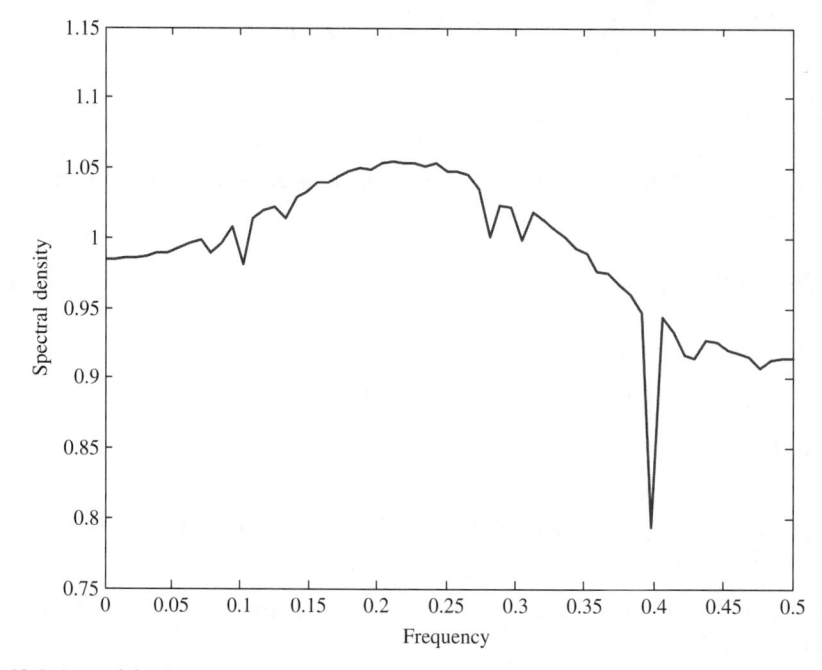

Figure 12.8 Spectral density.

where

$$C_{kj} = 2\cos(2\pi t_k f_j) \tag{12.60}$$

With the use of equation (12.57), the autocorrelation function can be written as

$$r_k = \int \left(\sum_{j=1}^{N} x_j C_{kj} \right) p(\vec{x}) d\vec{x} \tag{12.61}$$

which has the same form as expected value constraints given by equation (12.43). Now given the prior by equation (12.53) and new information given by equation (12.61), the minimum cross-entropy-based posterior estimate of $p(\vec{x})$ becomes

$$p(\vec{x}) = q(\vec{x}) \exp\left[-\lambda_0 - \sum_{k=0}^{M} \lambda_k \sum_{j=1}^{N} x_j C_{kj} \right] \tag{12.62}$$

where the Lagrange multipliers are determined by equation (12.61) and the normalization condition. Substitution of equation (12.53) in equation (12.62) and eliminating the zeroth Lagrange multiplier by equation (12.10), one gets

$$p(\vec{x}) = \prod_{j=1}^{N} \left(u_j + \frac{1}{G_j} \right) \exp\left[-(u_j + \frac{1}{G_j}) x_j \right] \tag{12.63}$$

where

$$u_j = \sum_{k=0}^{M} \lambda_k C_{kj} \tag{12.64}$$

Hence the posterior distribution is a Gaussian distribution. Now the posterior estimate of the power spectrum is given as

$$T_j = \int x_j p(\vec{x}) d\vec{x} = \frac{G_j}{1 + G_k u_j} \tag{12.65}$$

or

$$T_j = \frac{1}{\dfrac{1}{G_j} + \sum_k \lambda_k C_{kj}} \tag{12.66}$$

in which the multipliers λ_k in u_j are chosen such that T_j satisfy the autocorrelation constraints given by equation (12.59).

Example 12.9: Determine the spectral density from equation (12.63), using observations from Example 12.1.

Solution: From the given data, $r_0 = 1.0000, r_1 = 0.0184$, and $r_2 = -0.0256$ from Example 12.7. The Lagrange multipliers computed from equation (12.66) are: $\lambda_1 = 1, \lambda_2 = -0.638$, and $\lambda_3 = 0.452$. With $G_j = P_k$ computed in Example 12.3, the spectral density is plotted in Figure 12.9.

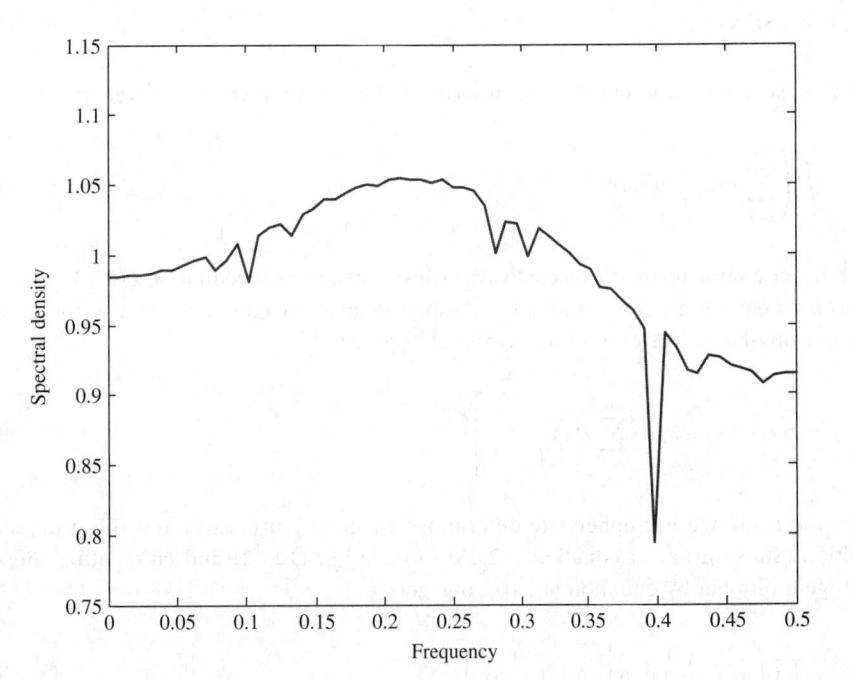

Figure 12.9 Spectral density.

12.4 Cross-entropy between input and output of linear filter

12.4.1 Given input signal PDF

Consider a linear filter with a characteristic function $Y(v_k)$. Suppose a signal with probability density function $q(\vec{x})$ is passed through the filter. Then the magnitude x_k of each power spectrum component is modified by the factor:

$$S_k = |Y(v_k)|^2 \tag{12.67}$$

where $v_k = w_k/2\pi$. The PDF of the signal output $p(\vec{x})$ and the PDF of the input signal $q(\vec{x})$ are related as

$$p(\vec{x}) = \frac{q\left(\dfrac{x_1}{S_1}, \dfrac{x_2}{S_2}, \ldots, \dfrac{x_n}{S_n}\right)}{S_1 S_2 \ldots S_n} \tag{12.68}$$

Equation (12.68) shows that the effect of the filter is one of linear coordinate transformation. The cross-entropy between input and output can be expressed as

$$
\begin{aligned}
H(p; q) &= \int p(\vec{x}) \log \frac{p(\vec{x})}{q(\vec{x})} d\vec{x} \\
&= \int q(y_1, y_2, \ldots, y_n) \log \frac{q(y_1, y_2, \ldots, y_n)}{q(y_1 S_1, y_2 S_2, \ldots, y_n S_n)} d\vec{y} - \sum_k \log(S_k)
\end{aligned} \tag{12.69}
$$

where $y_k = x_k/S_k$. Equation (12.69) holds for any input signal $q(\vec{x})$.

Uniform input PDF

If the input signal has uniform PDF and second the input PDF is exponential. For the uniform PDF of the filter input, the first term in equation (12.69) vanishes and the cross-entropy becomes

$$H(p, q) = -\sum_k \log(S_k) \tag{12.70}$$

Equation (12.70) is the negative of the Burg entropy. This means that minimizing equation (12.70) is equivalent to the Burg entropy minimization or cross-entropy minimization reduces to entropy maximization if the prior is uniform.

Exponential input PDF

The filter input is assumed to have an exponential PDF as equation (12.24) which is Gaussian in terms of the spectral amplitudes a_k and b_k. Then, cross-entropy becomes

$$H(p, q) = -\int q(\vec{y}) \sum_k \frac{(y_k - y_k S_k)}{P_k} d\vec{y} - \sum_k \log(S_k) \tag{12.71}$$

or

$$H(p, q) = -\sum_k [1 - S_k + \log(S_k)] \tag{12.72}$$

because

$$\int y_k q(\vec{y}) d\vec{y} = P_k \tag{12.73}$$

Equation (12.72) does not depend on the P_k values.

12.4.2 Given prior power spectrum

No prior power spectrum estimate is given

Equation (12.51) can also be derived using a filtering argument. The power spectrum of the output signal is given as $Q_k = |Y(f_k)|^2$, if a white signal $q(x)$ is passed through a linear filter having a characteristic function Y. If $p(\vec{x})$ is the output signal then Q_k must satisfy equation (12.40):

$$r_k = \sum_{j=1}^{n} Q_j C_{jk} \tag{12.74}$$

Equation (12.74) suggests that the linear filter is designed with the minimum cross-entropy between input and output. This change in information from the prior yet is still accounting for the given information, since cross-entropy is a measure of information required to transform the prior into the posterior. If the prior $q(\vec{x})$ is white uniform, the cross-entropy between the input and output of the filter is as equation (12.74).

$$H(p, q) = -\sum_{k} \log(Q_k) \tag{12.75}$$

Hence, equation (12.75) is maximized subject to constraints given by equation (12.73). Cross-entropy maximization results in

$$-\frac{1}{Q_k} + \sum_{r=1}^{m} \lambda_r C_{rk} = 0 \tag{12.76}$$

Solution in equation (12.76) is the same as equation (12.36). It may be noted that minimization of equation (12.75), subject to equation (12.74), is equivalent to maximizing the Burg entropy subject to autocorrelation function samples, indicating the equivalence between MESA and MCESA for uniform priors.

Example 12.10: Determine the spectral density using observations from Example 12.1.

Solution: The Lagrange multipliers computed from equation (12.76) are the same as those computed in Example 12.6, which are: $\lambda_1 = 1$, $\lambda_2 = -0.675$, and $\lambda_3 = 0.483$. Thus, the spectral density is also the same as that in Example 12.6.

A prior power spectrum estimate is given

Equation (12.56b) can also be derived using a filtering argument. Suppose a linear filter with characteristic function $Y(w)$ is subjected to a signal with power spectrum P_k. The output power spectrum will be $Q_k = P_k S_k$, where $S_k = |Y(w)|^2$. If the output power spectrum is the new estimate then Q_k must satisfy equation (12.74). For a previous estimate P_k, we design a filter with minimum cross-entropy between input and output given that input probability density function satisfies $\overline{x_k} = P_k$ and the output spectrum satisfies equation (12.60).

If the input density is of the exponential form given by equation (12.53), the cross-entropy between input and output is expressed by equation (12.72). Hence S_k is chosen by minimizing equation (12.72), subject to the constraints given by equation (12.74) re-written as

$$r_j = \sum_{k=1}^{n} P_k S_k C_{jk} \tag{12.77}$$

Minimization of cross-entropy leads to

$$1 - \frac{1}{S_k} + \sum_{r=1}^{m} \lambda_k P_k C_{rk} = 0 \tag{12.78}$$

Solution of equation (12.78) for S_k and computation of $Q_k = P_k S_k$ leads to equations (12.56).

Given spectral powers

Equation (12.65) can be interpreted in terms of linear filtering. As discussed before, when a process with the prior PDF $q(\vec{x})$ is passed through a linear filter with characteristic function $Y(f)$ then the magnitude of each power spectrum component is modified by the factor $A_j = |Y(f_j)|^2$ from G_j to $G_j A_j$ This suggests that the posterior estimate $p(\vec{x})$ can be produced by passing the prior estimate $q(\vec{x})$ through the linear filter with $A_j = (1 + G_j u_j)^{-1}$. Then, choosing the minimum cross-entropy-based posterior given by equation (12.63) is equivalent to designing a linear filter output PDF satisfies equation (12.59) and has the smallest cross-entropy given the prior $q(\vec{x})$. Because the posterior $p(\vec{x})$ has the form:

$$p(\vec{x}) = \prod_{j}^{N} \frac{1}{T_j} \exp\left(-\frac{1}{T_j}\right) \tag{12.79}$$

and prior is given by equation (12.53), the normalized cross-entropy can be expressed as

$$H^*(p, q) = \frac{H(p, q)}{N} = \frac{1}{N} \sum_{j=1}^{N} N\left(\frac{T_j}{G_j} - \log \frac{T_j}{G_j} - 1\right) \tag{12.80}$$

Given spectral powers: another way using a filter

The filtering result can be analyzed in another way. When the input process has a Gaussian PDF with the form given by equation (12.53), then the cross-entropy equation (12.57) becomes

$$H(p, q) = -\sum_{j} \log(A_j) - \int q(\vec{y}) d\vec{y} \sum_{j} (y_j - y_j A_j)/G_j \tag{12.81}$$

or

$$H(p, q) = \sum_{j} [A_j - \log(A_j) - 1] \tag{12.82}$$

Since

$$G_j = y_j q(\vec{y}) d\vec{y} \tag{12.83}$$

Equation (12.82) is independent of G_k.

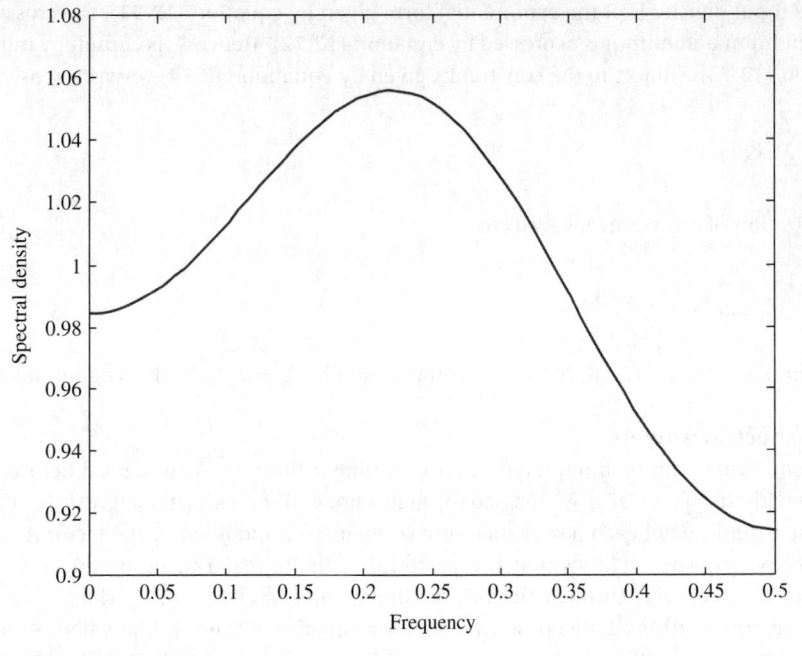

Figure 12.10 Spectral density.

Now suppose equation (12.82) is to be minimized subject to autocorrelation constraints given by equation (12.59) which can be re-written as

$$r_k = \sum_{j=1}^{N} G_j A_j C_{kj} \tag{12.84}$$

This yields

$$A_j = \frac{1}{1 + G_j u_j} \tag{12.85}$$

so that $T_j = G_j A_j$ yields equation (12.66). Equation (12.82) is another expression of equation (12.80).

Example 12.11: Determine the spectral density using observations from Example 12.1.

Solution: From equations (12.84) and (12.85), with given $r_0 = 1.0000$, $r_1 = 0.0184$, and $r_2 = -0.0256$, $u_1 = 1.025$, $\lambda_2 = -0.734$, and $\lambda_3 = 0.366$ are obtained. Then the spectral density through the filter is plotted in Figure 12.10.

12.5 Comparison

When the information is given in the form of autocorrelation functions, the minimum cross-entropy signal PDF is given by equation (12.53) where Q_k are the posterior estimates of the signal power spectrum. The form of Q_k depends on the form of the prior. If the prior is uniform white (UW) then Q_k is given by equation (12.55) as

$$Q_k^{(1)} = \frac{1}{\sum_r \lambda_r^{(UW)} C_{rk}} \tag{12.86}$$

If the prior is Gaussian white (GW) then Q_k is given by equation (12.83) with $P_k = p$ for all k:

$$Q_k^{(2)} = \frac{1}{(1/P) + \sum_r \lambda_r^{(GW)} C_{rk}} \tag{12.87}$$

where P is the maximum value of the expected power per unit frequency.

If the prior is Gaussian nonwhite (GNW), then Q_k is given by equation (12.83):

$$Q_k^{(3)} = \frac{1}{(1/P_k) + \sum_r \lambda_r^{(GNW)} C_{rk}} \tag{12.88}$$

where P_k are prior estimates of the power spectrum. In equations (12.86) to (12.88), the m Lagrange multipliers are determined by satisfying the requirement that $Q_k^{(i)}$ satisfy the autocorrelation constraints:

$$r_j = \sum_{k=1}^{n} Q_k^{(i)} C_{jk}, \ j = 1, 2, \ldots, m; i = 1, 2, 3 \tag{12.89}$$

First $Q_k^{(1)}$ and $Q_k^{(2)}$ are compared. If it is assumed that one of the sample autocorrelations, say r_1 is for zero lag ($t_1 = 0$), then equation (12.41) yields $C_{1k} = 2$ for all k. Equation (12.87) can then be written as

$$Q_k^{(2)} = \frac{1}{\sum_r \lambda_r C_{rk}} \tag{12.90}$$

where $\lambda_r = \lambda_k^{(i)}$ for $r = 1, 2, \ldots, m$, and $\lambda_1 = \lambda_1^{(1)} + 1/(2P)$. Comparison with equation (12.65) shows that $Q_k^{(1)} = Q_k^{(2)}$ for all k. It is thus concluded that a uniform white prior and a Gaussian white prior lead to the same posterior power spectrum estimate, provided one of the autocorrelation samples is for zero lag. This makes sense because the zero lag sample is nothing but the total expected power per discrete frequency, and the Gaussian white prior results from the minimization of the cross-entropy with respect to uniform white prior given the information on the total expected power per unit frequency. However, equations (12.65) and (12.66) are not equivalent if there is nonzero autocorrelation sample. To visualize this, assume there is only one correlation sample $r_1 = \sum Q_k C_{1k}$ with $t_1 \neq 0$. If $Q_k^{(1)} = Q_k^{(2)}$ holds for all $k = 1, 2, \ldots, n$, then

$$\lambda_1^{(1)} C_{1k} = P^{-1} + \lambda_2^{(2)} C_{1k} \tag{12.91}$$

would be true for all k. Here $\lambda_1^{(1)}$ and $\lambda_2^{(2)}$ are constants and C_{1k} varies with k, since $t_1 \neq 0$. This suggests that equation (12.91) and therefore $Q_k^{(1)} = Q_k^{(2)}$ would not hold for all k.

Now equations (12.76) and (12.78) are compared. Let there be a single autocorrelation sample $r_1 = \sum_k Q_k C_{1k}$ that may or may not be a zero-lag sample. If $Q_k^{(2)} = Q_k^{(3)}$ were to hold for all $k = 1, 2, \ldots, n$, then

$$P^{-1} + \lambda_1^{(2)} C_{1k} = P_k^{-1} \lambda_1^{(3)} C_{1k} \tag{12.92}$$

would have to be satisfied for all k. Here $\lambda_1^{(2)}$ and $\lambda_1^{(3)}$ are constant and P_k^{-1} varies independent of C_{1k} equation (12.92) cannot be true for all k whether or not C_{1k} is constant (zero lag sample).

12.6 Towards efficient algorithms

Equations (12.86) to (12.88) yield cross-entropy spectral estimates, based on autocorrelation samples. First, the Lagrange multipliers λ_k must be determined, which can be achieved by substituting any of these equations into equation (12.89), and solving the m resulting equations. However, there is no mention of the number and spacing of either frequencies or autocorrelation samples. This, therefore, may not be the best way to obtain the actual power spectrum values.

It is reasonable to assume that the frequencies w_k are equally spaced $w_k = k\Delta w$ at the Nyquist rate, $t_r = 1/(2n\Delta w)$ In this case coefficients C_{rk} become

$$C_{ik} = 2\cos(\pi rk/n) \tag{12.93}$$

The Lagrange multipliers are determined by inserting equation (12.93) in equation (12.74) and solving the resulting system of equations:

$$r_k = \sum_{k=1}^{n} P_k C_{rk} \left(1 + P_k \sum_{j=1}^{m} \lambda jC_{jk} \right)^{-1} \tag{12.94}$$

Expanding equation (12.94) and neglecting terms of $O(\lambda_j^2)$ and higher, if λ_j are close to zero, the result is

$$r_k = \sum_k \left(P_k C_{rk} - P_k^2 C_{rk} \sum_j \lambda_j C_{jk} \right) \tag{12.95}$$

Note the prior spectral estimates P_k and the power autocorrelation samples r_k' are related as $r_k' = \sum_k P_k C_{rk}$. Then equation (12.95) can be written as

$$\sum_{j=1}^{m} \lambda_j dr_j = r_r - r_r' \tag{12.96}$$

where

$$dr_j = \sum_{k=1}^{n} P_k^2 C_{rk} C_{jk} \tag{12.97}$$

Thus, when the autocorrelation function changes gradually, equation (12.94) reduces to a set of linear equations (12.96) whether or not equation (12.93) holds.

12.7 General method for minimum cross-entropy spectral estimation

Tzannes et al. (1985) introduced a new form of MCESA based on the minimization of a cross-entropy function entailing spectral densities rather than probability density functions. Let $S(w)$ be a prior estimate (guess) of spectrum and $T(w)$ the desired spectrum. In order to determine the optimum $T(w)$, $S(w)$ is normalized to have an area of one and the cross-entropy is minimized:

$$H = \int T(w) \log \frac{T(w)}{S(w)} dw \tag{12.98}$$

The constraints for minimization are

$$\frac{R(\tau)}{R(0)} = \int_0^\infty T(w) \exp(iw\tau) dw, \quad \tau = 1, 1, 2, \ldots, N \tag{12.99}$$

which in noncomplex form is:

$$\frac{R(\tau)}{2R(0)} = \int_0^\infty T(w) \cos(w\tau) dw, \quad \tau = 1, 1, 2, \ldots, N \tag{12.100}$$

where factor $2R(0)$ serves the normalization purpose. This is the autocorrelation function-power spectrum Fourier transform pair.

Minimization of H, subject to equation (12.100), is done using the method of Lagrange multipliers, resulting in

$$T(w) = S(w) \exp\left[-1 - \sum_{k=0}^N \lambda_k \cos wk\right] \tag{12.101}$$

The values of the Lagrange multipliers are determined using the known autocorrelation function values in equation (12.100). If $S(w)$ is uniform white noise then the method reverts to maximizing

$$H(w) = \int S(w) \log S(w) dw \tag{12.102}$$

and it becomes equivalent to MESA.

References

Burg, J.P. (1967). Maximum entropy spectral analysis. Paper presented at 37th Annual Meeting of Society of Exploration Geophysicists, Oklahoma City, Oklahoma.

Burg, J.P. (1969). A new analysis technique for time series data. Paper presented at NATO Advanced Institute on Signal Processing with Emphasis on Underwater Acoustics.

Burg, J.P. (1972a). The relationship maximum entropy spectra and maximum likelihood spectra. *Geophysics*, Vol. 37, No. 2, pp. 375–6.

Burg, J.P. (1972b). The relationship between maximum entropy spectra and maximum likelihood spectra. in: *Modern Spectral Analysis*, edited by D.G. Childers, pp. 130–1, M.S.A.

Burg, J.P. (1975). Maximum entropy spectral analysis. Unpublished Ph.D. dissertation, Stanford University, Palo Alto, California.

Dalezios, N.R. and Tyraskis, P.A. (1999). Maximum entropy spectra for regional precipitation analysis and forecasting. *Journal of Hydrology*, Vol. 109, pp. 25–42.

Fougere, P.F., Zawalick, E.J. and Radoski, H.R. (1976). Spontaneous line splitting in maximum entropy power spectrum analysis. *Physics of the Earth and Planetary Interiors*, Vol. 12, pp. 201–7.

Johnson, R.W. and Shore, J.E. (1983). Minimum cross-entropy spectral analysis of multiple signals. *IEEE Transactions on Acoustics, Speech, and Signal Processing*, Vol. ASSP-31, No. 3, pp. 574–82.

Kapur, J.N. (1999). *Maximum Entropy Models in Science and Engineering*. Wiley Eastern, New Delhi, India.

Kapur, J.N. and Kesavan, H.K. (1997). *Generalized Maximum Entropy Principle (with Applications)*. Sandford Educational Press, University of Waterloo, Waterloo, Canada.

Kapur, J.N. and Kesavan, H.K. (1992). *Entropy Optimization Principles with Applications*. 409 p., Academic Press, New York.

Padmanabhan, G. and Rao, A.R. (1988). Maximum entropy spectral analysis of hydrologic data. *Water Resources Research*, Vol. 24, No. 9, pp. 1519–33.

Shore, J.E. (1979). Minimum cross-entropy spectral analysis. NRL Memorandum Report 3921, 32 pp., Naval Research Laboratory, Washington, D.C.

Shore, J.E. (1981). Minimum cross-entropy spectral analysis. *IEEE Transactions on Acoustics, Speech, and Signal Processing*, Vol. ASSP-291, pp. 230–7.

Shore, J.E. (1983). Axiomatic derivation of the principle of maximum entropy and the principle of minimum cross entropy. *IEEE Transactions on Information Theory*, Vol. IT-26, pp. 26–37.

Ulrych, T.J. and Clayton, R.W. (1976). Time series modeling and maximum entropy. *Physics of the Earth and Planetary Interiors*, Vol. 12, pp. 189–99.

Additional References

Brockett, P.L., Charnes, A. and Paick, K.H. (1986). Computation of minimum cross-entropy spectral estimates: An unconstrained dual convex programming method. *IEEE Transactions on Information Theory*, Vol. IT-32, No. 2, pp. 236–41.

Burr, R.L. and Lytle, D.W. (1989). Minimum cross-entropy spectral analysis of time-varying biological signals. *IEEE Transactions on Systems, Man, and Cybernetics*, Vol. 18, No. 5, pp. 802–7.

13

Evaluation and Design of Sampling and Measurement Networks

Hydrometric data are required for efficient planning, design, operation, and management of virtually all water resources systems, including water supply reservoirs, dams, levees, recreation and fisheries facilities, flood control structures, hydroelectric power plants, irrigation schemes, urban and rural drainage, and wastewater treatment facilities. Many studies have applied the entropy theory to assess and optimize data collection networks (e.g., water quality, rainfall, stream flow, hydrometric, elevation data, landscape, etc.). Decision making in water resources project design and evaluation is closely linked to the amount of hydrologic information available. If enough accurate and relevant information is available, the likelihood of an under-design or over-design is reduced. Thus, economic losses can be minimized, resulting in an overall increase in the benefit/cost ratio. However, it is not always easy to quantitatively define the optimum level of information needed for planning, design, and development of a specific project in a watershed. This is largely due to the difficulty in developing cost and benefit functions of hydrologic information. This then leads to the difficulty in achieving an optimum balance between the economic risk due to inadequate information and the cost of a network capable of providing the required information. This chapter presents basic entropy-related considerations needed for the evaluation and design of monitoring networks.

13.1 Design considerations

A methodology for data collection network design must take into account the information of each gaging station or potential gaging station in the network. A station with a higher information content would generally be given a higher priority over other stations with lower information content. The information content of a station must, however, be balanced with site-specific uses and users of the data collected at a station. For example, a station which is used by one user might be given a lower priority than a station that has diverse uses. Burn and Goulter (1991) developed a data collection network design framework considering such issues.

In general, a framework for network design or evaluation considers a range of factors, including: 1) objectives of sampling, 2) variables to be sampled, 3) locations of measurement

Entropy Theory and its Application in Environmental and Water Engineering, First Edition. Vijay P. Singh.
© 2013 John Wiley & Sons, Ltd. Published 2013 by John Wiley & Sons, Ltd.

stations, 4) frequency of sampling, 5) duration of sampling, 6) uses and users of data, and 7) socio-economic considerations. Effective monitoring is also related to these factors. Evaluation of a network has two modes: 1) number of gages and their location (space evaluation), and 2) time interval for measurement (time evaluation). The information in one mode may be supplemented by the other with appropriate transfer mechanisms and by cross-correlation structure (space-time tradeoff). Space-time evaluation of networks should not be considered as fixed but should periodically be subject to revision and is evolutionary. Uslu and Tanriover (1979) analyzed the entropy concept for the delineation of optimum sampling intervals in data collection systems, both in space and time.

All designs, whether of the network or the monitoring program, must be cost effective in gathering data and cost efficient in obtaining information from data. These two requirements call for evaluating the performance of a network. Such an evaluation must consider benefits of monitoring with respect to the objectives of monitoring and the cost, both marginal and average, of obtaining those benefits. Sometimes it is the budget that controls the network and monitoring program design. Then, the network problem reduces to one of obtaining the greatest benefit (most information) for the available budget.

13.2 Information-related approaches

Besides the entropy theory, there are many approaches that have been used in monitoring network analysis (Mishra and Coulibaly, 2009). These may include information variance; transfer function variance; correlation function; simulation; economic-worth of data approach; decision theory and Bayesian analysis; linear estimation techniques, such as Theissen polygons and spline surface fitting; kriging; square grid technique; amongst others. The first three approaches are briefly discussed here.

13.2.1 Information variance

One measure of information is variance denoting the mean of the squared differences between measured and true values of statistics, of say, for example, mean areal rainfall. The higher the variance the greater the measurement error; thus, more gages are needed in the area and vice versa. The information variance approach involves space-time tradeoff explicitly, that is, the lack of data in one domain (say, time) may be compensated for by extrapolation of similar data in the other domain (say, space). The decision whether to continue or discontinue a gage is based on the relative lack of information. Both space and time analyses use some common measure of information.

To illustrate the use of variance as a measure, consider an estimate of the mean rainfall depth over the watershed under consideration which is obtained from measurements of rain gages positioned at specified locations in space. The mean squared error of this estimate and the mean depth is the variance which decreases with increasing number of rain gages. For a fixed rain gage in time, variance is the mean squared error between measured rainfall depth and time-averaged rainfall depth at the same rain gage. Let x_{it} be rainfall depth at rain gage i for time t, n be the total number of rain gages, and T be the length of record (restricted to the number of months or years that rainfall network has been in operation). Then, the space-time mean rainfall depth is defined as

$$\bar{x} = \frac{1}{nT} \sum_{i=1}^{n} \sum_{t=1}^{T} x_{it} \tag{13.1}$$

for all rain gages in the watershed and for all available records. The corresponding space–time variance can be defined as

$$S_x^2 = \frac{1}{nT} \sum_{i=1}^{n} \sum_{t=1}^{T} (x_{it} - \bar{x})^2 \tag{13.2}$$

When evaluating a rainfall network by variance it is assumed that 1) the rainfall process is random in time and space; 2) the rainfall process is weakly stationary; and 3) the covariance function is separable in both time and space. If the space-time mean rainfall depth, obtained from equation (13.1) is denoted as $\bar{x}$, then the variance of $\bar{x}$, as a measure of the network efficiency, is

$$\text{var}(\bar{x}) = \text{var}\left[\frac{1}{nT} \sum_{i=1}^{n} \sum_{t=1}^{T} x_{it} \right] \tag{13.3a}$$

This variance can be expressed as the product of the variance of a rainfall record of one rain gage, σ^2 (point variance), and temporal and spatial reduction factors, $F(T)$ and $G(n)$ depending only on sampling in time and space, respectively (Rodriguez-Iturbe and Mejia, 1974). The point variance (σ^2) is equal for all rain gauges because of the assumed stationarity. Then,

$$\text{var}(\bar{x}) = \sigma^2 F(T) G(n) \tag{13.3b}$$

where $G(n)$ is the spatial reduction factor, and $F(T)$ is the temporal reduction factor. For sampling in space, Rodriguez-Iturbe and Mejia (1974) expressed $G(n)$ as

$$G(n) = \frac{1}{n^2} \{ n + n(n-1) E[r(d)] \} \tag{13.4a}$$

where d is the distance between rain gages and $r(d)$ is the spatial correlation function. A derivation of equation (13.4), using POME, is given by Krstanovic and Singh (1988); it can be written as

$$G(n) = \frac{1}{n^2} \left\{ n + n(n-1) \exp\left(\frac{C_v^2}{2} - 1 \right) \left[1 - \Phi\left(\frac{C_v^2 - 1}{\sqrt{2} C_v} \right) \right] \right\} \tag{13.4b}$$

where C_v is the coefficient of variation of the distance among existing rain gages and $\Phi(.)$ is the error function. It is found that for $C_v > 5.6$, $G(n)$ becomes almost constant regardless of the value of n. A realistic range of C_v is (0.1, 4.6), virtually covering a broad range of different values of means and variances of distances.

For sampling in time, Rodriguez-Iturbe and Mejia (1974) described a procedure that was modified by Krstanovic and Singh (1988) for which the temporal reduction $F(T)$ can be written as

$$F(T) = \frac{1}{T^2} \left[\sum_{k=1}^{T} \rho(0) + 2 \sum_{k=1}^{T-1} \sum_{j=k+1}^{T} \rho(j-k) \right] \tag{13.5a}$$

where $\rho(k)$ denotes the autocorrelation function (ACF) at lag k. Equation (13.5a) is rewritten as

$$F(T) = \frac{1}{T} + \frac{2}{T^2}\left[\sum_{t=1}^{T-1}(T-t)\rho(t)\right] \tag{13.5b}$$

Thus, the temporal reduction factor depends on the length of record T and the magnitude of autocorrelation function (ACF) $\rho(t)$. The shorter the record, the smaller the reduction [$F(T)$ approaches 1]; the longer the record is the more dependency is within ACF; $F(T)$ differs for each rainfall record.

Spatial reduction factor $G(n)$, temporal reduction factor $F(T)$, and total reduction in variance $F(T) \times G(n)$ are now computed. For spatial reduction factor, it is sufficient to compute the coefficient of variation, C_v, of the distances among existing rain gages. If one increases or decreases the number of rain gages such that C_v does not change, then the corresponding $G(n)$ is variable. For the temporal reduction factor, equation (13.5b) is used to compute $F(T)$ for each rain gage. The total reduction in the variance of rainfall record is obtained by multiplying $G(n)$ with $F(T)$. However, two disadvantages of the variance analysis should be noted. First, because of nearby values of the total reduction in variance, it is difficult to determine exactly the number of rain gages necessary for every sampling interval. Second, this analysis does not consider location of the chosen rain gages in the watershed.

Another way to assess the adequacy of an existing rain gage network is by computing the assigned percentage of error (ε) in the estimation of mean value which is a function of the number of gages as:

$$n = \left(\frac{C_v}{\varepsilon}\right)^2 \tag{13.6}$$

where n is the optimum number of gages, and C_v is the coefficient of variation of the rainfall values of the gages. If there are m rain gages in the watershed recording $R_1, R_2, \ldots, R_m$ values of rainfall for a fixed time interval, then C_v is computed as

$$C_v = \frac{100S}{\overline{R}} \tag{13.7a}$$

where

$$\overline{R} = \frac{1}{m}\sum_{i=1}^{m}R_i \tag{13.7b}$$

and S is the standard deviation of R:

$$S = \left[\frac{\sum_{i=1}^{m}(R_i - \overline{R})^2}{m-1}\right]^{1/2} \tag{13.8}$$

13.2.2 Transfer function variance

The transfer function variance approach employs spectral density function (SPF) that measures the distribution of variance of the variable, say rainfall, over the range of its inherent frequencies. A higher SPF value at a certain frequency makes a greater contribution to the variance from the component of that frequency. An advantage of using transfer-function variance is that it reduces all the information concerning space-time dependence to a functional form.

13.2.3 Correlation

Another measure of information is cross-correlation amongst records from nearby sites. Cross-correlation helps with space-time tradeoff and regional data collection. The correlation coefficient r_{xy} is calculated for each pair of stations as:

$$r_{xy} = \frac{Cov_{xy}}{S_x S_y} \tag{13.9a}$$

where Cov_{xy} is the covariance between the random variables X and Y; S_x and S_y are the standard deviation of variables X and Y, respectively. Cov_{xy} can be obtained as

$$Cov_{xy} = \frac{\sum_{i=1}^{n}(x_i - \bar{x})(y_i - \bar{y})}{n-1} \tag{13.9b}$$

where $\bar{x}$ and $\bar{y}$ are the means of variable X and Y, respectively. Plotting the correlation coefficient with distance between gages and assuming an acceptable value of the correlation coefficient, one can determine the location of gages as well as the number of gages. The correlation coefficient may be expressed as a function of distance using an exponential form:

$$r(d) = e^{-d/\lambda} \tag{13.10}$$

where d is the distance between two gages (or points), and λ is the correlation length (the inverse of the correlation length is the correlation decay rate).

13.3 Entropy measures

13.3.1 Marginal entropy, joint entropy, conditional entropy and transinformation

Entropy is used for measuring the information content of observations at a gaging station. The measures of information are: marginal entropy, conditional entropy, joint entropy, and transinformation which have been discussed in Chapter 2. The information observed at different sites (gaging stations) can be inferred, to some extent, from the observations at other sites. The information transferred among information emitters (predictor stations) and the information receivers (predicted or predictant stations) can be measured by transinformation or mutual information.

Mutual information is used for measuring the inferred information or equivalently information transmission. Entropy and mutual information have advantages over other measures of information. They provide a quantitative measure of 1) the information at a station, 2) the information transferred and information lost during transmission, and 3) a description of the relationships among stations based on their information transmission characteristics.

Let the data being collected at a station correspond to a random variable X. Then, the marginal entropy, $H(X)$, can be defined as the potential information of the variable; this is also the information of that gaging station. The joint entropy $H(X, Y)$ is the total information content contained in both X and Y, that is, it is the sum of marginal entropy of one of the stations and the uncertainty that remains in the other station when a certain amount of information that it can convey is already present in the first station. The multidimensional joint entropy for n gages (with measurement record denoted as $i_j, i = 1, 2, \ldots, N_i; j = 1, 2, \ldots, n; N_i$ is the length of record or the number of values of the i-th station.) in a watershed which represents the common uncertainty of their measured data sets can be expressed as:

$$H[X_1, X_2, \ldots, X_n] = -\sum_{i_1}^{N_1} \cdots \sum_{i_{n+1}}^{N_m} p(i_1, i_2, \ldots, i_n) \log[p(i_1, i_2, \ldots, i_n)] \tag{13.11}$$

For two gaging stations, X and Y, the conditional entropy $H(X|Y)$ is a measure of the information content of X which is not contained in Y, or entropy of X given the knowledge of Y or the amount of information that still remains in X even if Y is known. The conditional entropy can be interpreted as the amount of lost information. The amount of uncertainty left in the central gage when the records of all other gages are known is expressed by the multivariate conditional entropy of the central gage conditioned on all other records. Similarly, the uncertainty left in the group of gages $(i_1, \ldots, i_n)$ when any new gage is added (i.e., the expansion of the existing gage network) can be defined as

$$H[(X_1, X_2, \ldots, X_n)|X_{n+1}] = -\sum_{i_1}^{N_1} \cdots \sum_{i_{n+1}}^{N_{m+1}} p(i_1, i_2, \ldots, i_n) \log[p(i_1, i_2, \ldots, i_n|i_{n+1})] \tag{13.12a}$$

or

$$H[(X_1, X_2, \ldots, X_n)|X_{n+1}] = H(X_1, X_2, \ldots, X_n, X_{n+1}) - H(X_{n+1}) \tag{13.12b}$$

The mutual entropy (information) between X and Y, also called transinformation $T(X, Y)$, is interpreted as the reduction in the original uncertainty of X, due to the knowledge of Y. It can also be defined as the information content of X which is also contained in Y. In other words, it is the difference between the total entropy and the sum of entropies of two stations. This is the information repeated in both X and Y, and defines the amount of uncertainty that can be reduced in one of the stations when the other station is known. Thus, the information transmitted from station X to station Y is represented by $T(X, Y)$. Corresponding to equations (13.12a) and (13.12b), the multidimensional transinformation between the n existing gages and the new (added) gage $(n + 1)$ can be defined as

$$T[(X_1, X_2, \ldots, X_n), X_{n+1}] = H(X_1, X_2, \ldots, X_n) - H[(X_1, X_2, \ldots, X_n), X_{n+1}] \tag{13.13}$$

$T(X, Y)$ is symmetric, that is, $T(X, Y) = T(Y, X)$, and is non-negative. A zero value occurs when two stations are statistically independent so that no information is mutually transferred, that is, $T(X, Y) = 0$ if X and Y are independent. When two stations are functionally dependent, the information at one site can be fully transmitted to another site with no loss of information at all. Subsequently, $T(X, Y) = T(X)$. For any other case, $0 \leq T(X, Y) \leq H(X)$. Larger values of T correspond to greater amounts of information transferred. Thus, T is an indicator of the capability of the information transmission and the degree of dependency of two stations.

13.3.2 Informational correlation coefficient

Transferability of information between hydrologic variables depends on the degree and the structure of interdependence of variables. The most likely measure of association between variables is the correlation coefficient. Its use is valid under the assumption of normality of variables and linearity of relationship between them. If variables are nonlinearly related, then either the variables have to be transformed to linearize the regression function or nonlinear regression has to be developed. For both linear and nonlinear types of interdependence, the correlation coefficient measures the amount of information that is actually transferred by the assumed regression relationship. If correlation coefficient is zero, it does not necessarily mean the absence of association between variables but may also be due to the inappropriate choice of the regression relation. The informational correlation coefficient and transinformation represent the extent of transferable information without assuming any particular type of interdependence. They also provide a means of judging the amount of information actually transferred by regression. The informational correlation coefficient R_0 (Linfoot, 1957) is a dimensionless measure of stochastic interdependence, varies between 0 and 1, and is expressed in terms of transinformation as

$$R_0 = \sqrt{1 - \exp(-2T_0)} \qquad (13.14)$$

It does not assume normality or any type of functional relationship between stations and therefore has an advantage over ordinary correlation coefficient. It reduces to the classical correlation coefficient when the normality and linearity assumptions are satisfied.

When the marginal and joint probability distributions of stations X and Y are approximated by their relative frequency distributions or when no particular probability distribution is assumed for the stations, then T_0 represents the upper limit of transferable information between stations. The informational correlation coefficient R_0 is a dimensionless measure of stochastic interdependence which varies between 0 and 1, and is a function of mutual information T_0 between stations. It does not assume normality or any type of functional relationship between stations and therefore has advantages over the ordinary correlation coefficient. It reduces to the classical correlation coefficient when the normality and linearity assumptions are satisfied. Although R_0 and T_0 do not provide any functional relationship between stations to transfer information, they serve as criteria for checking the validity of assumed types of dependence and probability distributions of the stations. Since T_0 represents the upper limit of transferable information, it can be used as a criterion for defining the amount of actually transferred information under the assumptions made. If T_1 represents the transinformation for any assumed type of relation between stations then the ratio

$$t_1 = \frac{T_0 - T_1}{T_0} = 1 - \frac{T_1}{T_0} \qquad (13.15)$$

measures the amount of nontransferred information, and $1 - t_1$ measures the amount of transferred information. Likewise R_0, and t_1 can be used as criteria to judge the validity of assumptions made about the dependence between stations. Entropy or transinformation does not provide any means of transferring information but provides a means for measuring transferable information. Thus, it can help improve the transfer of information by regression analysis.

If the actual amount of information transferred by regression analysis is far below the transferable information defined by entropy measures, then one can attempt to improve the regression analysis. This can be accomplished by 1) selecting marginal and joint distributions that better fit the data, 2) searching for better relationships between stations, 3) analyzing the effect of autocorrelation of each station upon interdependence and regression, and 4) analyzing the effect of lag cross correlation upon interdependence and information transfer.

13.3.3 Isoinformation

The coefficient of nontransferred information, defined by equation (13.14), as a measure of the non-transferred information as the percentage of the total transferable information is employed. Here T_0 is the total transferable information (not necessarily achieved by the network), and T_1 is the measured transinformation between X_1 and X_2. To illustrate, assume that X_1 is associated with the rainfall record of the central rain gage in the region, and $X_2 = X_i (i = 2, \ldots, n)$ associated with rainfall record of any other rain gage. Then,

$$T_1 = T(X_1, X_i) = H(X_1) - H(X_1 | X_i) \ (i = 2, \ldots, n) \tag{13.16a}$$

The value of T_0 can be expressed as

$$T_0 = H(X_1) \tag{13.16b}$$

Thus, the coefficient of nontransferred information is

$$t_1 = \frac{H(X_1 | X_i)}{H(X_1)}, \qquad 0 \leq t_1 \leq 1 \tag{13.17}$$

Similarly, $1 - t_1$ defines the coefficient of transferred information, or transferred information as a fraction of the total transferable information defined by equation (13.14).

By computing the coefficient of transferred information for all rain gages, the isoinformation contours can be constructed. These contours will be the lines of equal transfer of information in the region (Krstanovic and Singh, 1988). The isoinformation contour of the bivariate rainfall record is the line of equal common information between any rain gage in the watershed and the other existing rain gages located in the watershed. In the selection process, the first chosen rain gage has the highest information content and this can be designated as central rain gage. Thus, it is convenient to choose that rain gage as the reference point in the construction of all isoinformation contours when choosing rain gages in order of descending importance. It is true that isoinformation contour can be constructed between any two rain gages, but it may not benefit the selection process.

13.3.4 Information transfer function

In a given watershed or area there are limited monitoring or gaging stations which define a discrete field. This field can be extended to a continuous field by considering each point in the watershed as a gaging station. In the continuous field the information transfer function is defined. Thus, the transmission of information of a given station changes successively in any direction. Let X define the central gage (or basic point) and Y another gage (or auxiliary point) in a certain direction, and let d_{xy} be the geometric distance between X and Y. Then the information transfer function (ITF) of X about Y can be defined (Zhang et al., 2011) as

$$ITF_X(d) = \frac{1}{(1 + ad)^b} \tag{13.18}$$

where a and b are parameters. If a and b are small, ITF_X will be large, indicating greater transmission of information from X. The values of a and b may vary from watershed to watershed. It is seen from equation (13.18) that $ITF_X(0) = 1$ and ITF_X tends to 0 as d tends to infinity. Further, the derivative of equation (13.18) will define the information transfer intensity (ITI) or rate:

$$ITI = \frac{d(ITF_X)}{dd} \leq 0 \qquad (13.19)$$

13.3.5 Information distance

Let L be the distance of station X to the boundary of the watershed along a given direction. Then, the information distance (ID_X) of X in this direction can be defined as

$$ID_X(L) = \left[2 \int_0^L sITF_X ds \right]^2 = \frac{1}{a^2} \left[\frac{(1 + aL)^{2-b} - 1}{2 - b} - \frac{(1 + aL)^{1-b} - 1}{1 - b} \right] \qquad (13.20)$$

where a and b are parameters. ID_X can be regarded as the average information transfer of X in the given direction. If $ITF_X = 1$, then $ID_X = 1$, implying the transfer of entire information to the boundary. On the other hand, if $L = 0$, then $ITF_X = 0$ and $ID_X = 0$. In this case station X does not transfer any information outward.

13.3.6 Information area

For a station X, information transfer occurs in all directions. It may, therefore, be appropriate to define the area influenced by X as the information field, abbreviated as IF_X of X. To mathematically formulate IF_X requires setting up an appropriate coordinate system. To that end, let the geometric location of X be a pole and one direction (e.g., the direction may be from east to west) be a polar axis. In this way, polar coordinates are defined. Let θ be the polar angle. Then one can relate parameters a and b in ITF to θ, that is, $a(\theta)$ and $b(\theta)$, and $ITF = ITF(\theta)$. Now the information field of X can be defined as

$$IF_X = \{IF_X(\theta),\ 0 \leq \theta \leq 2\pi\} \qquad (13.21)$$

Assuming ITF as a continuous function of θ and d, the information area IA_X can be written as

$$IA_X = \frac{1}{2} \int_0^{2\pi} (ID_X)^2 d\theta \qquad (13.22)$$

which measures the total region of influence of X with regard to its nearby region.

13.3.7 Application to rainfall networks

The use of entropy measures is now illustrated by applying to rainfall networks. A rainfall network should be designed such that the space-time variability of rainfall is sampled optimally or accurately. The accuracy should be sufficient enough so as to reduce the standard error of areal average rainfall to a negligible level. The standard error depends on the variability of the rainfall field, including rainfall variability and intermittency. The main objective of a rain gage network design is to select an optimum number of rain gages and their locations such that the desired amount of information on rain through rain data is obtained. Other considerations of importance in the network design may include the obtaining of an adequate record length prior to the utilization of data, the development of a mechanism for transferring information from gaged to ungaged locations whenever needed, and the estimation of the probable magnitude of error arising from network density, distribution, and record length.

Krstanovic and Singh (1992a, b) used the marginal entropy measure to draw contour maps of the rainfall network in Louisiana and evaluated the network according to the entropy map. Lee and Ellis (1997) compared kriging and the maximum entropy estimator for spatial interpolation and their subsequent use in optimizing monitoring networks. Husain (1989) and Bueso et al. (1999) used the entropy theory to illustrate a framework for spatial sampling of a monitoring network. Ozkul et al. (2000) presented a methodology using the entropy theory for assessing water quality-monitoring networks. Their study was a follow up of earlier work by Harmancioglu and Alpaslan (1992).

Evaluation of rainfall networks in time

Let X be a random rainfall variable, such as depth over a time interval (a day, week, month or year) and has a record length N. X has N different values x_i, ($i = 1, \ldots, N$) each with a probability $p(x_i) = p_i$, ($i = 1, \ldots, N$) (i.e., probability of that rainfall depth value to occur). Then, the marginal entropy of X measures the uncertainty associated with the realization of x_i from X. Here X can take on different duration rainfall depth values. For example, $X_1 = \{$(amount of rainfall during day $1 = x_1$), (amount of rainfall during day $2 = x_2$), $\ldots$, (amount of rainfall during day $N = x_n$)$\}$; or $X_2 = \{$(total rainfall during first two days $= x_1$), (total rainfall during second two days $= x_2$), $\ldots$, (total rainfall during last two days of the record $N = x_n$)$\}$; $X_7 = \{$(total rainfall in week 1 of the record $= x_1$), (total rainfall in week 2 of the record $= x_2$), $\ldots$, (total rainfall of the last week in the record $= x_n$)$\}$; $X_{30} = \{$(total rainfall in month $1 = x_1$), $\ldots$, (total rainfall of the last month $= x_n$)$\}$; $X_{year} = \{$(total rainfall in year $1 = x_1$), $\ldots$, (total rainfall in year $N = x_n$)$\}$.

The rainfall depth record of an i-th rain gage is denoted by X_1, and its values as time series. Using a power or log-transformation, these values may be converted to a normal distribution. The multivariate normal distribution contains autocovariance matrix S_{ai}, whose elements are autocovariances of i-th rain gage record; and the number of rain gages n is replaced by $m + 1$ (dimension of the autocovariance matrix where m denotes the number of lags) (Krstanovic and Singh, 1988). When using the real-world data, each rain gage record $X_j (j = 1, \ldots, n)$ is used to compute $H(x_0)$, $H(x_0|x_1)$, $H(x_0|x_1, x_2)$, and so on, where x_i denotes the record with i-th serial dependency, or the record dependent on the i previous records at $t - \Delta t, t - 2\Delta t, t - i\Delta t$, where Δt is the sampling interval.

Under the assumption that analysis of each rain gage is independent of all other rain gages, rainfall record or time series of each rain gage is investigated. The objective is to determine the variation in entropy for various sampling intervals. For each rain gage, the conditional entropies are computed: $H(x_0), H(x_0|x_1), H(x_0|x_1, x_2), \ldots, H(x_0|x_1, x_2, \ldots, x_m)$, where x_0 denotes the value at 0-serial dependency (each rainfall record does not know about its past values), x_1 denotes values with first serial dependency (each rainfall record remembers its immediate predecessor), and so on.

At 0-th lag ($m = 0$) the value of entropy is the highest, since the rainfall record at each lag is considered independent of all other lags. With the introduction of information at the first lag, at the first two lags, and so on, the values of entropy decrease. The conditional entropies are computed for all rain gages for daily, two-day, weekly, monthly, and yearly time intervals. Serial dependencies are considered until $m = 10$ lags except for yearly evaluation. In general, the greatest entropy reduction occurs at the first lag. For short sampling intervals (daily, two-day) this reduction can be significant. For longer sampling intervals, the reduction at the first lag is even higher. Thus, the lag-dependency decreases in longer sampling intervals, while it still exists in short intervals. Entropy values for longer sampling intervals are lower than those for shorter ones, again because of the higher rainfall fluctuations in smaller sampling intervals. This trend is observed until the annual sample. Because of the short record length, calculations are usually not made for time intervals greater than one year.

Evaluation of rainfall networks in space

Bivariate Case: Let X_1 and X_2 be two random variables associated with two different rainfall records, S_1 and S_2, respectively, and $p(x_1, x_2) = p_{i,j}$ be the joint probability of a particular combination of the rainfall records at rain gages 1 and 2. Then the entropy of that joint probability of records expresses the total amount of uncertainty associated with realizations of X_1 and X_2 of rain gages 1 and 2. The relation between the mutual entropy $H(X_1, X_2)$ and marginal entropies $H(X_1)$ and $H(X_2)$ can be defined on two sets of data, S_1 and S_2, and it can be shown that the mutual entropy of the two sets of rainfall depth values will be at most equal to the sum of the uncertainties or marginal entropies of the individual sets of rainfall depth values. The completely independent sets of depth values have unrelated uncertainties and the maximum mutual entropy $H(X_1, X_2)$. The magnitude of the dependency among sets of depths is expressed as a difference $H(X_1, X_2) - [H(X_1) + H(X_2)]$. This then depends on the sampling interval of individual rainfall records (time dependency), distances among rain gages 1 and 2, and the orographic characteristics of the area (mountainous, plains, etc.). Thus, all of these external factors should be included within the data.

For evaluation of dependent/independent rainfall records, the following questions must be addressed: (1) How is the rainfall uncertainty reduced at rain gage 2 when the rainfall record at rain gage 1 is already known? (2) What is the common information among two rain gages? (3) Knowing the records of rain gages 1 and 2, how much uncertainty is still left in the rainfall network? To answer the first question, the conditional entropy $H(X_2|X_1)$ is defined as the uncertainty still left in the rainfall depth X_2 when the first depth X_1 becomes known. The joint entropy of X_1 and X_2 [$H(X_1, X_2)$], is reduced by the knowledge of X_1 expressed by [$H(X_1)$]. For no reduction in uncertainty, X_1 and X_2 are independent. To answer the second question, transinformation of X_1 and X_2, $T(X_1, X_2)$, is defined as the amount of information repeated in both X's.

Multivariate case: For spatial evaluation, rainfall records of several rain gages in the watershed are used. Let n be the number of rain gages, each with X_1 and record length $N_i(i = 1, \dots, n)$ and S_c the cross correlation matrix of $n \times n$ dimensions. The use of the multivariate normal PDF yields an expression for multivariate entropy (Harmancioglu and Yevjevich, 1985; Krstanovic and Singh, 1988). The essential condition for the existence of $H(X)$ is the positive-definiteness of the cross correlation matrix S_c. This enables nonsingular determinant $|S_c|$ and the existence of S_c^{-1}.

When applying entropy for univariate, bivariate, and multivariate cases, it is simpler to use univariate, bivariate, and multivariate normal distributions than other distributions, such as gamma, Pearson type or Weibull. Normal distributions are chosen because of the complexity involved in the application of entropy with other distributions. The normal distribution requires (Guiasu, 1977) replacement of discrete probabilities by a continuous probability distribution function (PDF); replacement of a system of events by a function space, where PDFs are defined; and replacement of the summation in the domain by integration throughout the domain. The multivariate normal distribution has been derived using the principle of maximum entropy (POME) in Chapter 5. In application of the normal distribution to the evaluation of rainfall networks in space, three cases can be distinguished: 1) Rainfall record measured at a rain gage is time independent, but may depend on other rain gages whose records are measured simultaneously. This corresponds to space evaluation. 2) Rainfall record measured at a rain gage is space independent of all other rain gages, but depends on its own past history. This corresponds to time evaluation. 3) Rainfall record is both space and time dependent. This corresponds to space-time evaluation. The space and time dependencies between rainfall records diminish with longer sampling time intervals (for example, daily rainfall records decrease dependencies among short-term rainfall events) and for rain gages within the smaller area.

Since the use of entropy involves various multivariate normal distributions, rainfall data must be normalized. The normalization can be accomplished by employing the Box-Cox transformation (Bras and Rodriguez-Iturbe, 1985). The data are transformed for every sampling interval and for different seasons (with annual values divided into six-month periods). For example, let X_{1A} and X_{1B} denote daily records of two rain gages A and B. Their transformed versions can be denoted as X'_{1A} and X'_{1B}. Similarly, X_{2A} and X_{2B} denote two-day values and their transformed versions as X'_{2A} and X'_{2B}. For a zero value of either X'_{2A} or X'_{2B} or $X'_{2A} \cup X'_{2B}$, the appropriate transformed value is used. For example, assume that the rainfall record of rain gage A contains the storm of July 4th, but that storm is only localized and rain gage B does not contain the same storm. Since only nonzero values are handled, the expected storm of July 4th at B is treated as "transformed zero" (the chosen Box-Cox transformation applied to zero values).

It is convenient that all rain gages have a common record length, that is, $N_1 = N_2 = \ldots = N_n$. This enables easy computation of different covariance matrices S_c. Records associated with $X_{j1} (j = 1, \ldots, n) =$ for each rain gage are constructed first, then $\bar{x}_2 = (X_i, X_j)$, $(i \neq j)$ for all rain gage pairs, $\bar{x}_3 = (X_i, X_j, X_k)$ for all rain gage triples, and so on, until $\overline{X}_{n-1} = (X_1, X_2, \ldots, X_{n-1})$ and $\overline{X}_n = (X_1, X_2, \ldots, X_n)$. For every vector, the corresponding joint entropy $[H(\overline{X}_1), H(\overline{X}_2), \ldots, H(\overline{X}_n)]$ is computed, with S_c varying from S_{c1} of 1×1 dimensions to S_{cn-1} of $(n-1) \times (n-1)$ dimensions. At each stage, for every combination of rain gages, the corresponding conditional entropies and transinformation are computed. Then, depending on these entropy values, either an additional rain gage is selected or the procedure is terminated.

To summarize, space evaluation involves the following steps:

1 Compute marginal entropies of all rain gages for every sampling interval. From the computed entropy values, the rain gage (S_1) with the highest uncertainty or entropy [i.e., $H(X_1)$] is found. This is regarded as the most important rain gage, and is designated as the central rain gage.

2 Compute conditional entropy of rain gage, S_1, with respect to all other rain gages (S_2 to S_n).

3 Find the rain gage that gives the lowest reduction in uncertainty or transinformation, or find $\min[H(X_1) - H(X_1|X_2)] = \min[T(X_1, X_2)]$. The rain gage that has the least common information with the first (central) rain gage is the rain gage of the second highest importance. Keep rain gages S_1 and S_2, and compute the conditional entropy of rain gages with respect to all other rain gages (S_3 to S_n) and find the rain gage that gives the minimum reduction in uncertainty, or $\min\{H(X_1, X_2) - H[(X_1, X_2)|X_3]\} = \min\{T[(X_1, X_2), X_3]\}$. The rain gage, S_3, is the third most important rain gage.

Repeat the procedure such that the j-th important rain gage is the one that gives $\min[H(X_1, X_2, \ldots, X_{j-1}) - H[(X_1, X_2, \ldots, X_{j-1})|X_j] = \min[T(X_1, X_2, \ldots, X_{j-1}), X_j]$. For a multivariate normal distribution, the transinformation can be expressed as $T[(X_1, X_2, \ldots, X_{j-1}), X_j] = -(1/2)\ln(1 - R^2)$ where R is multiple correlation coefficient. Thus, the rain gage having the smaller multiple correlation with other rain gages is always chosen.

To determine the number of essential rain gages, the coefficient of nontransferred informa-tion t_i is computed. If at step i, $t_{i+1} \geq t_i$ occurs, then the nontransferred information of the $i+1$ rain gages is greater than the non-transferred information of the i-th rain gage. Then, a rain gage is added. If the worth of keeping "i" rain gages is higher than the worth of "$i+1$" rain gages, then the other nonchosen rain gages are discontinued. The discontinuation depends on the measurement and the sampling interval of the rainfall record. If rain gages collect daily data, then daily evaluation should be most useful. If rain gages measure only long-term rainfall depths (monthly or yearly), then monthly and yearly evaluations should be performed.

Two other alternatives may be employed to choose rain gages: 1) maximum joint entropy $H(X_1, X_2, \ldots, X_i)$, and 2) the maximum weighted entropy: $[H(X_1|X_i) + H(X_2|X_i) + \ldots + H(X_{i-1}|X_i)]/k$. Both alternatives were examined by Krstanovic and Singh (1988) and produced results comparable to those by the maximum conditional entropy.

For space evaluation of rainfall networks, isocorrelation lines of rain gages with respect to a central rain gage can be plotted. In a similar vein, isoinformation contours or the lines of the coefficient of nontransferred information t_i can be plotted, where t_i is computed. Thus, $1 - t_i$ measures only the information transferred from the central rain gage to any other rain gage. Contours of isoinformation encompass central rain gages and higher concentration of isoinformation contours surrounding the central rain gages apparently require more rain gages in the watershed to transfer that information, as is the case for shorter sampling intervals. If the sampling interval is longer, the isoinformation contours are less densely concentrated around the central rain gage. By comparing isoinformation contours of the same season for various sampling intervals, the effect of sampling interval on the selection of rain gages can be examined. A possible relationship between consecutively selected rain gages and the distance among them can also be examined.

Rainfall records of several rain gages: multivariate case

Let $X_1, X_2, \ldots, X_n$ denote rainfall depths corresponding to different rainfall data sets S_1, S_2, S_n; $p(X_1, X_2, \ldots, X_n) = p_{i1} \ldots, p_{in}$ the joint probability; $p(x_m|x_1, \ldots, x_n) = p(i_m|i_1, \ldots, i_n)$ the conditional probability record of the m-th rain gage when all other rainfall depth data sets are known and $p[(x_1, \ldots, x_n)|x_{n+1}] = p[(i_1, \ldots, i_n)|i_{n+1}]$ the conditional probability of the n rainfall data sets (of n rain gages) when the rainfall data set of the $(n+1)$st rain gage is added. The multidimensional joint entropy for n rain gages in a watershed represents the common uncertainty of their measured rainfall data sets. The uncertainty left in the central rain gage when the records of all other rain gages are known is expressed by the multivariate conditional entropy of the central rain gage conditioned on all other records. Similarly, the uncertainty left in the group of rain gages $(i_1, \ldots, i_n)$ when any new rain gage is added (i.e., the expansion of the existing rainfall network) can be defined. Correspondingly, the multidimensional transinformation between the n existing rain gages and the new (added) rain gage $(n+1)$ can be defined.

Now constructing a network by consecutively adding rain gages one by one for a specified watershed and sampling interval is considered. At any step, this adding can be terminated, depending upon the consecutive coefficients of transferred information $1-t_i$. At the first step, only the central rain gage is chosen and

$$H(X_1) = H(X_{central}) \tag{13.23a}$$

In choosing the second rain gage,

$$t_1 = \frac{H(X_1|X_j)}{H(X_1)} \quad \text{where } j \neq 1 \tag{13.23b}$$

is computed. In choosing the i-th rain gage.

$$t_{i-1} = \frac{H((X_1, \ldots, X_{i-1})|X_i)}{H(X_1, \ldots, X_{i-1})} \tag{13.23c}$$

is computed.

The maximum possible transformation T_0 varies from one step to the other, and is always equal to the joint entropy of the rain gages already chosen before that step. For example, in the second step $T_0 = H(X_1)$, and in the i-th step $T_0 = H(X_1, \ldots, X_{i-1})$. The rain gages are worth adding as long as the transferred information decreases significantly, that is,

$$1 \geq t_1 > t_2 > \ldots > t_i \tag{13.24}$$

If at any step $t_{i+1} \geq t_i$, then the new rain gage being added at the $i+1$ step has repetitive information and is not worth adding.

If $t_i > t_{i+1}$, the new $(i+1)$st rain gage contains new information. The magnitude of the difference between t_{i+1} and t_i is always in the $(0,1)$ range. The higher the magnitude, the more new information is added; the lower the magnitude the less new information is added. For the latter, other considerations (economic, rain gage access, etc.) might terminate the addition of new rain gages before the $t_i = t_{i+1}$ point is reached. The significance of the magnitude corresponds to the significance in a multiple correlation coefficient among i and $i+1$ variables (r_i and r_{i+1}), assuming a certain functional relationship among these variables. However, it also depends on the number of existing rain gages. For example, for 10 rain gages in a watershed, the added information content is significant if $t_i - t_{i+1} \geq 0.10$ (or the average expected information increases from the first rain gage to the tenth rain gage).

At each step during the rain gage selection process, the coefficient of nontransferred information t_i between the already chosen rain gage and each of the remaining rain gages $j(j = i, \ldots, n)$ is computed. Then, isoinformation contours can be constructed. The isoinformation contour of the multivariate rainfall record (at i-th step) is the line of equal common information between already chosen (i) rain gages and any not-chosen existing rain gage (j) located in the region.

13.4 Directional information transfer index

Although transinformation indicates the dependence of two variables, it is not a good index of dependence, since its upper bound varies from site to site (it varies from 0 to marginal entropy H). Therefore, the original definition of T can be normalized by dividing by the joint entropy. Yang and Burn (1994) normalized T as

$$\frac{T}{H} = DIT = \frac{(H - H_{Lost})}{H} = 1 - \frac{H_{Lost}}{H} \tag{13.25}$$

The ratio of T by H is called directional information transfer (DIT) index. Mogheir and Singh (2002) called it as Information Transfer Index (ITI). The physical meaning of DIT is the fraction of information transferred from one site to another. DIT varies from zero to unity when T varies from zero to H. The zero value of DIT corresponds to the case where sites are independent and therefore no information is transmitted. A value of unity for DIT corresponds to the case where sites are fully dependent and no information is lost. Any other value of DIT between zero and one corresponds to a case between fully dependent and fully independent.

DIT is not symmetrical, since $DIT_{XY} = T/H(X)$ will not, in general, be equal to $DIT_{YX} = T/H(Y)$. DIT_{YX} describes the fractional information inferred by station X about station Y, whereas DIT_{YX} describes the fractional information inferred by station Y about station X. Between two stations, the station with higher DIT should be given higher priority because of its higher capability of inferring (predicting) the information at other sites.

DIT can be applied to the regionalization of the network. If both DIT_{XY} and DIT_{YX} are high, the two related stations can be arranged in the same group, since they are strongly dependent and information can be mutually transferred between them. If neither of the DIT values is high, they should be kept in separate groups. If DIT_{XY} is high, station Y whose information can be predicted by X can join station X, if station Y does not belong to another group; otherwise it stays in its own group. The predictor station X cannot join station Y's group, since if it were discontinued the information at that site would be lost.

DIT can be both a measure of information transmission capability and an indicator of the dependency of a station. This is an indicator of the information connection. In the station selection process a predicted station should be removed first, because it recovers information efficiently but does not predict it efficiently. When all remaining stations in the group have strong mutual connections with each other (i.e., both DIT_{XY} and DIT_{YX} are high), they can be further selected based on a criterion, designated as $SDIT_i$, defined as

$$SDIT_i = \sum_{j=1, j \neq i}^{m} DIT_{ij} \tag{13.26}$$

where DIT_{ij} is the information inferred by station i about station j, and $m = $ the number of stations in the group. The station in each group with the highest value of $SDIT$, in comparison with members in the group, should be retained in the network.

13.4.1 Kernel estimation

In order to be able to use transinformation and DIT, the probability density function of the variable being sampled must be determined. To that end, a nonparametric estimation method can be employed. Non-parametric estimation does not describe a probability density function by a formula and parameters but rather by a set of point values everywhere in the domain. If the values of the density function are known everywhere then the function is known numerically. This method of describing the density function is known as non-parametric method. The method can be described as follows:

For a random variable X, let $x_1, x_2, \dots, x_N$ be a sample of observations, independently and identically distributed with PDF $f(x)$. If $f(x)$ is the derivative of $F(x)$ and is continuous, then at a given point x, it can be estimated by a kernel estimator as:

$$f_N(x) = \int_{-\infty}^{\infty} \frac{1}{h} K\left(\frac{x - x_j}{h}\right) dF_N \left(\frac{x - x_j}{N}\right) \approx \frac{1}{Nh} \sum_{j=1}^{N} K\left(\frac{x - x_j}{h}\right) \tag{13.27}$$

where $f_N(x)$ is the kernel estimator of $f(x)$; function $K(.)$ is called kernel; h is a positive number and is a smoothing factor for the kernel, a function of the sample size and sample values; and $F_N(x)$ is an estimate of CDF of X. Parzen (1962) showed that $f_N(x)$ is an unbiased estimate of $f(x)$ under certain conditions.

For a multidimensional case where $X = X(x_1, x_2, \ldots, x_N)$, Cacoullos (1966) has shown that at a given point $X^O = X^O(x_1, x_2, \ldots, x_N)$ the kernel estimator has the form:

$$f_N(X^O) = \frac{1}{N} \frac{1}{h_1 h_2 \ldots hp} \sum_{j=1}^{N} K\left(\frac{x_1 - x_{j1}}{h_1}, \frac{x_2 - x_{j2}}{h_2}, \ldots, \frac{x_p - x_{jp}}{h_p}\right) \tag{13.28}$$

where $X^j = X^j(x_{j1}, x_{j2}, \ldots, x_{jp})$ is the j-th observation of X which is a p-dimensional variable.

The components of X: $\{X_1, X_2, \ldots, X_p\}$ could be mutually dependent or independent, but observations of each component are still assumed to be independent and identically distributed. A simple approximation of this equation is the product function of the form:

$$K\left(\frac{x_1 - x_{j1}}{h_1}, \frac{x_2 - x_{j2}}{h_2}, \ldots, \frac{x_p - x_{jp}}{h_p}\right) = \prod_{i=1}^{p} K_i\left(\frac{x_i - x_{ji}}{h_i}\right) \tag{13.29}$$

For practical purposes, it can be assumed that

$$h_1 = h_2 = \ldots = h_p = h \tag{13.30}$$

According to Adamowski (1989) the choice of a kernel is not crucial for the estimation. However, the selection of h is crucial, because it affects both the bias and the mean square error of the estimator. Some common forms of kernels are presented by Parzen (1962) and Wertz (1978) and are as follows. h is frequently expressed as a function of $f(x)$ which is not known. A widely used method is cross-validation maximum likelihood which minimizes the integral mean square error of $f_N(x)$. h is selected to meet

$$Max\ L(h) = \prod_{j=1}^{N} f_N(x_j, h) = \prod_{j=1}^{N} \frac{1}{h^p} \sum_{i=1, i \neq j}^{N} K(x_j, h) \tag{13.31}$$

where $f_N(x_j, h)$ is the estimated density value at x_j but with x_j removed.

In practice, hydrologic variables are non-negative. Therefore, the original variables are logarithmically transformed and a Gaussian kernel is chosen to form the estimator:

$$K(x, y) = K_1(x)K_2(y) = \frac{1}{2\pi}\left[\exp\left(-\frac{x^2}{2}\right)\right]\left[\exp\left(-\frac{y^2}{2}\right)\right]$$
$$= \frac{1}{2\pi}\left\{\exp\left[-\frac{(x^2 + y^2)}{2}\right]\right\} \tag{13.32}$$

where $x = (x^* - x_j)/h$; $y = (y^* - y_j)/h$; (x_j, y_j) is an observation point; and (x^*, y^*) is the coordinate of any point in the field. All x^* and y^* (or x_j and y_j) are logarithmically transformed. The optimal h-values are derived from equation (13.31). Then $f_N(x, y)$ becomes

$$f_N(x, y) = \frac{1}{Nh^2} \sum_{i=1}^{N} \frac{1}{2\pi}\left\{\exp\left[-\frac{(x^2 + y^2)}{2}\right]\right\} \tag{13.33}$$

Table 13.1 Two-dimensional contingency table (frequency).

		1	2	3	...	u	Total
				y			
	1	f_{11}	f_{12}	f_{13}	...	f_{1u}	$f_{1.}$
	2	f_{21}	f_{22}	f_{23}	...	f_{2u}	$f_{2.}$
x	3	f_{31}	f_{32}	f_{33}	...	f_{3u}	$f_{3.}$
	.	.	.	.	.	.	.
	.	.	.	.	.	.	.
	.	.	.	.	.	.	.
	v	f_{v1}	f_{v2}	f_{v3}	...	f_{vu}	$f_{v.}$
	Total	$f_{.1}$	$f_{.2}$	$f_{.3}$	...	$f_{.u}$	f_x or f_y

13.4.2 Application to groundwater quality networks

Information measures, including transinformation, information transfer index and correlation coefficient, have been applied to describe the spatial variability of spatially correlated groundwater quality data. These measures can be calculated using two types of approaches: discrete and analytical. The discrete approach employs the contingency table and the analytical approach employs a functional form of the probability density function, such as normal probability density function. Most studies have employed the analytical approach which presumes knowledge of the probability distributions of the random variables under study. The problem of not knowing the probability distributions can, however, be circumvented if a discrete approach is adopted. For the discrete approach Transinformation (T) and Information Transfer Index (ITI) can be employed to describe the spatial variability of data which is spatially correlated and which fits the normal distribution function. The Transinformation Model (T-Model) is a relation between mutual information measure, specifically T, and the distance between wells. Transinformation is useful and comparable with correlation to characterize the spatial variability of data which is correlated with distance. Mogheir and Singh (2002) used the entropy theory to evaluate and assess a groundwater monitoring network by means of marginal entropy contour maps.

To calculate information measures, the joint or conditional probability is needed, and one way is to use a contingency table. An example of a two-dimensional contingency table is given in Table 13.1 To construct a contingency table, let the random variable X have a range of values consisting of v categories (class intervals), while the random variable Y is assumed to have u categories (class intervals). The cell density or the joint frequency for (i, j) is denoted by f_{ij}, $i = 1, 2, \ldots, v$; $j = 1, 2, \ldots, u$, where the first subscript refers to the row and the second subscript to the column. The marginal frequencies are denoted by $f_{i.}$ and $f_{.j}$ for the row and the column values of the variables, respectively. Construction of a two-dimensional contingency table is illustrated using an example.

Example 13.1: The time series of the chloride data measured for two wells (H-9 and H-8), as shown in Figure 13.1 selected from the Gaza Strip groundwater quality monitoring network, are presented in Table 13.2 (Mogheir and Singh, 2002). Illustrate the construction of a two-dimensional contingency table for this field data.

Solution: The construction of a two-dimensional contingency table involves the following steps:1) Each data set is subdivided into class intervals. 2) To fill the first table (frequency table),

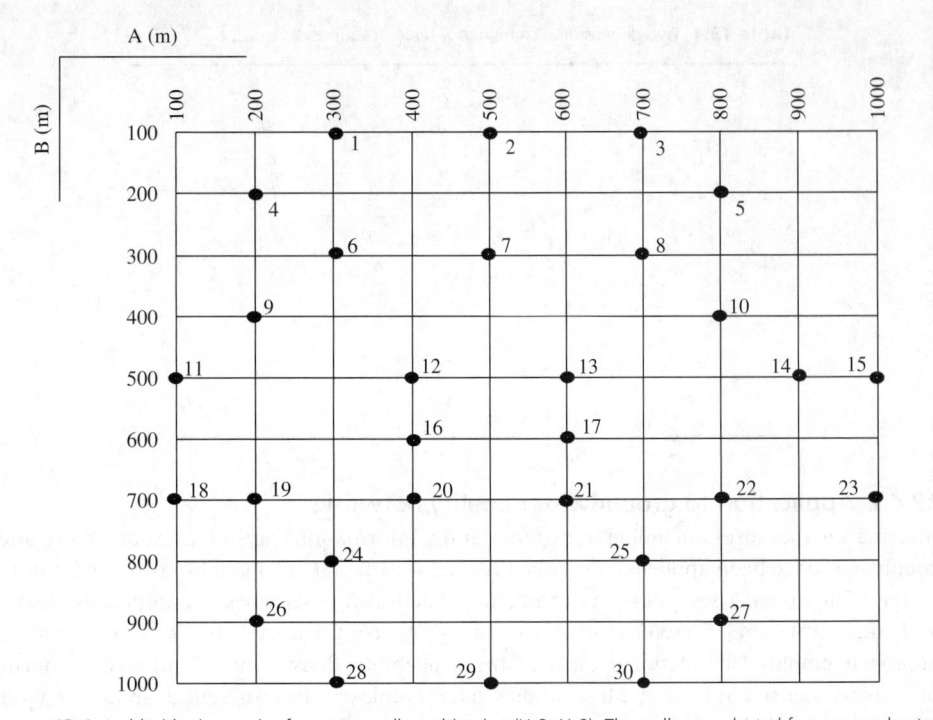

Figure 13.1 A chloride time series for a two-well combination (H-8, H-9). The wells are selected from groundwater quality monitoring network in the middle part of Gaza Strip.

Table 13.2 Chloride data for well H-9 and H-8.

Date	H-9 (Cl mg/l)	H-8 (Cl mg/l)	Date	H-9 (Cl mg/l)	H-8 (Cl mg/l)	Date	H-9 (Cl mg/l)	H-8 (Cl mg/l)
01-05-1972	644	427	20-03-1980	721	532	27-01-1990	868	700
28-10-1972	679	413	16-09-1980	749	546	26-07-1990	854	707
26-04-1973	721	483	15-03-1981	756	518	22-01-1991	840	770
23-10-1973	805	497	11-09-1981	861	525	21-07-1991	845	784
21-04-1974	693	483	10-03-1982	840	602	17-01-1992	819	770
18-10-1974	805	497	06-09-1982	861	630	15-07-1992	826	784
16-04-1975	693	504	05-03-1983	959	630	11-01-1993	819	805
13-10-1975	679	518	01-09-1983	882	644	10-07-1993	819	805
10-04-1976	721	511	28-02-1984	854	651	06-01-1994	819	784
07-10-1976	805	553	26-08-1984	868	658	05-07-1994	819	784
05-04-1977	658	630	22-02-1985	854	665	01-01-1995	819	777
02-10-1977	756	497	21-08-1985	868	644	30-06-1995	763	791
31-03-1978	735	504	17-02-1986	833	651	27-12-1995	714	777
27-09-1978	756	497	16-08-1986	868	658	24-06-1996	767	829
26-03-1979	735	504	12-02-1987	770	721	21-12-1996	739	921
22-09-1979	728	525	11-08-1987	868	658	19-06-1997	752	822
			07-02-1988	854	721			
			05-08-1988	819	707			
			01-02-1989	840	784			
			31-07-1989	819	707			

Table 13.3 Absolute frequency contingency table for H-8 and H-9 combinations.

	410<Cl<495	495<Cl<580	580<Cl<665	665<Cl<750	750<Cl<835	835<Cl<920	Marginal H-9
640<Cl<695	3	1	2	2	0	0	8
695<Cl<750	2	6	1	1	1	0	11
750<Cl<805	1	0	0	4	6	1	12
805<Cl<860	0	0	1	4	1	0	6
860<Cl<915	0	1	3	10	0	0	14
915<Cl<970	0	1	0	0	0	0	1
Marginal H-8	6	9	7	21	8	1	52

Table 13.4 Joint probability (contingency) table for H-8 and H-9 combinations.

	410<Cl<495	495<Cl<580	580<Cl<665	665<Cl<750	750<Cl<835	835<Cl<920	Marginal H-9
640<Cl<695	0.058	0.019	0.038	0.038	0.000	0.000	0.154
695<Cl<750	0.038	0.115	0.019	0.019	0.019	0.000	0.212
750<Cl<805	0.019	0.000	0.000	0.077	0.115	0.019	0.231
805<Cl<860	0.000	0.000	0.019	0.077	0.019	0.000	0.115
860<Cl<915	0.000	0.019	0.058	0.192	0.000	0.000	0.269
915<Cl<970	0.000	0.019	0.000	0.000	0.000	0.000	0.019
Marginal H-8	0.115	0.173	0.135	0.404	0.154	0.019	1.000

the rows or the columns are kept constant and the shared data of the other well are counted, as shown in Table 13.3) The joint probability table is constructed by dividing each count by the total number of the recorded data of one well, as shown in Table 13.4.

Note that $T(x,y) = 0$ if X and Y are independent. Transinformation is an indicator of the capability of information transmission. Normalizing transinformation, an Information Transfer Index (ITI) is defined [see equation (13.25)], which then indicates the standardized information transferred from one site to another. For both transinformation and correlation models, the geometrical distance (d) between two wells can be calculated as

$$d = \sqrt{(A_1 - A_2)^2 + (B_1 - B_2)^2} \qquad (13.34)$$

where A_1 and B_1 are the coordinates of well 1, A_2 and B_2 are the coordinates of well 2, and d is the distance between wells 1 and 2.

Smoothing of discrete values: Since the discrete T values may exhibit a scatter when plotted against the distance between wells, a smoothing method, such as axis transformation (e.g., logarithm transformation), moving average or exponentially weighted moving average (Berthouex and Brown, 1994), is employed. Mogheir and Singh (2002) employed the moving average method to smooth the T data using a 100 m distance interval, which is the distance between wells in the hypothetical network used. For 0 distance, transinformation T_0 Is assumed as the average of the marginal entropies of the wells. They also applied the moving average method to smooth the lognormal T, ITI and correlation values.

Analytical approach to transinformation: Assuming that the data is reduced such that the mean of the data is 0 and the data fits the normal probability distribution function, the marginal entropy, as a measure of information, is computed analytically using the expression (Lubbe, 1996):

$$H(x) = \ln(S_x) + 1.419 \qquad (13.35)$$

where S_x is the standard deviation for random variable x, and the values of $T(x, y)$ from the correlation coefficient (r_{xy}) as (Lubbe, 1996; Kapur and Kesavan, 1992):

$$T(x, y) = -0.5 \ln (1 - r_{xy}^2) \tag{13.36}$$

Example 13.2: Derive equation (13.36).

Solution: The marginal entropy for a variable X can be computed analytically for the normal distribution (Lubbe, 1996) as:

$$H(x) = \frac{1}{2} \ln \sigma_x^2 + \frac{1}{2} \ln (2\pi e) \text{ or } H(x) = \ln \sigma_x + 1.419$$

Similarly, for variable Y the marginal entropy is

$$H(y) = \frac{1}{2} \ln \sigma_y^2 + \frac{1}{2} \ln (2\pi e)$$

where σ_y is the standard deviation of variable Y.

The joint probability density $f(x, y)$ for variable X and Y, where $-\infty < x < +\infty, -\infty < y < +\infty$ and considering two-dimensional Gaussian distribution with mean for both X and Y equal to 0, can be expressed as (Kapur and Kesavan, 1992):

$$f(x, y) = \frac{1}{2\pi |C|^{\frac{1}{2}}} \exp \left\{ -\frac{1}{2} [x \; y] \; C^{-1} \begin{bmatrix} x \\ y \end{bmatrix} \right\}$$

where

$$C = \begin{bmatrix} \sigma_x^2 & \rho_{xy}\sigma_x\sigma_y \\ \rho_{yx}\sigma_x\sigma_y & \sigma_y^2 \end{bmatrix}$$

ρ_{xy} is the correlation coefficient between variable X and Y and can be calculated as

$$\rho_{xy} = \frac{\sigma_{xy}}{\sigma_x \sigma_y}$$

where σ_{xy} is the covariance between the variable X and Y.

For continuous variables X and Y with joint probability density $f(x, y)$, the joint entropy is equal to

$$H(x, y) = \ln (2\pi e) + \frac{1}{2} \ln |C|$$

For computing the transinformation $T(x, y)$ for two random variables, X and Y, the following expression can be used (e.g., Jessop, 1995):

$$T(x, y) = \frac{1}{2} \ln \sigma_x^2 + \frac{1}{2} \ln (2\pi e) + \frac{1}{2} \ln \sigma_y^2 + \frac{1}{2} \ln (2\pi e) - \ln (2\pi e) - \frac{1}{2} \ln |C|$$

which can be simplified as

$$T(x, y) = \frac{1}{2} \ln (\sigma_x^2 \sigma_y^2) - \frac{1}{2} \ln |C| = \frac{1}{2} \ln (\sigma_x^2 \sigma_y^2) - \frac{1}{2} \ln \left(\sigma_x^2 \sigma_y^2 \begin{vmatrix} 1 & \rho_{xy} \\ \rho_{yx} & 1 \end{vmatrix} \right)$$

$$= \frac{1}{2} \ln (\sigma_x^2 \sigma_y^2) - \frac{1}{2} \ln (\sigma_x^2 \sigma_y^2) - \frac{1}{2} \ln \begin{vmatrix} 1 & \rho_{xy} \\ \rho_{yx} & 1 \end{vmatrix}$$

Therefore, $T(x, y)$ can be obtained as

$$T(x, y) = -\frac{1}{2} \ln (1 - \rho_{xy}^2)$$

The use of sample correlation coefficient r_{xy} yields

$$T(x, y) = -\frac{1}{2} \ln (1 - r_{xy}^2)$$

The correlation is represented by equation (13.10) and therefore, the analytical T-Model can be expressed as

$$T(d) = -0.5 \ln [1 - (e^{-d/\lambda})^2]$$

One can investigate the sensitivity of the discrete transinformation (T) values to the factors that influence its behavior. The factors include the size of generated data and the number of class intervals. Different sizes of data can be used to compute the discrete T-values (200, 300, 400, and 500). The number of class intervals can be the same for all the different data sizes. It can be shown that the larger the data size the less the difference between discrete T values and analytically derived T values. This indicates that the discrete T-values are sensitive to the size of the data available for analysis, as in the case of actual groundwater data where the data is limited in time or is incomplete. In order to evaluate the importance of the class interval, for a given data size, different class intervals can be analyzed. Normally, when the class interval decreases the discrete T values come closer to the analytically derived T-values.

Usually, the correlation method is used to characterize the spatial variability (linear dependency) of many types of data in different fields (e.g., Cressie, 1990). It is noted that the data is correlated by distance, which means that the smaller the distance the higher the correlation. The transinformation can also be used to represent the spatial variability of data. It can be shown that there is a relation between transinformation and distance. The closely spaced wells have a higher value for T than the ones that are further apart. The T values become essentially constant as the distance increases. That may be because there is still mutual information that can be transferred, even for a long distance. Thus, transinformation can be used to represent the spatial dependency.

13.5 Total correlation

The total correlation can be used to quantify the amount of information shared by all gages (or variables) at the same time and hence provides an alternative way of examining multivariate dependency. It provides a direct and effective way of assessing such kind of repeated information and can be considered as a measure of general dependence, including both linear and nonlinear relationship, among multiple variables. The total correlation for N variables $(X_1, X_2, \ldots, X_N)$ can be defined as (McGill, 1954; Watanabe, 1960) as

$$C(X_1, X_2, \ldots, X_N) = \sum_{i=1}^{N} H(X_i) - H(X_1, X_2, \ldots, X_N) \tag{13.37}$$

Equation (13.37) shows that the total correlation is always positive, because the sum of marginal entropies of the N variables will be greater than their joint entropy. It is symmetric with respect to its arguments. If these N variables as well as their combinations are independent, then C will be zero. A large value of C may imply either a strong dependency amongst a few variables or a relatively small dependency amongst all of them. If $N = 2$, equation (13.37) will reduce to the usual transinformation T. equation (13.37) involves two components: marginal entropies and joint entropy.

The total correlation can be computed without resorting to the computation of multivariate entropy or multivariate probabilities. This can be accomplished by recalling the grouping property of total correlation and accordingly a systematic grouping of bivariate entropies. According to this property, if a new variable is formed by the union of two or more variables such that the marginal entropy of a new variable is equal to the joint entropy of the two forming variables, then the total correlation of the original variables is obtained by summing the marginal entropies of the new variables (Kraskov et al., 2003). It is seen that if the number of variables is two then transinformation becomes a special case of total correlation. Thus, total correlation is a multivariate extension of bivariate transinformation. From equation (13.37), the multivariate joint entropy can be expressed as

$$H(X_1, X_2, \ldots, X_n) = \sum_{i=1}^{n} H(X_i) - C(X_1, X_2, \ldots, X_n) \tag{13.38}$$

Although the total correlation concept has been widely used in medicine, neurology, psychology, clustering, significant feature selection and genetics (Jakulin and Bratko, 2004; Fass, 2006), there appears to have been limited application of total correlation in hydrology and water resources. Krstanovic and Singh (2002a, b) used it for evaluating multivariate $(N > 2)$ dependence, where it has been assumed that the random variables follow normal or lognormal distribution, which is not always the case for hydrologic variables, such as discharge, rainfall intensity, and so on. Furthermore, in network design, most analyses about the dependence of time series belonging to different potential hydrometric stations have been restricted up to bivariate analysis in terms of transinformation or directional information transfer index rather than multivariate analysis. A major reason is the difficulty of estimating multivariate probability distributions and the limited availability of data. However, often there is a need to evaluate the total amount of information/uncertainty duplicated by several stations under consideration. Recently Alfonso (2010) used total correlation to evaluate the performance of different optimal water level monitoring stations.

The total correlation can be computed directly using its grouping property. For trivariate total correlation $C(X_1, X_2, X_3)$, the grouping property can be expressed as $C(X_1, X_2, X_3) = C(X_1, X_2) + C(X_{1:2}, X_3)$ (Kraskov et al. 2005).

$$C(X_1, X_2, X_3) = C(X_1, X_2) + C(X_{1:2}, X_3) \tag{13.39}$$

where $X_{1:2}$ denotes the grouped variable formed by grouping X_1 and X_2. Using this property sequentially, the multivariate total correlation can be computed recursively as:

$$
\begin{aligned}
C(X_1, X_2, X_3, \ldots, X_N) &= C(X_1, X_2, X_{3 \to N}) \\
&= C(X_1, X_2) + C(X_{1:2}, X_{3 \to N}) \\
&= C(X_1, X_2) + C(X_{1:2}, X_3, X_{4 \to N}) \\
&= C(X_1, X_2) + C(X_{1:2}, X_3) + C(X_{1:3}, X_{4 \to N}) \\
&\cdots \cdots \\
&= C(X_1, X_2) + C(X_{1:2}, X_3) + C(X_{1:3}, X_4) + \ldots C(X_{1:N-1}, X_N) \\
&= \sum_{i=1}^{N-1} C(X_{1:i}, X_{i+1})
\end{aligned}
\tag{13.40}
$$

Notation $X_{1:i}$ represents the merged variable of $X_1, X_2, \ldots, X_i$. equation (13.40) shows that the total correlation is finally factorized as a summation of bivariate total correlation which is just the transinformation. In other words, the grouping property can reduce the dimension of multivariate total correlation and thus the estimation of multivariate probability distribution can be avoided.

13.6 Maximum information minimum redundancy (MIMR)

The foregoing discussion is extended to more than two dimensions without assuming any specific distribution, which means computations are based on a nonparametric method based on the available data only. This is based on the maximization of the amount of effective information retained by the selected optimal gaging stations and minimization of the amount of redundant information due to the dependence among the selected stations. Furthermore, a mechanism which reflects the preference of different decision makers may be built in the selection criterion by introducing two weights, effective information weight and redundancy weight. Importantly, it is easy to extend the criterion to cover more design considerations, such as cost and benefit of hydraulic information obtained from a network.

Let there be N potential candidate hydrometric stations located in the area of interest, $X_1, X_2, X_3, \ldots, X_N$; for example, the area may be a watershed, river, canal, estuary, or pipeline. It is assumed that for each candidate station there are some years of records about the hydrometric variable of interest denoted by X, such as discharge, rainfall amount, sediment discharge, and a water quality constituent. Let S denote the set of hydrometric stations already selected for the network and its elements are denoted by $X_{S_1}, X_{S_2}, X_{S_3}, \ldots, X_{S_k}, X_{S_1}, X_{S_2}, \ldots, X_{S_k}$, where S_i can be $1, 2, \ldots, k$ or only some of them. Similarly, let F denote the set of candidate stations to be selected and similarly its elements are denoted by $X_{F_1}, X_{F_2}, X_{F_3}, \ldots, X_{F_m}, X_{F_1}, X_{F_2}, \ldots, X_{F_m}$, where F_i can be $1, 2, \ldots, m$ or only some of them. The summation of k and m is equal to N, the total number of potential candidate stations. The amount of effective information retained by S can be modeled in terms of joint entropy and transinformation as

$$
H(X_{S_1}, X_{S_2}, \ldots, X_{S_k}) + \sum_{i=1}^{m} T(X_{S_1:S_k}; X_{F_i})
\tag{13.41}
$$

where $X_{S_1:S_k}$ denotes the merged time series of $X_{S_1}, X_{S_2}, X_{S_3}, \ldots, X_{S_k}$ such that its marginal entropy is the same as the multivariate joint entropy of $X_{S_1}, X_{S_2}, X_{S_3}, \ldots, X_{S_k}$. In other words, the merged variable $X_{S_1:S_k}$ contains the same amount of information as that retained by all of its individual members $X_{S_1}, X_{S_2}, X_{S_3}, \dots, X_{S_k}$. The same notation will be used to denote merged variables, for example, $X_{A:B}$ denotes the merged variable of those variables whose subscripts are A and B.

The effective information contains two parts. The first part is the joint entropy of the selected stations, measuring the total but not duplicated amount of information which can be obtained from the selected stations. The second part is the summation of transferred information from the group of already selected stations to each individual station which is still in the candidate set, respectively. To illustrate, consider $T(X_{S_1:S_k}, X_{F_i})$ which is the transinformation of $X_{S_1:S_k}$ and X_{F_i} and quantifies the common information shared by these two variables. When doing network design, it should be kept in mind that the major function of a hydrometric network is to monitor the hydrometric variables of interest and to make prediction. Therefore, the predictive ability of the network should not be neglected in the design. $T(X_{S_1:S_k}, X_{F_i})$ is a quantitative measure of the amount of information about the unselected station X_{F_i} which can be inferred from the selected stations. In other words, it is a measure of the predictive ability of the selected stations.

Husain (1987) and Al-Zahrani and Husain (1998) considered the predictive ability of an optimal network. However, the effective information equation (13.41) differs from the one they used in two respects. The first is that the predictive ability is measured in equation (13.41) considering the selected stations as a whole group containing the same amount of information as that of all its elements rather than treating them separately. This predictive ability measure can successfully filter the duplicated information (redundancy) of the selected stations. Second, the multivariate joint entropy is used to quantify the total information rather than using the summation of marginal entropies in which the duplicated information is summed again and again.

The effective information can also be expressed as

$$H(X_{S_1}, X_{S_2}, \dots, X_{S_k}) + T(X_{S_1:S_k}; X_{F_1:F_m}) \tag{13.42}$$

It contains two parts: total effective information part and predictive ability part. In this definition the unselected stations are also treated as a whole group. Transinformation $T(X_{S_1:S_k}, X_{F_1:F_m})$ is the amount of information about the unselected group that can be inferred from the selected group.

Another key point worthy of consideration in network design is the redundant information among selected stations. Such redundancy means the selected stations are not fully and effectively used, since a lot of information obtained from the network may be overlapping. In other words, some of the stations are not necessary and therefore the network is not an economical one. Even worse, the redundancy may deteriorate the predictive ability of the network even though the same amount of information can be obtained from a redundant network as that obtained from a minimum redundant one considering that no redundancy is impossible in real practice.

The total correlation of the selected stations can measure the redundancy among them, that is, $C(X_{S_1}, X_{S_2}, \dots, X_{S_k})$. The total correlation of already selected stations measures the common information shared by any combination of these stations unlike the interaction which only measures the information shared by all of these stations. Interaction information is sensitive to the newly added stations; in other words, it may change significantly from positive to negative or from a large value to a small value, if a new station is added to the network. To understand it, there may be a large amount of duplicated information between two stations;

however, when adding a new station, these three stations may have no simultaneous common information. In this sense the total correlation is a more reliable measure of redundancy than interaction information.

13.6.1 Optimization

An informative hydrometric network should provide as much information as possible and at the same time constrain the redundant information as much as possible. This kind of maximum information and minimum redundancy network can be determined as:

$$\begin{cases} \max : H(X_{S_1}, X_{S_2}, \ldots X_{S_k}) + \sum_{i=1}^{m} T(X_{S_1:S_k}; X_{F_i}) \\ \min : C(X_{S_1}, X_{S_2}, \ldots, X_{S_k}) \end{cases} \tag{13.43a}$$

or

$$\begin{cases} \max : H(X_{S_1}, X_{S_2}, \ldots X_{S_k}) + \sum_{i=1}^{m} T(X_{S_1:S_k}; X_{F_i:Fm}) \\ \min : C(X_{S_1}, X_{S_2}, \ldots, X_{S_k}) \end{cases} \tag{13.43b}$$

This constitutes a multi-objective optimization problem which can be reduced to a single objective optimization problem by recalling that both the effective information part and the redundancy part have the same unit. The two objectives can therefore be unified as

$$Max : w_1 [H(X_{S_1}, X_{S_2}, \ldots, X_{S_k}) + \sum_{i=1}^{m} T(X_{S_1:S_k}; X_{F_i})] - w_2 C(X_{S_1}, X_{S_2}, \ldots, X_{S_k}) \tag{13.44a}$$

$$Max : \lambda_1 [H(X_{S_1}, X_{S_2}, \ldots, X_{S_k}) + \sum_{i=1}^{m} T(X_{S_1:S_k}; X_{F_i})] - \lambda_2 C(X_{S_1}, X_{S_2}, \ldots, X_{S_k})$$

or

$$Max : w_1 [H(X_{S_1}, X_{S_2}, \ldots, X_{S_k}) + T(X_{S_1:S_k}; X_{F_1:Fm})] - w_2 C(X_{S_1}, X_{S_2}, \ldots, X_{S_k}) \tag{13.44b}$$

where w_1 and w_2, whose summation is 1, are the information weight and redundancy weight, respectively, since sometimes the decision maker needs a trade-off between the informativeness and redundancy of the hydrometric network.

One can unify the information and redundancy objectives as

$$\max : \frac{H(X_{S_1}, X_{S_2}, \ldots, X_{S_k}) + \sum_{i=11}^{m} T(X_{S_1:S_k}; X_{F_i})}{wC(X_{S_1}, X_{S_2}, \ldots, X_{S_k})} \tag{13.45a}$$

or

$$\max : \frac{H(X_{S_1}, X_{S_2}, \ldots, X_{S_k}) + T(X_{S_1:S_k}; X_{X_{F_1:Fm}})}{wC(X_{S_1}, X_{S_2}, \ldots, X_{S_k})} \tag{13.45b}$$

where w is the coefficient that makes a trade-off between information ability and redundancy of the network. Any of the two methods unifying information and redundancy objectives can be adopted in the hydrometric network design.

13.6.2 Selection procedure

Using the MIMR criterion, the selection procedure for a hydrometric network design entails the following steps:

1 Collect the hydrometric data for variable(s) of interest, for example, hourly, daily, weekly, or monthly water levels for each of the potential candidate stations. Then, discretize the continuous time series in such a way that each of the records is labeled by $1, 2, 3, \ldots, b$, where b is the number of bins used in the histogram. Therefore, the continuous time series will become a discrete time series.

2 Calculate marginal entropies for all candidate stations.

3 Identify the station having the maximum marginal entropy, and designate it as the central station.

4 Update the S set in which stations already selected are saved and the F set in which all unselected candidate stations are saved.

5 Select the next station from the F set by the MIMR criterion. In this step scan all the unselected stations and locate the station which can maximize the unified objective function [equation (13.44a) or equation (13.44b)]. In order to compare, the multivariate joint entropy $H(X_{S1}, X_{S2}, \ldots, X_{Sk})$, the mutual information between one grouped variable or station to another single variable or station $T(X_{S_1:S_k}, X_{F_i})$ or between two grouped variables $T(X_{S_1:S_k}, X_{F_1:F_m})$, and the total correlation $C(X_{S1}, X_{S2}, \ldots, X_{Sk})$ should be computed. It may be noted that all of these terms involve multiple variables as soon as the second important station is selected.

6 Repeat steps 4 and 5 recursively until the expected number of stations has been selected. The convergence of the selection can be determined by the ratio of joint entropy of selected stations to that of all potential candidate stations. If the ratio is over a threshold, such as 0.95, or the ratio will not change significantly which means no additional station still in the candidate set can provide significant amount of new information. These steps show that if no convergence threshold is provided then all the potential candidate stations will be ranked in descending order. This may be helpful when determining the station with the least importance or area or stations with the least degree of importance. These steps illustrate a forward selection procedure.

A hydrometric network can also be optimized in a backward manner, in which the criterion should be changed to minimum reduced information and maximum reduced redundancy. This criterion is also based on the principle of MIMR. The reduced information and redundancy can be quantified by the difference between joint entropies, total correlations of stations before and after one station is deleted, respectively. At the same time it should guarantee that the information of the station to be deleted can be inferred from the left-out station set as much as possible. A pseudo-code displaying the forward MIMR selection procedure is presented in Table 13.5. This algorithm is designed for ranking all the potential candidate stations such that it can provide the decision maker as much information as possible. That is why the convergence judgment appears outside of the selection loop.

In the selection procedure, one may use the simple histogram method to estimate the probability distribution of the hydrometric variable of interest. One can also use more sophisticated methods for probability density estimation, such as kernel density estimation, cross-entropy density estimation, and so on. However, the histogram method is sufficient if a proper bin width or number of bins is used (Scott, 1979). Minimizing the integrated mean square error (IMSE), Scott(1979) proposed an asymptotical optimal choice for bin width h_{opt} as

$$h_{opt} = \left(\frac{6}{\int [df(x)/dx]^2 dx} \right)^{\frac{1}{3}} n^{\frac{1}{3}} \tag{13.46}$$

Table 13.5 Selection procedure.

1: $F \leftarrow$ potential hydrometric station set $S \leftarrow$ empty set	Initialize candidate set F and empty set S
2: $optBins \leftarrow$ equation (13.48)	Determine the optimal bin number
3: Discretize the continuous time series and re-label them by 1, 2, ... $optBins$	
4: $TotalInfor \leftarrow H(F)$	Compute the total information of all the potential stations

5: For $i = 1 : N$
         $H(x_i) \leftarrow$ equation(13.1)
     End
	Compute the marginal entropy of each potential
	     station

6: $s_1 \leftarrow$ arg max$_i$ $H(x_i)$	Select the first center station
7: $F \leftarrow F_{-s_1}$ $S \leftarrow S_{+s_1}$	Update F and S for the first time

8: For $i = 2 : N$
         $m \leftarrow$ length(F)
         $n \leftarrow$ length(S)
         For $k = 1 : m$
             $infor_{s+s_k} \leftarrow$ First part of equation (13.43a)
             $redun_{s+s_k} \leftarrow$ Second part of equation (13.43a)
             $MIMR_{s+s_k} \leftarrow$ equation (13.44a) or (13.44b)
         End
         $s_k \leftarrow$ arg max$_{s_k}$ $MIMR_{s+s_k}$
         $F \leftarrow F_{-s_k}$
         $S \leftarrow S_{+s_k}$
     End
	Sequentially select station from the undated
	     candidate set according to MIMR criterion
	     Update the candidate set and already selected
	     set successively

9: For $i = 1 : N$
         $partialTotalInfor \leftarrow H(S_{1:i})$
         $pct \leftarrow partialTotalInfor/TotalInfor$
         If $|pct - Threshold| <$ eps
             $S_{final} \leftarrow S_{1:i}$
             return
         End
     End
	Determine the final optimal station set S_{final}
	     according to the information fraction of the
	     selected set to the total information

where $f(x)$ is the true underlying density and n is the random sample size. In practice the true density is unknown. Tukey (1977) suggested using the Gaussian density as a reference standard. Substituting the Gaussian density function in the asymptotical optimal bin width equation (13.46), one can have

$$h_{opt} = 2\left(\frac{1}{3}\right)^{1/3} \pi^{1/6} \sigma n^{-1/3} \tag{13.47}$$

where σ is the standard deviation of X. Replacing the underlying true variance σ^2 by in $h_{opt} = 2\frac{1}{3}^{1/3}\pi^{1/6}\sigma n^{-1/3}$ the sampling variance s^2, Scott's data-based choice for the optimal bin width can be expressed as

$$h_{opt}^* = 3.49sn^{-1/3} \tag{13.48}$$

Scott (1979) used this equation to estimate the probability density for several heavy-tailed non-Gaussian distributions and concluded that it produced satisfactory results. Compared to a kernel estimator, the histogram method, due to its slower convergence speed to the underlying density, is less sensitive to the choice of the smoothing parameter, such as the kernel band width and the bin width.

Given a continuous time series, one can compute the optimal bin width, the number of bins and the endpoints for each discretization interval. For example, if the optimal number of bins is b, then the empirical i/b quantiles can be computed, where $i = 0, 1, 2, \ldots, b$. These different i/b quantiles $q_{i/b}$ are just the thresholds used to re-label the continuous time series following the procedure that all the records of the time series falling in the interval $[q_{k-1}, q_{k/b}]$, where $k = 1, 2, \ldots, b$, will be labeled as k and this is repeated for each interval.

For the computation of other high dimensional information quantities, the key is to merge two discrete random samples into a single one such that the marginal entropy of the merged sample is equal to the joint entropy of the two original samples. In other words, the amount of information will be invariant before and after agglomeration. Two discrete or categorical random samples X_1 and X_2 can be merged in such a way placing a unique value for every combination of the corresponding records in X_1 and X_2 (Kraskov, 2003, Alfonso et al. 2010). For instance, if $X_1 = [1\ 2\ 1\ 2\ 1\ 3\ 3]$ and $X_2 = [1\ 2\ 2\ 2\ 1\ 3\ 2]$, then one of the options to merge X_1 and X_2 is by putting all the corresponding digits of X_1 and X_2 together, that is, [11 22 12 22 11 33 32]. [11 22 12 22 11 33 32]. However, this option has a serious defect in that it will cause the problem of "out of memory" as the number of merged samples increases especially when dealing with large samples. Alfonso (2010) suggested an alternative which can also accomplish the same purpose.

First, a new sample is created following the previously described method. Then, pick out the unique values and rank them in ascending order to obtain a new sample called as unique ranked sequence sample, that is, [11 12 22 32 33]. Denoting the length of this new sample as l, access the location index for each element of the previously merged sample in the unique ranked sequence sample. Finally, a new digit is assigned to replace each of the original merged elements by subtracting its location index from l and then adding 1. Figure 13.2 illustrates the entire agglomeration method. From this figure it is seen that two steps are involved. The first step is to weld the corresponding records directly. The second step is to re-label them according to their categories or classes. Results of computation in Figure 13.2 show that this agglomeration method keeps the amount of information retained by the composite samples invariant. However, this conclusion will not hold for continuous random samples. That is why the histogram method is used to estimate the probability distribution of the hydrometeorological variable of interest. This data agglomeration method can also be used for more than two random samples. Further it is symmetric with respect to its composites, which means one can merge several samples in any sequence one wants.

Using the variable agglomeration method, the joint entropy $H(X_{S1}, X_{S2}, \ldots, X_{Sk})$, transinformation between a grouped variable and a single variable $T(X_{S_1:S_k}, X_{F_i})$, and the transinformation between two grouped variables $T(X_{S_1:S_k}, X_{F_1:F_m})$ can be computed. This requires merging multiple variables sequentially to form a new variable without changing the information retained by them first; thus, all computations can be reduced to a univariate or bivariate case.

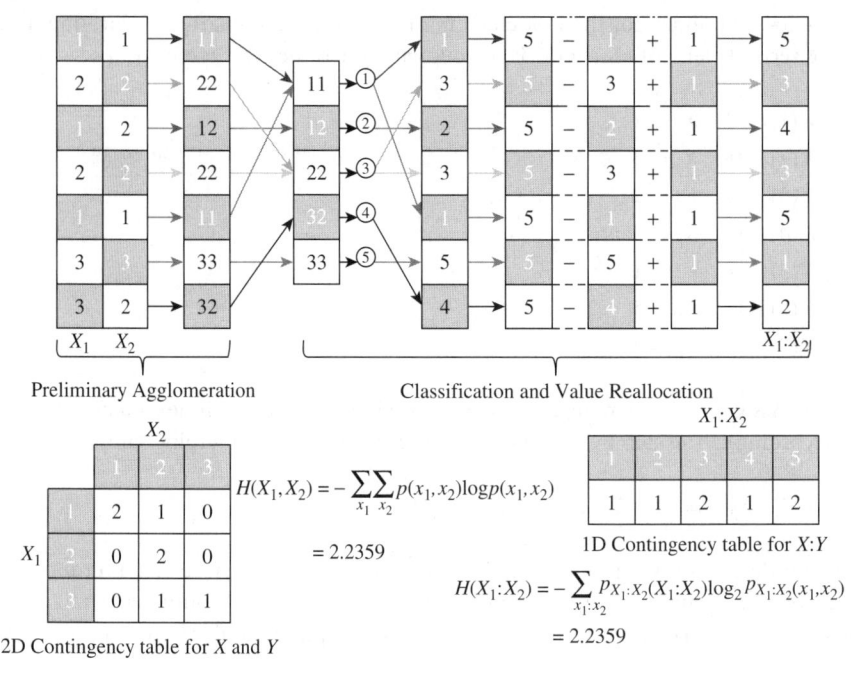

Preliminary Agglomeration   Classification and Value Reallocation

$$H(X_1, X_2) = -\sum_{x_1}\sum_{x_2} p(x_1, x_2)\log p(x_1, x_2)$$

$$= 2.2359$$

2D Contingency table for X and Y

1D Contingency table for $X:Y$

$$H(X_1:X_2) = -\sum_{x_1:x_2} p_{X_1:X_2}(X_1:X_2)\log_2 p_{X_1:X_2}(x_1,x_2)$$

$$= 2.2359$$

Figure 13.2 Schematic illustration for agglomeration of discrete samples.

Example 13.3: Considering three random samples, $X_1 = [1, 2, 1, 2, 1, 3, 3]$, $X_2 = [1, 2, 2, 2, 1, 3, 2]$ and $X_3 = [1, 1, 2, 2, 1, 3, 3]$, apply the variable merging approach. Compute entropy and show that it satisfies the law of association and communication, that is, the information retained by these samples keeps invariant before and after merging.

Solution: The step by step merging procedure is discussed as follows: Following these steps, the three random samples can be merged sequentially.

Step 1: Let us first merge X_1 and X_2 together. Create a new sample X from X_1 and X_2 by direct welding approach:

$$X = [1 \quad 2 \quad 1 \quad 2 \quad 1 \quad 3 \quad 3] \oplus [1 \quad 2 \quad 2 \quad 2 \quad 1 \quad 3 \quad 2]$$
$$= [11 \quad 22 \quad 12 \quad 22 \quad 11 \quad 33 \quad 32]$$

Step 2: Pick out the unique values in X and rank them in ascending order that results in a ranked sample X_r. To do this, first we pick out unique values in X. The result is [11 22 12 33 32]. Then we rank this sample in ascending order and obtain the ranked sample $X_r = [11 \quad 12 \quad 22 \quad 32 \quad 33]$.

Step 3: Access the location index of each element of X in the ranked sample X_r. For convenience and illustrative purposes, we restate the direct welding sample and the ranked sample:

$$X = [11 \quad 22 \quad 12 \quad 22 \quad 11 \quad 33 \quad 32]$$
$$X_r = [11 \quad 12 \quad 22 \quad 32 \quad 33]$$

The location index for the first element of X in X_r is 1. Similarly, the location index for the second element of X in X_r is 3. Following the same way, we have

$$X(1) \quad @ \quad X_r = 1$$
$$X(2) \quad @ \quad X_r = 3$$
$$X(3) \quad @ \quad X_r = 2$$
$$X(4) \quad @ \quad X_r = 3$$
$$X(5) \quad @ \quad X_r = 1$$
$$X(6) \quad @ \quad X_r = 5$$
$$X(7) \quad @ \quad X_r = 4$$

Step 4: Assign each element of X a new label as its location index obtained in step 3. According to the step by step merging procedure, the new merging sample is:

$$< X_1, X_2 > = [1 \quad 3 \quad 2 \quad 3 \quad 1 \quad 5 \quad 4]$$

Now in order to verify if the information content is invariant before and after merging, the joint entropy of X_1 and X_2 is first computed. To do that, the joint contingency table needs to be constructed first. The joint contingency tables for frequency and relative frequency are

		X_2		
		1	2	3
X_1	1	2	1	0
	2	0	2	0
	3	0	1	1

		X_2		
		1	2	3
X_1	1	2/7	1/7	0
	2	0	2/7	0
	3	0	1/7	1/7

Then the joint entropy is computed as

$$H(X_1, X_2) = -\frac{2}{7} \log_2 \left(\frac{2}{7} \right) - \frac{1}{7} \log_2 \left(\frac{1}{7} \right) - \frac{2}{7} \log_2 \left(\frac{2}{7} \right) - \frac{1}{7} \log_2 \left(\frac{1}{7} \right) - \frac{1}{7} \log_2 \left(\frac{1}{7} \right)$$
$$= 2.2359 \text{ bits}$$

Now for computing the marginal entropy of the merged variable $< X_1, X_2 >$, the frequency and relative frequency tables of this variable are noted:

$<X_1, X_2>$				
1	2	3	4	5
2	1	2	1	1

$$<X_1, X_2>$$

1	2	3	4	5
2/7	1/7	2/7	1/7	1/7

The marginal entropy of X is computed as

$$H(<X_1, X_2>) = -\frac{2}{7}\log_2\left(\frac{2}{7}\right) - \frac{1}{7}\log_2\left(\frac{1}{7}\right) - \frac{2}{7}\log_2\left(\frac{2}{7}\right) - \frac{1}{7}\log_2\left(\frac{1}{7}\right) - \frac{1}{7}\log_2\left(\frac{1}{7}\right)$$
$$= 2.2359 \text{ bits}$$

It is now apparent that before and after merging the information content keeps invariant.

Now the new variable $<X_1, X_2>$ and X_3 together are merged. Following the step by step procedure, $<X_1, X_2>$ and X_3 are first directly welded element by element as

$$X' = [1 \quad 3 \quad 2 \quad 3 \quad 1 \quad 5 \quad 4] \oplus [1 \quad 1 \quad 2 \quad 2 \quad 1 \quad 3 \quad 3]$$
$$= [11 \quad 31 \quad 22 \quad 32 \quad 11 \quad 53 \quad 43]$$

Then the unique values of X' are picked out and they are sorted in ascending order in order to obtain the ranked sample as $X'_r = [11 \quad 22 \quad 31 \quad 32 \quad 43 \quad 53]$. Then is accessed the location index for each element of the direct welding sample X' in the ranked sample X'_r as

$$
\begin{array}{lllll}
X'(1) & @ & X'_r & = & 1 \\
X'(2) & @ & X'_r & = & 3 \\
X'(3) & @ & X'_r & = & 2 \\
X'(4) & @ & X'_r & = & 4 \\
X'(5) & @ & X'_r & = & 1 \\
X'(6) & @ & X'_r & = & 6 \\
X'(7) & @ & X'_r & = & 5 \\
\end{array}
$$

Finally, the new merged sample is obtained as

$$<<X_1, X_2>, X_3> = [1 \quad 3 \quad 2 \quad 4 \quad 1 \quad 6 \quad 5]$$

Following what has been done above, it can also be verified that the information retained by $<X_1, X_2>$ and X_3 is equal to that retained by $<<X_1, X_2>, X_3>$. It can also be verified that the joint entropy of the original three samples is equal to the marginal entropy of the merged sample. This same way can be applied to sequentially merge any number of samples together. Therefore, the estimation of multivariate probability distribution estimation can be successfully bypassed. The total correlation can also be computed by the aid of its grouping property and variable merging method.

Example 13.4: Now consider an example explaining how to merge discrete (or categorical) variables step by step. Consider continuous stream flow observations. For the convenience of clear exposition, here a very small data set is used, as shown in Table 13.6. The selected data are monthly stream flow observations in the year of 2009 for three gages located along the upper main stem of the Brazos River. Address the following: 1) how to discretize a continuous time series, 2) how to merge discrete variables, 3) important properties of variable merging

Table 13.6 Observations of monthly stream flow in 2009 at three gages located along the main stem of the Brazos River.

Gage*	Jan.	Feb.	Mar.	Apr.	May.	Jun.	Jul.	Aug.	Sep.	Oct.	Nov.	Dec.
ST01	59.2	123.2	114.2	127.9	121.1	113.9	162.7	237.8	110.6	91.1	104.7	54.9
ST02	56.7	140.7	99.0	113.8	121.1	113.8	162.0	242.3	122.6	147.1	126.0	63.8
ST03	57.0	131.5	133.5	131.2	151.9	121.4	205.6	327.6	311	395.4	251.7	93.5

*ST01: USGS 08088610, ST02: USGS 08089000, ST03: USGS 08090800

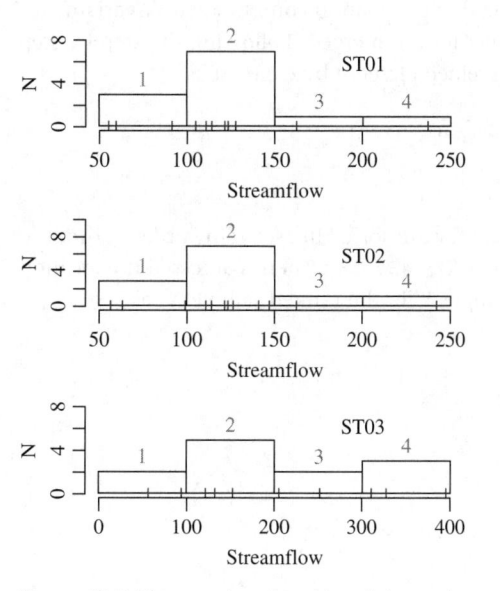

Figure 13.3 Histogram based partition of the continuous stream flow observations for the selected stations (The blue rugs indicate the number of observations falling in each interval and the red integers are the interval labels).

operator, 4) how to compute high dimensional joint entropy, and 5) total correlation via variable merging.

Solution: First, the continuous time series is discretized. The simplest and most commonly used method is the one of histogram partition. The continuous time series can be discretized (or labeled) as follows. First, determine the number of bins, assumed as *bins*. Then, compute the empirical quantiles $q(i/bins)$, where $i = 0, 1, 2, \ldots, bins$. These empirical quantiles are the thresholds to label the continuous time series data. Finally, all data falling in the interval $[q((k-1)/bins), q(k/bins)]$, where $k = 1, 2, \ldots, bins.$, are labeled as k. After discretization, a continuous random sample becomes as a discrete (or categorical) one. Figure 13.3 illustrates the histogram based partition of stream flow observations tabulated in Table 13.6. The resulting discrete (or categorical) samples for each station are listed in Table 13.7.

Taking ST01 as an example, we explain how to obtain Table 13.7 from Table 13.6. Divide the whole range of the stream flow observations into four intervals, that is, $[50, 100), [100, 150), [150, 200), [200, 250)$. At station ST01, the monthly stream flow observation in January is 59.2 m^3/s, falling in the first interval. Therefore, it is labeled as 1. The

Table 13.7 Discretized observations of monthly stream flow in 2009 at three gages located along the main stem of the Brazos River.

Gage*	Jan.	Feb.	Mar.	Apr.	May.	Jun.	Jul.	Aug.	Sep.	Oct.	Nov.	Dec.
ST01	1	2	2	2	2	2	3	4	2	1	2	1
ST02	1	2	1	2	2	2	3	4	2	2	2	1
ST03	1	2	2	2	2	2	3	4	4	4	3	1

*ST01: USGS 08088610, ST02: USGS 08089000, ST03: USGS 08090800

observation in February is 123.2 m³/s, falling in the second interval and is therefore labeled as 2. Following this way, the stream flow observations can be discretized. In some cases, it is possible to work with data which have been discretized or they are categorical variables. In this case, the data discretization procedure is not necessary.

Now the basic idea for variable merging lies in creating a new variable X such that the information retained by it is equal to that retained by the original variables, say $X_1, X_2, \ldots, X_n$. The simplest way of discrete variable merging is directly welding the corresponding digits together. Using $< \cdot >$ to denote the variable merging operator, the merged variable of ST01 and ST02 can be calculated as:

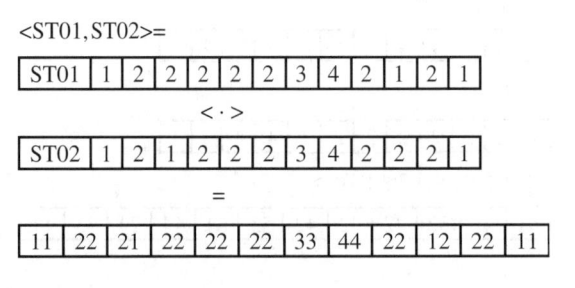

One problem is, however, associated with the above direct welding approach. Assume there are 10 variables (or stations) and assume the first entry in the discretized observation table is the same value of 1 for each variable, then direct welding would yield 1111111111, which is a very huge value. The similar thing may happen to other entries. These effects, acting together, may cause "out of memory." To avoid this problem, after direct welding we re-label the results. For example, the merged ST01 and ST02 can be re-labeled as

<ST01,ST02>=

| 11 | 22 | 21 | 22 | 22 | 22 | 33 | 44 | 22 | 12 | 22 | 11 |

=

| 1 | 4 | 3 | 4 | 4 | 4 | 5 | 6 | 4 | 2 | 4 | 1 |

Actually the re-labeling procedure can be done in many different ways. Here a simple approach is employed. In the following, we describe the re-label procedure. First, look into the direct welding sample for the two stations:

| 11 | 22 | 21 | 22 | 22 | 22 | 33 | 44 | 22 | 12 | 22 | 11 |

Then pick out the unique values:

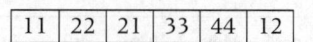

| 11 | 22 | 21 | 33 | 44 | 12 |

Then arrange the unique values in ascending order:

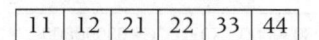

| 11 | 12 | 21 | 22 | 33 | 44 |

Then use the location index of the sorted unique value sample to re-label the direct welding sample. For example, the location index of 11 in the sorted sample is 1 then all elements in the direct welding sample are re-labeled as 1, as will be the case for the first and last elements in the direct welding sample above. Similarly, integer 12 locates at the second place of the sorted sample then all elements in the direct welding sample will be labeled as 2.

Now we want to merge $<ST01, ST02>$ and ST03 together. In the following, the above explained discrete variable merging procedure is displayed step by step.

Step 1: Direct welding

$<ST01, ST02>$:

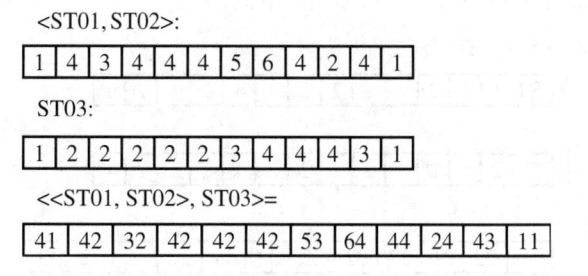

| 1 | 4 | 3 | 4 | 4 | 4 | 5 | 6 | 4 | 2 | 4 | 1 |

ST03:

| 1 | 2 | 2 | 2 | 2 | 2 | 3 | 4 | 4 | 4 | 3 | 1 |

$<<ST01, ST02>, ST03>=$

| 41 | 42 | 32 | 42 | 42 | 42 | 53 | 64 | 44 | 24 | 43 | 11 |

Step 2: Pick up the unique values in the merged sample

Unique values in the direct merging sample $<< ST01, ST02 >, ST03 >$ are:

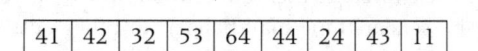

| 41 | 42 | 32 | 53 | 64 | 44 | 24 | 43 | 11 |

Step 3: Sort the unique values in ascending order

The sorted unique value sample is

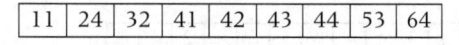

| 11 | 24 | 32 | 41 | 42 | 43 | 44 | 53 | 64 |

Step 4: Re-label the direct welding sample

The location indices for the elements of the direct welding sample in the sorted unique value sample are used to re-label the direct welding sample elements. For example, the first and second elements in the direct welding sample are 41 and 42, respectively, whose location indices are 4 and 5, respectively. Therefore, the first and second elements are re-labeled as 4 and 5, respectively. Repeating the same procedure for all the other elements, the direct welding sample can be re-labeled by small integers. The final result $<< ST01, ST02 >, ST03 >$ is

4	5	3	5	5	5	8	9	7	2	6	1

It is easy to verify that the information content keeps invariant before and after merging. In the following we verify this statement. The joint contingency tables can be constructed using the data in Table 13.7:

		STO1			
		[50, 100)	[100, 150)	[150, 200)	[200, 250)
ST02	[50, 100)	2	1	0	0
	[100, 150)	1	6	0	0
	[150, 200)	0	0	1	0
	[200, 250)	0	0	0	1

Then the relative frequency table is obtained by dividing each element in the contingency table by the summation of its elements:

		STO1			
		[50, 100)	[100, 150)	[150, 200)	[200, 250)
ST02	[50,100)	0.167	0.083	0	0
	[100,150)	0.083	0.5	0	0
	[150, 200)	0	0	0.083	0
	[200, 250)	0	0	0	0.083

Now the joint entropy, $H(ST01, ST02)$ is calculated as

$$H(ST01, ST02) = -\sum_{i=1}^{4}\sum_{j=1}^{4} p_{ij} \ln(p_{ij}) = -0.167 \ln(0.167) - 0.083 \ln(0.083)$$

$$- 0.083 \ln(0.083) - 0.5 \ln(0.5) - 0.083 \ln(0.083) - 0.083 \ln(0.083)$$

$$= 1.4735 \text{ nats}$$

On the other hand, the merged variable of ST01 and ST02 is

1	4	3	4	4	4	5	6	4	2	4	1

From this the one-dimensional contingency table is constructed as

	Label 1	Label 2	Label 3	Label 4	Label 5	Label 6
<ST01,ST02>	2	1	1	6	1	1

and the relative frequency table is

<ST01, ST02>	Label 1	Label 2	Label 3	Label 4	Label 5	Label 6
	0.167	0.083	0.083	5	0.083	0.083

Using the definition of marginal entropy, $H(<\text{ST01}, \text{ST02}>)$ can be computed as

$$H(<\text{ST01}, \text{ST02}>) = -\sum_{i=1}^{6} p_i \ln(p_i) = -0.167 \ln(0.167) - 0.083 \ln(0.083)$$
$$- 0.083 \ln(0.083) - 0.5 \ln(0.5) - 0.083 \ln(0.083) - 0.083 \ln(0.083)$$
$$= 1.4735 \text{ nats}$$

Obviously, before and after merging the information content is invariant.

Now important properties of variable merging are briefly examined. The variable merging approach satisfies the law of association and commutation in terms of information content. Considering the merging of three variables as an example, the following equalities are satisfied according to the law of association and commutation:

$$H(< \text{ST01}, \text{ST02}, \text{ST03} >) = H(<< \text{ST01}, \text{ST02} >, \text{ST03} >)$$
$$= H(< \text{ST01}, < \text{ST02}, \text{ST03} >>)$$
$$= H(<< \text{ST01}, \text{ST03} >, \text{ST02} >)$$

The above equalities can be easily verified using selected data set:

$$H(<< \text{ST01}, \text{ST02} >, \text{ST03} >) = H([1, 6, 3, 6, 6, 6, 7, 8, 4, 2, 5, 1]) = 1.9073 \text{ nats}$$
$$H(< \text{ST01}, < \text{ST02}, \text{ST03} >>) = H([1, 3, 2, 3, 3, 3, 7, 8, 5, 6, 4, 1]) = 1.9073 \text{ nats}$$
$$H(<< \text{ST01}, \text{ST03} >, \text{ST02} >) = H([1, 3, 4, 3, 3, 3, 7, 8, 6, 2, 5, 1]) = 1.9073 \text{ nats}$$

Now high dimensional entropy terms are computed. The multivariate joint entropy, $H(X_1, X_2, \ldots, X_n)$, can be computed by sequentially applying the variable merging approach, that is,

$$H(X_1, X_2, \ldots, X_n) = H(< X_1, X_2 >, X_3, \ldots, X_n)$$
$$= H(<< X_1, X_2 >, X_3 >, \ldots, X_n)$$
$$= H(< \ldots << X_1, X_2 >, X_3 >, \ldots X_{n-1} >, X_n)$$
$$\ldots \ldots$$
$$= H(< \ldots < \ldots << X_1, X_2 >, X_3 >, \ldots, X_{n-1} >, X_n >)$$

Since variable merging operator satisfies the law of association and commutation in terms of information content, one can neglect the merging sequence.

For the computation of total correlation, two different ways are available. One is the shortcut formula:

$$C(X_1, X_2, \ldots, X_n) = \sum_{i=1}^{n} H(X_i) - H(X_1, X_2, \ldots, X_n)$$

First the joint entropy and marginal entropies are computed. Then, the above equation is directly applied to calculate the total correlation.

Another one is to exploit the grouping property [*Kraskov* et al., 2005] and variable merging as shown in the following:

$$
\begin{aligned}
C(X_1, X_2, \ldots, X_n) &= C(X_1, X_2) + C(<X_1, X_2>, X_3, X_4, \ldots, X_n) \\
&= C(X_1, X_2) + C(<X_1, X_2>, X_3) + C(<X_1, X_2, X_3>, X_4, X_5, \ldots, X_n) \\
&\cdots \cdots \\
&= C(X_1, X_2) + C(<X_1, X_2>, X_3) + C(<X_1, X_2, X_3>, X_4) \\
&\quad + \cdots + C(<X_1, X_2, \ldots, X_{n-1}>, X_n)
\end{aligned}
$$

Since total correlation at the bivariate level reduces to transinformation, the above equation indicates that n-dimensional total correlation is finally factorized as a summation of traditional transinformation values.

Questions

Q.13.1 Take monthly discharge data for several gaging stations along a river. Compute marginal entropy of monthly discharge at each gaging station and plot it as function of distance between gaging stations. What do you conclude from this plot? Discuss it.

Q.13.2 For the discharge data in **Q.13.1**, compute transinformation of monthly discharge. Are all gaging stations needed? What is the redundant information? What can be said about increasing or decreasing the number of gaging stations?

Q.13.3 For the gaging stations in **Q.13.1**, take daily discharge data and compute transinformation of daily discharge. Are all stations needed?

Q.13.4 For the gaging stations in **Q.13.1**, take weekly discharge data and compute transinformation of weekly discharge. Are all stations needed?

Q.13.5 For the gaging stations in **Q.13.1**, take three-month discharge data and compute transinformation of three-month discharge. Are all stations needed?

Q.13.6 For the gaging stations in **Q.13.1**, take yearly discharge data and compute transinformation of yearly discharge. Are all stations needed?

Q.13.7 Plot transinformation of two gages against different time periods, such as day, week, month, three-month, and year. Do this for several pairs of stations? What does the plot tell?

Q.13.8 Plot the number of gages obtained in **Q.13.1–13.6** against the time period. What does the plot tell?

Q.13.9 Consider a drainage basin which has a number of rainfall measuring stations. Obtain annual rainfall values for each gaging station. Using annual rainfall values, determine marginal entropy at each station and also compute transinformation amongst stations. Comment on the adequacy of the rain gage network.

Q.13.10 Consider the same basin and the rain gage network as in **Q.13.9**. Now obtain monthly rainfall values and compute marginal entropy as well transinformation. Comment on the adequacy of the rain gage network. How does the adequacy change with reduced time interval. Which stations are necessary and which are not?

Q.13.11 For the gaging stations in **Q.13.10**, take daily rainfall data and compute transinformation of daily rainfall. Are all stations needed?

Q.13.12 For the gaging stations in **Q.13.10**, take weekly rainfall data and compute transinformation of weekly rainfall. Are all stations needed?

Q.13.13 For the gaging stations in **Q.13.10**, take three-month rainfall data and compute transinformation of three-month rainfall. Are all stations needed?

Q.13.14 For the gaging stations in **Q.13.10**, take yearly rainfall data and compute transinformation of yearly rainfall. Are all stations needed?

Q.13.15 Plot transinformation of two gages against different time periods, such as day, week, month, three-month, and year. Do this for several pairs of stations? What does the plot tell?

Q.13.16 Plot the number of rain gages obtained in **Q.13.10–13.15** against the time period. What does the lot tell?

Q.13.17 Can entropy be employed for designing a monitoring network? If yes, then how? Can entropy be employed for evaluating the adequacy of an existing network? If yes, how?

References

Adamowski, K. (1989). A Monte Carlo comparison of parametric and nonparametric estimation of flood frequencies. *Journal of Hydrology*, Vol. 108, pp. 295–309.

Alfonso, L. (2010). Optimisation of monitoring networks for water distribution systems. Ph.D. Dissertation, Delft University of Technology, Delft, The Netherlands.

Alfonso, L., Price, R. and Lobbrecht, A., 2010. Information theory-based approach for location of monitoring water level gauges. *Water Resources Research*, Vol. 46, W03528, doi:10.1029/2009WR008101.

Al-Zahrani, M. and Husain, T. (1998). An algorithm for designing a precipitation network in the south-western region of Saudi Arabia. *Journal of Hydrology*, Vol. 295, No. 3-4, pp. 205–16.

Berthouex, P.M. and Brown, L.C. (1994). *Statistics for Environmental Engineers*. CRC Press, Inc: Florida; 335 pp.

Bras, R.L. and Rodriguez-Iturbe, I. (1985). *Random Functions and Hydrology*. Addison Wesley, Reading, Massachusetts.

Bueso, M.C., Angulo, J.M., Cruz-Sanjulian, J. and Carcia-Arostegui, J.L. (1999). Optimal spatial sampling design in a multivariate framework. *Mathematical Geology*, Vol. 31, No. 5, pp. 507–25.

Burn, D.H. and Goulter, I.C. (1991). An approach to the rationalization of streamflow data collection networks. *Journal of Hydrology*, Vol. 122, pp. 71–91.

Cacoullos, T. (1966). Estimation of a multivariate density. *Inst. Stat., Math. Ann.*, Vol. 18, pp. 179–89.

Cressie, N.C. (1990). *Statistics for Spatial Data, Revised Edition*. Wiley: New York; 900 pp.

Fass, D.M. (2006). *Human Sensitivity to Mutual Information*. Rutgers, The State University of New Jersey, New Brunswcks, New Jersey.

Guiasu, S. (1977). *Information Theory with Applications*. McGraw Hill Book Company, London, U.K.

Harmancioglu, N.B. (1981). Measuring the information content of hydrological processes by the entropy concept. *Journal of Civil Engineering Faculty of Ege University*, Special Issue: Centennial of Ataturk's Birth, Izmir, pp. 13–40, Turkey.

Harmancioglu, N.B. (1984). Entropy concept as used in determination of optimum sampling intervals. Proceedings of Hydrosoft '84, International Conference on Hydraulic Engineering Software, Portoroz, Yugoslavia, pp. 6–99 to 6–110.

Harmancioglu, N.B. and Alpaslan, N. (1992). Water quality monitoring network design. *Water Resources Bulletin*, Vol. 28, No. 1, pp. 179–92.

Harmancioglu, N.B., Fistikoglu, O., Ozkul, S.D., Singh, V.P. and Alpaslan, M.N. (1999). *Water Quality Monitoring Network Design*. Kluwer Academic Publishers: Boston; 299 pp.

Harmancioglu, N.B. and Yevjevich, V. (1985). Transfer of hydrologic information along rivers fed by karstified limestones. Proceedings of the Ankara-Antalya Symposium on Karst Water Resources, IAHS Publication 61, pp. 151–31.

Harmancioglu, N.B. and Yevjevich, V. (1987). Transfer of hydrologic information among river points. *Journal of Hydrology*, Vol. 91, pp. 103–18.

Husain, T. (1987). Hydrologic network design formulation. *Canadian Water Resources Journal*, Vol. 12, No. 1, pp. 44–63.

Husain, T. (1989). Hydrologic uncertainty measure and network design. *Water Resources Bulletin*, Vol. 25, No.3, pp. 527–34.

Husain, T., Ukayli, M.A. and Khan, H.U. (1986). Meteorological network expansion using information decay concept. *Journal of Atmospheric and Oceanic Technology, AMA*. Vol. 3, No. 1, pp. 27–37.

Jakulin, A. and Bratko, I., 2003. Quantifying and visualizing attribute interactions. Arxiv preprint cs.AI/0308002.

Jakulin, A. and Bratko, I., 2004. Testing the significance of attribute interactions. ACM International Conference Proceedings Series.

Jessop, A. (1995). *Informed Assessments, An Introduction to Information, Entropy and Statistics*. Ellis Horwood: New York; 366 pp.

Kapur, J.N. and Kesavan, H.K. (1992). *Entropy Optimisation Principles with Applications*. Academic Press Inc.: San Diego; 408 pp.

Kraskov, A., Stogbauer, H., Andrzejak, R.G. and Grassberger, P. (2003). Hierarchical clustering based on mutual information. ArXiv preprint q-bio. QM/0311039.

Krstanovic, P.F. and Singh, V.P. (1988). Application of entropy theory to multivariate hydrologic analysis. Technical Reports WRR8 and WRR9, Department of Civil Engineering, Louisiana State University, Baton Rouge, Louisiana.

Krstanovic, P.F. and Singh, V.P. (1992a). Evaluation of rainfall networks using entropy: 1. Theoretical development. *Water Resources Management*, Vol. 6, pp. 279–93.

Krstanovic, P.F. and Singh, V.P. (1992b). Evaluation of rainfall networks using entropy: 1. Application. *Water Resources Management*, Vol. 6, pp. 295–314.

Lathi, B.P. (1969). *An Introduction to Random Signals and Communication Theory*. International Textbook Company, Scanton, Pennsylvania.

Lee Y. and Ellis, J.H. (1997). On the equivalence of Kriging and maximum entropy estimators. *Mathematical Geology*, Vol. 29, No. 1, pp. 131–52.

Linfoot, E.H. (1957). An informational measure of correlation. *Information and Control*, Vol. 1, pp. 85–9.

Lubbe, C.A. (1996). *Information Theory*. Cambridge University Press: Cambridge; 350 pp.

McGill, W.J., 1954. Multivariate information transmission. *Psychometrika*, Vol. 19, pp. 97–116.

Mishra, A.K. and Coulibaly, P. (2009). Developments in hydrometric network design: a review. *Reviews of Geophysics*, Vol. 47, RG2001, doi:10.1029/2007RG000243.

Mogheir, Y. and Singh, V.P. (2002). Application of information theory to groundwater quality monitoring networks. *Water Resources Management*, Vol. 16, No. 1, pp. 37–49.

Ozkul, S., Harmancioglu, N.B. and Singh, V.P. (2000). Entropy-based assessment of water quality monitoring networks. *Journal of Hydrologic Engineering*, ASCE, Vol. 5, No. 1, pp. 90–100.

Parzen, E. (1962). On estimation of a probability density function and mode. *Annals of Mathematical Statistics*, Vol. 33, pp. 1065–76.

Rodriguez-Iturbe, I. and Mejia, J.M. (1974). The design of rainfall networks in time and space. *Water Resources Research*, Vol. 10, No. 4, pp. 713–28.

Scott, D.W. (1979). On optimal and data-based histograms. *Biometrica*, Vol. 66, No. 3, pp. 605–10.

Singh, V.P. (1998). *Entropy-Based Parameter Estimation in Hydrology*. Kluwer Academic Publishers: Boston.

Tukey, J. W. (1977). *Exploratory Data Analysis*. Reading, Mass: Addison-Wesley.

Uslu, O. and Tanriover, A. (1979). Measuring the information content of hydrological process. Proceedings of the First National Congress on Hydrology. Istanbul, pp. 437–43.

Watanabe, S. (1960). Information theoretical analysis of multivariate correlation. *IBM Journal of Research and development*, Vol. 6.

Wertz, W. (1979). *Statistical Density Estimation: a survey*. Vandenhoeck & Ruprecht, Gottingen.

Yang, Y. and Burn, D.H. (1994). An entropy approach to data collection network design. *Journal of Hydrology*, Vol. 157, pp. 307–24.

Zhang, J., Wang, H. and Singh, V.P. (2011). Information entropy of a rainfall network in China. Chapter 2 in: *Modeling Risk Management in Sustainable Construction*, edited by D.D. Wu and Y. Zhu, pp. 11–20, Springer, Berlin.

Additional Reading

Bras, R.L. and Rodriguez-Iturbe, I. (1976). Network design for the estimation of areal mean of rainfall event. *Water resources Research*, Vol. 12, pp. 1185–95.

Altiparmak, F. and Dengiz, B. (2009). A cross entropy approach to design of reliable networks. *European Journal of Operational Research*, Vol. 199, pp. 542–52.

Caselton, W.F. and Husain, T. (1980). Hydrologic networks: Information transinformation. *Journal of Water Resources Planning and Management*, ASCE, Vol. 106, No. WR2, pp. 503–29.

Chapman, T.G. (1986). Entropy as a measure of hydrologic data uncertainty and model performance. *Journal of Hydrology*, Vol. 85, pp. 111–26.

Chen, Y.C., Wei, C. and Yeh, H.C. (2008). Rainfall network design using kriging and entropy. *Hydrological Processes*, Vol. 22, pp. 340–6.

Harmancioglu, N.B. (1997). The need for integrated approaches to environmental data management. In: *Integrated Approaches to Environmental Data Management Systems*, edited by N.B. Harmancioglu, M.N. Alsplan, S.D. Ozkul and V.P. Singh, Kluwer Academic Publishers, NATO ASI Series, 2, Environment, Vol. 31, pp. 3–14.

Harmancioglu, N.B., Alpaslan, N. and Singh, V.P. (1992a). *Application of the Entropy Concept in Design of Water Quality Monitoring Networks. Entropy and Energy Dissipation in Water Resources*, edited by V.P. Singh and M. Fiorentino, Kluwer Academic Publishers, Dordrecht, The Netherlands, pp. 283–302.

Harmancioglu, N.B., Alpaslan, N., and Singh, V.P. (1994). *Assessment of Entropy Principle as Applied to Water Quality Monitoring Network Design. Time Series Analysis in Hydrology and Environmental Engineering*, Vol. 3, edited by K.W. Hipel, A.I. McLeopd, U.S. Panu, and V.P. Singh, pp. 135–48, Kluwer Academic Publishers, Dordrecht, The Netherlands.

Harmancioglu, N.B., Alpaslan, N. M. and Singh, V.P. (1998a). Needs for Environmental Data Management. Chapter 1, pp. 1–12, in *Environmental Data Management*, edited by Harmancioglu, N.B., Singh, V.P. and M.N. Alpaslan, Kluwer Academic Publishers, Dordrecht, The Netherlands.

Harmancioglu, N.B., Alkan, A., Singh, V.P. and Alpaslan, N. (1996). Entropy-Based Approaches to Assessment of Monitoring Networks. Proceedings, IAHR International Symposium on Stochastic Hydraulics, edited by Tickle, Goulter, I.C., Xu, C.C., Wasimi, S.A., and Bouchart, F., pp. 183–90.

Harmancioglu, N.B., Alpaslan, M.N., Whitfield, P., Singh, V.P., Literathy, P., Mikhailov, N. and Fiorentino, M. (1998b) Assessment of Water Quality Monitoring Networks-Design and Redesign. Final Report to NATO, Brussels, Belgium.

Harmancioglu, N.B., Alpaslan, N., Ozkul, S. D. and Singh, V.P. (1997). Integrated Approach to Environmental Data Management Systems. NATO ASI Series, Vol. 31, Kluwer Academic Publishers, Dordrecht, The Netherlands, 546 pp.

Harmancioglu, N.B., Fistikoglu, O. and Singh, V.P. (1998). Modeling of Environmental Processes. Chapter IX, pp. 213–242, in *Environmental Data Management*, edited by Harmancioglu, N.B., Singh, V.P. and M.N. Alpaslan, Kluwer Academic Publishers, Dordrecht, The Netherlands.

Harmancioglu, N.B. and Singh, V.P. (1990). Design of Water Quality Networks. Technical Report WRR14, 63 pp., Water Resources Program, Department of Civil Engineering, Louisiana State University, Baton Rouge, Louisiana.

Harmancioglu, N.B. and Singh, V.P. (1991). *An Information Based Approach to Monitoring and Evaluation of Water Quality Data. Advances in Water Resources Technology*, pp. 377–386, edited by G. Tsakiris, A. A. Balkema, Rotterdam, The Netherlands.

Harmancioglu, N.B. and Singh, V.P. (1998). Entropy in Environmental and Water Resources. pp. 225–41, Chapter in *Encyclopedia of Hydrology and Water Resources*, edited by D.R. Herschy, Kluwer Acdemic Publishers, Dordrecht, The Netherlands.

Harmancioglu, N.B. and Singh, V.P. (1999). *On Redesign of Water Quality Networks*. in: *Environmental Modeling*, edited by V.P. Singh, I.L. Seo and J.H. Sonu, pp. 47–60, Water Resources Publications, Littleton, Colorado.

Harmancioglu, N.B. and Singh, V.P. (2002). Data Accuracy and Validation. in *Encyclopedia of Life Support Systems*, edited by A. Sydow, EOLSS Publishers Co., Ltd., Oxford, U. K.

Harmancioglu, N.B., Singh, V.P. and Alpaslan, N. (1992b). *Versatile Uses of the Entropy Concept in Water Resources. Entropy and Energy Dissipation in Water Resources*, edited by V.P. Singh and M. Fiorentino, Kluwer Academic Publishers, Dordrecht, The Netherlands, pp. 91–118.

Harmancioglu, N.B., Singh, V.P. and Alpaslan, N. (1992c). Design of Water Quality Monitoring Networks. Chapter 8, pp. 267–96, in *Geomechanics and Water Engineering in Environmental Management*, edited by R. N. Chowdhury, A. A. Balkema Publishers, Rotterdam, The Netherlands.

Harmancioglu, N. B., Singh, V.P. and Alpaslan, N., editors (1997). *Environmental Data Management*. Kluwer Academic Publishers, Dordrecht, The Netherlands, 298 pp.

Husain, T. (1979). Shannon's information theory in hydrologic design and estimation. unpublished Ph.D., University of British Columbia, Vancouver, Canada.

Husain, T. and Khan, H.U. (1983). Shannon's entropy concept in optimum air monitoring network design. *The Science of the Total Environment*, Vol. 30, pp. 181–90.

Husain, T. and Ukayli, M.A. (1983). Meteorological network expansion for Saudi Arabia. *Journal of Research in Atmosphere*, Vol. 16, pp. 281–94.

Krstanovic, P.F. and Singh, V.P. (1992). Transfer of information in monthly rainfall series of San Jose, California. in: *Entropy and Energy Dissipation in Water Resources*, edited by V.P. Singh and M. Fiorentino, Kluwer Academic Publishers, Dordrecht, The Netherlands, pp. 155–74.

Markus, M., Knapp, H.V. and Tasker, G.D. (2003). Entropy and generalized least square methods in assessment of the regional value of stream flow gages. *Journal of Hydrology*, Vol. 283, pp. 107–21.

Masoumi, F. and Kerachian, R. (2010). Optimal design of groundwater quality monitoring networks: a case study. *Environmental Monitoring Assessment*, Vol. 161, pp. 247–57.

Mogheir, Y. and Singh, V.P. (2002a). Quantification of information for ground water quality monitoring networks. *Water Resources Management*, Vol.16, pp. 37–49.

Mogheir, Y. and Singh, V.P. (2002b). Specification of information needs for groundwater resources management and planning in a developing country: Gaza Strip case study. in *Ground Water Hydrology*, edited by M. M. Sherif, V.P. Singh, and M. Al-Rashid, A. A. Balkema, Rotterdam, The Netherlands, pp. 3–20.

Mogheir, Y., de Lima, J.L.M.P. and Singh, V.P. (2003a). Assessment of spatial structure of groundwater quality variables based on the entropy theory. *Hydrology and Earth System Sciences*, Vol. 7, No. 5, pp. 707–21.

Mogheir, Y., de Lima, J.L.M.P. and Singh, V.P. (2003b). Applying the entropy theory for characterizing the spatial structure of groundwater regionalized variables (EC and Chloride). Proceedings, Environment 2010: Situation and Perspectives for the European Union, May 6-10, Porto, Portugal.

Mogheir, Y., Singh, V.P. and de Lima, J.L.M.P. (2003c). Redesigning the Gaza Strip groundwater quality monitoring using entropy. in: *Ground Water Pollution*, edited by V.P. Singh and R.N. Yadava, pp. 315–31, Allied Publishers Pvt. Limited, New Delhi, India.

Mogheir, Y., de Lima, J.L.M.P and Singh, V.P. (2004a). Characterizing the spatial variability of groundwater quality using the entropy theory: 1. Synthetic data. *Hydrological Processes*, Vol. 18, pp. 2165–79.

Mogheir, Y., de Lima, J.L.M.P and Singh, V.P. (2004b). Characterizing the spatial variability of groundwater quality using the entropy theory: 2. Case study from Gaza Strip. *Hydrological Processes*, Vol. 18, pp. 2579–90.

Mogheir, Y., de Lima, J.L.M.P. and Singh, V.P. (2004c). Influence of data errors on groundwater quality monitoring network assessment and redesign. Proceedings, EWRA Symposium on Water Resources Management: Risks and Challenges for the 21st Century, held September 2-4, 2004, in Izmir, Turkey.

Mogheir, Y., de Lima, J.L.M.P. and Singh, V.P. (2005). Assessment of informativeness of groundwater monitoring in developing regions (*Gaza Strip Case Study*). *Water Resources Management*, Vol. 19, pp. 737–57.

Mogheir, Y., de Lima, J.L.M.P. and V.P. Singh (2009). Entropy and multi-objective based approach for groundwater quality monitoring network assessment and redesign. *Water Resources Management*, Vol. 23, pp. 1603–20.

Mogheir, Y., Singh, V.P. and de Lima, J.L.M.P. (2006). Spatial assessment and redesign of a groundwater quality monitoring network using entropy theory, Gaza Strip, Palestine. *Hydrogeoloogy Journal*, Vol. 14, pp. 700–12.

Moss, M.E. (1982). Concepts and techniques in hydrological network design. Operational Hydrology Report No. 19, 30 pp., World Meteorological Organization, Geneva, Switzerland.

Moon, Y.I., Rajagopalan, B. and Lall, U. (1995). Estimation of mutual information using kernel density estimators. *Physical Review* E, Vol. 52, No. 3, pp. 2318–21.

Sanders, T., Ward, R.C., Loftis, J.C., Steele, T.D., Adrian, D.D. and Yevjevich, V. (1983). Design of networks for monitoring water quality. Water Resources Publications, Littleton, Colorado.

Sarlak, N. and Sorman, A.U. (2006). Evaluation and selection of streamflow network stations using entropy methods. *Turkish Journal of Engineering and Environmental Science*, Vol. 30, pp. 91–100.

Stol, P.T. (1981a). Rainfall interstation correlation functions: 1. An analytic approach. *Journal of Hydrology*, Vol. 50, pp. 45–71.

Stol, P.T. (1981b). Rainfall interstation correlation functions: 1Application to three storm models with the percentage of dry days as a new parameter. *Journal of Hydrology*, Vol. 50, pp. 73–104.

Yoo, C., Jung, K. and Lee, J. (2008). Evaluation of rain gage network using entropy theory: comparison of mixed and continuous distribution function applications. *Journal of Hydrologic Engineering*, Vol. 13, No. 4, pp. 226–35.

14   Selection of Variables and Models

When investigating a water resources or environmental system experimentally or otherwise, the question arises: Which variables should be selected for measurement or in modeling? For example, for conducting experiments on river morphology, variables of flow (depth and velocity), geometry (cross-section, wetted perimeter and width), slope, sediment size, meander length, sinuosity, bed forms, and erosion and deposition should be measured. Other variables can be derived from measurements of these variables. On the other hand, there can be a situation where measurements on a lot of variables are available. The question then arises: Which variables are relevant and should be selected to describe the system? For assessing the hazard potential of debris flow, variables pertaining to different aspects of hydrology, topography, and geology are incorporated in a geographic information system (Lin et al., 2002; Rupert et al., 2003; Chen et al., 2007). It is quite possible that different investigators employ different observed variables, even if there is some consensus on debris flow assessment. The question arises: Which variables are most important and should be employed in debris flow assessment?

Another situation frequently encountered is one of selecting a model from amongst different models. For example, there is a multitude of models for the assessment of suspended sediment load and bed load of a river. Which model should one select for a particular river or how can different models be ranked? The same applies to models of flow routing, erosion around bridge piers, hydraulic geometry, velocity distribution, and so on. This chapter discusses the principle of minimum entropy for variable or model selection (Tseng, 2006).

14.1 Methods for selection

There exist several methods for model or variable selection, such as P-values, Bayesian, and Kullback-Leibler distance method. The P-values method (Raftery et al., 1997) is restricted to two models. For selecting a model it compares the probability of the model, given a null model and experimental data sets, to a threshold value determined from the same data sets. Determination of the threshold value is somewhat ad hoc.

The Bayesian method utilizes the Bayes theorem, generates prior distributions based on some prior modeling rules and updates beliefs and uncertainty about models. Then the Bayes

Entropy Theory and its Application in Environmental and Water Engineering, First Edition. Vijay P. Singh.
© 2013 John Wiley & Sons, Ltd. Published 2013 by John Wiley & Sons, Ltd.

factor, defined by the ratio of posterior distributions of different models, is computed for choosing a model. This is also referred to as the Bayesian Information Criterion (BIC) and is one of the most popular selection criteria (Forbes and Peyrard, 2003; Weiss, 1995). The drawback, however, is that it requires prior information to be generated somewhat ad hoc.

14.2 Kullback-Leibler (KL) distance

The Kullback-Leibler (KL) distance, relative entropy or cross-entropy, has also been employed for model selection (Bollander and Weigend, 1994; Dupuis and Robert, 2003). The KL distance measures the distance between a model and a reference prior for a system of interest. The models are ranked based on the KL distance. A model having the largest value of distance is preferred. Dupuis and Robert (2003) employed the KL distance for variable selection where the distance between the full model corresponding to the complete set of variables and sub-models or approximations corresponding to subsets of variables was evaluated. The information on the full model is given a priori. A sub-model whose KL distance reaches a threshold is preferred.

Let there be M models given by probability distributions $P^m = \{p_j^m\}$, $m = 1, 2, \ldots, M$ where m denotes the m-th model. Let there be a reference distribution $Q = \{q_j\}$ of a model $P = \{p_j\}$ Then, a preference measure can be constructed using the scalar relative or cross-entropy as

$$H(P^m|Q) = \sum_j p_j^m \ln\left[\frac{p_j^m}{q_j}\right] \tag{14.1}$$

$H(P^m|Q)$, given by equation (14.1), measures the difference between model p^m and the reference distribution Q (Tseng, 2006). A small value of $H(P^m|Q)$ would imply a small difference between p^m and Q.

Consider, for a moment, that the reference distribution is the true or real distribution of the system Q_{real}. Then a model p^m yielding the minimum $H(P^m|Q)$ will be the most preferred distribution. However, in real world Q_{real} is not known. Therefore, Tseng (2006) proposed the use of a uniform distribution function Q_{uni}, computing $H(P^m|Q_{uni})$ for each model, and then ranking all candidate models. It may now be noted that a model p^m with a minimum $H(P^m|Q_{uni})$ provides minimum information about the system. This is because the uniform distribution is the most uncertain distribution carrying little useful information and hence the minimum $H(P^m|Q_{uni})$ would suggest that the two distributions are identical.

On the other hand, p^m is other than a uniform distribution, meaning it is codified with some information and would lead to a larger $H(P^m|Q_{uni})$. Arranging the values of $H(P^m|Q_{uni})$ in descending order would indicate the preference of models. The model p^m that is farther away from Q_{uni} carries more relevant information about the system and a larger relative entropy is more preferable.

14.3 Variable selection

Consider a regression model $P(X)$ containing N variables, $X_1, X_2, \ldots, X_N$, that describes the behavior of a system. These variables may be determined by experimentation and may be correlated with each other and may or may not be the crucial characteristics of the system. One may model the system with various combinations of these variables. The question then arises: Which combination of the variables best describes the system or which variables are more important? This question was addressed by Tseng (2006).

For a set of N variables, $X = \{X_1, X_2, \ldots, X_N\}$, there will be $(2^N - 2)$ combinations or subsets of variables, $X_{s_i} \in X$. Each subset of variables X_{s_i} leads to a sub-model with $P^s(X_{s_i})$. Then equation (14.1) can be recast as

$$H(P_s|Q_{uni}) = \sum P_s(X_{s_i}) \ln \left[\frac{P_s(X_{s_i})}{Q_{uni}} \right]$$

$$= -H(P_s) + \ln Q_{uni} \tag{14.2}$$

where sub-model $P_s(X_{s_i})$ contains n_i variables, and

$$H(P_s) = - \sum_{X_{s_i} \in X} P_s(X_{s_i}) \ln P_s(X_{s_i}) \tag{14.3}$$

Quantity $\ln Q_{uni}$ is constant. Models or sub-models can be ranked in the decreasing order of $H(P^m|Q_{uni})$ which is identical to the decreasing order of $H(P_s)$. Then, variables can be selected in accordance with the preference of sub-models. It may be noted that the maximum entropy (ME) method can correctly rank variables so long as they can be codified in a sub-model.

14.4 Transitivity

Consider three models 1, 2, and 3, with distributions P_1, P_2, and P_3, respectively. If model P_1 is preferred to P_2 and model P_2 is preferred to model P_3, then P_1 is preferred to P_3. This is referred to as transitivity property. To each distribution, entropy $H(P)$ can be assigned. If P_1 is preferred to P_2, then $H(P_1) < H(P_2)$. The functional form of $H(P)$ can be determined using POME. Increasing relative entropy $H(P^m|Q_{uni})$ indicates increasing preference of models.

14.5 Logit model

Consider N variables $X = \{X_1, X_2, \ldots, X_N\}$ to which the logistic regression (in short logit) model (Johnson and Albert, 1999; Dupuis and Robert, 2003) can be applied. This model is a statistical model that predicts the probability of occurrence. Examples include the probability of flooding after a heavy rainfall or the probability of wildfire occurring during a drought. Rupert et al. (2008) have applied the logistic regression to predict the probability of debris flows in areas burned by wildfires in California. Similar to multiple linear regression, logistic regression also determines relations between one dependent variable and several independent variables, but there is one important difference. The logistic regression determines the probability of a positive binomial outcome, for example, flood did or did not occur after a rainfall or a wildfire did or did not occur during a drought or beach erosion did or did not occur during a hurricane or debris flow did or did not occur after a heavy storm or landslide did or did not occur after a snowstorm. The multiple linear regression determines the dependent variable continuously.

The logit model can be expressed as

$$
y(x) = \frac{\exp\left(\sum\limits_{i=1}^{N} a_i x_i\right)}{1 + \exp\left(\sum\limits_{i=1}^{N} a_i x_i\right)} \tag{14.4}
$$

which relates Y (y is a specific value), the system response or dependent variable to independent variables or covariates X_i (x_i is a specific value); $i = 1, 2, \ldots, N$; x is specific value of X; and a_i are constants. Let a normalizing constant be defined as

$$
Z = \sum\limits_{X} \frac{\exp\left(\sum\limits_{i=1}^{N} a_i x_i\right)}{1 + \exp\left(\sum\limits_{i=1}^{N} a_i x_i\right)} \tag{14.5}
$$

Now the probability distribution of the system response or output of a given subset of the variables $X = \{X_i, i = 1, 2, \ldots, N\}$ is defined as

$$
P(X) = \frac{y(X)}{Z} = \frac{1}{Z} \frac{\exp\left(\sum\limits_{i=1}^{N} a_i x_i\right)}{1 + \exp\left(\sum\limits_{i=1}^{N} a_i x_i\right)} \tag{14.6}
$$

Using equation (14.6) the entropy of $P(X_{s_i})$ related to the subset of variables can be computed, which then gives the rank of the sub-model. Constants a_i can be determined by fitting the logit model to experimental measurements by the method of maximum likelihood estimation (Johnson and Albert, 1999) or any other method, such as the least square method. Equation (14.6) determines the predictive success of the logit model (Kleinbaum, 1994; Hosmer and Lemeshow, 2000). For each independent variable, a p-value can be computed that reflects the statistical significance of that variable for the overall logit model. For example, a p-value of 0.1 would indicate a significance level of 90% and a p-value of 0.05 would indicate a significance level of 95%.

Example 14.1: Data on chemical analyses of brine (in ppm) for oil-field waters obtained from drillstem tests of three carbonate rock units, namely Ellenburger Dolomite and Grayburg Dolomite = Unit G and Viola Limestone, from Texas and Oklahoma are given in Table 14.1. These are extracted from Davis (2002). Chen et al. (2007) have also employed these data. Brines recovered during tests from wells may contain signatures of compositional characteristics that may provide clues to the origin or depositional environment of their source rocks. Thus, the first column in Table 14.1 indicates whether or not brine samples belong to a specific carbonate unit, that is, Grayburg Dolomite (briefly designated here as Unit G) and other columns state the six chemical ions. Thus, the values in the first column represent the dependent variable and those in the other columns represent independent variables, with no (meaning 0) and yes (meaning 1). Clearly the dependent variable in this case is a binary one. The objective is to determine the number of combinations in which independent variables can be selected to determine if the associated rock source is unit G or not to which the water belonged. Each combination leads to one specific model. Compute the relative entropy of each model.

Table 14.1 Data on chemical analysis of brines (in ppm).

Unit G (Y)	HCO$_3$ (X_1)	SO$_4$ (X_2)	Cl (X_3)	Ca (X_4)	Mg (X_5)	Na (X_1)
0	10.4	30	967.1	95.9	53.7	857.7
0	6.2	29.6	1174.9	111.7	43.9	1054.7
0	2.1	11.4	2387.1	348.3	119.3	1932.4
0	8.5	22.5	2186.1	339.6	73.6	1803.4
0	6.7	32.8	2015.5	287.6	75.1	1691.8
0	3.8	18.9	2175.8	340.4	63.8	1793.9
0	1.5	16.5	2367	412	95.8	1872.5
1	25.6	0	134.7	12.7	7.1	134.7
1	12	104.6	3163.8	95.6	90.1	3093.9
1	9	104	1342.6	104.9	160.2	1190.1
1	13.7	103.3	2151.6	103.7	70	2054.6
1	16.6	92.3	905.1	91.5	50.9	871.4
1	14.1	80.1	554.8	118.9	62.3	472.4
0	1.3	10.4	3399.5	532.3	235.6	2642.5
0	3.6	5.2	974.5	147.5	69	768.1
0	0.8	9.8	1430.2	295.7	118.4	1027.1
0	1.8	25.6	183.2	35.4	13.5	161.5
0	8.8	3.4	289.9	32.8	22.4	225.2
0	6.3	16.7	360.9	41.9	24	318.1

Then, select the combinations that have minimum Shannon entropy or maximum relative entropy values.

Solution: The first column in Table 14.1 means that brine samples belong (denoted as "1") or do not belong (denoted as "0") to Unit G and defines the values of the dependent variable. The rest of the columns are the six chemical ions (in ppm) and these constitute values of the six independent variables. Since the dependent variable Y is binary with outcome as 0 or 1, it is advisable to use the logistic regression. Here X_i, $i = 1, 2, \ldots, 6$, denote the independent variables representing chemical ions. For six candidate variables, the number of possible combinations is $2^6 - 2 = 62$, as shown in Table 14.2 (shortened) where "1" denotes

Table 14.2 Sixty-two possible combinations or sub-models.

Sub-model	X_1	X_2	X_3	X_4	X_5	X_6
1	0	0	0	0	0	1
2	0	0	0	0	1	0
3	0	0	0	0	1	1
4	0	0	0	1	0	0
5	0	0	0	1	0	1
6	0	0	0	1	1	0
7	0	0	0	1	1	1
–	–	–	–	–	–	–
59	1	1	1	0	1	1
60	1	1	1	1	0	0
61	1	1	1	1	0	1
62	1	1	1	1	1	0

Table 14.3 Coefficients a_i for 62 sub-models.

Sub-model	a_0	a_1	a_2	a_3	a_4	a_4	a_5
1	−0.8836	0	0	0	0	0	0.0001
2	−0.6608	0	0	0	0	−0.0015	0
3	−0.7821	0	0	0	0	−0.0042	0.0003
4	0.7977	0	0	0	−0.0110	0	0
5	−0.1775	0	0	0	−0.0255	0	0.0028
6	−1.2227	0	0	0	−0.0963	0.1998	0
7	−1.2482	0	0	0	−0.1016	0.2159	−0.0004
−	−	−	−	−	−	−	−
60	−43.6896	2.1458	0.3441	−0.0029	0.0366	0	0
61	−46.3787	2.3127	0.3405	0.0874	−0.0688	0	−0.0888
62	−46.6605	2.2533	0.2447	−0.0012	0.0188	0.0906	0

that the variable is included and "0" means that the variable is excluded. Each combination of variables corresponds to a sub-model.

First, the logit model is applied to relate the response Y to the covariates (X_i) using equation (14.4). To that end, coefficients a_i are determined by fitting the logit model to the data in Table 14.1 by the maximum likelihood estimation (MLE) which can be done in Matlab by the mnrfit function and results are shown in Table 14.3.

It is important to check how good these coefficient values are. One way to judge the goodness is to use each sub-model with the corresponding coefficients determined above and compute the dependent variable. For example, for sub-model 1, equation (14.4) becomes:

$$y(x) = \frac{\exp(-0.8836 + 0.0001x_6)}{1 + \exp(-0.8836 + 0.0001x_6)}$$

Then, the correlation coefficient between calculated and observed values of Y is computed for all 62 sub-models, as shown in Table 14.4.

From Table 14.4, it is seen that the calculated coefficients for each sub-model are satisfactory, because for some sub-models (such as sub-models 56, 57, etc.), the correlation coefficient values are 1.00. It may be noted for some sub-models (like sub-model 1) the correlation coefficient values are very low, pointing to the need for variable selection. Now, in order to apply equation (14.2), the model-computed y values need to be converted into probabilities, which is done using equation (14.6). When applying equation (14.2) the prior is selected as a uniform distribution. For each sub-model, there are n_i variables. Since, $\ln Q_{uni}$ is constant, the order of ranking given by increasing $H[p_s|Q_{uni}]$ is identical to ranking by decreasing $H[p_s]$. Now, the relative entropy of each sub-model is calculated as shown in Table 14.5 (shortened).

From Table 14.5, it is seen that the combinations that have minimum entropy values are as follows:

Sub-model	X_1	X_2	X_3	X_4	X_5	X_6	Entropy
59	1	1	1	0	1	1	1.79183823

Table 14.4 The coefficient of correlation between calculated and observed values of Y.

Sub-model	Correlation coefficient
1	0.04
2	0.03
3	0.10
4	0.43
5	0.65
6	0.74
7	0.74
8	0.08
9	0.73
–	–
–	–
56	1.00
57	1.00
58	1.00
59	1.00
60	1.00
61	1.00
62	1.00

Table 14.5 Relative entropy and entropy of each sub-model.

Sub-model	X_1	X_2	X_3	X_4	X_5	X_6	Relative entropy	Entropy
59	1	1	1	0	1	1	1.15260075	1.79183823
61	1	1	1	1	0	1	1.15260069	1.79183829
55	1	1	0	1	1	1	1.15260062	1.79183836
62	1	1	1	1	1	0	1.15260062	1.79183836
54	1	1	0	1	1	0	1.15259823	1.79184075
51	1	1	0	0	1	1	1.15259721	1.79184177
58	1	1	1	0	1	0	1.15259683	1.79184215
57	1	1	1	0	0	1	1.15259657	1.79184241
50	1	1	0	0	1	0	1.15259502	1.79184396
53	1	1	0	1	0	1	1.15259245	1.79184653
60	1	1	1	1	0	0	1.15259197	1.79184701
47	1	0	1	1	1	1	1.15259010	1.79184888
52	1	1	0	1	0	0	1.15258930	1.79184968
56	1	1	1	0	0	0	1.15258427	1.79185471
49	1	1	0	0	0	1	1.15258230	1.79185668
48	1	1	0	0	0	0	1.15258160	1.79185738
33	1	0	0	0	0	1	0.89989192	2.04454706
40	1	0	1	0	0	0	0.89629479	2.04814419
–	–	–	–	–	–	–	–	–
–	–	–	–	–	–	–	–	–
45	1	0	1	1	0	1	0.50137038	2.44306860
46	1	0	1	1	1	0	0.48144050	2.46299847

Table 14.6 $P(x)$ each sub-model.

$y(x)$	$Z = \mathrm{sum}[y(x)]$	$P(x)$
2.96E − 05	6.00E + 00	4.94E − 06
9.14E − 10	6.00E + 00	1.52E − 10
1.78E − 11	6.00E + 00	2.96E − 12
8.51E − 06	6.00E + 00	1.42E − 06
9.96E − 07	6.00E + 00	1.66E − 07
3.85E − 11	6.00E + 00	6.42E − 12
7.25E − 12	6.00E + 00	1.21E − 12
1.00	6.00E + 00	0.17
1.00	6.00E + 00	0.17
1.00	6.00E + 00	0.17
1.00	6.00E + 00	0.17
1.00	6.00E + 00	0.17
1.00	6.00E + 00	0.17
7.57E − 07	6.00E + 00	1.26E − 07
1.57E − 13	6.00E + 00	2.61E − 14
7.42E − 13	6.00E + 00	1.24E − 13
9.38E − 16	6.00E + 00	1.56E − 16
7.53E − 11	6.00E + 00	1.25E − 11
5.97E − 12	6.00E + 00	9.95E − 13

Example 14.2: For Example 14.1, prepare a matrix (a shortened one) of selected combinations of variables and entropy. Use 1 if the variable is included and 0 if it is not.

Solution: From the calculations in Example 14.1, sub-model 59 is considered as an example. First, substituting the corresponding model coefficients into equation (14.4), we get:

$$y(x) = \frac{\exp(-46.61 + 2.27 \times x_1 + 0.27 \times x_2 + 0.02 \times x_3 + 0.07 \times x_5 - 0.02 \times x_6)}{1 + \exp(-46.61 + 2.27 \times x_1 + 0.27 \times x_2 + 0.02 \times x_3 + 0.07 \times x_5 - 0.02 \times x_6)}$$

For each sub-model, $y(x)$ is calculated for each observed dataset and then all the $y(x)$ values are summed up in order to get Z using equation (14.5). Then, $P(x)$ is computed for each sub-model using equation (14.6), as shown in Table 14.6.

Then, using equation (14.3), the entropy of each sub-model is computed. For example, for sub-model 59, it is done as:

$$H(P_s) = - \sum_{X_{S_i} \in X} P_s(X_{S_i}) \, \ln P_s(X_{S_i})$$

$$= -4.94 \times 10^{-6} \times \ln(4.94 \times 10^{-6}) - 1.52 \times 10^{-10} \times \ln(1.52 \times 10^{-10}) - \cdots$$

$$= 1.79183823 \text{ Napier}$$

Finally, the selected combinations of variables and entropy are obtained, as shown in Table 14.7 (Note: "1" denotes the variable is included and "0" means the variable is excluded.)

Example 14.3: From calculations in Example 14.2, count the number of times each variable appears and then discuss the significance of the variables and models selected.

Solution: From the entropy value calculated for each sub-model in Table 14.7 in Example 14.2, it is determined that the entropy values of the first 16 sub-models are very close and very small. Hence, these 16 sub-models are included in Table 14.8.

Table 14.7 Entropy of each sub-model.

Sub-model	X_1	X_2	X_3	X_4	X_5	X_6	Entropy
59	1	1	1	0	1	1	1.79183823
61	1	1	1	1	0	1	1.79183829
55	1	1	0	1	1	1	1.79183836
62	1	1	1	1	1	0	1.79183836
54	1	1	0	1	1	0	1.79184075
51	1	1	0	0	1	1	1.79184177
58	1	1	1	0	1	0	1.79184215
57	1	1	1	0	0	1	1.79184241
50	1	1	0	0	1	0	1.79184396
53	1	1	0	1	0	1	1.79184653
60	1	1	1	1	0	0	1.79184701
47	1	0	1	1	1	1	1.79184888
52	1	1	0	1	0	0	1.79184968
56	1	1	1	0	0	0	1.79185471
49	1	1	0	0	0	1	1.79185668
48	1	1	0	0	0	0	1.79185738
33	1	0	0	0	0	1	2.04454706
40	1	0	1	0	0	0	2.04814419
–	–	–	–	–	–	–	–
46	1	0	1	1	1	0	2.46299847

Table 14.8 Entropy of 16 selected sub-models.

Sub-model	X_1	X_2	X_3	X_4	X_5	X_6	Entropy
59	1	1	1	0	1	1	1.79183823
61	1	1	1	1	0	1	1.79183829
55	1	1	0	1	1	1	1.79183836
62	1	1	1	1	1	0	1.79183836
54	1	1	0	1	1	0	1.79184075
51	1	1	0	0	1	1	1.79184177
58	1	1	1	0	1	0	1.79184215
57	1	1	1	0	0	1	1.79184241
50	1	1	0	0	1	0	1.79184396
53	1	1	0	1	0	1	1.79184653
60	1	1	1	1	0	0	1.79184701
47	1	0	1	1	1	1	1.79184888
52	1	1	0	1	0	0	1.79184968
56	1	1	1	0	0	0	1.79185471
49	1	1	0	0	0	1	1.79185668
48	1	1	0	0	0	0	1.79185738
Number of times	16	15	8	8	8	8	

The last row of Table 14.8 shows the number of times each variable appears. X_1 appears 16 times, X_2 appears 15 times, X_3 appears 8 times, X_4 appears 8 times, X_5 appears 8 times, X_6 appears 8 times. Based on the number of times each variable appears, X_1 and X_2 are more important than the other four variables (i.e., X_3, X_4, X_5 and X_6). Therefore, although sub-model 59's entropy is minimum, however, if one just chooses X_1 and X_2 in the model, that corresponds to sub-model 48, it can be seen from Table 14.8 that sub-model 48's entropy is also very small.

Example 14.4: Using the results from Example 14.3, evaluate if the models can be further simplified.

Solution: Based on the number of times X_1 and X_2 appear in Example 14.3, it seems that the model can be further simplified by just including X_1 and X_2. This can be shown as follows. The coefficients for sub-model 48 are:

Sub-model	a_0	a_1	a_2	a_3	a_4	a_5	a_6
48	−40.2965	2.0083	0.3194	0.0000	0.0000	0.0000	0.0000

For this sub-model, the logit model can be written as

$$y(x) = \frac{\exp(-40.2965 + 2.0083 \times x_1 + 0.3194 \times x_2)}{1 + \exp(-40.2965 + 2.0083 \times x_1 + 0.3194 \times x_2)}$$

Now, $y(x)$ is calculated and compared with observed $y(x)$. Then, the coefficient of correlation between calculated and observed y values is calculated, as shown in Table 14.9.

From Table 14.9, the coefficient of correlation of the calculated and observed values of Y is 1.00. Therefore, instead of using sub-model 59 (the entropy of sub-model 59 is minimum), one can further simplify sub-model 59 by including only variables X_1 and X_2 in the model and excluding four other variables. It is already shown that this model (i.e., sub-model 48) is

Table 14.9 Coefficient of correlation between calculated and observed y values for sub-model 48.

Calculated y by sub-model 48	Observed y	Correlation coefficient
5.39E − 05	0	
1.03E − 08	0	
8.18E − 15	0	
1.08E − 07	0	
7.82E − 08	0	1.00
2.73E − 12	0	
1.25E − 14	0	
1.00	1	
1.00	1	
1.00	1	
1.00	1	
1.00	1	
1.00	1	
1.19E − 15	0	
2.29E − 14	0	
3.60E − 16	0	
4.18E − 13	0	
4.43E − 10	0	
2.05E − 10	0	

accurate enough. Sub-model 48's entropy is also small and this model's predicted y values are basically the same as observed y values.

Example 14.5: Water quality data for Hawkesbury River at Richmond, New South Wales, Australia, are available as shown in Table 14.9, where Y indicates if the river is suitable for swimming. Here Y is the dependent variable and water quality constituents are independent variables. The objective is to determine the number of combinations in which independent variables can be selected to determine if the river is suitable for swimming. Each combination leads to one specific model. Compute the relative entropy of each model. Then, select the combinations that have minimum Shannon entropy or maximum relative entropy values.

Solution: Since the dependent variable Y is a binary outcome as 0 or 1, it is advisable to use logistic regression. It should be noted that the five independent variables X_i have different units, and therefore, they are standardized first. In Matlab, the zscore function can be used to standardize the independent variables. For the five candidate variables, the number of combinations or models is $2^5 - 2 = 32 - 2 = 30$. These possible combinations are shown in Table 14.11, where "1" denotes the variable is included and "0" means the variable is excluded.

First, the logit model is applied to relate the response Y to the covariates (X_i) using equation (14.4). To that end, coefficients a_i are determined by fitting the logit model to the data in

Table 14.10 The data I use in this assignment.

Y	EC (mS/cm) X_1	DO (%) X_2	Temp (Deg.C) X_3	NTU X_4	pH X_5
1	0.34	93.9	24.3	2.16	7.98
0	0.31	126.55	25.7	1.7	9.03
0	0.31	98.62	19.2	2.42	7.91
0	0.24	101.94	17.1	3.18	7.94
0	0.309	111.42	12.9	2.72	7.87
0	0.264	97.22	10.2	5.29	7.89
1	0.178	79.85	12.9	3.1	7.56
0	0.231	101.29	14.3	3.7	7.82
0	0.18	88.19	15	4	7.54
1	0.215	69.19	20.9	5.39	7.18
1	0.277	78.05	22.7	46.3	7.28
1	0.178	67.55	26.4	3.63	7.69
1	0.371	103.93	27.6	1.36	8.49
1	0.318	84.15	24.7	1.23	7.61
0	0.315	98.22	20.8	1.65	9.2
0	0.321	120.84	21.3	1.78	8.12
0	0.28	107.89	14.7	2.15	7.75
0	0.269	100.07	11.1	2.17	7.75
1	0.181	98.72	12.6	2.34	7.54
1	0.165	100.61	15.4	4.03	7.52
1	0.173	100.4	17.9	1.93	8.03
1	0.285	71.96	23.2	3.03	7.4
1	0.367	89.79	30.4	7.78	7.83
1	0.323	91.41	26.5	1.34	8.02
0	0.202	102.26	24.5	3.11	7.73
1	0.212	94.58	24	2	7.51
1	0.248	111.05	21.3	1.94	7.96

Table 14.11 Thirty possible sub-models.

Sub-model	X_1	X_2	X_3	X_4	X_5
1	0	0	0	0	1
2	0	0	0	1	0
3	0	0	0	1	1
4	0	0	1	0	0
5	0	0	1	0	1
6	0	0	1	1	0
7	0	0	1	1	1
–					
25	1	1	0	0	1
26	1	1	0	1	0
27	1	1	0	1	1
28	1	1	1	0	0
29	1	1	1	0	1
30	1	1	1	1	0

Table 14.10 by the method of maximum likelihood estimation (MLE) which can be done in Matlab by the mnrfit function and results are shown in Table 14.12.

It is important to check how good these coefficient values are. One way to judge the goodness is use each sub-model with the corresponding coefficients determined above and

Table 14.12 Coefficients a_i for 30 sub-models.

Sub-model	a_0	a_1	a_2	a_4	a_5	a_6
1	0.1916	0	0	0	0	−0.9914
2	0.3139	0	0	0	0.8265	0
3	0.2282	0	0	0	0.3011	−0.9085
–	–	–	–	–	–	–
5	0.4501	0	0	1.7254	0	−1.8814
6	0.5527	0	0	1.0308	1.6895	0
7	0.4323	0	0	1.7407	−0.1245	−1.9194
9	0.4018	0	−1.7148	0	0	−0.2478
10	0.3987	0	−1.8360	0	−0.0263	0
11	0.3966	0	−1.7437	0	−0.0894	−0.2590
12	0.6111	0	−1.8012	1.1004	0	0
13	0.7514	0	−1.3367	1.5285	0	−1.0237
14	0.6080	0	−1.8078	1.1007	−0.0288	0
15	0.7184	0	−1.3901	1.5556	−0.3416	−1.0772
–	–	–	–	–	–	–
21	0.4500	−0.7140	0	2.1195	0	−1.5212
22	0.3540	−1.2520	0	1.8210	0.6436	0
23	0.4431	−0.7128	0	2.1282	−0.0585	−1.5406
–	–	–	–	–	–	–
26	0.3991	0.0188	−1.8371	0	−0.0269	0
27	0.3997	0.1924	−1.7156	0	−0.1289	−0.3880
28	0.5536	−0.9749	−1.6406	1.7927	0	0
29	0.6717	−0.6793	−1.3088	1.8816	0	−0.6984
30	0.5494	−0.9765	−1.6524	1.7954	−0.0446	0

Table 14.13 Coefficient of correlation (ρ) between calculated and observed y values for 30 sub-models.

Sub-model	ρ	Sub-model	ρ	Sub-model	ρ	Sub-model	ρ
1	0.36	9	0.56	17	0.37	25	0.56
2	0.18	10	0.55	18	0.21	26	0.55
3	0.37	11	0.56	19	0.37	27	0.56
4	0.43	12	0.64	20	0.61	28	0.70
5	0.66	13	0.69	21	0.71	29	0.72
6	0.46	14	0.64	22	0.61	30	0.70
7	0.66	15	0.69	23	0.71		
8	0.55	16	0.11	24	0.55		

compute the dependent variable. For example, for sub-model 1, equation (14.4) becomes:

$$y(\vec{x}) = \frac{\exp(0.1916 - 0.9914x_6)}{1 + \exp(0.1916 - 0.9914x_6)}$$

Then, the correlation coefficient between calculated and observed values of Y is computed for all 30 sub-models, as shown in Table 14.13.

From Table 14.12, it is seen that the calculated a coefficients for sub-models should be good, because for some sub-models (such as sub-models 29, 21, etc.), the correlation coefficients of calculated and observed y are greater than 0.7. Now the relative entropy of each sub-model is calculated, as shown in Table 14.14. For calculating the relative entropy, the following formula is used: relative entropy $= \ln 27 - H(P_s)$, where $H(P_s)$ is the Shannon entropy.

From Table 14.14, it is seen that the combination that has minimum entropy value is:

Sub-model	X_1	X_2	X_3	X_4	X_5	Entropy
29	1	1	1	0	1	3.0441

Example 14.6: For Example 14.5, prepare a matrix (a shortened one) of selected combinations of variables and entropy. Use 1 if the variable is included and 0 if it is not.

Solution: From the calculations in Example 14.5, sub-model 29 is considered as an example. Substituting the corresponding model coefficients into equation (14.4), we get:

$$y(\vec{x}) = \frac{\exp(0.6717 - 0.6793 \times x_1 - 1.3088 \times x_2 + 1.8816 \times x_3 - 0.6984 \times x_5)}{1 + \exp(0.6717 - 0.6793 \times x_1 - 1.3088 \times x_2 + 1.8816 \times x_3 - 0.6984 \times x_5)}$$

For each sub-model, $y(x)$ is calculated for each observation data set and then all y values are summed up in order to get Z according using equation (14.5). Then, $P(x)$ is computed for each sub-model, as shown in Table 14.15.

As an example, the entropy of sub-model 29 is calculated as follows:

$$H(P_s) = -\sum P_s(x_{s_i}) \log p_s(x_{s_i})$$
$$= -0.05 \times \log(0.05) - 0.01 \times \log(0.01) - \dots .$$
$$= 3.0441 \text{ Napier}$$

Table 14.14 Relative entropy and entropy of each sub-model.

Sub-model	X_1	X_2	X_3	X_4	X_5	Relative entropy	Entropy
29	1	1	1	0	1	0.2517	3.0441
30	1	1	1	1	0	0.2480	3.0478
28	1	1	1	0	0	0.2475	3.0483
15	0	1	1	1	1	0.2350	3.0609
13	0	1	1	0	1	0.2311	3.0648
23	1	0	1	1	1	0.2249	3.0709
21	1	0	1	0	1	0.2244	3.0715
7	0	0	1	1	1	0.2066	3.0892
5	0	0	1	0	1	0.2054	3.0905
14	0	1	1	1	0	0.2049	3.0910
12	0	1	1	0	0	0.2046	3.0912
22	1	0	1	1	0	0.1634	3.1325
20	1	0	1	0	0	0.1597	3.1361
27	1	1	0	1	1	0.1506	3.1452
11	0	1	0	1	1	0.1500	3.1459
25	1	1	0	0	1	0.1493	3.1465
9	0	1	0	0	1	0.1491	3.1468
10	0	1	0	1	0	0.1469	3.1489
26	1	1	0	1	0	0.1468	3.1490
8	0	1	0	0	0	0.1467	3.1492
24	1	1	0	0	0	0.1466	3.1493
6	0	0	1	1	0	0.0888	3.2070
4	0	0	1	0	0	0.0811	3.2147
17	1	0	0	0	1	0.0795	3.2163
19	1	0	0	1	1	0.0778	3.2180
1	0	0	0	0	1	0.0743	3.2215
3	0	0	0	1	1	0.0732	3.2226
18	1	0	0	1	0	0.0168	3.2790
2	0	0	0	1	0	0.0114	3.2844
16	1	0	0	0	0	0.0049	3.2909

Similarly, entropy is calculated for other sub-models. Finally, the selected combinations of variables and entropy are shown in Table 14.16 (Note: "1" denotes the variable is included and "0" means the variable is excluded).

Example 14.7: From calculations in Example 14.5, count the number of times each variable appears and then discuss the significance of the variables and models selected.

Solution: From the already calculated entropy values of sub-models in Example 14.6, it is found that the entropy values of 11 sub-models are close and are very small. Hence, the following shortened table including the first 11 sub-models is shown as Table 14.17.

The last row of the table shows the number of times each variable appears. Thus, X_1 appears 5 times, X_2 appears 7 times, X_3 appears 11 times, X_4 appears 5 times, and X_5 appears 7 times. Based on the number of times each variable appears, X_3 is the most important variable, X_2 and X_5 are also important variables, while X_1 and X_4 are not very important in this sense. Therefore, although sub-model 29's entropy is minimum, however, if we just choose X_3, X_2 and X_5 in the model, that corresponds to sub-model 13, it can be seen from Table 14.16 that sub-model 13's entropy is also very small.

Table 14.15 $P(x)$ each sub-model 29.

$y(x)$	$Z = \text{sum}(y(x))$	$P(x)$
0.78	15.00	0.05
0.08	15.00	0.01
0.40	15.00	0.03
0.33	15.00	0.02
0.03	15.00	0.00
0.06	15.00	0.00
0.76	15.00	0.05
0.21	15.00	0.01
0.75	15.00	0.05
0.99	15.00	0.07
0.98	15.00	0.07
1.00	15.00	0.07
0.59	15.00	0.04
0.96	15.00	0.06
0.13	15.00	0.01
0.10	15.00	0.01
0.10	15.00	0.01
0.07	15.00	0.00
0.34	15.00	0.02
0.57	15.00	0.04
0.57	15.00	0.04
0.99	15.00	0.07
0.97	15.00	0.06
0.91	15.00	0.06
0.92	15.00	0.06
0.96	15.00	0.06
0.44	15.00	0.03

Example 14.8: Using the results from Example 14.7, evaluate if the models can be further simplified.

Solution: Based on Example 14.7, the model can be further simplified by just including X_3, X_2 and X_5 variables because of the number of times they appear. This is verified as follows.

The a coefficients for sub-model 13 are as follows:

Sub-model	a_0	a_1	a_2	a_3	a_4	a_5
13	0.7514	0	−1.3367	1.5285	0	−1.0237

Accordingly, this sub-model is expressed as follows:

$$y(\vec{x}) = \frac{\exp(0.7514 - 1.3367 \times x_2 + 1.5285 \times x_3 - 1.0237 \times x_5)}{1 + \exp(0.7514 - 1.3367 \times x_2 + 1.5285 \times x_3 - 1.0237 \times x_5)}$$

Using this formula, the y values are calculated and compared with the observed y values by calculating the correlation coefficient as shown in Table 14.18.

Table 14.16 Entropy of each sub-model.

Sub-model	X_1	X_2	X_3	X_4	X_5	Entropy
29	1	1	1	0	1	3.0441
30	1	1	1	1	0	3.0478
28	1	1	1	0	0	3.0483
15	0	1	1	1	1	3.0609
13	0	1	1	0	1	3.0648
23	1	0	1	1	1	3.0709
21	1	0	1	0	1	3.0715
7	0	0	1	1	1	3.0892
5	0	0	1	0	1	3.0905
14	0	1	1	1	0	3.0910
12	0	1	1	0	0	3.0912
22	1	0	1	1	0	3.1325
20	1	0	1	0	0	3.1361
27	1	1	0	1	1	3.1452
11	0	1	0	1	1	3.1459
25	1	1	0	0	1	3.1465
9	0	1	0	0	1	3.1468
10	0	1	0	1	0	3.1489
26	1	1	0	1	0	3.1490
8	0	1	0	0	0	3.1492
24	1	1	0	0	0	3.1493
6	0	0	1	1	0	3.2070
4	0	0	1	0	0	3.2147
17	1	0	0	0	1	3.2163
19	1	0	0	1	1	3.2180
1	0	0	0	0	1	3.2215
3	0	0	0	1	1	3.2226
18	1	0	0	1	0	3.2790
2	0	0	0	1	0	3.2844
16	1	0	0	0	0	3.2909

From Table 14.18, the coefficient of correlation of the calculated and observed y values is 0.69. Therefore, instead of using sub-model 29 (the entropy of sub-model 29 is minimum and the correlation coefficient of calculated and observed y values for sub-model 29 is 0.72), it appears that sub-model 29 can be simplified by just including only variable X_2, X_3 and X_5 in the model and excluding the other two variables.

14.6 Risk and vulnerability assessment

Consider the case of debris flow but the methodology can be extended to other hazard areas as well, including floods, droughts, groundwater contamination, landslides, earthquakes, volcanic eruptions, tsunamis, coastal hazards, ecological hazards, and so on. Debris flow results from interactions between manmade systems and natural processes. Risk assessment may entail three levels of assessment: 1) hazard assessment: determination of magnitude and frequency of occurrence of that magnitude; 2) vulnerability assessment: evaluation of exposed population and property in a given area; and 3) risk assessment incorporating both hazard and vulnerability.

Table 14.17 Entropy of 11 sub-models.

Sub-model	X_1	X_2	X_3	X_4	X_5	Entropy
29	1	1	1	0	1	3.0441
30	1	1	1	1	0	3.0478
28	1	1	1	0	0	3.0483
15	0	1	1	1	1	3.0609
13	0	1	1	0	1	3.0648
23	1	0	1	1	1	3.0709
21	1	0	1	0	1	3.0715
7	0	0	1	1	1	3.0892
5	0	0	1	0	1	3.0905
14	0	1	1	1	0	3.0910
12	0	1	1	0	0	3.0912
Counts of times	5	7	11	5	7	

Table 14.18 Comparison between calculated and observed values.

Calculated y by sub-model 13	Observed y	Correlation coefficient
0.86	1	
0.04	0	
0.55	0	
0.32	0	
0.07	0	
0.11	0	
0.74	1	
0.23	0	
0.70	0	
0.99	1	
0.99	1	
1.00	1	
0.66	1	
0.98	1	
0.10	0	
0.14	0	
0.18	0	
0.14	0	
0.32	1	
0.46	1	
0.35	1	
0.99	1	
0.99	1	
0.93	1	
0.84	0	
0.94	1	
0.37	1	0.69

　　Risk defines a measure of the expected loss of lives, damage to property, and disruption in socio-economic activity in a given area for a reference period of time. Risk entails both hazard (L) and vulnerability (V) and is scaled from 0 (0%) to 1 (100%) (Alexander, 1991). The hazard is comprised of two components: magnitude and the associated frequency of occurrence and indicates the potential threat. The hazard of an event is measured on a scale of 0 (0%) to 1 (100%) (Deyle et al., 1998). Vulnerability measures the potential total maximum losses due to a hazardous event in a given area for a reference period of time, and is also measured on a scale of 0 (0%) to 1 (100%). Thus, it can be noted that vulnerability represents the potential for maximum loss or harm and risk represents the probability that some or all of that vulnerability will occur due to the event. This means that risk is always less than or equal to vulnerability.

14.6.1 Hazard assessment

As mentioned earlier, two components of an event define hazard: magnitude reflecting the size of the event and frequency reflecting the time period between events of that magnitude. For modeling purposes, both are considered as random variables. For many processes, such as debris flow and geologic and geomorphic hazards, there is a general relationship between magnitude and frequency and this relationship is defined by a probability distribution, such as Pearson type III distribution or any other standard distribution. In general, the number of past occurrences in a given time period constitutes the basis for determining the time between occurrences and hence the expected probability.

For the purpose of simplicity, the relation between magnitude and frequency can be assumed to follow an exponential function:

$$F = a \exp(-aM) \tag{14.7}$$

where M is the magnitude of the event, say debris flow, F is the frequency of the event (say debris flow), and a is parameter to be determined from measurements of M and F. By definition, parameter a is the reciprocal of the mean of M.

The hazard (L) due to an event (debris flow) can be expressed by the product of magnitude and frequency as

$$L = MF \tag{14.8}$$

Combining equations (14.7) and (14.8) the hazard can be expressed as

$$L = Ma \exp(-aM) \tag{14.9}$$

Hazard L can be calculated from the area under the M-F curve by noting that

$$dL = FdM \tag{14.10}$$

where DL is the incremental hazard or loss and dM is the incremental size or magnitude. Therefore, integrating equation (14.10) one gets

$$L = \int_0^M FdM \tag{14.11}$$

or

$$L = \int_0^M FdM = 1 - \exp(aM), \quad a > 0 \tag{14.12}$$

where L ranges from 0 to 1 and M varies from 0 to ∞. If parameter a is known from observations, then equation (14.12) determines hazard L for a given magnitude M.

In many cases empirical approaches are employed to quantify the hazard value. To that end, the magnitude of an event can be expressed in a multiplicative form as

$$y = a \prod_{i=1}^{N} x_i^{b_i} \tag{14.13}$$

where y represents the magnitude, x_i is the i-th independent variable, N is the number of independent variables, a is parameter, and b_i is i-th exponent associated with x_i. This y value is an estimate of M or another independent variable. Then, this can be transformed to hazard potential h on a scale of 0 to 1 as

$$h = 1 - \exp(-ky) \tag{14.14}$$

where k is parameter which can be determined from observations. Combining equations (14.13) and (14.14), one obtains

$$h = 1 - Ka \exp\left(-\prod_{i=1}^{N} x_i^{b_i}\right) \tag{14.15}$$

14.6.2 Vulnerability assessment

The overall vulnerability depends on four factors: physical, economic, environmental, and social. The physical vulnerability entails losses due to damage to physical infrastructure and damage. The economic vulnerability includes losses of individual and collective assets. The environmental vulnerability entails degradation of water, air, and land resources. Social vulnerability includes impact on population and social structure (size, density, education, health, age, and wealth). It may then be assumed that the vulnerability V_1, representing the potential maximum property loss (say in U.S. dollars), can be expressed as the sum of fixed asset value (P), annual GDP (G), and total value of land resources (E):

$$V_1 = P + G + E \tag{14.16}$$

where

$$E = \sum_{i=1}^{n} B_i A_i \tag{14.17}$$

in which A_i is i-th the land area, B_i is i-th base line price per unit area of land, and n is the number of land areas. The property value can be made dimensionless on a scale of 0 to 1 as:

$$FV_1 = \frac{1}{1 + \exp(-c_1 \log V_1 - c_2)} \tag{14.18}$$

where c_1 and c_2 are constants, and FV_1 is the transformed value of V_1.

The social vulnerability can be expressed as

$$V_2 = \frac{1}{3(d_0 + d_1 + d_2)D} \tag{14.19}$$

where V_2 is the potential maximum loss of life (person/km^2), d_i, $i = 0$, 1, and 2, are constants, and D is the population density. This can be transformed to a scale of 0 to 1 as

$$FV_2 = 1 - \exp(-gV_2) \tag{14.20}$$

where g is the reciprocal of the mean value of V_2.

The vulnerability can be determined by summing FV_1 and FV_2. It can be argued that V increases with increasing FV_1 and FV_2 but the rate decreases. The regional vulnerability model can be expressed as

$$V = \left[\frac{1}{2}(FV_1 + FV_2) \right]^{0.5} \tag{14.21}$$

where V is scaled from 0 to 1. Note both FV_1 and FV_2 are weighted equally but can be weighted differently if need be.

14.6.3 Risk assessment and ranking

Risk entails both hazard (L) and vulnerability (V). For practical applications, policy formulation and decision making, it may be useful to express risk in discrete levels. Liu and Lei (2003) suggested five classes for both hazard and vulnerability: very low (0–02), low (0.2–0.4), moderate (0.4–0.6), high (0.6–0.8) and very high (0.8–1.0). Likewise, risk can be expressed in five classes with the same description but different ranges of values: very low (0.00–0.04), low (0.04–0.16), moderate (0.16–0.36), high (0.36–0.64), and very high (0.64–1.00).

One can graph hazard versus vulnerability and then construct risk curves. The curves separating risk classes would be hyperbolas represented by: $V = 0.04/L$, $V = 0.16/L$, $V = 0.36/L$, and $V = 0.64/L$. Now the probability of each of the five levels of risk needs to be computed. This probability is the proportion of area within the region of the L versus V graph. Lin and Lei (2003) showed that the probability value was: 16.9% for very low, 28.4% for low, 27.5% for moderate, 19.8% for high, and 7.4% for very high.

Questions

Q.14.1 Select a set of monthly data on pan evaporation, relative humidity, air temperature (mean, minimum, and maximum), wind speed, number of cloudy days, number of sunshine hours, and solar radiation for a station. Here monthly evaporation is the dependent variable, and others are independent variables. Determine the number of combinations in which variables can be selected. Each combination leads to one specific model for predicting monthly evaporation. Compute the relative entropy of each model. Then, select the combinations that have minimum entropy values.

Q.14.2 Prepare a matrix (a shortened one) of selected combinations of variables and entropy from Q.14.1. You can use 1 if the variable is included and 0 if it is not.

Q.14.3 In the shortened matrix in Q.14.2, count the number of times each variable appears and then discuss the significance of the variables and models selected.

Q.14.4 Evaluate if the models in Q.14.3 can be further simplified.

Q.14.5 Obtain for a number of watersheds a set of geomorphological data, including drainage area, density, slope, length, width, compact factor, vegetative cover, and soil type. Also obtain data on 50-year peak discharge values for these watersheds. The objective is to determine a model for predicting 50-year peak discharge using the geomorphological characteristics. Determine the number of combinations in which variables can be selected. Each combination leads to one specific model for predicting monthly evaporation. Compute the relative entropy of each model. Then, select the combinations that have minimum entropy values.

Q.14.6 Prepare a matrix (a shortened one) of selected combinations of variables and entropy from Q.14.5. You can use 1 if the variable is included and 0 if it is not.

Q.14.7 In the shortened matrix in Q.14.6, count the number of times each variable appears and then discuss the significance of the variables and models selected.

Q.14.8 Evaluate if the models in Q.14.7 can be further simplified.

Q.14.9 Obtain a set of data for debris flow, including gully number, gully occurrence, gully length, drainage basin area, basin slope, form factor, and area of landslide. The objective is to determine the occurrence of debris flow as a function of these variables. Determine the number of combinations in which variables can be selected. Each combination leads to one specific model for predicting monthly evaporation. Compute the relative entropy of each model. Then, select the combinations that have minimum entropy values. For debris flow data, refer to Chen et al. (2007) and the references therein.

Q.14.10 Prepare a matrix (a shortened one) of selected combinations of variables and entropy from Q.14.9. You can use 1 if the variable is included and 0 if it is not.

Q.14.11 In the shortened matrix in Q.14.10 count the number of times each variable appears and then discuss the significance of the variables and models selected.

Q.14.12 Evaluate if the models in Q.14.11 can be further simplified.

References

Alexander, D.E. (1991). Natural disasters: a framework for research and detaching. *Disasters*, Vol. 14, No. 3, pp. 209–26.

Bollander, B.V. and Weigend, A.E. (1994). Selecting input variables using mutual information and nonparametric density estimation. Proceedings of the 1994 International Symposium on Artificial Neural Networks, pp. 42–50.

Chen, C.C., Tseng, C.Y. and Dong, J.Y. (2007). New entropy-based method for variables selection and its application to the debris-flow hazard assessment. *Engineering Geology*, Vol. 94, pp. 19–26.

Davis, J.C. (2002). *Statistics and Data Analysis in Geology*. 3rd Edition, Wiley, New York.

Deyle, R.E., Fench, S.P., Olshansky, R.B. and Patterson, R.G. (1998). Hazard assessment: the factual basis for planning and mitigation. In: R.J. Burby, editor, *Cooperating with Nature: Confronting Natural Hazards with Land Use Planning for Sustainable Communities*. Joseph Henry Press, Washington, D.C., pp. 119–66.

Dupuis, J.A. and Robert, C.P. (2003). Variable selection in qualitative models via an entropic explanatory power. *Journal of Statistical Planning and Inference*, Vol. 111, pp. 77–94.

Forbes, F. and Peyrard, N. (2003). Hidden Markov random field model selection criteria based on mean filed-like approximations. *IEEE transactions: Pattern Analysis and Machine Intelligence*, Vol. 25, No. 9, pp. 1089–101.

Hosmer, D.W. and Lemeshow, S. (2000). *Applied Logistic Regression*. John Wiley, 375 p., New York.

Johnson, V.E. and Albert, J.H. (1999). *Ordinal Data Modeling*. Springer, New York.

Kleinbaum, D.G. (1994). *Logistic Regression, a Self-learning Text*. Springer Verlag, 292 p., New York.

Lin, P.S., Hung, J.Y. and Yang, J.C. (2002). Assessing debris flow hazard in a watershed in Taiwan. *Engineering Geology*, Vol. 66, pp. 295–313.

Liu, X. and Lei, J. (2003). A method for assessing regional debris flow risk: An application in Zhaotong of Yunnan province (WS China). *Geomorphology*, Vol. 52, pp. 181–91.

Reftery, A.E., Madigan, D. and Hoeting, J.A. (1997). Bayesian model averaging for regression models. *Journal of American Statistical Association*, Vol. 92, pp. 179–91.

Rupert, M.G., Cannon, S.H. and Gartner, J.E. (2003). Using logistic regression to predict the probability of debris flows occurring In areas recently burned by wildfires. Open File Report OF 03–500, US Geological Survey.

Tseng, C.Y. (2006). Entropic criterion for model selection. *Physica A*, Vol. 370, pp. 530–8.

Weiss, R.E. (1995). The influence of variable selection: a Bayesian diagnostic perspective. *Journal of American Statistical Association*, Vol. 90, pp. 619–25.

Additional Reading

Christensen, R. (1997). *Log-linear Models and Logistic Regression.* 2nd Edition, Springer, New York.

Gelfand, A. and Dey, D. (1994). Bayesian model choice: asymptotics and exact calculations. *Journal of Royal Statistical Society, B.*, Vol. 56, pp. 501–14.

Goutics, C. and Robert, C.P. (1998). Model choice in generalized linear models: a Bayesian approach via Kullback-Leibler projections. *Biometrika*, Vol. 85, pp. 29–37.

Kieseppa, I.A. (2000). Statistical model selection criteria and Bayesianism. *Philosophy of Science*, Vol. 68, pp. 5141–52.

Li, L., Cook, R.D. and Nachtsheim, C.J. (2005). Model-free variable selection. *Journal of Royal Statistical Society, B*, Vol. 67, pp. 285–99.

Madigan, D. and Raftety, A.E. (1994). Model selection and accounting for model uncertainty in graphical models using Occam's window. *Journal of American Statistical Association*, Vol. 89, pp. 1535–46.

Ostroff, A.G. (1967). Comparison of some formation water classification systems. Bulletin, *American Association of Petroleum Geologists*, Vol. 51, pp. 404–16.

Raftery, A.E. (1985). Bayesian model selection in social research. *Sociological Methodology*, Vol. 25, pp. 111–63.

Tseng, C.Y. and Chen, C.C. (2011). Entropic component analysis and its application in geologic data. Computers and Geosciences, doi: 10.1016/j.cageo.2010.11.016.

15 Neural Networks

A human brain consists of a large number of highly connected elements (approximately 10^{11}), called neurons. A neuron has three principal components: the dendrites, the cell body, and the axon, as shown in Figure 15.1. The dendrites are tree-like receptive networks of nerve filters that carry electrical circuits (or signals) into the cell body. The cell body effectively sums and thresholds these incoming signals. The axon is a single long fiber that carries the signal from the cell body to other neurons. The point of contact between the axon of one cell and a dendrite of another cell is called synapse. The arrangement of neurons and the strengths of individual synapses determine the function of a neural network.

In implementation of an artificial neural network (ANN) neurons are considered as elements and are simple abstractions of biological neurons. The principal attribute of an artificial neural network (ANN) is the distribution of knowledge through connections amongst a large number of neurons. These neurons are arranged in several distinct layers, including the sensory layer, hidden layers, and output layer or motor control layer, as shown in Figure 15.2. The sensory layer represents the interfacing layer on the input side; the output layer is on the output side; and hidden layers are in between or intermediate. All neurons may perform the same type of input-output operation or different layers of neurons may perform different kinds of input-output operation or transfer functions. In this manner ANNs have the ability to generalize input-output mapping from a limited set of training examples or data.

15.1 Single neuron

Consider a single-input neuron as shown in Figure 15.3, in which input I is multiplied by a scalar weight w and then wI is sent to the summer. The other input is 1, is multiplied by a bias b and then sent to the summer. The output from the summer, s, often referred to as the net input, goes into a transfer function (sometimes called activation function) f which produces the scalar neuron output O. Relating to the biological jargon, weight w corresponds to the strength of a synapse, the summer and the transfer function represent the cell body and the neuron output O represents the signal on the axon. The neuron output is computed as

$$O = f(s), s = wI + b \tag{15.1}$$

Entropy Theory and its Application in Environmental and Water Engineering, First Edition. Vijay P. Singh.
© 2013 John Wiley & Sons, Ltd. Published 2013 by John Wiley & Sons, Ltd.

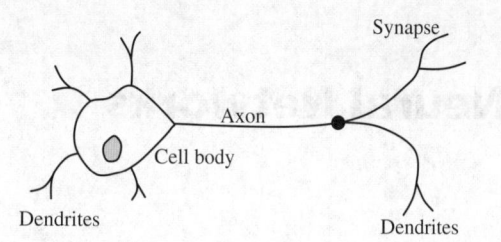

Figure 15.1 A neuron.

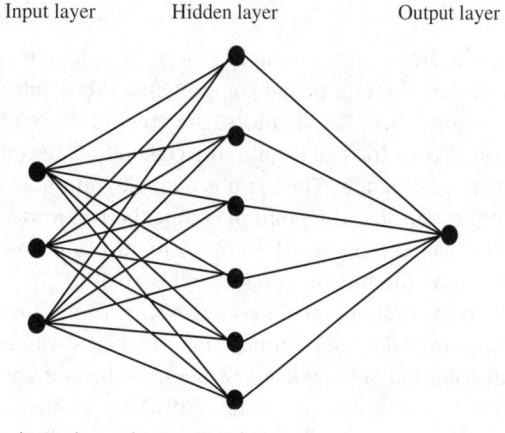

Figure 15.2 A simple three-layered neural network architecture.

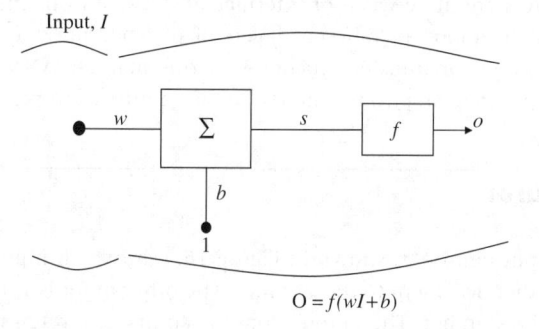

$$O = f(wI + b)$$

Figure 15.3 A single input neuron.

The actual output depends on the form of f chosen. The bias is like a weight and has a constant input of 1 (or -1). It can be omitted if it is not desired in a particular neuron. Both w and b are adjustable parameters of the neuron.

Similarly, one can construct a multiple-input single neuron as shown in Figure 15.4, in which input is a vector comprising a number of inputs to the neuron: $I_i, i = 1, 2, \dots, M$, M is the number of inputs. The weight assigned to each input is expressed as: $w_{1,i}, i = 1, 2, \dots, M$. Thus, the net input can be expressed as:

$$s = w_{11}I_1 + w_{12}I_2 + \dots + w_{1M}I_M + b \tag{15.2}$$

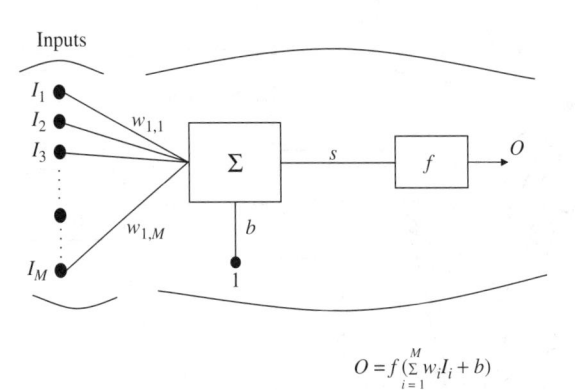

$$O = f(\sum_{i=1}^{M} w_i I_i + b)$$

Figure 15.4 A multiple input single neuron.

and the neuron output as

$$O = f\left(\sum_{i=1}^{M} w_i I_i + b_i\right) \tag{15.3}$$

In a similar manner one can construct neural network architectures, as shown in Figure 15.5. In a neural network, a layer is comprised of the weight, the summer, the bias, the transfer function, and the output. One can have layers of neurons, as shown in Figure 15.5. A single neuron is capable of only a linear mapping, but a layered network of neurons with multiple hidden layers yields any desired mapping. This is the principal reason for the popularity of ANNs. An example is shown in Figure 15.6.

All neurons in a layer are connected to all the neurons in the adjacent layers. The weight value indicates the connection strength between neurons from adjacent layers and functions as a signal multiplier on the corresponding link. The input to the neuron is the linear summation of all the incoming signals on the various connection links. The net summation is compared to a threshold, often referred to as bias. The difference due to comparison drives the output function, called activation function which produces an output signal. The most common output functions (activation functions) are sigmoid and hyperbolic tangent functions.

In pattern recognition, layered networks are also called multilayer perceptron (MLP) networks. Burr (1988), Lippmann (1987), among others, have shown that two hidden layers suffice to represent decision boundaries of any complexity in a piecewise linear manner. Two layers are not always needed for a particular decision region. The first hidden layer is called the partitioning layer that divides the entire feature space into several regions (The input data that represent a pattern are referred to as the measurement of feature vector). The function that a pattern recognition system performs is the mapping of the input feature vector into one of the various decision classes. The connotation of a linear input-output mapping then is that the decision boundaries are linear. The second hidden layer is referred to as the *ANDing* layer that performs *ANDing* of partitioned regions to yield convex decision regions for each class. The output layer is called the *ORing* layer that logically combines the results of the previous layer to produce disjoint regions of arbitrary shape with holes and concavities if needed.

Neural networks are self-organizing systems. The information theory offers a way to determine the efficiency of information representation by tree networks and the limitations in the reliable transmission of information over a network or a communication channel.

Inputs

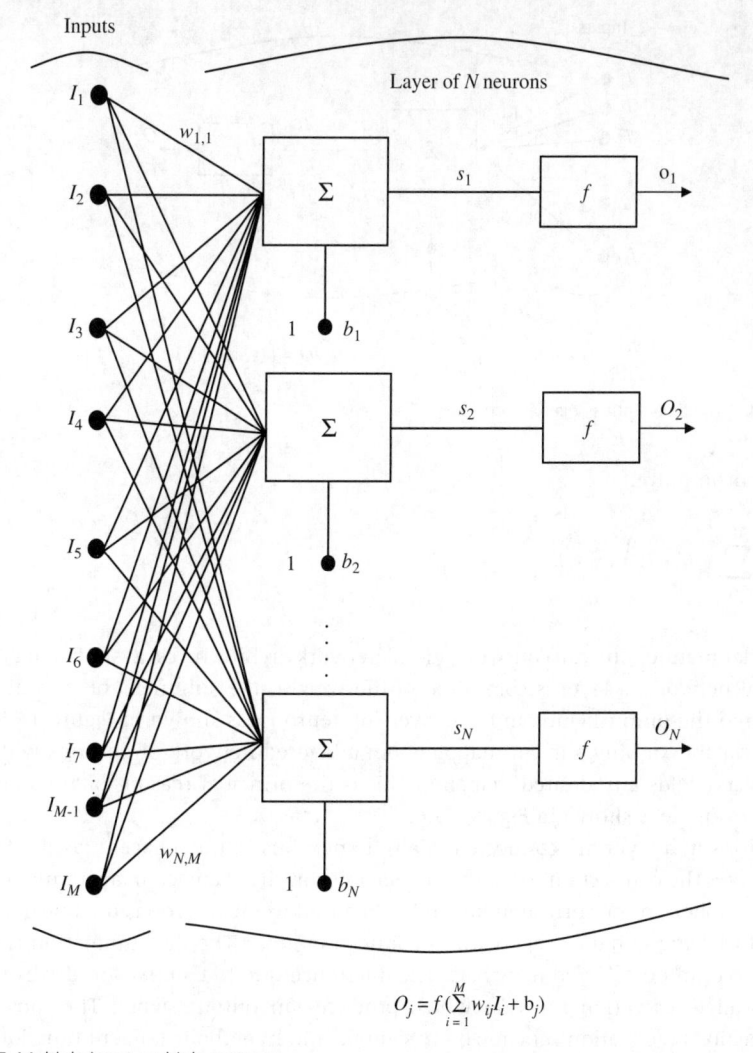

$$O_j = f\left(\sum_{i=1}^{M} w_{ij}I_i + b_j\right)$$

Figure 15.5 Multiple input multiple neurons.

The theory permits derivation of the principle of maximum information preservation which represents a principle of self-organization. It also allows us to compute ideal bounds on the optimal representation and transmission of information bearing signals or neural networks.

Example 15.1: The perceptron, as shown in Figure 15.7, represents feed forward neural networks and is a classifier that maps its input x to an output value $f(x)$, expressed as:

$$f(x) = \begin{cases} 1 & if \ w \cdot x + b > 0.5 \\ 0 & else \end{cases}$$

where w is a vector of real-valued weights and $\bullet$ is the dot product (which computes a weighted sum), and b is the 'bias', a constant term that does not depend on any input value. Show the calculations in tabular form?

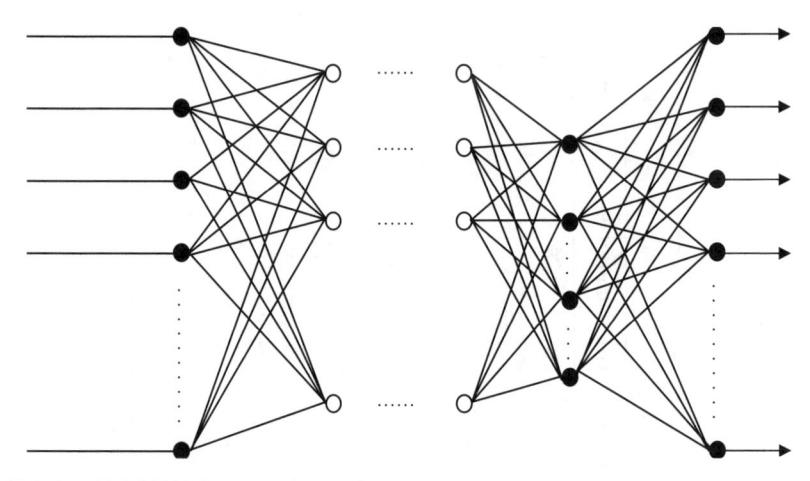

Figure 15.6 A multiple hidden layer neural network.

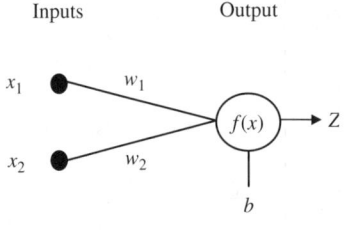

Figure 15.7 Perceptron configuration.

Solution: A perceptron has a number of external input links, one internal input (called a bias), a threshold, and one output link. Table 15.1 shows the calculations required to perform from input layer to output layer. The threshold value is given in the first column as 0.5. A linear learning rate that is a parameter of learning phase is selected as 0.1 (column 2). The measured sensor values x_1 and x_2, and desired output values, Z, are included in columns 3, 4, and 5. Initial weights are assigned, first, arbitrarily as 0.2 and 0.5 in columns 6 and 7. After the second iteration the initial weight takes on the final weight of the previous iteration. The C_1 and C_2 values are calculated by multiplying columns 3 and 6 and columns 4 and 7, respectively. Column 10 is for the sum of C_1 and C_2. Column 11 applies the function given in the example. If the summation is greater than the threshold value of 0.5 then the output equals 1, otherwise it is 0. The difference between calculated and measured outputs is shown in column 12. In order to obtain new weights for the next step, a correction factor that is the multiplication of learning rate and error amount is calculated in column 13. Finally, updated weights are obtained by adding the correction factor to old weights.

15.2 Neural network training

Training of a single-layer neural network is relatively simple. There are several methods to train layered networks, such as supervised learning (Sethi, 1990), Boltzmann learning (Ackley et al. 1985), counter-propagation (Hecht-Nielsen, 1987), and Madaline Rule-II (Widrow et al. 1988). Popular amongst these methods of training of layered networks is supervised learning.

Table 15.1 Calculation for Example 15.1.

Input					Initial		Calculated Output		Sum	Network	Error	Correction	Final	
Threshold	Learning Rate	Sensor values		Desired output	Weights								Weights	
TH	LR	x_1	x_2	Z	w_1	w_2	C_1	C_2	S	N	E	R	w_1	w_2
1	2	3	4	5	6	7	8	9	10	11	12	13	14	15
							$x_1 \times w_1$	$x_2 \times w_2$	$C_1 + C_2$	IF(S > TH, 1, 0)	Z − N	LR × E	R + w_1	R + w_2
0.5	0.1	1	0	1	0.2	0.5	0.2	0	0.2	0	1	0.1	0.3	0.6
0.5	0.1	1	1	1	0.3	0.6	0.3	0.6	0.9	1	0	0	0.3	0.6
0.5	0.1	1	0	1	0.3	0.6	0.3	0	0.3	0	1	0.1	0.4	0.7
0.5	0.1	0	0	0	0.4	0.7	0	0	0	0	0	0	0.4	0.7
0.5	0.1	1	1	1	0.4	0.7	0.4	0.7	1.1	1	0	0	0.4	0.7
0.5	0.1	1	0	1	0.4	0.7	0.4	0	0.4	0	1	0.1	0.5	0.8
0.5	0.1	0	0	0	0.5	0.8	0	0	0	0	0	0	0.5	0.8
0.5	0.1	0	1	1	0.5	0.8	0	0.8	0.8	1	0	0	0.5	0.8
0.5	0.1	1	1	1	0.5	0.8	0.5	0.8	1.3	1	0	0	0.5	0.8
0.5	0.1	1	0	1	0.5	0.8	0.5	0	0.5	0	1	0.1	0.6	0.9
0.5	0.1	0	1	1	0.6	0.9	0	0.9	0.9	1	0	0	0.6	0.9
0.5	0.1	1	1	1	0.6	0.9	0.6	0.9	1.5	1	0	0	0.6	0.9

In this way, the network is provided with examples of input-output mapping pairs from which the network learns. The learning is reflected through the modification of connection strengths or weights. This process of learning occurs continuously until the mapping present in the examples is achieved. One of the key issues in the learning process is one of credit assignment, that is, what should be the desired output of the neurons in the hidden layers during training? This issue has been addressed by propagating back the error in the output layer to the internal layers. This is often referred to as back propagation algorithm (Rumelhart et al. 1986). It minimizes the error at the output layer and is a gradient descent method. One of the main drawbacks of these training methods is that they do not specify the number of neurons needed in the hidden layers. This number significantly affects the learning rate and the overall classification performance, and is therefore an important parameter.

Back propagation algorithm (BPA) consists of mainly two activities: Forward pass and backward pass. In the forward pass, the activity is propagated from the input layer to hidden layers to output layer. In the backward pass, the activity is propagated from the output layer to hidden layers to input layer. The connection weights v_{ij} are adjusted based upon the following equation:

$$v_{ij}^{new} = v_{ij}^{old} - \eta \frac{\partial E(v_{ij})}{\partial v_{ij}} \tag{15.4}$$

where η is the learning rate and E is the error function that is defined as:

$$E = \frac{1}{2} \sum_{k=1}^{p} \sum_{j=1}^{q} (y_{kj} - t_{kj})^2 \tag{15.5}$$

$$\frac{\partial E}{\partial v_{ij}} = \sum_{k=1}^{p} \sum_{j=1}^{q} (t_{kj} - z_{kj})(-1)f(v_{ij})\left[1 - f(v_{ij})\right]y_j \tag{15.6}$$

where p is the number of training patterns, q is the number of output nodes, y_{ki} is the model output, and t_{ki} is the target output.

Example 15.2: Consider the network with two inputs and one output, as shown in Figure 15.8. Update the weights between hidden and output layers.

Solution: For the forward pass,

$$net_3 = (3)(1) + (4)(0) + (1)(1) = 4$$
$$net_4 = (1)(6) + (0)(5) + (1)(-6) = 0$$

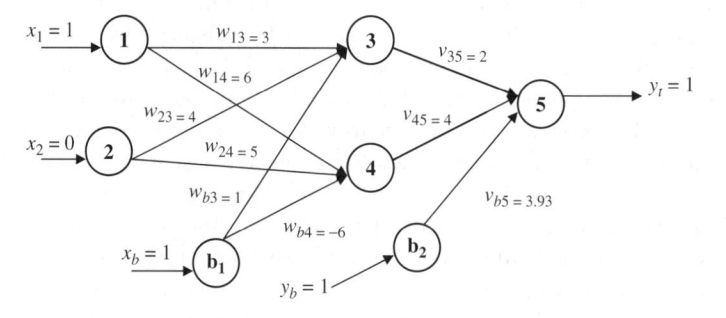

Figure 15.8 Network for Example 15.2.

$$y_3 = \frac{1}{1 + e^{-4}} = 0.982$$

$$y_4 = \frac{1}{1 + e^{-1}} = 0.50$$

$$net_5 = (0.982)(2) + (0.50)(4) + (1)(-3.93) = 0.04$$

$$y_4 = \frac{1}{1 + e^{-0.04}} = 0.51$$

$$E = 1 - 0.51 = 0.49$$

For the backward pass,

$$\text{Let} \quad \delta_{oj} = \sum_{k=1}^{p} \sum_{j=1}^{q} (t_{kj} - z_{kj})(-1)f'(w_{ij})$$

$$\delta_5 = y_5(1 - y_5)(y_5 - y_t) = 0.51(1 - 0.51)(-0.49) = -0.1225$$

$$v_{35}^{new} = v_{35}^{old} - \eta \partial_5 y_3 = 2 - (0.1)(-0.1225)(0.982) = 2.012$$

$$v_{45}^{new} = v_{45}^{old} - \eta \partial_5 y_4 = 4 - (0.1)(-0.1225)(0.50) = 4.012$$

$$v_{05}^{new} = v_{05}^{old} - \eta \partial_5 y_b = -3.93 - (0.1)(-0.1225)(1) = -3.9078$$

Other weights can be obtained in a similar fashion.

15.3 Principle of maximum information preservation

Linsker (1988) proposed a principle of self-organization which states that the development of synaptic connections of a multilayered neural network occurs by maximizing the amount of information that is preserved when signals are transformed at each processing stage of the network, subject to certain constraints. Before formalizing this principle, consider a two-layer neural network in which the input or source layer has forward connections to the output (target) layer that can have lateral connections, as shown in Figure 15.9. The network is required to process incoming signals in a self-organized manner. The input layers can have a number of activities represented by vector $\vec{X}$ whose elements are $x_1, x_2, \ldots, x_N$. Similarly, the output layer can have a number of resulting activities represented by a vector $\vec{Y}$ whose elements are $y_1, y_2, \ldots, y_M$. Linsker (1987, 1988) stated the principle of maximum information preservation (POMIP) as: "The transformation of a vector $\vec{X}$ observed in the input layer of a neural network to a vector $\vec{Y}$ produced in the output layer jointly maximizes information about the activities in the input layer. The parameter to be maximized is the

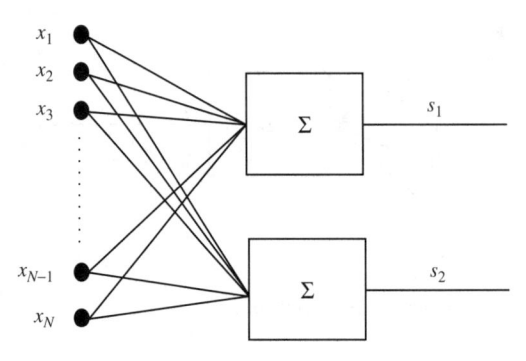

Figure 15.9 A two-layered network.

average information between the input vector $\vec{X}$ and the output vector $\vec{Y}$, in the presence of processing noise." This principle can be considered as the neural network counterpart of the concept of channel capacity which defines the limit on the rate of information transmission through a communication channel. POMIP is independent of the learning tool employed for its implementation.

15.4 A single neuron corrupted by processing noise

Consider a simple network of an input layer and an output layer. The input layer comprises N source nodes from which a single neuron receives input, as shown in Figure 15.10. The output of this neuron is corrupted by noise during processing and let this processing noise be denoted as ε. The source nodes are not equal and therefore let the synaptic weight of an i-th source node in the input layer to the neuron in the output layer be denoted as $w_i, i = 1, 2, \ldots, N$. For simplicity it is assumed that the output Y of the neuron is a Gaussian random variable with variance σ_y^2; the processing noise ε is also a Gaussian random variable with zero mean and variance σ_ε^2; and the processing noise is not correlated with any of the input components, that is,

$$E[\varepsilon x_i] = 0 \text{ for all } i \tag{15.7}$$

Let the output of the neuron be written as

$$Y = \sum_{i=1}^{N} w_i x_i + \varepsilon \tag{15.8}$$

The objective here is to derive the transinformation or the average mutual information $T(\vec{X}, Y)$ between the output of the neuron in the output layer, Y, and the input vector, $\vec{X}$. From the definition of mutual information (or trasninformation),

$$T(Y, \vec{X}) = H(Y) - H(Y|\vec{X}) \tag{15.9}$$

From equation (15.8) it is seen that the PDF of Y, given the input vector $\vec{X}$ is equal to the PDF of w plus the PDF of ε. Therefore, the conditional entropy $H(Y|\vec{X})$ is the information that the output neuron transmits about processing noise ε, rather than about the input signal vector

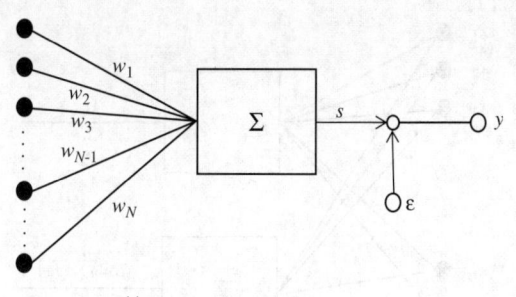

Figure 15.10 A single neuron corrupted by processing noise.

$\vec{X}$. That is,

$$H(Y|\vec{X}) = H(\varepsilon) \tag{15.10}$$

Substituting equation (15.10) in equation (15.9) one obtains

$$T(Y, \vec{X}) = H(Y) - H(\varepsilon) \tag{15.11}$$

Since Y is Gaussian distributed,

$$H(Y) = \frac{1}{2}\left[1 + \ln\left(2\pi\sigma_y^2\right)\right] \tag{15.12}$$

Similarly, one can write for ε assuming it to be Gaussian distributed as

$$H(\varepsilon) = \frac{1}{2}\left[1 + \ln\left(2\pi\sigma_\varepsilon^2\right)\right] \tag{15.13}$$

Inserting equations (15.12) and (15.13) in equation (15.11) one gets

$$T(Y, \vec{X}) = \frac{1}{2}\ln\left(\frac{\sigma_y^2}{\sigma_\varepsilon^2}\right) \tag{15.14}$$

where σ_y^2 depends on σ_ε^2.

The ratio $\sigma_y^2/\sigma_\varepsilon^2$ can be regarded as a signal-to-noise ratio. If the noise variance is fixed then equation (15.14) shows that $T(Y, \vec{X})$ is maximized by maximization of σ_y^2. Thus, it is concluded that under certain conditions maximizing the output variance of a neural network maximizes the average mutual information between the output signal of that neuron and its input. This conclusion however does not always hold.

Example 15.3: The Xinanjiang Rainfall-Runoff Model is a conceptual rainfall-runoff forecasting tool that was designed for humid and semi-humid regions and is based on the concept of runoff formation on repletion of storage, that is, runoff is not produced until the soil moisture content of the aeration zone reaches field storage capacity and thereafter runoff equals rainfall excess without further loss. In its simplest form the model comprises a single equation:

$$R = P - (W_m - W_0) + W_m\left[\left(1 - \frac{W_0}{W_m}\right)^{\frac{1}{1+b}} - \frac{p}{(1+b)W_m}\right]^{1+b} \tag{15.15}$$

in which $R =$ the runoff; $P =$ the effective precipitation; $W_m =$ the maximum field storage capacity; $W_0 =$ the initial field storage capacity; and b is an exponent that represents "nonuniform spatial distribution," that is, the nonuniform distribution of surface conditions, including factors, such as topography, geology, soil type, and vegetation coverage. How do we establish a neural network model by using the aforementioned variables?

Solution: Following the methodology proposed by Abrahart and See (2007), the question can be treated as follows:

Dividing by W_m converts equation (15.15) to a nondimensional form as:

$$\frac{R}{W_m} = \frac{P}{W_m} - \left(1 - \frac{W_0}{W_m}\right) + \left[\left(1 - \frac{W_0}{W_m}\right)^{\frac{1}{1+b}} - \frac{P}{(1+b)W_m}\right]^{1+b} \tag{15.16}$$

Defining $C_1 = \frac{P}{W_m}; C_2 = \frac{W_0}{W_m}; C_3 = \frac{R}{W_m}$ and substituting into equation (15.16) lead to:

$$C_3 = C_1 + C_2 - 1 + \left[(1 - C_2)^{\frac{1}{1+b}} - \frac{C_1}{(1+b)}\right]^{1+b} \tag{15.17}$$

The input values for the effective precipitation [P], maximum soil water storage [W_m] and the curve fitting exponent [b] are taken as random samples. A linear scaling is applied to the input drivers and output responses with each variable standardized to a fixed range [0, 1]. The following steps can be applied to obtain model results:

1 Preparation of input and output data: The model should be tested with a data set that is independent from the data set for which model parameters are determined. The model is developed and tested on a dataset of random input variables comprising 5000 records split into two equal groups: 2500 cases for training purposes; 2500 cases for split sample testing operations.

2 Configuration of network: The model architecture compromises three layers. There are three neurons represented by three input variables in the input layer, six hidden units positioned in one hidden layer and one neuron corresponding to final output in the output layer, as shown in Figure 15.11.

3 Selection of transfer functions for hidden layer and output layers: Each processing unit in the hidden layer and output layer contains a logistic transfer function, that is, sigmoid curve. Each weighted connection and processing unit bias is assigned an initial random setting in the range of ± 1.

4 Training of the system: The neural network (NN) feed forward models are trained using "back propagation," each run produces a "Back Propagation Neural Network" (BPNN). Training material is presented in random order and the training program is stopped at 10,000 epochs. Training parameters are set for automatic adjustment over the period; the learning rate is set to decrease from 0.8 to 0.2.

A strong agreement between expected and predicted runoff values is found when C_1, C_2 and b are used to predict C_3. However, dropping the C_1 input variable leads to poor agreement. Thus, the precipitation input variable C_1 is found to be the main driver in a deterministic process without which the original model would be illogical and nonsensical. Dropping the soil moisture condition input variable C_2 causes the model to be less accurate, but the model still retains some physical properties. In the case of dropping the curve fitting exponent input driver b, the revised model continues to produce a rational output. However the results are not as good as in the first case.

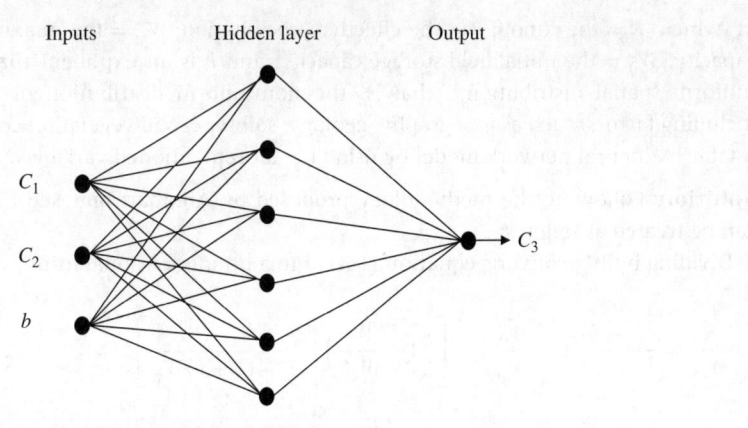

Figure 15.11 Neural network configuration for predicting output C_3.

15.5 A single neuron corrupted by additive input noise

Consider a simple neural network comprising an input layer consisting of N source nodes and an output layer, and a single neuron processing the input signals into an output signal. The noise in the output of the neuron in the output layer originates at the input ends of the synapses connected to the neuron, as shown in Figure 15.12. The output can be expressed as

$$y = \sum_{i=1}^{N} w_i(x_i + \varepsilon_i) \tag{15.18}$$

where each ε_i is assumed to be an independent Gaussian random variable with zero mean and common variance σ_ε^2. Equation (15.18) can be expressed as

$$y = \sum_{i=1}^{N} w_i x_i + \sum_{i=1}^{N} w_i \varepsilon_i = \sum_{i=1}^{N} w_i x_i + \upsilon \tag{15.19}$$

where

$$\upsilon = \sum_{i=1}^{N} w_i \varepsilon_i \tag{15.20}$$

Equation (15.19) is similar to equation (15.8). The quantity υ defines the composite component noise with variance equal to the sum of variances of individual components:

$$\sigma_\upsilon^2 = \left[\sum_{i=1}^{N} w_i\right]^2 \sigma_\varepsilon^2 \tag{15.21}$$

Similar to the preceding case, it is assumed that output Y of the neuron in the output layer is Gaussian distributed with variance σ_y^2. The conditional entropy $H(Y|\vec{X})$ is given as

$$H(Y|\vec{X}) = H(\upsilon) = \frac{1}{2}(1 + 2\pi\sigma_\upsilon^2) = \frac{1}{2}\left[1 + 2\pi\sigma_\varepsilon^2\left(\sum_{i=1}^{N} w_i\right)^2\right] \tag{15.22}$$

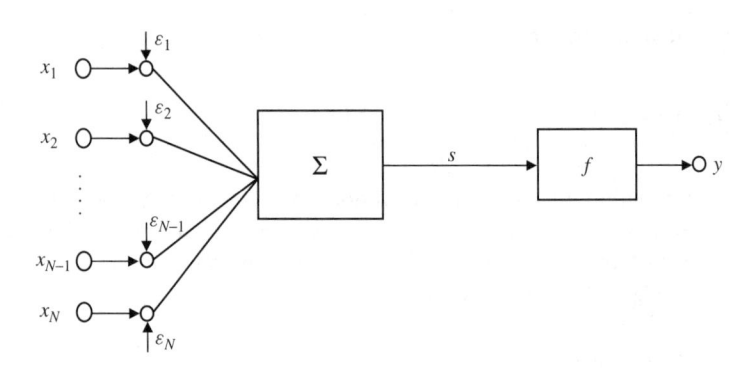

Figure 15.12 A single neuron corrupted by additive input noise.

Using equations (15.12), (15.21) and (15.22) in equation (15.11), one gets

$$T(Y, \vec{X}) = \frac{1}{2} \ln \left[\frac{\sigma_y^2}{\sigma_\varepsilon^2 \left(\sum\limits_{i=1}^{N} w_i \right)^2} \right] \tag{15.23}$$

Under the constraint that σ_ε^2 is constant, $T(Y, \vec{X})$ is maximized by maximizing the ratio of σ_y^2 to $\left(\sum\limits_{i=1}^{N} w_i \right)^2$.

Example 15.4: A rating curve relates sediment discharge or concentration to stream discharge, which can be used to estimate sediment loads from stream flow records. The sediment rating curve generally represents a functional relationship of the form:

$$S = aQ^b + \varepsilon_i$$

where Q is stream discharge, S is either suspended sediment concentration or yield, and ε_i is white noise. Values of a and b for a particular stream are determined from data using a linear regression between (log S) and (log Q). How can a simple ANN model represent this rating curve?

Solution: Daily data for a year or 365 days are used to train the ANN models and 365 daily data were used for testing. The input combinations used in this application to estimate suspended sediment values are (i) $Q(t)$; (ii) $Q(t)$ and $Q(t-1)$; (iii) $Q(t)$ and $S(t-1)$; and (iv) $Q(t)$ and $S(t-1)$, where $Q(t)$ and $S(t)$ represent, respectively, stream flow and sediment concentration at day t. To set up the model a MATLAB code is used. An ANN structure ANN(2, 5, 1) that consists of two input nodes, five nodes in hidden layers and one output node is used. In this case the input layer contains the current and one antecedent sediment $[Q(t), S(t-1)]$ and the output layer consists of the unique sediment concentration value at day t. The R^2 value for the testing period is 0.876. For the sediment rating curve, it is 0.852.

The weight and bias matrix of the trained network are given in columns 1, 2 and 3 of Table 15.2 and in Figure 15.13. Assume that $Q(t) = x_1 = 25.47$ and $S(t-1) = x_2 = 3.81$ is given. Output of the network can be obtained as follows:

1 Sum the weighted inputs, that is,

$$Nod_j = \sum_{i=1}^{N} (W_{ij}x_i) + b_j$$

where Nod_j = summation for the j-th hidden node; N = total number of input nodes; W_{ij} = connection weight of i-th input and j-th hidden node; x_i = input at the i-th input node; and b_j = bias value at the j-th hidden node. Calculations are shown in column 4.

2 Transform the weighted input (tansig function is used):

$$out_j = 2/(1 + \exp(-2Nod_j)) - 1$$

where out_j = output from the j-th hidden node (column 5).

3 Sum the hidden node outputs:

$$Nod_k = \sum_{J=1}^{NH} (W_{jk}out_j) + b_k$$

where Nod_k = summation for the k-th output node; NH = total number of hidden nodes; out_j = output from the j-th output node; W_{jk} = connection weight between the j-th hidden and k-th output node (column 6); and b_k = bias at the k-th output node (column 7). Since there is a one node in the output layer, summation of hidden node outputs produces one result which is equal to 3.491.

4 Transform the weighted sum (purelin function is used):

$$out_k = Nod_k$$

where out_k = output at the k-th output node. In this manner, the output obtained is 3.491.

Example 15.5: The longitudinal dispersion coefficient (K) in the literature is predicted using empirical models that relate the coefficient to channel physical characteristics (channel width, W and channel sinuosity, σ) and flow characteristics (flow depth, H, flow velocity, U, and shear velocity, $u*$). For example, Seo and Cheong (1998) derived the following equation to

Table 15.2 A sample calculation for given network.

| Nodes J | W_{1j} | W_{2j} | b_i | Nod_j | out_j | W_{jk} | b_k | Nod_k |
	1	2	3	4	5	6	7	8
1	17.705	13.816	−581.888	−660.195	−1.000	1.361	0.589	−0.772
2	−0.437	0.951	13.891	20.282	1.000	1.795	0.589	2.384
3	−71.837	−10.395	1969.257	2069.231	1.000	−0.424	0.589	0.165
4	−11.980	−1.596	307.722	304.232	1.000	−0.770	0.589	−0.181
5	0.076	−0.774	−0.249	−1.524	−0.909	−1.436	0.589	1.895
							Sum =	3.491

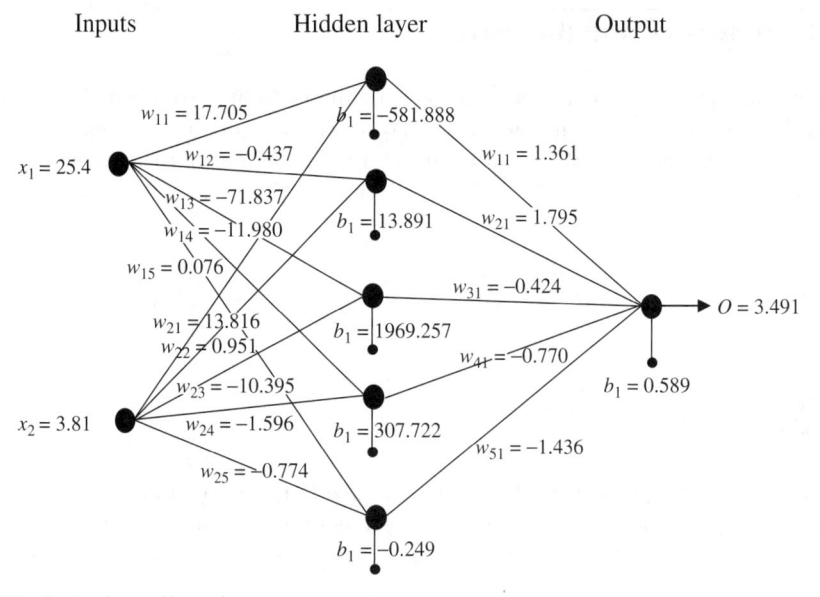

Inputs   Hidden layer   Output

Figure 15.13 Weights and bias of a trained network.

predict the dispersion coefficient in natural streams:

$$K = 5.915 \left(Hu_*\right) \left(\frac{B}{H}\right)^{0.62} \left(\frac{U}{u_*}\right)^{1.428}$$

where K is the dispersion coefficient, H is flow depth, U is flow velocity, B is channel width, and u_* is shear velocity. Tayfur and Singh (2005) compiled 72 sets of data from 30 different rivers. Outline the steps to predict K using an ANN model and predict K just using discharge data.

Solution:

Step 1: Randomly partition 70% and 30% of data, respectively, for the training and testing sets, paying attention to make sure that both sets have similar order of magnitude of mean, maximum and minimum values.

Step 2: Construct a three-layer ANN, the input layer having four neurons for B, H, U, u_* and output layer with one neuron for K. Start $2N + 1$ (where N is number of input layer neurons) thumb rule for the number of hidden layer neurons. That is nine hidden layer neurons.

Step 3: Assign random values for the connection weights and bias nodes. As an input, assign -1 for the bias nodes. Employ the learning rate of 0.1 and choose the sigmoid activation function.

Step 4: Start the iteration until the error levels off.

Step 5: After training, employ the same network to test the remaining 30% the data.

Step 6: Using several error measures, test the performance of the model.

Yes, K can be predicted using only flow discharge data. We can compute $Q = HUB$. Then, we can construct the 1-3-1 network and follow the steps outlined above.

15.6 Redundancy and diversity

Consider a network of two neurons each receiving inputs from N-source nodes in the input layer and producing outputs in the output layer. The outputs are corrupted during the processing of input signals as shown in Figure 15.13. The outputs of the two neurons can be expressed as

$$y_1 = \sum_{i=1}^{N} w_{1i} x_i + \varepsilon_1 \tag{15.24}$$

and

$$y_2 = \sum_{i=1}^{N} w_{2i} x_i + \varepsilon_2 \tag{15.25}$$

where w_{1i} are the synaptic weights from the input layer to neuron 1 in the output layer, and w_{2i} are the synaptic weights from the input layer to neuron 2 in the output layer.

In order to determine the average mutual information $T(Y, \vec{X})$ the following assumptions are invoked:

1 The noise terms ε_1 and ε_2 are uncorrelated:

$$E[\varepsilon_1, \varepsilon_2] = 0 \tag{15.26}$$

2 The noise terms ε_1 and ε_2 are Gaussian distributed with zero mean and common variance σ_ε^2.
3 Each noise term is uncorrelated with the input signals:

$$E[\varepsilon_i, X_i] = 0, \quad \text{for all } i = 1, 2, \text{ and } j = 1, 2, \ldots, N \tag{15.27}$$

4 The output signals Y_1 and Y_2 are each Gaussian distributed with zero mean.

The input is a vector defined as: $\vec{X} = [X_1, X_2, \ldots, X_N]^T$, and output is a vector defined as $Y = [Y_1, Y_2]^T$, and the noise vector is defined as $[\varepsilon_1, \varepsilon_2]^T$, where T is transpose. For the average mutual information, equation (15.9) holds for vectorial quantities as well:

$$T(\vec{X}, \vec{Y}) = H(\vec{Y}) - H(Y|\vec{X}) \tag{15.28}$$

Using the same rationale as before, $H(Y|\vec{X})$ can be expressed as

$$H(\vec{Y}|\vec{X}) = H(\vec{\varepsilon}) \tag{15.29}$$

$H(\vec{\varepsilon})$ can be written as

$$H(\varepsilon) = H(\varepsilon_1, \varepsilon_2) = H(\varepsilon_1) + H(\varepsilon_2) = 1 + \ln(2\pi\sigma_\varepsilon^2) \tag{15.30}$$

because ε_1 and ε_2 are statistically independent and equation (15.13) is invoked.

The entropy of the output vector Y can be written as

$$H(\vec{Y}) = H(Y_1, Y_2) = -\int_{-\infty}^{\infty} \int_{-\infty}^{\infty} f(y_1, y_2) \ln[f(y_1, y_2)] dy_1 dy_2 \tag{15.31}$$

where $f(y_1, y_2)$ is the joint PDF of Y_1 and Y_2 which are correlated because each depends on the same input signals. If R denotes the correlation matrix of the output vector then

$$R = E[YY^T] = \begin{bmatrix} r_{11} & r_{12} \\ r_{21} & r_{22} \end{bmatrix} \tag{15.32}$$

where

$$r_{ij} = E[Y_i Y_j], \quad i, j = 1, 2 \tag{15.33}$$

Using equations (15.24) and (15.25) and the assumptions in equations (15.26) and (15.27), individual elements of the correlation matrix can be written as

$$r_{11} = \sigma_{y_1}^2 + \sigma_\varepsilon^2 \tag{15.34}$$

$$r_{12} = r_{21} = \sigma_{y_1} \sigma_{y_2} \rho_{y_1 y_2} \tag{15.35}$$

$$r_{22} = \sigma_{y_2}^2 + \sigma_\varepsilon^2 \tag{15.36}$$

where $\rho_{y_1 y_2}$ is the correlation coefficient of output signals Y_1 and Y_2, and $\sigma_{y_1}^2$ and $\sigma_{y_2}^2$ are the variances of Y_1 and Y_2. These are all in the absence of noise.

To determine $H(Y)$, one can consider an N-dimensional Gaussian distribution discussed in Chapter 5, defined as

$$f(Y) = \frac{1}{2\pi \det(R)} \exp[-\frac{1}{2} Y^T R^{-1} Y] \tag{15.37}$$

where $\det(R)$ is the determinant of matrix R, and R^{-1} is the inverse of R. $H(Y)$ can be written as

$$H(Y) = \ln[(2\pi e)^{N/2} \det(R)] \tag{15.38}$$

In the present case, $N = 2$ and Y is composed of Y_1 and Y_2. Equation (15.38) reduces to

$$H(Y) = 1 + \ln[2\pi \det(R)] \tag{15.39}$$

Inserting equations (15.30) and (15.39) in equation (15.28), one obtains

$$T(Y, \vec{X}) = \ln[\frac{\det(R)}{\sigma_\varepsilon^2}] \tag{15.40}$$

For a fixed variance σ_ε^2, maximizing the determinant of R one obtains the maximum $T(Y, \vec{X})$. From equations (15.34) and (15.36), the determinant of R can be expressed as

$$\det(R) = r_{11} r_{22} - r_{12} r_{21}$$
$$= \sigma_\varepsilon^4 + \sigma_\varepsilon^2 (\sigma_{y_1}^2 + \sigma_{y_2}^2) + \sigma_{y_1}^2 \sigma_{y_2}^2 + (1 - \rho_{y_1 y_2}^2) \tag{15.41}$$

Depending on the value of noise variance σ_ε^2, two cases can be identified. First, if σ_ε^2 is large then the third term on the right side can be ignored. In this case maximization of $\det(R)$

depends on the maximization of $\sigma_{y_1}^2 + \sigma_{y_2}^2$. This can be got by maximizing either $\sigma_{y_1}^2$ or $\sigma_{y_2}^2$ separately in the absence of noise.

In the presence of noise, variances become $\sigma_{y_1}^2 + \sigma_{\varepsilon}^2$ and $\sigma_{y_2}^2 + \sigma_{\varepsilon}^2$. In accordance with the principle of maximum information preservation, $T(Y, \vec{X})$ can be maximized by maximizing the variance of either Y_1 or Y_2 for a fixed noise variance.

Second, if the noise variance is small, then the third term in equation (15.41), $\sigma_{y_1}^2 \sigma_{y_2}^2$ $(1 - \rho_{y_1 y_2})$, becomes important. $T(Y, \vec{X})$ can then be maximized by doing an optimal trade off between two options: 1) keeping $\sigma_{y_1}^2$ and $\sigma_{y_2}^2$ large; or 2) making $\rho_{y_1 y_2}^2$ small or making Y_1 and Y_2 independent.

From the above analysis one concludes: 1) If there is a high noise level, the two output neurons yield the same linear combination of inputs, provided there is only one such combination of inputs that produces a response with maximum variance. This then suggests that this case favors the redundancy of response. 2) If the noise level is low, then the two output neurons yield different combinations of inputs, even though such an option may produce a smaller output variance. This means that this case favors the diversity of response.

15.7 Decision trees and entropy nets

A decision tree represents a hierarchy of stages through which a decision in a given situation is reached. Depending on their construction, decision trees can be classified into what is referred to as classifiers. Decision tree classifiers are also called hierarchical classifiers. These classifiers accommodate the underlying distribution of input data whatever that may be, and are capable of generating arbitrarily complex decision boundaries as learnt from a set of training vectors. In decision trees, learning is noninteractive or single step. One advantage is that all examples are considered at the same time to formulate hypotheses. On the other hand, in neural networks learning is incremental, more akin to human learning, where hypotheses are continuously improved as a result of more and more training examples.

Decision tree classifiers and layered neural networks have two significant differences. First, tree classifiers are sequential wherein decisions are made in a sequence of steps, whereas layered ANNs have massive parallelism. Second, tree classifiers have limited generalization capabilities, whereas layered ANNs do not suffer from this limitation. Nevertheless, similarities between tree classifiers and layered networks can be exploited for the development and design of neural networks for classification.

A decision tree can be described by an order of set nodes which are root, internal, terminal or leaf. Each of the internal nodes is connected with a decision function of one or more features. The terminal or leaf nodes of the decision tree are connected with actions or decisions that the system is expected to perform. Most commonly used tree form is binary, but one can have an m-ary decision tree where there are m descendants for every node. One can also find an equivalent binary tree for any m-ary decision tree.

A decision is made within a space called decision space. Indeed this is the solution domain. A decision tree partitions this space in a hierarchical manner. In the case of a binary decision tree, beginning with the root node, each successive internal node divides its associated region into two half spaces where the node decision function defines the dividing hyperplane. Figure 15.14a shows a binary decision tree and the corresponding hierarchical partitioning introduced by the tree. Traversing the tree from the root node to one of leaf nodes using the unknown pattern vector charts the course of classification. The response obtained by the unknown pattern constitutes the class or label connected to the leaf node that is reached

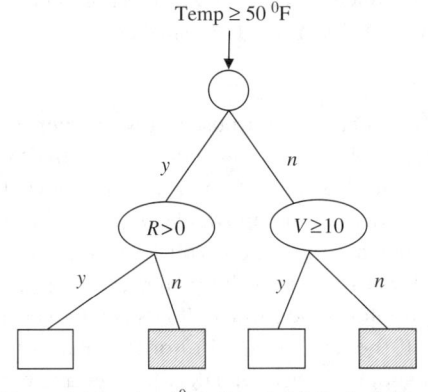

(Temp = temperature (^{0}F), R = rainfall (inches), V = wind velocity (miles/h)

Figure 15.14a   An example of a decision tree. (square boxes denote terminal nodes).

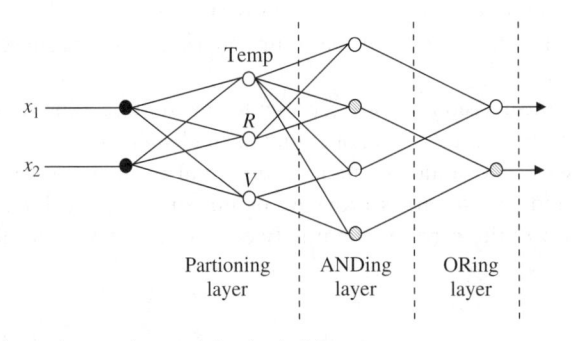

Figure 15.14b   Three-layered mapped network for the decision tree.

by the unknown vector. When a particular path is followed from the root node to the leaf node, it is implied that the conditions along that path are satisfied. Looking at the tree, each path implements an *AND* operation on a set of half spaces. It is likely that two or more leaf nodes lead to the same decision, then the corresponding paths form an OR relationship. Now comparing a layered neural network for classification, it also implements *ANDing* of hyperplanes followed by *ORing* in the output layer. This leads to the conclusion that a decision tree and a layered network are equivalent in terms of input-output mapping. In other words, a decision tree can be re-constructed as a layered network following certain rules, which can be stated as follows (Sethi, 1990):

1 The number of neurons in the first layer of the layered network, called the partitioning layer, is the same as the number of internal nodes of the decision tree. Each of these neurons implements one of the decision functions of internal nodes.

2 All leaf nodes have a corresponding neuron in the second hidden layer, called the *ANDing* layer, where the *ANDing* is implemented.

3 The number of neurons in the output layer is the same as the number of distinct classes or actions. In other words the *ORing* of those paths that result in the same decision is implemented in this layer.

4 The hierarchy of the tree is implemented by the connections between the neurons from the partitioning layer and the neurons from the *ANDing* layer.

This is illustrated by an example.

Example 15.6: Consider a case of a multipurpose reservoir where the water level (WL) is the primary determinant. The reservoir is used for flood control, hydropower generation, water supply, irrigation, and recreation. Depending upon the water level the reservoir is used and operated in different ways. The maximum allowable water level is 30 m. If the water level is above 30 m, any additional increase in the reservoir level is not accommodated, because the storage space is needed for flood storage. If the water level is equal to or more than 25 m, water is released to create space for flood storage. If it is equal to or more than 20 m it is used for hydropower generation. If the level is equal to or less than below 15 m, water is stored primarily for domestic water supply. Construct a decision tree and construct a layered mapped network for the decision tree.

Solution: The decision tree for this example is shown in Figure 15.15. The corresponding layered mapped network is shown in Figure 15.16. The following steps can be followed to constitute the decision tree and corresponding layered network:

1 There are four internal nodes of the decision tree which are $WL \geq 30, WL \geq 25, WL \geq 20,$ and $WL \geq 15$. So it should be four nodes in the partitioning layer of the three-layered network.

2 Five leaf nodes in the decision tree are represented by five nodes in the ANDing layer of the mapped network. In this layer ANDing operation is implemented.

3 Two distinct decisions whether demand (flood control, hydropower generation, water supply, irrigation, and recreation) is provided or not should stand for two neurons in the output (O Ring) layer of the correspondent network. For instance, according to the situation

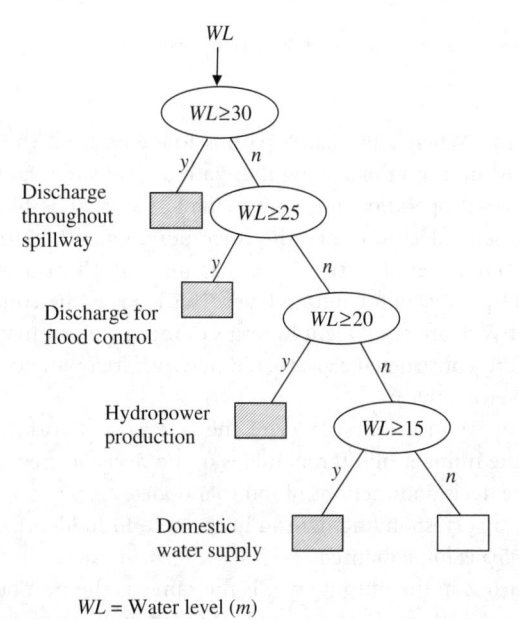

WL = Water level (*m*)

Figure 15.15 Decision tree for Example 15.6.

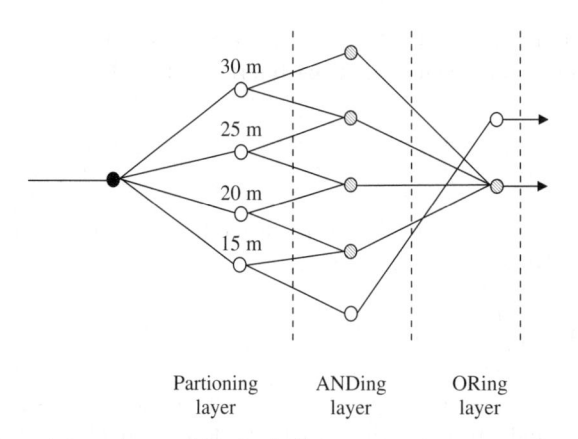

Partioning layer   ANDing layer   ORing layer

Figure 15.16 Three-layered mapped network for the decision tree.

that WL exceeds 30 m level or not, there are two options: discharge throughout the spillway or accommodate for flood control.

4 Neurons from the partitioning layer and the neurons from the *ANDing* layer are connected to each other by the implementation of the hierarchy of the tree.

This example shows a straightforward mapping of a decision tree into a layered network of neurons. However, the question arises: Is the number of neurons in the partitioning layer optimum? The mapping rules do not answer this question, although the tree-to-network mapping does specify the number of neurons in each of the three layers of the neural network. It may be noted that in the network so obtained except for one neuron in the partitioning layer that corresponds to the root node of the decision tree, the remainder of the neurons do not have connections with all the neurons in the adjacent layers. This results in fewer connections which are an advantage. This network is driven by the mutual information of the data-driven tree generation scheme, and can therefore be called an entropy net.

When designing a tree there are essentially three steps involved. First, the hierarchical ordering and the choice of node decision functions need to be defined. This can be accomplished by defining a goodness measure in terms of mutual information. Consider a variable X and its measurement defined by x. Let $x = y$ define the partitioning of the one-dimensional feature space into two outcomes x_1 and x_2: values of measurement x greater than y as x_1 and values of x less than y as x_2. There are two classes here and two outcomes. Then the amount of mutual information about the pattern classes of event X can be expressed as

$$H(C; X) = \sum_{i=1}^{2} \sum_{j=1}^{2} p(c_i, x_j) \log_2 [p(c_i|x_j)|p(c_j)]$$ (15.42)

where C denotes the set of pattern classes and $p(.)$'s denote the various probabilities. Clearly the threshold pays a pivotal role here, and therefore the choice of it is critical. One would want to select the value of y such that X yields the maximum information. This means that the value of y that maximizes equation (15.42) should be selected over all possible values of y. Thus, the value of average mutual information gain (AMIG) can serve as a basis for evaluating the goodness of partitioning. As new data become available and new knowledge is gained, the neural network can be modified by modifying the connection strengths or weights.

Second, a decision needs to be made as to when a node is to be labeled as a terminal node. Third, a decision rule is to be spelled out at each terminal node.

Questions

Q.15.1 An event-based hydrograph can be predicted using the following expression (Moramarco et al., 2005):

$$Q_d(t) = \alpha \frac{A_d(t)}{A_u(t - T_L)} Q_u(t - T_L) + \beta$$

where Q_u is upstream discharge; Q_d is downstream discharge; A_d and A_u are effective downstream and upstream cross sectional flow areas obtained from the observed stages, respectively; T_L are wave travel time depending on the wave celerity, c; and α and β are model parameters. Given the following events observed in Tiber River basin, construct an ANN to predict flow rate at a downstream station. Explain step-by-step. Note that Moramarco et al. (2006) measured data every half an hour, assuming that you have the data. Main characteristics of observed flood events at stations on the Tiber River are given as:

Date	Santa Lucia			Ponte Felcino			
	Q_b (m^3/s)	Q_p (m^3/s)	V $(10^6 m^3)$	Q_b (m^3/s)	Q_p (m^3/s)	V $(10^6 m^3)$	T_L (h)
Jan. 1994	35.6	108	19	50.8	241	34.7	3.0
May 1995*	4.2	71.0	10.3	8.8	138.7	19.1	4.0
Jan. 2003	24	58	13.5	50	218	40.9	3.5
Feb. 2004*	22	91	7.4	55	276	27.3	3.5

Q_b = base flow; Q_p = peak discharge; V = direct runoff volume; T_L = travel time
* Used for ANN model training

Q.15.2 For the following network, update weights between output and inner layers and inner and input layers.

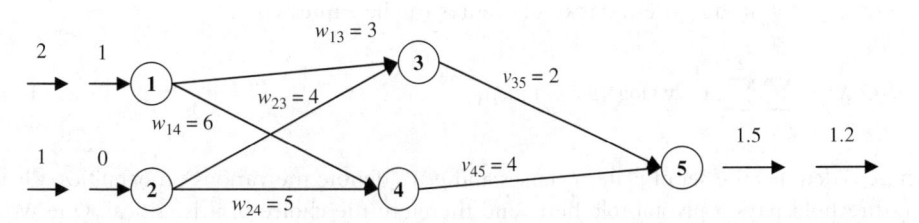

Q.15.3 In order to monitor an earth-fill dam body in terms of anomaly seepage, piezometers can be placed on the dam body. The levels in the piezometers show the pressure in the dam body. Swiatek (2002) monitored the Jaziorsky dam body seepage path using four piezometers plus the water levels in the upper and lower reservoirs. Swiatek monitored levels from 1995 to 2002 every other week. Can you develop ANN to predict seepage path. Explain step-by-step.

Q.15.4 Obtain stage discharge data for a gaging site on a river. Then develop an ANN model for the rating curve for this site. Compare this model with the standard rating curve.

Q.15.5 Obtain monthly rainfall and runoff volume data for a small watershed and develop an ANN model and compare it with a water balance model.

Q.15.6 Obtain data on event basis for rainfall, runoff and erosion. The objective is to predict sediment yield. Develop an ANN model and compare it with regression model or another model.

Q.15.7 Using the SCS-CN method, determine curve number for different types of land uses, soil moisture and soil type. Now develop an ANN model and compare the two.

Q.15.8 Obtain data for monthly pan evaporation and relevant hydrometeorological variables. Develop an ANN model to predict monthly evaporation and compare it with a standard evaporation estimation method.

Q.15.9 Obtain data on infiltration capacity rate for a watershed. Develop man ANN model and compare with a standard infiltration equation.

References

Abrahart, R. J. and See, L. M. (2007). Neural network modelling of non-linear hydrological relationships. *Hydrol. Earth Syst. Sci.*, 11, 1563–79.

Ackley, D.H., Hinton, G.E. and Sejnowski, T.J. (1985). A learning algorithm for Boltzmann machines. *Cognitive Science*, Vol. 9, pp. 147–69.

Burr, D.J. (1988). Experiments on neural net recognition of spoken and written text. *IEEE Transactions: Acoustic, Speech, Signal Processing*, Vol. 36, pp. 1162–8.

Haykin, S. (1994). *Neural Networks: A Comprehensive Foundation*. Macmillan College Publishing Company, New York.

Hecht-Nielsen, R. (1987). Counterpropagation networks. *Applied Optimization*, Vol. 36, pp. 4979–84.

Linsker, R. (1987). Towards an organizing principle for perception: Hebbian synapses and the principle of neural encoding. IBM Research Report RC12820, IBM Research, Yorktown Heights, New York.

Linsker, R. (1988a). Self generalization in a perceptual network. *Computer*, Vol. 21, pp. 105–17.

Linsker, R. (1988b). Towards an organizing principle for a layered perceptual network. In: *Neural Information Processing Systems*, edited by D.Z. Anderson, pp. 485–94, American Institute of Physics, New York.

Linsker, R. (1989a). An application of the principle of maximum information preservation to linear systems. *Advances in Neural Information Processing Systems*, Vol. 1, pp. 186–94.

Linsker, R. (1989b). How to generate ordered maps by maximizing the mutual information between input and output signals. *Neural Computation*, Vol. 1, pp. 402–11.

Linsker, R. (1990a). Perceptual neural organization: Some approaches based on network models and information theory. *Annual Review of Neuroscience*, Vol. 13, pp. 257–81.

Linsker, R. (1990b). Self-organization in a perceptual system: How network models and information may shed light on neural organization. In: *Connectionist Modeling and Brain Function: The Developing Interface*, edited by S.J. Hanson and C. Roldon, Chapter 10, pp. 351–92, MIT Press, Cambridge.

Lippmann, R.P. (1987). An introduction to computing with neural nets. *IEEE ASSP Management*, pp. 4–22.

Moramarco, T., Barbetta, S., Melone, F. and Singh, V. P. (2006). A real-time stage Muskingum forecasting model for a site without rating curve. *Hydrological Sciences Journal*, Vol. 51, No. 1, pp. 66–82.

Rumelhart, D.E., Hinton, G.E. and Williams, R.J. (1986). Learning internal representation by error propagation. In *Parallel Distributed Processing: Explorations in the Microstructure of Cognition*, Vol. 1: Foundations, edited by D.E. Rummelhart and J.L. McClelland, M.I.T. Press, Cambridge, Massachusetts.

Seo, I.W. and Cheong, T.S. (1998). Predicting longitudinal dispersion coefficient in natural streams. *Journal of Hydraulic Engineering*, ASCE, Vol. 124, No. 1, pp. 25–32.

Sethi, I.K. (1990). Entropy nets: From decision trees to neural networks. Proceedings of the IEEE, Vol. 78, No. 10, pp. 1605–13.

Swiatek, D. (2002). Application of filtration numerical model for estimation of river embankments antiseepage protections efficiency. *IMGW Research Papers*, No. 13 – Series: Water Engineering, Warsaw.

Tayfur, G. and Singh, V.P. (2005). Predicting longitudinal dispersion coefficient in natural streams by artificial neural network. *Journal of Hydraulic Engineering*, ASCE, Vol. 131, No. 11, pp. 991–1000.

Tayfur, G., Swiatek, D., Wita, A., and Singh, V.P. (2005). Case study: Finite element method and artificial neural network models for flow through Jeziorsko Earthfill Dam in Poland. *Journal of Hydrologic Engineering*, Vol. 131, No. 6, pp. 431–40.

Widrow, B., Winter, R.G. and Baxter, R.A. (1988). Layered neural nets for pattern recognition. *IEEE Transactions on Acoustic, Speech, Signal Processing*, Vol. 36, pp. 1109–18.

16 System Complexity

Hydrology, environmental and ecological sciences, and water engineering entail a range of systems from simple to complex. Complexity or simplicity is often viewed intuitively in terms of mathematical structure. The effect of complexity in both artificial and natural systems needs to be quantified by objective and operationally meaningful measures of system complexity, and these measures should supplement intuitive and quantitative measures of complexity. In recent years, entropy has been used to quantify complexity in dynamical systems by Bates and Shepard (1993), in hydrologic time series by Englehardt et al. (2009), in physics by Grassberger (1986), in ecosystems by Lange (1999), and in biology by Jimenez-Montano et al. (2002). The objective of this chapter is to discuss complexity and related aspects using entropy. Ferdinand (1974) developed a complexity measure for network systems using the principle of maximum entropy (POME), Cornacchio (1977) extended this measure, and Kapur (1983) developed additional measures. This chapter draws from the works of these authors.

16.1 Ferdinand's measure of complexity

Consider a system which has measurable properties. Any measurable property of the system can be called a system observable, as for example, the values of a function of the number of system defects or errors. The observable, its values and the function can also be referred to as a system observable. For example, if there are n defects or errors and f represents a system observable, then function $f(n)$ represents the value of the observable.

Now consider a system that has a number of defects denoted by $n \geq 0$; here the number of defects n is regarded as a random variable. Let $N \geq 0$ denote the maximum number of defects that can occur. This means that $0 \leq n \leq N$. The probability that n defects occur in the system is denoted as $0 \leq p(n) \leq 1$, that is, $p(n)$ is the probability distribution of n. The Shannon entropy of $p(n)$ or n can be expressed as

$$H(n) = -\sum_{n=0}^{N} p(n) \ln p(n), \quad p(n) \geq 0, n = 0, 1, 2, \ldots, N \tag{16.1}$$

The objective is to determine the least-biased probability distribution, $p(n)$, in accordance with the principle of maximum entropy (POME), subject to specified constraints. Determination of

Entropy Theory and its Application in Environmental and Water Engineering, First Edition. Vijay P. Singh.
© 2013 John Wiley & Sons, Ltd. Published 2013 by John Wiley & Sons, Ltd.

entropy-based distributions has been discussed in Chapter 4, but it will be revisited for the sake of completeness.

16.1.1 Specification of constraints

The probability distribution $p(n)$ must satisfy

$$\sum_{n=0}^{N} p(n) = 1 \tag{16.2}$$

Another constraint that $p(n)$ must satisfy can be specified in terms of the mean number of defects, m, that can occur in the system:

$$\sum_{n=0}^{N} np(n) = m \tag{16.3}$$

Thus, the canonical distribution $p(n)$ is specified by two parameters m and N. For purposes of simplicity, this distribution forms the basis for the analysis of system complexity in this chapter. It may, however, be remarked that depending on the system under consideration, other constraints can be specified that the probability distribution must satisfy.

16.1.2 Maximization of entropy

The Shannon entropy, given by equation (16.1), can be maximized, using the method of Lagrange multipliers, subject to the constraints expressed by equations (16.2) and (16.3). To that end, the Lagrangean function L can be expressed as

$$L = -\sum_{n=0}^{N} p(n) \ln p(n) - (\lambda_0 - 1) \left(\sum_{n=0}^{N} p(n) - 1 \right) - \lambda_1 \left(\sum_{n=0}^{N} np(n) - m \right) \tag{16.4}$$

where λ_0 and λ_1 are the Lagrange multipliers. Differentiating equation (16.4) and recalling the Euler-Lagrange calculus of variation, one gets

$$\frac{\partial L}{\partial p} = 0 = -\ln p(n) - \lambda_0 - \lambda_1 n \tag{16.5}$$

Equation (16.5) yields the least-biased entropy-based probability distribution of n:

$$p(n) = \exp(-\lambda_0 - \lambda_1 n) \tag{16.6}$$

Equation (16.6) describes the probability of n defects occurring in the system. It has two unknowns, λ_0 and λ_1, which can be determined using equations (16.2) and (16.3). It goes without saying that equation (16.6) satisfies constraint equations (16.2) and (16.3).

16.1.3 Determination of Lagrange multipliers

Inserting equation (16.6) in equation (16.2), one obtains

$$\exp(-\lambda_0) \sum_{n=0}^{N} \exp(-\lambda_1 n) = 1 \tag{16.7}$$

Similarly, substituting equation (16.6) in equation (16.3), one gets

$$\sum_{n=0}^{N} n \exp\left(-\lambda_0 - \lambda_1 n\right) = m \tag{16.8}$$

Equations (16.7) and (16.8) can be solved for λ_0 and λ_1. Combining equations (16.7) and (16.8), one obtains

$$m = \frac{\displaystyle\sum_{n=0}^{N} n \exp\left(-\lambda_1 n\right)}{\displaystyle\sum_{n=0}^{N} \exp\left(-\lambda_1 n\right)} \tag{16.9}$$

The only unknown in equation (16.9) is λ_1. Equation (16.9) can be solved for λ_1 in terms of m and then substituting it in equation (16.7) one gets λ_0.

16.1.4 Partition function

The partition function provides a link between the mechanical properties and thermodynamic properties of a system. It describes the partitioning of the system among different energy levels in the equilibrium distribution. Thus, important mechanical properties of the system are expressed in terms of the derivatives of the logarithm of the partition function. A high value of entropy is a measure of low degree of information. Thus, POME postulates that an equilibrium distribution corresponds to the condition of maximum ignorance or lowest degree of information for a given average number of elements distributed in the system.

Let $\alpha = \exp(-\lambda_1)$. Parameter α is used as a measure of complexity and is called the coefficient of complexity. Then, equation (16.7) can be written in terms of α as

$$\exp\left(\lambda_0\right) = Z(\alpha, N) = \sum_{n=0}^{N} \exp\left(-\lambda_1 n\right) = \sum_{n=0}^{N} \alpha^n \tag{16.10}$$

Term $Z(\alpha, N)$ or $\exp(\lambda_0)$ is called the partition function and is defined by equation (16.10). It can be written as

$$\exp\left(\lambda_0\right) = Z(\alpha, N) = \sum_{n=0}^{N} \alpha^n = \frac{1 - \alpha^{N=1}}{1 - \alpha}, \quad \alpha < 1 \tag{16.11}$$

$$= \frac{\alpha^{N+1} - 1}{\alpha - 1}, \quad \alpha \geq 1 \tag{16.12}$$

From equation (16.10), α is plotted as a function of λ_1 as shown in Figure 16.1.

Figure 16.2 plots the partition function as a function of N.

Equation (16.9) can be written in terms of α as

$$m = \frac{\displaystyle\sum_{n=0}^{N} n \alpha^n}{\displaystyle\sum_{n=0}^{N} \alpha^n} \tag{16.13}$$

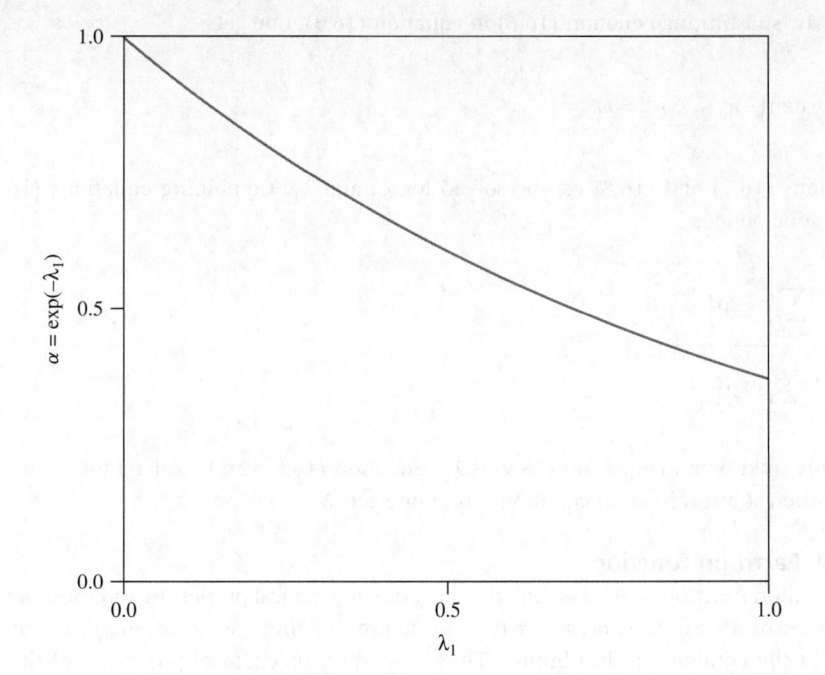

Figure 16.1 Plot of α as a function of λ_1.

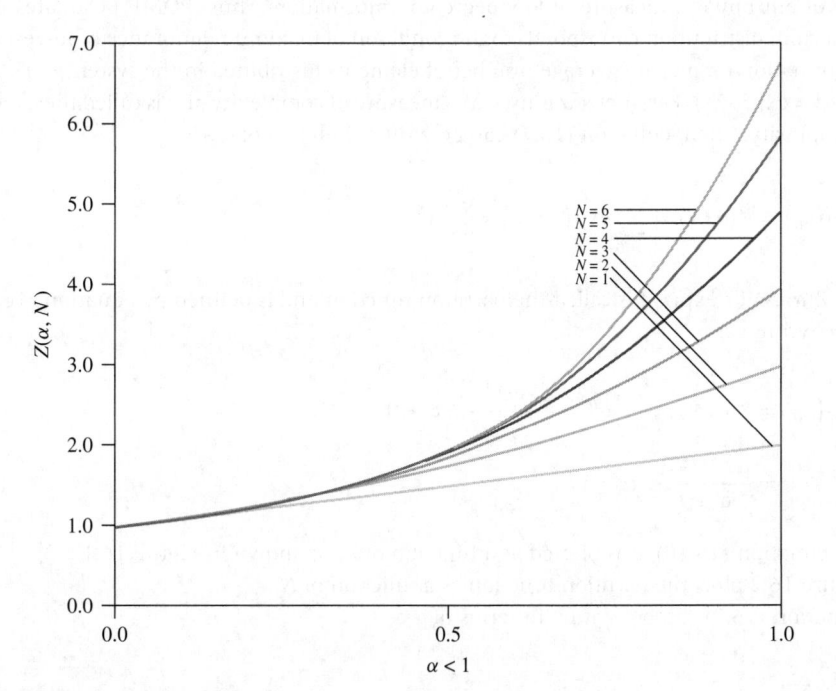

Figure 16.2a Partition function as a function of α for different values of N.

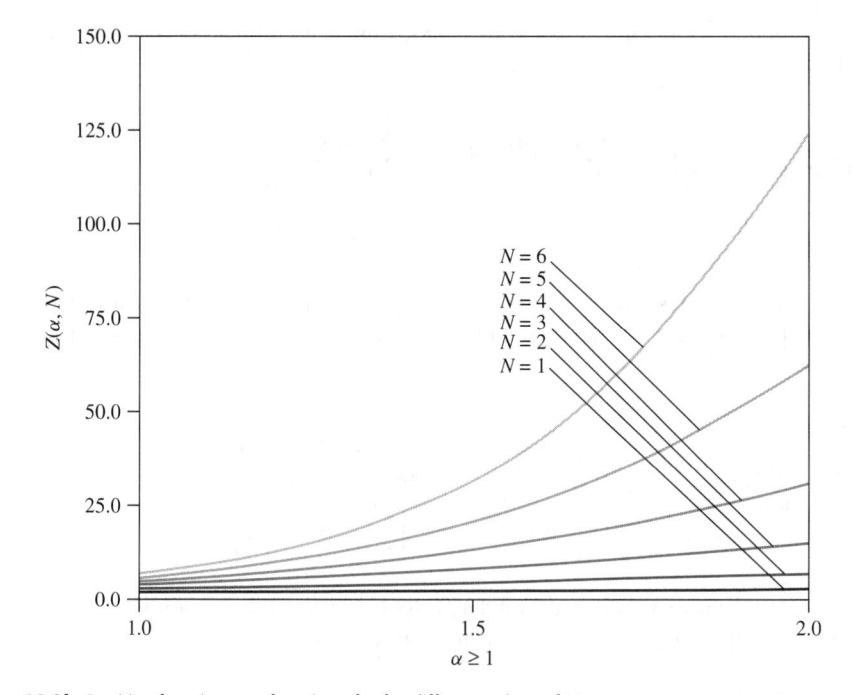

Figure 16.2b Partition function as a function of α for different values of N.

Taking the logarithm of equation (16.10), one obtains

$$\lambda_0 = \ln Z(\alpha, N) = \ln \left(\sum_{n=0}^{N} \alpha^n \right) \tag{16.14}$$

Differentiating equation (16.14) with respect to α one obtains

$$\frac{d\lambda_0}{d\alpha} = \frac{d \ln Z(\alpha, N)}{d\alpha} = \frac{\displaystyle\sum_{n=0}^{N} n\alpha^{n-1}}{\displaystyle\sum_{n=0}^{N} \alpha^n} \tag{16.15a}$$

Multiplying both sides of equation (16.15a) by α, one gets

$$\alpha \frac{d \ln Z(\alpha, N)}{d\alpha} = \frac{\displaystyle\sum_{n=0}^{N} n\alpha^n}{\displaystyle\sum_{n=0}^{N} \alpha^n} = m \tag{16.15b}$$

Now one needs to determine if equation (16.15b) has only one solution and if that solution is positive.

Writing equation (16.15b) in terms of λ_1 and noting that $d\alpha = -\exp(-\lambda_1)d\lambda_1 = -\alpha d\lambda_1$, one gets

$$\frac{d\ln Z[\exp(-\lambda_1), N]}{d\lambda_1} = -\frac{\sum\limits_{n=0}^{N} n\exp(-\lambda_1 n)}{\sum\limits_{n=0}^{N}\exp(-\lambda_1 n)} = -\frac{\sum\limits_{n=0}^{N} n\exp(-\lambda_0 - \lambda_1 n)}{\sum\limits_{n=0}^{N}\exp(-\lambda_0 - \lambda_1 n)} = -m \qquad (16.16)$$

Equation (16.16) states that the derivative of $\ln Z[\exp(-\lambda_1), N)]/d\lambda_1$ is always negative regardless of the value of λ_1. Differentiating equation (16.16) again with respect to λ_1, one obtains

$$\frac{d^2\ln Z[\exp(-\lambda_1), N]}{d\lambda_1^2} = \frac{\sum\limits_{n=0}^{N} n^2\exp(-\lambda_1 n)}{\sum\limits_{n=0}^{N}\exp(-\lambda_1 n)} - \frac{\left[\sum\limits_{n=0}^{N} n\exp(-\lambda_1 n)\right]^2}{\left[\sum\limits_{n=0}^{N}\exp(-\lambda_1 n)\right]^2} > 0 \qquad (16.17)$$

Equation (16.17) shows that the second derivative is positive if λ_1 is positive.

Furthermore, using equation (16.1) it can be shown that

$$\frac{\partial^2 H(p)}{\partial p(n)\,\partial p(r)} = \begin{cases} 0 & \text{if } n \neq r \\ -\dfrac{1}{p(n)} & \text{if } n \neq r \end{cases} \qquad (16.18)$$

Equation (16.18) shows that the matrix of $H(p)$ is negative definite and therefore $H(p)$ is maximized.

The probability distribution given by equation (16.6) can now be expressed with the use of equations (16.11) and (16.12) as

$$p(n) = \exp(-\lambda_0)\exp(-\lambda_1 n)$$

$$= \alpha^n \frac{1-\alpha}{1-\alpha^{N+1}}, \quad \alpha < 1 \qquad (16.19a)$$

$$= \alpha^n \frac{\alpha-1}{\alpha^{N+1}-1}, \quad \alpha \geq 1 \qquad (16.19b)$$

Equation (16.19) gives the probability of having n defects in a system with the maximum number of defects defined by N, and maximizes the Shannon entropy. Since $p(n)$ denotes the probability distribution of n defects in the system, it can be labeled as the defect probability distribution. Figure 16.3 plots the distribution.

16.1.5 Analysis of complexity

Now consider the mean number of defects in the system as a function of α and N which can be expressed from equation (16.15b). First, recalling equations (16.11) and (16.12),

$$\sum_{n=0}^{N} \alpha^n = \frac{1-\alpha^{N+1}}{1-\alpha}, \quad \alpha < 1 \qquad (16.20a)$$

$$= \frac{\alpha^{N+1}-1}{\alpha-1}, \quad \alpha \geq 1 \qquad (16.20b)$$

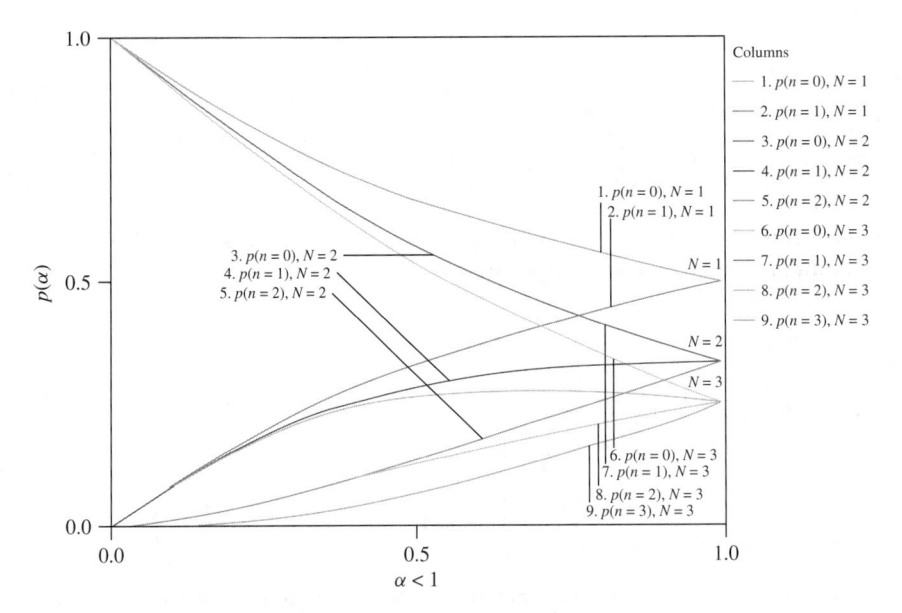

Figure 16.3a Probability distribution of defects for different values of N.

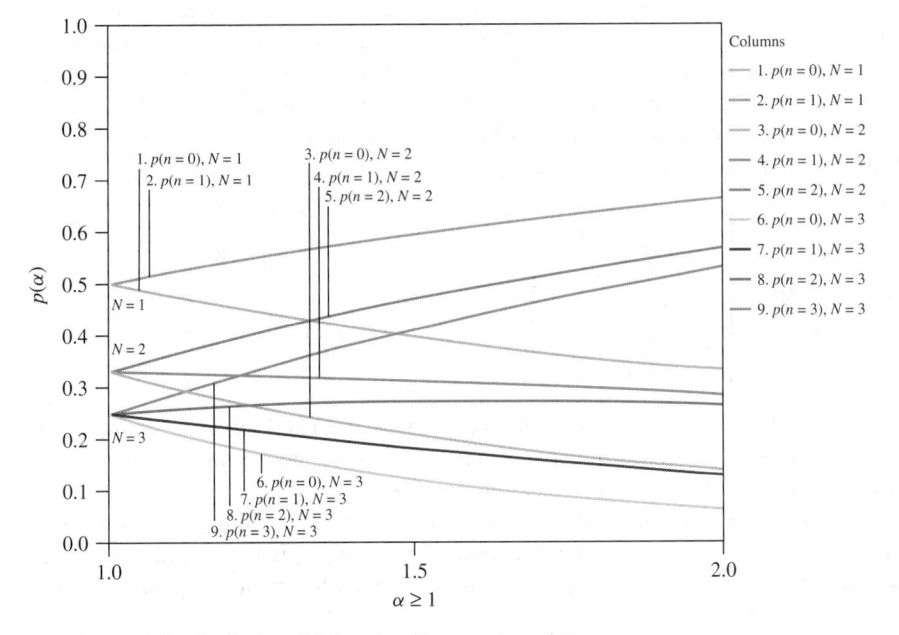

Figure 16.3b Probability distribution of defects for different values of N.

Second,

$$\sum_{n=0}^{N} n\alpha^n = \frac{\alpha\left(1 - \alpha^N\right)}{(1 - \alpha)^2} - N\frac{\alpha^{N+1}}{1 - \alpha}, \quad \alpha < 1 \tag{16.21a}$$

$$= \frac{\alpha\left(1 - \alpha^N\right)}{(\alpha - 1)^2} - N\frac{\alpha^{N+1}}{(\alpha - 1)}, \quad \alpha \geq 1 \tag{16.21b}$$

Therefore, using equation (16.16b), and taking advantage of equations (16.20) and (16.21), the mean number of defects can be expressed as

$$E(\alpha, N) = \alpha\frac{d}{d\alpha}\ln Z(\alpha, N)$$

$$= \frac{\alpha}{1 - \alpha} - \frac{(N + 1)\alpha^{N+1}}{1 - \alpha^{N+1}}, \quad \alpha < 1 \tag{16.22a}$$

$$= N - \left[\frac{\alpha}{\alpha - 1} + \frac{(N + 1)\alpha^{N+1}}{\alpha^{N+1} - 1}\right], \quad \alpha \geq 1 \tag{16.22b}$$

$$= N - \left[\frac{\beta}{1 - \beta} - \frac{(N + 1)\beta^{N+1}}{1 - \beta^{N+1}}\right], \quad \beta \leq 1 \tag{16.22c}$$

where $\beta = 1/\alpha$. For purposes of graphical illustration, a definition of $\beta = 1/\alpha$ comes in handy.

Now consider the behavior of $E(\alpha, N)$ as shown in Figure 16.4. Function $E(\alpha, N)$ is a monotonic function of α and for $\alpha > 0$ it is a monotonic function of N. Since quantity α is characterized as a coefficient of complexity, a system can be considered as simple if $\alpha = 0$; in this case $E(\alpha, N) = 0$. On the other hand, a system can be viewed as perfectly complex if $\alpha \rightarrow \infty$; in this case $E(\infty, N) = N$. Thus, α controls the number of defects expected in the system or the degree of complexity. It may be stated that it is also possible to define a threshold value of α, say α_0, corresponding to which the system is simple.

Now consider $E(\alpha, N)$ when $N \rightarrow \infty$.

$$\lim_{N \to \infty} E(\alpha, N) = \begin{cases} \dfrac{\alpha}{1 - \alpha}, & \alpha < 1 \\[2mm] \dfrac{1}{2N}, & \alpha = 1 \\[2mm] N - \dfrac{\beta}{1 - \beta}, & \beta > 1 \end{cases} \tag{16.23}$$

The point $\alpha = 1$ corresponds to a critical point in the system and serves to separate the region where the system is simple from the region where it is complex in which the number of expected defects tends to approach its ultimate value of N.

Figure 16.5 plots normalized $E(\alpha, N)/N$ against α for $\alpha < 1$ and $\alpha \geq 1$. When $N \rightarrow \infty$, the normalized function becomes a step function. Comparing equation (16.22) and (16.23), the difference $|E(\alpha, \infty) - E(\alpha, N)|$ is exponentially small even for moderately large N, that is, for all practical purposes:

$$E(\alpha, N) \approx E(\alpha, \infty) \quad \text{for small } \alpha \tag{16.24}$$

This is shown in Figure 16.5 which plots $E(\alpha, N)$ against α.

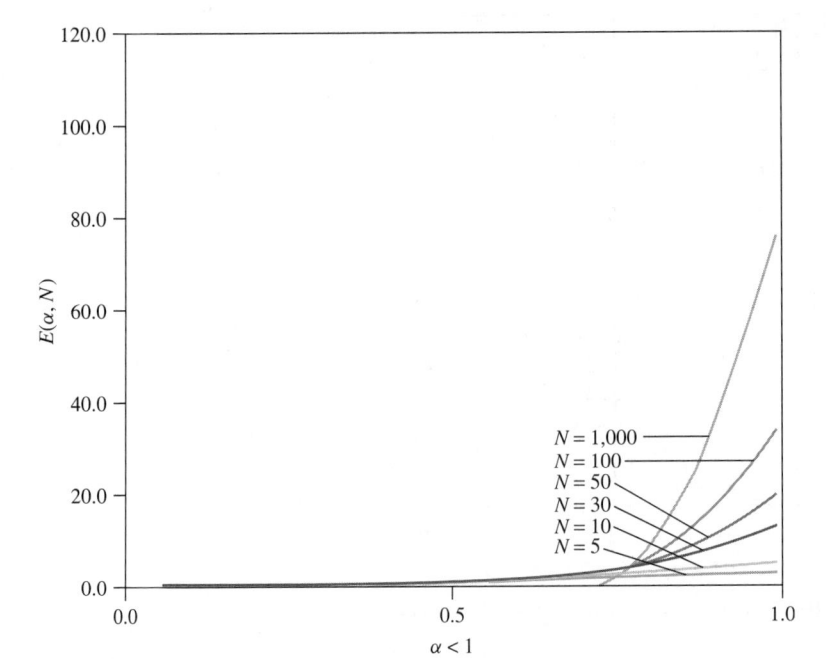

Figure 16.4a Behavior of $E(\alpha, N)$ as function of α for various values of N.

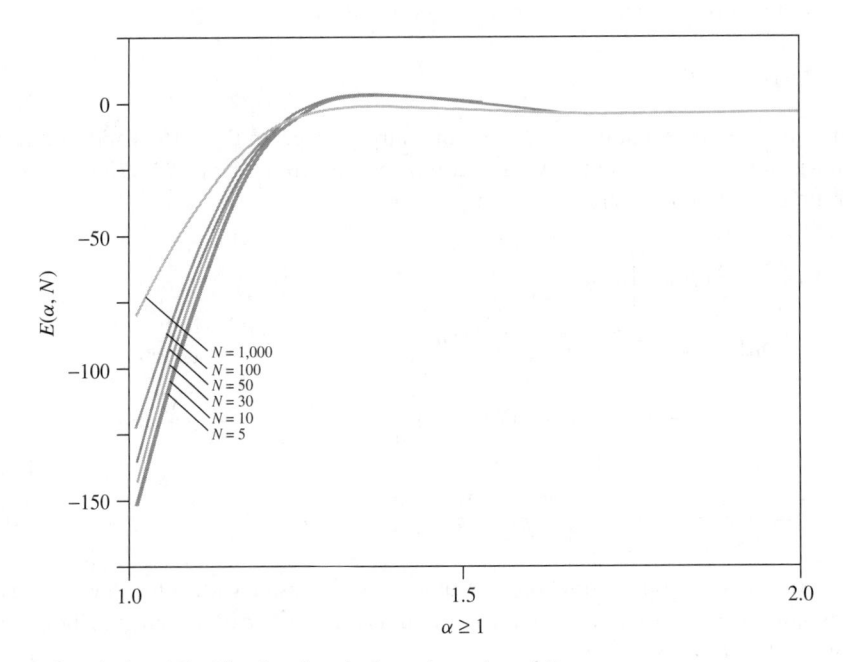

Figure 16.4b Behavior of $E(\alpha, N)$ as function of α for various values of N .

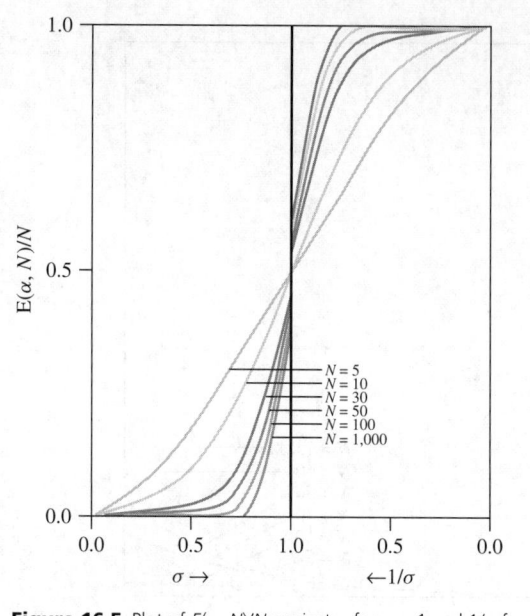

Figure 16.5 Plot of $E(\alpha, N)/N$ against α for $\alpha < 1$ and $1/\alpha$ for $\alpha > 1$.

16.1.6 Maximum entropy

The Shannon entropy of equation (16.6) can be written as

$$H(n) = \lambda_0 + \lambda_1 E[n] = \lambda_0 + \lambda_1 m \tag{16.25}$$

The entropy given by equation (16.25) is a measure of defect in the system and can therefore be characterized as a defect entropy. The entropy as a function of α and N can be obtained by substituting equation (16.19) in equation (16.1) as

$$H(\alpha, N) = -\sum_{n=1}^{N} p(n) \ln\left[\alpha^n \frac{1-\alpha}{1-\alpha^{N+1}}\right]$$

$$= \ln\left[1 - \alpha^{N+1}\right] - \ln(1 - \alpha) - \left[\frac{\alpha}{1-\alpha} - \frac{(N+1)\alpha^{N+1}}{1-\alpha^{N+1}}\right]\ln\alpha, \ \alpha < 1 \tag{16.26a}$$

$$= \ln\left[\alpha^{N+1} - 1\right] - \ln(\alpha - 1) - \left[N - \left(\frac{\alpha}{\alpha-1} + \frac{(N+1)\alpha^{N+1}}{\alpha^{N+1}-1}\right)\right]\ln\alpha, \ \alpha \geq 1 \tag{16.26b}$$

$$= \ln\left[\frac{1-\beta^{N+1}}{1-\beta}\right] - \left[\frac{\beta}{1-\beta} - \frac{(N+1)\beta^{N+1}}{1-\beta^{N+1}}\right]\ln\beta, \beta \leq 1 \tag{16.26c}$$

Entropy $H(\alpha, N)$ is a positive function, monotonically increasing with N for all values of $\alpha \geq 0$. It is a monotonic increasing function for $\alpha < 1$, and a monotonic decreasing function of $\alpha > 1$, as shown in Figure 16.6.

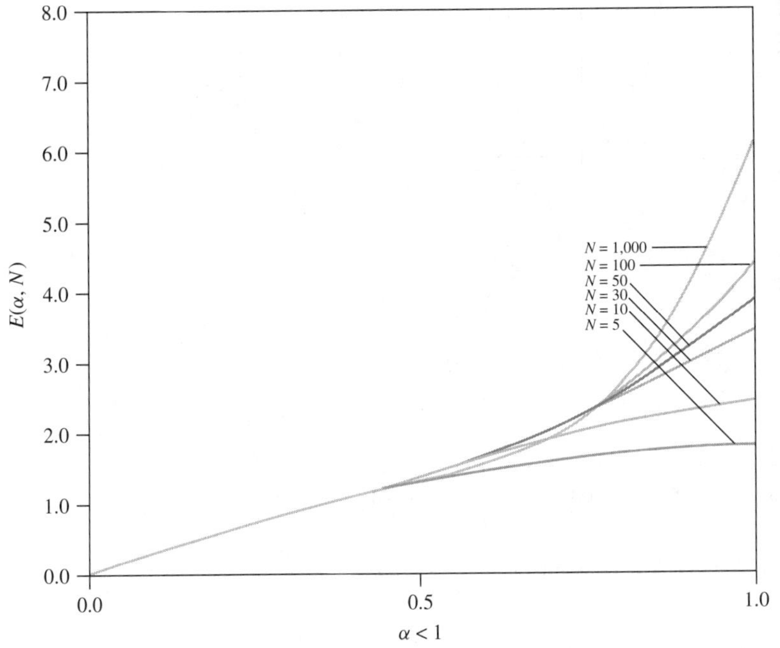

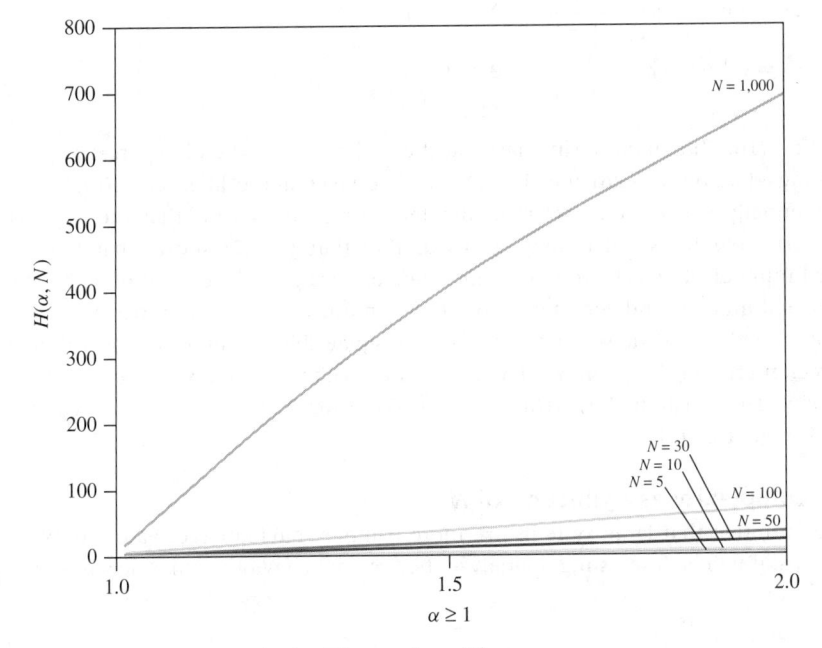

Figure 16.6a $H(\alpha, N)$ as a function of α for different values of N.

Figure 16.6b $H(\alpha, N)$ as a function of α for different values of N.

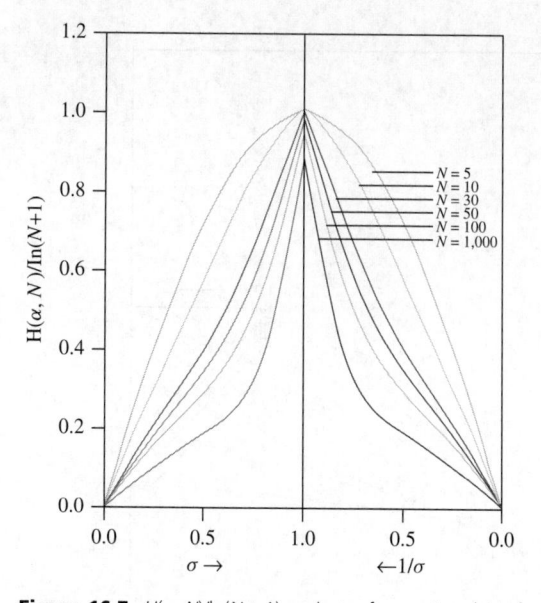

Figure 16.7 $H(\alpha, N)/\ln(N + 1)$ against α for $\alpha < 1$ and $1/\alpha$ for $\alpha > 1$.

In the limit when $N \to \infty$,

$$H(\alpha, \infty) = -\ln(1 - \alpha) - \frac{\alpha \ln \alpha}{1 - \alpha}, \quad 0 \leq \alpha < 1$$

$$= -\ln(1 - \beta) - \frac{\beta \ln \beta}{1 - \beta}, \quad 0 \leq \beta < 1 \tag{16.27}$$

Equation (16.27) exhibits a singularity at the critical $\alpha = 1$, when it diverges as $-\ln|1 - \alpha|$. Comparing equations (16.26) and (16.27), it is observed that the difference $|H(\alpha,N) - H(\alpha,\infty)|$ is exponentially small even if N is moderately large. This means that the approximation $H(\alpha,N) \sim H(\alpha,\infty)$ for small α may be used. Equation (16.22) shows that for $\alpha < 1$, the expected number of defects in the system tends to become independent of N, but for $\alpha > 1$, the expected number of defects tends to the maximum value N. These two extremes are not common in hydrology and water engineering. It is plausible that of most practical importance is the region around the point $\alpha = 1$ which is the dividing boundary between simplicity and complexity. The function $H(\alpha, N)/\ln(N + 1)$ is plotted against α for $\alpha < 1$ and against $\beta = 1/\alpha$ for $\alpha > 1$ in Figure 16.7.

16.1.7 Complexity as a function of *N*

It seems intuitive that large systems are more complex having larger values of N, and vice versa. Thus, it will be interesting to analyze the system behavior when α depends on N. Let

$$\alpha = 1 - h = 1 - \frac{\sigma}{N} \tag{16.28}$$

where σ is a parameter, and $h = \sigma/N$, such that when $h \to 0$, σ remains finite. Focusing on the region around the point $\alpha = 1$, which is of most interest, interesting results are obtained.

Expanding equation (16.28) in Taylor series at $\alpha = 1$, one gets

$$E[\sigma, N] \approx \frac{\frac{1}{2}N\left(1 - \frac{1}{3}\sigma\right)}{1 - \frac{1}{2}\sigma + \frac{\sigma^2}{6}} \tag{16.29}$$

$$H(\alpha, N) \cong \ln(N + 1) \tag{16.30}$$

This behavior is obtained if the coefficient of complexity is defined as an exponential function:

$$\alpha = \exp(-\sigma/N) \tag{16.31}$$

Thus, the system is simple if $\sigma > 0$ so that $\alpha < 1$. The complexity increases with N when $\sigma > 0$. One can now obtain the following for $\sigma > 0$:

$$Z\left[\exp(-\sigma/N), N\right] = \frac{N}{\sigma}\left[1 - \exp(-\sigma)\right] = \frac{1}{2}\left[1 + \exp(-\sigma)\right] + O\left(\frac{\sigma}{N}\right) \tag{16.32}$$

$$E\left[\exp(-\sigma/N), N\right] = NA(\sigma) - B(\sigma) + O\left(\frac{\sigma}{N}\right) \tag{16.33}$$

$$H\left[\exp(-\sigma/N), N\right] = \ln N + C(\sigma) + O\left(\frac{\sigma}{N}\right) \tag{16.34}$$

where

$$A(\sigma) = \frac{1}{\sigma} - \frac{1}{\exp(\sigma) - 1} \tag{16.35}$$

$$B(\sigma) = \frac{1}{2} + \frac{1}{\exp(\sigma) - 1} - \frac{\sigma \exp(\sigma)}{\left[\exp(\sigma) - 1\right]^2} \tag{16.36}$$

$$C(\sigma) = \sigma A(\sigma) + \ln\left\{\frac{1}{\sigma}\left[1 - \exp(-\sigma)\right]\right\} \tag{16.37}$$

The mean number of defects becomes an extensive property of the maximum number of defects N and the tendency of faults is proportional to $\ln N$.
    For $\alpha \geq 1$,

$$\alpha = \exp(-\sigma/N), \quad \sigma \geq 0 \tag{16.38}$$

Then,

$$E\left[\exp(-\sigma/N, N)\right] = N[1 - A(\sigma)] + B(\sigma) + O\left(\frac{\sigma}{N}\right) \tag{16.39}$$

$$H\left[\exp(-\sigma/N), N\right] = \ln N + C(\sigma) + O\left(\frac{\sigma}{N}\right) \tag{16.40}$$

Because $B(\sigma)$ is bounded,

$$B(0) = 0 \leq B(\sigma) \leq \frac{1}{2} = B(\infty) \tag{16.41}$$

Thus, in the vicinity of the region $\alpha = 1$ and for large N,

$$E[\exp(-\sigma/N), N] \approx NA(\sigma), \quad \sigma > 0$$
$$\cong N[1 - A(\sigma), \quad \sigma \leq 0 \tag{16.42}$$

Figure 16.8 displays functions $A(\sigma)$, $1 - A(\sigma)$, $B(\sigma)$, and $C(\sigma)$ against σ for $\sigma < 1$ and against $1/\sigma$ for $\sigma \geq 1$.

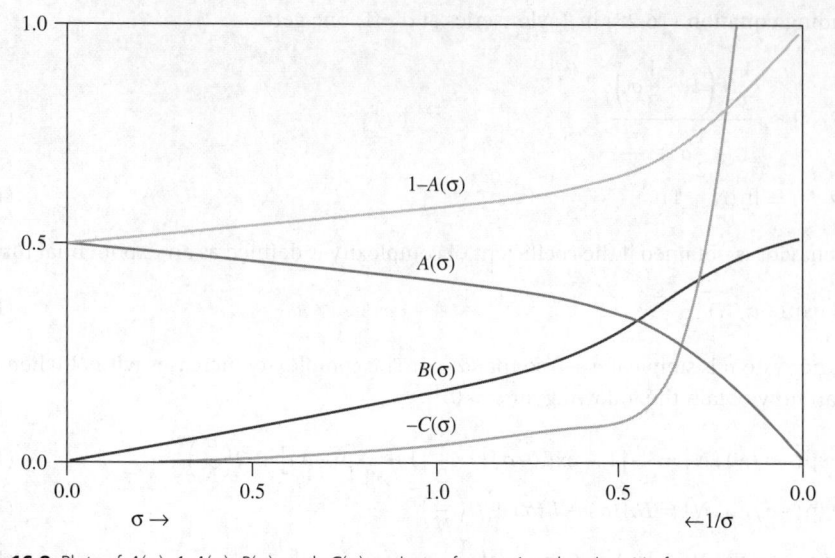

Figure 16.8 Plots of $A(\sigma)$, 1-$A(\sigma)$, $B(\sigma)$, and -$C(\sigma)$ against σ for $\sigma < 1$ and against $1/\sigma$ for $\sigma \geq 1$.

16.2 Kapur's complexity analysis

Kapur (1983) has given a solution for λ_1 as follows. Let

$$\alpha_0 = \exp(-\lambda_0), \quad \alpha_1 = \exp(-\lambda_1), \quad \text{and} \quad \frac{d\alpha}{\alpha} = -d\lambda_1 \tag{16.43}$$

Then, equations (16.6) to (16.8) become, respectively,

$$p(n) = \alpha_0 \alpha_1^n \tag{16.44}$$

$$\alpha_0 \sum_{n=0}^{N} \alpha_1^n = 1 \tag{16.45}$$

$$\alpha_0 \sum_{n=0}^{N} n\alpha_1^n = m \tag{16.46}$$

With the use of equations (16.45) and (16.46), α_1 is determined as

$$f(\alpha) = m + \alpha_1 (m - 1) + \alpha_1^2 (m - 2) + \ldots + \alpha_1^N (m - N) = 0 \tag{16.47a}$$

so that

$$f(0) = m > 0, \quad f(1) = (N + 1)\left(m - \frac{1}{2}N\right), \quad f(\infty) < 0 \tag{16.47b}$$

It is assumed that $m < N$. If $m = N$, the probability distribution is $(0, 0, \ldots, 0, 1)$ which is a minimum entropy distribution with zero entropy. It then follows that

$$\alpha_1 \leq 1 \quad \text{as } m \leq \frac{1}{2}N$$

$$\alpha_1 \geq 1 \quad \text{as } m \geq \frac{1}{2}N \tag{16.48}$$

This leads to three cases and with the use of equation (16.48), one obtains:

1 If $m < N/2$, $\alpha_1 < 1$, the probability p_n decreases as n increases.

2 If $m = N/2$, $\alpha_1 = 1$, the probability distribution is uniform, i.e., $p_i = 1/N$, $p_1 = p_2 = \ldots = p_N$.

3 If $m > N/2$, $\alpha_1 > 1$, the probability p_n increases as n increases.

The maximum entropy can be derived from equations (16.1) and (16.43) to (16.46) as:

$$H = -\sum_{n=0}^{N} p_n \ln\left(\alpha_0 \alpha_1^n\right) = -\ln \alpha_0 - \ln \alpha_1 \sum_{n=0}^{N} n p_n$$

$$= -\ln \alpha_0 - m \ln \alpha_1 = \ln(1 + \alpha_1 + \alpha_1^2 + \ldots + \alpha_1^N) - \frac{\alpha_1^2 + 2\alpha_2^2 + \ldots + N\alpha_1^N}{1 + \alpha_1 + \ldots + \alpha_1^n} \quad (16.49)$$

When $\alpha_1 = 1$, it can be shown that

$$\frac{dH}{d\alpha_1} = 0, \quad \frac{d^2 H}{d\alpha_1^2} < 0 \quad (16.50)$$

Hence, H is maximum when $\alpha_1 = 1$ and its maximum value is $\ln (N = 1)$.

Kapur (1983) noted that the first maximization is over probability distributions with fixed number of defects and the second maximization is for variation of number of defects. H is a defect entropy (DE) and depends on m and is maximum when $m = N/2$, $\alpha_1 = 1$, and the probability distribution is uniform. DE is less than $\ln (N = 1)$ when $\sigma > 1$ or $\sigma < 1$. Thus, the mean number of defects serves as a criterion of system complexity. If the mean is more than $N/2$, the system is complex, and it is simple if the mean is less than half. In this sense, it is more a criterion of system defectiveness than of complexity. Kapur (1983) proposed another measure of complexity of the system as the degree of departure of the probability distribution of the number of defects from the uniform distribution. For a uniform distribution the entropy is $\ln (N = 1)$. With increasing information about the system in terms of moments, the system becomes more structured and more complex and its defect entropy becomes smaller. Let complexity and simplicity be defined, respectively, as C and SI:

$$C = 1 - \frac{DE}{\ln(N+1)} = 1 - \frac{H_{max}}{\ln(N+1)} = 1 - SI \quad (16.51)$$

such that $0 \leq C \leq 1$ and $C = 1$ if and only if the system is complex. Likewise, $0 \leq SI \leq 1$, and $SI = 1$, if and only if the system is simple.

If the only constraint is defined as mean, then from equation (16.49),

$$C = \frac{1}{\ln(N+1)} \left[\ln(N+1) - \ln\left(1 + \alpha_1 + \ldots + \alpha_1^N\right) + \frac{\alpha_1 + 2\alpha_1^2 + \ldots + N\alpha_1^N}{1 + \alpha_1 + \ldots + \alpha_1^N} \ln \alpha_1 \right]$$

$$(16.52)$$

If $\alpha_1 = 1$, equation (16.52) yields $C = 0$ and $\alpha_1 \geq 1$, equation (16.52) yields $C > 0$. Hence, the system is characterized as complex when the mean departs from $N/2$. In this way, the measure of simplicity is the normalized defect entropy. This suggests that the system will have maximum simplicity, subject to the structure imposed by the constraints.

16.3 Cornacchio's generalized complexity measures

Complexity measures, formulated above, depend on the specified constraints and are, therefore, not absolutely necessary. Cornacchio (1977) introduced the concept of conjugate coefficients of complexity for determining the effect of additional constraints or a priori information. The fundamental system characteristic is system error or defect, measured by the number of defects or errors. This number (n) is assumed to be a random variable.

Relationships amongst the coefficient of complexity, expected number of defects, and defect entropy are derived from the canonical distribution. Cornacchio (1977) extended these relationships by specifying additional constraints on the number of defects or errors, such as the delay time through a large network, execution time in a sequence of statements in a computer program, the maximal flow in a directed network, half life of drug concentration in a biological system, and so on. These constraints that encode information may be given in a form other than the equality constraints or bounds. The existence of these constraints may be inferred from the discrepancy between predicted and observed values of the expected number of errors. The enlarged system of constraints represents all the information about the system properties which can be expressed as functions of the number of system defects or errors.

Let $R(1 \leq R \leq N - 1)$ be the equality constraints expressed as

$$\sum_{n=0}^{N} p(n) f_r(n) = \overline{f_r} = F_r, \ r = 1, 2, \ldots, R \tag{16.53}$$

and K be additional inequality constraints defined as

$$\sum_{n=0}^{N} p(n) g_r(n) \leq \overline{G_k}, \ k = 1, 2, \ldots, K \tag{16.54}$$

where $f_r(n)$ is the r-th function of n, $r = 0, 1, 2, \ldots, R$; $R =$ the maximum number of constraints; $\overline{f_r} = F_r$ is the average value of the r-th constraint; $g_k(n)$ is the k-th function of n, $k = 1, 2, \ldots, K$; $K =$ the maximum number of inequality constraints; and $\overline{G_k}$ represent known bounds on the expected values of K system attributes $g_1, g_2, \ldots, g_K$ whose expected values are defined as

$$G_k = \overline{g_k(n)} = \sum_{n=0}^{N} p(n) g_k(n), \ k = 1, 2, \ldots, K \tag{16.55}$$

Of course, $p(n) \geq 0, n = 0, 1, 2, \ldots, N$ and $\sum_{n=0}^{N} p(n) = 1$. This formulation has considered R (equality constraints) and K (inequality constraints) on system observables.

To analyze the effect of additional constraints on system complexity, we first consider the case where the additional constraints are expressed only in the form of equality constraints, that is, $R \neq 1$. Thus, the problem of governing the canonical distribution is formulated as

$$\text{maximize } H(p_0, p_1, p_2, \ldots, p_N) = -\sum_{n=0}^{N} p(n) \log p(n) \tag{16.56}$$

subject to equations (16.53) and (16.54). Cornacchio (1977) defined the coefficient of complexity conjugated to the system observable f_r as

$$\alpha_r = \exp(-\lambda_r), \ r = 1, 2, \ldots, R \tag{16.57}$$

where λ_r is the Lagrange multiplier associated with the r-th constraint. $\alpha_1, \alpha_2, \ldots, \alpha_R$ are called conjugate coefficients of complexity. In particular, α_1 is designed as the characterizing coefficient of complexity. All these coefficients are functions of $\alpha_1, \alpha_2, \ldots, \alpha_R$. For the total probability, the constraint given by equation (16.57) is $\alpha_0 = \exp(-\lambda_0)$.

Let $P = (p_0, p_1, \ldots, p_N)$ and $\lambda = (\lambda_0, \lambda_1, \ldots \lambda_R)$. The Lagrangean $L(P, \lambda)$ can be expressed as

$$L(P, \lambda) = -\sum_{n=0}^{N} p(n) \log p(n) - (\lambda_0 - 1) \left[\sum_{n=0}^{N} p(n) - 1 \right] - \sum_{r=1}^{R} \lambda_r [\sum_{n=0}^{N} f_r(n) p(n) - \overline{f_r}] \qquad (16.58)$$

For a stationary point of L, P and λ must satisfy

$$\nabla_\lambda L(P, \lambda) = 0 \qquad (16.59)$$
$$\nabla_p L(P, \lambda) = 0 \qquad (16.60)$$

where ∇_λ represents the gradient of L with respect to λ and ∇_p represents the gradient of L with respect to P. Therefore,

$$\frac{\partial L(P, \lambda)}{\partial P(n)} = 0, \quad n = 1, 2, \ldots, N \qquad (16.61)$$

which leads to

$$\ln p(n) = -\lambda_0 - \sum_{r=1}^{R} \lambda_r f_r(n), \quad n = 0, 1, 2, \ldots, N \qquad (16.62)$$

or

$$p(n) = \exp[-\lambda_0 - \sum_{r=1}^{R} \lambda_r f_r(n)], \quad n = 0, 1, 2, \ldots, N \qquad (16.63)$$

$$p(n) = \exp(-\lambda_0) \prod_{r=1}^{R} [\exp(-\lambda_r)]^{f_r(n)}, \quad n = 0, 1, 2, \ldots, N \qquad (16.64a)$$

$$= \alpha_0 \prod_{r=1}^{R} \alpha_r^{f_r(n)} \qquad (16.64b)$$

Equation (16.64) represents the canonical distribution $P = \{p(1), p(2), \ldots, p(N)\}$ in terms of the Lagrange multipliers $\{\lambda_r\}$ and system observables $\{f_r\}$.

Now we obtain the relationship between the Lagrange multipliers $\{\lambda_r\}$ and system observables $\{f_r\}$. To that end, substitution of equation (16.64) in equation (16.1) yields

$$\exp(-\lambda_0) \sum_{n=0}^{N} \prod_{r=1}^{R} [\exp(-\lambda_r)]^{f_r(n)} = 1 \qquad (16.65a)$$

or

$$\alpha_0 \sum_{n=0}^{N} \prod_{r=1}^{R} \left[\alpha_r^{f_r(n)} \right] = 1 \qquad (16.65b)$$

Equation (16.65) yields the partition function:

$$\exp(-\lambda_0) = Z^{-1}(\lambda_1, \lambda_2, \dots, \lambda_R) \tag{16.66}$$

where

$$\exp(\lambda_0) = Z(\lambda_1, \lambda_2, \dots, \lambda_R) = \sum_{n=0}^{N} \prod_{r=1}^{R} \left[\exp(-\lambda_r) \right]^{f_r(n)} \tag{16.67a}$$

or

$$\alpha_0 = \sum_{n=0}^{N} f_r(n) \prod_{r=1}^{R} \alpha_r^{f_r(n)} = F_r, r = 1, 2, \dots, R \tag{16.67b}$$

The canonical distribution can be expressed in terms of the partition function as

$$p(n) = \frac{1}{Z(\lambda_1, \lambda_2, \dots, \lambda_R)} \prod_{r=1}^{R} \left[\exp(-\lambda_r) \right]^{f_r(n)}, \quad n = 0, 1, 2, \dots, N \tag{16.68}$$

Substitution of equation (16.68) in equation (16.53) yields

$$Z^{-1}(\lambda_1, \lambda_2, \dots, \lambda_R) \sum_{n=0}^{N} f_r(n) \prod_{r=1}^{R} \left[\exp(-\lambda_r) \right]^{f_r(n)} = F_r, \quad r = 1, 2, 3, \dots, R \tag{16.69}$$

Equation (16.68) can be expressed in terms of conjugate complexities α_r by substituting $\alpha_r = \exp(-\lambda_r)$ as

$$p(n) = Z^{-1}(\lambda_1, \lambda_2, \dots, \lambda_R) \prod_{r=1}^{R} (\alpha_r)^{f_r(n)}, \quad n = 0, 1, 2, \dots, N \tag{16.70}$$

Likewise, equation (16.69) can be expressed as

$$Z^{-1}(\lambda_1, \lambda_2, \dots, \lambda_R) \sum_{n=0}^{N} f_r(n) \prod_{r=1}^{R} [\alpha_r]^{f_r(n)} = F_r, \quad r = 1, 2, 3, \dots, R \tag{16.71a}$$

or

$$Z(\alpha_1, \alpha_2, \dots, \alpha_R) = \sum_{n=0}^{N} \prod_{r=1}^{R} [\alpha_r]^{f_r(n)} \tag{16.71b}$$

where $Z(\alpha_1, \dots, \alpha_2, \dots, \alpha_R)$ is in reality $Z(-\ln \alpha_1, -\ln \alpha_2, \dots, -\ln \alpha_R)$. Equation (16.71b) shows the dependence of the partition function on conjugate complexities.

Equations (16.69) and (16.71b) can be employed in one of two ways. First, given $F_r, r = 1, 2, \dots, R, \lambda_1, \lambda_2, \dots, \lambda_R$ and then $\alpha_1, \alpha_2, \dots, \alpha_R$ can be determined by solving equation (16.69) or (16.71b), respectively. On the other hand, given $\lambda_1, \lambda_2, \dots, \lambda_R$ or $\alpha_1, \alpha_2, \dots, \alpha_R$, $F_1, F_2, \dots, F_R$ can be determined by solving equation (16.69) or (16.71b), respectively.

Differentiating equation (16.67) with respect to λ_r, one gets

$$\frac{\partial Z}{\partial \lambda_r} = -\sum_{n=0}^{N} f_r(n) \prod_{s=1}^{R} \exp\left[-\lambda_s f_s(n)\right] \exp\left[-\lambda_r f(n)\right] \qquad (16.72a)$$

or

$$\frac{\partial Z}{\partial \lambda_r} = -\sum_{n=0}^{N} f_r(n) \prod_{r=1}^{R} \exp\left[-\lambda_r f(n)\right], \ \ r = 1, 2, \ldots, R \qquad (16.72b)$$

From equation (16.72), it is seen that equation (16.71b) can be written as

$$\frac{\partial \ln Z(\lambda_1, \lambda_2, \lambda_R)}{\partial \lambda_r} = -F_r, \ r = 1, 2, \ldots, R \qquad (16.73)$$

Noting that

$$\frac{\partial Z(\lambda_1, \lambda_2, \lambda_R)}{\partial \lambda_r} = \frac{\partial Z(\alpha_1, \alpha_2, \ldots, \alpha_R)}{\partial \alpha_r} \frac{\partial \alpha_r}{\partial \lambda_r} = -\alpha_r \frac{\partial Z(\alpha_1, \alpha_2, \ldots, \alpha_R)}{\partial \alpha_r} \qquad (16.74)$$

one can write

$$\alpha_r \frac{\partial Z(\alpha_1, \alpha_2, \ldots, \alpha_R)}{\partial \alpha_r} = F_r, \ r = 1, 2, 3, \ldots, R \qquad (16.75)$$

Now the defect entropy $H(\alpha_1, \alpha_2, \ldots, \alpha_R)$ can be expressed with the use of equations (16.70) and (16.71b) as

$$H(\alpha_1, \alpha_2, \ldots, \alpha_R) = \ln Z(\alpha_1, \alpha_2, \ldots, \alpha_R) - \sum_{r=1}^{N} F_r \ln \alpha_r \qquad (16.76)$$

which is in terms of the expected values of observables and conjugate complexities.

Equation (16.76) shows that if $F_r, r = 1, 2, \ldots, R$, are known then $\alpha_r, r = 1, 2, \ldots, R$, are known and α_r can be determined from equation (16.71b) and then H is determined. From equations (16.71b) and (16.76),

$$\frac{\sum_{n=0}^{N} f_r(n) \alpha_1^{f_1(n)} \alpha_2^{f_2(n)} \ldots \alpha_R^{f_R(n)}}{\sum \alpha_1^{f_1(n)} \alpha_2^{f_2(n)} \ldots \alpha_R^{f_R(n)}} = a_r, r = 1, 2 \ldots, R \qquad (16.77)$$

Let $a_1 = N/2$. Then

$$\frac{\sum_{n=0}^{N} n \alpha_1^{f_1(n)} \alpha_2^{f_2(n)} \ldots \alpha_R^{f_R(n)}}{\sum \alpha_1^{f_1(n)} \alpha_2^{f_2(n)} \ldots \alpha_R^{f_R(n)}} = \frac{N}{2} \qquad (16.78)$$

Equation (16.77) expresses α_1 as a function of $\alpha_2, \alpha_3, \ldots, \alpha_R$. Cornacchio (1977) defined a μ-transition point, $(\alpha_1, \alpha_2, \ldots, \alpha_R)$, in the μ-dimensional Euclidean space, such that

$$E\big[f_1\,(n)\big] = E[n] = F'\big[\alpha_1\,(\alpha_2, \alpha_3, \ldots, \alpha_R), \alpha_2, \alpha_3, \ldots, \alpha_R\big] = \frac{1}{2}N \tag{16.79}$$

$$H = H\big[\alpha_1\,(\alpha_2, \alpha_3, \ldots, \alpha_R), \alpha_2, \alpha_3, \ldots, \alpha_R\big] = \ln(N = 1) \tag{16.80}$$

Equations (16.79) and (16.80) yield

$$E(n) = F'\,(\alpha_1) = \frac{N}{2} \tag{16.81}$$

$$H(\alpha_1) = \ln(N + 1) \tag{16.82}$$

where α_1, that is, the transition point $\alpha_1, \alpha_2, \ldots, \alpha_R)$ for $R = 1$ reduces to α_1.

16.3.1 Special case: $R = 1$

This indicates that the only constraint is the expected number of system errors or defects. Then the partition function from equation (16.67) reduces to

$$Z(\lambda_1, N) = \sum_{n=0}^{N} \exp[-\alpha_1)]^n = \sum_{n=0}^{N} \alpha_1^n = \frac{1 - \alpha_1^{N=1}}{1 - \alpha_1} \tag{16.83}$$

From equation (16.64), the canonical distribution becomes

$$p(n) = \frac{1 - \alpha_1^{N+1}}{1 - \alpha_1}\alpha_1^n, \; n = 0, 1, 2, \ldots, N \tag{16.84}$$

The expected number of system defects, $F'(\alpha_1; N)$ can be obtained using equation (16.69) as

$$F'\,(\alpha_1; N) = Z^{-1}\,(\alpha_1; N) \sum_{n=0}^{N} n\alpha_1^n \tag{16.85}$$

which reduces to

$$F'\,(\alpha_1; N) = \frac{\alpha_1}{1 - \alpha_1} - (N + 1)\,\frac{\alpha_1^{N+1}}{1 - \alpha_1^{N+1}} \tag{16.86}$$

The defect entropy then is obtained from equation (16.86) as

$$H(\alpha_1; N) = \ln\left[\frac{1 - \alpha_1^{N+1}}{1 - \alpha_1}\right] - \left[\frac{\alpha_1}{1 - \alpha_1} - \frac{(N + 1)\alpha_1^{N+1}}{1 - \alpha_1^{N+1}}\right]\ln\,\alpha_1 \tag{16.87}$$

These results concur with those derived by Ferdinand (1974).

16.3.2 Analysis of complexity: non-unique K-transition points and conditional complexity

The preceding discussion shows that if the only constraint for the canonical distribution is the expected number of system errors then with $\alpha_1 = 1$, the mean number of errors F^1 and the defect entropy H take on the values $N/2$ and $\ln\,(N = 1)$, respectively. Furthermore, $\alpha_1 = 1$ can be considered as a critical point or a transition point. This point separates the region in

α_1-space where the system can be classified as simple from the region where the system can be classified as complex. This point is also called the characterizing coefficient of complexity, that is, the coefficient conjugate to the system observable given by the number of defects. When additional constraints are specified, that is, $K > 1$, Cornacchio (1977) introduced the concept of K-transition point. To consider additional constraints, a K-transition point, $(\alpha_1, \alpha_2, \ldots, \alpha_K)$ is defined

$$F^1\left(\alpha_1, \alpha_2, \ldots, \alpha_K; N\right) = \frac{1}{2}N \qquad (16.88)$$

$$H\left(\alpha_1, \alpha_2, \ldots, \alpha_K; N\right) = \ln(N+1) \qquad (16.89)$$

Noting that any K-transition point $\left(\alpha_1^*, \alpha_2^*, \ldots, \alpha_K^*\right)$ defines a one-transition point, the transition point α_1^* will be related to $\alpha_2^*, \ldots, \alpha_K^*$ with the proviso that $F^1 = \frac{1}{2}N$ and $H = \ln(N+1)$. This one-transition point for α_1 will depend on the values of $\alpha_2, \ldots, \alpha_K$, i.e., $\alpha_1 = \alpha_1\left(\alpha_2, \ldots, \alpha_K\right)$ such that

$$F^1\left[\alpha_1\left(\alpha_2, \ldots, \alpha_K\right), \alpha_2, \alpha_3, \ldots \alpha_K\right] = \frac{1}{2}N \qquad (16.90)$$

$$H[\alpha_1\left(\alpha_2, \ldots, \alpha_K\right), \alpha_2, \ldots, \alpha_K) = \ln(N+1) \qquad (16.91)$$

This can be summarized that the one-transition point employed for classifying the system complexity is conditional upon the values of $K - 1$ conjugate coefficients of complexity defined by K-transition points. In other words, the system complexity depends not only on N but also on the assumed knowledge regarding other system observables.

Consider a set of observables f_k and arbitrary M. Then point $(1, 1, \ldots, 1)$ satisfies

$$F^1(1, 1, \ldots, 1) = \frac{1}{2}N \qquad (16.92)$$

$$H(1, 1, \ldots, 1) = \ln(N+1) \qquad (16.93)$$

From equations (16.67) and (16.71),

$$Z(1, 1, \ldots, 1; N) = \sum_{n=0}^{N}\prod_{r=1}^{R}[\alpha_r]^{f_r(n)} = \sum_{n=1}^{M}(1) = N = 1 \qquad (16.94)$$

$$F^1(1, 1, \ldots, 1; N) = Z^{-1}(1, 1, \ldots, 1; N)\sum_{n=0}^{N}f_n(1)\prod_{r=1}^{R}[1]^{f_n(1)} \qquad (16.95)$$

Inserting $f_n^1 = n$ and observing that

$$\sum_{n=0}^{N}n\prod_{k=1}^{K}(1)^{f_n 1} = \sum_{n=0}^{N}n = \frac{1}{2}N(N+1) \qquad (16.96)$$

it is found that

$$F^1(1, 1, \ldots, 1; N) = \frac{1}{2}N \qquad (16.97)$$

The defect entropy can be computed as

$$H(1, 1, 1, \ldots, 1; N) = \ln Z(1, 1, \ldots, 1; N) - \sum_{k=1}^{K}F^1(1, 1, \ldots, N)\ln(1) \qquad (16.98a)$$

That is,

$$H(1, 1, \ldots 1; N) = \ln(N+1) \tag{16.98b}$$

Consider the case $K = 2$, that is, there is one more constraint besides the mean number of defects. Let the additional information be expressed as

$$f_n^2 = N - n, \quad n = 0, 1, 2, \ldots, N \tag{16.99}$$

This constraint can represent, for example, the maximal flow in a network, where the maximal flow may decrease linearly with the number of defects. If there are no defects then $n = 0$ and the largest maximal flow $f^2 = N$ is attained. On the other hand, if the maximum number of defects is $n = M$, then the maximum flow goes to zero.

In this case, all the points given by $\alpha_1 = \alpha_2$ result in

$$F^1(\alpha_1, \alpha_2; N) = \frac{1}{2}N \tag{16.100}$$

$$H(\alpha_1, \alpha_2; N) = \ln(N+1) \tag{16.101}$$

Thus, all points $(\alpha_1, \alpha_2) = (\alpha, \alpha)$ for arbitrary α are two-transition points.

Using equations (16.77) through (16.71),

$$Z(\alpha_1, \alpha_2; N) = \alpha_2^N \sum_{n=0}^{N} \left(\frac{\alpha_1}{\alpha_2}\right)^n \tag{16.102}$$

$$F^1(\alpha_1, \alpha_2; N) = \alpha_2^N Z^{-1}(\alpha_1, \alpha_2; N) \sum_{n=0}^{N} n \left(\frac{\alpha_1}{\alpha_2}\right)^n \tag{16.103}$$

and

$$F^2(\alpha_1, \alpha_2; N) = N\alpha_2^N Z^{-1}(\alpha_1, \alpha_2; N) \sum_{n=0}^{N} \left(\frac{\alpha_1}{\alpha_2}\right)^n - \alpha_2^N Z^{-1} \sum_{n=0}^{N} n \left(\frac{\alpha_1}{\alpha_2}\right)^n \tag{16.104}$$

Substituting the expression for the partition function, one can express

$$F^2(\alpha_1, \alpha_2; N) = N - F^1(\alpha_1, \alpha_2; N) \tag{16.105}$$

The defect entropy $H(\alpha_1, \alpha_2; N)$ can now be expressed as

$$H(\alpha_1, \alpha_2; N) = \ln Z(\alpha_1, \alpha_2; N) - F^1(\alpha_1, \alpha_2; N) \ln \alpha_1 - F^2(\alpha_1, \alpha_2; N) \ln \alpha_2 \tag{16.106}$$

Using the expression for the partition function Z as well as for F^2 one gets

$$H(\alpha_1, \alpha_2; N) = \ln \sum_{n=0}^{N} \left(\frac{\alpha_1}{\alpha_2}\right)^n - F^1(\alpha_1, \alpha_2; N) \ln \frac{\alpha_1}{\alpha_2} \tag{16.107}$$

For any point (α, α), where α is arbitrary,

$$Z(\alpha, \alpha; N) = (N+1)\alpha^N \tag{16.108}$$

$$H(\alpha, \alpha; N) = \ln(N+1) \tag{16.109}$$

It can now be stated that all points (α_1, α_2) in the one-dimensional manifold specified by $\alpha_1 = \alpha_2$ are two-transition points for arbitrary N.

The simplicity can now defined in general terms from equation (16.87) as

$$
C = \frac{1}{\ln(N+1)}
$$

$$
\times \left\{ \left[\ln(N+1) - \ln \sum_{n=0}^{N} \alpha_1^{f_1(n)} \alpha_2^{f_R(n)} + \sum_{r=1}^{R} \left[\frac{\sum_{n=0}^{N} f_r(n)\alpha_1^{f_1(n)} + \ldots + \alpha_R^{f_R(N)}}{\sum_{n=0}^{N} \alpha_1^{f_1(n)} \alpha_2^{f_2(n)} \ldots \alpha_R^{f_R(n)}} \right] \ln \alpha_r \right] \right\} \quad (16.110)
$$

16.4 Kapur's simplification

For handling inequality constraints, Kapur (1983) showed that the programming technique proposed by Cornacchio (1977) is not necessary. To that end, suppose

$$
a_{r1} \leq \sum_{n=0}^{N} p_n g_r(n) \leq a_{r2}, \quad r = 1, 2, \ldots, K \quad (16.111)
$$

$H_{\max}$ is a concave function of $a_1, a_2, \ldots, a_K$:

$$
a_{r1} \leq a_r \leq a_{r2}, \quad r = 1, 2, \ldots, K \quad (16.112)
$$

This can be shown as follows.

$$
\frac{\partial H_{\max}}{\partial \alpha_r} = \lambda_r = \ln \frac{1}{\alpha_r} \quad (16.113)
$$

so that

$$
\frac{\partial H_{\max}}{\partial a_r} \geq 0, \alpha_r \leq 1; \quad \frac{\partial H_{\max}}{\partial a_r} \leq 0, \ \alpha_r \geq 1 \quad (16.114)
$$

16.5 Kapur's measure

If

$$
\alpha_1 = \alpha_2 = \ldots = \alpha_R = 1 \quad (16.115)
$$

then equations (16.44), (16.45), (16.46) and (16.47) lead to

$$
\alpha_0(N+1) = 1 \quad (16.116)
$$

$$
\alpha_0 \sum_{n=0}^{N} f_1(n) = \alpha_0 \sum_{n=0}^{N} n = \frac{N}{2} = \alpha_1 \quad (16.117)
$$

$$
\alpha_0 \sum_{n=0}^{N} f_r(n) = \alpha_r, \quad r = 1, 2, 3, \ldots, R \quad (16.118)
$$

$$
H_{\max} = -\ln \alpha_0 = \ln(N+1) \quad (16.119)
$$

Both equations (16.117) and (16.119) are satisfied. Kapur (1983) showed that equation (16.115) is the only case when that can happen and hence there can at most be one μ-transition point. This can be shown as follows. H_{max} is a concave function of $a_1, a_2, \ldots, a_R$ and its local maximum is a global maximum which occurs when

$$a_r = \frac{1}{N+1} \sum_{n=0}^{N} f_r(n), \quad r-1, 2, \ldots, R \tag{16.120}$$

The maximum of H, $H_{max} = \ln(N+1)$, occurs when $\alpha_1 = N/2$, and other values of $\alpha_1, \alpha_2, \ldots, \alpha_R$ are given by equation (16.116). If any of $\alpha_1, \alpha_2, \ldots, \alpha_R$ is different from that given by equation (16.120) then H_{max} will be less than $\ln(N+1)$ and equation (16.119) will not be satisfied.

16.6 Hypothesis testing

Let a system be satisfactory if $m \leq a_1$ and the variance of the number of defects $\leq a_2$. If $p_0, p_1, \ldots, p_N$ is the probability distribution, then in W trials, $Wp_0, Wp_1, \ldots, Wp_N$ are the expected frequencies. Consider $q_0, q_1, \ldots, q_N$ as observed frequencies. Then

$$\chi^2 = \sum_{n=0}^{N} \frac{(Wp_n - q_n)^2}{Wq_n} \tag{16.121}$$

For three constraints, the number of degrees of freedom

$$\upsilon = (N+1) - 3 = N - 2 \tag{16.122}$$

$P = \{p_1, p_2, \ldots, p_N\}$ is the maximum entropy-based distribution. From the χ^2 tables, one tests if P is different from $Q = \{q_1, q_2, \ldots, q_N\}$. In this case, no special form of the probability distribution is assumed.

16.7 Other complexity measures

Many time series can be replaced by symbolic strings using a binary alphabet in which the mean content is taken as normal content, a value higher than normal is represented by 1 and a value lower than normal by 0. This has been suggested by Lange (1999) and Wolf (1999). Now the length of a word, L, can be defined as a group or series of L consecutive symbols and the strings of symbols have 2^L possible words. For example, if the word length is 2, then there are two consecutive symbols and there are $2^2 = 4$ possible words: 00, 11, 10, and 01. Each word characterizes the state of the system. The change in the words starting from two consecutive observations, say 00 to 01, defines the transition from state 00 to state 01. Consider a string 11001 and the word length is two. The number of possible words is $2^5 = 32$. The first word is 11 and the shift from it to the second word 10 represents the transition from state 11 to state 10, and the shift from the second state 10 to the third state 00 represents the transition from state 10 to state 00, and so on.

Empirical probabilities can be considered, depending on the word. Let the word be of length L. There are three possibilities: 1) probability p_i for word i to occur in the symbolic string, $i = 1, 2, \ldots, 2^L$; 2) probability p_{ij} for the sequence of words i and j to occur, $i = 1, 2, \ldots, 2^L$, and $j = 1, 2, \ldots, 2^L$; and 3) $p_{j|i}$ the conditional probability of word j occurring after word i, $i = 1, 2, \ldots, 2^L; j = 1, 2, \ldots, 2^L$. Now the Shannon entropy can be defined for words of length L as

$$H(L) = -\sum_{i=1}^{2^L} p_i \log p_i \tag{16.123}$$

The metric entropy can be defined by dividing the Shannon entropy $H(L)$ by the word length L, that is, $H(L)/L$. It indicates the extent of disorder in the sequence of symbols, increasing to a maximum value of one when the random sequences of words are uniformly distributed.

The mean information content $H_m(L)$ can be defined as

$$H_m(L) = -\sum_{i,j}^{2^L} p_{ij} \log p_{j|i} \tag{16.124}$$

This is analogous to conditional entropy and is a measure of the (additional) information to be gained on average for the whole symbol sequence from the knowledge of the next symbol.

Now the complexity measures can be defined. Let the next information gain be defined as the difference between information gain and loss. Then the fluctuation complexity σ_c^2 is defined as the mean square deviation of the information gain:

$$\sigma_c^2 = \sum_{i,j}^{2^L} p_{ij} \left[\log \left(\frac{p_i}{p_j} \right) \right]^2 \tag{16.125}$$

A higher value of σ_c^2 (i.e., more the net information gain is fluctuating in the string) would lead to higher fluctuating complexity.

An effective complexity measure C_{em} measures the total minimum amount of information that must be stored at any time for the optimal prediction of the next symbol. It can be calculated as

$$C_{em} = \sum_{i,j}^{2^L} p_{ij} \log \left(\frac{p_{j|i}}{p_i} \right) \tag{16.126}$$

Bates and Shepard (1993) presented complexity measures for analyzing deterministic dynamical systems using information fluctuation. Their idea is that complex behavior lies between extremes of order and disorder. Consider a system in state i with probability p_i. Conditioning on this state, the system transits to state j (forward) with transitional probability $p_{i \to j}$ which can be estimated if the system dynamics were known. Let p_{ij} denote the probability that a transition from i to state j occurs. Then one can write

$$p_{ij} = p_i p_{i \to j} \tag{16.127}$$

If the transition is backward, that is the system is presently in state j and the prior state was i with probability $p_{i \leftarrow j}$ then

$$p_{ij} = p_{i \leftarrow j} p_j \tag{16.128}$$

It may also be noted that

$$p_j \sum_i p_{ij} = \sum_i p_{ji} \tag{16.129}$$

where the summation is over all the states. This leads to

$$p_j = \sum_i p_i p_{i \to j} \tag{16.130}$$

From the Shannon entropy, the information needed to specify the state i of the system can be expressed as

$$H_i = -\log_2(p_i) \tag{16.131}$$

The mean of this information yields the Shannon entropy as

$$H = \sum_{i=1}^{N} p_i H_i = -\sum_{i=1}^{N} p_i \log_2(p_i) \tag{16.132}$$

where N is the number of states.

The information gain G_{ij} due to the transition from state i to state j is expressed as

$$G_{ij} = -\log_2 p_{i \to j} \tag{16.133}$$

Likewise, the information loss is defined as

$$L_{ij} = -\log_2 p_{j \to i} \tag{16.134}$$

where the system goes from state j to state i. The difference between equations (16.133) and (16.135) yields the net information gain:

$$g_{ij} = G_{ij} - L_{ij} = \log_2\left(\frac{p_{j \to i}}{p_{i \to j}}\right) = \log_2\left(\frac{p_i}{p_j}\right) \tag{16.135}$$

Averaging over all transition states, equation (16.135) can be written as

$$\bar{g} = \sum_{ij} p_{ij} g_{ij} \tag{16.136}$$

It can be shown that the average gain will equal the average loss and equation (16.130) will vanish. However, the mean square deviations $\sigma_g{}^2$ of g_{ij} will not and can be written as

$$\sigma_g^2 = E\left[g - E(g)\right]^2 = E[g^2] = \sum_{i,j}^{N} p_{ij}\left[\log_2 \frac{p_i}{p_j}\right]^2 \tag{16.137}$$

This value reflects the fluctuation occurring in the system as it transitions from one state to another, and is termed fluctuation complexity. The fluctuation can be positive or negative, where positive would imply a net storage of information. If a system simultaneously gains and loses information then its net information storage capacity $g_{ij} = 0$. A system with zero entropy means that $G_{ij} = L_{ij} = 0$, and $\sigma_g = 0$. The net information gain can be expressed using equation (16.135) recast as

$$g_{ij} = \log\frac{p_i}{p_j} = H_j - H_i \tag{16.138}$$

Equation (16.138) shows that the cumulative net information gain for a sequence of states depends only on the initial and final states and is independent of the path. For sequences of transitions, the cumulative net information gain H_{in} needed to reach from state i to state n can be expressed as

$$H_{in} = g_{ij} = g_{jk} = \cdots = g_{lm} = g_{mn} = \log\frac{p_i}{p_n} = H_n - H_i \tag{16.139}$$

The quantity H_{in} can be thought of as information potential and H_n-H_i can be construed as a measure of the rarity of state n relative to state i or information force. It can also be shown that the variance of H_{in} is the same as the variance of H. In other words the fluctuation in H_{in} is the same as in H. Pachepsky et al. (2006) employed the above concepts to measure complexity of simulated soil water fluxes.

Questions

Q.16.1 Observations have been made on the leaks in a water supply system on a monthly basis for a number of years. Each month the number of leaks varies. Consider the number of leaks or defects in water supply system as a random variable. The number of leaks is used to describe if the water supply system is complex or simple. Let the maximum number of leaks or defects be 10 and the average number be 5. Compute the probability distribution of the number of leaks. Compute the Lagrange multipliers λ_0 and λ_1. Plot the probability distribution against the number of leaks.

Q.16.2 What is the probability of the occurrence of 1, 2, 3, 4, 5, 6, and 8 leaks in Q.16.1?

Q.16.3 Compute the coefficient of complexity α and also β in Q.16.1.

Q.16.4 Compute the defect entropy in Q.16.1.

Q.16.5 Compute the simplicity as well as complexity for Q.16.1.

Q.16.6 Consider the number of erroneous or missing values of rainfall in rain gage measurements in a watershed, that is, the number of missing values at a gage is a random variable. Some gages have more missing values than others. Consider the maximum number of missing values as 20 and the average number of missing values is 10. Compute the probability distribution of the number of leaks. Compute the Lagrange multipliers λ_0 and λ_1. Plot the probability distribution against the number of leaks.

Q.16.7 What is the probability of the occurrence of 1, 2, 3, 4, 5, 6, and 8 missing values in Q.16.6?

Q.16.8 Compute the coefficient of complexity α and also β in Q.16.6.

Q.16.9 Compute the defect entropy in Q.16.6.

Q.16.10 Compute the simplicity as well as complexity for Q.16.6.

References

Bates, J.E. and Shepard, H.K. (1993). Measuring complexity using information fluctuation. *Physical Letters A*, Vol. 172, No.6, pp. 416–25.

Cornacchio, J.V. (1977). Maximum-entropy complexity measures. *International Journal of Systems*, Vol. 3, N. 4, pp. 215–25.

Englehardt, S., Matyssek, R. and Huwe, B. (2009). Complexity and information propagation in hydrologic time series of mountain forest catchments. *European Journal of Forestry Research*, Vol. 128, pp. 621–31.

Ferdinand, A.E. (1974). A theory of system complexity. *International Journal of General Systems*, Vol. 1, No. 1, pp. 19–33.

Grassberger, P. (1986). Toward a quantitative theory of self-generated complexity. *International Journal of Theoretical Physics*, Vol. 25, pp. 907–38.

Jimenez-Montano, M.A., Ebeling, W., Pohl, T. and Rapp, P.E. (2002). Entropy and complexity of finite sequences as fluctuating quantities. *BioSystems*, Vol. 64, pp. 23–32.

Kapur, J.N. (1983). On maximum-entropy complexity measures. *International Journal of General Systems*, Vol. 9, pp. 95–102.

Lange, H. (1999). Time series analysis of ecosystem variables with complexity measures. *International Journal of Complex Systems*, Manuscript #250, New England Complex Systems Institute, Cambridge, Massachusetts.

Pachepsky, Y., Guber, A., Jacques, D., Simunek, J., van Genuchten, M.T., Nicholson, T. and Cady, R. (2006). Information content and complexity of simulated soil water fluxes. Geoderma. Vol. 134, pp. 253–266.

Wolf, F. (1999). Berechnung von information und komplexitat von zeitreihen-analyse des wasserhaushaltes von bewaldeten einzugsgebieten. Bayreuth. Forum Okol. Vol. 65, pp. 164 S.

Additional References

Abu-Mostafa, Y.S. (1986). The complexity of information extraction. *IEEE Transactions on Information Theory*, Vol. IT-32, No. 4, pp. 513–25.

Cady, R. (2006). Information content and complexity of simulated soil water fluxes. *Geoderma*, Vol. 134, pp. 253–66.

Clement, T.P. (2010). Complexities in hindcasting models-when should we say enough is enough? *Groundwater*, Vol. 48, doi:10.1111/j.1745-6584.2010.00765.x

Costa, M., Goldberger, A.L. and Peng, C.-K. (2002). Multiscale entropy analysis of complex physiologic time series. *Physical Review Letters*, Vol. 89, No. 6, 06812, pp.1-4.

Ferdinand, A.E. (1969). A statistical mechanics approach to systems analysis. IBM TR 21–348, York Heights, New York.

Ferdinand, A.E. (1970). A statistical mechanical approach to systems analysis. *IBM Journal of Research and Development*, Vol. 14, No. 5, pp. 539–47.

Ferdinand, A.E. (1993). *Systems, Software, and Quality Engineering: Applying Defect Behavior Theory to Programming*. 416 p., Van Nostrand Reinhold, New York.

Pincus, S. and Singer, B. (1996). Randomness and degrees of irregularity. Proceedings, National Academy of Science, Vol. 93, pp. 2083–8.

Pincus, S.M. (1995). Quantifying complexity and regularity of neurobiological systems. *Methods in Neurosciences*, Vol. 28, pp. 336–63.

Pincus, S. (1995). Approximate entropy (ApEn) as a complexity measure. *Chaos*, Vol. 5, No. 1, pp. 110–17.

Author Index

Subject Index

Entropy Theory and its Application in Environmental and Water Engineering, First Edition. Vijay P. Singh.
© 2013 John Wiley & Sons, Ltd. Published 2013 by John Wiley & Sons, Ltd.